EARTH SCIENCE

EARTH SCIENCE
VOLUME IV

WEATHER, WATER, AND THE ATMOSPHERE

EDITOR
JAMES A. WOODHEAD
Occidental College

EDITORIAL BOARD

DENNIS G. BAKER, VOLUME IV
University of Michigan

ANITA BAKER-BLOCKER, VOLUME IV
Applied Meteorological Services

DAVID K. ELLIOTT, VOLUME II
Northern Arizona University

RENÉ DE HON, VOLUME I
University of Louisiana at Monroe

CHARLES W. ROGERS, VOLUMES III & V
Southwestern Oklahoma State University

SALEM PRESS, INC.
Pasadena, California   Hackensack, New Jersey

MANAGING EDITOR: Christina J. Moose
PROJECT DEVELOPMENT: Robert McClenaghan
MANUSCRIPT EDITORS: Doug Long, Amy Allison
ACQUISITIONS EDITOR: Mark Rehn
RESEARCH SUPERVISOR: Jeffry Jensen
PHOTOGRAPH EDITOR: Philip Bader
ASSISTANT EDITOR: Andrea E. Miller
INDEXERS: Melanie Watkins, Lois Smith
RESEARCH ASSISTANT: Jeffrey Stephens
PRODUCTION EDITOR: Cynthia Beres
PAGE DESIGN AND LAYOUT: James Hutson
ADDITIONAL LAYOUT: William Zimmerman
GRAPHICS: Electronic Illustrators Group

Library of Congress Cataloging-Publication Data

Earth science / editor, James A. Woodhead.
   p. cm.
Expands and updates Magill's survey of science: earth science series.
Includes bibliographical references and indexes.
Contents: v. 1. The physics and chemistry of earth — v. 2. The earth's surface and history — v. 3. Earth materials and earth resources — v. 4. Weather, water, and the atmosphere — v. 5. Planetology and earth from space.
ISBN 0-89356-000-6 (set : alk. paper) — ISBN 0-89356-001-4 (v. 1 : alk paper) —
ISBN 0-89356-002-2 (v. 2 : alk. paper) — ISBN 0-89356-003-0 (v. 3 : alk. paper) —
ISBN 0-89356-004-9 (v. 4 : alk. paper) — ISBN 0-89356-005-7 (v. 5 : alk. paper)
1. Earth sciences. I. Woodhead, James A. II. Magill's survey of science. Earth science series.

QE28 .E12 2001
550—dc 21

00-059567

First Printing

CONTENTS

EARTH SCIENCE

1
EARTH'S ATMOSPHERE

ACID RAIN AND ACID DEPOSITION

Acid rain is rain that is more acid than natural rain as a result of reactions with pollutive acidic gases, such as sulfur dioxide and nitric oxides. Lakes, forests, soils, and human structures in the eastern part of the United States and southeastern Canada have been damaged by acid rain and deposition of sulfuric and nitric acid aerosol on terrestrial objects.

PRINCIPAL TERMS

ACIDITY: the concentration of hydrogen ions in a solution

ALKALINE: having a pH greater than 7

BICARBONATE: an ion formed of one hydrogen atom, one carbon atom, and three oxygen atoms that is very effective in natural waters at neutralizing hydrogen ions and reducing acidity

LIMESTONE: a rock containing calcium carbonate that reacts easily with acid rain and tends to neutralize it

NEUTRALIZATION: the removal of hydrogen ions from solution by their reaction with other ions, which lowers the acidity of the solution

NITRIC ACID: an acid formed in rain from nitric oxide gases in the air; it often contributes to acid rain

NITRIC OXIDE GASES: gases formed by a combination of nitrogen and oxygen, particularly nitrogen dioxide and nitric oxide

pH: a measure of the acidity of a solution; the lower the pH, the greater the concentration of hydrogen ions and the more acid the solution

SULFUR DIOXIDE: a gas formed by the combination of sulfur and two oxygen atoms

SULFURIC ACID: an acid formed in rain from sulfur dioxide gas in the air; it is the primary acid in acid rain

DEFINITION AND CAUSE

Acid rain is rain that is more acid than it would naturally be, usually because it has reacted with acidic pollutive gases. The acidity of rain is measured in pH units. Pure water, which is neutral, has a pH of 7; any solution with a pH greater than 7 is basic, and any solution with a pH less than 7 is acidic. The lower the pH, the more hydrogen ions and the more acid the solution. The natural acidity of rain is determined by its reaction with carbon dioxide gas in the atmosphere, a reaction that produces carbonic acid. Carbonic acid partly dissociates to produce hydrogen ions and bicarbonate ions. As a result of this reaction, pure rain should be moderately acidic, with a pH of 5.7. Any rain with a pH less than 5.7 is called "acid rain" and has reacted with acidic atmospheric gases other than carbon dioxide, such as sulfur dioxide, which produces sulfuric acid in rain, and nitrogen dioxide, which produces nitric acid in rain. In some extreme cases, acid rain has been analyzed with a pH as low as 2.4, which is as acidic as vinegar. In addition to acid rain, there is "dry deposi-

tion" from the atmosphere of acidic nitrate sulfate particles and sulfuric and nitric acid aerosols, which occurs without rain. The acidic particles are trapped by vegetation or settle out, and the gases are taken up by vegetation. "Acid deposition" usually refers to dry deposition of acids.

Acid rain was first recognized in Scandinavia in the early 1950's. It was discovered that acid rain (with a pH from 4 to 5) came from winter air masses that were carrying pollution from industrial areas in Central and Western Europe into Scandinavia. Rain became more acid over the next twenty years, and the area of Europe receiving acid rains increased. By the mid-1970's, most of northwestern Europe was receiving acid rain with a pH of less than 4.6. As a result of the discovery of acid rain in Europe, scientists began measuring the acidity of North American rain. Initially, around 1960, acid rain was concentrated in a bull's-eye-shaped area over New York, Pennsylvania, and New England. By 1980, however, most of the United States east of the Mississippi River and southeastern Canada was receiving acid rain (pH

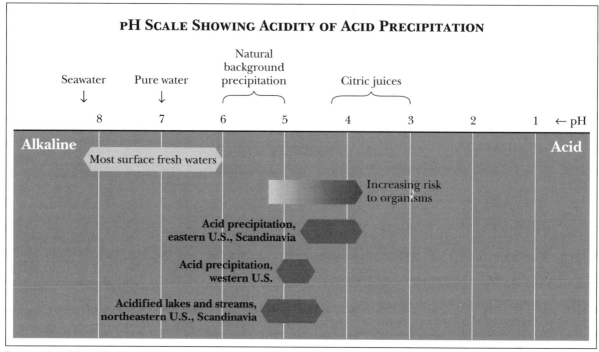

pH SCALE SHOWING ACIDITY OF ACID PRECIPITATION

SOURCE: Adapted from John Harte, "Acid Rain," in *The Energy-Environment Connection,* edited by Jack M. Hollander, 1992.

NOTE: The acid precipitation pH ranges given correspond to volume-weighted annual averages of weekly samples.

less than 5.0), and the central bull's-eye was receiving very acid rain (pH less than 4.2). The greatest increase in acidity of U.S. rain was in the southeastern United States.

The primary cause of acidity in U.S. and European rains is sulfuric acid, which comes from pollutive sulfur dioxide gas produced by the burning of sulfur-containing fossil fuels—particularly coal, but also oil and gas—by power plants and industry. In the United States, much of the sulfur dioxide gas is produced in the industrial area of the Midwest; however, sulfur dioxide gas, and the resulting sulfuric acid, can be transported for a distance of 800 kilometers to the northeast by the prevailing winds in the atmosphere before being rained out as acid rain in the northeastern United States and southeastern Canada. To reduce the acidity of rain in the East, then, the emissions of sulfur dioxide gas in the Midwest would have to be reduced. Another source of sulfur dioxide gas is smelters that process ores, such as the Sudbury, Ontario, smelter, located north of Lake Huron in Canada, which is one of the largest sulfur sources in the world. This smelter, which has a very high smokestack, spreads sulfuric acid aerosol over a large area hundreds of kilometers downwind from it. The original intent of building high smokestacks was to reduce local air pollution, but the net effect has been to spread the pollution over large areas. Acid rains are even found in Alaska, where sulfuric acid particles have been transported from the coterminous United States.

Nitric acid is a secondary cause of acid rain (contributing about 30 percent of the acidity), but one that is increasing. Nitric acid comes from the nitrogen oxide gases, nitrogen dioxide, and nitric oxide, which are produced by fossil fuel burning. In contrast to sulfur dioxide, 40 percent of the pollutive nitrogen oxide gas comes from vehicles and most of the rest from power and heating. The production of nitrogen oxide therefore tends to be concentrated in urban areas. Nitric acid is an important component of acid rain in Los Angeles, for example, because air pollution from vehicles tends to be trapped in this area.

Some acid rain results from natural causes. Reduced sulfur gases, such as hydrogen sulfide and dimethyl sulfide, are produced by organic matter decay and converted to sulfur dioxide and sulfuric acid in the atmosphere. This process results in naturally acid rain. Volcanoes are another natural source of sulfur dioxide gas. Nevertheless, about 75 percent of the sulfur dioxide gas produced in the United States comes from the burning of fossil fuels. Naturally acid rain (with a pH less than 5.5) is uncommon, being chiefly known in remote areas such as the Amazon basin and some oceanic areas.

There are natural factors that work to reduce or neutralize the acidity of rain in certain areas. Windblown dust (particularly that containing limestone) tends to make rains in arid areas of the western United States less acid, with a pH of 6 or more. Also, the presence of ammonia gas, which is produced in agricultural areas by animal wastes, organic matter decomposition, and fertilizers, will reduce the acidity of rain.

EFFECTS OF ACID RAIN AND ACID DEPOSITION

The detrimental effects from dry deposition and acid rain include the corrosion of human-made structures and buildings, changes in soils, increases in the acidity of lakes, and biological effects, particularly in high-altitude forests. The corrosive effects of airborne acids are particularly obvious on limestone, a rock composed of calcium carbonate, which reacts easily with acid rain. In many New England cemeteries, tombstones made of marble (metamorphosed or heated limestone) have been badly corroded, although older tombstones made of slate, which is less affected by acid rain, are intact. Limestone buildings are similarly corroded.

The effect of acid rain on soils depends on their composition. Alkaline soils, which contain limestone, easily neutralize acid rain. Even in soils that do not contain limestone, several processes operate to neutralize acid rain. Cation exchange occurs, whereby the hydrogen ions from the rain are exchanged for metal ions, such as calcium or magnesium, on the surface of clays and other minerals. This exchange removes hydrogen ions from soil solutions and makes the solutions less acid. Another neutralization process involves the release of soil aluminum into solution and the accompanying uptake of hydrogen ions. This process occurs by dissolution of aluminum bound to clays and organic compounds. Frozen soils and sandy soils containing mostly quartz, which does not react with acid rain, have little ability to neutralize acid rain.

Lakes in certain areas have become acidic (with a pH less than 5) from the deposition of acid rain. Lakes in granitic terrain are most affected by acid rain. Areas with many acid lakes include the Adirondack Mountains in New York, the Pocono Mountains in Pennsylvania, the Upper Peninsula of Michigan, Ontario, Nova Scotia, and Scandinavia. Generally, the deposition of very acid rain (with a pH less than 4.6) over a long period is required. Lake waters that have a tendency to become acidified initially have little ability to neutralize acid rain because they are low in bicarbonate ions, which come dominantly from limestone. Such lakes are described as being poorly buffered. (Buffering is the resistance to change in pH upon the addition of acid.) The soil in the drainage area surrounding acid lakes does not neutralize acid rain adequately before it reaches the lake because of a lack of limestone and clay minerals or because the soil cover is thin or lacking altogether. In addition, some lakes, although not usually acid, may have periods of acidity during the runoff of snowmelt, which collects acid precipitation stored in the snow. This runoff gives a sudden large pulse of very acid water to the lake. In certain areas, such as Florida, acid lakes result at least partly from causes other than acid rain, such as organic acids from the decay of vegetation in poorly drained areas and nitric acid from fertilizer runoff. The gradual acidification of lakes results in the death of their fish populations because of reproductive failure, as well as other changes in the organisms living in the lake. A reduction in the number of species occurs at all levels of the food chain. In some cases, snowmelt acidity can cause a massive, instantaneous fish kill in lakes.

Rivers are also known to become acid. Eastern U.S. rivers show high concentrations of sulfate and a low pH in cases where the soil cannot neutralize the acid rain that it receives. Certain acid rivers are caused by acid drainage from mine dumps rather than by acid rain. Organic-rich acid rivers, which are naturally acid, are found in the eastern United States coastal plain and in the Amazon basin. These rivers have naturally high concentrations of dissolved organic acids and are very dilute.

Acid rain and acid deposition are implicated in the decline and death of certain forests, particularly evergreen forests at high elevations. These forests receive very acid precipitation from the base of clouds at the mountaintops. It is thought that acid rain does not actually kill forests but rather provides a stress, causing them to become less resistant and die of other causes. The actual stress provided by acid rain is still being studied. Possible stresses include loss of soil nutrients and leaching of nutrients from the leaves, destruction of soil microorganisms, and increased susceptibility to cold winters.

Efforts have been made to reduce the acidity of rain, particularly by controlling sulfur emissions. Power plants have been required to reduce the sulfur content of coal that they burn; lower-sulfur coal produces less sulfur dioxide. The sulfate concentrations in rain in the northeastern United States have been reduced by this method. The nitrogen oxide emissions from cars have also been reduced. In some cases, acid lakes have been treated with limestone to neutralize their acidity temporarily, but the only permanent solution is a reduction in the acidity of the rain that they receive.

STUDY OF ACID RAIN AND ACID DEPOSITION

The acidity of rain can be measured directly by a pH meter. A pair of electrodes is inserted into a solution, and the electrical potential, or voltage, is measured between them. This voltage is a measure of the concentration of hydrogen ions in the solution—the acidity. Networks have been constructed to collect rain samples over large geographical areas. The acidity of rain over the whole year must be measured because pH varies between rainfalls, both seasonally and depending on whether the air masses that produce the rain have passed over significant sources of pollution. The pH of rain is also measured over a period of years. In addition to the concentration of hydrogen ions, the concentration of other ions, such as sulfate (from sulfuric acid) and nitrate (from nitric acid), is measured in the rain samples. That gives evidence of the source of the acidity—that is, which proportion is attributable to sulfuric acid and which to nitric acid. The pH levels of samples collected over a large geographical area are plotted on a map, and contours are drawn through equal values of the pH. Such maps show which areas are receiving the most acid rain. The amount

of sulfate and nitrate being deposited by rain is also plotted separately. Meteorologists also have information about the path followed by a storm system as it moves across the country. Such atmospheric systems transport pollutive gases from one area to another. Combining depositional patterns on maps with information about the path followed by a storm shows where the gases in rain may be coming from and suggests sources of the acidity.

Computers have been used to predict where acid rain will fall and how acid it will be, given the sources and amounts of sulfur and nitrogen emissions, particularly from power plants and smelters, and the weather patterns. Predictions of this type require a detailed knowledge of the atmospheric chemistry by which sulfur dioxide is converted to sulfuric acid and the oxides of nitrogen are converted to nitric acid. This type of modeling is necessary to predict how much reduction in the acidity of rain in a distant area will result from a given reduction in a power plant sulfur source, for example.

Studies of the effects of acidity on soils have been conducted. Laboratory experiments can be done to see how soil clays and other minerals react to the addition of acid rain, including which chemical species are taken up and released. In addition, soil solutions and minerals are collected and analyzed from actual field areas affected by acid rain. Ideally, this analysis should be done over a period of time to determine whether any changes in the soil solution chemistry are occurring. From a knowledge of the soil chemistry, it is possible to predict how long a soil can receive acid rain before it loses the ability to neutralize the acidity.

The acidity and chemical composition of lakes in various areas have been measured over long periods, and their fish and other biota have been sampled. That enables scientists to see increases in lake acidity and to correlate the increases with changes in the populations of fish and biota. In some areas, lakes have been artificially acidified so that the changes in their chemistry and biological populations can be observed. Apparently, acid lake water inhibits the reproduction of fish and other creatures in addition to destroying organisms that they use for food. Computer models of acid rain falling on susceptible drainage areas of lakes have been made in order to predict how the drainage area reacts to acid rain and how much reduction in acid rain would be necessary to lower the acidity of the lake to the point where it would support fish. In badly affected areas such as the Adirondacks, it may be necessary to reduce the acidity by one-half.

To study forest decline, surveys of present forest conditions are compared with historical records for the same areas. For example, in high-elevation areas in New England and the Adirondacks, more than one-half of the red spruce died between 1965 and 1990. Tree rings, which record annual growth, show reduced growth in certain forests. It is known that acid rain causes changes in the soil, such as the release of aluminum, which is toxic to root tissues and so prevents the uptake of essential nutrients. In addition, acid rain causes the loss from the soil of certain nutrients, such as sodium, calcium, and magnesium. Another effect of acid rain is the reduction of soil microorganisms. Yet, since acid rain from nitric acid contains nitrogen, a plant nutrient, it may fertilize the soil if there is a deficiency of soil nitrogen. One problem in studying forests receiving acid rain is determining which of the many changes occurring are contributing most to forest damage. It is often difficult to distinguish between the stresses of acid rain and other stresses, such as drought, cold, and insects. Field studies are being made that involve artificially acidifying forests to determine which mechanisms are important.

Elizabeth K. Berner

CROSS-REFERENCES

BIBLIOGRAPHY

Berner, Elizabeth K., and Robert A. Berner. *The Global Water Cycle: Geochemistry and Environment.* Englewood Cliffs, N.J.: Prentice-Hall, 1986. This book includes a number of sections on acid rain and its effects. Chapter 3 describes the formation and global distribution of sulfur and nitrogen gases, from both natural and man-made causes. Acid rain is also extensively discussed. Chapter 5 describes acid rivers, including those that result from sulfate pollution and those that result from natural organic acidity. Chapter 6 discusses acid lakes and how they form. Designed for college freshmen.

Cowling, E. B. "Acid Precipitation in Historical Perspective." *Environmental Science and Technology* 16 (February, 1982): A110-A123. This article traces the historical development of the study of acid rain. It has a chronological listing of the various important discoveries in the study of acid rain and excellent references to original articles. It also describes research programs designed to study the deposition of acid precipitation and its effects on forests, soils, and lakes. Suitable for the general reader.

Glass, N. R., et al. "Effects of Acid Precipitation." *Environmental Science and Technology* 16, no. 3 (1982): A162-A169. This article summarizes research on the effects of acid precipitation on lakes, soils, and crops. For example, efforts have been made to predict which areas in the eastern United States are sensitive to acid precipitation. Includes a number of references and is written on a college level.

Johnson, Arthur H., and T. G. Siccama. "Acid Deposition and Forest Decline." *Environmental Science and Technology* 17, no. 7 (1983): A294-A305. A summary of the evidence concerning the relationship between acid deposition and forest decline. It points out that a clear cause-and-effect relationship does not exist but that acid precipitation may provide a stress to forests. Readable by college students.

Klaassen, Ger. *Acid Rain and Environmental Degradation: The Economics of Emission Trading.* Brookfield, Vt.: Edward Elgar, 1996. An close examination of the pros and cons of the industrial practice of trading the right to emit acid-rain-causing pollutants. Illustrations, index, and fifteen pages of bibliographical references.

Likens, Gene E., R. F. Wright, J. F. Galloway, and T. F. Butler. "Acid Rain." *Scientific American* 241 (October, 1979): 43-51. This article is intended for the general reader. It has an excellent discussion of what acid rain is and how it was recognized. It discusses the areas receiving acid rain in the United States and Europe, changes in the acidity of rain over time, and how they correlate with changes in sulfur and nitrogen emissions. The chemistry of acid lakes is also described.

McCormick, John. *Acid Earth: The Politics of Acid Pollution.* 3d ed. London: Earthscan, 1997. A look at the causes and environmental consequences of acid rain and acid deposition, including efforts to curb the drift of acid-rain-causing gases across state and national boundaries. Appropriate for the high school reader. Illustrations, maps, index, and references.

Mohnen, Volker A. "The Challenge of Acid Rain." *Scientific American* 259 (August, 1988): 30-38. A discussion of the formation of acid rain and how it is possible to predict where acid deposition will occur from known sulfur emissions. The effects of acid rain on soils, forests, and lakes are covered. Ways of changing power-plant technology to reduce sulfur emissions are also evaluated. For a general audience.

National Research Council. *Acid Deposition: Atmospheric Processes in Eastern North America.* Washington, D.C.: National Academy Press, 1983. This text covers the relationships between emissions of acid-forming gases and their deposition as acid rain. Current and past research of the subject is summarized. The area covered in this article is particularly important in discussions of the United States-Canadian acid rain problem.

Rose, John, ed. *Acid Rain: Current Situation and Remedies.* Langhorne, Pa.: Gordon and Breach Science Publishers, 1994. This college-level textbook provides a clear description of the causes of acid rain, the techniques used to determine acid rain levels, and possible solutions to the acid rain and air pollution problems. A good introduction to the field. Illustrations, bibliographical references, and index.

Sommerville, Richard C. J. *The Forgiving Air: Understanding Environmental Change.* Berkeley: University of California Press, 1996. This thorough and easy-to-understand book focuses on the various consequences of air pollution, including the depletion of the ozone layer, climatic changes, the greenhouse effect, and acid rain. Intended as an introduction for the layperson. Bibliography and index.

Wyman, Richard L., ed. *Global Climate Change and Life on Earth.* New York: Chapman and Hall, 1991. A collection of articles on various aspects of climate changes and the effects of these changes.

AIR POLLUTION

Air pollution is generated from both natural and human-made sources. Natural sources include pollen from plants, gases and particulate matter from volcanoes, and windblown dust. Human-made sources include industrial and automobile emissions and airborne particles associated with human-induced abrasion.

PRINCIPAL TERMS

ACID RAIN: precipitation having relatively high levels of acidity

ATMOSPHERE: a mixture of gases surrounding the Earth

CARBON DIOXIDE: a natural component of the atmosphere; it is a good absorber of sensible heat energy

GREENHOUSE EFFECT: the environmental situation that results from the trapping of sensible heat energy in the atmosphere by atmospheric gases

INVERSION: an unusual atmospheric condition in which temperature in the troposphere increases with altitude

OXIDES OF NITROGEN: gases formed when molecular nitrogen is heated with air during combustion

OXIDES OF SULFUR: gases formed when fuels containing sulfur are burned

OZONE: a gas molecule composed of three atoms of oxygen; there is a zone in the stratosphere where ozone is highly concentrated

PHOTOCHEMICAL OXIDANTS: pollutants formed in air by primary pollutants undergoing a complex series of reactions

PHOTOCHEMICAL REACTION: a chemical reaction occurring in polluted air through the action of sunlight on pollutant gases to synthesize new gases

EARTH'S ATMOSPHERE

Air pollution results from the addition of gases, solids, and liquids to the atmosphere. The concentration of pollutants depends on prevailing atmospheric conditions as well as emission rates. Once pollutants are put into the atmosphere, it is impossible to control them to any significant degree. Thus emissions at the local level contribute to regional and global air pollution problems, such as smog and photochemical oxidants, acid precipitation, the depletion of the ozone layer, and global warming associated with the greenhouse effect. Although there are many air pollutants, the major ones are usually associated with burning, particularly the burning of coal and oil products. They are hydrocarbons, oxides of sulfur, oxides of nitrogen, carbon dioxide, carbon monoxide, photochemical oxidants, and particulate matter.

The atmosphere is a mixture of gases, aerosols, and particulate matter surrounding the Earth. The concentration of some of the gases in clean air is fairly constant both spatially and temporally. Consequently, these gases are referred to as stable or permanent gases. Nitrogen and oxygen, the two most abundant of the permanent gases, account for nearly 99 percent of the total atmosphere by volume: nitrogen 78 percent and oxygen 21 percent. The gases that experience noticeable temporal and spatial variations are termed variable gases. The two most abundant of these are water vapor and carbon dioxide. The average concentration of carbon dioxide is about 0.034 percent. It varies seasonally in response to the growth cycle of plants, daily in response to plant photosynthesis, and spatially in response to the burning of fossil fuels. Water vapor is also highly variable. Some variable gases have natural origins and tend to have relatively high concentrations in urban areas. They are methane, carbon monoxide, sulfur dioxide, nitrogen dioxide, ozone, ammonia, and hydrogen sulfide.

The atmosphere is stratified based on its vertical temperature gradient. From the surface up, the major layers are the troposphere, the stratosphere, the mesosphere, and the thermosphere. The troposphere contains the bulk of the atmo-

NATIONAL AMBIENT AIR QUALITY STANDARDS FOR CRITERIA POLLUTANTS

Pollutant	Averaging Time	Pollutant Level	Effects on Health
Carbon monoxide: colorless, odorless, tasteless gas; it is primarily the result of incomplete combustion; in urban areas the major sources are motor vehicle emissions and wood burning.	1-hour 8-hour	35 ppm 9 ppm	The body is deprived of oxygen; central nervous system affected; decreased exercise capacity; headaches; individuals suffering from angina, other cardiovascular disease; those with pulmonary disease, anemic persons, pregnant women and their unborn children are especially susceptible.
Ozone: highly reactive gas, the main component of smog.	1-hour 8-hour	0.120 ppm 0.080 ppm	Impaired mechanical function of the lungs; may induce respiratory symptoms in individuals with asthma, emphysema, or reduced lung function; decreased athletic performance; headache; potentially reduced immune system capacity; irritant to mucous membranes of eyes and throat.
Particulate matter {lt} 10 microns (PM10): tiny particles of solid or semisolid material found in the atmosphere.	24-hour Annual arithmetic mean	150 ug/m3 50 ug/m3	Reduced lung function; aggravation of respiratory ailments; long-term risk of increased cancer rates or development of respiratory problems.
Particulate matter {lt} 2.5 microns (PM2.5): fine particles of solid or semisolid material found in the atmosphere.	24-hour Annual arithmetic mean	65 ug/m3 15 up/m3	Same as PM10 above.
Lead: attached to inhalable particulate matter; primary source is motor vehicles that burn unleaded gasoline and re-entrainment of contaminated soil.	Calendar quarter	1.5 ug/m3	Impaired production of hemoglobin; intestinal cramps; peripheral nerve paralysis; anemia; severe fatigue.
Sulfur dioxide: colorless gas with a pungent odor.	3-hour 24-hour Annual arithmetic mean	0.5 ppm 0.14 ppm 0.03 ppm	Aggravation of respiratory tract and impairment of pulmonary functions; increased risk of asthma attacks.
Nitrogen dioxide: gas contributing to photochemical smog production and emitted from combustion sources.	Annual arithmetic mean	0.053 ppm	Increased respiratory problems; mild symptomatic effects in asthmatics; increased susceptibility to respiratory infections.

NOTES: ppm equals parts per million and ug/m3 equals micrograms per cubic meter.
SOURCE: United States Environmental Protection Agency (EPA); URL http://www.epa.gov.

spheric gases and, under normal conditions, is characterized by a fairly uniform temperature decline from the surface upward. The uppermost limit of the troposphere is called the tropopause, a transition zone between the troposphere and stratosphere where temperatures stabilize with increasing altitude. The troposphere extends up to about 10 kilometers. The next layer encountered is the stratosphere, which extends from about 12.5 kilometers up to about 45 kilometers above the surface. In its lower layer, the temperature gradient is somewhat stable. At an elevation of about 30 kilometers, however, the temperature starts to increase. Located within the stratosphere about 24 to 32 kilometers above the Earth's surface is a zone with a relatively high concentration of ozone, a form of oxygen with three atoms of oxygen. This zone is called the ozone layer. It is important because ozone is a good absorber of ultraviolet rays emitted by the Sun.

The two uppermost layers, the mesosphere and the thermosphere, have a distinctive temperature gradient. In the mesosphere, temperatures decline steadily with altitude, a condition that continues until its transition zone with the thermosphere, called the mesopause, is reached. At latitudes of 50 degrees north and higher, mesospheric clouds known as noctilucent clouds are sometimes seen during summer. These clouds may be anthropogenic in origin. The last layer, the thermosphere, slowly gives way to outer space and has no defined upper limit.

ATMOSPHERIC INVERSIONS AND SMOG

Vertical and horizontal mixing of air is necessary to dilute pollutants in the atmosphere. Under normal conditions, temperatures decline with altitude in the troposphere. This decline in temperature with altitude is referred to as the thermal or environmental lapse rate. The warmer air near the surface rises, mixes with the air above it, and is dispersed upward by winds. This dilution process is important in reducing the concentration of pollution near the surface. Conversely, the vertical mixing of air is inhibited when the temperature profile in the troposphere inverts, developing a temperature or thermal inversion. When an inversion exists, cooler air near the surface, which is heavier than the warmer air above it, does not rise. Pollutants then accumulate below the warmer air.

Conditions for temperature inversions develop when the Earth readily radiates heat energy from its surface on clear nights or when air subsides and warms adiabatically from compression. On cool, clear nights, the Earth readily radiates heat energy to space, causing the surface to cool. Air near the surface is, in turn, cooled by conduction, while the air above it is still fairly warm. This is referred to as a radiation inversion. Radiation inversions are common during the fall of the year and are usually short-lived, since the Sun heats the air near the surface when it rises, causing the inversion to dissipate as the day advances. The less frequent but more persistent subsidence-type inversions occur when air subsides in high-pressure systems or in valleys, where air descends along adjacent mountain slopes. These episodes may last for days, allowing pollutants to concentrate to excessive levels, causing eye irritation, respiratory distress, reduced visibility, corrosion of materials, and soiling of clothes.

The atmosphere has inherent self-cleaning mechanisms. Pollutants are removed from

Smog over New York City is visible in an inversion layer. (© William E. Ferguson)

the atmosphere through fallout from gravitational settling, through rainout in condensation processes, through washout as waterdrops and snowflakes fall to Earth, and through chemical conversion. Solar radiation, winds, and atmospheric moisture are important meteorological factors in these removal processes.

Chemical reactions between two or more substances in the atmosphere produce secondary pollutants—those created from other pollutants that are released directly from identifiable sources. Smog is a product of such reactions. Stability in the atmosphere that accompanies inversions provides favorable conditions for smog to develop. Smog is caused by chemical reactions between sulfur oxides, hydrocarbons, and oxides of nitrogen. Smog that is characterized by sulfur oxides is called sulfurous smog and is associated with burning fuels high in sulfur content. This type of smog is common in undeveloped countries. Photochemical smog develops when oxides of nitrogen and hydrocarbons undergo photochemical reactions and produce ozone and other chemical oxidizers. Sunlight catalyzes the reactions, and automobile exhaust is a primary source. This is the type of smog for which Los Angeles is noted.

ACID PRECIPITATION

While smog is a relatively localized phenomenon closely associated with urban areas, acid precipitation is more widespread. Its effects are found in national parks, agricultural regions, and forested areas as well as urban centers. Acid rain develops when oxides of sulfur and oxides of nitrogen combine with water vapor in the atmosphere to form sulfuric and nitric acids that fall back to Earth during the precipitation processes. Once released into the atmosphere, these compounds can travel hundreds of kilometers before returning to Earth in precipitation or as dry particulates. They can travel from 1,000 to 2,000 kilometers over three to five days. This long-range transport allows time for chemical reactions to convert pollutant gases into components of acid precipitation. Evidence suggests that the pH values in precipitation have been dropping, becoming more acidic, for some years. Wet precipitation is not the only way pollutants find their way to the surface. Diffusion and settling enable acidic gases and particles to find their way to the ground even under dry con-

ditions. It is now widely accepted that both wet and dry deposition can be traced to human activity.

Much evidence has been gathered indicating the damaging effects of acid precipitation. Some of the effects are damage to wildlife in lakes and rivers, reduction of forest productivity, damage to agricultural crops, and deterioration of human-made materials. Acid precipitation is also suspected of helping release heavy metals from soils and pipelines into drinking water supplies. It has different impacts on different ecological systems and is most damaging to aquatic ecosystems. Acidity in precipitation at a given time depends not only on the quality of pollutants being produced but also on the prevailing and immediate atmospheric conditions. Stagnant air, resulting from upper-level inversions, tends to cause higher levels of acidity. Furthermore, prevailing and local atmospheric systems are associated with the spread of acid precipitation over broader areas. Higher smokestacks simply disperse the pollutants over larger areas, increasing their residence time in the atmosphere.

OZONE DEPLETION AND GLOBAL WARMING

It is now realized that the impact of air pollution is more far-reaching than the troposphere. Evidence suggests that some pollutants are making their way up to the stratosphere and are causing the ozone layer to break down. Even though ozone constitutes a very small portion of the atmosphere (about one part per million), it absorbs ultraviolet rays from the Sun, preventing them from reaching the Earth. Research findings suggest that there has been a breakdown of the ozone shield in the Antarctic, where a hole has been identified in the ozone layer; the ozone layer also appears to be getting thinner over the Arctic. While early laboratory studies showed that oxides of nitrogen could attack ozone, attention later focused on chlorofluorocarbons (CFCs) as responsible for the decline in ozone. These compounds were used as coolants in refrigerators and air conditioners, propellants in aerosol sprays, agents for producing foam, and cleansers for electronic products. Being inert gases, they do not degrade readily in the troposphere. These substances eventually make their way into the stratosphere. Laboratory studies have shown that when the CFC molecules come in contact with ozone, they break down into more reactive gases, such as chlorine. Since these gases tend to linger in the atmosphere for many years, it is believed that even if people were to discontinue using CFCs, the ozone layer would continue to disintegrate for years to come.

Further evidence suggests that CFCs not only destroy the ozone but also trap heat energy radiated from the ground and contribute to heating the atmosphere. The trapping of sensible heat energy in the atmosphere by gases is called the greenhouse effect. The most important gas contributing to the greenhouse effect, however, is not a chlorofluorocarbon but carbon dioxide. Carbon dioxide moves in a continuous cycle throughout the environment. It provides a link between the organic and inorganic components of the environment. Reacting with water and solar energy in plants, it forms chemical energy that is passed through the food chain.

Columns at the famous Parthenon in Greece show black crusts caused by air pollution. (U.S. Geological Survey)

Carbon dioxide is given off by plants and animals to the atmosphere during respiration. When plant and animal remains decay, carbon dioxide is passed to the atmosphere and hydrosphere through the weathering of rocks, volcanic activity, and the burning of fossil fuels. When fossil fuels are burned, the natural processes are short-circuited, releasing large amounts of carbon dioxide directly into the atmosphere.

About 0.03 percent of the total atmosphere is carbon dioxide. Molecules of carbon dioxide in the atmosphere absorb infrared radiation, acting like a greenhouse. While it is transparent to shortwave radiation from the Sun, carbon dioxide absorbs strongly in the sensible heat or longwave radiation band. It is hypothesized that an increase in atmospheric carbon dioxide causes a decrease in outgoing longwave radiation and thus an increase in the atmospheric temperature.

The consequences of rising global temperatures will greatly alter the Earth's surface. As the atmosphere warms, polar ice and glaciers will begin to melt, and the rising oceans could drown many of the world's coastal regions, devastating low-lying countries. As shorelines rise, saltwater intrusion will contaminate drinking-water supplies of many cities worldwide. Agricultural regions of the middle-latitude countries will migrate farther northward, increasing the length of the growing seasons in Canada and Russia.

STUDY OF AIR POLLUTION

Methods of studying air pollution include controlled laboratory experiments, simulations in fluid-modeling facilities, computer simulations, and mathematical modeling. Controlled laboratory experiments are conducted in laboratories where gases are mixed to determine how they react. Laboratory experiments usually do not provide the definitive answer to what is actually occurring in the ambient environment because many variables cannot be replicated. These studies suggest what should be further studied and monitored in the natural environment. Some laboratory studies have simulated atmospheric conditions in a controlled environment, such as a biosphere where the impact of pollution on plants can be determined by introducing the pollutants at various levels. Simulation studies may also include gathering data from fluid-modeling facilities, where the environment is replicated using miniature models and atmospheric conditions are controlled. These studies often contribute to an understanding of the dispersion and deposition of air pollutants.

Monitoring the atmosphere is an essential component of air pollution studies involving computer simulations and mathematical modeling. These types of studies rely largely on data sources or values from the ambient environment and are constrained by difficulties of measuring ambient levels. Sometimes, vessels containing samples of air are collected and returned to the laboratory for analysis, but continuous monitoring devices that are placed in the ambient environment are more common.

Many of the monitoring devices involve a colorimetric or photometric technique. Air to be analyzed is pumped through a chamber and allowed to react with chemicals. The reaction product is then pumped through a chamber and allowed to react with chemicals. This reaction product is then pumped into a photometer, where the concentration of light-absorbing substance is indicated by the light intensity that reaches the photometer. Particulate matter is measured by fairly simple collectors that may involve adhesive coated paper or pumping air through filters and measuring the increase in weight resulting from the trapped particles. Another method involves passing a known volume of air through filter paper and measuring the intensity of light passing through it. The intensity of light indicates the scattering and absorptive properties of aerosols; it is expressed as a coefficient. Instruments may be located at the surface, mounted on airplanes, or allowed to ascend in balloons. Acidity in precipitation is determined by standard measures of acidity using a pH indicator.

SIGNIFICANCE

Industry continues to downplay the threat of pollution to the atmosphere and disputes the extent of damage caused by pollution. Yet as greater amounts of pollutants are released into the atmosphere, it becomes increasingly difficult to control their levels and reverse any resulting damage. The depletion of the ozone layer, acid rain, and increasing levels of smog are all very real and threatening manifestations of air pollution. Efforts to protect the atmosphere must be made on a worldwide basis, since gases cannot be confined to political boundaries. Air pollution may be lowered by reducing emissions or by extracting pollutants from the atmosphere through natural means. Thus any plans to reduce air pollution should center on one or both of these approaches.

Jasper L. Harris

CROSS-REFERENCES

Acid Rain and Acid Deposition, 1802; Atmosphere's Evolution, 1816; Atmosphere's Global Circulation, 1823; Atmosphere's Structure and Thermodynamics, 1828; Atmospheric Inversion Layers, 1836; Auroras, 1841; Carbonates, 1182; Cosmic Rays and Background Radiation, 1846; Earth-Sun Relations, 1851; Fallout, 1857; Freshwater Chemistry, 405; Global Warming, 1862; Greenhouse Effect, 1867; Hydrologic Cycle, 2045; Nuclear Winter, 1874; Ozone Depletion and Ozone Holes, 1879; Precipitation, 2050; Radon Gas, 1886; Rain Forests and the Atmosphere, 1890; Soil Chemistry, 1509; Surface Water, 2066; Weathering and Erosion, 2380.

BIBLIOGRAPHY

Bandy, A. R., ed. *The Chemistry of the Atmosphere: Oxidents and Oxidation in the Earth's Atmosphere.* Cambridge, England: Royal Society of Chemistry, 1995. This collection of essays and lectures first presented at the Priestley Conference in association with the Royal Society of Chemistry and the American Chemical Society at Bucknell University. The articles cover such topics as atmospheric chemistry, and the ozone layer, and the causes and remedies of air pollution. These papers can be somewhat technical at times, making them suitable for the reader with some backgound in the subject.

Brimblecombe, Peter. *Air Composition and Chemistry.* 2d ed. New York: Cambridge University Press, 1996. This college-level textbook provides a thorough account of atmospheric chemistry and the techniques and protocol involved in determining the makeup and properties of air. Although some background in chemistry and mathematics may be useful, the book as a whole is written in an understandable manner that will allow the layperson to obtain a grasp of the subject. Illustrations, maps, bibliography, and index.

Brodine, Virginia, ed. *Air Pollution.* New York: Harcourt Brace Jovanovich, 1972. A review of the social issues relevant to air pollution. It also addresses the ways in which air pollution affects the atmosphere and human health. An excellent reference for the general reader.

Commission on College Geographers. *Air Pollution.* Washington, D.C.: Association of American Geographers, 1968. A resource document designed to assist in teaching the fundamental concepts of air pollution.

Hopler, Paul. *Atmosphere: Weather, Climate, and Pollution.* New York: Cambridge University Press, 1994. This brief book offers a wonderful introduction to the study of the Earth's atmosphere and its components. Chapters explain the causes and effects of global warming, ozone depletion, acid rain, and climatic change. Illustrations, color maps, and index.

Ostmann, Robert. *Acid Rain: A Plague upon the Waters.* Minneapolis, Minn.: Dillon Press, 1982. This book provides insight into the political history and development of the acid rain problem. The effects of acid rain on society are addressed from the perspective of a concerned citizen.

Shaw, Robert W. "Air Pollution by Particles." *Scientific American* 257 (August, 1978): 96-103. An excellent article on tracing and measuring air pollution sources. A good source for those interested in air pollution monitoring processes. Applications of spectroscopy as a research tool in air pollution are addressed.

Stolarski, Richard S. "The Antarctic Ozone Hole." *Scientific American* 258 (January, 1988): 30-36. An excellent review of research efforts pertaining to ozone depletion in the Antarctic. A very understandable explanation of the chemical reactions associated with CFCs and ozone depletion. An argument is also presented for natural processes as a major contributor to the ozone hole.

Williamson, Samuel J. *Fundamentals of Air Pollution.* Reading, Mass.: Addison-Wesley, 1973. A good explanation of the physical aspects of air pollution. The approach is interdisciplinary, drawing from meteorology, chemistry, physics, engineering, medicine, and the social sciences. Some attention is given to the political aspects of air pollution problems.

Wyman, Richard L., ed. *Global Climate Change and Life on Earth.* New York: Chapman and Hall, 1991. A collection of articles on various aspects of climate change and the effects of these changes.

ATMOSPHERE'S EVOLUTION

The chemical composition of the atmosphere has changed significantly over the 4.6-billion-year history of the Earth. The composition of the atmosphere has been controlled by a number of processes, including the "outgassing" of gases or volatiles originally trapped in the Earth's interior during its formation; the geochemical cycling of carbon, nitrogen, hydrogen, and oxygen compounds between the surface, the ocean, and the atmosphere; and the origin and evolution of life.

PRINCIPAL TERMS

CHEMICAL EVOLUTION: the synthesis of amino acids and other complex organic molecules—the precursors of living systems—by the action of atmospheric lightning and solar ultraviolet radiation on atmospheric gases

PHOTOSYNTHESIS: the utilization of carbon dioxide and water vapor by chlorophyll-containing organisms in the presence of sunlight to metabolically produce carbohydrates used by the plant for food; oxygen is a by-product in the photosynthesis process

PRIMORDIAL SOLAR NEBULA: an interstellar cloud of gas and dust that condensed under gravity to form the Sun, the Moon, Earth, the other planets and their satellites, asteroids, comets, and meteors some 4.6 billion years ago.

SOLAR ULTRAVIOLET RADIATION: biologically lethal solar radiation in the spectral interval between approximately 0.1 and 0.3 micron (1 micron = 0.0001 centimeter)

VOLATILE OUTGASSING: the release of the gases or volatiles, such as argon, water vapor, carbon dioxide, and nitrogen, trapped within the Earth's interior during its formation

VOLATILE OUTGASSING

Some 4.6 billion years ago, a cloud of interstellar gas and dust, called the primordial solar nebula, began to condense under the influence of gravity. This condensation led to the formation of the Sun, the Moon, the Earth, the other planets and their satellites, asteroids, meteors, and comets. The primordial solar nebula was composed almost entirely of hydrogen gas, with a smaller amount of helium, still smaller amounts of carbon, nitrogen, and oxygen, and still smaller amounts of the rest of the elements of the periodic table. About the time that the newly formed Earth reached its approximate present mass, gases that were originally trapped in the Earth's interior were released through the surface, forming a gravitationally bound atmosphere. (It is believed that the atmospheres of the other terrestrial planets, Mars and Venus, also formed in this manner.) The release of these gases is called volatile outgassing. The period of extensive volatile outgassing may have lasted for many tens of millions of years. The outgassed volatiles or gases had roughly the same chemical composition as do present-day volcanic emissions: 80 percent water vapor by volume, 10 percent carbon dioxide by volume, 5 percent sulfur dioxide by volume, 1 percent nitrogen by volume, and smaller amounts of hydrogen, carbon monoxide, sulfur, chlorine, and argon.

The water vapor that outgassed from the interior soon reached its saturation point, which is controlled by the atmospheric temperature and pressure. Once the saturation point was reached, the atmosphere could not hold any additional gaseous water vapor. Any new outgassed water vapor that entered the atmosphere would have precipitated out of the atmosphere in the form of liquid water. The equivalent of several cubic kilometers of liquid water released from the Earth's interior in gaseous form precipitated out of the atmosphere and formed the Earth's vast oceans. Only small amounts of water vapor remained in the atmosphere—ranging from a fraction of a percent to several percent by volume, depending on atmospheric temperature, season, and latitude.

The outgassed atmospheric carbon dioxide, being very water soluble, readily dissolved into the newly formed oceans and formed carbonic acid.

In the oceans, carbonic acid formed ions of hydrogen, bicarbonate, and carbonate. The carbonate ions reacted with ions of calcium and magnesium in the ocean water, forming carbonate rocks, which precipitated out of the ocean and accumulated as seafloor carbonate sediments. Most of the outgassed atmospheric carbon dioxide formed carbonates, leaving only trace amounts of gaseous carbon dioxide in the atmosphere (about 0.035 percent by volume). Sulfur dioxide, the third most abundant component of volatile outgassing, was chemically transformed into other sulfur compounds and sulfates in the atmosphere. Eventually, the sulfates formed atmospheric aerosols and diffused out of the atmosphere onto the surface.

The fourth most abundant outgassed compound, nitrogen, is chemically inert in the atmosphere and thus was not chemically transformed, as was sulfur dioxide. Unlike carbon dioxide, nitrogen is relatively insoluble in water and, unlike water vapor, does not condense out of the atmosphere. For these reasons, nitrogen built up in the atmosphere to become its major constituent (78.08 percent by volume). Therefore, outgassed volatiles led to the formation of the Earth's atmosphere, oceans, and carbonate rocks.

CHEMICAL EVOLUTION

The molecules of nitrogen, carbon dioxide, and water vapor in the early atmosphere were acted upon by solar ultraviolet radiation and atmospheric lightning. In the process, molecules of formaldehyde and hydrogen cyanide were chemically synthesized in the early atmosphere. These molecules precipitated and diffused out of the atmosphere into the oceans. In the oceans, formaldehyde and hydrogen cyanide entered into chemical reactions, called polymerization reactions, which eventually led to the chemical synthesis of amino acids, the building blocks of living systems. The synthesis of amino acids from nitrogen, carbon dioxide, and water vapor in the atmosphere is called chemical evolution. Chemical evolution preceded and provided the material for biological evolution.

For many years, it was thought that the early atmosphere was composed of ammonia, methane, and hydrogen rather than of carbon dioxide, nitrogen, and water vapor. Experiments show, however, that ammonia and methane are chemically unstable and are readily destroyed by both solar ultraviolet radiation and chemical reaction with the hydroxyl radical, which is formed from water vapor. In addition, ammonia is very water soluble and is readily removed from the atmosphere by precipitation. Hydrogen, the lightest element, is readily lost from a planet by gravitational escape. Thus, an early atmosphere composed of methane, ammonia, and hydrogen would be very short lived, unless these gases were produced at a rate comparable to their destruction or loss rates. Today, methane and ammonia are very minor components of the atmosphere—methane at a concentration of 1.7 parts per million by volume and ammonia at a concentration of 1 part per billion by volume. Both gases are produced by microbial activity at the Earth's surface, and methane is released during coal mining and oil production. Clearly, microbial activity and microbes were nonexistent during the earliest history of the planet. The atmospheres of the outer gas giant planets—Jupiter, Saturn, Uranus, and Neptune—all contain appreciable quantities of hydrogen, methane, and ammonia. It is believed that the atmospheres of these planets, unlike the atmospheres of the terrestrial planets—Earth, Venus, and Mars—are captured remnants of the primordial solar nebula. Because of the outer planets' great distance from the Sun and their very low temperatures, hydrogen, methane, and ammonia are stable and long-lived constituents of their atmospheres. This is not true of hydrogen, methane, and ammonia in the Earth's atmosphere.

Some have suggested that at the time of its formation, the Earth may have also captured a remnant of the primordial solar nebula as its very first atmosphere. Such a captured primordial solar nebula atmosphere would have been composed of mostly hydrogen (about 90 percent) and helium (about 10 percent), the two major elements of the nebula. Even if such an atmosphere had surrounded the very young Earth, it would have been very short lived. As the young Sun went through its T Tauri phase of evolution, very strong solar winds (the supersonic flow of protons and electrons from the Sun) would have quickly dissipated this remnant atmosphere. In addition, there is no geochemical evidence to suggest that the early Earth ever possessed a primordial solar nebula remnant atmosphere.

PLANETARY ATMOSPHERES: COMPARATIVE DATA

	Mercury	*Venus*	*Earth*	*Mars*
Surface pressure (bars)	~10^{-15} bars	92 bars	1.014 bars	6.1-9 millibars
Surface density (kg/m³)	—	~65	1.217	~0.020
Avg. temperature (Kelvin)	440	737	288	~210
Scale height (kilometers)	—	15.9	8.5	11.1
Wind speeds (meters/second)	—	0.3-1.0	up to 100	2-30
Composition[a]				
Ammonia	—	—	—	—
Argon	tr?	70 ppm	9430 ppm	1.6%
Carbon dioxide	tr?	96.5%	350 ppm	95.32%
Carbon monoxide	—	17 ppm	—	—
Ethane	—	—	—	—
Helium	6%	12 ppm	5.24 ppm	—
Hydrogen	22%	—	0.55 ppm	—
Hydrogen chloride	—	tr	—	—
Hydrogen deuteride	—	—	—	0.85 ppm
Hydrogen fluoride	—	tr	—	—
Krytpon	tr?	—	1.14 ppm	0.3 ppm
Methane	—	—	—	1.7 ppm
Neon	tr?	7 ppm	18.18 ppm	2.5 ppm
Nitrogen	tr?	3.5%	78.084%	2.7%
Nitrogen oxide	—	—	—	100 ppm
Oxygen	42%	—	20.946	0.13%
Potassium	0.5%	—	—	—
Sodium	29%	—	—	—
Sulfur dioxide	—	150 ppm	—	—
Water	tr?	20 ppm	1%	210 ppm
Xenon	tr?	—	0.08 ppm	—

NOTES:

a. Composition: % = percentages; ppm = parts per million; tr = trace amounts.

b. Percentage composition of Pluto's atmosphere unknown; it is surmised to be mainly nitrogen and methane.

SOURCE: Data are from the National Space Science Data Center, NASA/Goddard Space Flight Center. URL: http://nssdc.gsfc.nasa.gov/planetary/factsheet.

EVOLUTION OF ATMOSPHERIC OXYGEN

There is microfossil evidence for the existence of fairly advanced anaerobic microbial living systems on the Earth by about 3.8 billion years ago. Photosynthesis evolved in one or more of these early microbial species. Through photosynthesis, the organism utilizes water vapor and carbon dioxide in the presence of sunlight and chlorophyll to form carbohydrates, used by the organism for food. During photosynthesis, oxygen is given off as a metabolic by-product. The production of oxygen by photosynthesis was a major event on the Earth and transformed the composition and chemistry of the early atmosphere. As a result of photosynthetic production, oxygen built up to become the second most abundant constituent of the atmosphere (20.90 percent by volume). It has been estimated that atmospheric oxygen reached only 1 percent of its present atmospheric level 2 billion years ago, 10 percent of its present atmospheric level about 550 million years ago (at the beginning of the Paleozoic), and its present atmospheric level as early as 400 million years ago.

The evolution of atmospheric oxygen had important implications for the evolution of life. The presence and buildup of oxygen led to the evolution of respiration and aerobic organisms. Accompanying and directly controlled by the buildup of atmospheric oxygen were the origin and evolution of atmospheric ozone, which is chemically formed

PLANETARY ATMOSPHERES: COMPARATIVE DATA

Jupiter	*Saturn*	*Uranus*	*Neptune*	*Pluto*[b]	
{gt}100 bars	{gt}100	{gt}100	{gt}100 bars	~3 microbar	Surface pressure (bars)
~0.16	~0.19	~0.42	~0.45	—	Surface density (kg/m^3)
~129	~97	~58	~58	~50	Avg. temperature (Kelvin)
27	59.5	27.7	~20	~60	Scale height (kilometers)
up to 150	up to 400	up to 200	up to 200		Wind speeds (meters/second)
					Composition[a]
260 ppm	125 ppm	—	—	—	Ammonia
—	—	—	—	—	Argon
—	—	—	—	—	Carbon dioxide
—	—	—	—	—	Carbon monoxide
5.8 ppm	7 ppm	—	1.5 ppm	—	Ethane
10.2%	3.25%	15.2%	19%	—	Helium
89.8%	96.3%	82.5%	80%	—	Hydrogen
—	—	—	—	—	Hydrogen chloride
28 ppm	110 ppm	~148 ppm	192 ppm	—	Hydrogen deuteride
—	—	—	—	—	Hydrogen fluoride
—	—	—	—	—	Krytpon
3000 ppm	4500 ppm	~2.3%	1.5%	✓	Methane
—	—	—	—	—	Neon
—	—	—	—	✓	Nitrogen
—	—	—	—	—	Nitrogen oxide
—	—	—	—	—	Oxygen
—	—	—	—	—	Potassium
—	—	—	—	—	Sodium
—	—	—	—	—	Sulfur dioxide
—	~4 ppm	—	—	—	Water
—	—	—	—	—	Xenon

NOTES:

a. Composition: % = percentages; ppm = parts per million; tr = trace amounts.

b. Percentage composition of Pluto's atmosphere unknown; it is surmised to be mainly nitrogen and methane.

SOURCE: Data are from the National Space Science Data Center, NASA/Goddard Space Flight Center. URL: http://nssdc.gsfc.nasa.gov/planetary/factsheet.

from oxygen. The evolution of atmospheric ozone resulted in the shielding of the Earth's surface from biologically lethal solar ultraviolet rays between 0.2 and 0.3 micron. The development of the atmospheric ozone layer and its accompanying shielding of the Earth's surface permitted early life to leave the safety of the oceans and go ashore for the first time in the history of the planet. Prior to the evolution of the atmospheric ozone layer, early life was restricted to a depth of several meters below the ocean surface. At this depth, the ocean water offered shielding from solar ultraviolet radiation. Theoretical computer calculations indicate that atmospheric ozone provided sufficient shielding from biologically lethal ultraviolet radiation for the colonization of the land once oxygen reached about one-tenth of its present atmospheric level.

VENUS AND MARS

Calculations indicate that the atmospheres of Venus and Mars also evolved as a consequence of the volatile outgassing of the same gases that led to the formation of Earth's atmosphere—water vapor, carbon dioxide, and nitrogen. In the case of Venus and Mars, however, the outgassed water vapor never existed in the form of liquid water on the surfaces of these planets. Because of Venus's closer distance to the Sun (108 million kilometers versus 150 million kilometers for Earth), its lower atmosphere was too hot to permit the outgassed water vapor to condense out of the atmosphere. Thus, the outgassed water vapor remained in gaseous form in the atmosphere and, over geological time, was broken apart by solar ultraviolet radiation to form hydrogen and oxygen. The very light hydrogen gas quickly escaped from the atmosphere of Venus, and the heavier oxygen combined with surface minerals to form a highly oxidized surface. In the absence of liquid water on the surface of Venus, the outgassed carbon dioxide remained in the atmosphere and built up to become the overwhelming constituent of the atmosphere of Venus (about 96 percent by volume). The outgassed nitrogen accumulated to make up about 4 percent by volume of the atmosphere of Venus. The carbon dioxide and nitrogen atmosphere of Venus is massive—its atmospheric surface pressure is about 90 atmospheres (the surface pressure of Earth's atmosphere is only 1 atmosphere). If the outgassed carbon dioxide in Earth's atmosphere did not leave via dissolution in the oceans and carbonate formation, its surface atmospheric pressure would be about 70 atmospheres, with carbon dioxide comprising about 98 to 99 percent of the atmosphere and nitrogen about 1 to 2 percent. Thus, the atmosphere of Earth would closely resemble that of Venus. The carbon dioxide-rich atmosphere of Venus causes a very significant greenhouse temperature enhancement, giving the surface of Venus a temperature of about 750 Kelvins, which is hot enough to melt lead. The surface temperature of Earth is only about 288 Kelvins.

Like Venus, Mars has an atmosphere composed primarily of carbon dioxide (about 95 percent by volume) and nitrogen (about 3 percent by volume). The total atmospheric surface pressure of Mars, however, is only about 7 millibars (1 atmosphere is equivalent to 1,013 millibars). There may be very large quantities of outgassed water in the form of ice or frost below the surface of Mars. In the absence of liquid water, the outgassed carbon dioxide remained in the atmosphere. The smaller mass of the atmosphere of Mars compared to the atmospheres of Venus and Earth may be attributable to the smaller mass of Mars and, therefore, the smaller mass available for the trapping of gases in the interior of Mars during its formation. In addition, the amounts of gases trapped in the interiors of Venus, Earth, and Mars during their formation apparently decreased with increasing distance from the Sun. Venus appears to have trapped the greatest amounts of gases and was the most volatile-rich planet. Earth trapped the next greatest amounts, and Mars trapped the smallest amounts.

STUDY OF EARTH'S ATMOSPHERE

Information about the origin, early history, and evolution of the Earth's atmosphere comes from a variety of sources. Information on the origin of Earth and other planets is based on theoretical computer simulations. These computer models simulate the collapse of the primordial solar nebula and the formation of the planets. Astronomical observations of what appears to be the collapse of interstellar gas clouds and the possible formation of planetary systems have provided new insights into the computer modeling of this phenomenon. Information about the origin, early history, and evolution of the atmosphere is based on theoretical computer models of volatile outgassing, the geochemical cycling of the outgassed volatiles, and the photochemistry/chemistry of the outgassed volatiles. The process of chemical evolution, which led to the synthesis of organic molecules of increasing complexity, the precursors of the first living systems on the early Earth, is studied in laboratory experiments. In laboratory experiments on chemical evolution, mixtures of gases simulating the Earth's early atmosphere are energized by solar ultraviolet radiation and atmospheric lightning. The resulting products are analyzed by chemical techniques. A key parameter affecting atmospheric photochemical reactions, chemical evolution, and the origin of life was the flux of solar ultraviolet radiation incident on the early Earth. Astronomical measurements of the ultraviolet emissions from young sunlike stars have provided important information about ultraviolet emissions from the Sun during the very early history of the atmosphere.

Geological and paleontological studies of the oldest rocks and the earliest fossil records have provided important information on the evolution of the atmosphere and the transition from an oxygen-deficient to an oxygen-sufficient atmosphere. Studies of the biogeochemical cycling of the elements have provided important insights into the later evolution of the atmosphere. Thus studies of the origin and evolution of the atmosphere are based on a broad cross section of science, involving astronomy, geology, geochemistry, geophysics, and biology as well as atmospheric chemistry.

SIGNIFICANCE

Studies of the origin and evolution of Earth's atmosphere have provided new insights into the processes and parameters responsible for global change. Understanding the history of the atmosphere provides better understanding of its future. Today, several global environment changes are of national and international concern, including the depletion of ozone in the stratosphere and global warming caused by the buildup of greenhouse gases. The study of the evolution of the atmosphere has provided new insights into the biogeochemical cycling of elements between the atmosphere, biosphere, land, and ocean. Understanding this cycling is a key to understanding environmental problems. Studies of the origin and evolution of the atmosphere have also provided new insights into the origin of life and the possibility of life outside the Earth.

Joel S. Levine

CROSS-REFERENCES

Acid Rain and Acid Deposition, 1802; Air Pollution, 1809; Atmosphere's Global Circulation, 1823; Atmosphere's Structure and Thermodynamics, 1828; Atmospheric Inversion Layers, 1836; Auroras, 1841; Climate, 1902; Cosmic Rays and Background Radiation, 1846; Earth-Sun Relations, 1851; Fallout, 1857; Global Warming, 1862; Greenhouse Effect, 1867; Nuclear Winter, 1874; Ozone Depletion and Ozone Holes, 1879; Radon Gas, 1886; Rain Forests and the Atmosphere, 1890.

BIBLIOGRAPHY

Bandy, A. R., ed. *The Chemistry of the Atmosphere: Oxidents and Oxidation in the Earth's Atmosphere.* Cambridge, England: Royal Society of Chemistry, 1995. This collection of essays and lectures first presented at the Priestely Conference in association with the Royal Society of Chemistry and the American Chemical Society at Bucknell University. The articles cover such topics as atmospheric chemistry, and the ozone layer, and the causes and remedies of air pollution. These papers can be somewhat technical at times, making them suitable for the reader with some backgound in the subject.

Brimblecombe, Peter. *Air Composition and Chemistry.* 2d ed. New York: Cambridge University Press, 1996. This college-level textbook provides a thorough account of atmospheric chemistry and the techniques and protocol involved in determining the make-up and properties of air. Although some background in chemistry and mathematics may be useful, the book as a whole is written in an understandable manner that will allow the layperson to obtain a grasp of the subject. Illustrations, maps, bibliography, and index.

Cloud, Preston. *Cosmos, Earth, and Man: A Short History of the Universe.* New Haven, Conn.: Yale University Press, 1978. A very readable, nontechnical account of cosmic evolution covering the evolution of stars, the Earth, life, and humankind. The volume defines various scientific terms such as elementary particles, chemical bonding, isotopes, periodic table, and mass. The author assumes that the reader does not have a scientific background, only an interest in our cosmic roots.

Henderson-Sellers, A. *The Origin and Evolution of Planetary Atmospheres.* Bristol, England: Adam Hilger, 1983. A technical treatment of the variation of the climate of the Earth over geological time and the processes and parameters that controlled it. The chapters in the book include the mechanisms for long-term climate change, the atmospheres of other planets, planetary climatology on shorter time scales, and the stability of planetary environments.

Holland, H. D. *The Chemical Evolution of the Atmosphere and Oceans.* Princeton, N.J.: Princeton University Press, 1984. A comprehensive and technical treatment of the geochemical cycling of elements over geological time and the coupling between the atmosphere, ocean, and surface. The book covers the origin of the solar system, the release and recycling of volatiles, the chemistry of the early atmosphere and ocean, the acid-base balance of the atmosphere-ocean-crust system, and carbonates and clays.

Levine, Joel S., ed. *The Photochemistry of Atmospheres: Earth, the Other Planets, and Comets.* Orlando, Fla.: Academic Press, 1985. A series of review papers dealing with the origin and evolution of the atmosphere, the origin of life, the atmospheres of Earth and other planets, and climate. The book contrasts the origin, evolution, composition, and chemistry of Earth's atmosphere with the atmospheres of the other planets. It contains two appendices that summarize all atmospheric photochemical and chemical processes.

Lewis, John S., and Ronald G. Prinn. *Planets and Their Atmospheres: Origin and Evolution.* New York: Academic Press, 1983. A comprehensive, textbook treatment of the formation of the planets and their atmospheres. This monograph begins with a detailed account of the origin and evolution of solid planets via coalescence and accretion in the primordial solar nebula and then discusses the surface geology and atmospheric composition of each planet.

Schopf, J. William, ed. *Earth's Earliest Biosphere: Its Origin and Evolution.* Princeton, N.J.: Princeton University Press, 1983. A comprehensive group of papers on such subjects as the early Earth, the oldest rocks, the origin of life, early life, and microfossils. Chapters include those on the oldest known rock record, prebiotic organic syntheses and the origin of life, Precambrian organic geochemistry, the transition from fermentation to anoxygenic photosynthesis, the development of an aerobic environment, and early microfossils. Very technical.

Smith, David E., et al. "The Global Topography of Mars and Implications for Surface Evolution." *Science* 284 (1995). The authors discuss the Martian ice caps and the presence of water ice on Mars.

Yung, Yuk Ling, and William B. DeMore. *Photochemistry of Planetary Atmospheres.* New York: Oxford University Press, 1999. This college-level text examines the atmospheric chemistry of all of the planets in the solar system, with a focus on photochemical processes. Intended for the advanced reader. Illustrations, bibliography, and index.

ATMOSPHERE'S GLOBAL CIRCULATION

The general circulation of Earth's atmosphere involves the large-scale movements of significant portions of that atmosphere. Variations in surface temperatures produce pressure gradients that combine with Coriolis force to circulate most of the air in the atmosphere. This involves the Hadley circulation, which moves air in the Northern and Southern Hemispheres in three huge convection cells each, and the Walker circulation along the equatorial belt, which produces El Niño when it oscillates.

PRINCIPAL TERMS

ANTICYCLONE: a general term for a high-pressure weather system that rotates clockwise in the Northern hemisphere and counterclockwise in the Southern hemisphere

CORIOLIS FORCE: a non-Newtonian force acting on a rotating coordinate system; on the Earth, this causes objects moving in the Northern Hemisphere to be deflected toward the right and objects moving in the Southern Hemisphere to be deflected toward the left

CYCLONE: a general term for a low-pressure weather system that rotates counterclockwise in the Northern Hemisphere and clockwise in the Southern Hemisphere

GEOSTROPHIC WIND: a wind resulting from the balance between a pressure gradient force and Coriolis force; the flow produces jet streams and is perpendicular to the pressure gradient force and the Coriolis force

PRESSURE GRADIENT FORCE: a wind producing force caused by a difference in pressure between two different locations

DRIVING FORCES

The general circulation of the atmosphere operates as a heat engine driven by the uneven distribution of solar energy on the planet. Atmospheric circulation transfers some of this energy from regions where it is abundant to regions where it is scarce. This energy transfer reduces the difference in surface temperature between the equatorial regions and the polar regions, between the oceans and the continents, and between continental interiors and coastal regions. The existence of an atmosphere, with its water vapor, carbon dioxide, and other greenhouse gases, keeps the average temperature at the surface of the Earth considerably warmer than it would otherwise be. The motions of this atmosphere cool the warmer regions and warm the cooler regions, smoothing out the extremes of temperatures.

The Sun is so far from Earth that its rays of light can be thought of as being parallel when they strike the planet. Because Earth is a sphere, the areas warmed by this light vary with latitude. During the equinoxes, the Sun is directly over the equator. A beam of sunlight with a square cross section, 1 meter on each side, will illuminate 1 square meter of Earth's surface at the equator. However, at a latitude of 45 degrees north (near Ottawa, Ontario, for instance), the same beam of sunlight will be spread out over 1.4 square meters. At a latitude of 60 degrees north (near Anchorage, Alaska), it will be spread out over 1.7 square meters.

The 23.5-degree tilt of Earth's axis shifts all of this during the course of a year. At the summer solstice, 45 degrees north has its 1 square meter of sunlight spread out over 1.07 square meters, and 60 degrees north has its spread out over 1.24 square meters. At the winter solstice, 45 degrees north has its 1 square meter of sunlight spread out over 2.73 square meters, while 60 degrees north has its spread out over 8.83 meters. The intensity of solar radiation at these two latitudes differs by a factor of 1.16 in the summer and 1.22 in spring and fall, but 3.23 in the winter. Other choices for latitudes would yield similar results. This helps to explain why the temperature contrasts between northern and southern states in the United States is so much greater during the winter months than during the summer ones. It also explains some of the seasonal differences in general atmospheric circulation.

MAJOR CONVECTION CELLS

Air above a warm surface is heated and expands. The expanded air is less dense than the air surrounding it, and it will rise, much like a hot air balloon. The rising air displaces air above it, pushing that air to the north or south with a pressure gradient force. To understand how temperature gradients at the surface produce pressure gradient forces aloft, consider two adjacent columns of air that initially contain the same amount of air, one of which is warmer than the other. The warmer one expands and reaches higher above the surface. Because the mass of air in both columns is initially the same, the air pressure at the surface is also the same beneath both columns. Next consider the air pressure halfway up the cooler column. This level is beneath one-half of the air in the cool column, but it is beneath more than one-half of the air in the warm column. The air pressure at this elevation in the warm column is therefore greater than the air pressure at this elevation in the cool column. This difference in pressure is a pressure gradient and will cause air to flow aloft from the warm column toward the cool column. As it does so, the total mass in the warm column will decrease, and the total mass in the cool column will increase; therefore, the air pressures at the surface will no longer remain the same. Beneath the cool column, the air pressure at the surface will be greater than beneath the warm column, and this pressure gradient will cause air to flow along the surface from the cool column toward the warm column. Eventually, a steady-state flow will result with an elevation that separates the two directions of air flow. Not surprisingly, above this elevation the total mass in each column will be the same.

If Earth were not rotating, displacements of the air initially above the equator would be toward the poles aloft and toward the equator at the surface; because it is rotating, however, an additional effect—Coriolis force—must be considered. A point on the equator is a distance of one Earth radius away from Earth's rotation axis, whereas a point at one of the poles is directly on the rotation axis. As a parcel of air moves to the north from the equator, it is moving closer to the rotation axis. It inherited a certain angular momentum from when it was on the equator, and conservation of this angular momentum requires it to move somewhat to the east. In contrast, a parcel moving to the south

in the Northern Hemisphere is moving farther from the rotation axis, and conservation of its angular momentum requires it to move a bit to the west. In either case, in the Northern Hemisphere, the Coriolis force causes objects to move to the right of the direction that they would normally be moving; in the Southern Hemisphere, it causes things to move to the left. This effect is only significant when there is very little frictional interaction between the moving object and Earth, but it is extremely important in the circulation of the oceans and atmosphere.

Air being displaced from its location over the equator will initially go due north or due south. Coriolis force deflects it more and more until, in the vicinity of 30 degrees north or south latitude, it will be moving due east. At high altitudes, this is a geostrophic wind, a wind from the west (called a westerly) resulting from a poleward directed pressure gradient force being deflected 90 degrees by Coriolis force. The fastest elements of this flow form the subtropical jet stream. No longer moving toward the poles, this high-altitude air is now considerably cooler and denser than it was when first heated at the surface of Earth. As a consequence, it descends. As it gets lower, the additional air pressure it experiences causes it to warm up. Bicycle pumps illustrate this effect, as they get hot when the air within them is pressurized.

The amount of water vapor that can be contained in air varies with temperature. This is commonly observed as water condenses out of cooler air at night to produce dew or frost. The air rising over the equator is initially warm and heavily laden with water. As it cools, this water condenses, producing the intense rainfall essential for equatorial rain forests. Losing its water warms the air, enhancing its ascent. Later, after moving to the northeast or southeast, this air descends, warming up so that it once again absorbs water vapor. The regions of Earth near 30-degree latitudes are characterized by intense evaporation that produces deserts such as the Sahara and the Kalahari in Africa.

Because the descending air at 30-degree latitudes is denser than average, air pressure at the surface beneath it will be higher than average. Similarly, the air pressure beneath the rising air column at the equator is lower than average. This difference in pressure causes winds to blow across the surface, from 30 degrees north or south latitude toward the equator. Coriolis force also affects this flow so that these winds are deflected; by the time they reach the equator, they are coming out of the east. These easterly winds are called the trade winds.

HADLEY CELLS

The movements of air already described connect to form two circulating cells, one in the Northern Hemisphere, and one in the Southern Hemisphere. These are called Hadley cells, named after George Hadley, who discovered them in 1735.

The convection responsible for the Hadley cells is not seen as clearly elsewhere on Earth. However, four additional cells, two in each hemisphere, have long been a part of the theoretical development of meteorology. One, called the Ferrel cell, after William Ferrel, who proposed it in 1856, also descends at about 30 degrees latitude and presumably ascends at latitudes of about 60 degrees. The other, also proposed by Ferrel and called the Polar cell, ascends at latitudes of about 60 degrees and descends at the poles. The surface winds produced by these cells return air to the region around 60 degrees. This returning air is deflected by Coriolis force, producing westerlies between 30 degrees and 60 degrees, and Polar easterlies at higher latitudes. Much of the general circulation of Earth's atmosphere can be explained by this six-cell model, and it continues to be used in many elementary meteorology and earth-science courses and textbooks. As a theoretical tool, it is useful and easily grasped, and it yields insights about global systems that are generally accurate. Air does descend at the poles and at 30 degrees latitude, and returns to the vicinity of 60 degrees latitude, but the situation there is not as simple as the rising heated air at the equator. The boundary between the Polar cell and the Ferrel cell, called the Polar front, has opposing surface winds, not converging ones; high-altitude winds over this front do not simply diverge as they do above the equator.

FRONTS AND CYCLONES

During World War I, dirigibles were used to bomb England. To avoid assisting this tactic, the global system of weather data gathering was suspended. Subsequently, Norway's fishing fleet was put at risk from the weather, and scientists there developed theoretical models to make up for the lack of distant weather data. Led by Vilhelm Bjerknes and Halvor Solberg, this group of meteorologists was called the Bergen school. Relying on more closely spaced but effectively synchronous data, they developed the concept of fronts and proposed the cyclone model for storm genesis and evolution.

The cyclone model grew out of observations of storm systems in the middle latitudes. Such a system has a low-pressure region near its center and a pattern of winds moving in a concentric fashion around this low. This is the result of surface winds trying to move into the low-pressure region but being deflected by Coriolis force. In the Northern Hemisphere, these cyclonic winds move around the low in a counterclockwise direction, whereas in the Southern Hemisphere, they move clockwise. Often, observations revealed a consistent pattern of temperature gradients, precipitation bands, and surface wind configurations. The entire storm system usually moved from west to east and evolved in similar ways from its initial genesis, to being fully developed with maximum winds, to fading out and disappearing. As this evolution occurred, the interactions between air masses of different temperatures and with different moisture content followed reasonably consistent patterns.

The meteorologists of the Bergen school saw that the shear zone between the westerlies of the warm, moist air to the south and the easterlies of the cold, dry air to the north was an important element in generating these storms. A line connecting the various cyclonic disturbances in the middle latitudes defined the Polar front—not as a simple, smooth surface, but one with major excursions to the north and south, much like a meandering river.

Air above the poles is cooler and more compressed than the warmer air in the middle latitudes. This produces a pressure gradient aloft that is directed toward the pole. This would cause poleward movement, except that Coriolis force deflects such a movement to the right, again producing a westerly geostrophic wind, the fastest part of which is called the Polar jet stream. This is the jet stream referred to on weather maps and in forecasts in the United States. As already described, the temperature gradients at the surface are greater at higher latitudes, and hence the pressure gradients and velocities of the Polar jet stream are greater than those for the subtropical jet stream. In addition, because the surface temperature gradients are greater in winter than in summer, the velocity and significance of the Polar jet stream are also greater in the winter.

As the Polar jet stream races around the globe at velocities of about 125 kilometers per hour in the winter and 60 kilometers per hour in the summer, instabilities develop that deflect it into a meandering path. As a meander develops, the range of temperatures over which it moves increases, causing higher winds and even greater meandering. Eventually, portions of its path may be nearly north-south, bringing warmer air to higher latitudes and cooler air to lower ones. This diminishes the temperature gradients, causing the meanders to shrink until nearly east-west flow is reestablished. Called Rossby waves, after Carl-Gustav Rossby, who described them in 1939, these meanders form a path that resembles a very blunt, rounded star with three to six points centered on the pole. Moving along such a path, the jet stream speeds up at some places and slows down at others. Speeding up decreases air pressure aloft, while slowing down increases it. Cyclonic disturbances tend to form beneath places where the air pressure aloft is reduced and to move along tracks that lie beneath the jet stream. As they evolve, fronts develop, precipitation occurs, and warm air is transported to higher altitudes and then to higher latitudes.

El Niño/Southern Oscillation

The surface components of the Hadley cells move toward the equator. Coriolis force deflects the flow, so that by the time these winds reach the equator, they are coming out of the east. Ocean currents, driven by their own geostrophic flows, are quite similar, with major east-to-west flows near the equator. The water moved by these ocean currents is warm surface water, and its transport to the west results in a buildup of such waters in the western part of the Pacific, and to some extent the Atlantic. With warmer sea-surface temperatures to the west and upwelling of cooler water on the eastern side of the Pacific, yet another temperature gradient exists of sufficient scale to influence general circulation patterns.

The convection cell in this case is called a Walker cell, named after Sir Gilbert Walker, who identified it in 1924. The conditions needed for its development are neither constant nor periodic.

Every three to five years, because of factors not yet well understood, this circulation breaks down. Generally coupled with a decrease in the strength of the trade winds, the Walker cell reverses its orientation: Instead of ascending air in the west, with sufficient rainfall to support equatorial rain forests in Indonesia, the air ascends over the eastern regions of the Pacific, bringing sometimes intense rainfall to regions that are otherwise deserts in Peru. Called El Niño by oceanographers and the Southern Oscillation by atmospheric scientists (hence often abbreviated as ENSO), this reversal has dramatic effects on weather patterns.

Usually lasting between twelve and eighteen months, the ENSO has been identified in historic and geologic data sets. Droughts in Africa, floods in the American West, and other phenomena appear to be related to the sea-surface temperature in the equatorial Pacific. Certainly one of the more interesting aspects of the ENSO is its effect on the other aspects of general atmospheric circulation.

Significance

Understanding general atmospheric circulation is essential for accurate, useful weather predictions. Knowledge of this circulation permits meteorologists to estimate how various parcels of air will move, how pressures will change, and how and where precipitation will occur. These estimates, in turn, permit them to project farther into the future and improve the accuracy of their predictions.

By recognizing what variables are most likely to alter the general atmospheric circulation, and being able to guess how these variables might have been different in the past, geologists and climatologists can make more informed models about ancient weather patterns and climate evolution. The Himalayan Mountains and Tibetan Plateau are comparatively recent features on the face of the Earth, for example, and their presence has dramatically altered circulation patterns. Better understanding of past climates will help assess the influence of anthropogenic inputs, such as carbon dioxide and chlorofluorocarbons (CFCs), and should serve to guide public policy.

Otto H. Muller

Cross-References

Acid Rain and Acid Deposition, 1802; Air Pollution, 1809; Atmosphere's Evolution, 1816; Atmo-

BIBLIOGRAPHY

Ahrens, C. Donald. *Meteorology Today: An Introduction to Weather, Climate, and the Environment.* 3d ed. St. Paul: West Publishing Company, 1988. A standard college textbook; the treatment of general atmospheric circulation is straightforward and easily grasped. Little physics and few equations are presented, but there are lists of terms and questions for thought.

Barry, Roger G., and Richard J. Chorley. *Atmosphere, Weather, and Climate.* 7th ed. London: Routledge, 1998. A classic text, first published in 1968, this book covers the subject of general atmospheric circulation in a very thorough, but not technically challenging, style. It presents a vast number of figures, generally black-and-white line drawings, to illustrate its points. Some physics is included, but nothing beyond the abilities of an advanced high school student.

James, Ian N. *Introduction to Circulating Atmospheres.* Cambridge: Cambridge University Press, 1994. Although many equations and technically challenging concepts are presented, the copious illustrations in this book can be grasped by most readers. In addition to developing the theoretical basis for much of terrestrial meteorology, the book gives considerable attention to atmospheres on other planets.

Lutgens, Frederick K., and Edward J. Tarbuck. *The Atmosphere.* 7th ed. Upper Saddle River, N.J.: Prentice Hall, 1998. A very readable textbook with a wealth of colored illustrations, including photographs and satellite images. Discussions are concise, easily understood, and internally consistent, although sometimes not as thorough as they might be.

Monmonier, Mark S. *Air Apparent: How Meteorologists Learned to Map, Predict, and Dramatize Weather.* Chicago: University of Chicago Press, 1999. This is an approachable treatment of meteorology cleverly presented as a history of weather cartography. While including discussions of television channels, Web sites, and NOAA radar and radio, it also traces much of the history of the subject since the end of the nineteenth century. It includes discussions of general circulation patterns throughout the text. This might make it less suitable as a reference text, but the clear writing style and wealth of information combine to make this book valuable reading for anyone interested in meteorology.

Wells, Neil. *The Atmosphere and Ocean: A Physical Introduction.* London: Taylor & Francis, 1986. The atmosphere and oceans are both fluids circulating on a rotating planet, and they influence each other intimately and profoundly. This book treats their interactions in a readable, yet thorough, manner. Although numerous quantitative concepts and equations are presented, many are illustrated with graphs or figures that make them understandable to readers with little technical background.

Williams, James Thaxter. *The History of Weather.* Commack, N.Y.: Nova Science Publishers, 1998. A very easy to read treatment of the development of meteorology, this little book provides insights into how the observational and theoretical approaches to understanding weather sometimes competed but generally complemented and augmented each other. Its treatment is qualitative, there are no illustrations, its organization is somewhat baffling, but it is entertaining and interesting.

ATMOSPHERE'S STRUCTURE AND THERMODYNAMICS

An atmosphere is a shell of gases that covers a planetary surface. Earth's current atmosphere is a complex, dynamic system that interacts closely with and controls the surface environment. Atmospheric thermodynamics involves the process by which energy from the Sun is absorbed and deposited on the Earth, including its oceans and atmosphere. Early life-forms substantially altered Earth's atmosphere, and humans continue to do so.

PRINCIPAL TERMS

CHLOROFLUOROMETHANES (CFMS): also called chlorofluorocarbons, these are chemicals in which chlorine and fluorine replace one or more of the hydrogen atoms in methane; in more complex variations, chlorine- and fluorine-based molecules are attached to the carbon base

COSMIC RAYS: high-energy atomic nuclei and subatomic particles, as distinct from electromagnetic radiation

ENVIRONMENTAL LAPSE RATE: the general temperature decrease within the troposphere; the rate is variable but averages approximately 6.5 degrees Celsius per kilometer

EXOSPHERE: the outermost layer of the Earth's atmosphere

GREENHOUSE EFFECT: a planetary phenomenon in which the atmosphere acts as a hothouse, trapping more heat radiation than it emits; the atmosphere traps solar energy until the mean surface temperature is raised several degrees

GROWING DEGREE-DAY INDEX: a measurement system that uses thermal principles to estimate the approximate date when crops will be ready for harvest

HETEROSPHERE: a major realm of the atmosphere in which the gases hydrogen and helium become predominant

HOMOSPHERE: a major realm of the atmosphere whose chemical makeup is similar to the sea-level proportions of nitrogen, oxygen, and trace gases; it overlaps the troposphere, stratosphere, and mesosphere

INFRARED RADIATION: electromagnetic radiation lower in energy than visible light but higher than radio and microwaves; generally beyond 770 nanometers in wavelength

IONOSPHERE: the ionized layer of gases in the Earth's atmosphere, occurring between the thermosphere and the exosphere (it starts at about 50-100 kilometers above the surface of the planet)

LATENT HEAT: the energy absorbed or released during a state of change

MESOSPHERE: the atmospheric layer above the stratosphere where temperature drops rapidly

NET RADIATIVE HEATING: the driving force for atmospheric thermodynamics, basically computed as the difference between solar heating and infrared cooling

PHOTODISSOCIATION: the splitting of molecules by light, generally in the ultraviolet spectrum

SENSIBLE HEAT: heat that can be felt or measured with a thermometer

STRATOSPHERE: the atmospheric zone above the troposphere that has the greatest ozone concentration and exhibits isothermal conditions followed by a gradual temperature increase

TEMPERATURE GRADIENT: the change in temperature with displacement in a given direction

TEMPERATURE INVERSION: an abnormal increase in air temperature with increasing elevation from Earth's surface

THERMOSPHERE: the atmospheric zone beyond the mesosphere in which temperature rises rapidly with increasing distance from the Earth's surface

TROPOSPHERE: the lowest atmospheric layer, marked by considerable turbulence and a decrease in temperature with increasing altitude

ULTRAVIOLET LIGHT: electromagnetic radiation higher in energy than visible light but lower than X rays; generally in the 310-110 nanometer wavelength range

ATMOSPHERIC CONTENT

"Atmosphere" usually refers to the layer of gases that covers Earth. Although most planets have atmospheres of some sort, Earth's atmosphere is unique among those known in this solar system in that it contains a substantial amount of oxygen.

The atmosphere of the Earth contains 78 percent nitrogen and 21 percent oxygen; the remainder contains 0.9 percent argon, 0.03 percent carbon dioxide, and traces of hydrogen, methane, nitrous oxide, and inert gases. In addition, the atmosphere carries varying amounts of water vapor and aerosol particles such as dust and volcanic ash, depending on local and global events. As early as 3.5 billion years ago, primitive algaelike life-forms emerged and fed on the carbon dioxide by using photosynthesis to metabolize carbon dioxide and water molecules and to form simple sugars that did not then exist. Photosynthesis over billions of years gradually brought about a drastic change in the gas ratios, resulting in the current carbon dioxide concentration of only 0.03 percent and free molecular oxygen concentration of 20 percent.

A DYNAMIC SYSTEM

Gases are compressible, and they absorb or transmit varying amounts of electromagnetic radiation; therefore, the atmosphere is not static but is instead a highly dynamic system. Phenomena such as weather and climate are short- and long-term events involving the exchange of energy and transport of mass within the atmosphere and with the solid Earth, liquid oceans, and space.

Links between solar activities (especially sunspot cycles) and weather and climate have been sought for decades, but little more than contradictory hints have been discovered. Interactions between the oceans and the air are more substantial: Winds drive waves and thus ocean currents; in return, the oceans act as a heat source or sink for the atmosphere. The most famous interaction is the El Niño/Southern Oscillation event. El Niño ("the child"; it usually happens around Christmas) is an upwelling of cold water off the Pacific coast of South America that occurs every two to seven years. Besides having a disastrous effect on the fishing industry, it is associated with changes in circulation and precipitation patterns in the atmosphere over the Pacific basin.

TEMPERATURE

The atmosphere has no clear upper boundary, but it generally is considered to extend to 300 kilometers, the altitude at which it responds more to electromagnetic effects and acts less like a fluid body. It can be described in three major characterizations: temperature, chemistry, and electrical activity.

Temperature changes with altitude and, as there is less overlying gas, with pressure and exposure to radiation. The lowest region of the atmosphere, enclosing virtually all life and weather on Earth, is the troposphere, extending to an altitude of 8-18 kilometers (*tropo* is Greek for "turn" and refers to the fact that this region turns with the solid earth). The bottom of the troposphere is the boundary layer, where the atmosphere interacts directly with the surface of the Earth. The boundary layer often is turbulent, as moving air masses (winds) encounter and flow around or over obstructions and exchange heat with the ground or with the water. Temperature and pressure in the troposphere decrease at about 2 degrees Celsius per kilometer until the top of the troposphere, the tropopause, is reached. Life becomes increasingly difficult to maintain with altitude (humans may require additional oxygen above 4 kilometers and must wear pressure suits above 10.6 kilometers). Some 75 percent of the gases and 90 percent of the water vapor in the atmosphere are contained below the tropopause.

At the tropopause, temperature reaches a minimum of about −50 degrees Celsius, then rises sharply as one enters the stratosphere to peak at about 15 degrees Celsius at the stratopause, about 50 kilometers high. Above this level, it declines again in the mesosphere to a low of −60 degrees Celsius at the mesopause, 85 kilometers high. Only 1 percent of the atmosphere is in the mesosphere and above; 99 percent lies below. Particles from atomic nuclei to meteors generally are destroyed in the mesosphere. Nuclei—better known as cosmic rays—encountering gas molecules will be shattered into secondary and tertiary particles. Most meteors are heated and evaporated by friction when they encounter the mesosphere. Above the mesosphere, the thermosphere (the hot atmosphere) extends to approximately 300 kilometers in altitude, and temperatures soar to between 500 and 2,000 degrees Celsius, depending on solar activity.

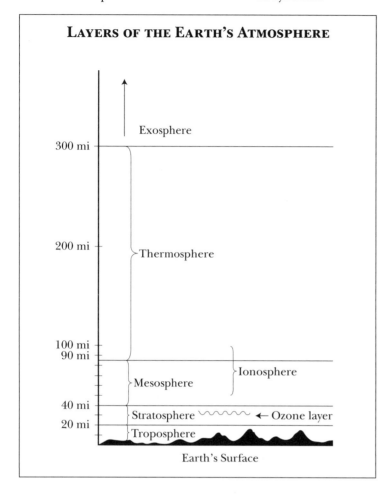

LAYERS OF THE EARTH'S ATMOSPHERE

Solar absorption is most concentrated on Earth's surface, particularly in tropical regions, whereas most of the infrared radiation going out to space originates in the middle troposphere and is more evenly distributed between equatorial and polar regions. Time-averaged temperature distribution is maintained by the system by transporting heat from regions where solar heating dominates over infrared cooling to regions where radiative cooling is able to dominate. In this way, the atmosphere transports heat from the ground to the upper troposphere, and the atmosphere and the ocean work together to transport heat toward the Arctic and Antarctic poles from the equatorial belt.

NET RADIATIVE HEATING

Net radiative heating, the difference between solar heating and infrared cooling, is the driving force for atmospheric thermodynamics. Solar radiation reaches the top of the atmosphere, where about 30 percent of it is reflected back into space. The remaining radiation is absorbed. About 70 percent of solar absorption happens at the surface, and about 40 percent of that leaves the surface as net infrared radiation. This serves to leave a net surface radiative heating of about 150 watts per square meter. On a global and annual average, the net radiative heating of Earth's surface is equivalent to the net radiative cooling of the atmosphere; normal thermodynamics within the atmosphere thus maintains an appropriate balance between the two. A prolonged and severe imbalance between these two systems would result in global warming, which could melt the polar ice caps and raise the elevations of the oceans.

Over the oceans, most of the net radiative heating of the surface is used in the process of water evaporation. Energy is removed from the surface as latent heat; when this vapor condenses, it deposits almost all of its latent heat into the air rather than into the condensed water, which re-

Because the atmosphere is so thin, however, the total heat present is minuscule. Finally, beyond the thermosphere is the exosphere (outer layer), which extends from 300 kilometers to the solar wind (also called the interplanetary medium).

SOLAR AND INFRARED RADIATION

Atmospheric thermodynamics can be described as the process by which energy from sunlight is absorbed and deposited on Earth's surface, in the oceans, and in the atmosphere. This energy must be returned to space in the form of infrared radiation in order for the planet to maintain a cyclic range of ambient temperatures within normal limits. Even though solar radiation and infrared radiation must be approximately balanced on a global level, they are frequently out of balance on a local level, accounting for the ambient weather changes seen on a day-by-day basis.

turns to the surface as precipitation. The result is that heat is transported from the surface to the air, where the potential for condensation takes place. On a global average, about 70 percent of the net radiative heating of the surface is removed by latent heat flux, and the remaining 30 percent leaves the surface by conduction of the sensible heat to the overlying air. The atmosphere thus experiences diabatic heating or cooling from four ongoing processes: latent heating, sensible heating at the surface, solar absorption, and infrared heating or cooling.

TEMPERATURE INVERSION

The general temperature decrease within the troposphere is called the "environmental lapse rate." The occurrence of shallow layers where temperatures actually reverse this normal pattern and get hotter with increasing height is known as a "temperature inversion." Temperature inversions are often seen in cities located at high altitudes with a low partial pressure of oxygen, and in basins surrounded by mountains that serve to block winds and trap industrial pollutants. A "greenhouse effect" inversion causes a reversal of an ecosystem's normal atmospheric temperature gradient, cooking harmful chemicals and enhancing their negative effects on organisms living below. A notable example is the strong eastern wind that blows toward Denver, Colorado, and traps a brown cloud of pollutants against the Rocky Mountains. This requires the daily broadcast of air-quality reports on radio and television and leads to the frequent calls for weak and elderly residents to stay indoors during hotter parts of the day. Trapped carbon monoxide, coming mainly from automobile tailpipes via the incomplete combustion of gasoline, combines quickly with hemoglobin, the oxygen-carrying compound in the blood in humans and animals. By taking up binding sites on the hemoglobin molecule, carbon monoxide impairs oxygen delivery to the tissues. Elderly persons with heart disease are at special risk through any restrictions in oxygen delivery.

The thermosphere, which does not have well-defined limits, exhibits another increase in temperature as a result of the absorption of very short-wavelength solar energy by atoms of oxygen and nitrogen. Although temperatures rise to values in excess of 1,000 degrees Celsius in this outermost layer, it is difficult to compare the temperature here with that seen on Earth's surface. Temperature is directly proportional to the average speed at which molecules are moving, and because gases within the thermosphere are moving at very high speeds, ambient temperature remains very high. However, the gases in this region are so sparse that only a minimal number of these fast-moving air molecules collide with foreign bodies, thus causing only a minimal amount of energy to be transferred. For this reason, the temperature of a satellite orbiting Earth in the thermosphere is determined by the amount of solar radiation it absorbs and not directly by the temperature of the surrounding environment. Thus, if an astronaut in the space shuttle were to expose his or her hand outside into ambient space, it would not feel hot.

CHEMISTRY

From a chemical standpoint, the atmosphere is divided into two major realms: the homosphere and the heterosphere. The homosphere—which overlaps the troposphere, stratosphere, and mesosphere—has a chemical makeup identical to sea-level proportions of nitrogen, oxygen, and trace gases, even though the absolute numbers of atoms and molecules drop sharply. With increasing altitude, some important differences start to appear. Ozone becomes an important constituent of the atmosphere in this realm. Ozone is formed in the stratosphere by short-wavelength ultraviolet sunlight splitting oxygen molecules (photodissociation). These free oxygen atoms then form ozone with oxygen molecules. Molecular nitrogen also is dissociated.

Although the stratosphere and mesosphere (sometimes treated together as the middle atmosphere) are quite tenuous compared to the troposphere, gases in them form an optically dense layer that absorbs or reflects short-wavelength ultraviolet and X-ray radiation that would be damaging to life on the surface. Ozone is especially important with regard to ultraviolet radiation. Nevertheless, ozone is quite fragile and can be destroyed by chlorofluoromethanes (Freon and related compounds) used in spray cans and refrigeration systems. Studies indicate that these gases migrate upward in the atmosphere and chemically remove thousands of times their own mass in ozone molecules before they are broken down after several decades or even centuries. The result is believed to have led to the formation of an ozone hole over the South Polar region. Atomic oxygen becomes more common in the mesosphere. In the heterosphere, the gas mixture changes drastically, and hydrogen and helium become dominant.

ELECTRICAL ACTIVITY

From an electrical standpoint, there are the neutral atmosphere and the ionosphere. The neutral atmosphere, below 50 kilometers, is largely devoid of electrical activity other than lightning, which might be regarded as localized "noise." Above 50 kilometers, atoms and molecules are ionized largely by sunlight (ultraviolet radiation in particular) and to a lesser extent by celestial X-ray sources, collisions with other atoms, and geomagnetic fields and currents. Although the ionosphere as a whole is electrically neutral, it comprises positive (ion) and negative (electron) elements that conduct currents and that respond to magnetic disturbances. The ionosphere starts at about 50-100 kilometers in altitude and extends outward to more than 900 kilometers as it gradually merges with the magnetosphere and its components. It is sometimes called the Heaviside layer after Oliver Heaviside (1850-1925), who predicted a layer of radio-reflection ionized gases. The ionosphere is divided into C, D, E, and F layers, which in turn are subdivided (F_1, for example).

The ionosphere is one of the most active regions of the atmosphere and one of the most responsive to changes in solar activity. Ions and electrons in the ionosphere form a mirrorlike layer that reflects radio waves. Radio waves are absorbed by the lower (D) region of the ionosphere (which also reaches down into the mesosphere). The D-layer dissipates at night in the absence of solar radiation, allowing radio waves to be reflected by the F-layer at higher altitudes, thus causing radio "skip." These effects vary at different wavelengths. Intense solar activity can alter the characteristics of the ionosphere and make it unreliable as a radio reflector, either through the input of high-energy radiation or by the injection of particles carried by the solar wind.

ATMOSPHERIC LIGHT DISPLAYS

Such particle injections would go unnoticed but appear as the aurora borealis and aurora australis (northern and southern lights, respectively). The Earth's magnetic field shields the planet from most charged radiation particles. At the polar regions, however, where the magnetic field lines are vertical (rising from the surface), the environment is magnetically open to space. Many charged particles from space or from the solar wind will be funneled into the polar regions. When the particles strike the atmosphere, they surrender their energy as light with spectral lines unique to the electrochemical interactions taking place. These auroral displays generally take place at 120-300 kilometers in altitude, with some occurring as high as 700 kilometers.

The aurora is the best-known atmospheric light display. Other "dayglow" and "nightglow" categories are caused by lithium, sodium, potassium, magnesium, and calcium at altitudes from about 60 to 200 kilometers. These metals may be deposited by meteors as they are evaporated on atmospheric entry. A layer of hydroxyl radicals causes an infrared glow at about 100 kilometers, and dull airglows are caused by poorly understood effects at 100-300 kilometers in altitude.

JET STREAMS AND WAVES

Although the principal division of the atmosphere is vertical, there are horizontal differences related to latitude and to weather. Two major phenomena which affect the atmosphere are the jet streams and waves. The jet stream is a high-speed river of air moving at about 10-20 kilometers in altitude and at 100-650 kilometers per hour. Its location plays a major role in the movements of larger air masses that make up weather fronts in the troposphere.

More than twenty wave phenomena take place in the atmosphere in response to different events. The three principal categories are gravity, Rossby, and acoustic. Gravity waves are not associated with relativity but with vertical oscillations of large air masses causing ripples, like a bottle bobbing in a pond. Rossby (or planetary) waves are associated with the wavelike distribution of weather systems. Acoustic waves are related to sound.

ATMOSPHERIC STUDY TOOLS

The earliest types of instruments used for atmospheric study remain among the most important. Barometers, thermometers, anemometers, and hygrometers provide the most immediate records of atmospheric change and warnings of impending events. Vertical profiles of atmospheric conditions are obtained by transporting such instruments up into the air on balloons and on suborbital rockets. The term "sounding rocket" comes from the earliest days of atmospheric study, when scientists were "sounding" the ocean of air just as they would the ocean of water: Small charges were attached to balloons, and the time sound took to reach the ground was a crude measure of atmospheric density. Balloon-borne instrument packages continued to be called radiosondes ("radio sounders").

Instrumentation carried aboard spacecraft is of a different nature. Many of the most revealing devices have been spectrometers of various types that analyze light reflected, emitted, or adsorbed by the atmosphere. Absorption of ultraviolet light by the atmosphere led to the discovery of the ozone hole; drops in absorption meant that ultraviolet light was passing through, rather than being returned to space (typically, such measurements also require observation of the solar ultraviolet output). Optical instruments usually are most effective when they view the atmosphere "edge-on" so as to increase the brightness of the signal (somewhat like viewing a soap bubble at the edges). Atmospheric studies can be difficult when viewing straight down, because the weak signals from airglow and other effects are washed out by the brighter glow of Earth or the stellar background. Special techniques can be employed. The U.S. space shuttle has twice carried sensors designed to monitor carbon monoxide pollution in the atmosphere. Gas cells containing carbon monoxide at different pressures acted as filters that blocked all signals but the wavelengths corresponding to carbon monoxide at the same pressure (that is, altitude) as that in the cell.

The most powerful tools used in studying the atmosphere have been the weather satellites, which observe the atmosphere from geostationary orbit (affording continuous views of half a hemisphere) and from lower polar orbits. Images from these satellites reveal the circulation of the atmosphere by the motion of clouds. Other sensors (called sounders) provide temperature profiles of the atmosphere at various altitudes.

The most extensive analyses of the atmosphere have been carried out by the Atmosphere Ex-

plorers, the Orbiting Geophysical Observatories, the Atmosphere Density Explorers, and the Dynamics Explorers. Operating in the upper reaches of the atmosphere, these spacecraft have delineated the structure and composition of the atmosphere and the changes it experiences with seasons and solar activity. The more sensitive chemical assays, however, have been conducted by instruments carried aboard the manned Spacelab 1 and 3 missions on the space shuttle. An Imaging Spectrometric Observatory carried on Spacelab 1 produced highly detailed emission spectra of the atmosphere between 80 and 100 kilometers in altitude. Atmospheric Trace Molecules Observed by Spectroscopy (ATMOS), on Spacelab 3, measured the altitude ranges of some thirty chemicals and identified five, such as methyl chloride and nitric acid, in the stratosphere, where previously they were only suspected.

COMPARATIVE PLANETOLOGY

Comparative planetology analyzes the differences and similarities between and among the planets. Earth, Venus, and Mars are used most often in comparative atmospheric studies. These three "terrestrial" planets are similar in size and in general chemistry but totally different in environment, largely because of their different atmospheres. Venus has a dense atmosphere composed largely of carbon dioxide and topped by clouds of sulfuric acid, which has led to surface temperatures of 900 degrees Celsius and to pressures ninety times greater than those of Earth. The circulation pattern, though, is unaltered by precipitation and oceans and thus can be used as a model in studying the Earth. Efforts to understand how Venus became a "runaway greenhouse" have suggested a similar scenario for Earth. Mars, in contrast, has a tenuous atmosphere composed of carbon dioxide and traces of water vapor and oxygen. Studies of Mars focus on how its climate and atmosphere evolved and whether it was once Earthlike.

ATMOSPHERIC ALTERATION BY LIFE-FORMS

The atmosphere as it currently exists is a relatively recent phenomenon brought about by the gradual alteration of the environment by life-forms. Awareness of this global alteration is helping humans understand the effects they are having on the environment over a relatively short timespan—essentially since the onset of the Industrial Revolution. The widespread use of fossil fuels and the burning of forests to clear land for agriculture have converted the carbon that plants spent billions of years converting into solid carbon compounds back into gaseous carbon dioxide. Furthermore, the plants that were "sinks," or absorbers, of carbon dioxide are available in lesser quantities to perform the ancient task of liberating oxygen. Sulfur compounds are naturally introduced by volcanoes, biological decay, and oceanic processes, but large quantities have been added by industrial processes, including coal burning. One product, sulfur dioxide, combines with water vapor at low altitudes to form sulfuric acid; at high altitudes it, too, can alter the ozone layer and the terrestrial radiation balance. In addition, the ratios of nitrogen compounds are altered by inefficient combustion and by widespread use of nitrogen-giving fertilizers; these products, too, have an adverse effect on ozone.

The immediate concern is not that the oxygen supply will be depleted—although that is a credible, long-term possibility—but that the increased amounts of carbon dioxide in the atmosphere will cause a greenhouse effect. In the greenhouse effect, long-wavelength radiation emitted by the soil or ground is absorbed by the atmosphere and then reemitted. Glass serves this purpose for a greenhouse; carbon dioxide has the same effect in Earth's atmosphere. Other human-made gasses that enhance the greenhouse effect are nitrous oxide, methane (which is also produced naturally), and chlorofluoromethanes (which also deplete ozone, thus allowing more radiation to enter). Because so little is known about the causes of different effects, there are uncertainties in predicting what will happen. It is expected, however, that increases in the carbon dioxide content of the atmosphere will raise global temperatures. This rise in temperature will shift weather patterns and will cause large portions of the polar ice caps to melt, thus flooding coastal regions.

AGRICULTURAL APPLICATIONS

There are many practical applications of the knowledge of thermodynamic data, one of which is regularly utilized by the agricultural industry to estimate the approximate date when crops will be ready for harvest. The growing degree-day index estimates the number of growing degree-days for a particular crop on any given day as the difference between the daily mean temperature and the minimum temperature required for growth of a particular crop. For example, the minimum growing temperature for corn is 50 degrees Fahrenheit (10 degrees Celsius), which means that on a day when the mean temperature is 75 degrees Fahrenheit, the number of growing degree-days for sweet corn is estimated at twenty-five. Starting with the onset of the growth season, the daily growing degree-day values are added. If two thousand growing degree-days are needed for corn to mature in a particular region, the corn should be ready to harvest when the accumulation reaches that figure.

Daniel G. Graetzer and Dave Dooling

CROSS-REFERENCES

BIBLIOGRAPHY

Ahrens, C. D. *Meteorology Today: An Introduction to Weather, Climate, and the Environment.* Minneapolis, Minn.: West Publishing, 1994. A useful text designed for college-level students taking an introductory course on atmospheric science.

American Meteorological Society. "Policy Statement of the American Meteorological Society of Global Climate Change." *Bulletin of the American Meteorological Society* 72 (1991): 57-59. A review of current knowledge and a call for more research into the nature of climate variability.

Burroughs, W. J. *Watching the World's Weather.* New York: Cambridge University Press, 1991. A very readable text with helpful explanations regarding weather changes in all parts of the world.

Campbell, I. M. *Energy and the Atmosphere: A Physical-Chemical Approach.* New York: John Wiley & Sons, 1986. An excellent text on the physical and chemical nature of the atmosphere and the effects of pollution.

Curry, Judith A., and Peter Webster. *Thermodynamics of Atmospheres and Oceans.* San Diego, Calif.: Academic Press, 1999. This book offers a look at the effects of the interaction between oceans and the atmosphere on weather patterns and climatic changes. Provides good insight into the role that atmospheric thermodynamics plays in meteorology. Illustrations, maps, and index.

Eagleman, J. R. *Air Pollution Meteorology.* Lenexa, Kans.: Trimedia Publishing, 1991. Describes the role of Earth's atmosphere in dissipating industrial air pollution, with regard to thermodynamic principles.

Garstang, Michael, and David R. Fitzjarrald. *Observations of Surface to Atmosphere Interactions in the Tropics.* New York: Oxford University Press, 1999. A look at the atmospheric and oceanic systems of tropical environments. The specific nature of this book makes it suitable for the reader with some background in meteorology. Illustrations and maps.

Grotjahn, R. *Global Atmospheric Circulations: Observations and Theories.* Oxford, England: Oxford University Press, 1993. An advanced text that provides a useful presentation of old concepts and new theories regarding atmospheric thermodynamics.

Lutgens, F. K., and E. J. Tarbuck. *The Atmosphere: An Introduction to Meteorology.* Englewood Cliffs, N.J.: Prentice-Hall, 1995. Contains excellent chapters on the atmosphere, meteorology, and weather patterns.

National Research Council. *Space Science in the Twenty-first Century: Imperatives for the Decades 1995-2015.* Washington, D.C.: National Academy Press, 1988. A National Academy of Sciences survey of the state of human knowledge of the atmosphere and the environment. Includes recommendations for missions to expand that knowledge.

Williams, J. *The "USA Today" Weather Book.* New York: Random House, 1992. An excellent guide to the interpretation of weather reports. Gives numerous applied examples of thermodynamic principles.

Yung, Yuk Ling, and William B. DeMore. *Photochemistry of Planetary Atmospheres.* New York: Oxford University Press, 1999. This college-level text examines the atmospheric chemistry of all of the planets in the solar system, with a focus on photochemical processes. Intended for the advanced reader. Illustrations, bibliography, and index.

ATMOSPHERIC INVERSION LAYERS

Temperature inversions occur frequently around the world. They are marked by a temperature increase with increasing altitude and occur most often in conjunction with a stagnating high pressure. Temperature inversions can be hazardous to humankind, and they frequently cause unpleasant atmospheric conditions.

PRINCIPAL TERMS

ADIABATIC: characterizing a process in which no heat is exchanged between a system and its surroundings.

ADVECTION: the process of transport of air solely by the horizontal motion of the atmosphere

AIR DRAINAGE: the flow of cold, dense air down slopes in response to gravity

ALBEDO: the percentage of reflectivity from the Earth environment, being greatest from white surfaces and decreasing toward black

LAPSE RATE: the rate of decrease in temperature with increasing height

NEGATIVE LAPSE RATE: an increase in temperature with increasing height

RADIATIONAL COOLING: the cooling of the Earth's surface and the layer of air immediately above it by a process of radiation and conduction

RADIOSONDE: a balloonborne instrument package for the simultaneous measurement and transmission of weather data

STAGNATING ANTICYCLONE: a high pressure that remains over an area for three to four days

TURBULENCE: an irregular, perhaps random, motion of air appearing as eddies

SUMMARY OF THE PHENOMENON

A temperature inversion is an increase in temperature with an increase in altitude above the Earth's surface in the troposhere, the lowest layer of the atmosphere. Normally, temperatures decrease with increasing altitude in the troposphere. To comprehend temperature inversions in the troposphere, it is first essential to understand how the troposphere is heated. The Sun is the source of energy for heating the troposphere, but the Sun does not heat this layer of the atmosphere directly. Only a minuscule amount of solar radiation is absorbed by the gases that make up the troposphere; rather, most of the energy used to heat this layer is derived from the surface of the Earth.

The Sun radiates mostly short wavelengths, and short wavelengths tend to pass through the atmosphere without much interference on the part of atmospheric gases. These gases, including variable gases such as water vapor and carbon dioxide, are transparent to shortwave energy; that is, these gases allow shortwave energy to pass through to the Earth's surface without much of the shortwave energy being absorbed. The shortwave energy is absorbed by the Earth's surface, and the Earth radiates longwave energy back toward the sky. The terrestrial radiation is intercepted by clouds and gases such as water vapor, carbon dioxide, and methane, which absorb the longwave energy and reradiate it back toward the Earth. In this way, there is a continual exchange of longwave energy between the Earth and atmosphere and, thus, heat is held within the troposphere. Therefore, the surface of the Earth is the direct source of heat energy for heating the troposphere, and temperatures decrease with height.

Temperature inversions occur largely because of the mobility of air. The gases that comprise the atmosphere are fluids and, therefore, behave as fluids. The air is capable of flowing in much the same way that water does. Fluids may have various densities in response to differing temperatures. The warmer the fluid's temperature, the less dense is the fluid. It is logical, therefore, to assume that the colder, denser air is near the bottom of the troposphere and that the warmer air overlies it. This is not the case, however, except when a temperature inversion is present. The vertical variation of temperature in the troposphere is called the lapse rate. A normal or positive lapse rate is present when the temperatures decrease with increasing altitude, while a temperature increase with height results in a negative lapse rate and indicates the presence of a temperature inversion. Temperature inversions are of two general kinds: ground inversions and upper air inversions.

Ground inversions occur frequently and commonly extend upward for 100 meters or more. They can develop from several different causes. The primary cause of ground temperature inversions is radiational cooling, which occurs at night. During the day, the ground is heated and builds up a store of energy. After the Sun has set, the Earth's surface continues to radiate heat energy, thereby cooling the ground surface. When two bodies—such as the atmosphere and the Earth—are in contact, heat energy is conducted from the warmer body to the cooler body. Thus, energy from the air is conducted into the ground, thereby cooling the air immediately above it. Air is a poor conductor of heat energy, so only about the first 100 meters experience a temperature decrease as a result of radiational cooling. Such nighttime-induced temperature inversions are more likely to occur on nights with clear skies than on nights with cloudy skies, as the amount of radiation lost by the Earth at night is reduced by cloud cover. Calm wind conditions or light breezes also are more conducive to developing a ground inversion, because strong winds cause air to whirl, and mixing of the air breaks up the inversion. In addition, ground inversions are more likely to develop and to last longer in the cold climates of the world because a snow or ice cover reflects sunlight, and the small amount of heat absorbed is utilized in the melting process, thus cooling the surface very rapidly and producing an inversion. Therefore, ground inversions are more frequent and longer-lasting in high latitudes, especially at the North and South poles, than in middle latitudes. Likewise, the fall and winter seasons in the Northern Hemisphere, when temperatures are coldest, are more likely times for ground inversions to develop and to be more long-lasting than are other times of the year.

Normally, ground inversions are local in character and are found most frequently in valleys or other topographic depressions. Thus, a second method of inversion formation is the result of a phenomenon called air drainage. On cold nights, over rolling topography, cold air, being denser than warm air, responds to the pull of gravity and slides downslope and collects in depressions. Continued cooling can cause inversions to extend over larger areas both vertically and horizontally, if the vertical cooling is great enough to extend above the summits of the rolling terrain. Evidence of the initial development of a ground inversion often is heavy dew or frost. Frequently, ground fogs occur in association with inversions because of cooling of the air. This is particularly true in an air drainage situation where fogs first appear in depressions in the surface.

A third way in which ground inversions are formed is from a warm air mass moving into a region. A warm current of air may move in over a cool ground surface or over a cooler layer of surface air. The lower portions of the air mass are cooled, and stable or nonturbulent conditions result, producing an inversion.

The frequency of ground inversions varies across the United States (frequency being expressed as a percentage of the total time a region has inversions). During winter, Southern California and Arizona have inversions about 55 percent of the time; the central United States experiences inversions about 45 percent of the time; and Utah, Wyoming, Montana, and other Rocky Mountain states also have frequent ground inversions. The southeastern United States also experiences frequent ground inversions, which occur about 45 percent of the time. Most of these winter inversions occur at night. Summer inversions are less frequent than winter inversions, but they do occur. The central United States has inversions only about 35 percent of the time in summer, and the East Coast has inversions only about 20 percent of the time.

Upper-air inversions are more persistent and last for longer periods of time, usually several days rather than a few hours. These inversions normally occur above the surface at heights of 50-2,000 meters. Inversions develop above the surface in the troposphere as a result of several physical processes. One phenomenon that leads to the formation of an upper-air inversion is the subsidence, or sinking, of air. Air that subsides in the troposphere is compressed because of increasing pressure at lower levels. Compression of air causes it to heat adiabatically (from within) at a lapse rate of 10 degrees Celsius per kilometer. One such region of subsiding air is in a high pressure area or anticyclone. Here the upper atmospheric layers descend farther and heat more, thus developing an upper-air inversion. Therefore, regions that experience a high frequency of high-pressure areas have numerous upper-air inversions.

A second means of producing an upper-air inversion is through the movement of air masses. An air mass is a body of air with uniform physical characteristics horizontally. Air mass source regions affecting North America consist of the Pacific high and the Bermuda-Azores (Atlantic) high at about 30 degrees north latitude, the Arctic high at the North Pole, and the Canadian high and the Aleutian low at about 60 degrees north latitude. Often warm, moist air from the Pacific and Atlantic highs invades the United States. At the same time, polar or Arctic continental air may move southward and encounter the warm, moist air. The colder air, being denser, will push beneath the warmer air, lifting it aloft. The result is an upper-air inversion positioned above the boundary of the two air masses.

A third phenomenon that causes an upper-air inversion is the daily heating and cooling cycle. Generally, an inversion can develop when cloudy conditions during the day interfere with heating the surface more than cloudy conditions interfere with cooling it at night. If insufficient heating occurs during the day to offset the night-time cooling, an upper-air inversion is left in the troposphere.

Still another upper-air inversion is called the trade wind inversion and occurs in the low latitudes lying between the subtropical high (Pacific and Atlantic highs) and the equatorial low. Here the inversion is formed by the subsiding of air in the Hadley cell (where air rises near the equator and sinks near 30 degrees latitude, producing the trade winds in each latitude).

METHODS OF STUDY

Every twelve hours, at midnight and at noon Greenwich mean time (the time at the observatory in Greenwich, England), hundreds of soundings of the atmosphere are taken. Soundings or readings of temperature and other weather elements are made by sending helium-filled balloons aloft. Attached to each balloon are a package of weather instruments and a radio transmitter called a radiosonde. Information on temperature, humidity, and pressure is sensed by the instruments and transmitted back to weather stations on the ground, where the data are recorded and plotted.

Over the oceans and sparsely settled regions of the world, where radiosondes are not sent aloft, weather data are obtained remotely from satellites. The data acquired from radiosondes and satellites are plotted on a special chart from which the meteorologist can make interpretations about the stability of the atmosphere and can determine the presence of temperature inversions and at what level they exist. When a temperature inversion is present, the atmosphere is said to be stable because the vertical movement of air is restricted. Once the temperatures received from the soundings are plotted on the special chart at the various elevations, a linear graph of temperature changes with height is generated. From the chart, the meteorologist can quickly interpret the rate of temperature change (lapse rate) for various layers in the troposphere. A temperature increase with height (negative lapse rate) is discerned easily on the chart, and the depth of the inversion can easily be seen.

The lines on the chart show how much rising air would decrease in temperature with increasing altitude in noncloudy air (with a dry adiabatic rate of 10 degrees Celsius per kilometer) and how much it will decrease in cloudy air (with a wet adiabatic rate of 6 degrees Celsius per kilometer). By comparing the temperature soundings with these latter lines, the meteorologist can determine just how stable or unstable particular layers of the troposphere are. For example, in a hypothetical situation, suppose that a meteorologist has identified four layers of air on the special chart. The first layer of the sounding had temperatures decreasing at the rate of 12 degrees Celsius per kilometer, which is greater than the dry rate mentioned above. With this rate of temperature decrease, the layer would be absolutely unstable, and air could rise readily within it. Hence, no temperature could exist in this layer. Suppose that the second layer had an increase in temperature of 2 degrees Celsius per kilometer. This increase indicates a negative lapse rate, and because 2 degrees Celsius per kilometer is much less than the wet rate, the layer would be considered absolutely stable and definitely indicative of the presence of a temperature inversion. Then suppose that the third layer has a temperature decrease of 8 degrees Celsius per kilometer. This rate is in between the dry and the wet adiabatic rates and is considered conditionally unstable, which means that air would have to be lifted to an elevation where clouds are forming to be able to rise on its own. Again, a temperature inversion is not present in this layer. Finally, assume that the top layer has a temperature decrease of 4 degrees Celsius per kilometer. This rate is less than the wet adiabatic rate and would be absolutely stable, but it would not have a temperature inversion within it because the temperature is decreasing even though it is absolutely stable. In other words, temperature inversions cause stable layers in the atmosphere, but not all stable layers contain temperature inversions. Thus the meteorologist collects, analyzes a vertical profile of the atmosphere, and detects temperature inversions in the atmosphere, which, in turn, contribute to a weather forecast.

CONTEXT

Air beneath an inversion will not rise because it is colder than the inversion layer. Likewise, air will not rise within a ground inversion layer because air gets warmer with height, and air has to get cooler with height for air currents to rise within it. Owing to this resistance to air rising, pollution is concentrated near the Earth's surface when an inversion is present. Pollution has always been present in the atmosphere naturally, but it began to worsen when humankind discovered fire. Burning coal causes pollution because much soot and sulfur dioxide is released to the atmosphere.

England's pollution problems culminated in the worst pollution disaster of all time in 1952. A temperature inversion formed over London on December 5, and fog developed in the humid air and gradually thickened over a five-day period. The people heated their homes more than usual to combat the wetness in the air. This additional burning of coal and oil added more soot and sulfur dioxide to the air, and four thousand people died as a direct result of the polluted air. Finally, in 1956, Great Britain passed a clean air act partly in response to this disaster.

The United States has had its own pollution disasters. On October 26-30, 1948, a high-pressure area stagnated over Donora, Pennsylvania, which had several industries that issued pollutants into the air. With the temperature inversion associated with the stagnating high pressure, the pollution thickened near ground level and could not be dispersed. By October 30, the situation had reached crisis proportions and people began to die. In all, seventeen people died; although that may seem to be a relatively small number, had Donora had the same population as London and had the ratio of deaths to population total carried over to a city that size, eight thousand people would have died.

Pollution disasters have occurred in other countries as well, and many countries have considerable pollution problems in their cities even though they have yet to suffer a disaster. An upper-air inversion is one of the reasons that Los Angeles has extreme air pollution levels, because the Pacific high area covers that part of the state for extended periods of time. The inversion layer associated with this high is lower than the mountains to the east, so the air and thus the pollution are entrapped above the city. This situation in effect places a "lid" over the city, and winds from the Pacific Ocean cannot carry away the pollutants because of the highlands to the east. In Florida, temperature inversions during the cold season often result in freezes that threaten the citrus crop. The freezes result from temperature inversions entrapping cold air at ground level. Other parts of the country experience such a threat to crops as well.

Fogs are quite frequently associated with temperature inversions. On calm, clear nights when radiational cooling is ongoing and the air is moist, radiation fog will form. This type of fog, which is usually quite patchy, is frequent in topographic depressions where air drainage results. Fog may also form when a warm, moist mass of air moves in over a cold ground surface. The air is cooled from below, an inversion is formed, and the water vapor in the air condenses. This type of fog is usually deeper and more extensive than is radiation fog. Fogs are a hazard to transportation.

Ralph D. Cross

CROSS-REFERENCES

ground Radiation, 1846; Earth-Sun Relations, 1851; Fallout, 1857; Global Warming, 1862; Greenhouse Effect, 1867; Nuclear Winter, 1874; Ozone Depletion and Ozone Holes, 1879; Radon Gas, 1886; Rain Forests and the Atmosphere, 1890; Skylab, 2827; Weather Forecasting, 1982; Wind, 1996.

BIBLIOGRAPHY

Ahrens, C. Donald. *Meteorology Today: An Introduction to Weather, Climate, and the Environment.* 3d ed. New York: West Publishing, 1988. Chapter 5, "Seasonal and Daily Temperature," provides a good background of temperature characteristics necessary to understanding temperature inversions. A good discussion of the various types of inversions is also included. A section entitled "Focus on a Special Topic" describes radiation inversions in detail. Suitable for entry-level college students.

Battan, Louis J. *Fundamentals of Meteorology.* Englewood Cliffs, N.J.: Prentice-Hall, 1984. Chapter 5 provides a fairly thorough, nontechnical analysis of atmospheric stability and vertical air motions. Included is a discussion of temperature inversions along with associated factors of stability, instability, temperature lapse rates, vertical motions, and turbulent diffusion. Diagrams showing temperature inversion profiles of both types in the atmosphere are included, as well as a table showing seasonal temperature inversion frequency.

Curry, Judith A., and Peter Webster. *Thermodynamics of Atmospheres and Oceans.* San Diego, Calif.: Academic Press, 1999. This book offers a look at the effects of the interaction between oceans and the atmosphere on weather patterns and climatic changes. Provides good insight into the role that atmospheric thermodynamics plays in meteorology. Illustrations, maps, and index.

Eagleman, Joe R. *Meteorology: The Atmosphere in Action.* New York: Van Nostrand Reinhold, 1980. Chapter 5 contains a reasonably good analysis of both ground and upper-air inversions. As in most meteorology books, much space is devoted to factors affecting stability in the atmosphere, an understanding of which is necessary to comprehend temperature inversions. The chapter also covers lapse rates, adiabatic rates, stability in saturated and unsaturated air, and applications of stability.

Gedzelman, Stanley D. *The Science and Wonders of the Atmosphere.* New York: John Wiley & Sons, 1980. Chapter 13 provides very good coverage of temperature inversions and other factors in addition to characteristics related to stability/instability in the atmosphere. The author demonstrates temperature inversions with the use of thermodynamic charts (a chart showing temperature, pressure, and adiabatic rates). He also discusses some actual soundings taken in the atmosphere.

Lutgens, Frederick K., and Edward J. Tarbuck. *The Atmosphere: An Introduction to Meteorology.* 7th ed. Upper Saddle River, N.J.: Prentice Hall, 1998. An excellent introduction and description of the atmosphere, meteorology, and weather patterns. A perfect textbook for the reader new to the study of these subjects. Color illustrations and maps.

Spiegel, Herbert J., and Arnold Gruber. *From Weather Vanes to Satellites: An Introduction to Meteorology.* New York: John Wiley & Sons, 1983. Chapter 3 presents the factors of energy exchange among Sun, Earth, and atmosphere necessary to understand temperature inversions. Chapter 4 covers temperature causes and characteristics and distributions in the atmosphere. Adiabatic rates (cooling rates of air irrespective of surrounding air) and lapse rates are described, an understanding of which is necessary to study temperature inversions. The chapters are well written, nontechnical, and easy to understand.

AUROROS

Auroras, the northern and southern lights, are caused by geomagnetic activity taking place in the Earth's atmosphere. By understanding auroras, scientists can gauge the effects of solar activities on the Earth's environment.

PRINCIPAL TERMS

ATMOSPHERE: the layer of gases that cloaks the surface of the Earth

GEOMAGNETISM: the external magnetic field generated by forces within the Earth; this force attracts materials having similar properties, inducing them to line up (point) along field lines of force

IONOSPHERE: the upper level of the atmosphere, which is ionized (electrically charged) by exposure to sunlight

SOLAR WIND: the stream of gases emitted by the Sun's surface

AURORAL OVAL

"Aurora" is a general term for the light produced by charged particles interacting with the upper reaches of the Earth's atmosphere. The term "aurora borealis" specifically refers to the northern dawn, or northern lights; "aurora australis" refers to the southern lights. The aurora appears in an oval girdling the Earth's geomagnetic poles, where the field lines are perpendicular to the surface. In this region, the Earth is not shielded from the space environment as it is at lower latitudes (where the field lines are almost parallel to the surface); thus, electrons and ions moving along magnetic field lines can strike the atmosphere directly. Normally, the auroral oval is located about 23 degrees from the north magnetic pole and 18 degrees from the south magnetic pole. Because the north magnetic pole is located in Greenland, the oval is offset toward Canada and away from Europe. Generally, the auroras appear at altitudes between 100 and 120 kilometers high, in sheets 1 to 10 kilometers thick and several thousand kilometers long.

The auroral oval is a product of the Earth's magnetic field and is driven by the Sun's output of charged particles. The oval can be enlarged as far north or south as 20 degrees latitude; its normal range is around 55 to 60 degrees. These variations in range and intensity have been correlated with sunspots, showing that solar activity is the engine that drives the aurora and other geomagnetic disturbances. Additionally, scientists usually describe auroral activity in terms of local time relative to the Sun rather than the geographic point over which it occurs. Thus, the Earth can be considered to be rotating beneath the auroral events (even though the shape of the oval remains skewed). The first indication that the aurora might be linked to solar activity came in 1859, when Richard Carrington observed an especially powerful solar flare in white light. A few hours later, he observed a strong auroral display and suspected that the two might be linked.

ELECTRON PRECIPITATION

Auroras are caused by electron precipitation: Electrons "rain" on the upper atmosphere from this field-aligned current. The analogy is limited, as rain falls under the influence of gravity, while electrons moving in a magnetic field do so in a helix wrapped around a field line, somewhat like the rifling of a gun barrel. The helix of electrons trapped in the Earth's magnetic field becomes more pronounced as they approach the poles, until finally their direction is reversed (at the "mirror point") and they are reflected back to the opposite pole. Motion back and forth is normal.

If the electrons are accelerated into the ionosphere, they encounter oxygen atoms (from molecules dissociated by sunlight) and nitrogen molecules. These collisions will release Bremsstrahlung (braking) X rays, which are absorbed by the atmosphere or radiated into space. The oxygen is ionized (and an electron freed) and radiates light when it is neutralized by a free electron. Nitrogen

The aurora borealis as seen near Anchorage, Alaska, in 1977. (Collection of Dr. Herbert Kroehl, NGDC)

tion, electric fields and currents move freely, although the net electrical charge is zero.

The structure of an aurora varies widely. Three major forms have been discerned: quiet (or homogeneous) arcs, rayed arcs, and diffuse patches. Homogeneous arcs appear as "curtains" or bands across the sky. They sometimes will occur as pairs or (rarely) sets of parallel arcs and have also been described as resembling ribbons of light. The lower edge of the arc will be sharply defined as it reaches a certain density level in the atmosphere, but the upper edge usually simply fades into space. Pulsating arcs vary rapidly in brightness. Also in the category of quiet arcs are diffuse luminous surfaces, which are like clouds and have no defined structure; they may also appear as a pulsating surface. Finally, the weakest homogeneous display is a feeble glow, which actually is the upper level of an auroral display just beyond the horizon.

Auroras with rays appear as shafts of light, usually in bundles. A rayed arc is similar to a homogeneous arc but comprises rays rather than evenly distributed light. The formation and dissipation of individual rays may produce the illusion that rays are moving along the length of a curtain. Among rayed arcs, the "drapery" most resembles a curtain and is most active in shape and color changes. If the viewer is directly below the zenith of an auroral event, then it appears as a corona, with parallel rays appearing to radiate from a central point. Drapery displays are often followed by flaming auroras that move toward the zenith.

either is excited and radiates when it returns to the "ground" state or is dissociated and excited.

The aurora has been compared to a television picture tube: The face of the tube has been relatively simple to understand, but the circuitry that drives the display remains elusive. As the atmosphere fades gradually into space, starting at about 60 kilometers above the surface of the Earth, it forms the ionosphere, which contains many free ions and electrons. Because many of these free atoms and molecules are also ionized by solar radia-

A controversial aspect of auroras is whether they produce any sound. Many observers, from antiquity, have reported "hearing" the aurora; however, sensitive sound-recording equipment has yet to capture this sound. This leaves open the question of whether the sound is a psychological perception, an electrostatic discharge, or some other phenomenon.

COLOR AND BRIGHTNESS

The colors of the aurora—pink, red, green, and blue-green—are distinct and correspond with a specific wavelength of light rather than being a continuous spectrum typical of a uniformly hot body (such as the Sun). Major emissions come from atomic oxygen (at wavelengths of 557.7 and 630 nanometers) and molecular nitrogen (391.4, 470, 650, and 680 nanometers). These emissions come from distinct altitudes. The green oxygen line (557.7 nanometers), which peaks at an altitude of 100 kilometers, is caused by an energy state that decays in 0.7 second. The red oxygen line (630 nanometers), which peaks at about 300 kilometers, comes from an energy state that decays in 200 seconds. While oxygen is energized to this level at lower altitudes, its energy will be lost to collisions with other gases long before it can decay naturally. From such comparisons, geophysicists were able to deduce some of the vertical structure of the atmosphere. X rays and ultraviolet light are also emitted but cannot be detected from the ground. The Dynamics Explorer 1 satellite has recorded the aurora at 130 nanometers in hundreds of images taken several Earth radii above the North Pole.

The brightness of the aurora can vary widely. Four levels of international brightness coefficients have been assigned, ranging from IBC I, which is comparable to the brightness of the Milky Way, to IBC IV, which equals the illumination received from a full Moon. Auroras usually are eighty times brighter in atomic oxygen than in ionized nitrogen molecules. Doppler shifting is commonly recorded in the spectra around 656.3 nanometers (hydrogen-alpha), indicating the motion of protons that are neutralized and reionized as they accelerate up or down the field lines. It is theorized that, as with many natural effects, only a small fraction (about 0.5 percent) of the energy that goes into the auroras actually produces light. The remainder goes into radio waves, ultraviolet rays, and X rays, and into heating the upper atmosphere.

Single images from the Dynamics Explorer satellite have confirmed the indication by ground-based camera chains that the aurora is uneven in density and brightness. One image, for example, shows that the auroras thin almost to extinction on the dayside but expand to several hundred kilometers in thickness between about 10:00 P.M. and 2:00 A.M. local time. "Theta" auroras have been recorded in which a straight auroral line crosses the oval in the center, giving the appearance of the Greek letter θ. This phenomenon may be caused by the splitting of the tail of the plasma sheet (which extends well into the tail of the magnetosphere) or by the solar wind's magnetic field when it has a direction opposite the Earth's.

SOLAR WIND

Imagery by the spin-scan auroral imagers aboard the Dynamics Explorer 1 satellite showed that auroral substorms start at midnight, local time, and expand around the oval. Observations of hundreds of substorms showed that they have the same generalized structure but that no two are identical. The satellite imager also showed expansions and contractions in the aurora in response to changes in the interplanetary magnetic field and in the solar wind. As the solar wind—which is simply a plasma—meets the Earth's magnetosphere, a shock wave is formed, and the wind is diverted around the Earth. This diversion compresses the Earth's magnetic field on the sunward side, while it extends like a comet's tail on the nightside. When the field of the solar wind is oriented toward the south, its field lines reconnect with the field lines of the Earth and allow protons (free hydrogen nuclei) and electrons to enter the magnetosphere (they are normally blocked when the field is oriented to the north).

Auroral activity is driven by the solar wind. When the magnetic field of the wind points north—aligned with the Earth's magnetic field—the auroral oval is relatively small, and its glow is hard to see. When the solar wind's magnetic field reverses direction, a substorm occurs. The oval starts to brighten within an hour, and bright curtains form within it. At its peak, the oval will be thinned toward the noon side and quite thick and active on the midnight side. As the storm subsides, about four hours after the field reversed (actually, as it starts to revert to normal), the aurora dims. Finally, a large, diffuse glow covering the pole may be left as the field becomes stronger in the northward direction.

The flow of the solar wind past the magnetosphere generates massive electrical currents, which flow mostly from one side of the magnetosphere to the other. Some of the currents, however, connect down the Earth's magnetic field, into and through the auroral oval. Because an electric current is caused by the flow of charged particles (electrons in this case), in the process, electrons are brought directly into the ionosphere around the poles. The primary currents enter around the morning side and exit around the evening side. Secondary currents flow in the opposite direction. Changes in the electrical potential of the magnetosphere, as when it is pumped up by particles arriving in the solar wind, will force the electrons through the mirror point; they are then accelerated deeper into the ionosphere. This auroral potential structure, as it is called, is thin but extends around the auroral oval for thousands of kilometers, even to the point of closing on itself.

E-Cross-B Drift and SAR Arcs

Electrojets also form in the auroras at low altitudes from an effect known as "E-cross-B drift" (written $E \times B$). At high altitudes, electrons and protons flow freely because there is a low gas density and no net current change. At lower altitudes, around 100 kilometers, the protons are slowed by collisions with atoms and molecules, but the electrons continue to move unopposed. The result is a pair of electrojets, eastward (evening) and westward (morning), which flow toward midnight, then cross the polar cap toward noon. These electrojets heat the ionosphere, especially during active solar periods, when the aurora is more intense.

This $E \times B$ drift in the auroral ovals appears to be a major source of plasma for the magnetosphere. It appears that the ions (which are positively charged) are accelerated upward along the same magnetic field lines where electrons (negatively charged) precipitate. Hydrogen, helium, oxygen, and nitrogen make up this ion flow. Each has the same total energy, so their paths vary according to mass. The net effect is that of an ion fountain blowing upward from the auroras then spread by a wind across the poles.

A little-known subset of the aurora is the subauroral red (SAR) arcs, which appear at the midlatitudes; the magnetic field lines on which they oc-

cur are different from those on which auroras appear. SAR arcs always emit at 660 nanometers (from oxygen atoms) and are dim and uncommon. Modern instrumentation has shown that the SAR arcs are a separate phenomenon from the polar auroras. These arcs may be caused by cold electrons in the plasmasphere interacting with plasma waves or with energetic ions. SAR arcs are believed to originate at approximately 19,000 to 26,000 kilometers altitude during especially strong geomagnetic storms, although the arcs themselves appear at altitudes around 400 kilometers as the energy from the storm leaks or is forced downward.

The aurora also appears in the radio spectrum. Studies in the twentieth century showed that the aurora could be sounded by radar at certain frequencies. Satellites in the 1970's started recording bursts of energy in the low end of the AM radio spectrum. This radiation is called auroral kilometric radiation (AKR), because its wavelength is up to 3 kilometers, reflected outward by the ionosphere. Such bursts can release 100 million to 1,000 million watts at a time, making them far more powerful than conventional broadcasts by humans. The bursts originate in a region of the sky about 6,400 to 18,000 kilometers high, in the evening sector of the auroral oval. Because the radiation is polarized, it is likely that AKR is caused directly by electrons spiraling along magnetic field lines in a natural mimic of free-electron lasers in the laboratory.

Dave Dooling

Cross-References

BIBLIOGRAPHY

Akasofu, Syun-Ichi. "The Dynamic Aurora." *Scientific American* 260 (May, 1989). A detailed, college-level treatment of the current understanding of auroras, written by the physicist who is generally accepted as the world expert.

Bone, Neil. *The Aurora: Sun-Earth Interactions.* 2d ed. New York: Wiley, 1996. This college-level text provides the perfect introduction to the origins, physics, and significance of auroras. Suitable for the reader with no prior knowledge in the field. Color illustrations, bibliographical references, and index.

Bryant, Duncan Alan. *Electron Acceleration in the Aurora and Beyond.* Philadelphia: Institute of Physics, 1999. A good look into the geophysics of auroras. Bryant explains the relationship between the electron and the aurora in a somewhat technical manner. This college text is recommended for the advanced reader with much interest in the field. Color illustrations, index, and bibliography.

Delobeau, Francis. *The Environment of the Earth.* New York: Springer-Verlag, 1971. A technical description of the terrestrial environment, written as a reference for space scientists. Although the work is dated by subsequent discoveries, its description of auroral chemistry is still valid.

Dooling, Dave. "Satellite Data Alters View on Earth-Space Environment." *Spaceflight* (July, 1987). An article focusing on the exploration of the magnetosphere by the Dynamics Explorer satellites, with details on auroral imaging and radiation.

Eather, Robert H. *Majestic Lights: The Aurora in Science.* Washington, D.C.: American Geophysical Union, 1980. A well-illustrated, informative booklet describing auroras through history and their modern scientific interpretation. Written for general audiences.

Lysak, Robert L., ed. *Auroral Plasma Dynamics.* Washington, D.C.: American Geophysical Union, 1993. A well-illustrated, informative booklet describing auroras and auroral plasma through history and their modern scientific interpretation. Written for general audiences.

Petrie, William. *Keoeeit: The Story of the Aurora Borealis.* Oxford: Pergamon Press, 1963. Highly detailed description of the history of auroras (largely from the Canadian and Eskimo point of view) and of the structure of auroral displays. Also tells much of the ground-based exploration of the aurora.

COSMIC RAYS AND BACKGROUND RADIATION

Cosmic rays are highly energetic protons and heavier atomic nuclei that continually rain down upon the Earth from space. Ranging in energies up to 10^{20} electronvolts (about 20 joules), cosmic rays are not well understood. Primary cosmic rays can produce a secondary "cosmic-ray air shower"—a plethora of secondary particles reaching the Earth's surface as a result of collisions of primary rays with atoms in the atmosphere.

PRINCIPAL TERMS

ALPHA PARTICLE: a helium 4 nucleus consisting of two protons and two neutrons

ELECTRONVOLT (eV): a unit of energy used for atomic and subatomic measurements; 1 eV is the kinetic energy acquired by an electron accelerated through a potential difference of 1 volt

FLUX: the number of particles striking a unit of surface area per unit of time

ISOTOPES: atoms of a given element with the same number of protons but different numbers of neutrons

NEUTRINOS: massless (or nearly massless) particles given off in certain types of nuclear reactions

PHOTON: the smallest energy packet of light for a given frequency; X rays and gamma rays are examples of high-energy photons

POSITRON: the antiparticle to the electron

PROTON: the nucleus of the hydrogen 1 atom; a proton carries one unit of charge, and the number of protons in the nucleus of an atom determines what element it is

COSMIC-RAY PARTICLES

Cosmic-ray particles are fast-moving subatomic particles that are continually striking the Earth's atmosphere. They are mostly highly energetic, completely ionized nuclei. About 90 percent of primary cosmic rays are protons, 9 percent are alpha particles (helium 4 nuclei), and the rest are nuclei of heavier elements and electrons (beta rays). Most cosmic rays are moving so fast that they must be described in relativistic terms—that is, their kinetic energies are comparable to or much greater than their rest mass energies. Sources of cosmic rays include the Sun, other sources believed to be in this galaxy (for example, supernovas), and other galaxies, particularly active galactic nuclei.

The flux of cosmic-ray particles decreases approximately exponentially with increasing energy. The highest-energy cosmic rays are in the range of 10^{20} electronvolts (100 million million million electronvolts, or approximately 20 joules). This energy is about eleven orders of magnitude greater than the rest mass-energy of the proton.

Cosmic rays are generally divided into two categories: primary and secondary. The primary rays strike the Earth's atmosphere; secondary rays are the resulting cascading showers of particles produced by collisions of the primary rays with atoms in the atmosphere. Primary rays have the following characteristics: Their intensity seems to be essentially constant; their flux and spectrum appear to be isotropic in space (the same in all directions); they are anomalous in composition; and their spectrum includes very energetic particles.

An important feature of cosmic rays is that their chemical abundance is significantly different from that of the Sun. Likewise, their chemical abundance is significantly different from that of the universe in general. Cosmic-ray composition is particularly rich in heavier elements. However, even with the lighter elements, cosmic rays contain about one million times as much lithium, beryllium, and boron relative to hydrogen as does the Sun.

Not only are cosmic rays individually energetic, but, given their high spatial particle density, they also represent a large fraction of the total energy associated with astrophysical phenomena. The energy density of cosmic rays is comparable to that of photons, interstellar magnetic fields, and the turbulent motion of interstellar material, each of which is approximately 1 electronvolt per cubic centimeter.

SOURCE AND EFFECT OF COSMIC RAYS

Since their discovery in 1911 by V. F. Hess, cosmic rays have presented scientists with an enigma. The fundamental questions with regard to cosmic rays—where do they come from, and how are they accelerated to such high energies—have yet to be fully understood. Although the Sun is a source of some of the lower-energy cosmic rays, it is clear that it cannot be responsible for the higher energy of the cosmic-ray spectrum. As yet, there has not been a satisfactory mechanism developed that can realistically account for the high acceleration necessary to account for the higher-energy rays. However, some promising models have recently been developed whereby cosmic rays may be repeatedly

accelerated by shock waves from violent astrophysical phenomena. Sources for these shock waves would include exploding galaxies and supernovas (exploding stars). It is well known that supernovas are rich in heavier elements. Indeed, this is presently thought to be the primary mechanism whereby heavier elements are synthesized.

The flux of primary cosmic rays incident upon the Earth's atmosphere is about 1 particle per square centimeter per second. However, the Earth's magnetic field effectively prohibits cosmic rays of less than 10^8 eV (100 million electronvolts) from reaching the surface, the radius of curvature of the path of the charged particles being sufficiently short that they are turned away by the Earth's magnetic field.

There is a measured latitude effect in the flux of cosmic rays reaching the Earth. This effect is caused by the Earth's magnetic field, which mani-

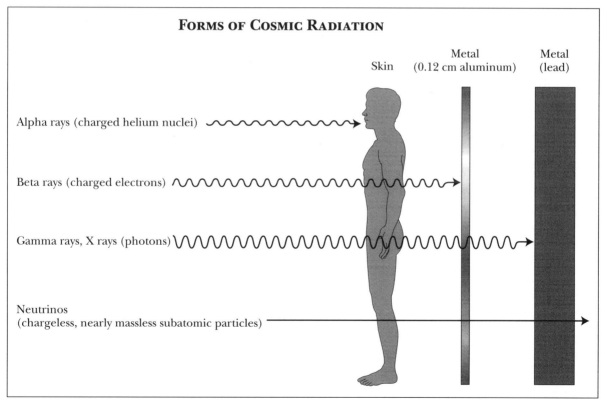

FORMS OF COSMIC RADIATION

Skin   Metal (0.12 cm aluminum)   Metal (lead)

Alpha rays (charged helium nuclei)

Beta rays (charged electrons)

Gamma rays, X rays (photons)

Neutrinos (chargeless, nearly massless subatomic particles)

Different forms of cosmic radiation can penetrate different forms of matter: Alpha rays cannot penetrate skin; beta rays can penetrate skin but not metal; gamma rays can penetrate both but are stopped by lead; and neutrinos, chargeless, nearly massless particles, can penetrate even lead, making them extremely difficult to detect. Although they interact very little with matter, neutrinos are believed to be produced in the nuclear reactions at the core of the Sun and other stars and may constitute a large portion of the "missing mass" of the universe.

fests a higher cutoff energy (the minimum kinetic energy required for the cosmic rays to reach the atmosphere) at the geomagnetic equator than at higher latitudes. For example, the energy cutoff for vertically arriving protons at the geomagnetic equator is about 15 gigaelectronvolts (GeV), whereas at geomagnetic latitude 50 degrees, the cutoff energy is 2.7 GeV.

Furthermore, there is a significant east-west effect in the flux of cosmic rays incident upon the Earth, also the result of the Earth's magnetic field. The cutoff energy for cosmic rays reaching the Earth's surface is less from the west (about 10 GeV) than from the east (about 60 GeV) for positive charges interacting in the Earth's magnetic field.

High-energy gamma-ray photons have also been observed in cosmic rays. Gamma rays and X rays produced by the interaction of primary cosmic rays with interstellar material yield information on the distribution and composition of matter in the galaxy.

Several point sources of X rays and gamma rays have been observed, for example, the Crab Nebula, Hercules X-1, and Cygnus X-3. These sources have been observed to emit gamma rays in the range of 10^{12} eV.

Intense gamma-ray burst events (about five per year) have also been observed. These bursts are characterized by an initial intense pulse of 0.1 second to 4 seconds duration followed by one or more pulses, the entire event taking place within 1 minute.

SHOCK WAVES AND REMOTE MATTER

Researchers have also begun to investigate the ability of shock waves to accelerate cosmic rays to some of the higher energies observed in the cosmic-ray spectrum. Shocks are common in the solar system. For example, "bow" shock waves are produced when the solar wind is deflected by planetary magnetic fields, from solar flares, and when fast solar-wind streams overtake slower streams (somewhat like the effects that produce tsunamis in the Earth's oceans). Some solar-wind particles attain high energies through this mechanism, which is generally referred to as "diffusive shock acceleration." Furthermore, mathematical models have shown that the shock waves generated by supernova explosions can dramatically accelerate charged particles.

The resulting acceleration from shock waves may also account for the anomalous component of cosmic rays. Thought to originate from neutral particles in the interstellar medium that become ionized via collisions with the solar wind or by solar photons, anomalous cosmic rays may then be further accelerated at their termination by the shock of the solar wind. However, this is still a very speculative model, and spacecraft have yet to detect the termination shock of the solar wind. Hence, its properties and effects are currently somewhat conjectural; the ultrahigh energies observed in the range of 10^{20} eV still defy a clear understanding as to their production and acceleration mechanisms.

The nuclear composition of cosmic rays enables scientists to sample matter that comes to the Earth from remote parts of the galaxy. Indeed, some cosmic rays are thought to be of extragalactic origin as well. Data on the composition of cosmic rays suggest that they may well have experienced nuclear conditions similar to those of the interstellar medium in the distant past. The cosmic-ray spectrum contains abundances of heavier nuclei in approximate proportion to their nuclear charge. For example, iron 56 (atomic number 26) is about twenty-six times more abundant than would otherwise be expected. Likewise, this observation extends to elements as heavy as lead and uranium. The lighter elements lithium, beryllium, and boron are present in cosmic rays by a factor of about 100,000 times greater than that found in typical universal abundance. In contrast, the relative proportions of nuclei of odd-even charge are similar to those of universal abundance for those nuclei from carbon to iron.

The extremely high proportions of lithium, beryllium, and boron can be accounted for by collisions of the more abundant heavier nuclei, carbon through iron, with protons—the overwhelmingly most populous nuclei present in the interstellar medium. If most cosmic rays are confined to the galactic disk, then this process, known as "spallation," can be used to calculate a mean galactic age of the cosmic rays of several million years.

One radioactive species, beryllium 10, is particularly suitable as a radioactive clock when compared with its stable neighbor, beryllium 9. The results of this comparison yield a mean age for galactic cosmic rays of greater than 20 million years.

The "leaky box" model of galactic cosmic rays assumes that spallation is not complete by the time cosmic rays reach the Earth and, hence, leads to the prediction that most cosmic rays leave the galactic disk at a relatively young age. The galactic magnetic field has an energy density slightly less than that of cosmic rays. Hence, it is thought that galactic gravity also plays a role in confining cosmic rays within the galactic halo.

Given the energy density of cosmic rays and the requisite confinement energies, it appears that the principal source of cosmic-ray energy is supernovas—events that are also the source for the synthesis of nuclei heavier than helium and lithium.

STUDY OF COSMIC RAYS

There is a variety of methods for observing cosmic rays: ground-based arrays of detectors both at low altitudes and on mountain tops, balloons carrying a detector payload, and rockets and satellites carrying payloads designed for cosmic-ray research.

Ground-based instruments can employ a variety of detection techniques. Some of the more traditional techniques use cloud chambers, scintillation detectors, and photomultiplier tubes. Usually, these instruments are networked into an extensive two-dimensional array (or, in some cases, with limited three-dimensional information). Computer modeling of the arrival times at the various detectors then can be used to reconstruct a shower front, yielding the approximate direction of origin of the shower. A more recent technique detects the Cherenkov radiation given off by fast-moving electrons through the Earth's atmosphere.

Ground-based detectors rarely observe a primary cosmic ray. Rather, they detect the shower of secondary particles produced by the interaction of the high-energy primary ray with atoms in the Earth's atmosphere. Primary particles with energies in excess of a few hundred million electron-volts (MeV) per nucleon will readily interact with atmospheric nuclei, mainly producing neutral pions, charged pions, protons, and neutrons. The neutral pions decay into gamma-ray photons that in turn produce electron-positron pairs in a cascading process known as an "extensive air shower." The charged pions decay into muons, some of which can be very penetrating. A case in point is the muon detector 1.6 kilometers underground at the Homestake Detector in South Dakota. The nucleons (neutrons and protons) produced by primary cosmic rays are sufficiently energetic to produce additional nucleons, resulting in neutron-proton cascades, which are also detected at ground-based array detectors. Production of these cascades is very sensitive to atmospheric conditions, in particular, temperature and pressure.

The use of rockets and satellites beginning in the 1960's enabled researchers to establish crucial characteristics of cosmic rays not directly discernible from ground-based detectors. In particular, scientists were able to determine the energy spectra and particle abundances of primary cosmic rays. For example, gamma-ray emissions from the plane of the galaxy were first detected by the OSO III satellite in 1967. The SAS-2 satellite then detected the diffuse gamma-ray background, and the COS-B satellite provided a detailed map of gamma-ray emissions in the galaxy. In addition, COS-B discovered twenty-five discrete gamma-ray sources, including 3C273 (a quasar) and the pulsars in the Crab and Vela supernova remnants. The Einstein X-Ray Observatory was launched in 1978, making high-resolution surveys of selected areas of the sky. The Einstein Observatory provided detailed images of many X-ray sources.

Stephen Huber

CROSS-REFERENCES

BIBLIOGRAPHY

Darrow, K. K. "Cosmic Radiation: Discoveries Reported in *The Physical Review*." In *The Physical Review: The First 100 Years*. New York: American Institute of Physics, 1996. The lengthy introduction provides an overview of the development of cosmic-ray physics. Includes an excellent bibliography for further investigation.

Davies, P., ed. *The New Physics*. New York: Cambridge University Press, 1990. Several chapters in this excellent book include discussion of cosmic rays. In particular, the chapter "The New Astrophysics," by Malcolm Longair, is strongly recommended.

Gaisser, Thomas K. *Cosmic Rays and Particle Physics*. New York: Cambridge University Press, 1990. Although this text can be somewhat technical at times, it is an excellent overview of the discipline of cosmic-ray astrophysics.

Gombosi, Tamaas I. *Physics of the Space Environment*. New York: Cambridge University Press, 1998. This college-level text explores the field of atmospheric physics to determine the makeup of the upper atmosphere and the space environment and how these forces affect the Earth's atmosphere and weather cycles. A focus is placed on solar winds and the magnetosphere. Intended for the advanced reader. Illustrations, index, and bibliography.

Murthy, R., P. V. Wolfendale, and A. W. Wolfendale. *Gamma-Ray Astronomy*. New York: Cambridge University Press, 1986. A good overview of gamma-ray (high-energy photon) astronomy.

Silberberg, R., C. H. Tsao, and J. R. Letaw. "Composition, Propagation, and Reacceleration of Cosmic Rays." In M. M. Shapiro et al., eds., *Particle Astrophysics and Cosmology*. Boston: Kluwer Academic Publishers, 1993. An excellent, though somewhat technical, discussion of cosmic-ray astrophysics.

EARTH-SUN RELATIONS

The relationship between the Sun and Earth controls life on this planet. Earth's "heat budget" is a result of many factors, including the effects of the atmosphere and of the oceans, but the phenomena of Earth rotation and revolution are primary. Earth motions also produce observable periodic changes in apparent Sun paths, perhaps most visible in the directions of sunrise and sunset.

PRINCIPAL TERMS

AXIAL TILT: a 23.5-degree tilt in the pole-to-pole line about which Earth rotates, relative to the plane of the ecliptic

CIRCLE OF ILLUMINATION: the circle on Earth's surface that bisects Earth and separates the sunlit half from the shadowed half

EARTH'S HEAT BUDGET: the balance between incoming short-wavelength solar radiation and outgoing long-wavelength infrared terrestrial radiation

ELLIPSE: the shape of Earth's orbit; rather than a circle with one center, the ellipse has two foci with the Sun located at one of the foci

GREENHOUSE EFFECT: the ability of Earth's atmosphere to allow short-wavelength solar radiation to pass in to Earth's surface but to retard the escape of long-wavelength heat into space

PATHS OF THE SUN: the apparent motions of the Sun as it tracks across the sky

PLANE OF THE ECLIPTIC: the plane in which Earth's orbit lies as it revolves around the Sun

SOLAR RADIATION: the energy given off by the Sun

SOLAR SYSTEM: the Sun and all the bodies that orbit it, including the nine planets and their satellites, plus numerous comets, asteroids, and meteoroids

SPECIFIC HEAT: the number of calories of heat required to raise the temperature of 1 gram of a substance 1 degree Celsius

SOLAR RADIATION

Earth-Sun relations are the dominant controls of life on Earth. The Sun is a star, and its radiation warms Earth and supplies the energy that supports life on the planet. Earth-Sun relations are the phenomena that determine the amount, duration, and distribution of solar radiation that is received by Earth. Earth motions of rotation and revolution cause day and night and the seasons, which serve to distribute solar radiation over Earth. Earth's atmosphere and oceans influence the reflection, absorption, and distribution of solar energy. The result of these interacting phenomena is an Earth heat budget that is hospitable and constant. Earth motions also result in a pattern of periodic changes that are observable in the apparent paths of the Sun.

The Sun radiates energy from every part of its spherical surface. Earth, 150 million kilometers away, receives a minute portion of the star's radiation, no more than one two-hundred-billionth, and yet Earth cannot tolerate full exposure to even that amount of radiation. The greatest amount of solar radiation is, in effect, wasted as far as Earth is concerned, radiating outward in all directions. That small amount of the Sun's energy that strikes Earth is Earth's energizer. It sustains life on Earth and drives the weather systems and the oceans' circulations. Solar radiation from the past has been preserved in the form of fossil fuels—coal, petroleum, and natural gas.

TEMPERATURE REGULATION

Perhaps the most remarkable aspects of Earth—remarkable because they are rare in the universe—are its moderate temperatures and the constancy of those temperatures. The adjectives "hot" and "cold" are frequently used in describing the weather. In relation to the temperatures that are found in the solar system, Earth is always moderate, and the words "hot" and "cold" better describe the other planets. The mean temperature of Earth is about 15 degrees Celsius; the absolute extremes recorded anywhere on Earth are 58 degrees in North Africa and −89 degrees in Antarctica. Few inhabitants of Earth will ever experience a temperature range of much over 55 degrees in a lifetime. Compare those temperatures with those of Earth's near neighbor, the Moon, where temperatures range from 127 to −173 degrees. Earth's sister planet Venus has a surface temperature of about 480 degrees. The outer planets of the solar system experience a permanent deep freeze, below −200 degrees. All the planets' temperatures are extreme in comparison with Earth's.

The most convincing proof of the moderate nature of Earth's temperature is the presence of the world's oceans. Water in the liquid state is extremely fragile and will exist only in a narrow temperature band, 0 to 100 degrees at Earth's atmospheric pressure. Water must be rare indeed in the universe; it is probably nonexistent elsewhere in the solar system, with the possible exception of trace amounts on Mars. Yet Earth has 71 percent of its surface covered with this rare and fragile substance. Almost 98 percent of Earth's water is in the liquid state. The polar ice caps contain 2 percent, and a minute portion is water vapor in the atmosphere at any time.

The factors that cause Earth to experience such a moderate and unchanging temperature are complex and interrelated. The Sun is the source of the energy. Yet being the right distance from the Sun cannot be the sole cause of Earth's moderate temperature; witness the Moon. Rather, the explanation has to do with Earth's atmosphere, the oceans, and—in particular—Earth-Sun relations, determined by rotation and revolution. The atmosphere protects Earth during both daylight and darkness. During daylight, the atmosphere blocks excessive amounts of short-wave solar radiation from reaching the surface. During darkness, the atmosphere retards the escape of long-wave infrared heat energy back into space and thus prevents excessive overnight cooling. The result is a moderation of temperatures. The oceans also have a pronounced effect on the heat budget of Earth. Water has the highest specific heat of any common substance. More heat is needed to raise the temperature of water than to raise the temperature of other materials. Summers and daylight periods are kept cooler by the water's abil-

FACTS ABOUT THE SUN

	Sun	Earth
Mass (10^{24} kg)	1,989,100	5.9736
Volume (10^{12} km^3)	1,412,000	1.083
Volumetric mean radius (km)	696,000	6371
Mean density (kg/m^3)	1408	5520
Surface gravity at equator (m/s^2)	274.0	9.78
Escape velocity (km/s)	617.7	11.2
Ellipticity	0.00005	0.0034
Absolute magnitude	+4.83	—
Luminosity (10^{24} J/s)	384.6	—
Mean distance from Earth (10^6 km)	149.6	—

THE SUN'S ATMOSPHERE

Surface gas pressure (top of photosphere)	0.868 millibar
Effective temperature	5778 Kelvins
Temperature at bottom of photosphere	6600 Kelvins
Temperature at top of photosphere	4400 Kelvins
Temperature at top of chromosphere	~30,000 Kelvins
Photosphere thickness	~400 km
Chromosphere thickness	~2500 km
Sunspot cycle	11.4 yrs.
Photosphere Composition:	
Hydrogen	90.965%
Helium	8.889%
Oxygen	774 ppm
Carbon	330 ppm
Neon	112 ppm
Nitrogen	102 ppm
Iron	43 ppm
Magnesium	35 ppm
Silicon	32 ppm
Sulfur	15 ppm

SOURCE: Data are from the National Aeronautics and Space Administration/Goddard Space Flight Center, National Space Science Data Center. URL: http://nssdc.gsfc.nasa.gov/planetary.

ity to absorb great amounts of solar energy without the water's temperature being raised significantly. During winter and during darkness, on the other hand, the water slowly gives up large amounts of heat without significant cooling of the water. Thus the oceans act as a huge temperature buffer and, similar to the atmosphere, add a moderating effect to temperature extremes.

Even more significant in controlling the temperature of Earth, however, are Earth's rotation and revolution. Rotation is Earth's turning on its axis (counterclockwise, if one's vantage point were above the North Pole). Rotation causes places on Earth to be alternately turned toward and away from the Sun. The effect is to prevent Earth from overheating or overcooling. If one side of Earth were continuously exposed to the Sun, the illuminated side would heat to hundreds of degrees, and the dark side would cool hundreds of degrees. The rotation of Mercury is such that a given point on the planet's surface is exposed to the Sun for eighty-eight Earth days and then is on the dark side for eighty-eight days. The resultant temperature extremes range from about 450 to −170 degrees Celsius. The other planets have various rates of rotation. If the rate of Earth rotation, once in twenty-four hours, were different, Earth's heat budget would be different. A slower rotation would result in greater extremes of daily temperatures.

EARTH REVOLUTION

One complete revolution, or orbit, of Earth around the Sun defines the time unit of one year. During a single revolution, Earth rotates on its axis 365¼ times; therefore, there are 365 days in a year, with an extra day every fourth year—leap year. Earth orbits the Sun in an elliptical path, or an ellipse, and the ellipse lies in a plane referred to as the plane of the ecliptic. The Sun is located at one of the two foci of the ellipse; thus Earth is nearer to the Sun at one position in the orbit than

An X-ray image of the Sun. (PhotoDisc)

at any other position. That position occurs on or about January 3 and is called perihelion, meaning "near the Sun." The Earth-Sun distance at perihelion is about 147,000,000 kilometers. At the opposite position on the elliptical orbit, called aphelion, on or about July 4, Earth is farthest from the Sun, 152,000,000 kilometers away.

This variation in Earth's distance from the Sun does alter the amount of solar radiation that is received by Earth, but it is not the cause of the seasons. Perihelion, when Earth is nearest to the Sun and seemingly when Earth would be the warmest, occurs during winter in the Northern Hemisphere, and aphelion occurs during the Northern Hemisphere's summer. The distance variations are out of phase with the Northern Hemisphere's seasons, then, but in phase with seasons in the Southern Hemisphere. In both cases, they modify the seasons but do not cause them. The cause of the seasons is the fact that Earth's axis of rotation is tilted 23.5 degrees from the vertical to the plane of the ecliptic. The tilt remains constant year-round in reference to space and the plane of the ecliptic, but because Earth revolves around the Sun, the axis in the Northern Hemisphere alternately tilts toward and away from the Sun. When the North Pole is tilted away from the Sun, the Southern Hemisphere receives more solar radiation than does the Northern Hemisphere. At the position in orbit where the northern tip of the axis is tilted directly away from the Sun, the Sun is directly overhead at the Tropic of Capricorn (23.5 degrees south latitude), and the circle of illumination is tangent to the Arctic Circle (66.5 degrees north latitude). This position in orbit is referred to as the December solstice and occurs on December 21 or 22. For the Northern Hemisphere, it is the "winter solstice," but it is the "summer solstice" for the Southern Hemisphere. Six months later, when the northern tip of the axis is tilted directly toward the Sun, the Northern Hemisphere in turn receives more solar radiation than does the Southern Hemisphere. This position in orbit, known as the June solstice, occurs on June 21 or 22. At this time, the Sun is directly overhead at the Tropic of Cancer (23.5 degrees north latitude), and the entire area north of the Arctic Circle is experiencing continuous daylight. Approximately halfway in the orbit between the two solstices occur the two equinoxes. The March equinox, occurring on March 20 or 21, is the vernal or spring equinox for the Northern Hemisphere, but it is the autumnal or fall equinox for the Southern Hemisphere. The September equinox, September 22 or 23, is the fall equinox for the Northern Hemisphere but the spring equinox for the Southern Hemisphere. On the two equinoxes, the Sun is directly overhead at the equator (0 degrees latitude), and the circle of illumination passes through both poles and bisects the parallels of all latitudes. Both hemispheres receive equal solar radiation on the equinoxes, but in March the Sun is moving northward, whereas in September it is moving southward. It is on these two dates only that daylight and darkness are equal. It is also only on these dates that the Sun rises due east and sets due west.

Between March and September, when the Sun's vertical rays are north of the equator, sunrise is north of east and sunset is north of west. This is true for all locations that would experience sunrise and sunset on a particular date, both Northern and Southern Hemisphere locations. (The places that would not experience sunrise and sunset are those areas near the poles that would be experiencing either continuous daylight or continuous darkness on that date.) Sunrise would be farthest north of east on the June solstice, and sunset would be farthest north of west on that same day. After the June solstice, sunrise

and sunset both migrate daily southward, being due east and due west on the September equinox. Between September and March, sunrise is south of east and sunset is south of west for all places that experience sunrise and sunset on a particular date, for both hemispheres. Sunrise is farthest south of east and sunset farthest south of west on the December solstice, after which they both begin a northward migration to repeat the pattern.

STUDY OF EARTH-SUN RELATIONS

Astronomers and astrophysicists study the Sun as they do any other star—that is, as having a life cycle that will eventually end in its death some 5 billion years hence, when it begins to exhaust its hydrogen fuel. Earth scientists, however, investigate the Earth-Sun system as though the Sun were everlasting, which, from the perspective of Earth-Sun relations, it is. Earth scientists are concerned principally with the effects of Earth's rotation and revolution on the amount of solar energy Earth receives, as well as its effects on the atmosphere and on the ocean. Earth scientists are interested in studying such solar phenomena as sunspots, prominences, and flares in order to ascertain whether they have any effects on Earth's weather and climate patterns. The cycles of solar phenomena have been compared to the records of tree growth rings. These solar phenomena are detected and photographed by filtering solar radiation so that light of a particular wavelength (color) can be viewed. Spectacular solar prominences have been captured on film.

Modern technology enables precise measurements over time of rotation and revolution in reference to the stars. Telescopic photography and the principle of parallax enable scientists to record Earth's position in space over time with great precision. Friction caused by tidal action, in turn caused by the Moon's gravity, may have caused an eons-long slowdown in rotation. Perhaps, over time, precise measurements will shed light on such matters. It has been determined that there are wobbles and gyrations in Earth's axis tilt. At present, the axis orientation is slowly moving away from the star Polaris and toward Vega. It is hypothesized that the ellipticity of the orbit undergoes cyclic changes. The positions in orbit where the solstices and equinoxes occur have been ascertained to be slowly changing. A changing relationship between the solstices and perihelion will doubtless alter the distribution of solar radiation between the hemispheres. The analysis of orbit changes offers a possible explanation for the ice ages.

Several areas of research relate to possible Earth-Sun relations in the past. Fossils, sedimentation rates, and extinction of species offer insight into heat budgets that prevailed on Earth in former eras. Hypotheses of the causes of the ice ages make reference to such factors as changes in solar output, changes in Earth output, changes in Earth's atmospheric makeup, a reconfiguration of ocean currents, and even the presence of volcanic dust. Satellite images of the oceans, on the various wavelengths of the electromagnetic spectrum, are analyzed to detect any slight temperature changes over time that may portend changes in Earth's environment. Sensitive instruments measure and record the intensity of sunlight in such areas as Antarctica. Satellite imagery also is providing an accurate record-base for areas of snow cover in polar regions. If Earth's heat budget should change, the areas and duration of snow cover will be among the earliest evidences to be detected. Intense research continues on changes in the composition of the atmosphere and their possible effects on the heat budget. One conclusion seems clear: Earth's heat budget is a product of many factors in Earth-Sun relations, not all of which are fully understood, and any changes in the budget are likely to be detrimental.

Significance

Life on Earth is profoundly dependent upon the relationship between Earth and the Sun. Fortunately, the Sun is a long-lived and constant radiating star. Yet continuous exposure to it would blister Earth. Rotation and revolution make Earth a rotisserie, slowly turning so as to expose all sides for a more even heat. The lengths of daylight and darkness and the changes of the seasons are phenomena that are beyond human control. It is to be hoped that they will not change, because life on Earth can tolerate very little change.

While humans cannot alter basic Earth-Sun relations, they apparently can have effects on other factors in Earth's heat budget. The atmosphere protects Earth from overheating by day and from overcooling at night. Earth's "greenhouse effect" is a result of the atmosphere's ability to trap solar radiation as heat during the day and retard its escape back into space at night, when the Sun is not above the horizon. A critical constituent gas that is responsible for the greenhouse effect is carbon dioxide. There is concern that a global warming resulting from an enhanced greenhouse effect may be underway because of additional carbon dioxide being added to the atmosphere through the burning of fossil fuels. This effect is exacerbated by deforestation. Another concern regarding changes in the atmosphere is that human-made pollutants, principally chlorofluorocarbons, may be depleting the gas ozone in the upper atmosphere. A reduction of ozone would allow greater amounts of harmful ultraviolet radiation to penetrate to Earth's surface.

The temperature of Earth is at a fine balance between the input of radiation from the Sun and the absorption of that solar radiation and subsequent reradiation of heat energy from Earth back into space. Life on the planet is dependent on this solar energy and the moderate, constant temperature that results. Changes could possibly occur in Earth's rotation, revolution, or axis tilt. Any such changes would be catastrophic. If any changes in Earth's heat budget occur, however, it is more likely that they will be as a result of changes in the atmosphere brought about by human actions.

John H. Corbet

Cross-References

Acid Rain and Acid Deposition, 1802; Air Pollution, 1809; Atmosphere's Evolution, 1816; Atmosphere's Global Circulation, 1823; Atmosphere's Structure and Thermodynamics, 1828; Atmospheric Inversion Layers, 1836; Auroras, 1841; Cosmic Rays and Background Radiation, 1846; Fallout, 1857; Global Warming, 1862; Greenhouse Effect, 1867; Nuclear Winter, 1874; Nucleosynthesis, 425; Ozone Depletion and Ozone Holes, 1879; Radiocarbon Dating, 537; Radon Gas, 1886; Rain Forests and the Atmosphere, 1890; Van Allen Radiation Belts, 1970.

BIBLIOGRAPHY

Ahrens, C. Donald. *Meteorology Today.* St. Paul, Minn.: West, 1982. This introductory college-level text on meteorology presents a thorough treatment of weather phenomena and explains the seasons and the effects of solar energy on the atmosphere. Written for students with little background in science or mathematics. Includes many illustrations.

Gabler, Robert E., Robert J. Sager, Sheila M. Brazier, and D. L. Wise. *Essentials of Physical Geography.* 3d ed. New York: Saunders College Publishing, 1987. A general introductory-level text on physical geography. Covers rotation, revolution, solar energy, and the elements of weather and climate. Well illustrated. For the general reader.

Gribbin, John R., and Simon Goodwin. *Empire of the Sun: Planets and Moons of the Solar System.* New York: New York University Press, 1998. A wonderful explorations of the planets, their atmospheres, and the space environment, this book provides a good introduction to astronomy for the reader without much background in the field. Color illustrations and index.

Harrison, Lucia Carolyn. *Sun, Earth, Time, and Man.* Chicago: Rand McNally, 1960. This book is considered the classic reference for earth-sun relations. Although dated, it is an excellent source of information and offers a remarkably extensive coverage of earth-sun relations. Most useful to the student who already has a working knowledge of earth-sun relations and wants to investigate further.

Jones, B. W., and Milton Keynes. *The Solar System.* Elmsford, N.Y.: Pergamon Press, 1984. A thorough discussion of the entire solar system from the life cycle of a star to the minor members of the system, including comets and meteoroids. Provides highly detailed information on each planet and its satellites.

Contains some highly technical data, but they are presented in a lucid manner.

Oberlander, Theodore M., and Robert A. Muller. *Essentials of Physical Geography Today.* 2d ed. New York: Random House, 1987. A general introductory text on physical geography, the book includes explanations of earth-sun relations and the energy balance of the Earth, atmosphere, and ocean systems. Well illustrated with maps and diagrams.

Pasachoff, Jay M. *Astronomy Now.* Philadelphia: W. B. Saunders, 1978. An introductory text in astronomy for students with no background in mathematics or physics. Well illustrated, it covers the planets, the solar system, and the galaxies. The portion on the structure and nature of the Sun is helpful.

Scott, Ralph C. *Physical Geography.* New York: John Wiley & Sons, 1988. A general introductory-level text on physical geography. Earth-sun relations are presented within an introduction to the study of the Earth's weather and climates. Clearly written and well illustrated. Recommended as an initial source.

Strahler, Arthur N., and Alan H. Strahler. *Modern Physical Geography.* 4th ed. New York: John Wiley & Sons, 1992. In this general college-level text on physical geography, earth-sun relations are discussed within the context of the study of weather and climate. Diagrams are well employed to explain the Earth's orbit, rotation, revolution, and axis tilt. Easy to read.

Tarbuck, Edward J., and Frederick K. Lutgens. *Earth: An Introduction to Physical Geology.* 6th ed. Upper Saddle River, N.J.: Prentice Hall, 1999. Several chapters in this introductory text deal with the solar system. The nature of solar activity and the motions of the Earth are explained. Well illustrated and highly accessible.

FALLOUT

Fallout consists of radioactive particles that settle out of the atmosphere. Nuclear explosions, nuclear reactor accidents, and improperly contained nuclear wastes are all possible sources of fallout particles. Fallout can carry hazardous radioactivity far from the site of its release.

PRINCIPAL TERMS

GROUND ZERO: the point on the ground that lies directly beneath a nuclear device when it explodes

HALF-LIFE: the time it takes for one-half of a sample of a radioactive isotope to decay

ISOTOPE: a variant form of an element having the same number of protons but a different number of neutrons from the typical form

RADIOACTIVITY: the spontaneous decay of an unstable nucleus

YIELD: the explosive power of a weapon, stated in terms of how many tons of dynamite would be required to produce the same explosive power

NUCLEAR EXPLOSIONS

Fallout is the name given to radioactive particles that rain down from the debris cloud of a nuclear explosion. The first nuclear weapons were fission bombs incorporating the radioactive isotopes uranium 235 or plutonium 239 (isotopes of an element are denoted by the total number of neutrons and protons in an atom). In the process called fission, neutrons striking nuclei of these particular isotopes cause these nuclei to fragment into two roughly equal pieces. Three hundred isotopes of thirty-six elements have been identified in fission fragments. Many of these isotopes are radioactive, and they make up the primary source of radioactive fallout.

In less than one-millionth of a second after detonation, a nuclear explosive changes from a solid material into an expanding fireball with a temperature of 100 million Kelvins and a pressure of 100 million atmospheres. As the boundary of the fireball expands outward, it vaporizes the remaining parts of the bomb. The fireball cools as it grows; eventually, it cools enough so that the bomb vapor condenses into the solid particles that will become fallout. The particles are radioactive because they incorporate the radioactive ashes of the nuclear explosive.

Eventually, the expansion of the fireball slows, and then the simple phenomenon of hot air rising lifts the fireball and its burden of fallout particles high into the atmosphere. For explosions with yields of less than 100 kilotons, the fireball will remain in the troposphere, the lowest layer of the atmosphere. Precipitation along with the natural settling of heavier particles will clear the air of fallout in a time span of hours to days, although some very small particles may remain aloft longer. This fallout is called "local fallout" and may occur near the site of the explosion (often called "ground zero") to as far as a few hundred kilometers distant. After the Hiroshima and Nagasaki explosions, smoky fires started by the bombs caused black rains to fall about thirty minutes after the blasts, and these rains brought down fallout with them.

FUSION AND FISSION

Modern nuclear weapons are generally three-stage devices that use fission, fusion, and fission in turn. In nuclear fusion, nuclei of lighter elements are fused to form nuclei of heavier elements. Typically, nuclei of hydrogen 2 (deuterium) and hydrogen 3 (tritium) are fused to form helium in a reaction that also produces high-energy neutrons. These neutrons can cause the common isotope of uranium, uranium 238, to fission. The detonation of a modern weapon's first stage causes plutonium to fission, and then the second stage uses the energy released by the plutonium to produce hydrogen fusion. In the third stage, neutrons from the hydrogen fusion cause uranium 238 to fission.

Since the fusion reaction products are mostly nonradioactive, a fusion weapon can be designed to produce more or less fallout than a fission weapon. To make a "clean" bomb, care must be taken in choosing the nonnuclear explosive materials, because the fusion-produced neutrons can make many materials become radioactive. Choosing materials that will be made radioactive by fusion neutrons, on the other hand, produces a "dirty" bomb.

The ultimate fallout weapon is perhaps the cobalt bomb, made by placing a blanket of cobalt around a fission-fusion weapon. Radioactive cobalt has a half-life of about five years, and it emits a very penetrating gamma ray. At one hour after detonation, a cobalt bomb's fallout radiation would be fifteen thousand times less intense than that of a conventional nuclear bomb; however, the conventional bomb's radioisotopes decay faster. After six months, the fallout levels from the two types of weapons would be equal; after five years, cobalt fallout radiation would be 150 times more intense. The detonation of thousands of cobalt bombs would constitute a "doomsday weapon," since it would make large areas uninhabitable for many years. Any nation launching such a doomsday attack would likely also suffer from it as radioactive cobalt dust blew onto its own territory.

In an air burst, the fireball does not touch the ground. For air bursts of weapons with yields between 100 kilotons and 1 megaton, less fallout is deposited locally, and more goes into the stratosphere as the explosive yield increases. Because missiles and bombs were not very accurate in the 1950's and 1960's, nuclear powers made weapons with large yields. For example, the Titan II missile carried a 10-megaton warhead, and the Soviet SS-18 missile carried either a 10-megaton or 20-megaton warhead. Such missiles need only to hit somewhere near their targets to destroy them. As accuracy improved, nuclear stockpiles were downsized to weapon yields of 100 to 700 kilotons. Perversely, these downsized weapons produce more local fallout than the larger weapons they replace.

For yields above 1 megaton, the fireball will punch its way up into the stratosphere; in the case of an air burst, the fireball will carry most of the radioactive fallout with it. Fine particles can remain in the stratosphere for years, circling the globe many times. Most of these particles will not cross the equator and so will remain in either the Northern Hemisphere or the Southern Hemisphere depending upon their point of origin. Fallout that takes several days to a few years to settle to the ground is called "global fallout." A fireball that touches the ground sucks up dirt and debris and mixes them with the radioactive bomb vapor. This greatly increases fallout. While an air burst causes the most widespread damage, the greatest damage at ground zero is caused by a ground burst, and therefore ground bursts would be used against hardened targets such as missile silos. For a 1-megaton ground burst, about half the fallout will be deposited locally, and the other half will rise into the stratosphere.

FALLOUT HAZARDS

The length of time it takes for fallout to reach the ground depends largely on the size of the particles and on how high they are lofted. Since radioactive nuclei decay, the longer they remain aloft, the weaker the radioactivity will be when the fallout reaches the ground. Half-lives of radioactive isotopes from nuclear weapons range from a fraction of a second to billions of years, but the total mix obeys a simple rule: For each factor-of-seven increase in time, the intensity of radioactivity decreases by approximately a factor of ten. For example, seven hours after a blast, the radioactivity will be only 10 percent of what it was one hour after the blast. Forty-nine hours after the blast, the radioactivity will be only 1 percent of what it was at one hour, and so on. After about six months, the radioactivity decreases even more rapidly.

Short-lived isotopes are intensely radioactive, but only for a brief time; long-lived isotopes are only weakly radioactive, but they are active for a long time. Those with intermediate half-lives are the most dangerous fallout isotopes. Two other important considerations are the type of radiation emitted and the human body's affinity for certain elements. Fallout radiation is usually gamma rays or beta particles. Gamma rays are similar to X rays, but they have higher energies and are more penetrating. Staying inside a frame house will reduce gamma radiation from fallout by a factor of two or three. A basement will reduce the radiation by a factor of ten to twenty, and a below-ground shelter covered with 1 meter of dirt will reduce the radiation by a factor of five thousand. Beta particles are electrons emitted by a radioactive nucleus. Heavy clothing can block much of the beta radiation. Beta emitters are most harmful if left on the skin for long periods or if they are ingested. If fallout dust is not rinsed off, people and animals may receive skin burns. Because of their cause, such burns are sometimes called "beta burns."

The most hazardous elements in fallout include the short-lived isotopes of iodine that emit both beta particles and gamma rays. Cows eating fodder contaminated with radioactive iodine pass it on in their milk. When humans consume this milk, the radioactive iodine is concentrated by the thyroid gland, which consequently may receive a very high radiation dose. Radioactive strontium is a beta-particle emitter and is chemically similar to calcium, so it is readily absorbed by the body. Radioactive cesium emits both beta particles and gamma rays. It is chemically similar to potassium, and so it is also readily taken up by the body. In the ground burst of a fusion weapon, neutrons will make some elements in the soil radioactive. The most hazardous of these include the short-lived isotopes of sodium, manganese, and silicon.

Consider an event that would qualify as a worst-case scenario for local fallout: a 1-megaton ground burst. Such a blast would form a crater at least 300 meters in diameter and 60 meters deep. Much of the crater material would be sucked into the rising fireball, while the rest would be flung into a wide ring around the crater. The first fallout would likely begin within minutes of the blast, as radioactive pebbles rained down near the stem of the mushroom cloud. Fallout of ever-finer particles would continue to rain out over the next hours and days in a region beginning near ground zero and extending downwind for one hundred kilometers or even a few hundred kilometers. Hot spots would exist in the fallout pattern where a lethal dose might be received within a few hours. As much as a few thousand square kilometers might receive enough radioactive fallout to be lethal to anyone remaining in the open for a week.

Radioactive fallout can also be produced by a nuclear reactor accident, as happened when the Soviet reactor at Chernobyl exploded in 1986. Because of operator error coupled with a poor design, two chemical explosions rocked the core of the Unit 4 reactor early in the morning of April 26. The explosions lifted the 1,000-ton cover plate from the reactor core, blasted radioactive particles more than 7 kilometers into the air, and started several fires in and around the reactor. Emissions from the fires formed a radioactive cloud 1.5 kilometers high at one point. As the fire burned, radioisotopes continued to be released over a period of several days. News of the accident did not reach the West until the morning of April 28, when workers contaminated by Chernobyl fallout set off radiation monitors as they entered the Forsmark Nuclear Plant north of Stockholm. It is estimated that all the radioactive xenon gas, about half the radioactive iodine and cesium, and from 3 percent to 5 percent of the remaining radioactive material in the core were released in the accident. Fallout was widely dispersed across all of Europe.

Charles W. Rogers

The Chernobyl nuclear power plant shortly after the 1986 accident that killed 31 people and sent highly radioactive material into the atmosphere, leading to fallout across western Europe as far west as the United Kingdom. A decade later, more than 40,000 people in the vicinity had been diagnosed with cancer. (AP/Wide World Photos)

BIBLIOGRAPHY

Ball, Howard. *Justice Downwind: America's Atomic Testing Program in the 1950's.* New York: Oxford University Press, 1986. The author uses newspaper accounts and recorded interviews to carefully trace changes in public attitudes toward atmospheric nuclear tests. He documents the exposure of the "downwinders" to fallout and traces their efforts to win redress through the courts and through Congress. He gives evidence for and against the existence of fallout-induced cancer and describes the settlement finally agreed to by the government.

Byrne, John, and Steven M. Hoffman, eds. *Governing the Atom: The Politics of Risk.* New Brunswick, N.J.: Transaction, 1996. This volume from the Energy and Environmental Policy Series researches the environmental and social aspects of the nuclear industry. Careful attention is paid to government policies that have been implemented to monitor safety and accident prevention measures.

Chernousenko, Vladimir M. *Chernobyl: Insight from the Inside.* Berlin: Springer-Verlag, 1991. An interesting collection of personal anecdotes, pictures, reports, evaluations, and recommendations. Begins with an account of the accident and carries through to the medical problems caused by fallout. The author faults the government for being more concerned with managing the news than with dealing with physical problems.

Chivian, Eric, Susanna Chivian, Robert J. Lifton, and John E. Mack, eds. *Last Aid: The Medical Dimensions of Nuclear War.* San Francisco: W. H. Freeman, 1982. Sponsored by the International Physicians for the Prevention of Nuclear War, this is a calm, factual report on the effects of nuclear weapons. It discusses fallout, including that which fell on the inhabitants of Rongelap Atoll and on the fishermen aboard the *Fortunate Dragon.* The appendix includes an extensive book list for suggested reading.

Fradkin, Philip L. *Fallout: An American Tragedy.* Tucson: University of Arizona Press, 1989. This somewhat graphic text looks at the health hazards and toxicology aftermath of the nuclear weapons testing that occured in Nevada and Utah. There is a strong focus on the victims of these test sites and the physiological effects they have undergone. Attention is also paid to the liability for nuclear testing damages.

Johnson, J. Christopher, et al., eds. *Mortality of Veteran Participants in the Crossroads Nuclear Test.* Washington, D.C.: National Academy Press, 1996. The essays in this volume describe Operation Crossroads and the effects of the testing that occurred under its auspices. Illustrations and bibliographical references.

Katz, Arthur M. *Life After Nuclear War: The Economic and Social Impacts of Nuclear Attacks on the United States.* Cambridge, Mass.: Ballinger, 1982. A classic work on the problems that would arise after a nuclear exchange. Katz concludes that in addition to the casualties caused, fallout would lead to denial of rescue services, fire control, and distribution of food and other supplies, would contaminate farmland, and would drive people to make irrational decisions as a consequence of fear.

National Research Council. *Exposure of the American People to Iodine-131 from Nevada Nuclear-Bomb Tests.* Washington, D.C.: National Academy Press, 1999. This report examines the effects of Iodine 131 on humans and the environment. There is quite a bit of information on radioisotopes, radiation toxicology, and iodine toxicology, as well as an assessment of the threat that nuclear weapons testing poses to public health.

Schell, Jonathan. *The Fate of the Earth*. New York: Alfred A. Knopf, 1982. An eloquent, even poetic treatment of how nuclear explosions and fallout might reduce life on Earth to a republic of insects and grass, both of which are relatively radiation tolerant. While Schell's predictions may be extreme, his concerns merit consideration.

Solomon, Fredric, and Robert Q. Marston, eds. *The Medical Implications of Nuclear War*. Washington, D.C.: National Academy Press, 1986. Sponsored by the National Academy of Sciences Institute of Medicine. Somewhat technical in places, but written for a general audience and includes a helpful glossary. An excellent, authoritative source of information on fallout and on the effects of radiation on humans.

GLOBAL WARMING

Global warming is the term applied to rising global air temperatures. This rise in temperature has the potential to cause drastic changes in climate and weather patterns worldwide.

PRINCIPAL TERMS

CLIMATE: the long-term weather patterns for a region, distinct from the day-to-day weather patterns

GREENHOUSE EFFECT: the trapping of heat in the atmosphere by certain gases

GREENHOUSE GAS: gases that absorb infrared radiation more efficiently than other gases; examples include carbon dioxide, methane, and water vapor

INFRARED RADIATION: electromagnetic radiation emitted by warm objects; it is frequently confused with heat but is actually a different phenomenon

GREENHOUSE EFFECT

"Global warming" is the term for the rise in the Earth's average temperature. Scientists know that the Earth's average global temperature has been rising since the beginning of the Industrial Revolution in the second half of the eighteenth century. Increases in air temperature could alter precipitation patterns, change growing seasons, result in coastal flooding, and turn some areas into deserts. Scientists do not know the cause or causes of this phenomenon but believe that it could be part of a normal climate cycle or be caused by natural events or the activities of humankind.

When the ground is heated by sunlight, it gives off the heat as infrared radiation. The atmosphere absorbs the infrared radiation and keeps it from escaping into space. This is called the "greenhouse effect" because it was once believed that the glass panes of greenhouses acted similarly to the atmosphere, capturing the infrared radiation given off by the Earth inside the greenhouse and not allowing it to pass through. Although it has subsequently been shown that greenhouses work by trapping the heated air and not allowing it to blow away, the name has stuck.

The atmosphere eventually releases its heat into space. The amount of heat stored in the atmosphere remains constant as long as the composition of gases in the atmosphere does not change. Some gases, including carbon dioxide, water vapor, and methane, store heat more efficiently than others and are called "greenhouse gases." If the composition of the atmosphere changes to include more of these greenhouse gases, the air retains more heat, and the atmosphere becomes warmer.

Global levels of greenhouse gases have been steadily increasing and in 1990 were more than 14 percent higher than they were in 1960. At the same time, the average global temperature has also been rising. Meteorological records show that from 1890 to the mid-1990's, the average global temperature rose by between 0.4 and 0.7 degrees Celsius. About 0.2 degrees Celsius of this temperature increase has occurred since 1950. In comparison, the difference between the average global temperature in the 1990's and in the last ice age is approximately 10 degrees Celsius, and it is estimated that a drop of as little as 4 to 5 degrees Celsius could trigger the formation of continental glaciers. Therefore, the rise in average temperature is significant and already beginning to cause some changes in the global climate. Documented changes include the melting of glaciers and rising sea levels as the ocean gets hotter and its waters expand. Measurements of plant activity indicate that the annual growing season has become approximately two weeks longer in the middle latitude regions.

EFFECTS OF GLOBAL WARMING

A common misunderstanding is that global warming simply means that winters will be less cold and summers will be hotter, while everything else will be basically the same. Actually, the Earth's weather system is very complicated, and higher global temperatures will result in significant changes in weather patterns. Most changes will be observed in the middle and upper latitudes, with equatorial regions witnessing fewer changes. The areas that will experience most of the changes include North America, Europe, and most of Asia. The Southern Hemisphere will experience less severe effects because it contains more water than the Northern Hemisphere does, and it takes more energy to heat water than land.

It is difficult to predict precisely what these changes will be, but observation of the changing climate and scientific studies allow researchers to make some rough estimates of the kinds of changes the Earth will experience. Summers will be hotter, with more severe heat waves. Because hot air holds more moisture than cool air, rain will fall less frequently in the summer. Droughts can be expected to be more common and more severe. Through the late 1980's and into the early 1990's, annual temperatures climbed higher and higher, and summer heat waves became more frequent. This is a particularly troubling problem in areas where homes are typically built without air conditioning. The air in a closed-up house during a heat wave can reach temperatures well over 40 degrees Celsius. Because these temperatures exceed people's normal body temperature, which is about 36.5 degrees Celsius, it becomes very difficult for the body to cool down. For this reason, heat waves are particularly dangerous for the very young or old and the sick.

More frequent heat waves will cause increases in the use of air conditioning, which requires more energy and will release additional greenhouse gases.

Global warming will also produce severe autumn rains. The overheated summer air will cool in the autumn and will no longer be able to hold all of the moisture it was storing. It will release the moisture as heavy rains, causing flooding. This

Two views of the same scene in Glacier National Park, Montana: in 1932 (top), showing an ice cave in Boulder Glacier, and in 1988 (bottom). At the current rate of melting, the glaciers in the park will disappear by 2050. (AP/Wide World Photos)

phenomenon has already been observed, but not for a period of time that is scientifically significant. It is difficult to tell the difference between long-term changes and short-term fluctuations with only a few years of observations.

These changing rain patterns—droughts and severe autumn storms—will certainly have an effect on the Earth's landscape. Some areas may be turned to deserts, and others may be transformed from plains to forests.

One of the strange aspects of global warming is that it is predicted to result in not only hotter periods during the summers but also colder periods during the winter. It takes energy to move large cold-air masses from the polar regions in winter, so it is possible that large winter storms will be colder, more violent, and more frequent. This pattern has been evident since about the mid-1970's. Winter storms have brought record low temperatures and enough snow to close cities for days. However, this time span is too short to determine whether this is a temporary phenomenon or a trend. It is possible that some smaller, less permanent event than global warming is responsible for the more severe winter storms. A shorter (although more severe) winter may create a surge in pest populations and diseases that are normally controlled by long winters.

Global warming will cause ocean levels to rise because water expands when heated and also because of the melting of glaciers on Greenland and Antarctica. The melting of the ice in the northern polar areas will not contribute to rising ocean levels because that ice, unlike the ice on Greenland and Antarctica, is already in the ocean. Just as the melting of the ice in a drink does not cause the level of the drink to rise, the melting of the northern oceanic ice sheets will not affect sea levels.

CAUSES OF GLOBAL WARMING

Although many scientists are convinced, based on the abundant evidence, that global warming is occurring, they are less sure of why. One significant factor, the rise in greenhouse gases, can be attributed to the activities of humankind. Burning forests to clear land and operating factories and automobiles produce carbon dioxide and water vapor. Livestock herds and rotting vegetation release methane, and fertilizers used on farms also

release greenhouse gases. Power plants that consume fossil fuels such as coal, oil, or natural gas release massive amounts of carbon dioxide into the air. However, no one knows whether humankind's activities are the real or only reason for the increase in global temperature. The last ice age ended very recently in geologic terms, and a number of changes are still taking place as the globe recovers from the presence of huge ice sheets. It is possible that the world's climate is still warming up from the last ice age. Volcanoes are another major source of greenhouse gases, and the level of volcanic activity has been increasing since approximately the beginning of the Industrial Revolution.

Global warming could also be part of a natural cyclical change in the climate. Evidence indicates that the Earth's climate varies between warmer and colder periods that last a few centuries. History provides many stories of dramatically changing climate, including one period from 1617 to 1650 that was so unusually cold that it is called the "Little Ice Age." Therefore, the Earth may be merely experiencing another cyclical change in its climate.

STUDY OF GLOBAL WARMING

The problem with studying global warming is that no one can be sure of its extent, and some scientists debate whether it actually exists. They argue that the observed changes may only be a warm phase of a climatic cycle that will make the Earth warmer for a few years, then cooler for a few years. If this is true, then the coming decades may see average temperatures leveling off or even dropping.

A major source of the confusion involves the way global warming is studied. It would seem easy to record temperatures for a number of years and then compare them. However, detailed records on the weather have not been kept for more than a few decades in many areas. Scientists are forced to rely on interpretations of historical accounts and the clues left in fossil records. The analysis of tree rings, sedimentary deposits, and even very old ice from deep within glaciers can provide data about the climate in the past.

In addition, the existing records must be reviewed carefully to identify local changes that may not reflect global ones. For example, as towns grow into cities, the temperature climbs simply because larger cities are warmer than smaller ones, a phenomenon known as the urban heat island effect. Measurements taken years ago in a more rural environment should be lower than those taken after the population around the measuring station increased. This problem can be overcome with balloons. By sending instruments high in the atmosphere on weather balloons, air temperatures can be measured without being affected by urbanization. Although data recorded this way show a consistent rise in global temperature, such measurements go back only a few decades.

Measurements of the level of greenhouse gases in the atmosphere are also affected by urbanization. As a small town becomes a city, levels can be expected to rise. However, recording stations located in regions far removed from cities and factories show an increase in the level of greenhouse gases. One station on the island of Hawaii has shown a rise in the amount of carbon dioxide present in the air since early 1958, with similar reports coming from stations in Point Barrow, Alaska, and Antarctica.

Another important variable in looking at global warming is sea surface temperature. Measurements can be skewed by local effects that have no impact on the global climate. One method used to make detailed measurements of seawater temperature is to broadcast a particular frequency of noise through the water and measure it at distant locations. The speed and frequency of the sound are affected by the temperature of the water.

However, more accurate and more global data are becoming available through the Mission to Planet Earth program of the National Aeronautics and Space Administration (NASA). The program places satellites in orbit to study the Earth and make a variety of detailed measurements, many of which are of factors that contribute to global warming. These data will help immensely in enabling scientists to understand the global climate and the changes it is undergoing.

Significance

The global climatic environment is very complicated, and many factors contribute to it. All these factors together produce the climate in which people live. For this reason, it is impossible to say which, if any, changes are attributable simply to a particular natural event or to the activities of humankind. However, at the same time, the activities of humankind are so large-scale and widespread that it is safe to conclude that they have definitely changed the world. Unfortunately, the effect of these changes is still not clear. People have generated massive amounts of greenhouse gases and released them into the air, but no one really knows the extent of their effect on the environment.

If scientists determine that humankind's activities are responsible for the observed changes in the environment, then people can identify what they must do to preserve the environment and take measures to change the way they live. People may eventually gain enough knowledge to plan the changes in the climate to occur in such a way as to provide maximum benefit to the environment and themselves. Even if researchers determine that global warming is caused by natural events beyond human control, people still must evaluate their role. Humankind's activities are likely to contribute to the problem, so altering behavior and ways of life might lessen the negative effects of global warming. For example, although global warming might not be caused by emissions of greenhouse gases, these emissions might worsen the droughts and severe rainstorms. If so, then by reducing emissions, people might be able to counteract some of the effects of the climatic changes.

If the changes in the climate approach what is expected based on scientific models of global warming, the impact on the environment and humanity will be staggering. The changes are such that virtually every person on Earth would be affected. If rain patterns change drastically, it is possible that the food supplies of the world will be sharply reduced. In some areas, if the average temperature increases, certain diseases associated with warmer climates will begin to affect the people. An increase of a few meters in the ocean level will inundate many of the world's coastal areas and the large cities found in those areas. In addition to these large-scale changes, changes in the weather will have a great impact on the day-to-day lives of the majority of the world's people. Any solution to the problem of global warming will also probably dramatically affect people's lives.

Christopher Keating

CROSS-REFERENCES

Acid Rain and Acid Deposition, 1802; Air Pollution, 1809; Atmosphere's Evolution, 1816; Atmosphere's Global Circulation, 1823; Atmosphere's Structure and Thermodynamics, 1828; Atmospheric Inversion Layers, 1836; Auroras, 1841; Cosmic Rays and Background Radiation, 1846; Earth-Sun Relations, 1851; Environmental Health, 1759; Fallout, 1857; Greenhouse Effect, 1867; Nuclear Power, 1663; Nuclear Waste Disposal, 1791; Nuclear Winter, 1874; Ozone Depletion and Ozone Holes, 1879; Radioactive Decay, 532; Radon Gas, 1886; Rain Forests and the Atmosphere, 1890.

BIBLIOGRAPHY

Abrahamson, Dean Edwin, ed. *The Challenge of Global Warming*. Washington, D.C.: Island Press, 1989. This detailed examination of what global warming is and how it occurs is suitable for all readers. Also discusses greenhouse gases and policy-making decisions regarding global warming.

Bach, Wilfrid, Jurgen Pankrath, and William Kellogg. "Man's Impact on Climate." Vol. 10 in *Developments in Atmospheric Science*. Amsterdam: Elsevier Scientific, 1979. A well-written but technical account of research into changes in the atmosphere and their effects. Suitable for more advanced readers.

Gates, David M. *Climate Change and Its Biological Consequences*. Sunderland, Mass.: Sinauer Associates, 1993. This excellent book examines the effects that climatic changes would have on life on Earth, discussing many of the models used in making predictions. Suitable for high school students and up.

Magill, Frank N., and Russell R. Tobias, eds. *USA in Space*. Pasadena, Calif.: Salem Press, 1996. This three-volume set consists of an extremely well-written series of articles on U.S. space activities. A number of pieces discuss data collection on the environment and detecting people's effects on the environment. Suitable for all levels.

Nance, John J. *What Goes Up: The Global Assault on Our Atmosphere*. New York: William Morrow, 1991. This is a clearly presented documentation of the investigation into the changes occurring in the atmosphere. Topics covered include global warming and the depletion of the ozone layer. Suitable for all readers.

Sommerville, Richard C. J. *The Forgiving Air: Understanding Environmental Change*. Berkeley: University of California Press, 1996. This thorough and easy-to-understand book focuses on the various consequences of air pollution, including the depletion of the ozone layer, climatic changes, the greenhouse effect, and acid rain. Intended as an introduction for the layperson. Bibliography and index.

Weiner, Jonathan. *Planet Earth*. Toronto: Bantam Books, 1986. This exceptionally well-written book is the companion book to the *Planet Earth* series developed by the Public Broadcasting Service. It covers many aspects of the planet and devotes a chapter to its climate. Suitable for all audiences.

Wyman, Richard L., ed. *Global Climate Change and Life on Earth*. New York: Chapman and Hall, 1991. A collection of articles on various aspects of climate changes and the effects of these changes.

GREENHOUSE EFFECT

Greenhouse gases absorb or trap infrared, or heat, energy emitted by the Earth's surface. The absorbed or trapped heat energy is then released or reemitted by the greenhouse gases, resulting in an additional heating of the Earth's surface. Atmospheric greenhouse gases include water vapor, carbon dioxide, methane, nitrous oxide, ozone, and a class of human-made molecules called chlorofluorocarbons. There is serious national and international concern that increasing atmospheric levels of these gases will lead to a global warming.

PRINCIPAL TERMS

EARTH-EMITTED RADIATION: the portion of the electromagnetic spectrum, from about 4 to 80 microns, in which the Earth emits about 99 percent of its radiation

EFFECTIVE TEMPERATURE: the temperature of a planet based solely on the amount of solar radiation that the planet's surface receives; the effective temperature of a planet does not include the greenhouse temperature enhancement effect

INFRARED RADIATION: the portion of the electromagnetic spectrum that extends from about 0.75 to 100 microns

MICRON: a unit used to measure the wavelength of electromagnetic radiation; one micron equals 0.0001 centimeter

SOLAR RADIATION: the portion of the electromagnetic spectrum, from about 0.15 to about 4 microns, in which the Sun emits about 99 percent of its total radiation

TROPOSPHERE: the lowest region of the Earth's atmosphere, which extends upward from the surface to between 12 and 15 kilometers; the troposphere contains about 85 percent of the total mass of the atmosphere, almost all the water vapor in the atmosphere, and most of the other greenhouse gases

VISIBLE RADIATION: the portion of the electromagnetic spectrum that extends from about 0.4 to 0.75 micron

SOLAR RADIATION

The Earth and the other planets receive almost all their energy in the form of electromagnetic radiation from the Sun. While the Sun emits radiation over the entire electromagnetic spectrum, from shortwave X rays to longwave radio waves, the bulk of the solar radiation is in the visible part of the electromagnetic spectrum, from about 0.15 to about 4 microns. The Sun radiates energy primarily in the visible part of the spectrum, as that is the spectral region of maximum emission for an object at a temperature of about 6,000 Kelvins, or about 5,727 degrees Celsius, which is the temperature of the Sun's surface, or "photosphere." According to the laws of physics governing the emission of electromagnetic radiation, an object at a temperature of 6,000 Kelvins emits most of its radiation at a wavelength of about 0.55 micron, which corresponds to visible radiation with a yellowish-white color. Hence, the Sun appears as a yellowish-white object in the sky.

The amount of solar radiation intercepted by the Earth and available for heating the planet depends on the product of the amount of solar radiation reaching the top of the planet and the area of the disk of the Earth illuminated by the Sun. At the Earth's distance from the Sun (about 150 million kilometers), the amount of solar radiation hitting the top of the Earth's atmosphere is about 1.4 million ergs per square centimeter per second. Not all this incoming solar radiation, however, is available for heating of the Earth's surface. A fraction of the incoming solar radiation is reflected back to space by clouds and by the surface itself. The fraction of incoming solar radiation reflected back to space is called the albedo of the planet. The albedo of the Earth is about 33 percent. Therefore, about 67 percent of the incoming solar radiation is available for heating the surface of the Earth. This incoming solar radiation is absorbed at the Earth's surface and heats the surface. The Earth's surface in turn emits its own radiation to balance the incoming solar radiation, which heats the surface. If the total amount of Earth-emitted radiation were greater than the incoming solar radiation, the Earth would cool off; however, if the total amount of Earth-emitted radiation were less than the incoming solar radiation, the Earth would heat up. Over a period of several years, Earth-emitted radiation just balances the incoming solar radiation. By equating Earth-emitted radiation, which depends on the temperature to which the surface is heated, to the incoming solar radiation, one can find the "effective" temperature of the Earth.

GREENHOUSE GASES

The Earth's effective temperature is about 253 Kelvins, or about −20 degrees Celsius. At this temperature, Earth-emitted radiation falls in the infrared part of the electromagnetic spectrum, between about 4 and 80 microns. Unlike the incoming solar radiation, however, which travels through the atmosphere without significant attenuation or loss from absorption by atmospheric gases, Earth-emitted infrared, or heat, energy is

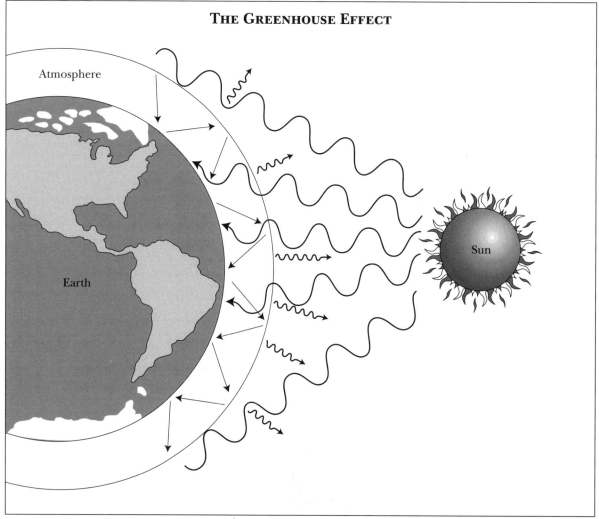

THE GREENHOUSE EFFECT

The greenhouse effect is aptly named: some heat from the Sun is reflected back into space (squiggled arrows), but some becomes trapped by Earth's atmosphere and re-radiates toward Earth (straight arrows), heating the planet just as heat is trapped inside a greenhouse.

absorbed by several atmospheric gases, called greenhouse gases. Greenhouse gases absorb Earth-emitted infrared, or heat, energy; however, these gases cannot continuously absorb infrared radiation. After a fraction of a second, the greenhouse gases release or reemit the absorbed or trapped infrared radiation in all directions. About 50 percent of the reemitted infrared radiation is directed in the upward direction, and about 50 percent is directed in the downward direction. The downward component of the reemitted infrared radiation is absorbed by the surface, with an additional heating effect. Hence, the surface of the planet is heated not only by incident solar radiation (0.3 to 5 microns) but also by Earth-emitted infrared radiation (4 to 80 microns) that was absorbed and then reemitted by atmospheric greenhouse gases. The additional heating of the Earth's surface by the reemitted infrared radiation heats the surface an additional 35 Kelvins, from the effective temperature of about 253 Kelvins to 288 Kelvins, or about 15 degrees Celsius—the average temperature of the Earth. The 35-Kelvin temperature enhancement is termed the greenhouse effect. It is this temperature enhancement that makes the Earth habitable for life.

While the greenhouse gases are only very minor constituents of the Earth's atmosphere, carbon dioxide is the major constituent of the atmosphere of Venus. Carbon dioxide comprises about 96 percent by volume of Venus's atmosphere, with nitrogen accounting for the remaining 4 percent. The surface pressure of the atmosphere of Venus is about 90 atmospheres, compared with 1 atmosphere for the Earth. As a result of the high percentage of carbon dioxide and the high surface pressure of the atmosphere, there is a very efficient and significant greenhouse effect on Venus. The greenhouse effect increases the temperature of Venus by about 450 Kelvins (177 degrees Celsius), from its effective temperature of about 244 Kelvins (−29 degrees Celsius)—close to the effective temperature of the Earth—to the measured surface temperature of Venus of about 700 Kelvins (427 degrees Celsius)—close to the melting point of lead.

GLOBAL WARMING

There is national and international concern that the buildup of greenhouse gases in the Earth's atmosphere will lead to an enhanced greenhouse effect and a global warming resulting in high temperatures in many regions and severe droughts in other regions. Such a global warming may lead to thermal expansion of the oceans, as ocean water expands in volume when heated. In addition, warming may result in a significant melting of the glacial ice and snow at the Earth's poles, producing an even greater volume of water in the world's oceans. The resulting increase in sea level could result in the flooding of many low-lying land areas.

The most important atmospheric greenhouse gases are water vapor, carbon dioxide, methane, nitrous oxide, tropospheric ozone, and a human-made family of gases termed "chlorofluorocarbons," of which CFC-11 and CFC-12 are the most abundant species. All the greenhouse gases are very minor constituents of the atmosphere, which is composed primarily of nitrogen (78.08 percent by volume), oxygen (20.95 percent), and argon (0.93 percent). Water vapor is a variable constituent of the atmosphere and ranges from a small fraction of a percent to several percent. The concentration of carbon dioxide is about 0.035 percent, or 350 parts per million by volume. The concentration of methane is about 1.7 parts per million by volume. The nitrous oxide concentration is about 0.31 part per million by volume. Ozone is a variable constituent of the troposphere and ranges from about 0.02 to 10 parts per million by volume. The atmospheric concentrations of CFC-12 and CFC-11 are only 0.00038 and 0.00023 part per million by volume, respectively.

All these greenhouse gases, with the exception of water vapor, are produced by human activities, such as the burning of fossil fuels and biomass (trees, vegetation, and agricultural stubble), the use of nitrogen fertilization on agricultural areas, and various industrial activities. Each of the greenhouse gases produced by human activities is increasing in concentration in the atmosphere with time. Carbon dioxide is increasing at a rate of about 0.3 percent per year; methane is increasing at a rate of about 1.1 percent per year; nitrous oxide is increasing at a rate of about 0.3 percent per year; and CFC-11 and CFC-12 are each increasing at a rate of about 5 percent per year. Estimates suggest that tropospheric ozone is also increasing with time, with increases of between 1 and 2 percent per year over North America and Europe during the 1970's and 1980's. Collectively, methane, nitrous oxide, tropospheric ozone, and chlorofluorocarbons are now estimated to trap about as much infrared radiation as does carbon dioxide alone. These gases absorb infrared radiation in the spectral region, from about 7 to 13 microns, known as the atmospheric window, where water vapor and carbon dioxide do not absorb. Water vapor is a strong absorber below about 7 microns, and carbon dioxide is a strong absorber above about 13 microns. If the rates of increase of the greenhouse gases persist with time, the greenhouse effect of the other gases, when added to that of carbon dioxide, will amount to an effective doubling of carbon dioxide levels by the year 2030, some fifty years sooner than the level of carbon dioxide alone is likely to double.

A doubling of carbon dioxide in the atmosphere would lead to a global temperature increase of about 1.2 Kelvins if there were no other changes in the climate system. A warming caused by the doubling of carbon dioxide would lead to more evaporation of water vapor from the ocean. In addition, a warming would permit the atmosphere to hold more water vapor, as the capacity for air to hold water vapor increases with increasing temperature. Because water vapor is itself a greenhouse gas, as its concentration increases in the atmosphere, the Earth would warm even further. The net result of these different feedback processes would be that for a doubling of atmospheric carbon dioxide, the surface temperature would increase about 4 Kelvins. It is estimated that the average surface temperature has increased globally by about 0.5 Kelvin over the twentieth century, although this temperature increase cannot unambiguously be attributed to the buildup of greenhouse gases over this period. Temperature records also indicate that the 1980's were the warmest decade on record up to that time.

GREENHOUSE GAS SOURCES

The sources of the greenhouse gases, with the exception of water vapor, are mostly human-initiated. Carbon dioxide results from the burning of fossil fuels and of living and dead biomass, such as deforestation burning in the tropics and the burning of agricultural stubble after the harvest. Methane is produced as a combustion product of bio-

mass burning, from natural gas leakage, and by the action of anaerobic bacteria in wetlands (such as rice paddies), in landfills, and in the stomachs of ruminants (such as cattle and sheep). Nitrous oxide is produced as a combustion product of biomass burning and by the action of nitrifying and denitrifying bacteria (which add and remove nitrogen) in natural, fertilized, and burned soils. Ozone results from atmospheric reactions involving methane, carbon monoxide, and oxides of nitrogen, which are produced by the burning of fossil fuel and by biomass burning. Chlorofluorocarbons (CFC-11 and CFC-12) are released into the atmosphere when they are used as refrigerants.

There are a number of other important factors in the climate scenario whose roles must be more fully understood before scientists can completely assess the impact of increasing concentrations of greenhouse gases on the future climate. Other climate factors that must be studied in more detail include the role of clouds and, especially, their impact on a warmer Earth; the role of the ocean, its heat capacity and circulation; and the variability of incoming solar radiation. The Earth's climate is a complex system affected by many processes and parameters. Experts need to learn more about the interplay of atmospheric greenhouse gases, clouds, the ocean, and solar variability before they may accurately assess the future climate.

STUDY OF THE GREENHOUSE EFFECT

Studies of the greenhouse effect and its impact on the climate of the Earth are multidisciplinary in nature and involve theoretical computer modeling, laboratory studies of the spectroscopic parameters and properties of greenhouse gases, atmospheric measurements of greenhouse gases, and aircraft and satellite measurements of parameters that control climate. Theoretical computer models of climate include zero-dimensional, one-dimensional, two-dimensional, and three-dimensional models. Zero-dimensional models give climate parameters that represent an average for the entire system, such as the mean temperature of the Earth's surface. One-dimensional models are used to study climate in either a horizontal (latitude) or vertical (altitude) direction. In these models, a latitude-dependent surface temperature is the climate parameter of major interest. The vertical one-dimensional model is known as a radiative-convective model and is used to study the effects of changes in concentrations of greenhouse gases on the surface temperature. Two-dimensional models involve characterizing the temperature variation as a function of latitude and altitude. The most complex climate model is the three-dimensional model, or the general circulation model (GCM). This model gives a complete description of climate as a function of latitude, longitude, and altitude.

Major uncertainties exist in the understanding of several key parameters in the theoretical modeling of climate. A major uncertainty in these models is the role of cloud feedback, particularly as the Earth heats up. Calculations indicate that as the temperature of the Earth increases, clouds will appear in greater quantity. Clouds both reflect incoming solar radiation and trap Earth-emitted thermal infrared radiation. Hence, clouds can enhance or decrease global warming. The radiative properties of clouds and how they will vary with a warmer Earth are poorly understood and therefore are not included in most theoretical climate models.

To understand the greenhouse effect of gases, the spectroscopic parameters and properties of greenhouse gases are measured in laboratory experiments. These studies provide information on the spectral location of the absorption of infrared radiation by these gases and the intensity of the absorption. These laboratory spectroscopic studies involve filling an absorption cell with a known concentration of a greenhouse gas and measuring the absorption of infrared radiation as a function of spectral wavelength using a scanning spectrometer.

To assess the impact of atmospheric greenhouse gases on the Earth's climate, scientists must obtain accurate measurements of those gases that are found at trace levels in the atmosphere—for example, at atmospheric concentrations of parts per million by volume, parts per billion by volume, and parts per trillion by volume. Measuring atmospheric greenhouse gases at these very low atmospheric levels involves a series of different analytical chemistry instruments. Carbon dioxide may be measured with a gas chromatograph equipped with a thermal conductivity detector or using nondispersive infrared instrumentation. Methane is measured with a gas chromatograph equipped with a flame ionization detector or using nondispersive infrared instrumentation. Nitrous ox-

ide, CFC-11, and CFC-12 are measured using a gas chromatograph equipped with an electron capture detector. Ozone may be measured using various ultraviolet and infrared absorption techniques. Water vapor is measured using a hygrometer. Aircraft and satellite measurements have provided very important data on several of the parameters that control climate, including the flux of incoming solar radiation, the flux of outgoing Earth-emitted thermal infrared radiation, the flux of solar radiation reflected by clouds, and the geographical and temporal variability of clouds and surface temperature. Earth-orbiting satellites have provided important information on these important climate parameters.

SIGNIFICANCE

The buildup in the atmosphere of greenhouse gases, such as carbon dioxide, methane, nitrous oxide, tropospheric ozone, CFC-11, and CFC-12, will result in a global warming of the Earth. Atmospheric greenhouse gases are increasing with time. These greenhouse gases result from a variety of human activities, including the burning of fossil fuels (carbon dioxide), the burning of living and dead biomass (carbon dioxide, methane, and nitrous oxide), and the application of nitrogen fertilizers and the burning of agricultural acreage and grasslands (nitrous oxide). Greenhouse gases are also produced from rice paddies, cattle, and sheep (methane) and from several industrial applications (CFC-11 and CFC-12).

A global warming of the Earth from the buildup of greenhouse gases in the atmosphere would have a significant impact on people's daily lives. The growing seasons in most regions would change, as would patterns of rainfall. One of the most important effects would be a predicted increase in the height of the world's oceans, which would result from the thermal expansion of seawater because of the Earth's high temperature (water is a compressible fluid and expands in volume when heated) and because of the melting of polar ice and snow as the Earth becomes warmer. An increase in sea level would result in the flooding of the world's low-lying land areas. It has been estimated that a global temperature increase of about 4 degrees Celsius might result in a 2-meter increase in the height of the world's oceans. Higher temperatures resulting from the buildup of atmospheric greenhouse gases would tax the world's air conditioning facilities, which require the burning of fossil fuels for their operation. Ironically, the burning of fossil fuels is a major source of atmospheric greenhouse gases.

It must be emphasized that there are fundamental uncertainties and deficiencies in scientists' understanding of climate and the processes and parameters that control it. More must be learned about the effect of clouds and oceans on climate and how these phenomena would affect the climate as the Earth begins to warm. Theoretical computer models of climate are not complete; more research is needed before the future climate of the Earth can be assessed with greater certainty.

Joel S. Levine

CROSS-REFERENCES

Acid Rain and Acid Deposition, 1802; Air Pollution, 1809; Atmosphere's Evolution, 1816; Atmosphere's Global Circulation, 1823; Atmosphere's Structure and Thermodynamics, 1828; Atmospheric Inversion Layers, 1836; Auroras, 1841; Climate, 1902; Cosmic Rays and Background Radiation, 1846; Earth Science Enterprise, 1752; Earth-Sun Relations, 1851; El Niño and La Niña, 1922; Fallout, 1857; Global Warming, 1862; Meteorological Satellites, 2786; Mission to Planet Earth, 2792; Nuclear Winter, 1874; Ozone Depletion and Ozone Holes, 1879; Paleoclimatology, 1131; Radon Gas, 1886; Rain Forests and the Atmosphere, 1890; Satellite Meteorology, 1945.

BIBLIOGRAPHY

Environmental Protection Agency. *The Greenhouse Effect: How It Can Change Our Lives*. EPA Journal 15 (January/February, 1989). Popular, nontechnical accounts of the impact of climate change on agriculture, forests, energy demand, and other areas in a special issue devoted to the greenhouse effect. The principles that control and regulate global climate, including the greenhouse effect, are presented in a simple and readable manner for a general audience. Well illustrated with photographs and charts.

Flavin, Christopher, Odil Tunali, and Jane A. Peterson, et al., eds. *Climate of Hope: New Strategies for Stabilizing the World's Atmosphere*. Washington, D.C.: Worldwatch Institute, 1996. This brief Worldwatch paper examines the causes and current state of the greenhouse effect and global warming. The paper also looks at fossil fuels and energy policies put into place that will help curb the climatic changes and lessen their severity. Illustrations and fifteen pages of references.

Goody, R. M., and J. C. G. Walker. *Atmospheres*. Englewood Cliffs, N.J.: Prentice-Hall, 1972. A good nontechnical description of solar and infrared radiation in planetary atmospheres and how the radiation balance controls the temperature of a planet. The chapters of the book include sections on the Sun and the planets, solar radiation and chemical change, atmospheric temperatures, winds of global scale, condensation and clouds, and the evolution of atmospheres.

Hansen, Joel E., and T. Takahashi, eds. *Climate Processes and Climate Sensitivity*. Geophysical Monograph 29. Washington, D.C.: American Geophysical Union, 1984. A collection of papers dealing with various aspects of the climate system, including atmosphere and ocean dynamics, the hydrologic cycle and clouds, albedo and radiation processes, polar ice, and ocean chemistry. Each paper was written by an expert in that particular area of climate research. The papers summarize what is known, along with the major uncertainties and deficiencies in understanding the processes and parameters that control climate.

Henderson-Sellers, A., ed. *Satellite Sensing of a Cloudy Atmosphere: Observing the Third Planet*. London: Taylor and Francis, 1984. The subtitle refers to the Earth being the third planet from the Sun. The subjects of the chapters in this volume, each written by an active researcher in the area, cover radiation and satellite sensors, the Earth's radiation budget and clouds, water and the photochemistry of the troposphere, vertical temperature sounding of the atmosphere, cloud identification and characterization from satellites, and the remote sensing of land, ocean, and ice from space.

Kondratsev, Kirill, and Arthur P. Cracknell. *Observing Global Climate Change*. London: Taylor and Francis, 1998. This book serves as an excellent reference for anyone interested in the greenhouse effect and climatic change. Suitable for the reader without background in environmental sciences. Illustrations, maps, index, and a fifty-page bibliography.

Levine, Joel S., ed. *The Photochemistry of Atmospheres: Earth, the Other Planets, and Comets*. Orlando, Fla.: Academic Press, 1985. A comprehensive textbook covering atmospheric composition, chemistry, and climate, the sources and sinks of greenhouse gases, and the climate modeling of the Earth. The chapter on climate includes discussions on zero-dimensional, one-dimensional, and three-dimensional (global circulation model) climate models and the underlying physical, radiative, and dynamic processes and parameters in each model.

National Research Council. *Changing Climate: Report of the Carbon Dioxide Assessment Committee*. Washington, D.C.: National Academy Press, 1983. Technical report that addresses the possible impacts of climate change on sea level, agriculture, plant growth, and society in general. Considers future carbon dioxide emissions from fossil fuels, the dissolution of carbon dioxide into the oceans, the biosphere storage of carbon dioxide, and the impact of increased carbon dioxide on climate, agriculture, and sea level.

Sommerville, Richard C. J. *The Forgiving Air: Understanding Environmental Change*. Berkeley: University of California Press, 1996. This thorough and easy-to-understand book focuses on the various consequences of air pollution, including the depletion of the ozone layer, climatic changes, the greenhouse effect, and acid rain. Intended as an introduction for the layperson. Bibliography and index.

Wyman, Richard L., ed. *Global Climate Change and Life on Earth*. New York: Chapman and Hall, 1991. A collection of articles on various aspects of climate change and the effects of these changes.

NUCLEAR WINTER

Nuclear winter is a model of the projected consequences of nuclear war, wherein smoke and dust concentrated in the upper atmosphere may drastically reduce surface temperature and light by blocking solar radiation over several seasons or years, leading to collapse of the food chain, mass starvation, and possible human extinction. This model has drastically altered thinking on the survivability of humanity following a major nuclear exchange.

PRINCIPAL TERMS

AEROSOL: a suspension of solid or liquid particles in a gas

GREENHOUSE EFFECT: various mechanisms for increasing the absorptive capacity of radiated energy by the atmosphere

THERMAL BUDGET: the balance of incoming and outgoing radiative energy in the Earth's ecosystem

EARTH'S THERMAL BUDGET

The theory of "nuclear winter" proposes that massive quantities of smoke and dust injected into the atmosphere by multiple nuclear detonations would drive temperatures on the Earth's surface below the freezing point of water for extended periods and would diminish sunlight reaching the planet's surface, creating a period of prolonged cold and darkness. Atmospheric circulation could distribute the smoke and dust veils from even a moderate-sized nuclear exchange to make the effects global, rather than confined only to the regions of the detonations. Secondary effects from radioactive fallout and toxic substances carried by the smoke and dust veils also would be extremely widespread. The combined effects of nuclear winter could be sufficient to cause the eventual extinction of plant and animal life and possibly of humanity itself.

The key factor in the nuclear-winter scenario is that the combined effects of smoke and dust injected into the atmosphere by multiple nuclear detonations could severely reduce the amount of sunlight reaching the surface of the Earth, throwing the "thermal budget" of the ecosystem into imbalance and sending surface temperatures well into the subfreezing range. Over time and across latitudes, the thermal budget of the Earth is more or less in balance, with the amount of thermal radiation incoming from the Sun roughly equal to the amount of energy radiated back into space. (Low latitudes characteristically receive more thermal radiation than they emit, and higher latitudes less, creating large-scale atmospheric and oceanic circulation patterns that balance the thermal budget and account for major weather patterns.)

Radiation from the Sun reaches the Earth mostly in the visible-light wavelengths of the electromagnetic spectrum, but thermal radiation emitted from the Earth's surface back into space is mostly in infrared wavelengths. If the Earth did not have an atmosphere, the thermal budget would balance out at a level below the freezing point of water. Fortunately, components of the atmosphere—primarily water vapor and carbon dioxide molecules—are excellent absorbers of solar radiation; consequently, they help to keep ambient temperatures on most of the Earth's surface above the freezing point of water, making possible a biological regime based on liquid water.

DIMINISHED SUNLIGHT

Nuclear winter is a kind of "anti-greenhouse" effect. The greenhouse effect is produced by increases in the amount of infrared-absorbing carbon dioxide in the atmosphere, increasing ambient temperatures at the Earth's surface by reducing the amount of infrared energy being emitted into space. In the nuclear-winter scenario, changes in the atmosphere reduce the amount of incoming solar radiation reaching the surface by absorbing greater quantities of solar radiation in the upper atmosphere, with no reduction in surface infrared radiation to maintain a thermal balance. Surface temperatures would drop significantly; since any appropriate atmospheric agent responsible for such a reduction would have to absorb primarily visible light, continuous darkness would result at the Earth's surface.

Depending upon the extent of a general nuclear exchange, from 100 million to 300 million metric tons of smoke and soot particles might be injected into the atmosphere, enough to diminish sunlight reaching the Earth's surface by more than 90 percent if it were distributed uniformly throughout the atmosphere. The smoke would come from immense firestorms generated by detonations over urban or large industrial centers, so powerful that they would generate hurricane-like convective winds similar to those experienced by the victims of the Allied incendiary bombing of Dresden, Germany, during World War II.

The convective energy of these firestorms would drive the smoke and soot clouds beyond the dense troposphere into the stratosphere. The higher these clouds are injected, the longer they are likely to persist, since smoke and dust are cleansed from the atmosphere mostly by rainfall, and there is little water vapor in the stratosphere. Research suggests that this material would remain aloft long enough to become distributed throughout the stratosphere, with global rather than merely regional repercussions. Firestorms in modern urban centers, burning not only structures and material life but also huge quantities of plastics, light metals, and a host of synthetic chemicals of still unknown potential, would be far more dangerous over much greater areas than anything previously experienced. Additional smoke accumulation would come from forests and grasslands ignited by nuclear explosions.

EFFECTS OF NUCLEAR WINTER

The horrific loss of human life from the immediate effects of a nuclear conflict remains in the nuclear-winter scenario, but the secondary, longer-term losses are even more sobering. Deprived of heat and light from the Sun, much of the Earth would experience a multiyear state of subfreezing temperatures and continuous darkness, even at midday. Deaths from cold and exposure, and extraordinary psychological stress, would continue to increase the fatality rate in the

aftermath of nuclear war. Over several seasons, the result would be collapse of the food chain, the destruction of most plant life and the animal life that feeds on it, and mass starvation. All remaining order would likely break down as survivors fought for food, water, and shelter.

Interruption of sunlight to the surface would result in breakdown of the process of photosynthesis in plants and, therefore, destruction of crops. Wild plant species would also be affected, and those susceptible to small changes in the environment would be driven to extinction. Large-scale extinction of vertebrate species dependent on plant matter for food would soon follow. At sea, the phytoplankton population that anchors the food chain for many large species would collapse; many of the larger species of aquatic life would face extinction as well.

Moderate-size thermonuclear explosions are capable of perforating the ozone layer of the Earth's atmosphere; a significant number of such explosions could lead to massive ozone depletion. In addition to radioactive fallout from the explosions themselves—a hazard now judged to be more serious than previously believed, as a result of the long-term suspension of matter in the atmosphere—survivors of a nuclear war could be exposed to harmful ultraviolet radiation from the Sun that is ordinarily blocked by atmospheric ozone.

The impact of radioactive fallout, once presumed to be localized in the vicinities of the nuclear explosions themselves, would become global and prolonged as these substances circulated in the upper atmosphere. Water and food sources hundreds or thousands of miles distant from actual detonations would be contaminated by atmospheric circulation of radioactive matter and toxic substances from burning cities. Human populations could decline, making the species vulnerable to eventual extinction.

Possible consequences of nuclear winter include long-term climatic disruption. For example, deposition of dust and soot over much of the Earth's surface could lower the albedo, or reflective value, of the surface, with unknown climatic consequences over time. Coastal areas—in contrast to continental interiors, where convection would be almost nonexistent for years—might suffer storms of unprecedented magnitude, as warm air currents from over the oceans collided with chilled air masses on land.

MAGNITUDE OF NUCLEAR WAR

Phenomena associated with the nuclear-winter scenario may vary considerably depending upon the magnitude of a nuclear exchange and the nature of selected targets; catastrophic results are possible far short of a full-scale thermonuclear war. In a "medium-range" exchange, for example, involving some 5,000 megatons of detonations, including a thousand ground bursts—the sort of exchange that might be envisioned should one superpower attempt to eliminate the first-strike capability of another—the resulting smoke and dust veils could create a global atmospheric inversion that would essentially stop wind currents near the surface. As the smoke and dust absorbed visible sunlight and blocked it from reaching the surface, the upper atmosphere would be heated as much as 30 to 80 degrees Celsius. As the ground cooled, these hot layers would rise and expand, blanketing the planet in an unbroken cloud of smoke and dust. Even several months after the exchange, the lowest levels of the atmosphere would receive only enough radiation to drive very weak convective currents. This temperature inversion would inhibit the rise of moist air from the surface and thereby inhibit cleansing rainfall, increasing the time that the smoke and dust veils would remain suspended in the atmosphere.

One of the most significant features of the nuclear-winter model is the ability of its climatic consequences to involve the tropics through atmospheric circulation. Tropical plant and animal species are far more vulnerable than others to even minor shifts in temperature or rainfall. Prolonged cold and darkness over the tropics could result in the extinction of many species on the planet. At the very least, destruction of the forest regions of the tropics, major absorbing agents for solar energy, would further alter the Earth's thermal balance toward long-term subfreezing conditions. Widespread destruction of forests would also lead to further large-scale fires.

The nuclear-winter model predicts much more severe radiation hazards, spread over wider areas, than those estimated by earlier projections. Most radioactive isotopes in a nuclear explosion condense onto aerosols and dust sucked in by the initial fireball. Inasmuch as the model predicts very large quantities of dust aerosols in the atmosphere for longer periods of time, the dangers of transport or dispersal of radioactive materials would multiply, particularly with respect to intermediate and prolonged fallout. Although not immediately lethal, this fallout could lead to chronic radiation sickness, depressed immune systems, genetic defects, and delayed death for millions from cancer or other radiation-induced causes.

STUDY OF NUCLEAR WINTER

Formulation of the nuclear-winter scenario, and its widespread discussion in the scientific, military, and political communities, resulted largely from the work of a research team appointed in 1982 by the National Academy of Sciences (NAS). The NAS requested that Richard B. Turco, Owen B. Toon, Thomas P. Ackerman, James B. Pollack, and Carl Sagan (a team subsequently known by their last initials as "TTAPS") investigate the possible effects of dust raised by nuclear detonations. The TTAPS team comprised scientists who already had investigated a variety of atmospheric effects of dust particles, including the consequences of volcanic explosions, the behavior of dust storms on Mars, and theories of mass extinction caused by asteroid or comet impacts in the geologic past.

Computer modeling represents the only feasible means of studying the consequences of a nuclear exchange. The TTAPS researchers had access to advanced models for atmospheric circulation, particle microphysics, and radiative-convective behavior. Also used were models of nuclear-exchange scenarios of varying size and geographical complexity. Other researchers, including Paul J. Crutzen at the Max Planck Institute for Chemistry in West Germany and John W. Birks at the University of Colorado, persuaded the TTAPS team to include in their research the possible effects of smoke from massive fires caused by nuclear explosions.

Prior to the TTAPS research, little was known about potential large-scale smoke-and-dust pollution except as the consequence of volcanic explosions. The TTAPS team was able to study firsthand results of the explosion of El Chichón in 1982, which generated a significant dust veil and possibly some short-term surface temperature reduction in some areas. The much larger explosion of

Krakatau in 1883 caused visible atmospheric effects around the world and, possibly, minor surface-temperature changes. Historically, several volcanic explosions far larger than Krakatau are known from the eighteenth and nineteenth centuries, some of them comparable in energy to several-multimegaton nuclear detonations. Yet none generated more than minor multiyear climatic effects.

The TTAPS team concluded that earlier attempts to judge the possible effects of nuclear explosions based on records and observations of volcanic explosions were misleading. No matter how large, volcanic events were episodic and localized. The clouds of soil-and-dust particles, or aerosols, created by these explosions consist mostly of relatively large and bright, light-scattering particles that can create dazzling displays but do not absorb much visible light. By including in their model the smoke and soot from fires generated by nuclear explosions, however, the TTAPS team identified the potential source of aerosols of very dark and very fine particles with average diameters smaller than the typical wavelength of infrared radiation. These smoke and soot particles would have a very high absorptive capacity for visible light but would not correspondingly increase infrared absorption in the atmosphere, thus bringing about surface darkness and temperature reduction.

One of the most unsettling results of the TTAPS research was to show that a full-scale nuclear exchange between the superpowers would not be needed to bring about the most serious nuclear-winter effects. Presuming that such an exchange would entail a total explosive yield of about 25,000 megatons, nuclear winter could be triggered by much smaller conflicts—in the range from 3,000 to 5,000 megatons, depending upon the nature of the targets. Prior to the 1963 international treaty banning atmospheric testing of nuclear weapons, scenarios of nuclear conflict centered on massive atmospheric explosions of thermonuclear weapons in the 20-to-40-megaton range, principally over cities and refinery areas. Many of these weapons were to have been delivered by strategic bombers.

Subsequent development of ballistic missiles using multiple re-entry warhead configurations of 1 to 5 megatons each, directed initially at the enemy's own silos, greatly increased the anticipated number of ground-level detonations in a nuclear war scenario, thereby increasing the expected amount of material injected into the atmosphere by the explosions. Ironically, the strategic use of much larger numbers of considerably smaller-yield warheads makes a nuclear-winter catastrophe even more likely, even if the nuclear exchange were limited to "surgical" strikes concentrated on missile silos and military installations.

Ronald W. Davis

Cross-References

Acid Rain and Acid Deposition, 1802; Air Pollution, 1809; Atmosphere's Evolution, 1816; Atmosphere's Global Circulation, 1823; Atmosphere's Structure and Thermodynamics, 1828; Atmospheric Inversion Layers, 1836; Auroras, 1841; Climate, 1902; Cosmic Rays and Background Radiation, 1846; Earth-Sun Relations, 1851; Fallout, 1857; Global Warming, 1862; Greenhouse Effect, 1867; Hydrologic Cycle, 2045; Ice Ages, 1111; Ocean-Atmosphere Interactions, 2123; Ozone Depletion and Ozone Holes, 1879; Paleoclimatology, 1131; Radon Gas, 1886; Rain Forests and the Atmosphere, 1890.

BIBLIOGRAPHY

Diermendjian, Diran. *"Nuclear Winter": A Brief Review and Comments on Some Recent Literature.* Santa Monica, Calif.: RAND Corporation, 1988. An independent study of the first few years of debate about nuclear winter and policy implications. Covers some of the research literature outside the United States.

Ehrlich, Paul R., et al. *The Cold and the Dark: The World After Nuclear War.* New York: W. W. Norton, 1984. A summary of the TTAPS research for the general public.

_____. "Long-Term Biological Consequences of the Nuclear War." *Science* 222 (1983): 1293-1300. Confirms and elaborates upon the TTAPS team projections of the dangers of nuclear winter to plants and animals (and, hence, human food supplies) and projects likely conditions over several different time intervals. Together with a TTAPS article preceding it in the issue, this article announced the nuclear-winter scenario, and debate about it, to the scientific community.

Fisher, David E. *Fire and Ice: The Greenhouse Effect, Ozone Depletion, and Nuclear Winter.* New York: Harper & Row, 1990. Juxtaposes the twin hazards of global warming through the greenhouse effect and possible global winter resulting from nuclear war. Calls for more environmental and foreign-policy responsibility to avoid these dangers.

Greene, Owen. *Nuclear Winter: The Evidence and the Risks.* Cambridge, England: Polity Press, 1985. An independent confirmation of the TTAPS research that had a significant impact in Europe.

Harwell, Mark. *Nuclear Winter: The Human and Environmental Consequences of Nuclear War.* New York: Springer-Verlag, 1984. General survey of nuclear-winter conditions, with particularly good treatment of the potential hazards of widespread distribution of toxic materials by fires in bombed cities.

Nuclear Winter and National Security: Implications for Future Policy. Maxwell Air Force Base, Ala.: Air University Press, 1986. Papers from a military symposium that concluded that publicizing the nuclear-winter scenario would lead to increased international pressure for nuclear disarmament, thus affecting the capability of the military to respond to threats. Reflects a certain unwillingness on the part of military planners to accept the predictions of the nuclear-winter model.

Sagan, Carl. "Nuclear Winter and Climatic Catastrophe: Some Policy Implications." *Foreign Affairs* 62 (Winter, 1983): 257-292. A major contribution toward introducing the nuclear-winter issue into foreign-policy debates among the informed public, by the most prominent science writer on the TTAPS team.

Sagan, Carl, and Richard Turco. *A Path Where No Man Thought: Nuclear Winter and the End of the Arms Race.* New York: Random House, 1990. Two members of the TTAPS team tell the story of the research program and the campaign to bring the results to the attention of political and military authorities. Interesting speculations on how the findings may have influenced geopolitical and military thinking of major powers other than the United States.

Turco, Richard P., et al. "The Climatic Effects of Nuclear War." *Scientific American* 251, no. 2 (August, 1984): 33-43. Extraordinarily important article for its introduction of the TTAPS research to the scientifically educated public.

_____. "Nuclear Winter: Global Consequences of Multiple Nuclear Explosions." *Science* 222 (1983): 1283-1292. The key article announcing the TTAPS team findings to the scientific community.

OZONE DEPLETION AND OZONE HOLES

One of the natural gases in the atmosphere is ozone, which absorbs ultraviolet radiation from the Sun. When chlorofluorocarbons, which are used as a primary refrigerant and in making plastic, are broken up, they release chlorine ions, which destroy molecules of ozone. Scientists are concerned with possible ultraviolet damage to all organisms if ozone depletion continues. The term "ozone hole" refers to the seasonal decrease in stratospheric ozone concentration occurring over Antarctica. Ozone-hole formation is evidence that human activities can significantly alter the composition of the atmosphere.

PRINCIPAL TERMS

CATALYST: a substance that increases the rate of a chemical reaction without itself being produced or destroyed

CHLOROFLUOROCARBON (CFC): a group of chemical compounds containing carbon, fluorine, and chlorine, used in air conditioners, refrigerators, fire extinguishers, and other applications

DOBSON SPECTROPHOTOMETER: a ground-based instrument for measuring the total column abundance of ozone at a particular geographic location

FOOD CHAIN: an arrangement of the organisms of an ecological community according to the order of predation in which each uses the next, usually lower, member as a food source

OZONE LAYER: a region in the lower stratosphere, centered about 25 kilometers above the surface of the Earth, which contains the highest concentration of ozone found in the atmosphere

OZONE: an elemental form of oxygen containing three atoms of oxygen per molecule

PHYTOPLANKTON: free-floating microscopic aquatic plants that convert sunlight into food for themselves and for other organisms in the food chain

POLAR STRATOSPHERIC CLOUDS: clouds of ice crystals formed at extremely low temperatures in the polar stratosphere

POLAR VORTEX: a closed atmospheric circulation pattern around the South Pole that exists during the winter and early spring; atmospheric mixing between the polar vortex and regions outside the vortex is slow

STRATOSPHERE: the region of the atmosphere between 10 and 50 kilometers above the surface of the Earth

TOTAL COLUMN ABUNDANCE OF OZONE: the total number of molecules of ozone above a 1-centimeter-square area of the Earth's surface

TOTAL OZONE MAPPING SPECTROMETER (TOMS): a space-based instrument for measuring the total column abundance of ozone globally

ULTRAVIOLET SOLAR RADIATION: radiation composed mainly of ultraviolet A (UV-A) and ultraviolet B (UV-B), which are categories based on their wavelengths

CONCENTRATION OF OZONE IN ATMOSPHERE

Ozone, although only a minor component of the atmosphere, plays a vital role in the survival of life on Earth. Ozone molecules absorb high-energy ultraviolet (UV) light from the Sun. Absorption of ultraviolet light in the ozone layer, a region of the stratosphere that contains the maximum concentration of ozone, prevents most such light from reaching the surface of the planet. If none of the Sun's ultraviolet radiation were blocked by the ozone layer, it would be difficult for most forms of life, including humans, to survive on land.

The concentration of ozone in the atmosphere is highly variable, changing with altitude, geographic location, time of day, time of year, and prevailing local atmospheric conditions. Long-term fluctuations in ozone concentration are also seen, some of which are related to the solar sunspot cycle. While long-term average ozone concentrations are relatively stable, short-term fluctuations of as much as 10 percent in total column abundance of ozone as a result of the natural variability in ozone concentration are often observed.

Beginning in the early 1970's, a new and unexpected decrease in stratospheric ozone concentration was first observed. The decrease was localized in geography to the Southern Polar region, and in time to early spring (which begins in October in the Southern Hemisphere). The initial decrease in ozone was small, but by 1980, decreases in total column abundance of ozone of as much as 30 percent were being seen, well outside the range of variation expected as a result of random fluctua-

tions. This seasonal destruction of stratospheric ozone above Antarctica, which by 1990 had reached 50 percent of the total column abundance of ozone, was soon given the label "ozone hole."

ROLE OF CFCs

While it was initially unclear whether formation of the Antarctic ozone hole stemmed from natural causes or from anthropogenic effects on the environment, extensive field studies combined with the results of laboratory experiments and computer modeling of the atmosphere quickly led to a consistent and detailed explanation for ozone-hole formation. The formation of the ozone hole has two principal causes: chemical reactions that occur generally throughout the stratosphere, and special conditions that exist in the Antarctic region.

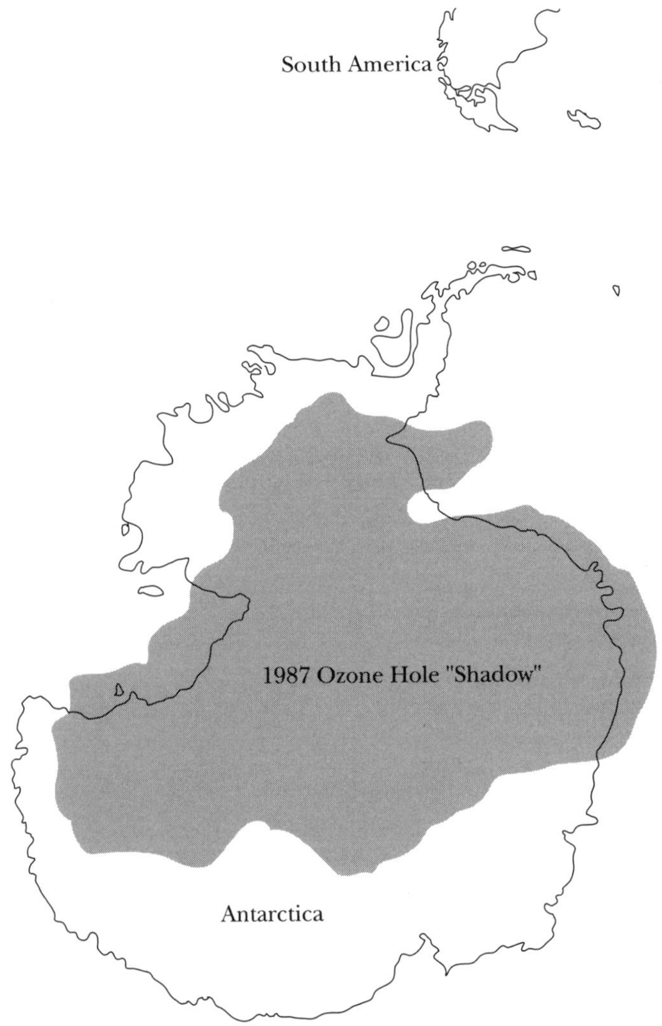

Under normal conditions, the concentration of ozone in the stratosphere is determined by a balance between reactions that remove ozone and those that produce ozone. The removal reactions are mainly catalytic chain reactions, in which trace atmospheric chemical species destroy ozone molecules without themselves being consumed. In such processes, it is possible for one chain carrier to remove many ozone molecules before being itself removed. The trace species involved in ozone removal include hydrogen oxides and nitrogen oxides, formed primarily by naturally occurring processes, and chlorine and bromine atoms and their corresponding oxides.

A major source of chlorine in the stratosphere is the decomposition of a class of compounds called chlorofluorocarbons (CFCs). Such compounds can be used in refrigeration and air condition-

THE OZONE HOLE, 1987

South America

1987 Ozone Hole "Shadow"

Antarctica

ing, as aerosol propellants, and as solvents. Chlorofluorocarbons are extremely stable in the lower atmosphere, with lifetimes of several decades. The main fate of chlorofluorocarbons is slow migration into the stratosphere, where they absorb ultraviolet light and release chlorine atoms. The chlorine atoms produced from the breakdown of chlorofluorocarbons in the stratosphere provide an additional catalytic process by which stratospheric ozone can be destroyed. A similar set of reactions involving a class of bromine-containing compounds called halons, used in some types of fire extinguishers, leads to additional ozone destruction. By 1986, the average global loss of stratospheric ozone caused by the release of chlorofluorocarbons, halons, and related compounds into the environment was estimated to be 2 percent.

ANTARCTIC CONDITIONS

While the decomposition and subsequent reaction of chlorofluorocarbons and other synthetic compounds explains the small general decline in ozone concentration observed in the stratosphere, additional processes are needed to account for the more massive seasonal ozone depletion observed above Antarctica. These processes involve a set of special conditions that in combination are unique to the stratosphere above Antarctica.

During daylight hours, a portion of the chlorine present in the stratosphere is tied up in the form of reservoir species, compounds such as hydrogen chloride and chlorine nitrate that do not react with ozone. This slows the rate of removal of ozone by chlorine. Processes that directly or indirectly involve absorption of sunlight transform reservoir species into ozone-destroying chlorine atoms. During the Antarctic winter, when sunlight is entirely absent, stratospheric chlorine is rapidly converted into reservoir species.

In the absence of additional chemical processes, the onset of spring in Antarctica and the return of sunlight would convert a portion of the reservoir compounds into reactive chlorine species and reestablish the balance between ozone-producing and ozone-destroying processes. However, the extremely low temperatures occurring in the stratosphere above Antarctica during the winter months leads to the formation of polar stratospheric clouds, which, because of the extremely low concentration of water vapor in the stratosphere, do not form during other seasons or outside the polar regions of the globe. The ice crystals that compose the clouds act as catalysts that convert reservoir species into diatomic chlorine and other gaseous chlorine compounds that, in the presence of sunlight, re-form ozone-destroying species. At the same time, the nitrogen oxides found in the reservoir species are converted into nitric acid, which remains attached to the ice crystals. As these ice crystals are slowly removed from the stratosphere by gravity, the potential for conversion of active forms of chlorine into reservoir species is greatly reduced. Because of this, when spring arrives, large amounts of ozone-destroying chlorine species are produced by the action of sunlight, and only a small fraction of this reactive chlorine is converted into reservoir species. The increased rate of ozone removal caused by the abundance of reactive chlorine present in the stratosphere leads to ozone depletion and formation of the ozone hole.

An additional process important in formation of the ozone hole is the unique air-circulation pattern in the stratosphere above Antarctica. During the winter and early spring, a vortex of winds circulates about the South Pole. This polar vortex minimizes movement of ozone and reservoir-forming compounds from other regions of the stratosphere. As this polar vortex breaks up in midspring, ozone concentrations in the Antarctic stratosphere return to normal levels, and the ozone hole gradually disappears.

ATMOSPHERIC OZONE STUDY AND INTERPRETATION

Researchers utilize a great diversity of devices and techniques in their study and interpretation of atmospheric ozone. One popular technique is the use of models. A good model is one that simulates the interrelationships and interactions of the various parts of the known system. The weakness of models is that, often, not enough is known to give an accurate picture of the total system or to make accurate predictions. Most modeling is done on computers. Scientists estimate how fast chemicals such as CFCs and nitrous oxide will be produced in the future and build a computer model of the way these chemicals react with ozone and with one another. From this model, it is possible to estimate future ozone levels at different altitudes and at different future dates.

Ozone instruments are being loaded onto ER-2, the NASA research aircraft. NASA's ER-2 aircraft is a modified U-2 reconnaissance plane that carries instruments up to 20 kilometers altitude for seven-hour flights to 80 degrees north latitude. (NASA/JPL/ Caltech)

Similar processes appear to be at work in the Arctic stratosphere, leading to ozone depletion, as in the Antarctic; however, the National Oceanic and Atmospheric Administration (NOAA) Aeronomy Laboratory in Boulder, Colorado, reported a discrepancy between observed ozone depletion and predicted levels, based on models that account accurately for the Antarctic depletions. This report suggests that some other mechanism is at work in the Arctic. Thus, good models can be very useful in studying new data. There are two models favored by most scientists in this area. Some scientists put forth a chemical model that says the depletion is caused by chemical events promoted by the presence of chlorofluorocarbons created by industrial processes. Acceptance of this model was promoted by the discovery of fluorine in the stratosphere. Fluorine does not naturally occur there, but it is related to CFCs. The other model assumes that the ozone hole was formed by dy-

namic air movement and mixing. This model best fits data gathered by ozone-sensing balloons that sample altitudes up to 30 kilometers and then radio the data back to Earth. Ozone depletion is confined to air between 12 and 20 kilometers. While the total ozone depletion is 35 percent, different strata showed various amounts of depletion from 70 to 90 percent. Surprisingly, about half the ozone was gone in twenty-five days. This finding does not fit the chemical model very well.

Besides ozone-sensing balloons, satellites are of much help. The National Aeronautics and Space Administration (NASA) obtains measurements with its Nimbus 7 satellite. Ozone measurements made by this satellite helped to develop flight plans for the specialized aircraft NASA also deploys in ozone studies. NASA's ER-2 aircraft is a modified U-2 reconnaissance plane that carries instruments up to 20 kilometers in altitude for seven-hour flights to 80 degrees north latitude. A

DC-8, operating during the same period, is able to survey the polar vortex, owing to its greater range. In addition, scientists utilize many meteorological techniques and instruments, including chemical analysis of gases by means of infrared spectroscopy, mass spectroscopy and gas spectroscopy combined, gas chromatography, and oceanographic analysis of planktonic life in the southern Atlantic, Pacific, and Indian oceans. As new research methods become available, they are applied to this essential study.

PUBLIC HEALTH CONCERNS

Atmospheric ozone provides a gauze of protection from the lethal effects of ultraviolet radiation from the Sun. This ability to absorb ultraviolet radiation protects all life-forms on the Earth's surface from excessive ultraviolet radiation, which destroys the life of plant and animal cells. Currently, between 10 and 30 percent of the Sun's ultraviolet B (UV-B) radiation reaches the Earth's surface. If ozone levels were to drop by 10 percent, the amount of UV-B radiation reaching the Earth would increase by 20 percent.

Present-day UV-B levels are responsible for the fading of paints and the yellowing of window glazing and for car finishes becoming chalky. These kinds of degradation will accelerate as the ozone layer is depleted. There could also be increased smog, urban air pollution, and a worsening of the problem of acid rain in cities. In humans, UV-B causes sunburn, snow blindness, skin cancer, cataracts, and excessive aging and wrinkling of skin. Skin cancer is the most common form of cancer—more than 400,000 new cases are reported every year in the United States alone. The National Academy of Sciences has estimated that each 1 percent decline in ozone would increase the incidence of skin cancer by 2 percent. Therefore, a 3 percent depletion in ozone would produce some 20,000 more cases of skin cancer in the United States every year.

ECOLOGICAL CONCERNS

Many other forms of life—from bacteria to forests and crops—are adversely affected by excessive radiation as well. Ultraviolet radiation affects plant growth by slowing photosynthesis and by delaying germination in many plants, including trees and crops. Scientists have a great concern for the organisms that live in the ocean and the effect ozone depletion may have on them. Phytoplankton, zooplankton, and krill (a shrimplike crustacean) could be greatly depleted if there were a drastic increase in ultraviolet A and B. The result would be a tremendous drop in the population of these free-floating organisms. These organisms are important because they are the beginning of the food chain. Phytoplankton use the energy of sunlight to convert inorganic compounds, such as phosphates, nitrates, and silicates, into organic plant matter. This process provides food for the next step in the food chain, the herbivorous zooplankton and krill. They, in turn, become the food for the next higher level of animals in the food chain. Initial studies of this food chain in the Antarctic suggest that elevated levels of ultraviolet radiation impair photosynthetic activity. Recent studies show that a fifteen-day exposure to UV-B levels 20 percent higher than normal can kill off all anchovy larvae down to a depth of 10 meters. There is also concern that ozone depletion may alter the food chain and even cause changes in the organism's genetic makeup. An increase in the ultraviolet radiation is likely to lower fish catches and upset marine ecology, which has already suffered damage from human-made pollution. On a worldwide basis, fish presently provides 18 percent of all the animal protein consumed.

INTERNATIONAL RESPONSE

The United Nations Environmental Program (UNEP) is working with governments, international organizations, and industry to develop a framework within which the international community can make decisions to minimize atmospheric changes and the effects they could have on the Earth. In 1977, UNEP convened a meeting of experts to draft the World Plan of Action on the Ozone Layer. The plan called for a program of research on the ozone layer and on what would happen if the layer were damaged. In addition, UNEP created a group of experts and government representatives who framed the Convention for the Protection of the Ozone Layer. This convention was adopted in Vienna in March, 1985, by twenty-one states and the European Economic Community and has subsequently been signed by many more states. The convention pledges states that sign to protect human health and the environment from the effects of ozone depletion. Action has already been taken to protect the ozone layer. Several countries have restricted the use of CFCs or the amounts produced. The United States banned the use of CFCs in aerosols in 1978. Some countries, such as Belgium and the Nordic countries, have in effect banned CFC production altogether. The group has also worked with governments on a Protocol to the Convention that required nations that signed to limit their production of CFCs. It is the hope and aim of these nations that such international cooperation will lead to a better global environment.

George K. Attwood and Jeffrey A. Joens

CROSS-REFERENCES

Acid Rain and Acid Deposition, 1802; Air Pollution, 1809; Atmosphere's Evolution, 1816; Atmosphere's Global Circulation, 1823; Atmosphere's Structure and Thermodynamics, 1828; Atmospheric Inversion Layers, 1836; Auroras, 1841; Climate, 1902; Cosmic Rays and Background Radiation, 1846; Earth-Sun Relations, 1851; Fallout, 1857; Global Warming, 1862; Greenhouse Effect, 1867; Nuclear Power, 1663; Radon Gas, 1886; Rain Forests and the Atmosphere, 1890; Volcanic Hazards, 798.

BIBLIOGRAPHY

Bast, Joseph L., Peter J. Hill, and Richard C. Rue. *Eco-Sanity: A Common Sense Guide to Environmentalism*. Lanham, Md.: Heartland Institute, 1994. Highly critical of claims that human activities have the potential to degrade the global environment significantly. Although the authors are selective in the information they present, the book represents an interesting minority opinion on issues of global change. Ozone depletion and the ozone hole are briefly discussed in chapter 4.

Cagin, Seth, and Philip Dray. *Between Earth and Sky: How CFCs Changed Our World and Endangered the Ozone Layer*. New York: Pantheon Books, 1993. A history of chlorofluorocarbons. Beginning with the discovery of CFCs in 1928, the book goes on to discuss the increasing use of CFCs during the following decades, the establishment of the link between CFCs and stratospheric ozone depletion, and the subsequent restrictions placed on the use and manufacture of CFCs. Chapters 16 through 21 focus on the ozone hole.

Firor, John. *The Changing Atmosphere: A Global Challenge*. New Haven, Conn.: Yale University Press, 1990. An interesting and readable introduction to atmospheric science and global change. Complex scientific research is presented in a clear and balanced manner. Both scientific issues and the corresponding political challenges are discussed.

Fisher, David E. *Fire and Ice: The Greenhouse Effect, Ozone Depletion, and Nuclear Winter*. New York: Harper & Row, 1990. A thorough discussion of the real and potential effects of human activities on the global environment. Chapter 12 gives a brief history of the events surrounding the discovery of the Antarctic ozone hole.

Graedel, T. E., and Paul J. Crutzen. *Atmospheric Change: An Earth System Perspective*. New York: W. H. Freeman, 1993. An introduction to atmospheric science by two well-known workers in the field, one of whom, Crutzen, shared the 1995 Nobel Prize in Chemistry for his work in atmospheric chemistry. While portions of the book require an understanding of college-level mathematics and chemistry, most of the book can be understood without such a background. The large number of figures and tables are of particular interest.

Roan, Susan. *Ozone Crisis: The Fifteen-Year Evolution of a Sudden Global Emergency*. New York: John Wiley & Sons, 1989. A detailed account of the scientific and political events associated with the discovery of the connection between chlorofluorocarbons and stratospheric ozone depletion. Chapters 8 through 16 focus on the evidence linking the ozone hole to increasing atmospheric concentrations of CFCs, and the regulation of CFCs that followed.

Rowland, F. Sherwood. "Stratospheric Ozone Depletion." *Annual Review of Physical Chemis-*

try 42 (1991): 731. Although this article may be difficult to find, it is well worth the effort. Rowland, cowinner of the 1995 Nobel Prize in Chemistry for his work on ozone depletion, has written a thorough but highly readable review of stratospheric ozone chemistry and the role of chlorofluorocarbons and related compounds in ozone depletion. While some background in chemistry is useful for full understanding of the material, the essential points of the article can be understood by the layperson.

Shell, E. R. "Weather Versus Chemicals." *The Atlantic* 259 (May, 1987): 27-31. A good source for discussion in fairly simple terms of the complex problems of the ozone hole and the widely recognized role of chlorofluorocarbons in this global issue. *The Atlantic* magazine offers thorough discussion of environmental issues on a popular reading level.

Somerville, Richard C. J. *The Forgiving Air: Understanding Environmental Change.* Los Angeles: University of California Press, 1996. A thorough investigation of the relationship between human activities and changes in the atmosphere and global climate. Written by a scientist involved in atmospheric research, but aimed at a nonscientific audience. Chapter 2 is a discussion of the ozone hole.

RADON GAS

The chemical element radon is a radioactive gas and the heaviest of the noble gases. It is produced by the radioactive decay of radium, which is a natural decay product of the uranium found in various types of rocks. Trace amounts of radon seep from rocks and soil into the atmosphere and can become a health hazard in sufficient concentrations.

PRINCIPAL TERMS

HALF-LIFE: the time required for one-half of the nuclei in a sample of a radioactive isotope to decay into another isotope by emitting various types of particles

ISOTOPES: different atoms of the same element with different numbers of neutrons in the nucleus

NOBLE GAS: any of the six elements of group O in the periodic table—including helium, neon, argon, krypton, xenon, and radon—that are monatomic with filled electron shells; they are often called inert gases since they usually are chemically inert

PICOCURIE (pCi): a unit of radioactivity corresponding to one-trillionth of that from 1 gram of radium (0.037 disintegration per second or 2.22 disintegrations per minute)

RADIOACTIVITY: the spontaneous and continuous emission from unstable atomic nuclei of high-energy particles and radiation, including helium nuclei (alpha particles), electrons (beta particles), and electromagnetic waves (gamma and X rays)

DISCOVERY AND PROPERTIES OF RADON GAS

The French physicist Antoine Henri Becquerel discovered radioactivity in 1896 when he accidentally detected spontaneous and continuous radiation emitted by uranium. One of his students, Polish scientist Marie Curie, found that thorium was also radioactive and that the energy of radioactivity was about 1 million times greater than the energy of chemical reactions. In 1898, Marie and her husband Pierre Curie discovered the radioactive elements polonium and radium by separating various components of the uranium ore pitchblende. Within a few years, radioactivity was found to consist of three components in increasing order of their ability to penetrate matter: helium nuclei (alpha particles), electrons (beta particles), and electromagnetic radiation (gamma and X rays).

In 1900, the German chemist Friedrich E. Dorn detected a radioactive gas given off in the decay of radium along with helium. This gas was originally called radium emanation but was later called radon. It was found to have a half-life of 3.82 days. Even before Dorn's discovery, British physicists R. B. Owens and Ernest Rutherford had observed in 1899 that some of the radioactivity of thorium compounds could be blown away in the form of a gas they called thoron, which was found to have a 51.5-second half-life. In 1904, Friedrich Giesel and André-Louis Debierne independently discovered another radioactive gas produced from actinium called actinon, which was found to have a 3.92-second half-life. After the development of the isotope concept by Rutherford and Frederick Soddy in 1912, it was eventually found that thoron and actinon were isotopes of radon, with atomic masses 220 for thoron (radon 220), 219 for actinon (radon 219), and 222 for radon (radon 222). Radon is now known to have at least seventeen artificial radioactive isotopes in addition to its three natural isotopes.

Radon is an odorless, tasteless, colorless gas nearly eight times heavier than air and more than one hundred times heavier than hydrogen. It has an atomic number of 86 (86-proton nucleus), and its isotopes have atomic mass numbers (protons plus neutrons) ranging from 204 to 224, all radioactive. It has a stable electronic configuration of eight electrons in its outer shell, accounting for its usual chemical inactivity, although it is not completely inert. In 1962, the compound radon difluoride was produced in the laboratory and is apparently more stable than other noble-gas compounds.

Radon is rare in nature because its isotopes all

have short half-lives and because its source, radium, is such a scarce element. Radon 222 is the longest lived of the radon isotopes and is an alpha-decay product of radium 226, which itself results from the decay of uranium 238 along with about one dozen other radioactive isotopes. Traces of radon are found in the atmosphere near the ground because of seepage from soil and rocks, most of which contain small uranium deposits that produce minute amounts of radium that continually decay into radon.

HEALTH EFFECTS OF RADON GAS

Because it is inert and has a relatively short half-life, radon in itself poses little danger. However, in the process of radioactive decay, it produces several daughters that are solids, some of which emit alpha particles that can be especially dangerous to the delicate tissues of the lungs. Radon 222 poses the greatest danger because it has the longest half-life and occurs most commonly in nature, making up about 40 percent of background radiation. It is a product of uranium 238, which makes up 99.3 percent of natural uranium and has a half-life of 4.5 billion years so that it is a virtually unending source of radioactivity in the crust of the Earth. Other radon isotopes are of less danger because of their short half-lives and less common occurrence, with the possible exception of thoron in some local areas where it occurs at higher than normal concentrations.

The decay products of radon 222 are solids, two of which emit high-energy alpha particles. These two radon daughters are polonium 218, with a half-life of 3.05 minutes, and polonium 214, with a half-life of 164 microseconds. Because of their electric charge, these isotopes readily attach to airborne particles. When the polonium is inhaled, it lodges in the lung and can cause damage to the lining of the lung by alpha radiation. Most of the damage is done in the bronchial tubes, which contain the precursor (stem) cells that are particularly sensitive to the cancer-causing effects of alpha radiation. The primary data relating to lung cancer deaths caused by radon exposure come from studies of underground uranium miners. These studies indicate that from 3 to 8 percent of the miners studied developed lung cancer above and beyond those cancers attributed to smoking and other causes.

In the 1970's, concern began to be expressed about radon contamination of indoor air, especially for homes constructed on or with waste rock or tailings associated with the mining and processing of uranium and phosphate ores with significant concentrations of radium 226. This led Congress to pass the Uranium Mill Tailings Act in 1978. Wider concern emerged in 1984 when a nuclear power engineer named Stanley Watras set off monitors at his job, detecting radioactivity that was traced to his exposure to high radon concentration in his home in Boyertown, Pennsylvania. Further studies indicated that radon levels in houses far removed from uranium tailings or phosphate ores were often as high as or higher than houses near such sites, especially in poorly sealed basements.

By the late 1980's, it was recognized that radon gas seeping into the foundations, basements, or piping of poorly ventilated buildings is a potentially serious health hazard. Radon levels are highest in well-insulated homes built over geological formations that contain uranium mineral deposits. Even though these levels might be significantly lower than those in underground mines, it was feared that long-term exposure to even moderate amounts of radon might greatly increase the risk of developing lung cancer. Radon is now thought to be the largest source of natural radiation exposure and the single most important cause of lung cancer among nonsmokers in the United States. Studies indicate that indoor radon exposure increases considerably in the presence of cigarette smoke, both primary and secondary (passive smoke), since radon daughters apparently bind more effectively with smoke particles in the air.

CONCENTRATIONS OF RADON GAS

In the United States, the concentrations of radon and its decay products are usually expressed in picocuries per liter (pCi/L). A picocurie is one-trillionth (10^{-12}) of a curie, where 1 curie equals 37 billion becquerels (Bq, or disintegrations per second). In the International System of Units, radon concentrations are expressed as becquerels per cubic meter, so that 1 picocurie per liter equals 37 becquerels per cubic meter. Average radon concentrations in outdoor air at ground level are about 0.20 picocurie per liter and range from less than 0.1 picocurie per liter to about 30 picocuries per liter. Radon dissolved in ground water ranges from about 100 to nearly 3 million picocuries per liter. Indoor air averages about 1.5 picocuries per liter, but local conditions can result in levels several orders of magnitude higher than these, especially in some single-family dwellings. The National Academy of Science (NAS) estimates that at least fifteen thousand fatal lung cancers per year are caused by indoor radon, and another estimate suggests an additional five thousand fatalities from increased radon exposure caused by passive smoking.

Various standards for indoor radon have been established by extrapolating down from levels as high as 30,000 picocuries per liter in uranium mines associated with lung cancer. The Environmental Protection Agency (EPA) has set a radon guideline of 4 picocuries per liter for remedial action in buildings, which they estimate could produce between one and five lung cancer deaths for every one hundred individuals. They project up to seventy-seven fatalities out of one hundred people exposed to levels of 200 picocuries per liter. These estimates assume seventy years in the dwelling with about 75 percent of time spent indoors. The International Council on Radiation Protection (ICRP) has set an indoor radon level of 8 picocuries per liter as unsafe, which is about seventeen atoms per minute of radon decaying in every liter of air. The EPA estimates that a level of 10 picocuries per liter has a lung cancer risk similar to smoking one pack of cigarettes per day. Since the 1970's, a radiation safety limit of about 100 picocuries per liter has been set for uranium mining.

The indoor contamination problem in the Boyertown area of southeastern Pennsylvania was found to have radon levels as high as 2,600 picocuries per liter. Boyertown lies on a geological formation called the Reading Prong, which extends east from Reading through three counties of Pennsylvania and into parts of New Jersey, New York, and New England, with bedrock minerals containing elevated levels of uranium and thorium. These conditions led to the monitoring of eighteen thousand homes by the EPA in conjunction with the Pennsylvania Department of Health and local utilities, which found radon levels in excess of the EPA's 4 picocuries per liter guideline for remedial action in 59 percent of the homes. In a nationwide EPA residential survey, average radon levels ranged from 0.1 picocurie per liter in Hawaii to 8.8 picocuries per liter in Iowa. Researchers estimate that 2 percent of U.S. homes

have radon levels in excess of the ICRP guideline of 8 picocuries per liter.

DETECTION AND REDUCTION OF INDOOR RADON GAS

Indoor radon levels are difficult to measure because of such factors as air movement, the effects of cigarette smoke, water tables, barometric pressure, and seasonal conditions, with higher readings in the summer than in the winter. Both active and passive testing devices can be used to test homes. Active devices include continuous radon monitors used by trained testers over several days. Passive radon kits, which are available at hardware stores, include charcoal canisters and alpha-track detectors. Sealed charcoal-filled canisters are exposed for two or three days and then sent to EPA-approved laboratories for analysis of their level of gamma radiation, which is related to the radon level. Alpha-track detectors contain plastic strips that must be suspended for two or three months before being sent for analysis based on counting microscopic alpha tracks.

Radon levels in residential structures can be reduced by various techniques. Site selection is important in avoiding the high radon contamination associated with highly permeable (porous) soils. These can be identified from soil maps prepared by the Soil Conservation Service. High-radium-content surface materials can be covered with soil that has low permeability and radium content. A 3.3-meter fill depth can reduce radon emanation rates by about 80 percent.

The choice of substructure is also important in radon reduction, with well-ventilated crawl spaces providing much lower radon levels than basements. Radon control in basements is aided by using good-quality concrete on top of an impermeable plastic barrier and a complete system of drainage tile around the perimeter. Radon can be reduced in existing basements by sealing floor and wall cracks, capping sumps, and venting the air from under the basement floor.

SIGNIFICANCE

Radon has both positive and negative implications, although there are some claims that the risks from radon have been overstated. Radon 222 is used in the treatment of some cancers. It can be collected by passing air through an aqueous solution of radium-226 salt or a porous solid containing radium 226 salt and then pumping off the accumulated radon every few days. It is then purified and compressed in small tubes, which can be inserted in the diseased tissue. The gas produces penetrating gamma radiation from the bismuth-214 decay product of radon and can be used for both radiotherapy and radiography.

Radon is also a useful tracer for groundwater and atmospheric mixing. It is used in studies of groundwater interaction with streams and rivers. A high radon concentration in groundwater that makes its way into a stream or river is a sensitive indicator of such local inputs. Since atmospheric radon concentrations decrease exponentially with altitude and are lower over water than land, radon can serve as an effective tracer in measuring atmospheric mixing.

Critics of the EPA's 4 picocuries per liter radon guideline have raised questions about extrapolating from statistical data on lung-cancer deaths among miners with radon exposures as high as 30,000 picocuries per liter. Studies have indicated no unusual incidence of lung-cancer deaths for U.S. uranium miners when exposures are below 12,000 picocuries per liter. A massive study (published in *Science* magazine on August 22, 1980) of two groups in China, one living in a high-radiation area and the other in a low-radiation area, showed no significant cancer-rate difference between the two groups. A 1996 study by the Finnish Center for Radiation and Nuclear Safety found no increased risk for residents exposed to as much as 2.5 times the EPA's guideline. Perhaps the nearly 0.5 billion dollars spent by Americans testing for radon and renovating their homes has been an overreaction to the natural background radiation humans have lived with for thousands of years.

Joseph L. Spradley

CROSS-REFERENCES

Acid Rain and Acid Deposition, 1802; Air Pollution, 1809; Atmosphere's Evolution, 1816; Atmosphere's Global Circulation, 1823; Atmosphere's Structure and Thermodynamics, 1828; Atmospheric Inversion Layers, 1836; Auroras, 1841; Cosmic Rays and Background Radiation, 1846; Earth-Sun Relations, 1851; Fallout, 1857; Global Warming, 1862; Greenhouse Effect, 1867; Nuclear Winter, 1874; Ozone Depletion and Ozone Holes, 1879; Rain Forests and the Atmosphere, 1890.

BIBLIOGRAPHY

Bolch, Ben W., and Harold Lyons. *Apocalypse Not: Science Economics and Environmentalism.* Washington, D.C.: Cato Institute, 1993. This book attempts to debunk many of the arguments used by environmentalists. Chapter 5 on "A Multibillion-Dollar Radon Scare" is a good summary and evaluation of the radon problem.

Brookins, Douglas G. *The Indoor Radon Problem.* New York: Columbia University Press, 1990. This comprehensive book on radon and its health effects, detection, and reduction, contains a good glossary and many tables, graphs, and references.

Cole, Leonard A. *The Element of Risk: The Politics of Radon.* Washington, D.C.: AAAS Press, 1993. This is an interesting discussion of the politics of the radon problem, with information on many of the officials and scientists involved in shaping policy.

Fang, Hsai-Yang. *Introduction to Environmental Geotechnology.* New York: CRC Press, 1997. Chapter 10 on "Radiation Effects on Water, Soil and Rock" in this college-level text on environmental geology has many tables, graphs, and diagrams on radon and reduction methods.

Gates, Alexander E., and Linda C. S. Gundersen, eds. *Geologic Controls on Radon.* Boulder, Colo.: Geological Society of America, 1992. The special papers in this series on the geology of radon are somewhat specialized, but the introductory paper on "Geology of Radon in the United States" has good information that is accessible to the nonspecialist.

Godish, Thad. *Indoor Air Pollution Control.* Chelsea, Mich.: Lewis, 1989. Chapter 1 contains information on the problem of radon, and chapter 2 discusses the control of radon. Includes useful tables, diagrams, and references.

RAIN FORESTS AND THE ATMOSPHERE

Rain forests are ecosystems noted for their high biodiversity and high rate of photosynthesis. The rapid deforestation of such areas is of great concern both because it may lead to the extinction of numerous species and because it may reduce the amount of photosynthesis occurring on the Earth. Since photosynthesis releases large amounts of oxygen into the air, a curtailment of the process may have negative effects on the global atmosphere.

PRINCIPAL TERMS

BIODIVERSITY: a measure of the variety of life occupying a particular ecosystem

CELLULAR RESPIRATION: a series of chemical reactions that occurs within mitochondria of cells that leads to the release of energy as glucose is oxidized

CHLOROPHYLL: green pigment found in plants and algae that absorbs light and makes photosynthesis possible

ECOSYSTEM: a unit of nature composed of an interacting system of both living and nonliving components

PHOTOSYNTHESIS: a series of chemical reactions that occurs in chloroplasts of certain cells and that leads to the formation of glucose

LIFE AND THE ATMOSPHERE

All living things on the Earth—plants, animals, and microorganisms—depend on the "sea" of air surrounding them, the atmosphere. Present are the more abundant, permanent gases such as nitrogen (78 percent) and oxygen (21 percent), as well as smaller, variable amounts of others such as water vapor and carbon dioxide. Organisms absorb and utilize this air as a source of raw materials while releasing into it by-products of their life activities.

Of the life processes of greatest importance, cellular respiration is the most universal. Cellular respiration, a series of chemical reactions beginning with glucose and occurring in cytoplasmic organelles called mitochondria, produces a chemical compound called adenosine triphosphate (ATP). This essential substance furnishes the energy cells need to move, to divide, and to synthesize chemical compounds—in essence, to perform all the activities necessary to sustain life. It should be emphasized that cellular respiration is not confined to animals but also occurs in plants, and it also occurs during both the day and the night. In order for the last of the series of chemical reactions to be completed, oxygen from the surrounding air (or water in the case of aquatic organisms) must be absorbed. The carbon dioxide that forms is released into the air.

For cellular respiration to occur, a supply of glucose (a simple carbohydrate compound) is required. Photosynthesis, an elaborate series of chemical reactions occurring in chloroplasts, produces glucose, an organic (carbon) compound with six carbon atoms. Energy present in light and trapped by the chlorophyll within the chloroplasts is required to drive photosynthesis. Therefore, the process occurs only in plants and related organisms such as algae, and only during the daytime. Carbon dioxide, required as a raw material, is absorbed from the air, while the resulting oxygen is released into the atmosphere. The exchange of both gases typically involves tiny openings in leaves called stomata.

OXYGEN CYCLE

Oxygen is required for the survival of the majority of microorganisms, as well as for all plants and animals. It is from the surrounding air that organisms obtain the oxygen used in cell respiration. Terrestrial animals absorb it from the air by means of lungs or other organs, such as spiracles in the case of insects. Those living in water absorb dissolved oxygen by means of their gills. Plants absorb oxygen through the epidermal coverings of their roots and stems and through the stomatal openings of their leaves.

The huge amounts of oxygen removed by organisms during respiration must be replaced in order to maintain a constant reservoir of oxygen in the atmosphere. There are two significant sources of oxygen. One involves water molecules of the atmosphere that undergo a process called photodissociation: Oxygen remains after the lighter hydrogen atoms are released from the molecule and escape into outer space.

The other source is photosynthesis. Chlorophyll-containing organisms release oxygen as they use light as the energy source to split water molecules in a process called photolysis. The hydrogen is transported to the terminal phase of photosynthesis called the Calvin cycle, where it is used as the hydrogen source necessary to produce and release molecules of the carbohydrate glucose. In the meantime, the oxygen from the split water is released into the surrounding air.

Early in the history of the Earth, before certain organisms evolved the cellular machinery necessary for photosynthesis, the amount of atmospheric oxygen was very low. As the number and sizes of photosynthetic organisms gradually increased, so did the levels of oxygen in the air. A plateau was reached several millions years ago as the rate of oxygen release and absorption reached an equilibrium.

Another form of oxygen is ozone. Unlike ordinary atmospheric oxygen in which each molecule contains two atoms, ozone molecules have three oxygen atoms each. Most ozone is found in the stratosphere at elevations between 10 and 50 kilometers. This layer of ozone is important primarily because it helps to protect life on the Earth from the harmful effects of ultraviolet radiation. It is therefore understandable that scientists, especially ecologists, are concerned as the amount of ozone has been reduced drastically over the last few decades. Already, an increase in the incidence of skin cancer in humans and a decrease in the efficiency of photosynthesis has been documented. Another concern related to ozone is that of an increase in ozone levels nearer to the ground, where

Aerial view of the Amazon rain forest. (PhotoDisc)

living things are harmed as a result. The formation of ozone from ordinary oxygen within the atmosphere is greatly accelerated by the presence of gaseous pollutants released from smokestacks of industrial complexes.

CARBON CYCLE

Carbon is an element of unusual importance to life because all forms of life are composed of organic (carbon-containing) molecules. In addition to carbohydrates already discussed, these include lipids (fats, oils, steroids, and waxes), proteins of endless variety, and nucleic acids. The ability of carbon to serve as the backbone of these molecules results from the ability of carbon atoms to form chemical bonds not only with other atoms such as those of oxygen, hydrogen, and nitrogen, but also with other carbon atoms.

Like oxygen, carbon cycles in a predictable manner between living things and the atmosphere. In photosynthesis, carbon is "fixed" as carbon dioxide in the air (or dissolved in water) is absorbed and converted into carbohydrates. Carbon cycles to animals as they feed on plants and algae. As both green and nongreen organisms respire, some of their carbohydrates are oxidized, releasing carbon dioxide into the air. Each organism must eventually die, after which decay processes, aided by microorganisms, returns the remainder of the carbon to the atmosphere.

Levels of atmospheric carbon dioxide have fluctuated gradually during past millennia, as revealed by the analysis of the gas trapped in air bubbles of ice from deep within the Earth. In general, levels were lower during glacial periods and higher during warmer ones. After the beginning of the Industrial Revolution in the nineteenth century, levels rose slowly until about 1950 and then much more rapidly afterward. The apparent cause has been the burning of increased amounts of fossil fuels required to supply the energy demands of the growing world population. The

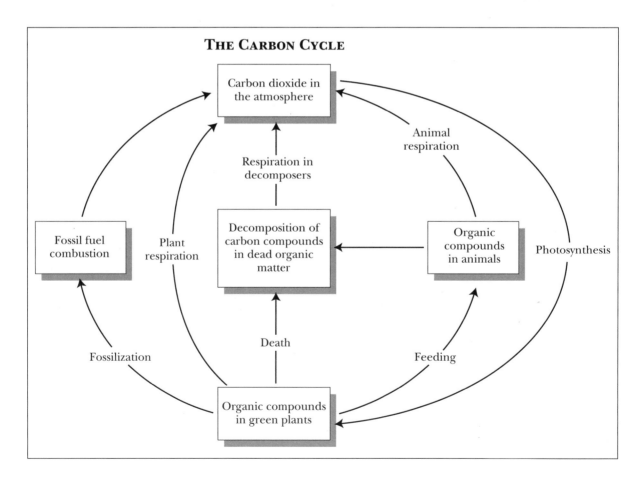

global warming that is now being experienced is believed by most scientists to be the cause of increased carbon dioxide levels. The "greenhouse effect" is the term given to the insulating effects of the atmosphere with increased amounts of carbon dioxide. The Earth's heat is lost to outer space less rapidly, thus increasing its average temperature.

In considering both the oxygen and carbon cycles, it should be realized that both elements—in addition to being cycled between organisms and the atmosphere—are stored in the Earth's crust on a long-term basis as rocks.

FOREST ECOSYSTEMS

The biotic (living) portions of all ecosystems include three ecological or functional categories: producers (plants and algae), consumers (animals), and decomposers (bacteria and fungi). Thus, the everyday activities of organisms of all the categories involve the constant exchange of oxygen and carbon dioxide between the organisms and the surrounding atmosphere.

Because they release huge quantities of oxygen during the day, producers deserve special attention. In both fresh and salt water, algae are the principal producers. On land, this role is played by a variety of grasses, other small plants, and trees. Forest ecosystems, dominated by trees but also harboring many other plants, are major systems that produce a disproportionate amount of the oxygen released into the atmosphere by terrestrial ecosystems.

Forests occupy all continents except for Antarctica. A common classification of forests recognizes these principal categories: coniferous (northern evergreen), temperate deciduous, and tropical evergreen, with many subcategories for each.

The designation "rain forest" refers to the subcategories of these types that receive an amount of rainfall well above the average. Included are tropical rain forests (the more widespread type) and temperate rain forests. Because of the ample moisture they receive, both types contain lush vegetation that produces and releases oxygen into the atmosphere on a larger scale than do other forests.

TROPICAL AND TEMPERATE RAIN FORESTS

Tropical rain forests occur at relatively low elevations in a band about the equator. The Amazon Basin of South America contains the largest contin-

Interior of a tropical rain forest in the Amazon Basin, Brazil. (© William E. Ferguson)

uous tropical rain forest. Other large expanses are located in western and central Africa, as well as the region from southeastern Asia to Australia. Smaller areas of tropical rain forests occur in Central America and on certain islands of the Caribbean Sea, the Pacific Ocean, and the Indian Ocean.

Seasonal changes within tropical rain forests are minimal. Temperatures, with a mean near 25 degrees Celsius, seldom vary more than 4 degrees Celsius. The amount of rainfall each year is at least 400 centimeters.

Tropical rain forests have the highest biodiversity of any terrestrial ecosystem. Included is a large number of species of flowering plants, insects, and animals. These organisms are arranged into horizontal layers, or strata. In fact, all forests are stratified, but not to the same degree as tropical rain forests. A mature tropical rain forest typically has five layers, beginning with the uppermost downward: an emergent layer (the tallest trees that project above the next layer), a canopy of tall trees, understory trees, shrubs, tall herbs and ferns, and low plants on the forest floor.

Several special life-forms are characteristic of

the plants of tropical rain forests. Epiphytes are plants such as orchids that are perched high in the branches of trees. Vines called lianas wrap themselves around trees. Most tall trees have trunks that are flared at their bases to form buttresses that help support them in the thin soil.

This brief description of tropical rain forests helps to explain their role in world photosynthesis and the related release of oxygen into the atmosphere. Because of the high rainfall and year-round warm temperatures, photosynthesis occurs throughout the year. As a result of their many layers of vegetation, the energy from sunlight as it passes downward is efficiently utilized. Furthermore, the huge amounts of oxygen released are available for use not only by the forests themselves but also, because of global air movement, by other ecosystems throughout the world. Because of this, tropical rain forests are often referred to as "the Earth's lungs."

Temperate rain forests are much less extensive than tropical rain forests. Temperate rain forests occur primarily along the Pacific Coast in a narrow band from southern Alaska to central California. Growing in this region is a coniferous forest, but one with warmer temperatures and a higher rainfall than those to the north and inland. Its rainfall of 65 to 400 centimeters per year is much less than that of a tropical rain forest but is supplemented in the summer by frequent heavy fogs. As a result, evaporation rates are greatly reduced.

Because of generally favorable climatic conditions, temperate rain forests, like tropical ones, support a lush vegetation. Also, the rate of photosynthesis and release of oxygen is higher than in most other world ecosystems.

SIGNIFICANCE

Ecologists and conservationists are greatly concerned about the massive destruction of rain forests. They are being cut and burned at a rapid rate to plant crops, to graze animals, and for their timber. Even though rain forests occupy less than 10 percent of the Earth's land surface, they produce a huge proportion of all oxygen released into the atmosphere. The ultimate effect of deforestation of these special ecosystems is yet to be seen.

Thomas E. Hemmerly

CROSS-REFERENCES

Acid Rain and Acid Deposition, 1802; Air Pollution, 1809; Atmosphere's Evolution, 1816; Atmosphere's Global Circulation, 1823; Atmosphere's Structure and Thermodynamics, 1828; Atmospheric Inversion Layers, 1836; Auroras, 1841; Cosmic Rays and Background Radiation, 1846; Earth's Crust, 14; Earth-Sun Relations, 1851; Environmental Health, 1759; Fallout, 1857; Global Warming, 1862; Greenhouse Effect, 1867; Landfills, 1774; Nuclear Winter, 1874; Ozone Depletion and Ozone Holes, 1879; Radioactive Decay, 532; Radioactive Minerals, 1255; Radon Gas, 1886; Soil Profiles, 1525.

BIBLIOGRAPHY

Barry, R. G., and R. J. Chorley. *Atmosphere, Weather, and Climate.* London: Methuen and Co., 1987. A good introduction to an understanding of weather, climate, and the atmosphere, written by recognized authorities.

Hunter, M. L. *Fundamentals of Conservation Biology.* Cambridge, Mass.: Blackwell Science, 1996. The primary theme of this book is biodiversity at all levels: species, ecosystem, and genetic.

Smith, R. L. *Ecology and Field Biology.* 5th ed. New York: Harper Collins, 1996. This leading textbook of ecology gives a good description of the structure and functioning of rain forests and other ecosystems.

Vankat, J. L. *The Natural Vegetation of North America: An Introduction.* New York: John Wiley and Sons, 1979. A highly useful overview of the ecosystems—including rain forests—of North America.

Walker, L. C. *Forests: A Naturalist's Guide to Trees and Forest Ecology.* New York: John Wiley and Sons, 1990. A fascinating guidebook for amateur naturalists.

Whitmore, T. C. *Tropical Rain Forests.* Oxford: Clarendon Press, 1990. Whitmore is concerned with all aspects of tropical rain forests: composition, climate, and the future. The book is scholarly but easy to understand.

2
WEATHER AND METEOROLOGY

ATMOSPHERIC AND OCEANIC OSCILLATIONS

The oceans' effect on developing weather patterns has long been known and taken into consideration in weather prediction. Atmospheric and oceanic patterns fluctuate over a one- to twenty-year course. These fluctuations, or oscillations, create major climate change, such as the well-known El Niño Southern Oscillation (ENSO) in the tropical Pacific, which alters weather across the globe.

PRINCIPAL TERMS

CONVEYOR BELT CURRENT: a large cycle of water movement that carries warm waters from the north Pacific westward across the Indian Ocean, around southern Africa, and into the Atlantic, where it warms the atmosphere, then returns at a deeper ocean level to rise and begin the process again

EKMAN SPIRAL: water movement in lower depths that occurs at a slower rate and in a different direction from surface water movement

SOLAR RADIATION: transfer of heat from the Sun to the Earth's surface, where it is absorbed and stored

TRADE WINDS: winds that blow steadily toward the equator; north of the equator, trade winds blow from the northeast, while south of the equator, they blow from the southeast

UPWELLING: the process by which colder, deeper ocean water rises to the surface and displaces surface water

AIR AND OCEAN INTERACTION

Currents in the water and in the air push heated water to cooler climates, where the heat is released, helping to regulate temperature on northern continents. As the heat is released, currents in the ocean and atmosphere return this cooled water to warmer climates, and the process repeats. These exchanges are an important way of recycling energy through the ocean-atmosphere system, which balances land temperatures and climate. Ocean currents, which keep water in constant motion, are affected by the Earth's rotation, the Sun's energy, wind, and the salt content and temperature of the ocean. Just as there are a series of air streams, pressures, and currents in the various levels of the atmosphere, the ocean mirrors the atmosphere with a similar network of circulation patterns and pressure zones.

Some currents are deep-sea streams, such as the Gulf Stream, that push through surrounding waters like rivers flow through land. By coupling the sciences of meteorology and oceanography, several major ocean currents have been detected using peak evaporation levels caused by warm, dry air in the subtropics. Strong evaporation throughout the year off the coast of South Africa revealed the presence of the Agulhas Current, strong evap-

oration off the eastern United States in January revealed the existence of the Gulf Stream, and strong evaporation throughout the year in the northeast Atlantic revealed the North Atlantic Drift. These currents all have generally higher surface temperatures in relation to overlying air—especially in winter—and are frequently accompanied by strong winds. Approximately 40 percent of total heat transport from the Southern to the Northern Hemisphere is caused by surface ocean currents. The Gulf Stream warms the western side of the Atlantic and spreads across Europe.

Another type of ocean current, called an upwelling, brings deeper, colder waters to the surface. This colder water replaces warmer surface waters, which are pulled away by strong trade winds. Upwellings carry with them nutrients that feed large fish crops. Such currents are an excellent example of how atmospheric currents and ocean currents interact: Without the trade winds, which pull surface waters away, colder water could not surface.

Ocean currents called Ekman spirals also demonstrate how the atmosphere affects surface waters. Winds drive surface water in the same direction that the wind blows. Water just below the surface moves more slowly and is slightly deflected to the right because of the Earth's gravitational rotation. Water at lower depths moves even more slowly, but in the opposite direction to the flow of surface water. This causes the water to flow in a downward motion called the Ekman spiral. Because of the Ekman spiral, surface water actually flows at a 90-degree angle to the wind flowing out to sea along coastlines, thus creating the opportunity for upwelling.

CLIMATE SHIFTS AND GLOBAL IMPACT

Changes in ocean temperature and air currents influence weather, and weather changes affect climate. As warm-water masses float toward the coasts, they bring with them moisture that causes rainfall. When air currents shift, as when strong trade winds die down or reverse course, areas that normally get little rain may be flooded, and areas that normally get a lot of rain may have droughts. In both cases, crops fail, possibly resulting in starvation. Other natural disasters such as fires, mudslides, and hurricanes may also be triggered. Further impact from climate shifts may cause ocean fisheries to decline. With marine food supplies disrupted, some fish also migrate to other locations. The warming waters of tropical regions cause bleaching of coral reefs, an already fragile and endangered ecosystem.

The best-known climate shift has occurred for centuries. It was named El Niño (Spanish for "Christ child" or "the little boy") by Peruvian fisherman because it occurs during the Christmas season. Peru is known for its anchovy fisheries, and El Niño hampers this important harvest. Normally, strong trade winds pull warmer surface waters away from Peru's coast so upwelling can occur. This upwelling, which brings rich nutrients from the lower ocean waters, feeds the anchovies, and Peruvian fisherman reap strong harvests. During El Niño years, which cycle approximately every seven years and may last up to two years, trade winds weaken, and warmer surface waters remain along Peru's coast. Upwellings do not take place, surface waters heat up, and the anchovy harvest suffers.

The effects of El Niño do not end with the Peruvian anchovy fisheries, however. The phenomenon triggers a climatic spasm that disrupts weather across the globe. Storms may rage across North America, causing floods and landslides. Drought conditions have been experienced in northeastern Brazil, southeastern Africa, and western Pacific islands, causing dust storms and wildfires. El Niño has also caused unusually wet springs in the eastern United States and elsewhere, which often leads to an increase in the number of encephalitis-bearing mosquitoes.

The 1982-1983 El Niño caused more than $8 billion in damage. The severe 1997-1998 El Niño caused more than $15 billion in damage. It was not until the 1982 El Niño that meteorologists and oceanographers began to seriously study the phenomenon and try to learn ways to predict both its approach and the severe weather it causes. Researchers learned that a fluctuation in wind pattern known as the Southern Oscillation—the same fluctuation that causes the trade winds to weaken—triggers El Niño. Researchers also learned that the ocean-current pattern is dependent upon the wind pattern; together, the climatic event is known as the El Niño Southern Oscillation (ENSO).

ENSO

Meteorologists and oceanographers have worked together to learn all they can about ENSO and other atmosphere-ocean oscillations around the globe. The ENSO pattern was the first they studied, and many weather forecasting models were created to aid the study of atmosphere-ocean oscillations. These computer models have also helped researchers study ENSO historical trends for the past five hundred years. Many scientists have come to believe that ENSO events may have contributed to plagues and other similar disasters throughout history.

ENSO patterns swing between the extremes of heated tropical waters associated with El Niño and a colder weather-front pattern known as La Niña, or "the little girl." This oscillation pattern takes about seven years to cycle, and each extreme (hot or cold) in the pattern may last up to two years. ENSO researchers have also linked the onset of El Niño with other climate events in surrounding ocean waters. Just before El Niño begins warming the Pacific Ocean, the tropical Indian Ocean warms. Warming then appears in the tropical Pacific, triggering ENSO, which causes warm winds to blow over South America. About nine months after this occurs, the circulation of the tropical Atlantic changes, and the waters there begin warming.

The study of the ENSO pattern has led researchers to discover a major ocean circulation pattern called the conveyor belt. Though this conveyor belt of water has always existed—it serves to connect the major bodies of water making up the global ocean—its significance was overlooked until oscillation patterns were studied. The conveyor belt current circulates heated ocean water from the north Pacific Ocean westward across the Indian Ocean, around southern Africa, and up into the Atlantic Ocean. As the surface waters cool in the Atlantic Ocean—thus heating eastern North America, Europe, and Ireland—they sink to a lower ocean level and return eastward across the Atlantic Ocean, around southern Africa, across the Indian Ocean, and back to the tropical waters of the Pacific Ocean.

Some researchers are concerned that if this conveyor belt current were to stop or slow down, heat would pile up in the Southern Hemisphere, while the Northern Hemisphere would experience a severe drop in temperature. These scientists claim that such an occurrence may have caused the last ice age. In an article for *Science* magazine, Richard A. Kerr wrote, "Climate records from ice and sediments show that subtle oscillations lasting roughly 1,600 years seem to have persisted through the Holocene." Research findings continue to stress the importance of this conveyor belt current to global climate. Temperatures of surface water change in direct relation to changes in this conveyor belt current. According to Kerr, "With a flow equaling that of one hundred Amazon Rivers, the conveyor delivers enough heat to the North Atlantic's northern half to equal 25 percent of the solar energy reaching the surface there."

NAO AND PADO

The study of the ENSO pattern and the use of computer modeling to forecast weather related to ENSO led researchers to discover other atmospheric-oceanic oscillations, such as the North Atlantic Oscillation (NAO), which fluctuates on a twenty-year time scale. The atmospheric behavior of NAO has long been known to meteorologists. It is typically a low-pressure, counterclockwise wind circulation that centers over Iceland. This weather pattern contrasts with a high-pressure, clockwise circulation near the Azores Islands off the coast of Portugal. Strong winds blow west to east between these two weather centers. NAO seesaws between

these two centers, using strong winds to drive heat from the Gulf Stream across Eurasia during high-index years. This extreme results in unusually mild winters there. At the other extreme, air pressure builds up over Iceland, weakening the warming winds and delivering a bitter winter to Europe and Greenland.

Oceanographers have concluded the ocean must be the cause of NAO. Since the ocean has a huge capacity for storing heat and reacts at a more ponderous rate, it must provide the "memory" for atmospheric patterns operating in the same mode year after year. The source of the oceanic oscillator is a pipeline of warm water fed by the Gulf Stream. It takes twenty years to complete one cycle, thus setting the timing for the long-term swings of NAO. Researchers are unsure, however, how the temperature of the waters in the pipeline actually trips NAO.

The discovery of the Pan-Atlantic Decadal Oscillation (PADO) resulted from a theory that tropical oscillations in the Atlantic actually extended beyond the tropical region. PADO covers an area of more than 11,000 kilometers extending from the southern Atlantic to Iceland. Meteorologists Shang-Ping Xie of Hokkaido University and Youichi Tanimoto of Tokyo Metropolitan University proposed the existence of PADO, claiming that it fluctuates on a ten- to fifteen-year time scale. PADO consists of east-west bands of water spanning the Atlantic Ocean. The bands alternate between warmer and cooler water and are accompanied by changes in atmospheric circulation. An oscillation to the other extreme reverses the temperature variance. The two researchers also believe that PADO is triggered by NAO.

The study of ENSO, NAO, and PADO has led many researchers to theorize that such oscillations are linked in a chain that circles the global ocean. Several scientists at the University of Washington in Seattle have proposed the existence of the Arctic Oscillation, which reverberates from the far northern Atlantic Ocean to the far northern Pacific Ocean. Kerr's *Science* article refers to this oscillation:

> Varying from day to day and decade to decade, the oscillation is a natural response of the atmosphere to the random jostlings of the ocean-atmosphere system, the way a drum has a natural tone when struck.

The Pacific Decadal Oscillation (PDO) is believed to be another link in the chain of atmosphere-ocean oscillations. This high-latitude oscillation was first noted in 1977 when the northern Pacific Ocean cooled dramatically. An atmospheric low-pressure center off the Aleutian Islands intensified and shifted eastward. This brought more frequent storms to the West Coast of North America and warmed Alaska, while Florida experienced periodic winter freezes. A multitude of other environmental changes also resulted. Researchers have been able to pinpoint other shifts occurring in 1947 and 1925 and have also identified close ties between PDO and ENSO events.

SIGNIFICANCE

The greatest impact of the study of atmosphere-ocean oscillations has been the joining of two separate fields of science. As meteorologists and oceanographers work together to research the manner in which atmospheric and oceanic patterns rely on each other and ultimately influence climate, they will become more dependent on coupled research.

Computer simulations—referred to as "oceans in a box"—will become more sophisticated as the two fields combine their research efforts. Climate scientists began using computer simulations in the mid-1980's to study winds in the atmosphere and watch how they stir up ocean currents, which alters pressure patterns in the atmosphere that feed back on the ocean again. Joint research will help these scientists learn more about the inner workings of atmosphere-ocean oscillations and their influence upon each other and the climate.

Lisa A. Wroble

CROSS-REFERENCES

Atmosphere's Global Circulation, 1823; Atmosphere's Structure and Thermodynamics, 1828; Climate, 1902; Clouds, 1909; Drought, 1916; El Niño and La Niña, 1922; Geochemical Cycle, 412; Global Atmospheric Research Program, 2763; Greenhouse Effect, 1867; Hurricanes, 1928; Lightning and Thunder, 1935; Monsoons, 1940; Satellite Meteorology, 1945; Seasons, 1951; Storms, 1956; Tornadoes, 1964; Van Allen Radiation Belts, 1970; Volcanoes: Climatic Effects, 1976; Weather Forecasting, 1982; Weather Forecasting: Numerical Weather Prediction, 1987; Weather Modification, 1992; Wind, 1996.

BIBLIOGRAPHY

Bigg, Grant R. *The Oceans and Climate.* Cambridge, England: Cambridge University Press, 1996. This thorough discussion of the ocean-atmosphere interaction was written by a noted professor of environmental sciences. Bigg details atmospheric and oceanic circulation patterns, and discusses their influence on meteorological developments. Several chapters are devoted to the direct influence of the atmosphere and the ocean on each other and to showing how this interaction influences major ocean-atmosphere oscillations.

Houghton, John. *Global Warming: The Complete Briefing.* 2d ed. Cambridge, England: Cambridge University Press, 1997. Houghton provides an overview of the evidence for global warming. Several chapters focus on the climate system and seasonal forecasting. The accessible coverage of the ENSO system includes discussions of how ENSO influences weather and of the potential for improved weather modeling based on research of ENSO.

Kerr, Richard A. "As the Oceans Switch, Climate Shifts." *Science* 281 (July 10, 1998): 157-158. A noted science and environmental writer discusses oscillations in the northern Pacific and North Atlantic that are suspected of shifting the climate in decade-long cycles. Kerr also describes emerging oscillations in nontropical waters and evidence of similar shifts occurring more than 12,000 years ago.

_____. "A New Driver for the Atlantic's Moods and Europe's Weather?" *Science* 275 (February 7, 1997): 754-755. Kerr discusses the seesaw pattern of NAO, comparing it to the greatly researched ENSO and showing how these two oscillations may impact each other.

Monastersky, Richard. "Spying on El Niño: The

Struggle to Predict the Pacific Prankster." *Science News* 152 (October 25, 1997): 268. Monastersky explains in accessible language what the ENSO pattern is, how scientists study it, and what methods are used to model ocean-atmosphere interactions so forecasts of the weather caused by the phenomenon will be more accurate.

_____. "The Case of Global Jitters: Even in Seemingly Stable Times, Climate Can Take an Abrupt Turn." *Science News* 149 (March 2, 1996): 140. Monastersky discusses evidence that ocean-atmosphere oscillations and erratic climate shifts caused by them have repeated for thousands of years.

National Oceanic and Atmospheric Administration (NOAA) Web site (http://www.pmel .noaa.gov/toga-tao/tao-tour.html). Researchers at NOAA and other research institutions work together to track and model current oscillations, including ENSO and NAO. More information about their findings and other ocean-atmosphere interactions are available through links from their Web site.

Weber, Michael L., and Judith A. Gradwohl. *The Wealth of Oceans*. New York: W. W. Norton, 1995. A thorough overview of the resources available from the ocean and the danger posed by human behavior. The book includes explanations of the current system and ocean-atmosphere interaction. Weber served on the staffs of the Center for Marine Conservation and the National Marine Fisheries Service, and Gradwohl is director of the Smithsonian Institution's Environmental Awareness Program and curator of their Ocean Planet traveling exhibit.

Woods Hole Oceanographic Institution (WHOI) Web site (http://www.whoi.edu/index.html). WHOI is the world leader in oceanography. The WHOI Web site contains general information about the ocean, research, education, and resources, including video animations. WHOI researchers work to understand the complexities of the ocean; atmosphere-ocean interaction is just one component of their research.

CLIMATE

Climate is the long-term total of atmospheric variations. Over shorter periods of time, climate is referred to as weather. Climate always refers to a specific geographical location and is determined by many factors, including wind belts, topography, elevation, barometric pressure and movement of air masses, amount of solar radiation available, proximity to oceanic influences, and planetary cycles.

PRINCIPAL TERMS

AIR MASS: a mass of air in the lower atmosphere that has generally uniform properties of temperature and moisture

ATMOSPHERE: the envelope of gases surrounding the Earth, consisting of five clearly defined regions

GENERAL CIRCULATION MODELS (GCMs): comprehensive, mathematical-numerical formulas, in climate studies, that attempt to express the basic dynamical equations thought to govern the large-scale behavior of the atmosphere and include the numerical methods and resolution used to solve those equations

GREENHOUSE EFFECT: a situation in which levels of carbon dioxide and water vapor in the atmosphere of a planet lead to a trapping of solar energy, which results in increased temperatures, commonly called "global warming"

PARAMETERIZATION: the arbitrary assignment of a value to physical processes which occur on scales too small to be resolved by a general circulation model

PRECIPITATION: phenomena such as rain, snow, and hail deposited on the Earth's surface from moisture in the atmosphere

SCIENCE OF CLIMATOLOGY

From early times, humans have attempted to classify climates. The ancient Greeks in the sixth century B.C.E. visualized the Earth as having three temperature zones based on the Sun's elevation above the horizon. (The Greek word *klima*, meaning "inclination," is the origin of the word "climate.") They called these three zones torrid, temperate, and frigid. This system did not consider the differences between climates over land and over water, the effects of topography, and other atmospheric elements such as precipitation.

Many people confuse weather with climate. The two phenomena are related, but weather is concerned with day-to-day atmospheric conditions, such as air temperature, precipitation, and wind. Climatology is concerned with the mean (average) physical state of the atmosphere, along with its statistical variations in both time and space over a period of many years. In addition to the description of climate, climatology includes the study of a wide range of practical matters determined by climate and the effects and consequences of climatic change. As a result, it has be-

come an interdisciplinary science, important to a wide range of other fields, such as geophysics, biology, oceanography, geography, geology, engineering, economics, statistics, solar system astronomy, and political and social sciences.

For centuries, the content of the atmosphere and the amount of solar radiation reaching the Earth's surface were fairly constant. Until the middle of the 1950's, it was assumed that the climates of the Earth also were relatively unchanging; scientists now understand that climate is never constant. Climate consists of the ocean, atmosphere, cryosphere (areas of permanent ice), land surface, and biomass. The various components of climate are coupled to each other in a nonlinear feedback process—that is, a change in one of these factors produces a feedback or effect on the other factors. Therefore, all or any of these processes can change the statistical state of the system called climate.

The science of climatology has developed along two main lines. Regional climatology studies the discrete and characteristic qualities of a particular region of the globe. The second approach of climatology is a physical analysis of the basic relationships existing among various atmospheric elements of temperature, precipitation, air pressure, and wind speed. A third branch now widely used, which originated in the 1960's, is called dynamical climatology. This branch uses models to simulate climate and climatic change based on the averaged forms of the basic equations of dynamic meteorology.

CLASSIFICATION OF CLIMATES

The classification of climates is of two types: genetic and empirical. A genetic system is based on air masses and global wind belts, which control climates. The empirical classification system utilizes the observed elements of climate, such as temperature. An empirical classification system called the Köppen system uses five categories: tropical forest climates; dry climates; warm, temperate rainy climates with mild winters; cold forest climates with severe winters; and polar climates. Another empirical classification system, which may be more useful in describing world climates, is based on decreasing temperature and increasing precipitation. This system of classification is convenient for use in many other sciences, as well. The first category is the desert climate, which has the highest temperatures and the lowest precipitation. The next category is the savanna, which may be either temperate or tropical. In either case, this climate is characterized by nearly treeless grassland. The steppes of Eastern Europe are an example of a temperate savanna, with plants adapted to very hot conditions and extremely limited water supply. The grassy plains of western Africa are typical tropical savanna. The next category or climate type is temperate and tropical Mediterranean climate. An example of temperate Mediterranean climate is the coast of southern California. Tropical Mediterranean climate is found typically in the northern region of Africa. The next climate type is the temperate and tropical rain forests. An example of a temperate rain forest is the Olympia Peninsula in Washington State. Brazil offers a good example of a tropical rain forest. The two last climate types are the coldest and have a great amount of precipitation in the form of snow. The first of these two, the taiga, is characterized by great forests of evergreen cone-bearing trees. Taigas occur only in the temperate zone or at great

elevation. Large portions of Canada are typical of this climate type. The last region is the tundra. This coldest climate type may have small amounts of precipitation. Again, this region is found in either the temperate zone or the highest elevations on Earth. The best example is northern Canada and most of Alaska.

FACTORS OF CLIMATE CHANGE

Climate is not a steady, unvarying cycle of weather; it fluctuates and varies over a period of time. One of the major factors causing these variations is the role of atmospheric circulation. The atmosphere extends out from the Earth's surface for hundreds of miles and consists of five clearly defined regions. The troposphere, the region closest to the Earth, extends about 11 miles above the Earth's surface. It is in this region that most weather phenomena affecting climate take place.

In the tropics there is intense solar heating, whereas in the polar regions there is little solar heating. These differences of heating and cooling of the atmosphere result in a large-scale global circulation of air that carries excess heat and moisture from the tropical areas of intense heating into the higher latitudes, where there is little excess heat and moisture. The great movements of air masses that are produced alter various local climates.

Global cloud patterns are also extremely important factors in regulating worldwide climate. Heavy cloud cover reduces the amount of solar radiation reaching the surface of the Earth, which can have a cooling effect. Yet cloud cover also acts as a kind of blanket to reduce the amount of heat normally lost from the ground. A clear sky in winter always results in colder temperatures at the surface than when there is cloud cover. Even so, only the solar heat that manages to reach the Earth's surface is able to be kept under this protective cover. Extended cloud cover will result in less solar heat reaching the Earth's surface, leading to a cooler climatic condition.

Modification of the climate of various parts of the globe is also strongly influenced by movement of ocean currents and global wind belts. An oceanic effect termed El Niño has made dramatic climate changes, possibly on a global scale; scientists are not sure of the initial cause of this phenomenon.

It is important to remember that climate has never been stable and unchanging. Many factors can cause changes in local and even global climate; when one factor becomes very active, it inevitably modifies the rate of change and influence of other factors affecting climate.

FACTORS RESULTING FROM HUMAN ACTIVITY

Some factors causing climate change are of natural origin, while many others result from human activity. The globe's tropical rain forests are vital to maintaining a proper carbon dioxide balance in the global atmosphere. Heavy deforestation in South America and Asia has endangered the natural means for maintaining the balance of atmospheric carbon dioxide. Industrial exhaust and automobile emissions also contribute heavily to the buildup of carbon dioxide in the atmosphere, re-

---

TOP WEATHER, WATER, AND CLIMATIC EVENTS OF THE TWENTIETH CENTURY

	Year
TOP U.S. EVENTS	
Galveston Hurricane	1900
Tri-state Tornado	1925
Great Okeechobee Hurricane/Flood	1928
Dust Bowl	1930's
Florida Keys Hurricane	1935
New England Hurricane	1938
Storm of the Century	1950
Hurricane Camille	1969
Super Tornado Outbreak	1974
New England Blizzard	1978
El Niño episode	1982-1983
Hurricane Andrew	1992
Great Midwest Flood	1993
Superstorm	1993
El Niño episode	1997-1998
Oklahoma/Kansas Tornado Outbreak	1999
TOP GLOBAL EVENTS	
India Droughts	1900, 1907, 1965-1967
China Droughts	1907, 1928-1930, 1936, 1941-1942
Sahel Droughts, Africa	1910-191914, 1940-1944, 1970-1985
China Typhoons	1912, 1922
Soviet Union Drought	1921-1922
Yangtze River Flood, China	1931
Great Smog of London	1952
Europe Storm Surge	1953
Great Iran Flood	1954
Typhoon Vera, Japan	1958
Bangladesh Cyclone	1970
North Vietnam Flood	1971
Iran Blizzard	1972
El Niño	1982-1983
Typhoon Thelma, Philippines	1991
Bangladesh Cyclone	1991
Hurricane Mitch, Honduras/Nicaragua	1998

NOTE: Chosen by NOAA scientists from among the world's most notable tornadoes, floods, hurricanes, climatic events and other weather phenomena, taking into account an event's magnitude, meteorological uniqueness, economic impact, and death toll.
SOURCE: National Oceanic and Atmospheric Administration.

sulting in measurable increases in the global temperature. Carbon dioxide absorbs heat reflected from the Earth's surface, creating a greenhouse effect that normally raises the global temperature to a life-supporting level. Even a small increase of 2 degrees Celsius, however, would have a potentially disastrous effect on climatic conditions. Global warming would mean that as the oceans warmed, they would expand; ocean temperature and the direction of upper-level winds are the main factors in the development of hurricanes. Some scientists argue that the warming of the oceans has already increased the frequency and the force of hurricanes. These storms result in loss of life, money, and property.

One of the main dangers of a warming climate is flooding from rising sea levels. Scientists have calculated that if the average temperature increases by about 3.7 degrees Celsius, the sea levels could increase by approximately 81 centimeters, enough to flood huge areas of unprotected coastal land. Nearly 30 percent of all human beings live within about 60 kilometers of a coastline. A rise in sea level of only 20 inches could have a profound effect, flooding many of the world's most important cities and ports.

Various forms of pollution over large urban areas such as Los Angeles, Tokyo, and London have caused local climate changes in these areas. Concentrations of ozone built up at ground level from industrial and automobile exhausts cause extreme eye and lung irritation. Oxides of nitrogen, carbon dioxide, sulfur dioxide, and particle-laden air, which can cause a depletion in the amount of total solar radiation, have altered climate conditions over local urban areas. Also, changes in surface characteristics—such as converting forests and prairies to agricultural fields, building cities, damming rivers to create lakes, and spilling oil at sea—all affect climate conditions. Usually, these surface changes appear to have only a minor global effect, though they do alter local conditions significantly.

There is now widespread national and international interest in how human activity is fundamentally altering the global climate. Organizations are seeking to inform the general public regarding society's role in maintaining a healthful climate. Climate Institute in Washington, D.C., sponsors world conferences where scientists and govern-ment leaders from many nations gather. The Climate Institute provides an interchange among climate researchers and analysts, policymakers, planners, and opinion makers.

DEPARTURES FROM THE "NORM"

For centuries, the content of the atmosphere and the amount of solar radiation reaching the Earth's surface have been fairly constant. Scientists now accept that climate is never constant, and that it is the departures from the expected "norm" that provide the greatest insight into climatic processes and have the greatest human impact.

The ice ages were a decided departure from the "norm." Drought in the United States during the 1920's and 1930's resulted in the Dust Bowl disaster, which disrupted human life and crop production in the middle region of the country. The impact of the devastating drought of the Sahel region in Africa led, in the early 1970's, to increased desertification. Loss of the anchovy fisheries along the Peruvian coast in the 1970's (probably the result of the oceanic fluctuations known as El Niño), severe droughts in the United States' farmbelt in the 1980's, and the large-scale flooding in Bangladesh all attest the fact that dramatic climatic fluctuations and changes are the true norm. It is now seen that climate modification may even result in microclimates as small as portions of a backyard garden.

DATA-GATHERING TOOLS

Because the science of climatology is the study of weather patterns for a geographic area over a long span of time, records of scientific data on the weather are valuable for interpreting climate trends. Because of the international agreements and standardized procedures concerning climate that were adopted in 1853, about 100 million observations have been taken from ships since then. The quantities recorded were sea-surface temperature, wind direction and speed, atmospheric pressure and temperature, the state of the sea, and cloudiness.

A great variety of instruments are used by meteorologists in gathering weather information. The net radiometer, pyranometer, and Campbell-Stokes sunshine recorder are used for measuring radiation and sunshine. The sling psychrometer is used to determine relative humidity. For remote sensing, a humidity gauge is used in which electrical transmission of the variation depends on the fact that the passage of electrical current across a chemically coated strip of plastic is proportional to the amount of moisture absorbed at its surface. This type is used in radiosondes for upper-air observations.

Measurement of wind velocity and direction employs a simple device called an anemometer. Anemometers have four-bladed propellers that are driven by the wind to record wind speed. When the propellers are mounted at right angles, an anemometer responds to air movement in three dimensions. Weather pilot balloons, or "pibals," are released and then tracked visually using a theodolite, or a right-angled telescopic transit that is mounted in a way to make possible the reading of both azimuthal (horizontal) and vertical angles. The progress of the balloon is then plotted minute by minute on a special board, and the direction and speed is computed for each altitude. This technique does not work, however, for upper-air wind observations when there is a low cloud cover. Under these conditions, rawin observation is used. Here, a metal radar target is attached to the balloon and tracked by a radar transceiver or radio theodolite until the balloon breaks or is out of range.

Precipitation is measured in several ways. Rain measurements are made using a rain gauge. An improved version is the tipping-bucket gauge, which has a small divided metal bucket mounted so that it will automatically tip and empty measured quantities of rainfall. This tipping closes electrical contacts that activate a recording pen. Another type of improved rain gauge is a weighing-type gauge. Here, as precipitation falls on a spring scale, a pen arm attached to the scale makes a continuous record on a clock chart. Snow depth is determined by averaging three or more typical measured depths. Where snow depths become very great, graduated rods may be installed in representative places so that the depth can be read directly from the snow surface. Heat and air pressure are measured by thermometers and barometers, most of which can record data automatically.

AERIAL PHOTOGRAPHS AND CORE SAMPLES

Specially equipped planes and satellites can obtain previously inaccessible data of value in determining climate changes. The National Aeronautics and Space Administration (NASA) has supplied a modified U-2 spy plane to carry instruments to 20 kilometers for seven-hour flights up to 80 degrees north latitude. A DC-8 is also used at the same time because of its extended range. The Nimbus 7 satellite is also employed. The U.S. satellite Landsat continually photographs the Earth's surface. These photographs are radioed back to Earth stations and are computer-enhanced to provide visual information about such things as the amount and type of vegetation, precipitation, and underground waterways. The enhanced photographs also provide information about changing conditions that may cause changes in local and/or global climates.

In order to gain information about the relationship between carbon dioxide and global temperature, scientists have drilled deep into Arctic and Antarctic ice and have obtained core samples of the polar ice pack. By studying the amount of carbon dioxide trapped in air bubbles in very old layers of ice pack, scientists have been able to chart the ebb and flow of carbon dioxide in the atmosphere. Fossilized plant tissues indicate how warm the air was during the same period as the bubbles in the core samples. This clue helps provide a picture of the warming and cooling trends that have occurred in very ancient times, and it also makes possible comparisons with current climate conditions.

GENERAL CIRCULATION MODELS

Climatologists have found modeling very useful. This is concerned primarily with general circulation models, or GCMs. The most important part of a GCM is the parameterization of a wide range of physical processes too small to be resolved by the model. These may include all of the turbulent fluxes that occur in the surface boundary layer as well as the occurrence of convection, cloudiness, and precipitation. These relatively small processes must be parameterized in terms of the variables which are resolved by GCMs. It is primarily this parameterization which makes GCMs different from each other and leads to variations between the results of different GCM models. The value of these models, however, depends mostly on the accuracy with which the model actually simulates the natural conditions. Also, assumptions sometimes must be made in modeling, and these assumptions can give a false picture under certain conditions. Models, therefore, are intensely tested against weather data as newer methods of observation and more data become available to ensure the accuracy of the models.

As newer methods for measurement, comparison, and prediction become available, the understanding of global climate change, and more accurate predictions, will lead to better understanding of human activities affecting the climate.

George K. Attwood

CROSS-REFERENCES

Air Pollution, 1809; Atmosphere's Global Circulation, 1823; Atmosphere's Structure and Thermodynamics, 1828; Atmospheric and Oceanic Oscillations, 1897; Atmospheric Inversion Layers, 1836; Clouds, 1909; Desertification, 1473; Drought, 1916; El Niño and La Niña, 1922; Floods, 2022; Greenhouse Effect, 1867; Hurricanes, 1928; Hydrologic Cycle, 2045; Ice Ages, 1111; Lightning and Thunder, 1935; Monsoons, 1940; Ozone Depletion and Ozone Holes, 1879; Precipitation, 2050; Satellite Meteorology, 1945; Seasons, 1951; Storms, 1956; Tornadoes, 1964; Van Allen Radiation Belts, 1970; Volcanoes: Climatic Effects, 1976; Weather Forecasting, 1982; Weather Forecasting: Numerical Weather Prediction, 1987; Weather Modification, 1992; Wind, 1996.

BIBLIOGRAPHY

Bolin, Bert, and B. R. Doos. *The Greenhouse Effect: Climatic Change and Ecosystems.* New York: John Wiley & Sons, 1986. Published with the support of the United Nations Environment Programme and the World Meteorological Organisation. Rich with information. Students can skip the many technical areas and glean the valuable original scientific information.

Curry, Judith A., and Peter Webster. *Thermodynamics of Atmospheres and Oceans.* San Diego, Calif.: Academic Press, 1999. This book offers a look at the effects of the interaction between oceans and the atmosphere on weather patterns and climatic changes. Provides good insight into the role that atmospheric thermodynamics play in meteorology. Illustrations, maps, and index.

Fletcher, J. O. "Climatology: Clues from Sea-Surface Records." *Nature* 310 (August 23, 1984): 630. A study of the large, sudden changes in globally averaged sea-surface temperature in terms of the effect on global climate.

Graham, N. E., and W. B. White. "The El Niño Cycle: A Natural Oscillator of the Pacific Ocean Atmosphere System." *Science* 240 (June 3, 1988): 1293-1302. The article discusses the oceanic phenomenon known as El Niño. The journal *Science* provides excellent scientific materials on the cutting edge of discovery.

Henderson-Sellers, A., and P. J. Robinson. *Contemporary Climatology.* New York: John Wiley & Sons, 1986. Well written and easy to follow. Gives a broad introduction to the science of climatology and its many aspects.

Hopler, Paul. *Atmosphere: Weather, Climate, and Pollution.* New York: Cambridge University Press, 1994. This brief book offers a wonderful introduction to the study of the Earth's atmosphere and its components. Chapters explain the causes and effects of global warming, ozone depletion, acid rain, and climatic change. Illustrations, color maps, and index.

Linden, Eugene. "Big Chill for the Greenhouse." *Time* 132 (October 31, 1988): 90. *Time* magazine frequently offers brief but accurate coverage of current scientific research and discoveries for the average reader. This article discusses the so-called greenhouse effect.

Lutgens, Frederick K., and Edward J. Tarbuck. *The Atmosphere: An Introduction to Meteorology.* 7th ed. Upper Saddle River, N.J.: Prentice Hall, 1998. An excellent introduction and description of the atmosphere, meteorology, and weather patterns. A perfect textbook for the reader new to the study of these subjects. Color illustrations and maps.

Shell, E. R. "Weather Versus Chemicals." *The Atlantic* 259 (May, 1987): 27-31. A discussion of the problems associated with the ozone hole. *The Atlantic* has thorough discussions of environmental issues on a popular reading level.

Stevens, William Kenneth. *The Change in the Weather: People, Weather, and the Science of Climate*. New York: Delacorte Press, 1999. Stevens describes various natural and human-induced causes of changes in the climate. Includes a twenty-page bibliography and an index.

CLOUDS

Clouds provide an indication of the current weather and a forecast of weather to come as well as information regarding climate and other aspects of the surface of the Earth and its atmosphere. They are also a resource for the investigation of the dynamic intermingling of solid, liquid, and gaseous substances.

PRINCIPAL TERMS

CIRRUS: trail or streak clouds, ranging from 5 to 13 kilometers above the ground, that are feathery or fibrous in appearance

CONDENSATION: the transformation from water vapor into a liquid when, subject to falling temperatures, droplets form (or condense) around small particles in the atmosphere

CONVECTION: the transmission of heat by mass transport within a substance; the motion of warm (less dense) air parcels rising as cooler (denser) parcels sink

CUMULUS: clouds with vertical development, or heap clouds, with bases ranging from ground level to 6 kilometers above the ground

RADIATION: the transfer of energy through a transparent medium, as occurs when the Sun warms the Earth

STRATUS: sheet or layer clouds, ranging from 2 to 6 kilometers above the ground (altostratus, or middle) or 0 to 2 kilometers above the ground (stratocumulus, or low)

SUPERSATURATION: a state in which the air's relative humidity exceeds 100 percent, the condition necessary for vapor to begin transformation to a liquid state

CLOUD FORMATION AND LEVEL

The formation of clouds is essentially a two-part process. The heat provided by the radiation of the Sun warms liquid water on the Earth's surface. This moisture evaporates and rises as an invisible vapor and cools. The atmosphere is filled with numerous particles that provide a surface to which water molecules adhere. These particles become nuclei for the formation of water droplets or ice crystals, depending on the temperature of the atmosphere. The particles are usually composed of combustion products, meteoritic dust, volcanic material, soil, or salt. When a sufficient number of droplets or crystals have formed, a cloud has come into existence. The size, shape, and growth patterns of the cloud will depend upon the moisture and the atmospheric energy available during its formation.

Although clouds appear to be relatively stable, they are always in motion. They rise or fall, depending on the temperature of the surrounding air, and move with the direction of the winds. Eventually, clouds either precipitate, in which case the water droplets (in the form of rain or snow) fall back to the Earth's surface, or evaporate, in which case the water returns to a vapor that remains in the atmosphere. Clouds have been divided into three levels: high, middle, and low. The upper limits of the high stage, which is called the tropopause, range from 8 kilometers in the polar regions to 18 kilometers in the tropics. In the temperate zones, which include most of the world's landmasses, the high stage ranges up to approximately 13 kilometers.

CLOUD TYPES

Cirrus clouds, the highest clouds, are composed of ice crystals. The air temperature of these clouds is usually below −25 degrees Celsius. They are formed when air rises slowly and steadily over a wide area. They are usually delicate in appearance, are white in color, and tend to occur in narrow bands. They are thin enough to permit stars or blue sky to be seen through them, and they are responsible for the ring or halo effect that appears to surround the Sun. They often signal large, slow-moving warm fronts and rain. They sometimes seem to be gathered in branching plumes or arcs with bristling ends. Since the color of a cloud depends on the location of the light source which illuminates it—usually the Sun—the position of the Sun at sunset or sunrise may cast these clouds in bright reds or yellow-oranges. A variant of this classification is the cirrocumulus, a cloud composed of ice crystals gathered in columns or prismatic shapes. The cirrocumulus usually is arranged in ripples or waves; it is the cloud that sailors describe when they speak of a "mackerel sky." A close relation is the cirrostratus, which resembles a transparent white veil. Suffusing the blue of the sky with a milky tone, it is composed of ice crystals shaped like cubes. It is the cloud that herdsmen describe as "mares' tails."

On the middle atmospheric level are found the altocumulus clouds. They signal an approaching cyclonic front. The altocumulus clouds reach an altitude of 6 kilometers, where they intersect the lower level of the cirrocumulus, and they tend to resemble the cirrocumulus at their uppermost reaches. Below this point, they tend to be larger and to have dark shadings on their lower boundaries. They are composed primarily of water drops and are formed by a slow lifting of an unstable layer of air. Typically, they look like large, somewhat flattened globules and are often arranged evenly in rows or waves. In the summer, they may appear as a group of small, turretlike shapes, and in that form they often precede thunderstorms. The other basic middle-level cloud is the altostratus. These clouds are formed in air that is ascending slowly over a wide area, and they often occur in complex systems between the higher cirrostratus and the low-level nimbostratus. They

are composed of both ice crystals and water droplets, with raindrops or snowflakes in the lower levels of the cloud. They are usually gray or bluish-gray, resembling a thick veil that is uniform in appearance, and often cover a substantial portion of the sky. There is a reciprocal relationship between altocumulus and altostratus clouds, in that a sheet of altocumulus clouds—particularly high, scaly ones—may be forced down and become transformed into altostratus. Conversely, as weather improves, altostratus may be altered toward altocumulus.

Clouds on the lowest levels generally are associated with the onset of precipitation. They often produce what is frequently called a leaden sky—one that is flat and "dirty" in appearance. These clouds range in altitude from 2 kilometers to the air just above the Earth's surface. The stratocumulus cloud may appear in sheets, patches, or layers close together, in which case the undersurface has a wavelike appearance. These clouds are formed by an irregular mixing of air currents over a broad area and suggest instability in the atmosphere. They are composed primarily of water droplets, and because they often have a marked vertical development, they can be confused with small cumulus clouds; however, they have a softer, less regular shape than cumulus clouds. Stratus clouds tend to be the lowest level of cloud formation. Sometimes resembling fog, they are usually an amorphous, gray layer, and while they do not produce steady rain, drizzle is not uncommon. They are formed by the lifting of a shallow, moist layer of air close to the ground and are composed almost exclusively of water droplets. They produce the dullest of visible sky conditions.

CUMULUS AND NIMBUS CLOUDS

The cumulus cloud is probably the cloud form most frequently depicted in artistic illustrations of the sky. In fair weather, cumulus clouds are separated into cottonlike puffs with distinctive outlines against a deep blue background. When they are few in number and almost purely white, they are a prominent feature of the most pleasant weather during the spring and summer; however, these clouds also have the potential to change into storm-making systems. They are initially formed by the ascent of warm air in separate masses or bubbles, the air rising with increasing velocity as

TYPES OF CLOUDS		
Name	*Altitude (km)*	*Altitude (miles)*
Altocumulus	2-7	6,500-23,000
Altostratus	2-7	6,500-23,000
Cirrocumulus	5-13.75	16,500-45,000
Cirrostratus	5-13.75	16,500-45,000
Cirrus	5-13.75	16,500-45,000
Cumulonimbus	to 2	to 6,500
Cumulus	to 2	to 6,500
Nimbostratus	2-7	6,500-23,000
Stratocumulus	to 2	to 6,500
Stratus	to 2	to 6,500

SOURCE: National Oceanic and Atmospheric Administration.

A cumulonimbus cloud. (Weather Stock)

the cloud takes shape. Their development is driven by convection currents, and while their base is usually about 1 kilometer above the land surface (or somewhat less over the ocean), they may grow vertically to an altitude of more than 7 kilometers; in the case of the cumulonimbus, the classic anvil-shaped thunderhead, they may reach an altitude of more than 15 kilometers. In the transition to cumulonimbus, the warm air rises rapidly into a cooler layer, and the convergence of radically varying temperatures produces a "boiling" at the tops of the clouds. When the rising air encounters strong winds at its upper reaches, it becomes flattened against the stable, colder layer of air that resists the convection from below. This shearing effect is what leads to the classic anvil shape of the thunderhead.

The cumulonimbus is the great thundercloud of the summer sky. It tends to be white, dense, and massive, with a dark base. Generally growing out of cumulus clouds, these are the tallest of conventional cloud forms, created by strong convection currents with updrafts of high velocity within the clouds. The thunderstorm is a product of considerable turmoil within a cumulonimbus cloud. Violent updrafts in the center of the cloud result in an increasing size of water droplets that alternately descend and rise within the cloud until they fall as some form of precipitation. The size of the drops that eventually fall is an indication of the altitude of the cloud as well as the severity of the storm; in their lower regions, cumulonimbus are composed of water droplets, but as they rise, ice crystals, snowflakes, and hail may form. The development of the cumulonimbus is generally cellular (or compartmentalized) so that there is usually some space between clouds of this type, although they may occur in a long "squall line" that is often the boundary between warm and cold airflows. Cumulonimbus clouds, because of the great energy involved in their production, often generate subsidiary clouds as well. On lower levels, the sky seems to take on a convoluted, disorganized appearance, while upper levels often feature exten-

Mammatocumulus clouds, which are often seen where tornadoes are developing.
(National Oceanic and Atmospheric Administration)

A cloud is essentially composed of water in a liquid state or of ice particles suspended in the air or both; however, other substances that may not technically be considered the components of clouds—essentially dry particles (lithometeors)—may be suspended in the atmosphere and gathered in cloudlike masses. Haze, for example, is a suspension of extremely fine particles invisible to the naked eye but sufficiently numerous to produce an opalescent effect. Clouds of smoke are usually composed of the products of combustion gathered in a dense, swirling mass that rises from its source. Particles of sand may be gathered by strong, turbulent winds into clouds that move near the surface of the Earth. One other type of cloud, the noctilucent, is a high-altitude form found in the mesosphere at 8 to 90 kilometers above the Earth's surface in northern latitudes. Its formation is somewhat mysterious, but one theory maintains that it is created by turbulence that carries water vapor to unusual heights during the arctic summer. The nuclei for the formation of these clouds may be deposited in the atmosphere by the residue of meteors. Noctilucent clouds are either pure white or blue-white and have pronounced band or wave structures. One theory maintains that this wave configuration is a result of gravity waves in the atmosphere, but the source of these gravity waves is unknown. Noctilucents resemble high cirrus clouds more than any other form.

On rare occasions, clouds have also been observed in the stratosphere. Their composition has not been determined, but at a range of 20 to 30 kilometers above the Earth's surface, nacreous or "mother-of-pearl" clouds—so designated because they may display a full spectrum of colors—have been seen in the sky over Scandinavia and Alaska during the winter.

sions that are like cirrus clouds. The forward edge of a cumulonimbus cloud often produces what looks like the front of a wave or rolling scud just under the base of the main cloud.

Although original attempts at cloud classification included the word "nimbus" to describe a separate order of low, rain-producing clouds, the term is now applied as a modifier. In addition to its application to the storm-producing version of the cumulus cloud, there are also nimbostratus clouds, or low, gray, dark clouds with a base close to the ground, which generally contain rain. They are formed by the steady ascent of air over a wide area, usually associated with the arrival of a frontal system. They differ from stratus clouds in that they are much darker and are composed of a mixture of ice crystals and water droplets. These clouds have a ragged base, highly variable in shape and often hard to discern in steady rain. Depending on the temperature, the base will be primarily snowflakes or raindrops.

OTHER CLOUD-RELATED PHENOMENA

In addition to the standard forms, other cloud-related phenomena may occur in the atmosphere.

STUDY OF CLOUDS

Recorded descriptions of the sky and weather-related phenomena date at least as far back as Aristotle's *Meteorologica* (c. 350 B.C.E.). The first formal system of cloud classification owes its development to the work of Luke Howard in London in the early nineteenth century. Howard proposed the use of the Latin names currently used to identify clouds. In 1897, C. T. R. Wilson's cloud chamber experiments simulated the cloud formation process in the laboratory. By 1925, radio wave observations were beginning to augment free balloon ascensions as a method of examining the atmosphere, and, in 1928, the first radiosonde apparatuses—a kind of electrical thermometer that transmits temperature, pressure, and humidity information to a receiver—were placed into operation. Immediately after World War II, high levels of atmospheric exploration became possible with rockets.

In 1960, the first meteorological satellite, Tiros 1, was launched by the United States. It transmitted pictures of cloud patterns back to the Earth, and whereas previously the motion of cloud systems could be developed only from separate

ground sightings, the satellite provided an overview of areas more than 1,000 kilometers wide. The use of time-lapse imaging permitted an observer to follow the birth and decay of cloud systems and, in conjunction with computer data storage, made it possible to develop a history of the entire range of cloud formation in the atmosphere.

There are seven basic factors that must be considered to determine the fundamental properties of cloud systems. They are, according to Horace Byers, the amount of sky covered, the direction from which the clouds are moving, their speed, the height of the cloud base, the height of the cloud top, the form of the cloud, and the constitution of the cloud. Each of these categories depends upon the location of the observing system. The amount of sky covered will vary considerably, depending on whether a ground observer is reporting or a satellite photograph is being analyzed against a schematic grid. The direction of cloud

Polar stratospheric, or "mother-of-pearl," clouds over Sweden, January, 2000. These rare clouds contain small droplets that, at the clouds' high altitude of 20 to 30 kilometers, act as prisms to separate incoming light into many colors. (NASA/JPL/Caltech)

motion also depends on the point of observation because relatively specific motion with regard to a fixed point is much easier to chart than the complex series of measurements necessary to describe the motion of a large-scale cloud formation. The upper and lower reaches of a cloud may be measured from balloons, from airplanes, by satellites, and in ground stations. The refinement of cloud investigation by satellite imaging has led to the consideration of what the eminent meteorologist Richard Scorer calls "messages," which include the fact that cyclonic patterns are more varied than had been realized, requiring the assimilation of much more data into an explanation of how large-scale cloud systems are formed and why they exist for a particular duration in time.

In addition to measurements that are essentially external, the inner mechanics of a cloud must be studied to determine its basic properties and potential behavior. The inner cloud physics that influence its development and its potential for precipitation involve the super-cooling and freezing of water (that is, water droplets that do not form ice crystals below 0 degrees Celsius), the origin and specific form of the ice nuclei that lead to the growth of a cloud and that depend on the rise and descent of particles within a cloud, the rate of collisions between particles in a cloud that influence the growth of droplets and crystals, and the reflecting and refracting properties of the components of a cloud that determine the color of a cloud in terms of both visual observation and radiophotometric measurement. Aside from spectographic observations, most cloud physics depend on an understanding of thermodynamics, or the effects of heat changes on the motion and elemental properties of particles in the atmosphere.

Leon Lewis

CROSS-REFERENCES

Atmospheric and Oceanic Oscillations, 1897; Climate, 1902; Deep Ocean Currents, 2107; Drought, 1916; El Niño and La Niña, 1922; Global Warming, 1862; Gulf Stream, 2112; Hurricanes, 1928; Lightning and Thunder, 1935; Monsoons, 1940; Ocean-Atmosphere Interactions, 2123; Satellite Meteorology, 1945; Seasons, 1951; Storms, 1956; Surface Ocean Currents, 2171; Tornadoes, 1964; Van Allen Radiation Belts, 1970; Volcanoes: Climatic Effects, 1976; Weather Forecasting, 1982; Weather Forecasting: Numerical Weather Prediction, 1987; Weather Modification, 1992; Wind, 1996; World Ocean Circulation Experiment, 2186.

BIBLIOGRAPHY

Ahrens, C. Donald. *Meteorology Today.* St. Paul, Minn.: West Publishing, 1999. An up-to-date, clearly written examination of the entire spectrum of meteorological phenomena that might be of interest to the nonspecialist. Cloud systems are discussed within the context of atmospheric science. Extensively illustrated. Suitable for college-level students.

Anthes, Richard, Hans Panofsky, John Cahir, and Albert Rango. *The Atmosphere.* Columbus, Ohio: Charles E. Merrill, 1957. A book for the casual as well as the sophisticated observer of the weather. Chapter 5 follows a weather system across the United States over four days, concentrating on changes in cloud systems. Includes a good chart of the atmosphere and a useful annotated bibliography.

Byers, Horace. *General Meteorology.* New York: McGraw-Hill, 1959. An older but still essentially accurate study. A good chapter on cloud observation includes an introduction to the science prior to satellites and detailed descriptions of cloud classifications.

Lee, Albert. *Weather Wisdom.* Garden City, N.Y.: Doubleday, 1976. An entertaining and informative discussion of the weather and its effects on human life. Includes a lively and lucid discussion of cloud systems, generously supported by quotations from poems about clouds.

Ludham, F. H., and R. S. Scorer. *Cloud Study: A Pictorial Guide.* London: John Murray, 1957. This book, nearly a classic in the field, has been described by Sverre Petterssen, former Director of Scientific Services for the U.S. Air Force Weather Service, as "a charmingly written, well-illustrated discussion, readily understood by nonspecialists." It provides an excellent basis for observation of cloud systems.

Miller, Albert, and Jack Thompson. *Elements of Meteorology.* Columbus, Ohio: Charles E. Merrill, 1970. A solid examination of cloud formation within the larger context of atmospheric phenomena. An excellent appendix has a descriptive explanation of each of the major cloud classifications, including a color photograph for each type.

Moran, Joseph M., and Michael Morgan. *Meteorology: The Atmosphere and the Science of Weather.* 2d ed. New York: Macmillan, 1989. One of the best current texts available for college-level students. Written by two longtime scholars in the field, the book has charts, photographs, and illustrations in addition to clear and informative discussions of weather phenomena.

Neiburger, Morris, James G. Edinger, and William D. Bonner. *Understanding Our Atmospheric Environment.* San Francisco: W. H. Freeman, 1973. A very clearly written, essentially introductory examination of how clouds are formed, with a good table of cloud classifications and photographs. Also contains a somewhat more advanced study of clouds and precipitation and a very useful table of units of measurement and conversion factors.

Parker, Sybil P., ed. *Meteorology Source Book.* New York: McGraw-Hill, 1988. Clear definitions and comprehensive illustrations make this a good source for basic information about meteorology. Different sections have been written by specialists in the field.

Petterssen, Sverre. *Introduction to Meteorology.* New York: McGraw-Hill, 1969. Petterssen, an eminent authority in the field, addresses himself "to readers who wish to make a first acquaintance with the atmosphere and its environs." In addition to a good introduction to basic cloud states, with a list of important terms (mostly derived from the Latin) describing variants in basic cloud types, there are several analytical chapters that consider development and change in cloud systems. Includes an annotated bibliography.

Rogers, R. R. *A Short Course in Cloud Physics.* Elmsford, N.Y.: Pergamon Press, 1976. For the student who wishes to go beyond the preliminary material. Some knowledge of mathematics and physics is required. See also H. R. Pruppacher and J. D. Klett, *Microphysics of Clouds and Precipitation* (1980).

Scorer, R. S. *Cloud Investigation by Satellite.* New York: John Wiley & Sons, 1986. The author, Emeritus Professor and Senior Research Fellow at Imperial College of Science and Technology, University of London, is one of the most distinguished environmental scholars in the world. Although this book is designed primarily for the specialist, it is written clearly enough for the layperson. Reading is enhanced by Scorer's penetrating and engaging style. There is a wonderful variety of photographs of cloud systems.

Shaw, Glenn E. *Clouds and Climatic Change.* Sausalito, Calif.: University Science Books, 1996. A good look at the physics of clouds, their evolutionary processes, and the effects those changes have on the climate. Illustrated, index, and bibliography.

Trefil, James. *Meditations at Sunset: A Scientist Looks at the Sky.* New York: Charles Scribner's Sons, 1988. Trefil's tone is clear, affable, and enthusiastic, and the chapter "When Clouds Go Bad" is a superb description of a storm—vivid and energetic.

Westerlund, Bengt E. *The Magellanic Clouds.* New York: Cambridge University Press, 1997. Westerlund provides a well-written account of the formation and life span of magellanic clouds. This book is easily understood by the nonscientist and scientist alike. Illustrations and index.

DROUGHT

Drought is an unusually long period of below-normal precipitation. It is a relative rather than an absolute condition, but the end result is a water shortage for an activity such as plant growth or for some group of people such as farmers.

IMPACT OF DROUGHT

Droughts have had enormous impacts on human societies since ancient times. The most obvious effect is crop and livestock failure, which have caused famine and death through thousands of years of human history. Drought has resulted in the demise of some ancient civilizations and, in some instances, the forced mass migration of large numbers of people. Water is so critical to all forms of life that a pronounced shortage could, and has, decimated whole populations.

The effects of drought remain profound. The dry conditions in the Great Plains of North America in the early 1930's in conjunction with extensive farming activities resulted in the creation of the Dust Bowl, which at one point covered more than 200,000 square kilometers, or an area about the size of Nebraska. During the early 1960's, a severe drought affected the Mid-Atlantic states. Parts of New Jersey had sixty consecutive months of below-normal precipitation, so depleting local water supplies that plans were actively considered to bring rail cars of water into Newark and other cities in the northern part of the state, since the reservoirs that usually supplied the region were practically dry. The Sahel region south of the Sahara in West Africa had a severe drought beginning in the late 1960's and continuing into the early 1970's, creating an enormous and negative impact on the local population, livestock, and vegetation. Hundreds of thousands of people starved, thousands of animals died, and many tribes were forced to migrate south to areas of more reliable precipitation.

DROUGHT CHARACTERISTICS

Almost all droughts occur when slow-moving air masses that are characterized by subsiding air movements dominate an area. Often, the air comes from continental interiors where the amount of moisture that is available for evaporation into the atmosphere is scarce. When these conditions occur, the potential for precipitation is low for a number of reasons. First, the initial status of the humidity in the air is low, since the continental air mass is distant from maritime (moist) influences. Second, air that subsides undergoes adiabatic heating at the rate of 10 degrees Celsius per 1,000 meters. The term "adiabatic" refers to a change of temperature within a gas (such as the atmosphere) that occurs as a result of compression (descending air) or expansion (rising air), without any gain or loss of heat from the outside. For example, assume that air at a temperature of 0 degrees Celsius is passing over the Sierra Nevada in eastern California at an elevation of 3,500 meters. As the air descends and reaches Reno, Nevada, at an elevation of 1,200 meters, the higher atmospheric pressure found at lower elevations results in compression and heating at the dry adiabatic rate of 10 degrees Celsius per 1,000 meters, yielding a temperature in the Reno area of 23 degrees Celsius. Thus, adiabatic heating from subsiding air masses results in a decline in relative humidity and an increase in moisture-holding capacity. In addition, the movement of air under these conditions is usually unfavorable for vertical uplift and the beginning of the condensation process. The final factor that reduces precipitation potential is the decrease in cloudiness and corresponding increase in sunshine, which in turn leads to an increase in potential evapotranspiration demands, which favor soil moisture loss.

Another characteristic associated with droughts is that once they have become established within a particular location, they appear to persist and even expand into nearby regions. This tendency is apparently related to positive feedback mechanisms. For example, the drying out of the soil influences air circulation and the amount of moisture that is then available for precipitation further downwind. At the same time, the atmospheric interactions that lead to unusual wind systems associated with droughts can induce surface-temperature variations that, in turn, lead to further development of the unusual circulation pattern. Thus, the process feeds on itself to make the drought last longer and intensify. The problem persists until a major change occurs in the circulation pattern in the atmosphere.

Many climatologists concur with the concept that precipitation is not the only factor associated with drought. Other factors that demand consideration include moisture supply, the amount of water in storage, and the demand generated by evapotranspiration. Although the literature in climatology is replete with information about the intensity, length, and environmental impacts of

---

A TIME LINE OF HISTORIC DROUGHTS

1270-1350	SOUTHWEST: A prolonged drought destroys Anasazi Indian culture.
1585-1587	NEW ENGLAND: A severe drought destroys the Roanoke colonies of English settlers in Virginia.
1887-1896	GREAT PLAINS: Droughts drive out many early settlers.
1899	INDIA: The lack of monsoons results in many deaths.
1910-1915	SAHEL, AFRICA: First in a series of recurring droughts.
1933-1936	GREAT PLAINS: Extensive droughts in the southern Great Plains destroy many farms and create the Dust Bowl during the worst U.S. drought in more than 300 years.
1968-1974	SAHEL, AFRICA: Intense period of drought; 22 million affected in four countries, 200,000-500,000 estimated dead, millions of livestock lost.
1977-1978	WESTERN UNITED STATES: Severe drought compromises agriculture.
1981-1986	AFRICA: Drought in 22 countries, including Angola, Botswana, Burkina Faso, Chad, Ethiopia, Kenya, Mali, Mauritania, Mozambique, Namibia, Niger, Somalia, South Africa, Sudan, Zambia, and Zimbabwe, results in 120 million people in 22 countries affected, several million forced to migrate, significant loss of life and of livestock.
1986-1988	MIDWEST: Many farmers are driven out of business by a drought.
1986-1992	SOUTHERN CALIFORNIA: Drought brings increased water prices, loss of water for agricultural production, water rationing.
1998	MIDWEST: Drought destroys crops in the southern part of the Midwest.
1999	LARGE PART OF UNITED STATES: Major drought strikes the Southeast, the Atlantic coast, and New England; billions of dollars in damage to crops.

drought events, the role of individual climatological factors that can increase or decrease the severity of a drought is not fully understood.

DROUGHT IDENTIFICATION

Research in drought identification has been changing over the years. Drought was once considered solely in terms of precipitation deficit. Although that lack of precipitation is still a key atmospheric component of drought, sophisticated techniques are now used to assess the deviation from normal levels of the total environmental moisture status. These techniques have enabled investigators to better understand the severity, length, and areal extent of drought events.

Drought has been defined in numerous ways. Some authorities consider it to be merely a period of below-normal precipitation, while others relate it to the dangers of forest fires. Drought is also said to occur when the yield from a specific agricultural crop or pasture is less than expected; it has also been defined as a period when soil moisture or groundwater reaches a critical level.

Drought was identified early in the twentieth century by the U.S. Weather Bureau as any period of twenty-one or more days when precipitation was 30 percent or more below normal. Subsequent examination of drought events that were identified by this method revealed that soil moisture reserves were often elevated during these events to the extent that there was sufficient water to support vegetation. It was also determined that the precipitation preceding the drought event was either ample or even heavy. Thus, it became apparent that precipitation should not be used as the sole measure to identify drought. Subsequent research has shown that the moisture status of an area is affected by additional factors.

Further developments in drought identification during the middle decades of the twentieth century began to focus on the moisture demands that are associated with the evapotranspiration of an area. Evaporation is the process by which water in liquid or solid form (snow and ice) at the Earth's surface is converted into water vapor and conveyed into the atmosphere. Transpiration refers to the loss of moisture by plants to the atmosphere. Although evaporation and transpiration can be studied and measured separately, it is convenient to consider them in applied climatological studies as a single process, evapotranspiration.

There are two ways to define evapotranspiration. The first is actual evapotranspiration, which is the actual or real rate of water-vapor return to the atmosphere from the Earth and vegetation; this process could also be called "water use." The second is potential evapotranspiration, which is the theoretical rate of water loss to the atmosphere if one assumes continuous plant cover and an unlimited supply of water. This process could also be called "water need," as it indicates the amount of soil water needed if plant growth is to be maximized. Procedures have been developed that enable one to calculate the potential evapotranspiration for any area as long as one has monthly mean temperature and precipitation values.

Some drought-identification studies have focused on agricultural drought, looking at the adequacy of soil moisture in the root zone for plant growth. The procedure involved the evaluation of precipitation, evapotranspiration, available soil moisture, and the water needs of plants. The goal of this research was to determine drought probability based on the number of days when soil moisture storage is reduced to zero.

Evapotranspiration was also used by the Forest Service of the U.S. Department of Agriculture when it developed a drought index to be used by fire-control managers. The purpose of the index was to provide a measure of flammability that could create forest fires. This index has limited applicability to nonforestry users, as it is not effective for showing drought as an indication of total environmental stress.

PALMER DROUGHT INDEX

One of the most widely adopted drought-identification techniques was developed by W. C. Palmer in 1965. The method, which became known as the Palmer Drought Index, defines drought as the period of time—usually measured in months or years—when the actual moisture supply at a given location is consistently less than the climatically anticipated or appropriate supply of moisture. The calculation of this index requires the determination of evapotranspiration, soil moisture loss, soil moisture recharge, surface runoff, and precipitation. The Palmer Drought Index values range from approximately +4.0 for an extremely wet moisture status class to −4.0 for extreme drought. Normal conditions have a value close to 0. Positive values indicate varying stages of abundant moisture, whereas negative values indicate varying stages of drought.

Although the Palmer Drought Index is recognized as an acceptable procedure for incorporating the role of potential evapotranspiration and soil moisture in magnifying or alleviating drought status, there have been some criticisms of its use. For example, the method produces a dimensionless parameter of drought status that cannot be directly compared with other environmental moisture variables, such as precipitation, which are measured in units (centimeters, millimeters) that are immediately recognizable. In addition, the index is not especially sensitive to short drought periods, which can affect agricultural productivity.

In order to address these shortcomings, other researchers use water-budget analysis to identify deviations in environmental moisture status. The procedure is similar to the Palmer method inasmuch as it incorporates the environmental parameters of precipitation, potential evapotranspiration, and soil moisture. However, the moisture status departure values are expressed in the same units as precipitation and are therefore dimensional. Drought classification using this index method ranges from approximately 25 millimeters for an above-normal moisture status class to −100 millimeters for extreme drought. The index would be close to 0 for normal conditions.

SIGNIFICANCE

Drought is invariably associated with some form of water shortage, yet many regions of the world have regularly occurring periods of dryness. Three different forms of dryness—perennial, seasonal, and intermittent—have a temporal dimension. Perennially dry areas include the major deserts of the world, such as the Sahara, Arabian, and Kalahari. Precipitation in these areas is not only very low but also very erratic. Seasonal dryness is associated with regions where the bulk of the annual precipitation comes during a few months of the year, leaving the rest of the year rainless. Intermittent dryness is associated with those instances where the overall precipitation is reduced in humid regions or where the rainy season in seasonally dry areas does not occur or is shortened.

The absence of precipitation when it is normally expected creates a major problem for people. For example, the absence of precipitation for one week in an area where daily precipitation is the norm would be considered a drought. In contrast, it would take two or more years without any rain in parts of Libya in North Africa for a drought to occur. In those areas that have one rainy season, a 50 percent reduction in precipitation would be considered a drought. In regions that have two rainy seasons, the failure of one could lead to drought conditions. Thus, the word "drought" is a relative term, since it has different meanings in different climatic regions.

User demands also influence drought definition. Distinctions are often made among climatological, agricultural, hydrologic, and socioeconomic drought. Climatological, or meteorological, drought occurs at irregular periods of time, usually lasting months or years, when the water supply in a region falls far below the levels that are typical for that particular climatic regime. The degree of dryness and the length of the dry period is used as the definition of drought. For example, drought in the United States has been defined as occurring when there is less than 2.5 millimeters of rain in forty-eight hours. In Great Britain, drought has been defined as occurring when there are fifteen consecutive days with less than 0.25 millimeters of rain for each day. In Bali in Indonesia, drought has been considered as occurring if there is no rain for six consecutive days.

Agricultural drought occurs when soil moisture becomes so low that plant growth is affected. Drought must be related to the water needs of the crops or animals in a particular place, since agricultural systems vary substantially. The degree of agricultural drought also depends on whether shallow-rooted or deep-rooted plants are affected. In addition, crops are more susceptible to the effects of drought at different stages of their development. For example, inadequate moisture in the subsoil in an early growth stage of a particular plant will have minimal impact on crop yield as long as there is adequate water available in the topsoil. However, if subsoil moisture deficits continue, then the yield loss could become substantial.

Hydrologic drought definitions are concerned with the effects of dry spells on surface flow and groundwater levels. The climatological factors associated with the drought are of lesser concern. Thus, a hydrological drought for a particular watershed is said to occur when the runoff falls below some arbitrary value. Hydrological droughts are often out of phase with climatological and agricultural droughts and are also basin-specific; that is, they pertain to certain watersheds.

Socioeconomic drought includes features of climatological, agricultural, and hydrological drought and is generally associated with the supply and demand of some type of economic good. For example, the interaction between farming (demand) and naturally occurring events (supply) can result in inadequate water for both plant and animal needs. Human activities, such as poor land-use practices, can also create a drought or make an existing drought worse. The Dust Bowl in the Great Plains and the Sahelian drought in West Africa provide ready examples of the symbiotic relationship between drought and human activities.

In a sense, droughts differ from other major geophysical events such as volcanic eruptions, floods, and earthquakes because they are actually nonevents: that is, they result from the absence of events (precipitation) that should normally occur. Droughts also differ from other geophysical events in that they often have no readily recognizable beginning and take some time to develop. In many instances, droughts are only recognized when plants start to wilt, wells and streams run dry, and reservoir shorelines recede.

There is wide variation in the duration and extent of droughts. The length of a drought cannot be predicted, since the irregular patterns of atmospheric circulation are still not fully known and predictable. A drought ends when the area receives sufficient precipitation and water levels rise in the wells and streams. Since the severity and areal extent of a drought cannot be predicted, all that is really known is that they are part of a large natural system and that they will continue to occur.

Robert M. Hordon

CROSS-REFERENCES

Atmosphere's Structure and Thermodynamics, 1828; Atmospheric and Oceanic Oscillations, 1897; Climate, 1902; Clouds, 1909; El Niño and La Niña, 1922; Hurricanes, 1928; Lightning and Thunder, 1935; Monsoons, 1940; Satellite Meteorology, 1945; Seasons, 1951; Storms, 1956; Tornadoes, 1964; Van Allen Radiation Belts, 1970; Volcanoes: Climatic Effects, 1976; Weather Forecasting, 1982; Weather Forecasting: Numerical Weather Prediction, 1987; Weather Modification, 1992; Wind, 1996.

BIBLIOGRAPHY

Bryson, Reid A., and Thomas J. Murray. *Climates of Hunger.* Madison: University of Wisconsin Press, 1977. An interesting account of the profound effect of climate on human societies going back to ancient times. The treatment is nonmathematical and is suitable for a wide range of readers. Major climate changes and droughts for various regions in the world are discussed in separate chapters. The material is suitable for senior-level high school students and above.

Climate, Drought, and Desertification. Geneva, Switzerland: World Meteorological Organization, 1997. This brief booklet, prepared by the World Meteorological Organization, deals with the climatic factors, such as drought, that lead to desertification worldwide. Includes color illustrations.

Dixon, Lloyd S., Nancy Y. Moore, and Ellen M. Pint. *Drought Management Policies and Economic Effects in Urban Areas of California, 1987-92.* Santa Monica, Calif.: RAND, 1996. This 131-page report examines the impacts of the 1987-1992 drought in California on urban and agricultural users. The purpose of the investigation was to assess the effects of the drought on a broad range of residential, commercial, industrial, and agricultural water users.

Fisher, R. J. *If Rain Doesn't Come: An Anthropological Study of Drought and Human Ecology in Western Rajasthan.* New Delhi: Manohar, 1997. This book focuses on the human ecology associated with India and other countries prone to drought. There is a lot of attention paid to the factors involved with drought and its durations, as well as drought relief tactics and evaluation of such programs. Illustrations, maps, and references.

Frederiksen, Harald D. *Drought Planning and Water Resources Implications in Water Resources Management.* Washington, D.C.: World Bank, 1992. This relatively short report contains two papers on drought planning and water-use efficiency and effectiveness. The papers deal with policy and program issues in water-resources management from the perspective of the World Bank.

Garcia, Rolando V., and Pierre Spitz. *Drought and Man: The Roots of Catastrophe.* New York: Pergamon Press, 1986. Part of a larger series, *Impact of Climate Change on the Character and Quality of Human Life,* sponsored by the International Federation of Institutes for Advanced Study. The topics in this volume include the food insecurity and social disjunctions that are caused by drought. Case studies include northeastern Brazil, Tanzania, and the Sahelian countries.

Karanth, Gopal Krishna. *Surviving Droughts.* Bombay: Himalaya, 1995. Karanth's brief book examines agricultural strategies used by farmers in the drought-prone Chitradurga district in Karantaka, India. There is a heavy focus on relief and prevention programs. Bibliographical references and index.

Mather, John R. *Drought Indices for Water Managers.* Newark: University of Delaware Center of Climatic Research, 1985. A useful monograph on the variety of drought indices that have been developed by researchers. Examples are provided of the different types of methods used by investigators.

Oliver, John E., and John J. Hidore. *Climatology.* Columbus, Ohio: Merrill, 1984. A thorough, well-written textbook that discusses all aspects of climatology. Contains numerous black-and-white illustrations and maps. Suitable for college-level students. Although quantitative measures are included, no particular mathematical background is necessary.

Russell, Clifford S., David G. Arey, and Robert W. Kates. *Drought and Water Supply.* Baltimore: Johns Hopkins University Press, 1970. An excellent, scholarly study of the losses experienced by municipal water-supply system customers in Massachusetts as a result of a 1962-1966 drought. The well-documented methodology discussed in the book has applicability to water systems in other areas.

Tannehill, Ivan R. *Drought: Its Causes and Effects.* Princeton, N.J.: Princeton University Press, 1947. A classic text on the climatology of droughts. Although the discussion is technical, it is nonmathematical and very readable and should appeal to the general adult audience.

U.S. Congress. House. *Effects of Drought on Agribusiness and Rural Economy.* 100th Congress, 2d session, 1988. This report contains the testimony of a variety of witnesses at a congressional hearing held July 13, 1988. In addition to the text and tables documenting the impacts of the drought of July, 1988, a map of the United States is included that illustrates the use of the Palmer Drought Index. The testimony also provides the flavor of a congressional hearing on drought issues.

Wilhite, Donald A., and William E. Easterling, with Deborah A. Wood, eds. *Planning for Drought: Toward a Reduction of Societal Vulnerability.* Boulder, Colo.: Westview Press, 1987. An extensive collection of thirty-seven short chapters on drought covering a wide range of topics from the climatological to the institutional. The different authors provide a good background to the many issues pertaining to drought, social impacts, governmental response, and adaptation and adjustment. Suitable for college-level students.

EL NIÑO AND LA NIÑA

El Niño Southern Oscillation (ENSO) is a series of linked atmospheric and oceanic events in which a reduction or reversal of the normal large-scale pressure systems acting in the equatorial regions of the Pacific Ocean results in dramatic changes in precipitation, currents, and wind patterns. Repercussions occur over wide areas from the interior of North America to India. Although not regularly periodic, El Niño phenomena generally occur every two to seven years.

PRINCIPAL TERMS

CONVECTION CELL: the path taken when an air parcel is heated in one location, expands, rises through the atmosphere, cools, contracts, descends through the atmosphere to arrive at a different location, and then moves along the surface to return to its original location

EL NIÑO: warm extreme of the Southern Oscillation cycle, with unusually warm surface water in the equatorial Pacific Ocean and subdued trade winds

ENSO: acronym for El Niño Southern Oscillation, used to denote the complete linked atmospheric/ocean phenomenon

LA NIÑA: the cold extreme of the Southern Oscillation cycle, with unusually cold surface water in the equatorial Pacific Ocean and enhanced trade winds

SOUTHERN OSCILLATION: an atmospheric "see-saw" that tilts between atmospheric pressure extremes at Tahiti and Darwin, Australia

THERMOCLINE: the depth interval at which the temperature of ocean water changes abruptly, separating warm surface water from cold, deep water

TRADE WINDS: winds blowing from the northeast in the Northern Hemisphere and the southeast in the Southern Hemisphere and converging at the Intertropical Convergence Zone, which wanders back and forth across the equator during the course of a normal year

GLOBAL WIND PATTERNS

El Niño is the term used for the oceanic aspects of one phase of a climatalogical phenomenon more generally referred to as an El Niño Southern Oscillation (ENSO) event. *El Niño* is a Spanish term meaning "the Christ child" that was used by Ecuadoran and Peruvian fishermen to refer to unusually warm surface currents that arrived—if they did at all—around Christmastime. ENSO events typically recur every two to seven years and involve nutrient upwellings and dramatic changes in sea surface temperatures and precipitation and wind patterns. Such environmental fluctuations influence and in some cases may control the ecology of large areas of the Earth where the effects of the ENSO events are most pronounced.

The large-scale wind patterns of the Earth result from the different amounts of solar energy received at the equator and at the poles. Generally, a low-pressure region of ascending air can be found at the equator. This air rises, cools, and releases its water vapor as rain, creating equatorial rain forests over land. High in the atmosphere, this dry air spreads out, eventually returning to the Earth near the latitudes of the tropics, between 23 degrees and 30 degrees north and south latitudes. Most of the Earth's large deserts are located in this latitude band. To finish its circuit, this air travels across the surface of the Earth toward the equator. Because of the Earth's rotation, however, its path is deflected. In the Northern Hemisphere, deflection is to the right, and in the Southern Hemisphere, it is to the left. Such a deflection causes the air to travel obliquely toward the west in both hemispheres, resulting in the system of winds that are known as the trade winds. (Most of the United States lies north of this area and is under the influence of a similar set of winds blowing in the opposite direction, the prevailing westerlies.)

In one pattern called Walker circulation, warm, wet air rises over Indonesia, spreads out, and then falls as cool, dry air on the eastern Pacific Ocean and the coast of Peru. The existence of this phenomenon became apparent to British scientist Sir Gilbert Walker as he collected data from Tahiti and Darwin, Australia, during the early part of the twentieth century. He found that the atmospheric pressures in these two locations varied together but out of phase with each other. That is, if the pressure rose in Darwin, it fell in Tahiti. Sometimes conditions were normal, with high-pressure zones over Tahiti and lows over Darwin, while at other times they were reversed. Walker called this phenomenon the Southern Oscillation and showed that many other meteorological phenomena from around the world seemed to be tied to it. His particular goal, predicting when the monsoons would fall in India, has yet to be achieved.

As the trade winds move across the Pacific Ocean, they drag along great quantities of water. This water is at the sea surface, is warmed by the Sun, and piles up in the western Pacific in the vicinity of Indonesia. The elevation of the sea there can be 40 centimeters higher than in the eastern equatorial Pacific. The temperature of the sea changes rather abruptly between the water that has been warmed by the Sun and the water beneath it. The depth at which the change occurs is called the thermocline. Because the density of water varies with temperature, this abrupt temperature change corresponds to a density contrast, and mixing of water across the thermocline does not readily occur. The piling up of warm surface water in the area east of Indonesia causes the thermocline to get deeper—as much as 200 meters deep. Off the coast of Peru, the thermocline is typically 50 meters deep. The difference of 40 centimeters of water at the surface is balanced by a difference of 150 meters on the thermocline because the density contrast across the thermocline, between the cold water and warm water, is much less than the density contrast between the water and air at the surface.

OCCURRENCE OF EL NIÑO AND LA NIÑA

Every few years, for reasons that are not well understood and are quite complex, the strength of the trade winds decreases. Often this is accompanied by an eastward spreading of the surface pool of warm water. These two phenomena are linked: Each can and probably does, to some extent, cause the other. As the trade winds collapse,

a major readjustment of the thermocline occurs. As the small surface slope dwindles, the much larger slope on the thermocline also disappears. As the thermocline rises, it generates Kelvin waves, which are large-scale gravity waves. These race across the Pacific Ocean, deepening the thermocline to the east as they go.

The deeper thermocline and warmer surface water reduce the effectiveness of upwelling in supplying nutrients to the surface water. This causes dramatic decreases in the primary productivity of the oceans off the coast of Peru and often spectacular die-offs in many species of fish and the seabirds and other animals that rely on them. During the 1972 El Niño, die-offs resulted in the collapse of the anchovy fishing industry, although its decline was exacerbated by questionable fishery management policies.

As the thermocline deepens and warmer surface waters move east, the area in the Pacific equatorial region where the ascent of warm, moisture-laden air produces intense precipitation shifts to the east. This area, called the Intertropical Convergence Zone, is generally found over the warmest water. During strong ENSO events, the Walker circulation may be completely reversed, with warm, moist air rising over coastal Peru, where cool, dry air normally falls. Suddenly, regions that have not received a drop of rainfall for years are inundated.

The paths and severity of many storms tracking across the continental United States are affected by the location of the Intertropical Convergence Zone. During ENSO events, the jet stream is often moved or disrupted by the heavy storm activity above the zone. Sometimes this results in more precipitation across certain sections of the United States and sometimes less, but typically the result is abnormal weather.

Eventually, El Niño weakens, and trade winds pick up. Cold water wells up from the depths to replace it, bringing nutrients to reestablish the high productivity of the waters. As warm surface water piles up in Indonesia, the thermocline there deepens again, and the Intertropical Convergence Zone moves toward its earlier location. If, as sometimes happens, the return of the trade winds cools the surface waters beyond their normal temperatures, a phase called La Niña begins.

Apparently, the ocean/atmosphere system os-

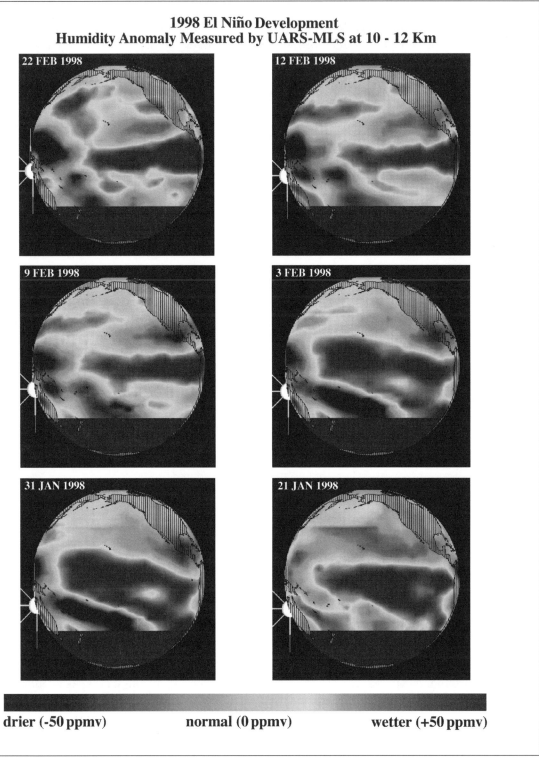

This series of images shows the development of water vapor over the Pacific Ocean during the El Niño event of January and February, 1998. Warmer water temperatures result in greater-than-normal water evaporation, warmer, moister air, and finally altered global weather patterns. (NASA/JPL/Caltech)

cillates between two states, La Niña and El Niño. What drives it from one state to the other remains a matter of considerable speculation. Although the complexities of the ocean/atmosphere system would seem more than adequate to produce almost any sequence of events, some researchers have suggested the involvement of yet another complex system, the Earth's tectonic engine. ENSO events may be correlated with eruptions of large equatorial volcanoes, which inject dust into the air and can have dramatic impact on the atmospheric components of the cycle.

There is a great deal of interest—scientific and general—in learning more about ENSO events. Computational capabilities and theoretical developments have progressed rapidly, and although scientists are not yet on the verge of understanding this complex interaction of air and water, they have begun to deploy water temperature sensing buoys across the Pacific to study it.

STUDY OF EL NIÑO

El Niño Southern Oscillation events are huge. They last longer than one year, involve massive amounts of air and water, and affect a substantial percentage of the Earth's surface. During the course of an ENSO event, the changes that have already occurred in the system influence the changes that will occur as it progresses. Winds affect sea surface temperatures, and sea surface temperatures affect the winds. Such feedback and interactions make it difficult to develop models with predictive capability. In the past, computer models used for atmospheric studies often oversimplified the ocean, and those used for oceanic studies oversimplified the atmosphere. ENSO events involve a tight coupling of the ocean and atmosphere, and any successful model will need to treat both with sophistication and detail.

Large computer models are being developed to try to study ENSO events. As they evolve, deficiencies in the available data set are being identified, and additional data are being acquired. Much of the necessary data are of the type normally gathered during almost any routine oceanographic study: temperatures, salinities, and surface wind velocities and directions. Researchers would like to have such data from before, during, and after an ENSO event. Because these events recur so frequently, almost any data must be collected within

a few years of an event. The difficulty lies in figuring out which data are important. Although there is agreement on what typically happens during an ENSO event, the specifics vary widely from event to event. Just when scientists think they understand it well enough to venture forth with a predictive model, nature comes along and shows them that their model is faulty.

Models, whether sophisticated computer models or simple physical models, can further the understanding of ENSO events. One simple physical model uses a clear Pyrex baking pan as the ocean basin; water as the deep, cool, ocean water; vegetable oil as the warmer, less dense, surface water; and a hair dryer (set to blow unheated air) as the trade winds. If the flow of air produced by the hair dryer is directed across the oil, the oil piles up at the far side of the baking pan. The surface of the oil will form a slight slope, with the higher elevation at the far side of the baking pan. A more pronounced, readily visible slope will form where the oil meets the water. Friction from the air produced by the hair dryer pushes surface oil toward the far side of the pan, but as it piles up, it increases the pressure on the water beneath it. This causes the water beneath the oil to move back toward the near side of the baking pan. Eventually a wedge of oil sits over the water, its upper slope maintained by the hair dryer "wind," and its lower slope positioned where the pressures at depth are the same. (The fraction of a centimeter of additional oil instead of air at the top surface is compensated by a considerably greater deflection of the oil-water boundary because the density contrast between oil and water is much less than that between oil and air.) This represents the normal state in the equatorial Pacific: a shallow thermocline off the coast of Peru; winds moving off the South American continent at Peru, preventing rainfall; a deep thermocline in Indonesia; and air moving from east to west across the ocean, becoming warm and moist, rising over Indonesia, and drenching it in rain.

In the model, an El Niño can be produced by turning off the hair dryer. The forces that maintained the slope on the upper surface of the oil are gone, so the oil sloshes down and its surface becomes horizontal. A wave forms where the oil meets the water, a boundary that corresponds to the thermocline. Such a wave is called an internal wave because it occurs between two layers of water with contrasting densities but in general is not visible at the surface. The wave races across the baking pan just as the waves race across the Pacific during an El Niño.

Although this physical model demonstrates many of the basics of an ENSO event, no physical model is capable of capturing the entire natural event because its scale is so enormous. The rotation of the Earth plays a very important role, and the nature of currents and waves as they move directly along the equator is also quite significant. Computer models can incorporate these effects, however, and developing such models is an active area of research.

Additional data on the ocean are also required. The Pacific Ocean between Indonesia and Peru has not been studied in enough detail to enable scientists to refine their models as much as they would like. The processes that occur during an ENSO event involve large areas and take place over a considerable period of time. Monitoring the sea to detect gradual changes in sea surface temperatures, depth of the thermocline, and surface wind velocity and direction is a complex, expensive project. As more data are acquired, understanding of ENSO events is certain to improve.

SIGNIFICANCE

The El Niño Southern Oscillation cycle is as much a part of the Earth's weather and climate as the cycle of seasons is. Just as people spend about one-quarter of the year in winter, people spend about one-quarter of their lives in an El Niño phase. Therefore, El Niño events are not truly anomalous. Unlike winter, however, the timing of the ENSO cycle is not regular. For example, from 1991 through 1994, three El Niños developed. Analysis of historic records shows that El Niños have occurred at nearly the same frequency since at least 1525, when record keeping began in Peru. In addition to historic records, evidence for prehistoric ENSO events is found in tree rings, flood deposits, ice cores, and other geological and biological record keepers. These, too, suggest that the frequency and severity of ENSO events has been fairly constant for at least several thousand years.

The severity of El Niños can vary remarkably. Substantial effort has been made to identify predictors that would permit an early estimate of the occurrence and severity of ENSO events. Several promising predictors of the timing or intensity have emerged but have been abandoned as additional events transpired in which the predictor was present but the expected development did not occur.

Sir Gilbert Walker's suggestion that local weather in widely separated parts of the planet might be closely related was ridiculed by some of his contemporaries. However, his belief has been borne out by the data and understanding acquired subsequent to his early work. ENSO-related changes can also be seen in the Indian and Atlantic Oceans. Some droughts in Africa and India and most droughts in Australia seem connected to ENSO events. For example, weak floods on the Nile River, records of which have been kept since 622 C.E., seem directly related to El Niño events.

Droughts and floods can probably never be prevented. However, many of the ensuing hardships could be avoided or reduced if they could be predicted a year or so in advance. People and governments in affected areas could take preventive measures: Drought-resistant or flood-resistant varieties of crops could be planted, dikes could be reinforced, deforestation could be halted, and water could be temporarily impounded. Before these measures are taken, however, considerable confidence in any prediction is required. Preparing for a flood may worsen the effects of a drought. Someday, however, it may be possible to predict ENSO events with the same degree of accuracy with which meteorologists predict the arrival of a significant winter storm.

Otto H. Muller

CROSS-REFERENCES

Atmosphere's Global Circulation, 1823; Atmospheric and Oceanic Oscillations, 1897; Climate, 1902; Clouds, 1909; Drought, 1916; Earth-Sun Relations, 1851; Global Warming, 1862; Greenhouse Effect, 1867; Hurricanes, 1928; Lightning and Thunder, 1935; Monsoons, 1940; Precipitation, 2050; Satellite Meteorology, 1945; Seasons, 1951; Storms, 1956; Tornadoes, 1964; Van Allen Radiation Belts, 1970; Volcanoes: Climatic Effects, 1976; Weather Forecasting, 1982; Weather Forecasting: Numerical Weather Prediction, 1987; Weather Modification, 1992; Wind, 1996.

BIBLIOGRAPHY

Davidson, Keay. "What's Wrong with the Weather? El Niño Strikes Again." *Earth*, June, 1995, 24-33. Reviews the anomalous 1991-1994 period when El Niños of various strengths occurred in three out of the four years. Effects in the Indian Ocean are also described. Includes some discussion of Rossby waves from earlier El Niños bouncing around the Pacific basin. No equations, several diagrams. Suitable for readers at the high school level.

Diaz, Henry F., and Vera Markgraf, eds. *El Niño: Historical and Paleoclimatic Aspects of the Southern Oscillation*. Cambridge, England: Cambridge University Press, 1992. A collection of twenty-two papers dealing with historical aspects of El Niño. These papers cover a variety of climate indicators from ice cores to tree rings to records of floods on the Nile River and seek to reveal the long-term characteristics of El Niños. Most have useful summaries and concluding sections. Suitable for college-level readers.

Knox, Pamela Naber. "A Current Catastrophe: El Niño." *Earth*, September, 1992, 31-37. Provides an easy-to-read overview of many aspects of El Niño events. Some historical perspectives are presented along with various theories on causes, including possible tie-ins with volcanic eruptions. No equations, several good diagrams. Suitable for high school readers.

Open University Oceanography Course Team. *Ocean Circulation*. Oxford, England: Pergamon Press, 1989. This carefully written textbook provides an introduction into the analysis and understanding of all sorts of circulation in the world's oceans. Intended for an undergraduate college-level reader, it provides the mathematical background needed to understand many of the aspects of geophysical fluid dynamics. Although treatment of El Niño is brief, it builds on many other discussions of topics such as Kelvin waves and Rossby waves, which are presented very well.

Philander, S. George. *El Niño, La Niña, and the Southern Oscillation*. San Diego, Calif.: Academic Press, 1990. Although it contains some quantitative material beyond the grasp of the average reader, this book covers its subjects thoroughly and with enough qualitative discussions to make it worthwhile. Plenty of diagrams back up the material. Includes visual representations of much of the math. Suitable for most college-level readers, particularly those with some background in science.

Ramage, Colin S. "El Niño." *Scientific American*, June, 1986, 76-83. This balanced, scientific paper avoids throwing all of the vagaries of weather into the El Niño basket and presents several different theories on what causes and sustains these events. It is careful to distinguish between the generally accepted model of El Niños and the actual course taken by any particular event. Several excellent diagrams and maps. Suitable for high school readers.

HURRICANES

Hurricanes are cyclonic storms that form over tropical oceans. A single storm can cover hundreds of thousands of square kilometers and has interior wind speeds of 65-230 knots (74-200 miles) per hour near its more tranquil eye. Destruction is caused by wind damage, as well as storm surge and subsequent flooding.

PRINCIPAL TERMS

CONDENSATION: the process in which water changes from a vapor state to a liquid state, releasing heat into the surrounding air; this process is the opposite of evaporation, which requires the input of heat

CORIOLIS FORCE: a force caused by the Earth's rotation in which a moving object (such as the wind) deflects to the right in the Northern Hemisphere and to the left in the Southern Hemisphere

TROPICAL CYCLONE: a storm that forms over tropical oceans and is characterized by extreme amounts of rain, a central area of calm air, and spinning winds that attain speeds of up to 300 kilometers per hour

TROPICAL DEPRESSION: a storm with wind speeds up to 63 kilometers per hour

TROPICAL STORM: a storm with winds of 64 to 118 kilometers per hour

VORTEX: a spinning mass of air or water that can reach high velocities.

ANATOMY OF A HURRICANE

Hurricanes are huge, swirling storm systems that can cover thousands—sometimes hundreds of thousands—of square kilometers. Often called the "greatest storms on Earth," hurricanes have sustained winds of at least 65 knots (74 miles) per hour with maximum wind speeds of 230 knots (200 miles) per hour. In the Western Hemisphere, these storms are called hurricanes. They are referred to as typhoons in the western Pacific, cyclones in the Indian Ocean, Willy Willys near Australia, and *baguious* in the Philippines. The swirling motion of these storms is counterclockwise in the Northern Hemisphere and clockwise in the Southern Hemisphere.

Hurricanes are as individual and unique as fingerprints. Their behavior is difficult to predict, even when satellites are used to track them. Scientists who have studied decades of hurricane data have found some patterns. First, hurricanes evolve in specific areas of the west Atlantic, east Pacific, south Pacific, western north Pacific, and north and south Indian Oceans. Second, they rarely move closer to the equator than 4 to 5 degrees latitude north or south, and no hurricane has ever crossed the equator. Third, they are more common during certain months of the year depending on their ocean of origin. In the Northern Hemisphere, hurricanes are common from May to September; in the Southern Hemisphere, the hurricane season ranges from December to May.

The reasons for these patterns are that hurricanes need warm surface waters, high humidity, and winds from the same direction at a constant speed in order to form. Cyclonic depressions can only develop in areas where the ocean temperatures are more than 24 degrees Celsius; the eye structure, which must be present in order to be classified as a hurricane, requires temperatures of 26 to 27 degrees Celsius to form. This means that hurricanes will rarely develop above 20-degree latitudes because the ocean temperatures are never warm enough to provide the heat energy needed for formation. In the Northern Hemisphere, the convergence of air that is ideal for hurricane development occurs above tropical waters when easterly moving waves develop in the trade winds. The region around the equator is called the "doldrums" because there is no wind flow. Hurricanes, needing wind to form, can be found as close as 4 to 5 degrees away from the equator. At these latitudes, the Coriolis effect—a deflecting force associated with the Earth's rotation—gives the winds the spin necessary to form hurricanes.

In hurricane formation, heat is extracted from the ocean, and warm, moist air begins to rise, forming clouds and causing instability in the upper atmosphere. As the air rises, it spirals inward toward the center of the system. This spiraling movement causes the seas to become turbulent, and large amounts of sea spray are captured and suspended in the rising air. This spray increases the rate of evaporation fueling the storm with water vapor.

As the vortex of wind, water vapor, and clouds spin at a faster and faster rate, the eye of the hurricane forms. The eye or center of the hurricane is a relatively calm area that has only light winds and fair weather. The most violent activity in the hurricane takes place in the area right around the eye called the eyewall. It is in the eyewall that the spiraling air rises and cools, while moisture condenses into droplets that form rainbands and clouds. The process of condensation releases latent heat, which causes the air to rise and generate more condensation. The air rises rapidly, creating an area of extremely low pressure close to the storm's center. The severity of a hurricane is often indicated by how low the pressure readings are in the central area of the hurricane.

EYE FORMATION

As the air moves higher, up to 15,000 meters, the air is propelled outward in an anticyclonic flow. However, some of the air is forced inward into the eye. The compression of air in the eye causes the temperature to rise. This warmer air can hold more moisture, and the water droplets in the central clouds will evaporate. As a result, the eye of the hurricane becomes nearly cloud-free.

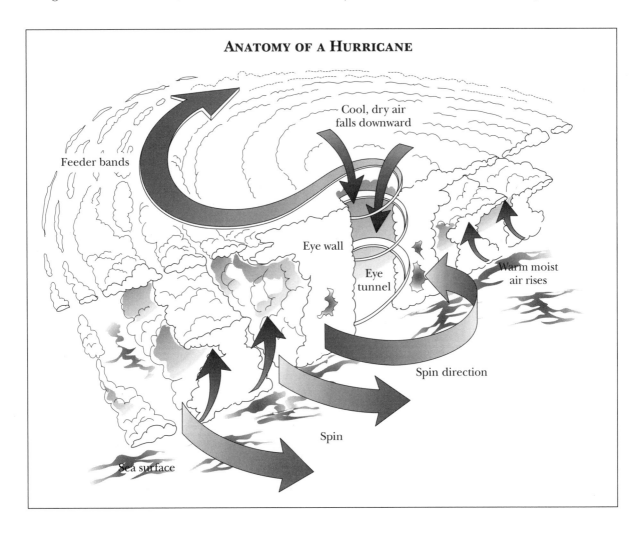

ANATOMY OF A HURRICANE

The temperature is much warmer in the eye, especially in the middle and upper levels, than outside of it. Therefore a large pressure differential develops across the eyewall, which establishes the violence of the storm. Waves of 15 to 20 meters are common in the open ocean because of winds around the eye. Winds in a hurricane are not symmetrical around the eye; when one faces the direction that the hurricane is moving, the strongest winds are usually to the right of the eye. The radius of hurricane-level winds (velocities of 120 kilometers per hour or more) can vary from 15 kilometers away from the eye in small hurricanes to 150 kilometers in large hurricanes. The strength of the wind decreases in relation to its distance from the eye.

A hurricane, with eye in center, as seen in a satellite image. (PhotoDisc)

Depending on the size of the eye, which can range from 5 to 65 kilometers in diameter, the calm period of blue skies and mild winds can last from a few minutes to hours when the storm hits. The calm is deceiving because it is not the end of the storm but a momentary lapse until the winds from the opposite direction hit.

Hurricane Damage

Hurricanes, and similar storms that are less intense, are classified by their central pressure and their sustained wind speed. Tropical depressions have wind speeds below 20-34 knots per hour, whereas tropical storms have wind speeds of 35-64 knots. To be classified as a hurricane, storms must have sustained winds of 65 knots (74 miles) per hour or higher.

The Saffir-Simpson scale further categorizes the storm intensity of hurricanes into five levels. Category 1 hurricanes are considered weak, with sustained winds of 63-83 knots (74-95 miles) per hour. They cause minimal damage to buildings but do damage unanchored mobile homes, shrub-bery, and trees. Normally they cause coastal road flooding and minor damage to piers. Category 2 hurricanes have winds of 83-95 knots (96-110 miles) per hour and can easily damage roofs, doors, and windows on buildings. They also cause substantial damage to trees, shrubs, mobile homes, and piers. Flooding of roads and low-lying areas normally occurs with this level of storm. Category 3 hurricanes are considered to be strong, with winds of 96-113 knots (111-130 miles) per hour. These storms destroy mobile homes and can cause structural damage to residences and utility buildings. Flooding from this level of hurricane can destroy small structures near the coast, while larger structures normally sustain damage by floating debris. There can be flooding 15 kilometers or more inland. Category 4 hurricanes are categorized as very strong, boasting winds of 114-135 knots (131-155 miles) per hour. These storms can cause major damage to lower floors of buildings and can cause major beach erosion. Residences often sustain roof structure failure and subse-

---

STORM CLASSIFICATIONS

Tropical Classification

Gale-force winds	>15 meters/second
Tropical depression	20-34 knots and a closed circulation
Tropical storm (named)	35-64 knots
Hurricane	65+ knots (74+ mph)

Saffir-Simpson Scale for Hurricanes

Category 1	63-83 knots (74-95 mph)
Category 2	83-95 knots (96-110 mph)
Category 3	96-113 knots (111-130 mph)
Category 4	114-135 knots (131-155 mph)
Category 5	>135 knots (>155 mph)

NOTES: 1 knot = 1 nautical mile/hour = 1.152 miles/hour = 1.85 kilometers/hour.
SOURCE: National Aeronautics and Space Administration, Office of Space Science, Planetary Data
   System. URL: http://atmos.nmsu.edu/jsdap/encyclopediawork.html

---

quent rain damage. Land lower than 3 meters above sea level can be flooded, which may lead to the massive evacuation of residential areas up to 10 kilometers inland. Category 5 hurricanes are devastating, classified by having sustained winds greater than 135 knots (155 miles) per hour. These can completely destroy roofs on residential and industrial buildings, and may destroy the buildings themselves. Structures less than 5 meters above sea level can sustain major damage to lower floors, and massive evacuations of residential areas 10 to 20 kilometers inland from the shoreline can be required.

Since wind speeds of about 43 knots (50 miles) per hour can break tree branches and cause some damage to structures, hurricane winds are considered to be among the most devastating natural disasters. The wind force applied to an object increases with the square of the wind speed. A building 30 meters long and 3 meters high that has 100-mile-per-hour hurricane speed winds blowing against it would have 40,000 pounds of force exerted against its walls. This is because a 100-mile-per-hour wind exerts a force of approximately 430 pounds per square meter. If the wind speed was 138 knots (160 miles) per hour, the force against the house would be over 1 million pounds per square meter. Rain blown by the wind also contributes to an increase of the pressure on buildings.

Hurricane Andrew, which hit the United States in 1992, caused wind gusts of at least 280 kilome-

ters per hour in south Florida. Hurricane Andrew caused an estimated $25 billion in damage, making it the most expensive hurricane to hit the United States.

STORM PATHS

Hurricanes can travel in very sporadic paths. Some will travel in a general curved path, while others will change course quite rapidly. They can reverse direction, zigzag, veer from the coast back to the ocean, intensify over water, stall, or return to the same area. The paths are affected by pressure systems of the surrounding atmosphere and the influence of prevailing winds and the Earth's rotation. They can also be influenced by the presence of high- and low-pressure systems on the land they invade. The high-pressure areas can act as barriers to the hurricanes; if a high is well-developed, its outward-spiraling flow will guide the hurricane around its edges. Low-pressure systems tend to attract the hurricane system toward it.

After forming in tropical water, hurricanes generally travel indirectly toward the poles until they lose their energy over cooler waters. An average hurricane can travel from 450 to 650 kilometers per day and more than 1,600 kilometers before it is downgraded to a tropical storm. Rarely, a hurricane can maintain the required wind speed to latitudes as high as 40 degrees.

STORM SURGE

The greatest cause of death and destruction in a hurricane comes from the rise of the sea from the storm surge. As the hurricane crosses the continental shelf and moves to the coast, the water level may increase 5 to 7 meters. This is caused by the drop in atmospheric pressure at sea level inside the hurricane, which allows the hurricane's force to pick up the sea while the winds in front of the hurricane will pile the water up against the coastline. This results in a wall of water that can be up to 7 meters tall and 75 to 150 kilometers wide. This wall of water can sweep across the coastline where the hurricane makes landfall. The combi-

DEADLIEST HURRICANES IN NORTH AMERICA

Hurricane	Year	Category	Deaths
Georgia/South Carolina	1881	Unknown	700
Louisiana	1893	Unknown	2,000
South Carolina/Georgia	1893	Unknown	1,000-2,000
Galveston, Texas	1900	4	8,000+
Florida	1906	2	164
Mississippi/Alabama/Pensacola	1906	3	134
Grand Isle, Louisiana	1909	4	350
Velasco, Texas	1909	3	41
Southwest Florida	1910	3	30
Galveston, Texas	1915	4	275
New Orleans, Louisiana	1915	4	275
Southwest Louisiana	1918	3	34
Florida Keys/South Texas	1919	4	600#
Miami/Mississippi/Alabama/Pensacola	1926	4	243
Lake Okeechobee, Florida	1928	4	1,836
San Felipe, Puerto Rico	1928	4	312
Freeport, Texas	1932	4	40
U.S. Virgin Islands, Puerto Rico	1932	2	225
South Texas	1933	3	40
Florida Keys	1935	5	408
New England	1938	3*	600
Southern California	1939	TS	45
Georgia/South and North Carolina	1940	2	50
Northeast U.S.	1944	3*	390@
SE Florida/Louisiana/Mississippi	1947	4	51
Carol (NE U.S.)	1954	3*	60
Hazel (South and North Carolina)	1954	4*	95
Diane (NE U.S.)	1955	1	184
Audrey (Louisiana/Texas)	1957	4	390
Donna (St. Thomas, VI)	1960	4	107
Donna (Florida/eastern U.S.)	1960	4	50
Carla (Texas)	1961	4	46
Hilda (Louisiana)	1964	3	38
Betsy (SE Florida/Louisiana)	1965	3	75
Camille (Mississippi/Louisiana)	1969	5	256
Agnes (NE U.S.)	1972	1	122
Eloise (Puerto Rico)	1975	TS	44
Alberto (NW Florida/Georgia/Alabama)	1994	TS	30
Mitch (Central America)	1998	5	11,000+

NOTES: Intensity is for time of landfall. The cyclones may have been stronger at other times.
+ = May actually been as high as 10,000 to 12,000.
# = More than 500 of these lost on ships at sea; 600-900 estimated deaths.
* = Moving more than 30 miles an hour.
@ = Some 344 of these lost on ships at sea.
TS = Only of Tropical Storm intensity.
Unknown = Intensity not sufficiently known to establish category.
SOURCE: National Oceanic and Atmospheric Administration.

nation of shallow shore water and a strong hurricane causes the highest surges of water.

If the storm surge arrives at the time of high tide, the water heights of the surge can increase nearly 1 meter. The height of the storm surge also depends upon the angle at which the storm hits the mainland. Hurricanes that make landfall at right angles to the coast will cause a higher storm surge than hurricanes that enter the coast at an oblique angle. Often, the slope or shape of the shore and ocean bottom can create a bottleneck effect and cause a higher storm surge.

Water weighs approximately 2,250 pounds per cubic meter and thus has considerable destructive power. The storm surge is responsible for 90 percent of the deaths in a hurricane. The pounding of the waves caused by the hurricane can easily demolish buildings. Storm surges cause severe erosion of beaches and coastal highways. Often, buildings that have survived hurricane winds have had their foundations eroded by the sea surge or have been demolished by the force of the pounding waves. Storm tides and waves in harbors can destroy ships. The salt water that inundates land can kill existing vegetation, and the residual salt left in the soil makes it difficult to grow new vegetation.

The most destructive hurricane in the twentieth century occurred in November of 1970 in what is now known as Bangladesh. Of the 500,000 people who were killed, most of them were victims of the storm surge. In 1985, the same area was hit by another hurricane, causing the deaths of 100,000 people. In 1989, Hurricane Hugo hit the mainland United States near Charleston, South Carolina. It blew in with sustained winds of more than 113 knots (130 miles) per hour and a 7-meter storm surge. The storm caused losses of up to $6 billion and caused damage in North Carolina, Puerto Rico, and the Virgin Islands.

RAINFALL AND FLOODING

The amount of rainfall received depends on the diameter of the rainband within the hurricane and the hurricane's speed. One typhoon in the Philippine Islands caused 185 centimeters of rain to fall in a twenty-four-hour period, a world record. Heavy rainfall can cause flash floods or river floods. Flash floods last from thirty minutes to four hours and are caused by heavy rainfall over a small area that has insufficient drainage. This excess water overflows streambeds, damaging bridges, underpasses, and low-lying areas. The strong currents in a flash flood can move cars off the road, wash out bridges, and erode roadbeds.

Floods in an existing river system develop more slowly. It can take two or three days after a hurricane hits before large rivers overflow their banks. River floods cover extensive areas, last one week or more, and destroy both property and crops. The flood waters eventually retreat, leaving buildings and residences full of mud and destroying furnishings. Rain driven by the wind in hurricanes can cause damage to buildings around windows, through cracks, and under shingles.

TORNADOES AND OTHER DANGERS

Hurricanes may also spawn tornadoes. The tornadoes associated with hurricanes are usually about one-half the size of tornadoes in the Midwest and are of a shorter duration. The area these tornadoes affect is small, usually 200 to 300 meters wide and about 2 kilometers long. Despite their small size, they can be very destructive. Tornadoes normally occur to the right of the direction of the hurricane's movement. Approximately 94 percent of tornadoes occur within 10 to 120 degrees from the hurricane eye and beyond the area of hurricane-force winds.

Tornadoes associated with hurricanes are most often observed in Florida, Cuba, and the Bahamas, as well as along the coasts of the Gulf of Mexico and the South Atlantic Ocean. In 1961, 26 tornadoes were associated with Hurricane Carla; Hurricane Beulah released 115 tornadoes in 1967. Hurricane Camille, one of the most lethal storms to hit the United States, created more than 100 tornadoes in 1969.

During almost every hurricane, snakes are driven from their natural habitats by high inflow of salt water. They are strong swimmers and can be found along roads, in buildings, and in high, dry places. Many people who are bitten by snakes have problems receiving medical attention for their bites because of breakdowns in communications and transportation.

Toby Stewart and Dion Stewart

CROSS-REFERENCES

Atmosphere's Global Circulation, 1823; Atmospheric and Oceanic Oscillations, 1897; Climate,

BIBLIOGRAPHY

Bryant, Edward A. *Natural Hazards*. Cambridge, England: Cambridge University Press, 1991. Bryant provides a sound scientific treatment for the educated layperson. Readers should have a basic understanding of mathematical principles to fully appreciate this book. Contains a glossary of terms.

Pielke, R. A., Jr., and R. A. Pielke, Sr. *Hurricanes: Their Nature and Impacts on Society*. New York: John Wiley & Sons, 1997. A very informative book written by a father and son meteorological team. The book focuses on the United States, integrating science and societal policies in response to these storms.

Robinson, Andrew. *Earth Shock: Hurricanes, Volcanoes, Earthquakes, Tornadoes, and Other Forces of Nature*. London: Thames and Hudson, 1993. This interesting and informative book provides a good mix of science, individual event summaries, and noteworthy facts and figures. It is written for high school students and adults with no background in science.

Tufty, Barbara. *1001 Questions Answered About Hurricanes, Tornadoes, and Other Natural Air Disasters*. New York: Dover, 1987. This book provides much more than just excellent answers to commonly asked questions. The text has a logical flow to it, in which questions raised by reading an answer become the next question to be answered. Excellent illustrations accompany the text.

LIGHTNING AND THUNDER

Lightning is the flash of light produced by discharge of electricity from a thundercloud. It is accompanied by intense heating and explosive expansion of air, producing the sonic boom called thunder. Lightning may cause property damage and death, but it also contributes to the nitrogen absorption in the soil necessary for plant life and may have even helped to start life on the Earth.

PRINCIPAL TERMS

BALL LIGHTNING: a rare form of lightning appearing as luminous balls of charged air

BEAD LIGHTNING: lightning that appears as a series of beads tied to a string

CONVECTION: vertical air circulation in which warm air rises and cool air sinks

CORONA DISCHARGE: a continuous electric discharge from highly charged pointed objects that produces a luminous greenish or bluish halo known as St. Elmo's Fire

CUMULONIMBUS: thunderstorm clouds that develop vertically by deep convection in the atmosphere and often form a top in the shape of an anvil

DART LEADERS: surges of electrons that follow the same intermittent ionized channel taken by the initial stepped leader of a lightning stroke

GRAUPEL: ice particles between 2 and 5 millimeters in diameter that form by a process of accretion in a cloud

IONOSPHERE: an electrified region of the upper atmosphere from about 80 to 800 kilometers above the Earth that contains a relatively high concentration of ions (charged atoms or molecules) and free electrons

RETURN STROKE: the luminous lightning stroke that propagates upward from the Earth to the base of a cloud as electrons surge down and as positive current flows to the cloud

STEPPED LEADER: an initial discharge of electrons that proceeds in a series of steps from the base of a thundercloud toward the ground

EARLY LIGHTNING STUDIES

The search for understanding about lightning has been long and circuitous, and even today there are conflicting theories. In the ancient civilizations, lightning and thunder were viewed as manifestations of the power and wrath of the gods. Lightning was the weapon of Zeus, the father of all the Greek gods. The Roman god Jupiter subdued monsters with thunderbolts and presented Venus to Vulcan as a bride in gratitude for forging thunderbolts. The book of Exodus in the Bible relates that before Moses climbed Mount Sinai to receive the Ten Commandments, "there was thunder and lightning, with a thick cloud over the mountain, and a very loud trumpet blast. Everyone in the camp trembled."

A scientific understanding of lightning began in about 1750 with the work of Benjamin Franklin. Three years earlier, he had experimented with the electric charging of pointed objects connected to the ground and found that such points could draw a charge off of a charged object when placed near it. In 1749, he wrote a letter to a friend in the English Royal Society, speculating on the electrification of clouds and suggesting that lightning is an electric discharge to the ground that favors trees, spires, chimneys, and masts. He also compared the snap of an electric spark with the thunder produced by lightning. In 1750, he wrote the Royal Society suggesting that a pointed metal rod extending several feet above a building and connected by a wire to the ground could protect the building from lightning damage, even drawing a charge silently from the cloud.

Along with his idea of the lightning rod, Franklin proposed an experiment for collecting electricity from clouds by placing a properly insulated metal rod above a high tower or steeple and drawing sparks from its lower end. The Comte De Buffon performed this experiment successfully in

France in May of 1752, but when the German scientist George W. Richmann repeated it in Russia in 1753, a bolt of lightning killed him. In October of 1752, Franklin performed his famous kite experiment with a pointed wire fastened to the top of the kite and a metal key at the lower end of the string, which was held by a silk ribbon for insulation. Standing under a shed to keep the silk ribbon dry, Franklin was able to draw sparks from the key and collected enough electric charge to show that it was the same as ordinary electricity. In experiments with a metal rod above his house, he found that the charge at the bottom of thunderclouds was usually negative but sometimes positive, suggesting the complexity of cloud charging.

THEORIES OF THUNDERCLOUD CHARGING

Lightning occurs when separate regions in a cumulonimbus thundercloud contain opposite electric charges. Several theories have been proposed to account for this state of affairs. The precipitation hypothesis, first proposed by the German physicists Julius Elster and Hans Geitel in 1885, assumed that raindrops and ice pellets in a thunderstorm are pulled down by gravity past smaller water droplets and ice crystals suspended in the cloud. The hypothesis suggested that collisions between these particles cause the larger particles to gain a negative charge (as charge transfers to shoes from a rug), and the smaller ones become positively charged. Updrafts within the cloud then sweep the smaller positive particles up near the top of the cloud, while the larger negative particles accumulate near the lower part of the cloud.

A second theory, the convection hypothesis, was developed independently by the French scientist Gaston Grenet in 1947 and by the American scientist Bernard Vonnegut in 1953. They assumed that cosmic rays from outer space ionize the air above a cloud, separating positive and negative charges, while corona discharges around sharp objects on the Earth's surface produce positive ions. Warm air then carries these positive ions upward by convection until they near the top of the cloud, where they attract the negative ions formed by cosmic rays. These negative ions enter the cloud and attach to water droplets and ice crystals, which are then carried by downdrafts at the edges of the cloud to the lower regions, charging the cloud negative at the bottom and positive at the top.

A theory developed in the 1980's combines the effects of convection and precipitation with the microphysics of ice particles and is able to explain Franklin's observations that the bottoms of clouds are sometimes positive. It assumes that charging occurs as millimeter-size ice particles called graupel fall through a region of supercooled droplets (below 0 degrees Celsius) and ice crystals. When these droplets collide with the graupel, they freeze on contact by releasing latent heat. This heat keeps the surface of the graupel warmer than the surrounding ice crystals. Several studies at American and British universities have shown that negative charge is transferred to graupel below a critical temperature of about −15 degrees Celsius. When the graupel and crystals come into contact, positive ions are transferred to the ice crystals, which are then carried by updrafts to the top parts of the cloud. The graupel and larger hailstones fall, depositing a strong negative charge near the critical temperature level in the cloud. Below this level, positive charge is picked up by the graupel, causing some weaker positive regions near the bottom of the cloud.

CLOUD-TO-GROUND LIGHTNING

The normal fair-weather atmosphere is characterized by a positively charged upper atmosphere, called the ionosphere, and a negative charge on the Earth's surface. As a thundercloud develops with strong convection updrafts from warmer surface regions, its strong negative charge repels electrons in the ground below, causing it to become positively charged relative to the cloud. The positive charge becomes most concentrated on objects that protrude above the ground, such as trees, poles, and buildings. When the difference in charge between the ground and the cloud builds up to an electric potential of about 1 million volts per meter, the air begins to ionize enough to overcome the air's resistance, and current begins to flow. High-speed camera studies have revealed that the resulting lightning stroke is a complex series of events.

In cloud-to-ground lightning, a local electric potential of more than 3 million volts per meter develops within the cloud along a path of some 50 meters. This causes a discharge of electrons to rush toward the base of the cloud and then toward the ground in a series of steps, carrying a negative current of a few hundred amperes downward. Each step in the discharge covers about 50 to 100 meters, with stops of about fifty microseconds between steps. This initial discharge is called the "stepped leader" and produces only a faint glow nearly invisible to the human eye. As the stepped leader approaches to within about 50 meters of the ground, a region of positive charge moves up into the air through any conducting object (usually elevated) to meet it.

When the stepped leader of electrons meets the upward-moving positive charge, a large number of electrons flows to the ground starting from the bottom of the ionized channel in a "return stroke" equivalent to an upward positive current of some 10,000 amperes. The luminous return stroke, several centimeters in diameter, generates heat and light in a bright flash lasting less than

two-tenths of a millisecond, too fast for the eye to resolve the motion. This leader-and-stroke process is often repeated in the same ionized channel several times at intervals of about two-hundredths of a second. These subsequent surges of electrons from the cloud, called "dart leaders," proceed downward more rapidly because of the lowered resistance of the path, each followed by a less energetic return stroke. The entire process lasts less than one-half of a second, too short to distinguish individual strokes. Some dart leaders that do not reach the ground appear as lighter forked flashes.

VARIED FORMS OF LIGHTNING AND THUNDER

Lightning and thunder take on a variety of shapes and forms. Sometimes a dart leader is met by a return stroke along a new conducting path from the ground, causing a forked lightning bolt that strikes the ground in more than one place. In fewer than 10 percent of cases, a positively

Multiple cloud-to-ground lightning strikes during a storm. (Weather Stock)

charged leader emanates from the cloud; even more rarely, a leader starts from tall objects on the ground with a return stroke from the cloud. If the wind moves the ionized channel fast enough between strokes, a broader streak called ribbon lightning can be produced, or a series of bright streaks called bead lightning results when the channel appears to break up. A rare phenomenon called ball lightning consists of an electric discharge in the form of a slowly moving basketball-size luminous sphere that has been variously observed to either explode or simply decay.

Studies have shown that about 80 percent of lightning occurs within clouds where it is hidden from view, and occasionally even strikes between clouds. Lightning inside of clouds can illuminate them with flashes of light called sheet lightning. Lightning from distant thunderstorms that can be seen but is too far away for its thunder to be heard is called heat lightning. Since 1989, three new rare types of lightning have been videotaped jumping from the tops of clouds into the atmosphere. These consist of bright bluish, cone-shaped bursts called blue jets that rise more than 20 kilometers from the cloud center, clusters of dim reddish bursts called red sprites rising to heights of more than 50 kilometers, and doughnut-shaped bursts of light called elves some 400 kilometers wide and 100 kilometers above the cloud tops.

Lightning usually produces thunder. A lightning discharge heats the air along the conducting path so quickly (a few microseconds) and to such a high temperature (several times that of the surface of the Sun) that the air expands explosively. This produces a shock wave of compressed air that decays into an acoustic wave within a couple of meters. Because sound travels almost 1 million times slower than light, thunder is heard after the lightning is seen. Sound travels at about 330 meters per second, so the distance to the flash can be estimated as 1 kilometer for every three seconds between the lightning flash and the thunder. Within about 100 meters from lightning, thunder sounds like a clap followed by a loud bang. At greater distances, thunder often rumbles as it emanates from sections of the lightning stroke (about 5 kilometers long) at different distances from the observer. Because of temperature differences and turbulence in the air, thunder is seldom heard beyond about 20 kilometers.

SIGNIFICANCE

Lightning has both destructive and beneficial effects. Estimates indicate that more than forty thousand thunderstorms and 8 million lightning flashes occur daily around the world. Tropical landmasses experience thunderstorms on more than one hundred days each year, while they rarely occur in Polar regions or dry climates. In the United States, the average number of thunderstorm days each year varies from ninety in Florida to forty in the Midwest and less than ten along the West Coast. Annually in the United States, lightning causes about one hundred deaths and starts about ten thousand forest fires, with timber losses and property damage of about 80 million dollars. Lightning can be most dangerous near elevated places and isolated trees. The best protection is secured inside automobiles and buildings, but out of contact with conducting surfaces.

Lightning is essential for life on the Earth by providing a source of plant fertilizer and ozone. Its electrical action combines nitrogen and oxygen in the atmosphere to form nitric oxide, which is then dissolved in precipitation and brought to the Earth as a nitrate. Hundreds of millions of tons of nitrates are produced by lightning each year and absorbed by the soil, where it helps to nourish plant life and food crops. Studies at the National Center for Atmospheric Research using a computer model called MOZART showed that lightning is probably the main source of ozone in the upper atmosphere, which protects life by blocking ultraviolet radiation from the Sun.

Lightning also helps to maintain the potential difference of some 300,000 volts between the Earth and the ionosphere, which reflects radiowaves used for long-distance communication. This voltage causes a fair-weather current of about 2,000 amperes to discharge the Earth (a few microamperes per square kilometer). Lightning balances this flow by returning electrons to the Earth. Lightning may have even been the spark that started life among organic molecules on Earth, as well as the source of the first fire used by humans for protection and survival.

Joseph L. Spradley

CROSS-REFERENCES

Atmosphere's Global Circulation, 1823; Atmospheric and Oceanic Oscillations, 1897; Climate,

BIBLIOGRAPHY

Ahrens, C. Donald. *Meteorology Today: An Introduction to Weather, Climate, and the Environment.* 6th ed. Pacific Grove, Calif.: Brooks/Cole, 2000. Chapter 15 in this college-level text has a good description of the current understanding of lightning and thunder with several good colored photographs and diagrams.

Barry, James Dale. *Ball Lightning and Bead Lightning: Extreme Forms of Atmospheric Electricity.* New York: Plenum Press, 1980. This book provides a comprehensive description of two unusual forms of lightning supported by numerous photographs and drawings and an extensive bibliography.

Few, Arthur A. "Thunder." *Scientific American* 233 (July, 1975): 80-90. This important article describes extensive research on lightning, with a special emphasis on its relation to thunder, including several illustrations and diagrams.

Golde, R. H., ed. *Lightning.* London: Academic Press, 1977. This authoritative book is volume 1 in the Physics of Lightning series, with chapters on lightning in history and research by experts in various fields. It includes many photographs and diagrams.

Viemeister, Peter E. *The Lightning Book.* Cambridge, Mass.: MIT Press, 1972. This is a readable description of the history of lightning studies, the nature of lightning, and how to protect yourself from it. Many illustrations and photographs.

Williams, Earle R. "The Electrification of Thunderstorms." *Scientific American* 259 (November, 1988): 88-99. This article is a well-illustrated discussion of the causes of charge separation in clouds and the lightning discharge.

MONSOONS

Monsoons are seasonal wind systems that reverse directions biannually and are crucial for the economic stability and agricultural productivity of affected geographic areas.

PRINCIPAL TERMS

AUSTRAL: referring to an object or occurrence that is located in the Southern Hemisphere or related to Australia

BOREAL: alluding to an item or event that is in the Northern Hemisphere

CONVECTION: heat transfer that occurs during the circulation of liquids or gases

GEOSTROPHIC: force that causes directional change because of the Earth's rotation

INTERTROPICAL CONVERGENCE ZONES (ITCZ): low-pressure areas where southern and northern trade winds meet

OROGRAPHY: study of mountains that incorporates assessment of how they influence and are affected by weather and other variables

OSCILLATION: varying behavior of an object such as an ocean wave changing direction or velocity

TROPOSPHERE: closest level of the atmosphere to Earth

TROUGH: immense area of low barometric pressure, usually in deserts or underwater

VORTEX: a vacuum formed in the center of rapidly whirling liquid or gas

ORIGINS OF MONSOONS

The term "monsoon" originated from the Arabic word *mauism*, meaning season, used by sailors to comment about changing winds above the Arabian Sea. Solar heat produces winds that shift to the north and south according to the Sun's position each season. Scientists cite geological evidence that suggests conditions favorable for monsoons began millions of years ago when the Indian subcontinent collided with the Asian plate, creating the Himalaya Mountains and the Tibetan Plateau. Warm and cool ocean water and landmasses affect atmospheric circulation, creating recurring wind systems that sweep over large regions. The majority of monsoons occur in tropical areas adjacent to the Indian Ocean, although seasonal winds also affect Africa, northeast Asia, Australia, and North and South America. Monsoons are described by geographical terminology; for example, the two predominant patterns are called the Asian-Australian and American monsoon systems. Monsoons are also referred to as boreal or austral, designating the hemisphere in which they occur.

For thousands of years, monsoons have been incorporated into the literature, folklore, and religious rituals of numerous cultures. The monsoon motif is a universal symbol for rebirth and fertility as well as devastation and death. Monsoons are paradoxical breezes because they are predictable to the extent that humans know that the monsoons probably will occur during specific seasons. However, the winds are often erratic, being delayed or appearing prematurely and precipitating extreme or minute amounts of rainfall or sometimes bypassing regions. Usually, between April and October, monsoon winds develop in the southwest, shifting direction to originate from the northeast between October and April. Monsoons are a prolonged series of winds and not restricted to a single storm.

Monsoons manipulate the planet's climate, often proving to be beneficial and occasionally detrimental. The winter monsoon blowing from land to sea is usually associated with dryness, while the summer monsoons storming from the sea onto land produce torrential, sustained rainfall that is vital for agricultural activity. Sufficient precipitation assures growth of ample crops to nourish domestic populations, to export for economic profits, and to provide employment for farm laborers. The absence of monsoon rains can result in famines and impoverishment.

MONSOON DYNAMICS

For several centuries, scientists have analyzed monsoons to determine the physical forces that generate the winds and regulate their behavior. Researchers have agreed on basic explanations regarding fundamental aspects of monsoons, such as their relationship to atmospheric and oceanic conditions and how monsoons tend to vary instead of conforming to exacting standards. Scholars continue to seek answers to complex questions about monsoons, especially concerning fluctuations displayed in time periods ranging from seasons to decades. Such information might enable meteorologists to predict possible occurrences and outcomes that could impact both local and global populations and economies. Researchers want to understand more about the onset of monsoons and their active and break periods thought to be caused by shifting troughs. The definite beginning of a monsoon is often disputed, with some scientists saying that increased humidity indicates the monsoon's start, while many say precipitation or formation of a vortex signals the beginning.

Monsoons are caused by sea and land breezes that create temperature and air-pressure differences between landmasses and bodies of water. Land is unable to absorb and maintain heat like water does, and the difference in land temperature relative to water heat instigates monsoons. Areas in low latitudes near the equator undergo circulatory and precipitation changes because of temperature deviations on adjacent continents and seas. The amount of solar radiation emitted in each season affects the temperature and air pressure over continents. During the Northern Hemisphere's summer, when it tilts toward the Sun, high-pressure systems move from the cooler ocean into the land's low-pressure area. As the landmass cools during winter, its high-pressure air mass moves from the land to the low-pressure air mass over the warmer ocean.

The jet stream moves south during winter and north in the summer, transporting air masses to and from monsoon regions. In winter, the Siberian high-pressure system causes air to circulate clockwise. Winds move from the northeast, down the Himalaya Mountains, and across land cooled during shortened days toward the sea, creating dry conditions and causing monsoon rains in Indonesia and Australia. By summer, the winds reverse direction because the land is heated by increased solar energy. Moving counterclockwise, the winds carry moisture from the sea in the southwest toward land. For example, the Somali jet stream near Africa moves across the equator to the Arabian Sea and alters the ocean current's direction, which causes deep water to rise and surface temperatures to drop. Humid winds shift into India, rising when they reach the Himalaya Mountains. As the winds are lifted over the Tibetan Plateau, the air cools sufficiently to become saturated, triggering thunderstorms and convectional rainfall.

Scientists have designated three monsoon circulation patterns. The lateral component indicates a monsoon that circulates across the equator. Transverse circulation moves from North Africa and the Middle East into southern Asia. The Walker circulation, also a transverse pattern, moves across the Pacific Ocean. The size, shape, coastal positions, and elevation of landmasses influence these patterns, and the shifting of the low-pressure Intertropical Convergence Zone (ITCZ) north and south can determine how monsoon systems move. These factors contribute to variations in monsoon intensity and duration. In areas such as Australia, monsoon winds do not rise over or descend from mountains, and high-pressure masses undergo geostrophic adjustment as they ascend over monsoon troughs. Orography influences the nature of monsoons globally. African and Australian monsoons tend to be weaker because they are not elevated as high as Asian winds. Because the Rocky Mountains and the Sierra Madre lift air masses somewhat like the Himalaya, some researchers claim that a monsoon circulation pattern occurs in North America, causing increased precipitation in Mexico and the southwestern United States every summer.

IMPACT ON CIVILIZATION

Monsoons permeate the civilizations of regions where they occur. Approximately 60 percent of people worldwide are economically dependent on the climate affected by the Asian-Australian monsoon system. Humans, animals, and plants rely on monsoons to provide essential moisture. The absence of monsoon rains can cause droughts and famine, killing millions of people by starvation. Any fluctuation in the monsoon cycle, whether within one season, year, or decade, can be detrimental. At least 2 billion people rely on monsoons to irrigate rice and wheat crops. In India, grain production is essential to feed the growing population that expands annually at a rate greater than agricultural yields increase. At least one-fourth of the Indian economy is agriculturally based, and 60 percent of laborers are engaged in agricultural employment. Monsoon rain is critical to maintain this balance.

Ironically, monsoons also flood areas and drown people, cause landslides that destroy communities, and inundate crops. Millions of acres often remain underwater for months. Several thousand people die annually during monsoons, and thousands more become homeless. Many people are reported missing after a monsoon deluge. Flood waters wash away unstable dams, buildings, and graves in cemeteries. Extreme humidity is stifling for most people. Humanitarian and relief agencies such as the International Red Cross and Red Crescent Society provide emergency food rations and shelter. Shipping along trade routes in monsoon regions is often both helped and interrupted by winds. Military combatants in Asian wars complain about obscured landscapes and incessant rain.

Folklore, proverbs, and prayers provide insights into personal experiences with monsoons. Natives and visitors to monsoon regions have documented their encounters with monsoons. Some tourists purposefully travel to Asia during monsoons because hotels and businesses are not crowded, prices are lower, and reservations are easier to secure. However, some visitors note the inconvenience of always carrying an umbrella, traveling monsoon-damaged roads, and encountering storm-related delays. Monsoon tours are offered for Arabs to escape the hot desert, and indoor parks with arcades, music, and entertainment have been built for monsoon tourists to enjoy. Astronomers can observe celestial phenomena such as eclipses because of clear skies during the winter monsoons. Oceanographers also consider the monsoons useful because nutrient-rich waters rise to the surface, allowing scientists to study how plants, animals, the sea, and the atmosphere exchange carbon dioxide.

MODELING MONSOONS

Because monsoons are essential for affected populations to thrive despite the wind's uncertain behavior, scientists have initiated programs of cooperative research regarding monsoons. They hope to collect sufficient data to develop computer models that can predict when monsoons will occur and how they will impact land masses. Such forecasting efforts have often frustrated researchers because of the monsoons' capricious nature. Speculation is primarily hindered by intraseasonal oscillations between the monsoons' active and passive precipitation phases. The Center for Ocean-Land-Atmosphere Studies (COLA) examines the relationship of the ocean and atmosphere and heat sources. While empirically based forecasts have often proved more reliable than modelled simulations during the twentieth century, scientists seek to perfect their experimental methods.

Researchers recognize that regional topographic differences and varying hydrodynamic situations impede monsoon modeling attempts. Incorporating information about variables such as the temperature of the sea surface and snow cover, scientists try to understand how these factors alter monsoon behavior. They also want to comprehend how the monsoons affect ocean and atmosphere interactions in addition to how the water-air pressure system relationship influences monsoons. Studies have been designed to expand knowledge about the role of convection in monsoons. Researchers are also interested in studying the monsoons' impact on climates inside and outside the monsoons' immediate zone. Scientists disagree whether monsoons are impacted by or affect El Niño; studies have been done to evaluate global precipitation data and drought conditions to determine any correlations.

Monsoon variability hinders forecasting and potential benefits to agriculture based on information concerning the onset of rainfall. For example, with advance knowledge, farmers could plant crops that require less water in case a weaker monsoon is forecast. Monsoons, however, do not always begin when expected, and variations might happen within the yearly cycle or fluctuate over several years. Using different global circulation models, researchers can comprehend the fundamental physical processes of monsoons, then apply this knowledge to create specific models representing a seasonal cycle based on their observations, statistics, and hypotheses. Interpreting results produced by these models compared with satellite data, researchers become aware of how monsoons vary according to the landmasses they traverse and according to unique oceanic and atmospheric conditions.

Realizing these factors are linked, researchers create models to consider numerous variables concurrently but realize that more sophisticated modeling of factors such as sea-surface temperature, solar radiation, water vapor, cloud cover, soil moisture, and terrain is necessary to improve prediction methods. Such models must accurately simulate a monsoon's average seasonal rainfall and behavior, taking into account how it varies within one season and one year. The models must also consider anomalies with other documented monsoons in that area to isolate how external conditions such as altered topography may affect the monsoon's internal dynamics.

FUTURE RESEARCH

Monsoons fascinate researchers who strive to acquire more complex understanding of the seasonal wind systems. Scientists hope to predict monsoons more precisely, because the winds impact global economies and populations. Meetings of international monsoon experts have been scheduled to share knowledge of monsoon cycles, coordinate research methods, and set future goals. Primarily, scientists want to explain why monsoons vary in behavior and how such variables as sea-surface temperature and location of land-based heat sources interact and influence the monsoons' fluctuations. Researchers also want to explore the relationship between monsoons and El Niño.

Scientists realize that to achieve accurate predictive techniques, full understanding of sea-surface temperature anomalies that affect monsoon circulation must be obtained through an analysis of oceanic processes that remain vague. Future models will assess these temperatures in different geographic regions. Models will also further evaluate water surface fluxes, land surface coverings of snow and vegetation, and global warming, all of which affect temperatures and alter monsoon cycles. Until such comprehension is attained through enhanced model simulations designed to analyze numerous dynamic variables simultaneously, predictions will be minimized, slowing endeavors to manage the winds, and monsoons will continue to dominate life both advantageously and harmfully in tropical zones.

Elizabeth D. Schafer

CROSS-REFERENCES

Atmosphere's Structure and Thermodynamics, 1828; Atmospheric and Oceanic Oscillations, 1897; Climate, 1902; Clouds, 1909; Drought, 1916; El Niño and La Niña, 1922; Hurricanes, 1928; Life's Origins, 1032; Lightning and Thunder, 1935; Ozone Depletion and Ozone Holes, 1879; Precipitation, 2050; Satellite Meteorology, 1945; Seasons, 1951; Storms, 1956; Tornadoes, 1964; Van Allen Radiation Belts, 1970; Volcanoes: Climatic Effects, 1976; Weather Forecasting, 1982; Weather Forecasting: Numerical Weather Prediction, 1987; Weather Modification, 1992; Wind, 1996.

BIBLIOGRAPHY

Bamzai, Anjuli S. *Relations Between Eurasian Snow Cover, Snow Depth, and the Indian Summer Monsoon.* Calverton, Md.: Center for Ocean-Land-Atmosphere Studies, 1998. Bamzai studies the relation between the Indian summer monsoon and the snow cover in the Himalaya Mountains and other nearby regions.

Chang, Chih-Pei, and Tiruvalam Natarajan Krishnamurti, eds. *Monsoon Meteorology.* Oxford Monographs on Geology and Geophysics 7. New York: Oxford University Press, 1987. A thorough technical text that divides information into sections about summer and winter monsoons; heating, topography, and clouds; and theoretical and modeling studies. Includes charts and references.

Cheung, Chan Chik. *Synoptic Patterns Associated with Wet and Dry Northerly Cold Surges of the Northeast Monsoon.* Hong Kong: Royal Observatory, 1997. This short, somewhat technical report discusses monsoon patterns in the Hong Kong region of China. Bibliography.

Douglas, Michael W., Robert A. Maddox, Kenneth W. Howard, and Sergio Reyes. "The

Mexican Monsoon." *Journal of Climate* 6 (August, 1993): 1665-1677. Discusses the phenomenon of summer rainfalls in North America, which resemble Asian monsoons. Illustrated with maps. Sources listed.

Fein, Jay S., and Pamela L. Stephens, eds. *Monsoons*. New York: Wiley, 1987. Comprehensive anthology that includes meteorological and geological data, in addition to a discussion of cultural responses to monsoons throughout history. Illustrations, bibliography, and index. Useful for high school students.

McCurry, Steve. *Monsoon*. New York: Thames and Hudson, 1988. A brief personal account of author's travels during a monsoon which includes historical and cultural insights. Mostly photographs documenting people and places affected by monsoons. Readers of all ages can use this book to supplement more scholarly texts.

Oshima, Harry T. *Strategic Processes in Monsoon Asia's Economic Development*. Baltimore: The Johns Hopkins University Press, 1993. Examines how climatic conditions caused by monsoons influence employment and other socioeconomic concerns. References and index. Accessible to high school readers.

Vesilind, Priit J. "Monsoons." *National Geographic* 166 (December, 1984): 712-747. Narrative supplemented with photographs by Steve McCurry that describes author's experiences during monsoons and explains the scientific principles of monsoons. Suitable for high school readers.

Webster, P. J., et al. "Monsoons: Processes, Predictability, and the Prospects for Prediction." *Journal of Geophysical Research* 103 (June 29, 1998): 14,451-14,510. This scholarly article explores potential methods to model variables to forecast future monsoons. Illustrations and references.

SATELLITE METEOROLOGY

Satellite meteorology, the study of atmospheric phenomena using satellite data, is an indispensable tool for forecasting weather and studying climate on a global scale.

PRINCIPAL TERMS

ACTIVE SENSOR: a sensor, such as a radar instrument, that illuminates a target with artificial radiation, which is reflected back to the sensor

ALBEDO: the percentage of incoming radiation that is diffusely reflected by a surface

EL NIÑO: a periodic anomalous warming of the Pacific waters off the coast of South America; part of a large-scale oceanic and atmospheric fluctuation that has global repercussions

GEOSYNCHRONOUS (GEOSTATIONARY): pertaining to a satellite that orbits about the Earth's equator at an altitude and speed such that it remains above the same point on the Earth

NEAR-POLAR ORBIT: an Earth orbit that lies in a plane that passes close to both the north and south poles

PASSIVE SENSORS: sensors that detect reflected or emitted electromagnetic radiation

RADIOMETER: an instrument that makes quantitative measurements of reflected or emitted electromagnetic radiation within a particular wavelength interval

SPATIAL RESOLUTION: the ability of a sensor to define closely spaced features

SUN-SYNCHRONOUS ORBIT: for an Earth satellite, a near-polar orbit at an altitude such that the satellite always passes over any given point on the Earth at the same local time

SYNTHETIC APERTURE RADAR (SAR): a spaceborne radar imaging system that uses the motion of the spacecraft in orbit to simulate a very long antenna

---

TYPES OF SATELLITES AND INSTRUMENTS

Satellite meteorology is the study of atmospheric phenomena, notably weather and weather conditions, using information gathered by artificial satellites. These satellites, unmanned spacecraft equipped with instruments that monitor cloud cover, snow, ice, temperatures, and other parameters, give scientists a continuous and up-to-date view of meteorological conditions and activity over a large area. Satellite data are an important tool not only for forecasting weather and tracking storms but also for observing climate change over time, monitoring ozone levels in the stratosphere, and studying other aspects of weather and climate on a global and ongoing basis.

The satellites from which meteorological measurements are made can be categorized by their orbits. Some weather satellites have a geosynchronous, or geostationary, orbit, meaning that they travel around the globe at an altitude and speed that keeps them above the same point along the equator. A near-polar, Sun-synchronous orbit, by contrast, is a north-south orbit that passes close to the Earth's poles and carries a satellite over any given location on the Earth's surface at the same local time. Geosynchronous satellites have better temporal resolution: They provide updated information for an area every thirty minutes, while near-polar, Sun-synchronous satellites may take anywhere from a few hours to several days to transmit updates. However, near-polar satellites have the higher spatial resolution—that is, they are better at providing images in which closely spaced features can be identified. Geosynchronous satellites provide images with comparatively poor spatial resolution because they must orbit at a greater altitude (at least 35,000 kilometers above the Earth's surface). Examples of geosynchronous meteorological satellites include the U.S. Geostationary Operational Environmental Satellite (GOES) series, Europe's Meteosat series, Russia's Geostationary Operational Meteorological Satellite (GOMS) series, and Japan's Geostationary Meteorological Satellite (GMS), or Himawan, series. Near-polar, Sun-synchronous meteorological satellites include the National Oceanic and Atmospheric Administration (NOAA) series of American satellites and Russia's Meteor series.

The various orbiting platforms carry different sets of instruments. The weather satellites launched in the 1960's and early 1970's included television camera systems as part of their instrument packages. Later satellites have relied instead on instruments such as specialized radiometers (instruments that measure the amounts of electromagnetic radiation within a specific wavelength range) and radar systems. Radiometers measure such parameters as surface, cloud, and atmospheric temperatures; atmospheric water vapor and cloud distribution; and scattered solar radiation. Radar-system measurements include satellite altitude and ocean-surface roughness. Television cameras and radiometers are examples of passive sensors, which record radiation reflected or emitted from clouds, landforms, or other objects below. Radar systems are active sensors, sending out signals and recording them as they are reflected back. The data collected by a satellite's sensors are transmitted via radio to ground stations. If a near-polar orbiter is not within transmitting distance of a ground station, its onboard data-collection system will store the information until the satellite passes within range.

WEATHER SATELLITES

The Advanced Television Infrared Observation Satellite, or TIROS, and Next-Generation (ATN) near-polar orbiting satellites (NOAA 8 through 14), a series of American weather satellites, have carried an array of sophisticated instruments. Two of these satellites, NOAA 12 and NOAA 14, remained in operation into the late 1990's. All the ATN series satellites have included an advanced, very high resolution radiometer (AVHRR), which detects specified wavelength intervals within the visible, near-infrared, and infrared ranges to generate information on sea-surface and cloud-top temperatures and ice and snow conditions; a TIROS operational vertical sounder (TOVS), which measures emissions within the visible, infrared, and microwave spectral bands to provide vertical profiles of the atmosphere's temperature, water vapor, and total ozone content from the Earth's surface to an altitude of 32 kilometers; and a solar proton monitor, which detects fluctuations in the Sun's energy output, particularly those re-

lated to sunspot (solar storm) activity. With the exception of NOAA 8 and NOAA 12, all these satellites have included the Earth radiation budget experiment (ERBE), which uses long-wave and short-wave radiometers to provide data pertaining to the Earth's albedo (the amount of incoming solar radiation that is diffusely reflected by the ground, water, ice, snow, clouds, airborne particles, or other surfaces). All but NOAA 8, NOAA 10, and NOAA 12 have carried the solar backscatter ultraviolet (SBUV) radiometer, which measures the vertical structure of ozone in the atmosphere by monitoring the ultraviolet radiation that the atmosphere scatters back into space. Of the ATN series of satellites, NOAA 12 is the only one that has not also carried the Search and Rescue Satellite Aided Tracking (SARSAT) system, which detects distress signals from downed aircraft and emergency beacons from ocean vessels, then relays the signals to special ground stations.

The satellites of the U.S. GOES series that remained in operation in the late 1990's (GOES 8 and GOES 9) are geosynchronous orbiters with their own distinctive instrumentation. Instrumentation aboard GOES 8 (also called GOES-EAST) and GOES 9 (or GOES-WEST) includes a sounder, which uses visible and infrared data to create vertical profiles of atmospheric temperature, moisture, carbon dioxide, and ozone; a five-band multispectral radiometer, which scans visible and infrared wavelengths to obtain sea-surface temperature readings, detects airborne dust and volcanic ash, and provides day and night images of cloud conditions, fog, fires, and volcanoes; and a search-and-rescue support system similar to the ones flown on the NOAA near-polar orbiters. The GOES satellites are also equipped with a space environment monitor (SEM), which uses a solar X-ray sensor, a magnetometer, an energetic particle sensor, and a high-energy proton alpha detector to monitor solar activity and the intensity of the Earth's magnetic field. The GOES-EAST satellite is positioned above the equator at an approximate longitude of 75 degrees west, while the GOES-WEST satellite orbits above the equator at approximately 135 degrees west longitude. These locations are ideal: GOES-EAST can provide the United States with images of storms approaching across the Atlantic Ocean during hurricane season (June through November); GOES-WEST can monitor the weather systems that move in from across the Pacific Ocean, which affect the United States during most of the year. However, when instrument malfunctions impair a satellite's ability to provide data, the National Weather Service can use small rocket engines aboard the satellites to reposition a more functional platform to provide the desired coverage until a replacement satellite can be launched. After the imaging system on GOES 6 failed in 1989, for example, GOES 7 was relocated several times to compensate.

RADAR AND SATELLITE DATA

While operational weather satellites have generally carried radiometers, radar instruments have been part of the instrument package aboard experimental satellites and space shuttle flights. One of the best-known orbiters using active sensors is Seasat, a short-lived experimental craft launched in 1978 to monitor the Earth's oceans. During its three months of operation, this earth-resources satellite provided a wealth of data for meteorological study. Its radar altimeter determined the height of the sea surface, from which data scientists derived measurements of winds, waves, and ocean currents. Its radar scatterometer yielded information on wave direction and size, measurements that in turn provided insights into wind speed and direction. The most sophisticated of Seasat's active sensors was a synthetic aperture radar (SAR), a radar imaging system that created a "synthetic aperture" of view by using the motion of the platform to simulate a very long antenna. Images of the ocean's surface were obtained from SAR data.

Satellite data have become indispensable to the meteorologist. From orbit, information is readily available from any location on the planet, regardless of its remoteness, inaccessibility, climatic inhospitality, or political affiliation. Satellites yield regular, repeated, and up-to-date coverage of areas at minimal cost. They make it possible to view large weather systems in their entirety, and facilitate meteorological observations on a regional or global basis. Satellites also provide a single data source for multiple locations, alleviating the problem of individual variance of calibration and accuracy that would be associated with separate ground-based observations for each location. However, it is important to note that ground-based stations can make more accurate and detailed observations of a small area; such details may be lost in a view from space. Important though it is to modern meteorology, the use of satellite data augments, rather than replaces, other methods of study.

WEATHER FORECASTING

Satellite meteorology provides a relatively inexpensive means for obtaining current and abundant information on temperature, pressure, moisture, and other atmospheric, terrestrial, and oceanic conditions that affect weather and climate. These data, collected in digital form, are readily processed digitally and integrated with other information. Through these ongoing observations from orbit, scientists gain insights into the short-term and long-term implications of major atmospheric phenomena.

Weather forecasting is the best-known application of satellite meteorology. Anyone who has watched a televised weather report is familiar with geostationary satellite images, a series of which is usually presented in quick succession to show the recent movement of major weather systems. Meteorologists use computers to process the vast amounts of data provided by satellites and other information sources, including ground-based stations, aircraft, ships, and buoys. Data processing yields such forecasting aids as atmospheric temperature and water-vapor profiles, enhanced and false-color images, and satellite-image "movies." Computer models of atmospheric behavior also assist the meteorologist in short-range and long-range forecasting.

Satellite images of cloud cover alone yield a wealth of information for the forecaster. By comparing imagery from visible and infrared spectral regions, meteorologists can identify cloud types, structure, and degree of organization, then make assumptions concerning associated weather conditions. For example, the tall cumulus clouds that produce thunderstorms appear bright in the visible range, as they are deep and thus readily reflect sunlight. These clouds show up in infrared images as areas of coldness, an indicator of the clouds' altitude. Clouds that appear bright in visible-range imagery but that register as warm (low-altitude) in infrared scans may be fog or low-lying clouds. Wispy, high-altitude cirrus clouds, which are not precipitation-bearing, appear cold in infrared images but may not show up at all in visible-range scans.

Weather satellites have proved particularly useful in the science of hurricane and typhoon prediction. These violent rotating tropical storms originate as small low-pressure cells over oceans, where coverage by conventional weather-monitoring methods is sparse. Before the advent of satellites, ships and aircraft were the sole source of information on weather at sea, and hurricanes and typhoons often escaped detection until they were dangerously close to populated coastal areas. Now, using images and data obtained from orbit, a meteorologist can track and study these storms from their inception through their development and final dissipation. With accurate storm tracking and ample advance warning, inhabitants at risk can evacuate areas threatened by wind and high water, thereby minimizing loss of life.

CLIMATE STUDIES

Meteorological satellites also provide scientists with a view of how human activity affects climate on a local, regional, and even global basis. Terrestrial surface-temperature measurements clearly show urban "heat islands," where cities consistently exhibit higher temperatures than those of the surrounding countryside. In images obtained from orbit, thunderstorms can be seen developing along the boundaries of areas of dense air pollution: The haze layer inhibits heating of the ground surface, leading to the unstable atmospheric conditions that produce rainfall. Satellite imagery has revealed that in sub-Saharan Africa, where the overgrazing of livestock owned by nomads has contributed substantially to the spread of desert areas, the resulting increase in albedo has led to a reduction in rainfall and a subsequent reinforcement of drought conditions. Studies of deforestation in tropical areas have incorporated satellite data in their efforts to determine whether replacing forests with agricultural land affects rainfall by reducing evaporation or altering albedo. Satellite data have also played a major role in the ongoing debates regarding how human activity has affected global temperature trends and the Earth's ozone layer.

Satellite meteorology is useful for monitoring the climatological effects of natural occurrences as well. The 1991 eruption of Mount Pinatubo in the Philippine Islands marked the first time that scientists were able to quantify the effects of a major volcanic eruption on global climate. Satellites equipped with ERBE instruments tracked the dissemination of the ash and sulfuric acid particles resulting from the violent eruption. The larger ash particles more readily drift down into the lower portion of the atmosphere, where they can be removed by precipitation; the smaller sulfuric acid particles can remain suspended in the stratosphere for several years. The ERBE instruments measured the amount of sunlight reflected by clouds, land surfaces, and particles suspended in the atmosphere and detected the contribution of suspended particles, clouds, and trace gases such as carbon dioxide to the amount of heat that the atmosphere retained. The eruption was found to have brought about a uniform cooling of the Earth, slowing the global-warming trend that had been observed since the 1980's.

Another natural phenomenon, El Niño, has also been the subject of satellite-based study. This periodic anomalous warming of the Pacific waters off the coast of South America is part of a large-scale oceanic and atmospheric fluctuation, in which atmospheric pressure declines over the eastern Pacific Ocean and rises over Australia and the Indian Ocean. These widespread pressure changes influence rainfall patterns around the world. Satellite measurement of sea-surface temperatures facilitates the early detection of El Niño conditions, and satellite observations on a global basis help scientists to discern the climatic patterns that make up this complex phenomenon.

Meteorological satellites have also been used to gather data pertaining to "solar weather." Using orbiting sensors that detect energetic particles from the Sun, scientists can monitor and predict sunspots and other solar activity. The ability to predict the increases in solar emissions that are associated with sunspots allows scientists to anticipate the resulting ionospheric conditions on the Earth, the effects of which include magnetic storms and disruption of radio transmissions.

Karen N. Kähler

CROSS-REFERENCES

Atmosphere's Global Circulation, 1823; Atmospheric and Oceanic Oscillations, 1897; Climate, 1902; Clouds, 1909; Drought, 1916; El Niño and

La Niña, 1922; Floods, 2022; Himalaya, 836; Hurricanes, 1928; Hydrologic Cycle, 2045; Indian Ocean, 2229; Lightning and Thunder, 1935; Monsoons, 1940; Ocean-Atmosphere Interactions, 2123; Paleoclimatology, 1131; Seasons, 1951; Storms, 1956; Surface Ocean Currents, 2171; Tornadoes, 1964; Van Allen Radiation Belts, 1970; Volcanoes: Climatic Effects, 1976; Weather Forecasting, 1982; Weather Forecasting: Numerical Weather Prediction, 1987; Weather Modification, 1992; Wind, 1996.

BIBLIOGRAPHY

Bader, M. J., et al., eds. *Images in Weather Forecasting: A Practical Guide for Interpreting Satellite and Radar Imagery.* New York: Cambridge University Press, 1995. This collection of essays written by leading meteorologists describes the equipment and techniques used in weather forecasting. Chapters focus on remote sensing, radar meteorology, and satellite meteorology. Color illustrations.

Barrett, E. C., and L. F. Curtis. *Introduction to Environmental Remote Sensing.* New York: John Wiley & Sons, 1976. The chapter on weather analysis and forecasting discusses satellite systems and data applications. The following chapter deals with the earth/atmosphere energy and radiation budgets and other aspects of global climatology. Each chapter includes references.

Burroughs, William James. *Watching the World's Weather.* Cambridge, England: Cambridge University Press, 1991. This useful text focuses on the importance of satellite meteorology to an understanding of weather and climate on a global scale. It deals not only with satellites and instrumentation but also with the essentials of meteorology. Includes satellite images, a glossary, a list of acronyms, and an annotated bibliography.

Collier, Christopher G. *Applications of Weather Radar Systems: A Guide to Uses of Radar Data in Meteorology and Hydrology.* 2d ed. New York: Wiley, 1996. Collier offers the reader a detailed look into scientific advancements regarding the tools used in meteorology and hydrology. Chapters focus on radar meteorology, precipitation measurement, hydrometerology, and weather forecasting.

Colwell, Robert N. *Manual of Remote Sensing.* 2d ed. Falls Church, Va.: American Society of Photogrammetry, 1983. Volume 1, *Theory, Instruments, and Techniques,* includes a chapter on meteorological satellites. Volume 2, *Interpretation and Applications,* contains a chapter dedicated to remote sensing of weather and climate. This thorough technical manual includes a glossary, copious references, and examples of satellite imagery.

Fishman, Jack, and Robert Kalish. *The Weather Revolution.* New York: Plenum Press, 1994. Provides a nontechnical explanation of the basics of meteorology and outlines the evolution of weather forecasting, before and since the advent of weather satellites. The chapter on the development of satellite meteorology discusses programs from TIROS through GOES.

Gurney, R. J., J. L. Foster, and C. L. Parkinson, eds. *Atlas of Satellite Observations Related to Global Change.* Cambridge, England: Cambridge University Press, 1993. Suitable for college-level readers, this volume includes articles on the stratosphere, the troposphere, the Earth's radiation balance, ocean-atmosphere coupling, and snow and ice cover. Illustrated with satellite imagery. An appendix describes selected satellites and sensors.

Hill, Janice. *Weather from Above.* Washington, D.C.: Smithsonian Institution Press, 1991. This overview of U.S. weather satellite programs, intended for a nonscientific audience, includes a glossary of acronyms, a chronological list of meteorological satellites, and suggestions for further reading. Illustrations include photographs of weather satellites, many from the collection of the National Air and Space Museum.

Lillesand, Thomas M., and Ralph W. Kiefer. *Remote Sensing and Image Interpretation.* 2d ed. New York: John Wiley & Sons, 1987. The chapter on Earth resource satellites includes a description of the U.S. NOAA and GOES series of satellites and the Air Force's Defense Meteorological Satellite Program. Earlier chapters provide detailed information on various scanning instruments.

Lutgens, Frederick K., and Edward J. Tarbuck. *The Atmosphere: An Introduction to Meteorology.* 7th ed. Upper Saddle River, N.J.: Prentice-Hall, 1998. An excellent introduction and description of the atmosphere, meteorology, and weather patterns. A perfect textbook for the reader new to the study of these subjects. Color illustrations and maps.

Monmonier, Mark S. *Air Apparent: How Meteorologists Learned to Map, Predict, and Dramatize Weather.* Chicago: University of Chicago Press, 1999. This college-level text looks at the satellites and radar systems used to collect meteorological data, as well as the techniques used to interpret that information. Color illustrations, index, and bibliographical references.

Stevens, William Kenneth. *The Change in the Weather: People, Weather, and the Science of Climate.* New York: Delacorte Press, 1999. Stevens describes various natural and human-induced causes of changes in the climate. Includes a twenty-page bibliography and an index.

Villmow, Jack R. "Application of Remote Sensing in Weather Analysis." In *Introduction to Remote Sensing of the Environment*, edited by Benjamin F. Richason, Jr. Dubuque, Iowa: Kendall/Hunt, 1978. This chapter includes case studies involving typhoons and hurricanes, snowfall, frontal systems, and other weather conditions as observed from orbit. Suitable for college-level readers.

SEASONS

Seasons are the two (fair and rainy) or four (spring, summer, autumn, and winter) periods of the year that are distinguished by specific types of weather. Many plant and animal life cycles and periods of activity are based on the seasons.

PRINCIPAL TERMS

AUTUMNAL EQUINOX: the day that the Sun shines directly over the equator and the season of autumn begins; in the Northern Hemisphere, the date is about September 21, and in the Southern Hemisphere, it occurs about March 21

EQUATOR: a line of latitude on the Earth that is halfway between the North and South Poles

MONSOON: a wind system that results in an annual cycle of fair weather followed by rainy weather

PERIHELION: the point in a planet's orbit at which it is closest to the Sun

SUMMER SOLSTICE: the day that summer begins; in the Northern Hemisphere, it is about June 21, when the Sun is directly over the Tropic of Cancer; in the Southern Hemisphere, it is about December 21, when the Sun is directly over the Tropic of Capricorn

TROPIC OF CANCER: a line of latitude 23.5 degrees north of the equator; this line is the latitude farthest north on the Earth where the noon Sun is ever directly overhead

TROPIC OF CAPRICORN: a line of latitude 23.5 degrees south of the equator; this line is the latitude farthest south on the Earth where the noon Sun is ever directly overhead

VERNAL EQUINOX: the day that the Sun shines directly over the equator and the season of spring begins; in the Northern Hemisphere, the date is about March 21, and in the Southern Hemisphere, it is about September 21

WINTER SOLSTICE: the day that winter begins; in the Northern Hemisphere, about December 21, and in the Southern Hemisphere, about June 21

ENERGY FROM THE SUN

The seasons are the natural, weather-related divisions of the year. In some tropical areas, the seasons may be classified merely as "rainy" and "dry." In the North and South Temperate Zones of the Earth, the four seasons are spring, summer, autumn, and winter.

The seasons and their weather patterns are all caused by energy from the Sun. The seasons change because, in places, this solar energy increases and decreases in an unending annual cycle. This cycle results from the Earth's axis being tilted at an angle of 23.5 degrees, an angle formed by the axis of the Earth and a line perpendicular to the plane formed by the Earth's orbit around the Sun.

The Earth makes a complete orbit around the Sun every 365 days, 5 hours, and 49 minutes. As the Earth revolves around the Sun, the direction of the Earth's axis does not appreciably change. Thus the north end of the axis tilts toward the Sun in June; six months later in December, when the Earth has traveled in its orbit to the other side of the Sun, the north end of the axis tilts away from the Sun.

Light is most intense when it strikes the Earth vertically, because its energy is then concentrated. Light falling at an oblique angle is spread out as it hits the Earth and is thus less effective in heating the Earth. When the North Pole tilts toward the Sun, more sunlight falls on the Northern Hemisphere than on the Southern Hemisphere. In addition, this sunlight is more direct; that is, it hits the Earth closer to a vertical direction. In fact, it is exactly vertical at the Tropic of Cancer. This results in two factors that together cause the warmer weather of summer: The Sun is higher in the sky than it is in autumn and winter, and the number of hours of daylight is increased. The higher the Sun is in the sky, the more concentrated is its heat and thus the warmer that part of the Earth becomes.

A common misconception is that summer occurs because the Earth is closer to the Sun. In fact, the Earth is closest to the Sun (at perihelion) about January 3 of each year, which is winter in the Northern Hemisphere. The Earth is farthest from the Sun about July 3. The orbit of the Earth is an ellipse (an elongated circle), but the shape of this ellipse is so nearly circular that its effect (that is, the variation in distance to the Sun) is not as important a factor in the Earth's weather as is the angle at which the sunlight hits the Earth. This angle is greatest in the Northern Hemisphere in June, when the Sun shines directly over the Tropic of Cancer, and greatest in the Southern Hemisphere in December when the Sun shines directly over the Tropic of Capricorn. It takes an enormous amount of solar energy to warm the Earth's atmosphere, lakes, oceans, and land, and thus the warmest part of summer does not occur on the summer solstice itself (June 21) but instead about one month later.

EFFECTS OF LATITUDE

Latitude affects seasonal change in two ways. One of these is the number of hours of daylight. As one goes from the equator (0 degrees latitude) to either of the poles, the latitude increases until one is at the poles, where the latitude is 90 degrees. On the day of the summer solstice (about June 21), daylight is at a maximum for areas north of the equator and at a minimum for areas south of the equator. The farther north one goes on that date, the longer the daylight. At 20 degrees north of the equator (approximately the latitude of

Mexico City), daylight on that date is 13 hours and 12 minutes long; at 40 degrees north (approximately the latitude of New York, Rome, or Beijing), it is 14 hours, 52 minutes; at 60 degrees north (approximately the latitude of Anchorage or Oslo), daylight is 18 hours and 27 minutes.

The opposite is true at the winter solstice. At 20 degrees north of the equator, daylight on that date is 10 hours and 48 minutes long; at 40 degrees north, it is 9 hours, 8 minutes; at 60 degrees north, daylight lasts only 5 hours and 33 minutes. Therefore, the closer one is to the poles, the more extreme the variations of daylight and darkness.

The other way that latitude affects seasonal change is the angle of incoming sunlight. As one goes north from the Tropic of Cancer (or south from the Tropic of Capricorn), the angle of incoming solar light decreases. Even though far northern areas have extremely long periods of daylight from May to July, the angle of sunlight is so low that the solar energy is spread very thin. Therefore, the Earth there does not receive much heat, and thus summer temperatures in the region never get very high.

Two other common misconceptions are that the Sun is straight overhead at noon every day and that it rises due east and sets due west every day. In fact, the Sun is never straight overhead at any part of any day in any location north of the Tropic of Cancer or south of the Tropic of Capricorn. The location of sunrise and sunset changes with the seasons. In June, the Sun rises in the northeast and sets in the northwest, while in December it rises in

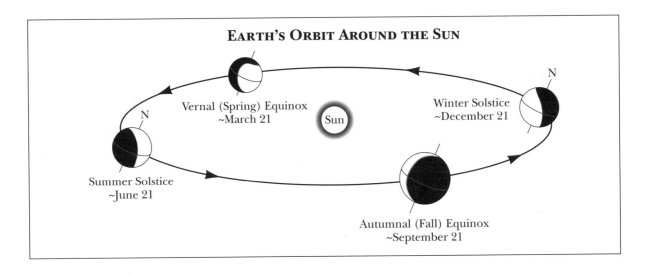

EARTH'S ORBIT AROUND THE SUN

Vernal (Spring) Equinox
~March 21

Sun

Winter Solstice
~December 21

Summer Solstice
~June 21

Autumnal (Fall) Equinox
~September 21

the southeast and sets in the southwest. The exact location is dependent upon one's latitude. The only days on which the Sun rises due east and sets due west are the vernal and autumnal equinoxes.

OTHER FACTORS AFFECTING SEASONS

Seasons are affected by other factors: altitude, nearby mountain ranges, ocean currents, and proximity to water. Mountains lose heat in a manner very similar to the way a person's fingers do on a cold day. Thus, the air at the top of a mountain is much cooler than at the bottom. A typical example is Mount Kilimanjaro in Tanzania, the bottom of which is located in a tropical rain forest but the top of which is covered by snow and ice.

The Gulf Stream in the Atlantic Ocean absorbs much solar energy as it passes through tropical areas near Florida. It then flows northeastward, carrying its warm water to Western Europe. Western Europe is thus much warmer than areas of Eastern North America at the same latitude. The same process occurs in the Pacific Ocean, where the Japan Current carries warm water to the western North American shore near the states of Oregon and Washington.

Mountain ranges often denude air currents of their moisture and deprive downwind areas of rainfall. The Cascade Mountains, for example, force moist Pacific air upward as it passes over them to the east. This causes the air to cool, water vapor to condense, and rain to fall on the windward side of the mountains. When the air gets to the other side of the mountain range, much moisture is gone, and the potential for rainfall is much lower. These factors cause western Oregon and western Washington to get much more rainfall than most places in North America but leave Idaho and Montana with very little.

Lakes and oceans heat up more slowly than land areas do. As a result, land areas near large bodies of water do not experience the extremes of heat and cold that areas farther from the water do. For example, downtown Chicago, which is at the Lake Michigan shoreline, does not get as cold on a winter night as Batavia, Illinois, which is 57 kilometers west of the Chicago lakefront. For the same reason, downtown Chicago does not get as hot on a summer afternoon as does Batavia.

Many areas near the tropics do not experience the four seasons already described but rather have an annual pattern of dry and wet seasons. In southern Asia, for example, the lower angle of the Sun in the winter causes temperatures to cool below the temperature of the nearby warm Indian Ocean. This causes winds to blow off the land toward the ocean and results in several months of fair or dry weather. By summer, the higher Sun has caused the continent to warm a little above the temperature of the ocean, and the wind pattern is reversed. Moist, warm air then flows from the ocean to the land, causing heavy rainfall; this wind system is called a "monsoon." The name also is given to the rains caused by this system.

STUDY OF THE SEASONS

The ability to predict the seasons accurately was important to early civilizations. Because of their dependence upon farming, ancient people had to know when a river would flood or when to plant their crops. Primitive people were able to note that the appearance of certain constellations, and the locations of sunrise and sunset, varied when warmer or cooler weather could be expected.

Several early civilizations, including those of Egypt, Babylon, India, and China, have left indirect evidence of astronomical writings as early as 2500 B.C.E. Stonehenge, built in southwest England about 1800 B.C.E., has a large stone that marks the direction of sunrise at the summer solstice. Medicine wheels made by Native Americans two thousand years ago also show the direction of the summer solstice sunrise.

A critical barrier in defining the seasons was the difficulty of determining the exact length of one year. For many centuries, astronomers and civil rulers tried to define the year based on a number of lunar months. The Egyptians, who needed a means to forecast the flooding of the Nile, may have been the earliest civilization to adopt a solar calendar. By the year 46 B.C.E., errors in the accepted calendar were so great that Julius Caesar redefined the year, entirely independently of the Moon, as 365 days, with one extra day added every four years. The Gregorian calendar, adopted in 1582 and still used today, corrected a small error in the Julian calendar by eliminating the extra day from would-be leap years that are divisible by one hundred but not by four hundred.

An understanding of the causes of seasonal change begins with an understanding of the shapes and relative motions of the Sun and the Earth. In the sixth century B.C.E., Pythagoras theorized that the Earth was a sphere resting in the center of the universe. Yet even though several of his students believed that the apparent movements of the Sun and the stars were caused by the Earth's movements, that explanation was rejected by many people, including Aristotle. The Greek astronomer Ptolemy confirmed Aristotle's geocentric views. His *Almagest* was the standard work on astronomy for 1,400 years.

In 1543, Nicolaus Copernicus published his explanation of the Sun-centered solar system and a new explanation for the changing of the seasons. His text shows the position of the tilted Earth in relation to the Sun at the two equinoxes and solstices.

Although Copernicus's heliocentric solar system and explanation for the seasons are commonly taught in schools today, his ideas did not win general acceptance for nearly two hundred years. Galileo Galilei's use of the telescope confirmed Copernicus's theories, as did Sir Isaac Newton's 1687 theory of gravitation. The last minor obstacle to universal acceptance of the Copernican system was the fact that the distant stars did not appear to move as the Earth revolved around the Sun. That opposition disappeared after Friedrich Bessel's 1838 discovery of stellar parallax, which confirmed that those stars do move, albeit by very small amounts.

SIGNIFICANCE

An understanding of the seasons is just as important to modern civilization as it was to ancient ones. Today, the need for an understanding of seasons is more complex but just as vital. A knowledge of seasons and the length of a growing season still affects decisions about which crops can be planted in a given area. Certain crops require longer growing seasons than others and must be grown in southern areas; some perennial flowers, such as tulips, require a cold dormant period and cannot be grown in areas without a cold winter season. Homeowners who purchase trees and other living plants from landscaping catalogs need to know about their seasonal climates in order to buy plants suited to local conditions. Knowing how high the Sun will be in the summer can help a homeowner or a landscape architect to decide in the spring where to put shade-producing trees so that they do not interfere with a Sun-loving vegetable garden.

Seasonal changes of the Sun can help the environmentally conscious architect to decide where to place windows and whether to design roof overhangs. For example, in summer, when most new houses and office buildings will likely be using air conditioning, midday sunlight streaming in windows can drive up electricity costs. To minimize this, homes and other buildings can be designed with roof overhangs that shade the windows from direct sunlight. In winter, when the buildings could use this sunlight to keep down heating costs, the Sun will be further south and thus lower in the sky. If well designed, a roof overhang can shade windows from the high summer Sun while allowing in light from the low winter Sun.

Kenneth J. Schoon

CROSS-REFERENCES

Atmospheric and Oceanic Oscillations, 1897; Climate, 1902; Clouds, 1909; Drought, 1916; Earth Resources Technology Satellite, 2754; El Niño and La Niña, 1922; Greenhouse Effect, 1867; Hurricanes, 1928; Infrared Spectra, 478; Lightning and Thunder, 1935; Monsoons, 1940; Mount Pinatubo, 761; Ozone Depletion and Ozone Holes, 1879; Remote Sensing and the Electromagnetic Spectrum, 2802; Satellite Meteorology, 1945; Storms, 1956; Tornadoes, 1964; Van Allen Radiation Belts, 1970; Volcanoes: Climatic Effects, 1976; Weather Forecasting, 1982; Weather Forecasting: Numerical Weather Prediction, 1987; Weather Modification, 1992; Wind, 1996.

BIBLIOGRAPHY

Armitage, A. *The World of Copernicus.* Wakefield, England: E. P. Publishing, 1972. Originally published in 1947 under the title *Sun, Stand Thou Still.* A description of the state of astronomy before and during the time of Copernicus. The last part deals with the conversion of scientific thought from belief in the geocentric solar system to acceptance of Copernicus's heliocentric system.

Asimov, Isaac. *The Clock We Live On.* Eau Claire, Wis.: E. M. Hale, 1965. The Earth's relation to the Sun and Moon form the basis for this detailed text about measuring seasons, years, and hours. Includes eighteen diagrams and an index.

Berry, A. *A Short History of Astronomy.* New York: Dover, 1898. Reprint, 1961. A comprehensive history of the science of astronomy. This 440-page book has excellent diagrams, including some reproduced from sixteenth century texts. Extremely well documented; includes both an index of names and a subject index.

Emiliani, Cesare. *Planet Earth: Cosmology, Geology, and the Evolution of Life and Environment.* Cambridge, England: Cambridge University Press, 1992. A large, comprehensive book containing basic information about matter and energy, many aspects of the physical and historical Earth, and a large section about the Earth's relationship to the universe. The last section is a brief history of the Earth sciences.

Fisher, Leonard Everett. *Calendar Art.* New York: Four Winds Press, 1987. A collection of stories about thirteen calendars developed by Central American, European, Middle Eastern, and East Asian civilizations. Each calendar is reproduced in its own language. Very good, in spite of an error in defining the vernal equinox. Diagrams are in blue and white.

Gay, Kathlyn. *Science in Ancient Greece.* New York: Franklin Watts, 1988. A book about the philosophers, astronomers, and mathematicians of ancient Greece and their concepts of the universe. The text is supported by black-and-white photographs and diagrams. Indexed, and has a small glossary.

Goudsmit, Samuel, and Robert Claiborne. *Time.* Alexandria, Va.: Time-Life Books, 1980. A comprehensive discussion of the elements of time, from the establishment of the calendar, to the building of clocks, to the theories of relativity and carbon-14 dating. Color and black-and-white photographs and diagrams.

Henes, Donna. *Celestially Auspicious Occasions: Seasons, Cycles, and Celebrations.* New York: Berkley, 1996. Henes' book looks at the history, rituals, and mythology surrounding seasons. Well illustrated and contains a bibliography.

Lambert, David, and the Diagram Group. *Field Guide to Geology.* New York: Facts on File, 1988. A profusely illustrated book about the Earth, its seasons, rocks, erosional forces, and geological history. Contains a list of "great" geologists (including Copernicus) and a list of geologic museums, mines, and spectacular geologic features. Indexed.

Montgomery, Sy. *Seasons of the Wild: A Year of Nature's Magic and Mysteries.* Buffalo, N.Y.: Firefly, 1995. Part of the Curious Naturalist series, Montgomery's book focuses on the seasonal changes that occur in North America throughout the course of one year. The author does an excellent job at presenting his observations and knowledge in a way that makes them interesting to readers who may not otherwise feel compelled to read about the natural history of seasons. Illustrations.

Neal, Harry Edward. *The Mystery of Time.* New York: Julian Messner, 1966. Especially useful for those interested in the historical development of the calendar. Contains sixteen pages of black-and-white prints. Bibliography and index.

Ronan, Colin. *The Practical Astronomer.* New York: Macmillan, 1981. A detailed description of basic astronomy along with ideas for experiments for the amateur astronomer. Very well illustrated with color and black-and-white photographs and diagrams. Glossary and index.

Stevens, William Kenneth. *The Change in the Weather: People, Weather, and the Science of Climate.* New York: Delacorte Press, 1999. Stevens describes various natural and human-induced causes of changes in the climate. Includes a twenty-page bibliography and an index.

STORMS

A storm is a weather phenomenon that has a specific structure and is associated often with whirling air and precipitation. Winds in a storm are often of high velocity and are frequently used to differentiate among storm stages.

WINTER STORMS

A storm is a disturbed state of the atmosphere that has an impact on the Earth's surface and that threatens potentially inclement and destructive weather. Some storms are confined to certain seasons or locations, and some occur anywhere. Among the storms in the atmosphere are winter storms, thunderstorms, tornadoes, and tropical cyclones. Winter storms can be quite severe, destructive, and life-threatening. Two types of winter storm are particularly catastrophic: the ice storm and the blizzard.

The term "ice storm" is given to rain that falls through the atmosphere in liquid form and freezes when the water comes in contact with a surface having a temperature of 0 degrees Celsius or lower. This form of precipitation is known also as freezing rain. In order for freezing rain to occur, the surface temperature of the ground must be below 0 degrees Celsius, and above-freezing temperatures must be present aloft. Snowflakes fall through the layer of warm air and are melted. The raindrops then are cooled as they pass through the colder air near the surface and freeze

quickly on impact with the freezing surface. Layers of ice, called glaze, coat streets, trees, automobiles, and power lines.

A blizzard is a strong wind, 56 kilometers per hour or higher, with temperatures lower than −7 degrees Celsius and enough snow to restrict visibility to less than 150 meters. Blizzards are associated with midlatitude cyclones. The type of cyclone most likely to produce blizzard conditions is one with a surface low pressure connected with a low pressure in the upper air at the level of the jet stream.

The area of heaviest snowfall in a cyclone is within about 200 kilometers north of the low-pressure center. The heavy snow results from moist air from the south turning counterclockwise around the low-pressure center. Farther to the south, usually along the cold front, sleet and freezing rain may occur. Sleet, like freezing rain, occurs when above-freezing temperatures are present aloft, only the raindrops freeze before reaching the surface. The strongest winds are behind the cold front, and blizzard conditions are most likely there.

(continued on page 1959)

A TIME LINE OF HISTORIC WINTER STORMS

Apr. 13, 1360	CHARTRES, FRANCE: A freak storm devastates the armies near the cathedral town of Chartres; lightning and huge hailstones strike the heavily armored English infantry and cavalry, killing people outright.
1643-1653	EUROPE: Severest winter since the Ice Age.
May 8, 1784	SOUTH CAROLINA: A deadly hailstorm hits the town of Winnsborough, Fairfield County.
July 13, 1788	FRANCE: A severe hailstorm damages French wheat crop.
June, 1816	NEW ENGLAND AND SOUTHERN UNITED STATES: Snow and temperatures just above freezing during one of the coldest summers on record, dubbed the Year Without a Summer.
Oct., 1846	SIERRA NEVADA: An early blizzard traps the Donner Party, some of whom finally turn to cannibalism in an effort to survive.
Nov. 17-21, 1798	NEW ENGLAND: Known as the Long Storm, this five-day blizzard covers New England with several feet of snow, resulting in hundreds dead.
May, 1853	INDIA: 84 people and 3,000 cattle die during a hailstorm. Another hailstorm of note occurs two years later, when, in 1855, particularly large hailstones fall on the town of Naini Tal, killing approximately 50.
Jan. 23, 1867	NEW YORK: The East River in New York City freezes.
Jan. 6-13, 1886	MIDWEST: A blizzard unexpectedly hits the region, particularly Iowa, killing 80.
Mar. 11-14, 1888	EASTERN UNITED STATES: The Great Blizzard of 1888 strikes; 400 die.
Apr. 30, 1888	MORADABAD, UTTAR PRADESH, INDIA: 246 people and 1,699 sheep and goats die during a hailstorm.
Feb. 7-8, 1891	MIDWEST: A blizzard affecting Iowa, Nebraska, South Dakota, Wyoming, Minnesota, and Wisconsin kills 23.
Nov. 26-27, 1898	NORTHEASTERN UNITED STATES: A blizzard in New York, Connecticut, and Massachusetts results in 455 estimated dead.
Feb. 10-14, 1899	U.S. EAST COAST: A blizzard introduces bitter-cold tempeartures across two-thirds of the United States, including Florida.
Jan. 27-29, 1922	U.S. EAST COAST: A blizzard particularly harmful to Washington, D.C., results in 108 dead, 133 injured.
Dec. 12, 1939	JAPAN: A blizzard kills an estimated 750.
Nov. 11-12, 1940	MIDWEST: A blizzard results in 157 dead.
Mar. 15-16, 1941	NORTH DAKOTA AND MINNESOTA: A blizzard leaves 151 dead.
Jan. 14, 1952	SIERRA NEVADA: A blizzard kills 26, leaves 226 stranded, and costs $4.25 million in damages.
Jan.-Feb., 1956	EUROPE: A blizzard from Siberia to the Mediterranean leaves 907 dead and costs $2 billion in damages.
Oct. 18, 1957	IRAQ: A hailstorm in Suleimaniyah, northern Iraq, kills 27 and leaves hundreds injured.
Feb. 17, 1962	GERMANY: Major storms blanket the country; 343 are killed.
Mar. 7, 1962	NEW ENGLAND: A blizzard rearranges the northeastern coastline, covering Ocean City, Delaware, in sand.
Jan. 29-31, 1966	EASTERN UNITED STATES: The worst blizzard in seventy years.

(continued)

Jan. 26-27, 1967	MIDWEST: Chicago residents experience the worst blizzard ever, with 24 inches of snow in 29 hours, 8 minutes.
Dec. 12-26, 1967	SOUTHWESTERN UNITED STATES: The region, especially New Mexico, endures a blizzard that leaves 51 dead.
Feb, 8-10, 1969	NORTHEASTERN UNITED STATES: The Lindsay Storm dumps 25 inches of snow on New York and Boston and dashes New York mayor John Lindsay's hopes for reelection when he is blamed for tardy snow removal.
Feb. 4-11, 1972	IRAN: Heavy snow falls on the warm-weather nation; thousands perish.
Dec. 1-2, 1974	MICHIGAN: Nineteen inches of snow falls on Detroit in the worst snowstorm in eighty-eight years.
Jan.-Mar., 1977	MIDWEST: Blizzards rage in and around Buffalo, which reports 160 inches of snow.
Jan. 25-26, 1978	MIDWEST: A major snowstorm strikes with 31 inches of snow and 18-foot drifts.
Jan. 29, 1978	SCOTLAND: 8 motorists die and 70 are rescued from a train during a blizzard.
Feb. 5-7, 1978	NEW ENGLAND: The worst blizzard in the region's history strikes the Northeast; eastern Massachusetts receives 50 inches of snow, and winds reach 110 miles per hour. All business stops for five days.
Jan. 12-14, 1979	MIDWEST: Blizzards yield 20 inches of snow, with temperatures at –20 degrees Fahrenheit; 100 die.
Feb. 18-22, 1979	EASTERN UNITED STATES: Snow blankets the District of Columbia during the President's Day Storm.
Mar. 1-2, 1980	EASTERN UNITED STATES: The mid-Atlantic region experiences a blizzard.
Feb., 1983	UNITED STATES: The Northeast and Mid-Atlantic are pummeled by a blizzard.
Feb. 5-28, 1984	GREAT PLAINS: A series of snowstorms strikes Colorado and Utah.
Mar. 29, 1984	EASTERN UNITED STATES: A snowstorm covers much of East Coast.
Oct., 1991	NORTHEASTERN UNITED STATES: The Halloween Storm is called the most extreme nor'easter in half a century.
Mar. 12-15, 1993	EASTERN UNITED STATES: From Maine to Florida, 270 are dead and 2.5 million homes without power as the result of a blizzard called the Superstorm of 1993.
May 5, 1995	TEXAS: A hailstorm in Dallas and Tarrant Counties kills 17 and causes several millions of dollars in damage.
Jan. 6-12, 1996	EAST COAST: Another big snowstorm leaves 154 dead, more than $1 billion in damages.
May 10-11, 1996	NEPAL: Mount Everest experiences wind speeds of 45 to 80 miles per hour and windchill factors of nearly –150 degrees Fahrenheit during a blizzard that kills 9 and causes severe frostbite in 4 others.
Aug., 1996	INDIA: Jammu and Kashmir endure a blizzard that kills approximately 239 pilgrims.
Jan. 1-4, 1999	U.S. MIDWEST AND EAST COAST: The Blizzard of '99 results in more than 100 dead, airline and rail traffic disrupted, hundreds of cars damaged in chain-reaction accidents, several hundred thousand homes without power.
Feb.-Mar., 1999	EUROPE: Heavy snowfall in the Alps triggers avalanches.

THUNDERSTORMS

A thunderstorm is a violent disturbance in the atmosphere and may occur along a front, especially a cold front. Associated with it is lightning, thunder, occasionally hail, and frequently heavy precipitation, although in dry climates precipitation at the surface may not occur. When temperatures in the atmosphere decrease rapidly with height, the atmosphere is unstable. Moreover, if the air is moist, a considerable amount of energy is stored inactively in the water vapor, and this energy will be released when the vapor changes to liquid water or ice. When this moist, unstable air is given an initial lift by unequal heating of the Earth's surface, a mountain range, or an advancing front, a rising air current is set in motion. As long as the rising air is less dense than the surrounding air, it will continue to rise. As the water vapor is condensed, air density is decreased, and a towering thunderhead cloud forms.

Conditions favorable for the formation of thunderstorms most often occur in warm, moist tropical air. For this reason, thunderstorms form most frequently in states bordering the Gulf of Mexico. Thunderstorms are divided into two basic types: local and organized thunderstorms. Local storms are isolated and scattered and usually short-lived. Normally they occur on warm summer afternoons near the time of the peak daily temperature. Organized thunderstorms are long-lived and occur over larger areas than do local storms. They form in rows called squall lines, along cold fronts, occasionally along warm fronts, and adjacent to mountain peaks.

The initial stage is the cumulus stage, in which the cloud is dominated by updrafts. Precipitation does not occur in this stage, but as the cloud gets larger, updrafts get stronger and more widespread. In the top of the cloud, where liquid water and ice crystals are abundant and where buoyancy is less, a downdraft is initiated. As soon as the downdraft starts, the second stage in the life cycle is reached, called the mature stage.

The most violent weather in the thunderstorm occurs during the mature stage. A strong updraft and downdraft exist, and heavy rain is produced in the downdraft side. Also associated with the downdraft are strong wind gusts at the surface. Depending on the strength of upper winds and downdraft currents, surface winds may be cool,

gentle breezes or strong blasts of air. Gradually, the downdraft spreads throughout the cloud, and the dissipating stage is reached. In this stage, the storm is characterized by weak downdrafts and light rain.

Hazards other than strong winds and heavy rain associated with thunderstorms are lightning and hail. Lightning occurs because of the separation of positive and negative charges within clouds, between clouds, and between the clouds and the ground. When the power of a charge is greater than the insulating effect of the air, a lightning stroke results. Thunder is caused by a rapid expansion of air as lightning, which is several times hotter than the Sun's surface, passes through it. Hail is formed as an ice crystal is buffeted about in a cloud and successive layers of supercooled water (liquid water below freezing) are added to the nucleus. The eventual size of a hailstone depends on the length of time and passage through the cloud.

TORNADOES AND HURRICANES

Tornadoes are also associated with severe thunderstorms. They are small, powerful storms usually less than 0.5 kilometer in diameter but, at times, may extend for 1 kilometer or more. A tornado may have the shape of a funnel, rope, or cylinder extending from the base of a thunderstorm cloud. The tone of the tornado depends on its background, the debris, and condensed moisture within it. A blue sky behind a tornado makes it appear dark, whereas intense rain behind it makes it look white. The funnel will appear dark when it is filled with debris and dust sucked up from the surface.

The motion of a tornado is variable, averaging 65 kilometers per hour, but tornadoes have been observed to travel as fast as 110 kilometers per hour. Inside, the storm winds, which almost always turn in a counterclockwise direction, may whirl around the center in excess of 500 kilometers per hour. Within a tornado are smaller, more intense vortices called suction vortices. There may be anywhere from one to three such vortices in a tornado. When there is more than one suction vortex, they rotate around a common center and account for total destruction in one area, while several meters away there is no damage.

Tornadoes occur in many parts of the world,

but the topography and pressure patterns of the central and eastern United States are especially suited to tornado formation. The Rocky Mountains block winds from the west. High pressure in the Atlantic at 30 degrees north latitude causes warm, moist air from the Gulf of Mexico to flow northward. The jet stream, flowing over the mountains, is colder and drier, making for an extreme vertical contrast in the two air masses. When unstable air rises, the contrast provides conditions for an explosive upward movement of air. Another aspect of the tornado is the low pressure in the center. Pressure here often approaches 800 millibars, whereas average sea-level pressure is 1,013 millibars.

Tornado occurrence follows the seasonal migration of the jet stream. In winter, the jet stream is near the Gulf of Mexico, and Gulf states are most prone to tornadoes. In spring, the southern plains states are most likely to experience tornadoes, and in late spring and summer, the northern plains and eastern United States are most prone. May is the peak month of occurrence, and Texas, because of its size, has more tornadoes than does any other state. When storms are averaged over area, however, Oklahoma ranks first.

Hurricanes are tropical cyclones that occur in the Atlantic Ocean and Gulf of Mexico. A hurricane goes through a four-stage development. Stage one is the tropical disturbance, where a low-pressure center has some clouds and precipitation but no enclosed isobars (lines connecting points of equal pressure) and only light winds. The second stage is the tropical depression, with lower pressure in the center and at least one enclosed isobar but with winds less than 60 kilometers per hour. Third is the tropical storm stage, with winds between 60 and 120 kilometers per hour around a low-pressure center with several enclosed isobars. The storm is given a name in this stage. Fourth is the tropical cyclone or hurricane stage, with pronounced rotation around a central core or eye. Winds are in excess of 120 kilometers per hour, circling as bands in toward the center. Hurricanes form over warm oceans and derive most of their energy from water with temperatures greater than 26.5 degrees Celsius. These storms form only in late summer and fall and outside 5 degrees latitude. They have no fronts and are smaller than a midlatitude cyclone,

while central pressures are lower and winds are stronger.

STUDY OF STORMS

One of the first major investigations into the nature of thunderstorms was undertaken in 1946-1947 during what was called the Thunderstorm Project. The project consisted of flying instrumented aircraft through thunderstorms to obtain various kinds of data. The study also was augmented by collection of data from instruments such as radar, radiosondes, and ground-based instruments. Much was learned about thunderstorm structure, internal activity, and life cycle from this investigation. It was largely through the Thunderstorm Project that the stages in a thunderstorm's life cycle were identified; that is, cumulus, mature, and dissipating stages were recognized. Also, aircraft are used to fly through hurricanes for data collecting.

One device used in data collection in the upper atmosphere is the rawinsonde, which is a package of weather instruments attached to a balloon that is sent aloft. The rawinsonde has instruments to record temperature and humidity, a reflector for collecting wind data by a ground-based radar, and a transmitter to send data to recorders in the weather station. Rawinsonde stations are spaced several hundred kilometers apart and are released only twice daily at 0000Z and 1200Z (the time in Greenwich, England). Thus, the rawinsonde is limited, because data collected in this manner are insufficient to make precise forecasts of severe thunderstorms or tornadoes hours in advance.

Another valuable meteorological tool is radar, which can be used to detect and observe storms hundreds of kilometers away from a station. A radar set sends out pulses of radio waves through an antenna that rebound from objects that include raindrops, cloud drops, ice crystals, and hail. A radio receiver intercepts the returning pulses between transmitted pulses, and these are recorded.

Doppler radar is being used increasingly as a forecasting tool, especially for severe thunderstorms and tornadoes. Doppler radar sends out continuous radio waves instead of pulses, so it can be used to measure speeds of objects moving toward or away from the radar antenna. Targets such as cloud drops, raindrops, and other liquid or solid particles reflect radio waves, so air movement can be discerned. Thus, air that is whirling within a storm can be detected. Doppler radar can identify and locate a tornado within a thunderstorm as much as fifteen to twenty minutes before the funnel touches the ground. A network of Doppler radars is in place in the United States.

Satellites are another useful tool for detecting and observing middle-latitude cyclones, thunderstorms, and hurricanes. Several meteorological satellites have been placed in orbit over the years, beginning with the television infrared observation satellites (TIROS). Satellites such as the applications technology satellites (ATS) and synchronous meteorological satellites (SMS) are in geosynchronous orbit, which means they make one revolution of the Earth in a twenty-four-hour period so that to an earthbound observer they appear to be stationary. These satellites sense images on and above the Earth and send coded information back to weather stations. The information is transformed into images that are frequently shown on television. Images are taken in both visible and infrared wavelengths. Both wavelengths can produce visible images, but infrared senses temperature differences of objects. These images provide information on midlatitude cyclones, thunderstorms, and hurricanes.

Significance

Storms occur somewhere in the country nearly every day. They can be quite destructive and often are life-threatening, although at times they can be beneficial. One of the most dangerous aspects of storms is lightning. Lightning deaths are single occurrences, however, and do not draw the attention that the often multiple deaths resulting from tornadoes or hurricanes do.

Thunderstorms, on an annual basis, can cause more cumulative damage than other storms simply because there are so many of them. In addition to lightning, thunderstorms produce heavy downpours, which can cause flooding. Hailstorms are also associated with thunderstorms. Hail destroys crops; damages buildings by breaking windows, tearing window screens, and perforating roofing shingles, causing roofs to leak; and dents automobile bodies. Strong gusts of wind on the downdraft side of a thunderstorm often do appreciable damage to vegetation and buildings.

Downbursts—air that rushes out of a thunderstorm downdraft and spreads out laterally near the surface—are particularly hazardous to aircraft taking off or landing and have caused several planes to crash.

Tornadoes are most often associated with severe thunderstorms, although they also occur in conjunction with hurricanes. They are difficult to forecast and appear quickly, without warning. The United States has far more tornadoes annually than does any other country, averaging between seven hundred and one thousand such storms each year. Deaths from tornadoes average more than one hundred per year, and damage is in the millions of dollars. Tornadoes migrate with the jet stream, and the month of May has more tornadoes than any other, although June has more days on which tornadoes occur.

Hurricanes are normally limited to late summer and fall, although they have occurred as late as December. They are not as intense as are tornadoes but, being larger, cause more damage. They are restricted to the Gulf of Mexico and the Atlantic coast and on occasion ravage the California coast from the Pacific. The most damaging aspect of a hurricane is the tidal surge that accompanies it. Surges have been as high as 75.5 meters (Hurricane Camille) and have carried floods inland for 6 kilometers or more. High-velocity winds also cause considerable damage, and the damage is more widespread than is surge damage.

Ice storms occur most often in the eastern and southern states. They are responsible for power outages when power lines break as the result of accumulated glaze. Ice storms are also very hazardous to transportation, especially automobiles. Blizzards occur in the northern United States. Blizzard winds may reach hurricane force, and because of drifting snow and low visibility, travelers are often stranded. Because of the cold temperatures, people lost in blizzards frequently freeze to death.

Ralph D. Cross

Cross-References

ning and Thunder, 1935; Monsoons, 1940; Satellite Meteorology, 1945; Seasons, 1951; Solar Power, 1674; Tornadoes, 1964; Van Allen Radiation Belts, 1970; Volcanoes: Climatic Effects, 1976; Weather Forecasting, 1982; Weather Forecasting: Numerical Weather Prediction, 1987; Weather Modification, 1992; Wind, 1996.

BIBLIOGRAPHY

Battan, Louis J. *Fundamentals of Meteorology.* 2d ed. Englewood Cliffs, N.J.: Prentice-Hall, 1984. A thorough, nontechnical treatise on severe storms is presented in chapter 11. All types of severe storms from thunderstorms to hurricanes are covered. Moreover, a discussion of storm forecasting is included, and particularly of note is the segment on the use of radar in storm forecasting. The chapter is well illustrated with maps, diagrams, and photographs.

Bluestein, Howard B. *Tornado Alley: Monster Storms of the Great Plains.* New York: Oxford University Press, 1999. Bluestein has compiled excellent color photographs that capture the power of storms and tornadoes in the American Midwest. He describes the hazards involved with these storms, as well as the factors that cause them.

Eagleman, Joe R. *Meteorology: The Atmosphere in Action.* New York: Van Nostrand Reinhold, 1980. A thorough coverage of storms, from frontal cyclones to tornadoes. The author devotes one-third of this meteorology book to the discussion of storms. Storm formation, development, and characteristics are detailed in a nontechnical fashion. Moreover, the excellent illustrations make the discussions vividly clear, so there is no problem understanding the material.

_____. *Severe and Unusual Weather.* 2d ed. Lenexa, Kansas: Trimedia, 1990. An excellent book dealing with severe weather of all kinds. The principles of storm formation are well presented, and descriptions of storms and how they form are clear and concise. The book is illustrated with photographs and explanatory diagrams. Suited to entry-level college students.

Erickson, Jon. *Violent Storms.* Blue Ridge Summit, Pa.: TAB Books, 1988. The title of the book is misleading, as only two chapters deal with severe storms. Another six chapters, however, cover subject matter closely related to storms and provide good background material. The book is written on a general level and should be understood easily by the layperson. Illustrated with diagrams and black-and-white and color photographs.

Gibilisco, Stan. *Violent Weather, Hurricanes, Tornadoes, and Storms.* Blue Ridge Summit, Pa.: TAB Books, 1984. An excellent book on storms of all kinds. The text is nontechnical but goes into considerable detail on storm types. The narrative is written in a style easily understood: suitable for the layperson, yet rigorous enough to be included on college reading lists. Illustrated with diagrams and black-and-white photographs.

MacGorman, Donald R., and W. David Rust. *The Electrical Nature of Storms.* New York: Oxford University Press, 1998. This college-level text describes the atmospheric conditions and electrical conditions that create thunderstorms, hailstorms, lightning, tornadoes, and hurricanes. The authors also discuss the ways in which violent weather affects the environment and humans.

Navarra, John Gabriel. *Atmosphere, Weather, and Climate: An Introduction to Meteorology.* Philadelphia: W. B. Saunders, 1979. A good coverage of atmospheric storms is presented in chapter 9. The writing is clear and easily comprehendible and nontechnical. Suited to the layperson as well as the college student. The author covers both storms and severe storms and includes discussions of phenomena associated with storms, such as lightning, thunder, and squalls.

Ross, Frank, Jr. *Storms and Man.* New York: Lothrop, Lee and Shepard, 1971. This book is suitable as a high-school-level introduction to storms. An introductory chapter covers the nature of storms, winds, and clouds. The next six chapters present a general analysis of storm systems, and a concluding chapter covers storm control. Lavishly illustrated with black-and-white photographs and a few diagrams.

Weisberg, Joseph S. *Meteorology: The Earth and Its Weather.* Boston: Houghton Mifflin, 1981. Chapter 8 presents a nontechnical analysis of major storm types. Also included is a discussion of waterspouts, whirlwinds, blizzards, and cold waves. The how and why of the formation and growth of storm types are described. In addition, the author includes, at the end of the chapter, lists of tips for surviving tornadoes, lightning storms, hurricanes, and winter storms. Suitable for entry-level college students.

TORNADOES

Tornadoes are small, violent, rotating storms that may produce devastating wind velocities of more than 400 kilometers per hour. A strong tornado can demolish well-built structures, and a person in the tornado's path should follow important safety precautions to avoid becoming a victim of the storm.

PRINCIPAL TERMS

COLD FRONT: the contact between two air masses when a bulge of cold, polar air surges southward into regions of warmer air

CORIOLIS EFFECT: a phenomenon in which, because of the Earth's rotation, objects in motion in the Northern Hemisphere appear to be deflected to the right; in the Southern Hemisphere, the deflection is to the left

CUMULONIMBUS CLOUD: the tall, billowy variety of cloud with precipitation falling

DUST DEVIL: a rotating column of rising air, made visible by the dust it contains; they are smaller and less destructive than tornadoes, having winds of less than 60 kilometers per hour

HURRICANE: a huge, tropical low-pressure storm system with sustained winds in excess of 118 kilometers per hour

SQUALL LINE: a line of vigorous thunderstorms created by a cold downdraft that spreads out ahead of a fast-moving cold front

UNSTABLE AIR: a condition that occurs when the air above rising air is unusually cool so that the rising air is warmer and accelerates upward

VORTEX: any rotating current in a fluid; there are vortices in water (whirlpools) and air (whirlwinds)

WATERSPOUT: a tornado over water; less violent and smaller waterspouts form in fair weather just as dust devils do over dry land

TORNADO FORMATION AND CLASSIFICATION

Tornadoes are relatively small, localized low-pressure areas associated with powerful cumulonimbus clouds. For its size, the tornado is the most violent of the whirlwinds. The "typical" tornado is 250 meters in diameter, with whirling winds of 240 kilometers per hour. The twisting funnel cloud travels at 65 kilometers per hour over the surface and lasts ten minutes, moving, in North America, along a northeasterly track. Very large, devastating tornadoes are relatively rare but have almost unbelievable destructive power. The Tri-State Tornado of March 18, 1925, touched down near Ellington, Missouri, at 1:00 P.M. and ripped a trail of havoc for 352 kilometers across southern Illinois; it finally broke up at 4:30 P.M. near Princeton, Indiana. The storm killed 695 people and injured 2,027. Damage, calculated in 1970 dollars, was $43 million. Compared with other tornadoes, it raced along the ground, averaging well over 100 kilometers per hour.

Tornadoes are most consistently associated with fast-moving cold fronts that sweep across the midsection of the United States, drawing warm, moist, tropical air from the Gulf of Mexico. The cold front is usually associated with a strong low-pressure storm system that rotates counterclockwise as it swirls across land in the prevailing westerly wind pattern. The counterclockwise rotation of the low-pressure system brings cold air in behind, which wedges underneath the warm Gulf air that is drawn in ahead of the storm center. When heavy, cold air wedges under the less dense warm, tropical air, the warm air is forced to rise. If the air is unstable, the cloud will accelerate upward, making a towering thundercloud. The upward surge stops only when the cloud has penetrated all the excessively cold upper air. If the cloud tops can penetrate above 11 kilometers, severe storms, including supercells that spawn tornadoes, are possible. These high cloud tops indicate unstable air at abnormal heights, and occasionally the unstable conditions can drive storm cloud tops to 20 kilometers or higher. A high zone of strong wind from the west (the "jet stream") tends to increase the chance for violence when associated with a storm system.

Tornadoes are classified by maximum wind velocity, which occurs on the skin of the spinning funnel. T. T. Fujita, of the University of Chicago, developed a scale from 1 to 5, or weakest to strongest. Tornadoes consisting of winds below 116 kilometers per hour are ranked "F0" (the F standing for Fujita) and incur slight damage (broken street signs and branches); F1 tornadoes have winds of

116 to 180 kilometers per hour and cause trees to snap and windows to break; F2 tornadoes have winds of 181 to 253 kilometers per hour and cause considerable damage, uprooting trees; F3 tornadoes have winds of 254 to 332 kilometers per hour and cause severe damage, flipping over cars or knocking down walls; F4 tornadoes, with winds of 333 to 419 kilometers per hour, are devastating, destroying frame houses; and finally, the strongest of tornadoes, with wind velocities above 419 kilometers per hour, are capable of destroying steel-reinforced buildings and are designated F5. More than one-half of all tornadoes reported are in the F0 to F1 range of intensity; however, nearly 70 percent of all fatalities are caused by tornadoes of F4 to F5 intensity. Only 2 percent of the storms are in the upper range. In the years from 1960 to 1985, the average number of tornado deaths in the United States was 88 per year.

Multiple tornadoes can occur when the weather conditions are ideal for severe weather. The worst twenty-four hours on record was April 3 to April 4, 1974, when a remarkable 148 tornadoes struck eleven states, centered on Kentucky. Canada also reported tornadoes during that episode. The swarm of funnels claimed 300 lives and left 5,500 injured.

OCCURRENCE AND DAMAGE

Tornadoes have been reported in each of the fifty states, but they are rare in Alaska and the mountainous portions of the western United States. From 1953 to 1976, only one tornado was reported in Alaska, and only fourteen from the whole state of Nevada. No doubt more twisters occurred in these sparsely inhabited states, but conditions generally are not favorable for tornadoes in these general areas. Central Oklahoma has the highest incidence of tornadoes, with more than nine twisters annually for every 26,000 square kilometers. For the whole state, the average is about fifty per year. The area from Fort Worth, Texas, to northwestern Nebraska has been dubbed Tornado Alley because the storms develop there so consistently. The four-state zone has more than five tornadoes per year for each 26,000 square kilometers. Other particularly vulnerable areas include the Texas Panhandle near Lubbock, northeast Colorado, parts of Iowa, most of Illinois and Indiana, most of the Florida peninsula, parts of

---

FUJITA TORNADO INTENSITY SCALE

CATEGORY F0: Gale tornado (40-72 mph); light damage. Some damage to chimneys; break branches off trees; push over shallow-rooted trees; damage to sign boards.

CATEGORY F1: Moderate tornado (73-112 mph); moderate damage. The lower limit is the beginning of hurricane wind speed; peel surface off roofs; mobile homes pushed off foundations or overturned; moving autos pushed off the roads.

CATEGORY F2: Significant tornado (113-157 mph); considerable damage. Roofs torn off frame houses; mobile homes demolished; boxcars pushed over; large trees snapped or uprooted; light-object missiles generated.

CATEGORY F3: Severe tornado (158-206 mph); Severe damage. Roofs and some walls torn off well-constructed houses; trains overturned; most trees in forest uprooted; heavy cars lifted off ground and thrown.

CATEGORY F4: Devastating tornado (207-260 mph); Devastating damage. Well-constructed houses leveled; structure with weak foundation blown off some distance; cars thrown and large missiles generated.

CATEGORY F5: Incredible tornado (261-318 mph); Incredible damage. Strong frame houses lifted off foundations and carried considerable distance to disintegrate; automobile sized missiles fly through the air in excess of 100 yards; trees debarked; incredible phenomena will occur.

SOURCE: National Oceanic and Atmospheric Administration; URL: http://www.outlook.noaa.gov/tornadoes.

Georgia, and most of southern Alabama. The United States accounts for 75 percent of all the global tornadoes.

Conditions that favor the formation of tornadoes include broad flatlands with no obstructions to the flow of surface wind; an elevation near sea level to allow a full thickness of the atmosphere for the development of towering clouds; a position on a large continent where very cold air from the north can be swept into a low-pressure storm system that has access to hot, humid tropical air to the south; a southward bulge of strong jet stream currents aloft; and springtime weather patterns that provide intense low-pressure systems that can penetrate rather close to the Gulf of Mexico coast of the United States. March through July is the peak season, with May the most tornado-prone

month. Winter tornadoes are mostly confined to the Gulf of Mexico coast, and the frequency moves north and swings toward Kansas as springtime progresses. By July and August, the area of tornado danger spreads through the northeastern states and into southern Canada. Oklahoma is in or near the worst areas most of the year. Most of the storms occur during the late afternoon, at the climax of daily heating, although a really violent frontal advance occasionally will generate night and early-morning tornadoes.

Some of the damage caused by tornadoes results from the rapid passage of low pressure. Most houses are built to withstand downward pressure from much water, snow, or wind against the structure, especially weight on the roof. When a tornado passes over a house, however, the low pressure above, countered with high pressure inside that cannot leak out quickly enough plus wind pressure under the eaves, causes the house to appear to "explode" from within. A rapid pressure drop of 10 percent would give the pressure inside a house a lifting force of nearly 1 metric ton per square meter. The roof is lifted slightly off the supporting walls, which, in turn, fall outward. The roof then drops back onto the interior of the house or blows away.

Whirlwinds can develop multiple vortices. Around the core of the funnel cloud, it is possible for high-speed "suction spots" to develop that might have wind velocities 100 kilometers per hour faster than the average velocity of the whirling funnel. Tornado paths observed in open ground often have a swirling pattern of streaks or scratches in the soil. On a larger scale, these suction traces match the paths of greatest destruction along the tornado path. Even in the smaller dust devils, multiple vor-

THE 25 DEADLIEST U.S. TORNADOES

Date	Location	Deaths
May 6, 1840	Natchez, Mississippi	317
Apr. 18, 1880	Marshfield, Missouri	99
Mar. 27, 1890	Louisville, Kentucky	76
May 27, 1896	St. Louis, Missouri	255
June 12, 1899	New Richmond, Wisconsin	117
May 18, 1902	Goliad, Texas	114
June 1, 1903	Gainesville, Holland, Georgia	98
May 10, 1905	Snyder, Oklahoma	97
Apr. 24, 1908	Amite, Louisiana/Purvis, Mississippi	143
Apr. 24, 1908	Natchez, Mississippi	91
Mar. 23, 1913	Omaha, Nebraska	103
June 26, 1917	Mattoon, Illinois	101
Apr. 20, 1920	Starkville, Mississippi	88
June 28, 1924	Lorain, Sandusky, Ohio	85
Mar. 18, 1925	Missouri, Illinois, Indiana	689
May 9, 1927	Poplar Bluff, Missouri	98
Sept. 29, 1927	St. Louis, Missouri	79
Apr. 5, 1936	Tupelo, Mississippi	216
Apr. 6, 1936	Gainesville, Georgia	203
June 23, 1944	Shinnston, West Virginia	100
Apr. 9, 1947	Woodward, Okalahomo	181
May 11, 1953	Waco, Texas	114
June 8, 1953	Flint, Michigan	115
June 9, 1953	Worcester, Massachusetts	90
May 25, 1955	Udall, Kansas	80

NOTE: "Deadliest" is defined by number of persons dead, which for earlier years should be regarded as estimates. Having occurred before the era of comprehensive damage surveys, some of these events may have been composed of multiple tornadoes along a damage path.
SOURCE: National Oceanic and Atmospheric Administration. URL: http://www.spc.noaa.gov/faq/tornado/killers.html

tices have been observed. Two or more small dust columns may rotate around the perimeter of the central column of the dust devil. Hurricane Celia, which struck near Corpus Christi, Texas, in 1971, approached landfall with only 145-kilometer-per-hour winds, but the core (eye) of the storm broke into multiple vortices at Corpus Christi, and the damage was more typical of winds of 250 kilometers per hour. The hurricane was expected to cause minor damage, but the center behaved more like a cluster of F2-scale tornadoes and caused extensive damage, even to well-built structures.

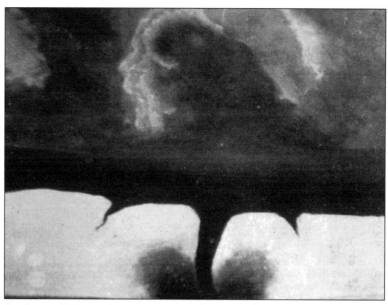

Southwest of Howard, South Dakota, on August 28, 1884, a tornado stirs the ground in this oldest known photograph of a tornado. (National Oceanic and Atmospheric Administration)

STUDY OF TORNADOES

Although meteorologists are quite successful at predicting the general region where tornadoes are likely to occur, they have not been so successful at measuring the wind velocities and air pressure in tornadoes. The storms are so small and so violent that it is nearly impossible to get instruments in the direct path and have the instruments able to survive passage of the funnel. The few weather stations that have been hit by tornadoes were destroyed or lost power to record the data. Barographs have recorded a drop in pressure of about 1 centimeter of mercury in about thirty seconds during the passage of tornadic storms.

Many of the best measurements of tornadoes are from indirect methods, such as calculating the wind velocity required to cause the kinds of destruction observed. Close estimations of wind velocity can be calculated from steel towers that have been toppled or from railroad boxcars that have been tipped over. Reinforced concrete grain silos have been ripped apart, and walls have collapsed in enough storms to get a large collection of approximate values. Surprisingly, a fairly reliable set of data involves the penetration of straw or splinters into wood surfaces. These projectiles are so small that they are quickly accelerated up to the velocity of the wind, and they are so common that most storms will hurl vast numbers of them at a variety of fixed targets. Experiments reported in 1976 provide many examples of the penetration of straw and toothpicks into all varieties of wood, both wet and dry. The data indicate that a velocity of 30 meters per second (108 kilometers per hour) is sufficient to drive a toothpick into soft pine. Broom straws need about twice as much velocity as do toothpicks to penetrate wood.

Meteorologists continue to improve their ability to forecast, locate, and track tornadoes. Space satellites, a worldwide network of manned weather stations and sophisticated computer systems, enable meteorologists to "see" weather as it develops. Balloons carry instrument packages aloft twice a day from about 90 of the 250 weather stations of the continental United States. The balloons give wind, humidity, temperature, and pressure data for all of the lower atmosphere where storms develop. Forecasters can determine which levels of the atmosphere are unstable and where the moist air is likely to be forced aloft with conditions that can generate violent storms. When dangerous storms begin to develop, Doppler radar is available in most parts of the nation. Doppler radar detects the wind component along the radar beam, then examines the pattern of the wind field to find locations of potential tornadoes. The radar helps to spot hail formation while other instruments monitor the location and frequency of lightning. All these clues to violent storms focus attention on the parts of storms that might spawn tornadoes. Critical information can be relayed to news services, which, in turn, can warn citizens who are in danger.

SIGNIFICANCE

There is no end to the documented stories of strange phenomena caused by tornadoes. Many instances are recorded where heavy boards have been punched through steel plate or metal pipes. Automobiles have been pounded and rolled into battered wads of twisted steel by the storms, and some autos have been thrown into upper stories of buildings. Wire fences have been ripped up and wound into prickly balls up to 16 meters in diameter. Strange objects, such as frogs, trash barrels, or blankets, may be picked up in one area by a tornado and deposited many kilometers away. A survivor of a tornado near Scottsbluff, Nebraska, reported seeing a head-sized boulder whirling around his car after the funnel engulfed the auto and its two occupants. After witnessing the flying boulder, the man was hurled from his car and nearly killed, regaining consciousness in a hospital. His passenger, also ejected from the car, died. The auto was destroyed and deposited in a nearby field.

The National Weather Service usually can predict severe weather regions several hours in the future, but the exact location of a tornado must wait for a visual sighting or the occurrence of tornado signature on Doppler radar. When threatening storms develop, it is wise to monitor local weather broadcasts and to keep a lookout for the characteristic funnel cloud. Often the twister is causing major damage long before the dust swirl on the ground ever connects with the descending visible funnel. Usually a tornado will travel toward the northeast along the ground at about 60 kilometers per hour. The storms do not alter course very much, and if the funnel appears to be heading toward the observer, it might be possible to leave the area of greatest danger by moving away in a direction perpendicular to the path of the storm. Attempting to escape a tornado, however, is often more dangerous than taking some protective measures, because auto drivers trying to flee tornadoes are often involved in serious accidents. Conditions of traffic, congestion, and available time should be considered before attempting to run from a twister. Those in buildings should try to get to lower floors and in narrow or confined inside corridors. Above all, windows should be avoided: Many injuries from tornadoes are from flying glass. Taking shelter under heavy tables or in a sturdy tub can prevent some injury from falling beams or masonry. Tornadoes tend to stay on or even above the ground surface, so a depression, pit, culvert, gutter, or ditch may provide safety. A deep storm cellar with a latched door provides excellent protection. A mobile home is one of the worst places for shelter from a tornado. Automobiles are not safe; cars are easily overturned and are often beaten into shapeless masses by tornadoes. Flying debris causes most of the injuries in tornadoes, and one's shelter should include protection from tumbling containers, planks, sections of fence, branches, splinters of glass, and other items that could be ripped loose by the storm.

Dell R. Foutz

CROSS-REFERENCES

BIBLIOGRAPHY

Battan, Louis J. *Fundamentals of Meteorology.* 2d ed. Englewood Cliffs, N.J.: Prentice-Hall, 1984. A standard beginning college text that assumes no background in mathematics. The book has nine pages on tornadoes, including several important illustrations. The index is thorough, as is the glossary and bibliography.

_____. *The Nature of Violent Storms.* Garden City, N.Y.: Doubleday, 1961. Reprint. Westport, Conn.: Greenwood Press, 1981. This pocket-sized paperback has twenty-four pages on tornadoes, plus nine black-and-white photographs of tornadoes and their aftermaths.

_____. *Weather in Your Life.* San Francisco: W. H. Freeman, 1983. The index for this 230-page paperback lists tornadoes on only eleven pages, but the author is an easily understood expert in the field. As a research professor of meteorology, the author has much experience to draw from and practical experience explaining weather subjects to beginners.

Bluestein, Howard B. *Tornado Alley: Monster Storms of the Great Plains.* New York: Oxford University Press, 1999. Bluestein has compiled excellent color photographs that capture the power of storms and tornadoes in the American Midwest. He describes the hazards involved with these storms, as well as the factors that cause them.

Crump, Donald J., ed. *Powers of Nature.* Washington, D.C.: National Geographic Society, 1978. This two hundred-page, colorful, special publication of the society deals with the more spectacular weather and geologic phenomena of the world. The fourteen pages on tornadoes include some remarkable photographs and a review of some of T. T. Fujita's research.

Lutgens, Frederick K., and Edward J. Tarbuck. *The Atmosphere: An Introduction to Meteorology.* 7th ed. Upper Saddle River, N.J.: Prentice Hall, 1998. A beginning college textbook. The authors have other introductory Earth science books that are very understandable for the beginner. This profusely illustrated book has twelve pages devoted to tornadoes. Contains a glossary and seven separate appendices to explain everything from metric conversions to the reading of daily weather charts.

MacGorman, Donald R., and W. David Rust. *The Electrical Nature of Storms.* New York: Oxford University Press, 1998. This college-level text describes the atmospheric conditions and electrical conditions that create thunderstorms, hailstorms, lightning, tornadoes, and hurricanes. The authors also discuss the ways in which violent weather affects the environment and humans.

National Research Council. *Severe Storms.* Washington, D.C.: National Academy of Sciences, 1977. This seventy-seven-page pamphlet of the Committee on Atmospheric Sciences is not particularly significant for tornado information except that it includes, in a pocket, a 50-by-40-centimeter chart of the 1974 outbreak that had 148 tornadoes. Prepared by T. T. Fujita, it is a colorful map that documents all the significant data observed from that stormy event.

Rosenfeld, Jeffrey O. *Eye of the Storm: Inside the World's Deadliest Hurricanes, Tornadoes, and Blizzards.* New York: Plenum Trade, 1999. This well-illustrated text is filled with accounts of storms and their powers of destruction. Rosenfeld focuses much attention on the aftermath of deadly storms but also does a nice job explaining the conditions that create weather systems of such power.

Weisberg, Joseph S. *Meteorology: The Earth and Its Weather.* Boston: Houghton Mifflin, 1981. This well-illustrated text for a broad subject has only seven pages on tornadoes, but the book is a standard for beginners. Indexed and contains several appendices, plus a bibliography. Several editions are available.

VAN ALLEN RADIATION BELTS

The Van Allen belts trap regions of high-energy charged particles that surround the Earth and that are contained within the Earth's magnetic field. The particles that make up the inner belt are energetic protons, while the outer belt consists mainly of electrons and is subject to day-night variations.

PRINCIPAL TERMS

ELECTRONS: negatively charged subatomic particles

MAGNETIC FIELD: magnetic lines of force that are projected from the Earth's interior and out into space

PLASMA: a state of matter consisting of ionized gasses

PROTONS: positively charged subatomic particles

CHARACTERISTICS

The Van Allen radiation belts consist of toroidal, or doughnut-shaped, structures that exist within the Earth's magnetic field. The inner belt begins at an altitude which varies between 250 kilometers and 1,200 kilometers, depending on latitude, and extends to an altitude of about 10,000 kilometers. This inner belt is made up of energetic protons. Within the inner belt is another belt consisting of ionized nitrogen, neon, and oxygen believed to have originated as neutral atoms in the interstellar medium that penetrated into the solar system. Before entering the Earth's magnetic field, these neutral atoms were ionized by solar ultraviolet radiation. The outer belt extends from the boundary of the inner belt to about 60,000 kilometers. During times of extensive solar activity, it may expand outward to more than 80,000 kilometers. The outer belt consists mainly of electrons.

The particles in the Van Allen belts spiral in a corkscrew-shaped path along the Earth's magnetic lines of force. The spirals are small compared to the scale of the Earth's magnetic field, and they curve to follow the field lines. As the lines of force converge toward a pole, the field becomes more intense. As a result, the particles travel in a tighter spiral as they approach the pole. Eventually, the converging tubes of magnetic force will cause the particles to be reflected back toward the equator and on to the opposite pole. The point of closest approach to a pole is the "mirror point" of the particle. The transit time between mirror points is about one second.

In addition to bouncing back and forth between mirror points, the particles undergo a slow lateral drift. The basic curvature of the path of a spiraling particle depends upon the local strength of the magnetic field. The particles in the inner belt have a tighter spiral, as the field nearest the Earth is stronger. Since there is a slight difference in the curvatures of the spirals in the inner and outer belts, the particles drift laterally. Because the charges of protons and electrons are opposite, they spiral in opposite directions. The drift direction is also opposite; the protons tend to drift westward and the electrons eastward. This drift leads to a uniformity of the radiation belts.

At the end of each path, some particles penetrate through the mirror point and descend into regions of higher atmospheric density. These particles interact with particles of atmospheric gases. Particles of radiation lose energy as a result of these collisions, and after a period of days or weeks, they are lost to the lower atmosphere, only to be replaced by more particles from the Sun.

A large influx of particles from the Sun will, in turn, cause large numbers of particles to be "dumped" into the upper atmosphere. The resulting interactions with oxygen and nitrogen atoms produce the colorful auroras. In the Northern Hemisphere, this phenomenon is called the "aurora borealis" or "northern lights"; in the Southern Hemisphere, the display is called the "aurora australis" or "southern lights."

The auroral displays are usually pink, blue, and green streaks or curtains of light. The emission of light is the result of collisions between particles of radiation from the Van Allen belts and atoms of gas in the atmosphere. When a particle of radiation strikes an atom of gas, the orbiting electrons of the atom absorb the energy of the collision. They then jump to a higher energy level. After remaining there for only a fraction of a second, the electrons fall back to the lowest energy level, or the "ground state." When they do this, a burst of light is released that represents the difference in energy between the excited state and the ground state.

FORMATION

The Van Allen radiation belts exist because of three natural phenomena: the solar wind, cosmic rays, and the Earth's magnetic field. As a result of the enormous heat in the Sun's upper atmosphere, atoms of gas are given enough energy to escape the Sun's gravity and move off into space. The solar wind, as this phenomenon has become known, consists of a plasma with a temperature near 100,000 Kelvin. A plasma contains charged particles, including protons, electrons, and ions of heavier elements, mainly helium. After the plasma escapes from the Sun's upper atmosphere, it flows outward into space. Since the plasma conducts heat well, the temperature remains high even after it has traveled a great distance from the Sun. The velocity of the solar wind increases as it expands radially outward. The speeds are near 300 kilometers per second at a distance of 30 solar radii from the Sun and nearly 400 kilometers per second at the distance of the Earth.

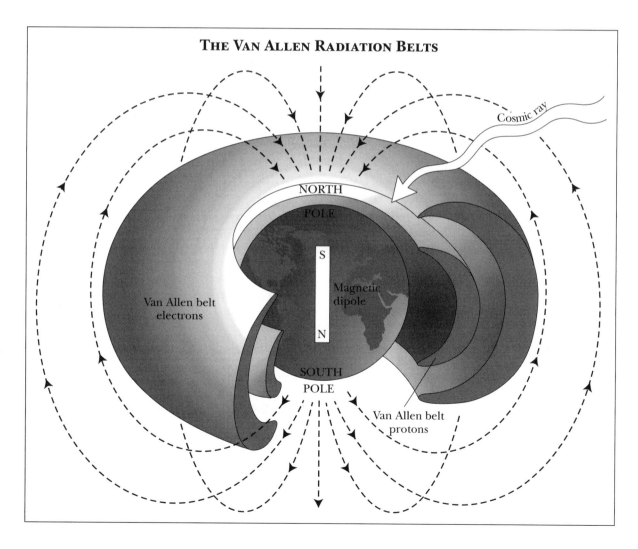

THE VAN ALLEN RADIATION BELTS

At the Earth's orbit, the density of the plasma is 5 particles per cubic centimeter at times of relatively little solar activity. This density changes considerably at times of peak solar activity. As the solar wind arrives at the Earth, it encounters the Earth's magnetic field. The edge of the magnetic field, the bowshock, deflects the solar-wind particles away from the Earth, but many protons and electrons penetrate into the magnetosphere where they become trapped. It is these trapped particles from the solar wind that make up the bulk of the Van Allen radiation belts.

A steady contribution to the inner radiation belts is made by cosmic rays. Cosmic rays consist of the nuclei of atoms and come to the Earth from two different sources. Some are hurled into space from the surface of the Sun during violent eruptions such as solar flares; others come from the depths of space.

Some cosmic rays have energies sufficient to penetrate the Earth's magnetosphere and reach the surface. Others collide with particles of atmospheric gas and release a shower of secondary radiation that is detected on Earth. The principles of the interaction of cosmic-ray particles with the Earth's magnetic field are the same as for the trapped particles of solar radiation.

The lifetimes of particles trapped in the Van Allen belts are highly variable. Under conditions of minimum solar activity, some particles remain trapped for months or years. Since the radiation belt density is fairly low, collisions that would send particles out of the belt are rare. The individual particle's lifetime is therefore determined by the height of its mirror points above the densest part of the atmosphere. The higher the mirror points, the less chance that the particle of radiation will collide with an atom of atmospheric gas.

The mechanism of loss and replenishment of particles in the radiation belt is not well understood. Several mechanisms work to distribute particles of various energies to different parts of the belts. The continuous supply of solar-wind particles from the Sun appears to be a necessity. Apparently, low-energy solar-wind particles enter the magnetosphere and are accelerated to energies necessary to become radiation-belt particles. Changes in Earth's magnetic field can excite the electrons in the outer belt to speeds approaching the speed of light.

STUDY OF RADIATION BELTS

Radiation in space consists of electromagnetic waves from the Sun; highly energetic protons, electrons, and atomic nuclei that make up the solar wind; cosmic rays both from the Sun and outer space; and trapped particles within the Van Allen radiation belts. While on the ground, we are protected from these forms of radiation by the Earth's atmosphere and its magnetic field. Astronauts flying in space however, have a much more significant risk. If a spacecraft is in orbit at an altitude lower than the altitude of the inner Van Allen belt, then there is no danger to the astronaut from the trapped particles of radiation. Although the radiation intensity of the inner belt is fairly constant, the intensity in the outer belt is much more sensitive to what is happening on the surface of the Sun. Activities on the Sun such as solar flares will result in a considerable fluctuation in the radiation intensity at this higher level. During periods of peak solar activity, the radiation in the outer belt could give an unprotected astronaut an exposure that could be fatal. The cabin of a spacecraft does offer some protection; however, when a particle strikes the spacecraft wall, it produces a shower of secondary radiation by a process known as "bremsstrahlung."

High-energy particles are capable of creating damage in human organisms. High-speed, high-energy particles pass through the body, colliding with and knocking electrons out of atoms that make up the cells. The result is chemical changes in the cells.

On the surface of the Earth, humans are subjected to a certain amount of background radiation. The cells of the body regularly absorb radiation, die, and regenerate. The body has acquired an immunity to this type of radiation. When the body is exposed to too much high-energy ionizing radiation, however, the cells may not be able to replace themselves because of damage to the nuclei that control the cell. Another possibility would be the uncontrolled reproduction of the cells and the spread of a cancer.

The best way to protect an astronaut from the effects of the Van Allen radiation belt is shielding, although shielding of the entire spacecraft may prove to be impractical because of the mass of the materials used in making the shield. It is generally more feasible to have a small portion of the spacecraft protected; the area must be large enough to provide a safe haven for the entire crew until the time of danger has passed. In August, 1972, a major solar flare occurred. A detailed study of the radiation released from that event showed that a shield of aluminum 0.2 meter thick would provide the necessary protection to keep astronaut exposure rates below the established limits.

During the Apollo missions, spacecraft trajectories were planned that avoided the regions of highest concentrations of radiation. Also, most of the missions were launched after the peak of the solar cycle. As a result of these precautions, the astronauts were exposed to about the same radiation dosages as someone would receive from a diagnostic X ray.

The inner Van Allen radiation belt interacts with the Earth's upper atmosphere in the polar regions to produce the aurora borealis and the aurora australis. Although the auroras are sometimes seen all over the globe, their regular and most frequent appearances are confined to the Arctic and Antarctic zones. The greatest frequency of the aurora in the Arctic region appears within a zone that lies from 15 to 30 degrees from the Earth's north magnetic pole.

In the region of the auroral zone, viewing an auroral display can be a nightly event. The development of an auroral exhibition may be described by the following sequence. After sunset, a faint arc becomes visible in the northern sky. The arc remains quiet for hours, with only a slight amount of movement. Suddenly, its lower border becomes more intense, and the arc breaks up into a series of parallel rays. The rays increase in color and intensity, and bundles of rays move along the arc. The arc then splits up into rays and draperies that fill the sky with dancing movements. The entire display usually lasts only minutes before the forms fade away. The faint arc again becomes visible in the northern sky. This display may be repeated several times during the night. A faint luminosity covers the northern sky after a particularly powerful auroral display. This phenomenon has given the aurora borealis the additional name the "northern dawn."

These colorful displays are produced when electrons and protons drop into the inner radiation belt and collide with atoms of atmospheric gases. The gas atoms are excited by these colli-

sions; when they return to their stable forms, visible light is emitted.

The trapped particles that make up the Van Allen belts are held in close proximity to the Earth by the Earth's magnetic field. Geophysical studies have concluded that the Earth's magnetism is generated within the liquid iron-and-nickel outer core rather than the solid inner core or the rock that makes up the mantle and the crust. The molten outer core flows at a rate of several kilometers per year in massive convective currents.

These currents, along with the rotation of the Earth, create electrical currents. The currents in turn, create the Earth's magnetic field. It has been found that other planets that have strong magnetic fields—Jupiter, for example—also have radiation belts similar to the Earth's Van Allen radiation belts.

SIGNIFICANCE

The history of the study of near-Earth radiation goes back to the early 1950's. During those years, scientists launched instrument packages into the upper atmosphere by using a combination of balloons and rockets. A balloon would lift a small rocket up to an altitude of about 25 kilometers, then the rocket would ignite and carry the package to an altitude of from 95 to 115 kilometers. The purpose of these tests was to monitor cosmic-ray intensities at high altitudes and latitudes. By doing so, scientists hoped to learn more about how cosmic rays are deflected by the Earth's magnetic field and absorbed in the atmosphere.

In 1953, a rocket launched into the northern auroral zone yielded a radiation count far greater than expected. It was proposed that since other instrument packages launched into regions both north and south of the zone revealed no anomalous data, there must be a connection between the high radiation count and the aurora. It was believed that the auroras were caused when showers of particles from the Sun entered the Earth's atmosphere along magnetic lines of force. During the International Geophysical Year (1957-1958), further tests were completed in the polar regions. The tests revealed that the radiation included energetic protons and electrons.

It was during the International Geophysical Year that a series of orbiting satellites were to be launched. It was decided that aboard these satellites would be placed radiation-detecting instruments. Before any American satellites could be orbited, the Soviet Union launched Sputnik 1; moreover, the first American launch attempts, using the Vanguard rocket, ended in a dismal failure. In January, 1958, the Jupiter-C rocket successfully carried Explorer 1 into orbit. The satellite carried with it a cosmic-ray detector.

The first reports from Explorer 1 indicated that radiation intensity increased with altitude. The data was somewhat incomplete, as it could only be taken when the satellite was near a tracking station. A later satellite, Explorer 3, carried a tiny magnetic tape recorder. This device could record data during an entire orbit and then send it upon receiving a radio command. The data from Explorer 3 confirmed that the radiation increased with altitude and eventually went to zero. Scientists eventually determined that the zero reading on their radiation counters was anomalous; the counters were actually being overloaded by radiation. The satellites had found a major new phenomenon: particles of radiation trapped in the Earth's magnetic field.

Data from satellites and lunar probes carrying more advanced instrument packages eventually revealed the existence of two belts of trapped radiation. Since their discovery in 1958, the radiation belts have been thoroughly mapped by space probes. Knowledge of the Van Allen belts is essential because of the effects of ionizing radiation.

Orbiting satellites are especially vulnerable to sudden changes in the Van Allen belts. The 1996 failure of Anik E1, a Canadian telecommunications satellite, was believed to have been caused by such a change. On May 19, 1998, a sudden change in the Earth's magnetic field caused electrons in the outer belt to accelerate to just under the speed of light, causing the Galaxy 4 telecommunications satellite to fail abruptly; as a result, 45 million pager customers lost service. Because of the energy of the electrons travelling at these speeds, heavy shielding would be required to completely protect sensitive satellites.

David W. Maguire

CROSS-REFERENCES

Atmospheric and Oceanic Oscillations, 1897; Climate, 1902; Clouds, 1909; Cosmic Rays and Background Radiation, 1846; Drought, 1916; Earth's Magnetic Field, 137; Earth's Magnetic Field: Origin, 144; El Niño and La Niña, 1922; Hurricanes, 1928; Lightning and Thunder, 1935; Magnetic Reversals, 161; Magnetic Stratigraphy, 167; Monsoons, 1940; Satellite Meteorology, 1945; Seasons, 1951; Storms, 1956; Tornadoes, 1964; Volcanoes: Climatic Effects, 1976; Weather Forecasting, 1982; Weather Forecasting: Numerical Weather Prediction, 1987; Weather Modification, 1992; Wind, 1996.

BIBLIOGRAPHY

Baugher, Joseph F. *The Space-Age Solar System.* New York: John Wiley & Sons, 1988. A well-written, well-illustrated volume on what has been learned about the solar system since the dawn of the space age. Suitable for the informed layperson.

Beatty, J. Kelly, and Andrew Chaikin. *The New Solar System.* Cambridge, Mass.: Sky Publishing, 1981. This well-illustrated, somewhat technical volume consists of a collection of essays by various experts in solar system research. Topics include the Sun, planets, moons, cornets, asteroids, and meteorites.

Bone, Neil. *The Aurora: Sun-Earth Interactions.* 2d ed. New York: Wiley, 1996. This college-level text provides the perfect introduction to the origins, physics, and significance of auroras. Suitable for the reader with no prior knowledge in the field. Color illustrations, bibliographical references, and index.

Bryant, Duncan Alan. *Electron Acceleration in the Aurora and Beyond.* Philadelphia: Institute of Physics, 1999. A good look into the geophysics of auroras. Bryant explains the relationship between the electron and the aurora in a somewhat technical manner. This college text is recommended for the advanced reader with much interest in the field. Color illustrations, index, and bibliography.

Damon, Thomas D. *Introduction to Space.* Malabar, Fla.: Orbit, 1989. A well-written, well-illustrated volume that covers various topics relevant to spaceflight. These topics include the history of spaceflight, propulsion systems, orbits, satellites, and living and working in space.

Harang, L. *The Aurorae.* New York: John Wiley & Sons, 1951. Contains detailed discussion of how the auroras form, their spectra, electromagnetic storms, and more. Well-illustrated with graphs and charts. The reader should have some basic knowledge of electromagnetism, spectroscopy, and advanced mathematics.

Plummer, Charles C., and David McGeary. *Physical Geology.* Dubuque, Iowa: William C. Brown, 1996. A general introduction to physical geology intended for use at the college-freshman level. Contains many tables, illustrations, and photographs; also includes a glossary and an index.

Shipman, Harry L. *Humans in Space.* New York: Plenum Press, 1989. A nonmathematical, very readable volume. The author deals with such subjects as living in space, space resources, and uses for advanced space technology.

Stine, G. Harry. *Handbook for Space Colonists.* New York: Holt, Rinehart, and Winston, 1985. A volume suitable for the general reader on the effects of spaceflight on humans. Such topics as space habitats, health and medicine, working in space, and social aspects of space living are discussed.

Tucker, R. H., et al. *Global Geophysics.* New York: American Elsevier, 1970. Deals with topics in geophysics such as geodesy, seismology, and geomagnetism. Well-illustrated; however, the author does make liberal use of advanced mathematics. Intended for the college-level physics student.

Van Allen, James A. "Radiation Belts Around the Earth." *Scientific American* 200, no. 3 (March, 1959). By the scientist who designed the experimentation that discovered the radiation belts that bear his name. Contains both a history of the study of radiation near the Earth and a technical description of the radiation belts.

Wentzel, Donat G. *The Restless Sun.* Washington, D.C.: Smithsonian Institution Press, 1989. A well-written volume intended for the layperson that describes, in nonmathematical terms, internal and external solar processes.

Zeilik, Michael, and Elske V. P. Smith. *Introductory Astronomy and Astrophysics.* 2d ed. New York: Saunders College Publishing, 1987. A text for an introductory college course on astrophysics. Well-illustrated with graphs, drawings, and photographs. Includes problem sets at the end of each chapter; the authors assume some knowledge of advanced mathematics.

VOLCANOES: CLIMATIC EFFECTS

Studies of volcanic activity confirm the existence of a significant relationship between the effects of an eruption and long-term climatic conditions.

PRINCIPAL TERMS

AEROSOL: an aggregate of dispersed gas particles suspended in the atmosphere for varying periods of time because of their small size

ATMOSPHERE: the thin layer of gases surrounding the Earth whose density decreases rapidly with height

CLIMATE: the sum total of the weather elements that characterize the average and extreme condition of the atmosphere over a long period of time at any one place or region of the Earth's surface

GREENHOUSE EFFECT: the retention of solar heat in the lower atmosphere caused by the water absorption and radiation of clouds and gases, creating an insulating effect similar to a greenhouse

OZONE LAYER: the band of gases that forms the protective layer in the stratosphere against ultraviolet rays

STRATOSPHERE: the atmospheric layer above the troposphere characterized by little or no temperature change with altitude

SULFUR DIOXIDE: a colorless, nonflammable, suffocating gas formed when sulfur burns

TROPOPAUSE: the layer of atmosphere separating the troposphere and the stratosphere

TROPOSPHERE: the lowest layer of atmosphere between the Earth's surface and the tropopause; most of the atmospheric turbulence and weather features occur here

ATMOSPHERIC ENVIRONMENT

Scientists have speculated for centuries on the impact of volcanic eruptions on the climate, and historical figures as early as Benjamin Franklin have noted the possible link between an eruption and climatic conditions. Two years after the explosion of Liki Fissure, Iceland, in 1783, Franklin suggested that the blue haze that lingered over the city of Paris during his stay there was caused by the Icelandic event. He believed the "fog," as he termed it, caused a decline in the temperature by absorbing portions of sunlight that otherwise would have reached the Earth's surface.

Technological advances such as satellite imagery have made it possible for scientists to pinpoint more precisely which types of matter injected by the volcano into the atmosphere have the greatest effect. Initially, researchers believed it was the amount of gas and ash hurled into the atmosphere that determined the impact. However, scientists have come to learn that it is the types of gases that are the critical determining factor. The molten rock of a volcano contains a multitude of gases that are released into the atmosphere before, during, and after an eruption. They range from water vapor to dense clouds of sulfur. Other gases include carbon dioxide, sulfur dioxide, hydrogen sulfide, hydrochloric acid, hydrogen, carbon monoxide, and hydrochloric acid, along with small amounts of other chemical compounds.

Exposure to the acid gases such as sulfur dioxide, hydrogen sulfide, and hydrochloric acid can be life threatening. One of the most serious hazards posed by a volcano is the large amount of carbon monoxide it can emit during an eruption. Since it is heavier than air, carbon dioxide tends to settle in lower elevations. As a result, numerous people residing near erupting volcanoes have been asphyxiated. When a volcano erupts, its immediate effect on the local area is evident, as it spews a great amount of ash, volcanic gases, and heat into the atmosphere. Violent eruptions usually are accompanied by thunderstorms, lightning, and torrential rains. In some cases, the intense heat can generate powerful whirlwinds, strong enough to topple nearby trees and destroy wildlife. Volcanic plumes also contain droplets of water in which acid gases have dissolved. These droplets eventually fall to the Earth in the form of acid rain, which can have a corrosive effect on a variety of metal objects. Long-term exposure to volcanic gases also can have a lethal effect on most varieties of vegetation.

Along with the research into the makeup of gases has come a greater awareness of the long-term effects on climatic patterns. The volcanic dust released into the lower atmosphere can create a temporary cooling effect by producing an ash cloud that blocks the sunlight. However, the ash particles are quickly washed out by the abundant supply of water and rain contained in the lower atmosphere. The majority of ash clouds stop rising at the troposphere, where they encounter increased air temperatures. Dust particles thrown into the stratosphere can linger for several weeks or months before they settle back to the Earth.

For many years scientists believed that a volcano's ash was the most important factor in climatic changes. They since have discovered that of all the ingredients tossed into the air, it is the volcanic gases that wield the greatest influence on climatic conditions. Specifically, the key factor is the conversion of the gases into sulfuric acid aerosols that are capable of remaining in the atmosphere for months. The droplets, though tiny in size, are able to reflect significant amounts of sunlight and have been detected as high up as 25 kilometers. They not only are able to cool the troposphere by absorbing infrared radiation, but also can warm the stratosphere by reflecting the solar rays back into space. The reflective effect is particularly strong when it occurs in normally cloudless areas. It has been estimated that the aerosols can increase the Earth's reflectivity by nearly 20 percent. Although aerosols eventually grow large enough to fall back to the Earth from their own weight, the process can take years because of the dry conditions in the stratosphere, which slows their growth. Aerosol cloud eruptions occur on an average of once every ten to twenty years, with major eruptions taking place at a rate of nearly one every one hundred years.

Only those volcanic eruptions that emit principal amounts of sulfur compounds have an effect on the global climate. Smaller eruptions can create atmospheric effects similar to the larger explosions if they release large concentrations of sulfur-rich elements directly into the stratosphere. Also, eruptions occurring at high latitudes have less impact on climatic conditions than those at lower latitudes, where air currents are greater. The aerosols travel more quickly around the globe when moving in an easterly or westerly direction. If they are traveling in a north-south direction, the movement is much slower, with many of the aerosols ending up confined for years in a zone surrounding the polar cap.

MEASUREMENT METHODS

There are two primary indexes that are used to measure the probable impact of volcanic activity. The first is the dust veil index (DVI), which is based on an estimate of the volume of material injected into the atmosphere, temperatures on the Earth's surface, and the amount of sunlight reaching the Earth's surface. This index is derived from observations made at mid-latitudes and thus is not entirely representative of the global scenario. The second index is the volcanic explosivity index, which is based on the magnitude, intensity, dispersion, and destructiveness of an eruption. Other measurements of impact include tree-ring records, ice-core readings, and solar radiation measurements.

In addition to the impact indexes, four primary methods are used to determine the volume and composition of gases that are emitted during large volcanic eruptions: an examination of the composition and amount of aerosol layers in polar ice

---

VOLCANIC EXPLOSIVITY INDICES OF THE DEADLIEST ERUPTIONS SINCE 1500 C.E.

Year	Location	VEI	Casualties
1586	Kelut, Indonesia	4	10,000
1783	Lakagigar (Laki), Iceland	4	9,000
1792	Unzen, Japan	3	15,000
1815	Tambora, Indonesia	7	92,000
1883	Krakatau, Indonesia	6	36,000
1902	Mount Pelée, Martinique	4	30,000
1985	Nevado del Ruiz, Colombia	3	25,000

SOURCE: U.S. Geological Survey, Cascades Volcano Observatory, Vancouver, Washington. URL: http://vulcan.wr.usgs.gov/ LivingWith/VolcanicFacts/deadly_eruptions.html

cores that are identified as being related to volcanic eruptions; a comparison of gases from eruptions and glass inclusions in crystals that had formed in the volcanic rock prior to the event; measurements of eruptions from satellite imagery; and measurements of volcanic aerosols from the Earth's surface. Scientists have had some success with using a laser instrument that sends out a pulsing light beam, which reflects back when it detects aerosols. This method enables researchers to construct a profile indicating the density and height of the aerosols.

Ice core readings are especially beneficial in providing a clear record of older eruptions. Atmospheric aerosols from nearly every historic event have shown up in deeply drilled polar ice. Since they tend to accumulate in layers, the ice cores offer clear annual records of climate and weather. As the aerosols fall to the Earth's surface over the poles, they begin to soil the surface snow with acid fallout in a process called sedimentation, in which thin layers of debris are formed. Over time, the snow becomes compacted with the glacial ice and can be detected through electrical conductivity measurements. Some ice core discoveries have provided records going back more than 100,000 years.

Satellite measurements have enabled scientists from the U.S. Geological Survey to determine the amounts and compositions of gases emitted by several active volcanoes in the United States. Satellite sensors detected up to 1 million metric tons of sulfur dioxide tossed into the stratosphere during the main eruption of Mount St. Helens in 1980. Of particular benefit are the satellite observations conducted by the National Aeronautical and Space Administration's (NASA's) Total Ozone Mapping Spectrometer (TOMS) instrument, which have helped to measure sulfur dioxide levels in the atmosphere following major events. TOMS was instrumental in tracking the band of sulfur dioxide across the Pacific produced by the eruption of Mount Pinatubo in the Philippines in 1991. Altogether, TOMS has made more than one hundred observations of volcanic events, including a major eruption of Chile's Cerro Hudson volcano in 1991. These measurements allow scientists to compare volcanic emissions of sulfur dioxide with injections of the gas from industrial plants and other human-based activities. Through comparative studies of volcanic activity, researchers are able to examine the effects of past and future eruptions with the aim of determining whether human or natural activities ultimately pose the greater threat to the environment.

HISTORICAL RECORD

Though less documented than modern eruptions, the larger historic events offer substantial proof for volcanic effects on climate for the simple reason they were bigger and left a more easily detectable trail of evidence.

By drilling into the seafloor south of Haiti and uncovering evidence of ash, researchers concluded that massive volcanic eruptions in the Caribbean Basin more than 55 million years ago created a sudden temperature inversion in the ocean waters that led to one of the most dramatic climatic changes in the Earth's history. Scientists discovered distinctly colored volcanic ash layers that were far different from the sediments located above and below them. The period when the ash layers were deposited corresponds with a time of rapid warming of the Earth. The presence of the ash indicates a gigantic eruption took place just as the warming began. Scientists believe that the dust and gases from the eruptions initially cooled the atmosphere, increasing the density of the sea water to the point where it sank into the deep ocean. The descending water, in turn, warmed the ocean floor and melted deposits of methane sediments, which then bubbled up into the air, creating a greenhouse effect that warmed the world. Evidence also indicates that the process resulted in the extinction of nearly one-half of all deep-sea animals, victims of asphyxiation by the suddenly warmer waters. Conversely, the evolution of new plant and animal species, including many primates and carnivores, was accelerated. Scientists already were aware that volcanic eruptions had occurred in the North Atlantic Ocean nearly 61 million years ago and believe that the Caribbean Basin event somehow may have acted in connection with these earlier eruptions.

In another major undersea discovery, scientists suspect that the islands of Tonga and Epi, located about 1,930 kilometers east of Australia, were the products of a massive eruption that took place around the year 1453. During the course of their research, scientists found that the entire stretch of seafloor separating the two islands was a crater more than 11 kilometers in width. They also uncovered charred vegetation that was carbon-dated between 1420 and 1475. To further narrow down the date of the eruption to 1453, they analyzed ice cores from Greenland and Antarctica; tree-ring records from California, Europe, and China; and reports of worldwide crop conditions during the period.

In 1815, the volcano Tambora erupted in Indonesia, precipitating one of the clearest examples of an eruption-induced global cooling event. The volcano emitted a massive column of solid material into the upper atmosphere. The aerosol veil extended to both hemispheres with effects that lasted well into the following year. The aftereffects were such that the year 1816 came to be known as "the year without a summer." In some regions of New England, up to 15 centimeters of snow fell in the month of June. There were numerous other reports of abnormally cool weather, including record low temperatures that forced people to wear coats and gloves in July. The average temperature in the Northern Hemisphere was reduced by as much as 0.5 degree Celsius, and parts of the United States and Canada experienced unusual summer frosts and loss of crops. In Europe, the unusually cold readings resulted in widespread famine, though at the time the connection with the volcanic veil went unrecognized. Researchers identified the Tambora eruption from evidence uncovered in ice cores in Antarctica and Greenland. The event coincides with the fact that the decade between 1810 and 1820 is considered perhaps the coldest on record.

The famous 1883 eruption of Krakatau in Indonesia marked perhaps the first time researchers became fully involved on a worldwide basis with a volcanically induced atmospheric event. Much of the global interest could be attributed to the advances in telegraphic communication, which enabled scientists to share their observations of the spectacular sunsets and other visible phenomena arising from the eruption. Measurements indicated that Krakatau generated a cloud of approximately 21 cubic kilometers of matter. Witnesses in the area recalled dramatic displays of lightning in the cloud veil and a strong odor of sulfur in the air. Researchers believe that the ash cast into the upper atmosphere by the eruption and the ensuing dust veil led to worldwide decreases in incom-

ing radiation. Mean annual global temperatures fell close to 0.5 degree Celsius in 1884, with the cooling period extending through the remainder of the 1880's. In 1884, there was a marked increase in the number of storms in the United States. Record snowfalls, an unusually high number of tornadoes, torrential rains, and severe flooding caused widespread damage. The abnormal atmospheric conditions attributed to Krakatau included brilliant sunsets and a blue or green tinge to the Sun, depending on which part of the globe the observation occurred. In 1888, the Royal Society of London published a volume that documented the eruption and formally established the connection between major volcanic activity and subsequent changes in worldwide atmospheric conditions.

By historic standards, the eruption of Mount Pinatubo may appear insignificant, but it stands as a watershed event in the ability of scientists to monitor the interaction of a volcanic cloud and the upper atmosphere. The eruption is believed to have sent nearly 20 billion kilograms of sulfur dioxide about 20 to 27 kilometers into the atmosphere, resulting in 30 billion kilograms of sulfuric acid aerosols. For several months the TOMS instrument tracked the sulfur cloud created by Pinatubo with images verifying that its particles circled the globe in about three weeks, forming an almost continuous band.

The Pinatubo eruption had a gigantic atmospheric impact. The year following its eruption turned out to be one of the coldest on record. Temperature measurements in the lower and middle atmospheres indicated a change of nearly 0.5 degree Celsius between 1991 and 1992. By 1994, readings revealed that the volcano's effect had waned and that global temperatures had returned to previous levels.

SIGNIFICANCE

Much has been learned and much remains a mystery concerning the chemical and physical processes that occur between a volcanic eruption and climatic change. It is an area of intense study because of the belief among scientists that the balance of the Earth's mild climate is dependent on the phenomenon of volcanism. The Earth's atmosphere essentially was developed through intermittent volcanic emissions of carbon dioxide and water vapor, along with nitrogen and possibly methane. There is no reason to believe that the relationship between the two forces has changed in any dramatic way.

Volcanic eruptions can be an agent for global cooling or global warming, depending on their interactions with other elements. The historical record indicates that when the atmospheric balance is threatened by natural or human forces, the consequences can be severe. To most people, fractions of degrees may not appear significant, but in the grand scheme of the environment, they can produce dramatic effects. During the ice ages, the global temperature was only about 5 degrees Celsius cooler than it was at the close of the twentieth century. Scientists believe that a global rise in temperature of as little as 3 degrees Celsius could bring about dramatic changes, including accelerated glacial melting, rising sea levels, additional severe storms, and droughts. The temperature increase during the twentieth century is considered evidence that the human production of greenhouse gases such as carbon dioxide is affecting the climate. However, it also is believed that multiple eruptions of large volcanoes over a long period of time can raise the carbon dioxide levels enough to cause substantial global warming. To add to the equation, studies also indicate a possible association between other volcanic vapors and the depletion of the ozone layer. A few researchers even have advanced the idea that there is a link between El Niño—the periodic warm ocean conditions that appear along the tropical west coast of South America—and volcanism. A number of eruptions have preceded El Niño in years past, leading to speculation that volcanic gases can trigger or strengthen the phenomenon.

Major events such as the Caribbean Basin volcanic eruption pose a special problem for the Earth's climatic equilibrium. The sudden warming of the deep ocean resulting from a series of large volcanic eruptions is a scenario that scientists believe could repeat itself, causing a major disruption of the Earth's atmospheric circulation. In attempting to trace the connections between volcanic activities, scientists have begun to think in global terms and to look upon their task as an interdisciplinary effort. In so doing, they have been able to make significant strides in developing the depth of understanding necessary to form reasonably accurate forecasts of future catastrophic events.

William Hoffman

CROSS-REFERENCES

Atmosphere's Global Circulation, 1823; Atmospheric and Oceanic Oscillations, 1897; Climate, 1902; Clouds, 1909; Drought, 1916; El Niño and La Niña, 1922; Hurricanes, 1928; Lightning and Thunder, 1935; Monsoons, 1940; Precipitation, 2050; Satellite Meteorology, 1945; Seasons, 1951; Storms, 1956; Tornadoes, 1964; Van Allen Radiation Belts, 1970; Weather Forecasting, 1982; Weather Forecasting: Numerical Weather Prediction, 1987; Weather Modification, 1992; Wind, 1996.

BIBLIOGRAPHY

Decker, Robert W., and Barbara Decker. *Volcanoes*. 3d ed. New York: Freeman, 1998. This introductory work on the study of volcanoes contains a concise chapter, with clear illustrations, on the connection between volcanic eruptions and climate.

Fisher, Richard V., Grant Heiken, and Jeffrey B. Hulen. *Volcanoes: Crucibles of Change*. Princeton, N.J.: Princeton University Press, 1997. The authors cover a long list of volcanoes, from Mount Vesuvius to Mount St. Helens. A segment on the effects of volcanic gases helps explain the complex interactions among gaseous, liquid, and solid volcanic matter; the surrounding atmosphere; and solar radiation.

Kuhn, Gerald G. "The Impact of Volcanic Eruptions on Worldwide Weather." *Twenty-first Century Science and Technology*, Winter, 1997-1998, 48-58. This well-documented article provides a comprehensive review of the evidence supporting a link between volcanic activity and climatic changes. Special attention is given to a wide range of weather-related abnormalities associated with the aftereffects of specific events.

Robinson, Andrew. *Earth Shock*. London: Thames and Hudson, 1993. The author offers an interdisciplinary look at the forces of nature, including volcanoes and their impact on the global climate. The work probes the interrelationships among the various natural disasters with a lucid style, buttressed by ample illustrations and charts.

Sparks, R. S. *Volcanic Plumes*. New York: Wiley, 1997. This work examines the basic geological principles and dynamics behind the volcanic plumes and their impact on the surface environment. How the plumes interact with the atmosphere and then disperse before depositing their debris and gases is the focal point of the text.

WEATHER FORECASTING

Weather systems are chaotic, thus limiting the degree of forecasting accuracy. The study of chaotic systems themselves and the development of innovative techniques give insight into weather predictability.

PRINCIPAL TERMS

APERIODIC: an irregularly occurring interval, such as found in most weather cycles, rendering them virtually unpredictable

CHAOS: an emerging scientific discipline that attempts to quantify and describe seemingly random, aperiodic events

NOWCASTING: a weather forecast for a specific area for an approaching weather system, made and disseminated within an hour or less

WEATHER ANALOGUE: an approach that uses the weather behavior of the past to predict what a current weather pattern will do in the future

DEVELOPMENT OF WEATHER FORECASTING

From the earliest days of recorded civilization, people have actively attempted to forecast the weather—and for good reason: The weather directly affects everyone. Predicting the weather has been a priority for reasons of simple comfort to protection of property and lives. Among uneducated people, weather forecasting was widely associated with lunar phases, rainbows, halos, and even the actions of animals and insects. With the advent of weather instruments such as thermometers, anemometers, hygrometers, and barometers, the science of meteorology evolved. The first recorded use of the term "weather forecast" was by Englishman Robert Fitzroy in 1860, a man with enough perception to avoid any nonscientific term such as "prophecy" or even one so bold as "prediction."

Record keeping and sharing of information across states, nations, and the world by telegraph and radio during and after World War I finally brought the science to a coherent body of knowledge. By the start of World War II, the invention of the high-altitude balloon for weather soundings (the radiosonde) enabled the tracking of upper air currents. The term "atmospheric science" began to encompass the grand, global scheme of weather forecasting. "Meteorology" technically accounts for the action in the lower atmosphere.

The earliest meaningful attempts at forecasting the weather on a scientific basis came with the advent of the barometer (a device that measures atmospheric pressure) and the realization that changes in the weather seemed to follow changes in the barometric pressure. It was also noted that the weather changed with the passage of air masses, which often dramatically altered the weather from what it had been. The observation was also made that the passage of air masses was often related to the barometric pressure. The accuracy of weather prediction increased with such observations.

FORECASTING METHODS

Prediction of the weather from present conditions to two days is called a short-range forecast. A medium-range forecast can be stretched out to only seven days. Beyond seven days is called a long-range forecast, of which thirty days are all that can be predicted with any statistical accuracy. A ninety-day outlook is given by the National Weather Service. Prediction method accuracy varies widely with the condition being predicted. Forecasting damaging weather such as local high winds, heavy snow, dangerous thunderstorms, or tornadoes beyond twelve hours is not possible. Very short-range forecasting for periods of one hour or less is a relatively new technique primarily used for issuing warnings. Such immediate forecasts are called nowcasts.

Satellites offer a wide view of the planet from orbits above the Earth. The advantage of these instruments is a global view of weather patterns and their movement, which offers a wide range of immediate forecasting options when related to ground weather observations. Radar offers a view of precipitation in most forms and is useful in forecasting imminent weather patterns. Doppler radar allows for pinpointing tornadoes as they are forming inside cloud formations.

The global weather scheme is plotted by powerful computers, which collect data from all over the globe and integrate them into a planetary weather picture. From this picture, forecasters adopt mathematical models to calculate what will happen to the system over time. The weather is predicted by supercomputers using a global scheme simulating the Earth's weather based on a vast grid of data points covering the globe at various resolutions. These forecasts are calculated at millions of operations per second. Yet forecasters have found that no single model or combination of models, using even the most powerful computers, gives an accurate prediction of weather schemes beyond one week. The reason for that is a condition called chaos.

Chaos accounts for the breakdown of forecasting accuracy over time. It was first related to weather patterns by Massachusetts Institute of Technology meteorologist Edward Lorenz in 1961. By 1963, Lorenz had constructed a computerized model of the atmosphere. Lorenz discovered that regardless of the point at which the computer begins calculating the equations for predicting the weather, every data point is potentially unstable, so that very small errors are magnified over time. That is called sensitive dependence on initial conditions, or the butterfly effect. The butterfly effect refers to the flapping of a butterfly's wings in, for example, Beijing, China, which initiates a tiny instability in the atmosphere, which is ultimately magnified into storm systems in New York; this rather fanciful analogue of the effect of chaos in weather systems is the real problem faced in weather forecasting. The future of weather forecasting probably lies in understanding and modeling chaotic systems. Such work is under way at the National Weather Service.

APPLICATIONS OF FORECASTING METHODS

The three ranges of weather forecasting—nowcasting (one hour or less), short-range forecasting (less than two days), and long-range forecasting (up to ninety days ahead)—all use different techniques for development of the forecast. Nowcasts are usually based on ongoing observations or predictions from radar for very specific locations. For example, a tornado warning, hurricane warning, or severe thunderstorm warning may be issued for a specific city based upon actual images of the storm bearing down. The only mathematical methods applied in these cases would be tracking equations based on speed and distance.

Short-range forecasting is much more complex; equations are developed from four different types of mathematical techniques. The National Weather Service uses combinations of all these equations in their computer analyses of the weather's short-term picture. These mathematical techniques are computations of physical parameters, computations of displacements of large weather systems, regression analysis, and statistical (time series) analyses. Computations of physical parameters are pure physical formulas: hydrodynamic, aerodynamic, thermodynamic, and classical physical laws. Examples are Sir Isaac Newton's equations for motion, rules for the conservation of mass, heat transfer equations, and the laws for gas states, all applied to dynamic (moving) weather systems. In computations of displacements of large weather systems, the rates of movement of well-defined frontal systems are calculated based on a series of equations developed for just these systems and their movements. By using a statistical analogy between past events and present conditions, a forecaster is able to relate past weather patterns to future conditions. Statistical (time series) regression analyses called model output statistics (MOS) enable the forecaster to make predictions of future weather events. These equations account for the possible variations in weather patterns over time and physical range of the weather system itself. In such forecasts, the weather is predicted to occur on a chance basis. For example, the forecast may give an area a 50 percent chance of measurable precipitation. All these parameters may be utilized in developing a single forecast.

Extended forecasting is developed by different methods from those employed in short-range forecasts. Meteorologists still use a battery of computerized tools available to them, but the effect of chaos (instability at every point in the atmosphere) is so great beyond two days that extending the computer tools used for short-range forecasting out many days is all but useless for specific areas. Yet there are seasonal weather patterns, worldwide circulations, oceanic currents, and historical norms that account for making long-range forecasts possible. These nearly cyclic conditions are called aperiodic and enable some degree of forecasting. In the study of chaos itself, there are predictable, recognizable elements: all the nuances of aperiodicity that ultimately lead to the chaotic state. It is precisely the investigation of these elements that atmospheric scientists are now pursuing.

One of the most elementary, subjective elements of such an approach actually surfaced in the early 1950's, a decade before chaos was quantitatively linked to the weather. In such an approach, called weather analogues or weather typing, weather charts of previous seasons were categorized and cataloged. Current charts were then compared with old charts to indicate possible future trends. In the days before a full understanding of the power of even minute atmospheric instabilities, these analyses proved of no use at all and were all but abandoned in favor of powerful computerized analyses.

In using computers to make long-range forecasts, meteorologists have noted that there are times when their automated models prove correct, often even surprisingly accurate. Chaos, again, proves to be the central antagonist in determining accuracy. Stability of the weather pattern determines the accuracy of the computerized forecast. In order to weed the good forecasts from the bad, atmospheric scientists have developed a method called the ensemble approach. In this technique, they predict the reliability of the forecast by testing the stability of the conditions: They run the same forecast at least ten times using a supercomputer, altering each run very slightly to mirror tiny instabilities in the atmosphere, and then compare results. If the end result is widely variable forecasts, then the forecasting reliability for that day is low. Using such techniques, atmospheric scientists can not only predict the ultimate accuracy of their forecasts but also decide on the degree of its reliability.

SIGNIFICANCE

From the obviously detrimental effect of storms to the periodic need for rain, the weather is a phenomenon of universal relevance. Predicting the weather is a science that has developed into a study involving individuals and governments and the pooling of an immense amount of information. Some forecasting ability has resulted from ordinary and practical application, while other methods are so complex that they are understood only by a very few highly skilled meteorologists in the scientific community.

Yet without regard for human progress and understanding, the deserts of equatorial Africa have been encroaching northward at tens of kilometers per year, rendering once fertile land into useless desert; hundreds of people have died daily from famine unfolding across the vast African savannah. The frequency of hurricanes may be changing along the Atlantic seaboard, threatening millions of seashore inhabitants. The sunbelt population in the United States is rapidly exploding along the path of hundreds of potentially deadly tornadoes. The world's atmospheric scientists are faced with the task of predicting the potential for the weather directly affecting the lives of millions. From giving imminent warnings to plotting the centimeter-by-centimeter alteration of long-term weather patterns, the task of prediction has significance for a great number of lives.

The discovery of the impact of chaos has delivered a stunning blow to the task at hand. Indeed, many scientists have declared the task of accurate weather forecasting to be impossible. Others continue the struggle to discern the meaning behind the science with instabilities at every point. The study of chaos itself has spread from the meteorologist's table in 1961 to nearly every other discipline.

Dennis Chamberland

CROSS-REFERENCES

soons, 1940; Ozone Depletion and Ozone Holes, 1879; Satellite Meteorology, 1945; Seasons, 1951; Storms, 1956; Tornadoes, 1964; Van Allen Radiation Belts, 1970; Volcanic Hazards, 798; Volcanoes: Climatic Effects, 1976; Weather Forecasting: Numerical Weather Prediction, 1987; Weather Modification, 1992; Wind, 1996.

BIBLIOGRAPHY

Bader, M. J., et al., eds. *Images in Weather Forecasting: A Practical Guide for Interpreting Satellite and Radar Imagery.* New York: Cambridge University Press, 1995. This collection of essays written by leading meteorologists describes the equipment and techniques used in weather forecasting. Chapters focus on remote sensing, radar meteorology, and satellite meteorology. Color illustrations.

Barrett, E. C. *Viewing Weather from Space.* New York: Frederick A. Praeger, 1967. This concise book provides a fine historical backdrop to the development of the weather satellite as an instrument for observing the weather from space. Also discusses the primary uses for the weather satellite as an instrument for predicting the weather. Illustrated with drawings and photographs.

Collier, Christopher G. *Applications of Weather Radar Systems: A Guide to Uses of Radar Data in Meteorology and Hydrology.* 2d ed. New York: Wiley, 1996. Collier offers the reader a detailed look into scientific advancements regarding the tools used in meteorology and hydrology. Chapters focus on radar meteorology, precipitation measurement, hydrometeorology, and weather forecasting.

Day, John A. *The Science of Weather.* Reading, Mass.: Addison-Wesley, 1966. This account of the science of weather observation and prediction details the art and technology of what goes into making predictions. Offers the standard textbook approach to descriptive meteorology but also an account of the developing use of satellites and computerized analyses. Illustrated throughout.

Dolan, Ed. *The Old Farmer's Almanac Book of Weather Lore.* Dublin, N.H.: Yankee Books, 1988. No accounting of modern weather forecasting can be told without its fabulously rich heritage of centuries of folklore and history. From the very source of folk meteorology itself, *The Farmer's Almanac*, comes this definitive treatment of the subject.

Gleick, James. *Chaos: Making a New Science.* New York: Penguin Books, 1988. This work details the birth of a new science: the study of chaos, which began with a meteorologist attempting to write a computer program to predict the weather. This book, a national best seller, discusses the recognition and development of the study of chaos. Offers a fine treatment of weather prediction as an example of chaos and why forecasting such a chaotic system may not be impossible as once thought. The book, written for a general audience, is illustrated and indexed.

Kerr, Richard A. "A New Way to Forecast Next Season's Climate." *Science* 244 (April 7, 1989): 30-31.

_____. "Telling Weathermen When to Worry." *Science* 244 (June 9, 1989): 1137. This pair of articles details current methods in weather forecasting and attempts to predict the long-range weather by computer analyses. They discuss the blend of forecasting by human subjective experience and computer estimates. The articles are also concerned with the accuracy of past forecasting methods and how the computer fares when compared with human judgments.

Monmonier, Mark S. *Air Apparent: How Meteorologists Learned to Map, Predict, and Dramatize Weather.* Chicago: University of Chicago Press, 1999. This college-level text looks at the satellites and radar systems used to collect meteorological data, as well as the techniques used to interpret that information. Color illustrations, index, and bibliographical references.

O'Neill, Gerard K. *2081: A Hopeful View of the Human Future.* New York: Simon & Schuster, 1981. This book describes what the world may be like in 2081 (under the best of circumstances) and how the pace of technology will determine the future. In one discussion, O'Neill describes weather forecasting's relationship to computer science. Illustrated. Written for the general public.

Thompson, Philip D. *Numerical Weather Analysis and Prediction.* New York: Macmillan, 1961. This book details the science of numerical analysis of the weather. Written for those with a good grasp of fairly complex mathematics. Lends a good insight into the requirements for mathematical modeling of the atmosphere and what supercomputers deal with when summarizing all the climatic data into a picture of tomorrow's weather.

WEATHER FORECASTING: NUMERICAL WEATHER PREDICTION

Most industrialized nations, including the United States, use numerical weather prediction techniques to formulate weather forecasts. Numerical weather prediction is based upon physical laws that are incorporated into a mathematical model; a solution is then determined using computer algorithms. A wide variety of data are used to initialize the numerical models, including observations from radiosondes, weather satellites, ground-based stations, airplanes, and buoys. Verification of short-range numerical weather prediction models shows a considerable improvement in skill over climatology. Numerical weather prediction models require large mainframe supercomputers, capable of millions of iterations, to formulate each forecast.

PRINCIPAL TERMS

CLIMATOLOGY: the scientific study of climate; the term "climatology" in forecasting refers to the statistical database of weather observed over a period of twenty or more years for a specific location

ENSEMBLE WEATHER FORECASTING: repeated use of a single model, run many times using slightly different initial data; the results of the model runs are pooled to create a single "ensemble" weather forecast

FORECAST VERIFICATION: comparison of predicted weather to observed weather conditions to determine forecasting skills

GLOBAL ATMOSPHERIC MODEL: based on a spherical coordinate system representing the entire world, these numerical models often use a reduced grid in which the longitudinal grid interval is fairly constant with respect to physical distance rather than angular distance

HEMISPHERIC MODEL: a numerical model that extends over the whole Northern or Southern Hemisphere

LONG-RANGE PREDICTION: a weather forecast for a future period greater than one week in advance for a specific region, often supplemented with climatological information

MESOSCALE MODEL: a weather forecast for an area of up to several hundred square kilometers on a time scale between one hour and twelve hours

NONDETERMINISM: chaotic, random events that cannot be predicted but that have a significant influence on development of weather systems

NOWCASTING: a short-term weather forecast usually for the prediction of rapidly changing, severe weather events

HISTORICAL DEVELOPMENT

The atmosphere is a very complex, chaotic natural system. Before the science of meteorology was developed, keen observers of the natural environment developed forecasting rules epitomized by such sayings as "Red sky at morning/ Sailors take warning/ Red sky at night/ Sailors delight." In the early nineteenth century, a systematic classification of clouds was introduced by Luke Howard—known as the father of British meteorology—and scientists began making routine daily weather observations at major cities and universities. A systematic study of the atmosphere as a chemical and physical entity began.

In 1904, Vilhelm Bjerknes brought the scientific community to understand that atmospheric motions were largely governed by the first law of thermodynamics, Newton's second law of motion, conservation of mass (the "continuity equation"), the equation of state, and conservation of water in all its forms. (These physical laws form what is now known as the "governing equations" for numerical weather prediction.) Bjerknes, writing that the fundamental governing equations constituted a determinate, nonlinear system, realized that the system had no analytic solution. He also recognized that available data to determine initial conditions were inadequate. In 1906, Bjerknes devised graphical methods to use in atmospheric physics. During the next decade, he adopted an approach to apply physics in a qualitative, as opposed to a numerical, technique to weather forecasting.

In 1916, Bjerknes began working at the Bergen Museum in Norway, a move that would be decisive in establishing meteorology as an applied science. In 1918, Bjerknes's son Jacob noticed distinctive features on weather maps that led him to publish an essay entitled "On the Structure of Moving Cyclones." This was followed by the development of the concept of "fronts" as a forecasting tool at the Bergen School in 1919. Following World War I, Norwegian and Swedish meteorologists educated at the Bergen School began using the theory of fronts operationally. Their ability to correctly predict severe weather events led other Europeans to adopt this frontal forecasting method. Enthusiasm for this empirical method blossomed and remains strong among the general public today.

Meanwhile, attempts to develop numerical weather prediction techniques languished. A method for numerically integrating the governing equations was published by L. F. Richardson in 1921. It contained several problems and errors. Following World War II, the development of mainframe computers allowed development of objective forecasting models incorporating many meteorological variables. By 1950, the first scientific results of a computer model, based on numerical integration of barotropic vorticity equations, was published. In the following decades, computer models grew increasingly sophisticated through the incorporation of larger and larger numbers of operations, iterations, and data input. Computer speeds increased dramatically, allowing use of sophisticated equations modeling atmospheric behavior. Larger models with relatively fine grids were developed.

Use of Supercomputers

Modern weather forecasting attempts to model processes in the atmosphere by representing the appropriate classical laws of physics in mathematical terms. Models use assemblages of approximations of physical equations (algorithms). Large (mainframe) supercomputers generate numerical forecasts using these algorithms and an array of observations. Most models have at least fifteen vertical levels and a grid with a mesh size less than 200 kilometers. The exact formulation of any particular model depends upon the amount of data being input, how long in advance, how detailed, and which variables are being forecast.

Historically, weather observations for numerical forecasts were taken mainly at 0000 Coordinated Universal Time (UTC), or Greenwich time, and 1200 UTC using ground stations and radiosondes (weather balloons). Now these observations are supplemented by asynchronous observations from infrared and microwave detectors on satellites, radar, pilot reports, and automatic weather observation stations (including ocean buoys). These new data sources have substantially enlarged the amount of input data. However, because these data are fed into the models as they run, the distinction between input data and model predictions has become blurred. Thus, a weather map showing a numerically generated forecast (as seen on the Internet or in a television presentation) may contain information from many data sources of varying accuracy taken at various times and locations.

Data from all available observations are interpolated to fit the model grid size. Every forecast incorporates inherent errors from the observations, data interpolations, model approximations, the instabilities of the mathematics, and the limitations of the computers used. Models can be separated by their time frame into nowcasts, short-term forecasts, medium-range forecasts, long-term forecasts, and outlooks. In addition, specialized forecast models for hurricanes and other specific types of severe weather have been developed. Because of hazards associated with severe, rapidly developing weather threats, there is increasing interest in developing specialized nowcast and short-range mesoscale (local) forecast models.

Many countries staff their own meteorological offices, using satellite images and local weather observations to develop computer models uniquely suited to their needs. Countries with specific needs and adequate resources develop models to suit their own unique needs. Japan, with its high population concentrations exposed to tropical storm threats, has directed its meteorological efforts toward mesoscale forecasting. The entire country was covered by a network of weather radars by 1971, and Japan supplements its ground-based observations with geostationary weather satellites. This maximizes the Japanese ability to predict heavy rains from tropical typhoons and to issue flash flood forecasts.

In 1999, the United States National Weather Service (NWS) completed a twelve-year modernization program that included 311 automatic weather observing systems and 120 new Nexrad Doppler radars. Observed weather and forecasts were displayed on state-of-the-art computer systems at NWS forecasting offices.

In the mid-1970's, the European Centre for Medium Range Forecasting (ECMWF) and the U.S. National Weather Service began utilizing high-speed supercomputers to generate and solve numerical weather models. Both of these entities have developed global atmospheric models for medium-range forecasts, producing a three- to six-day forecast. The numerical methods used in these global models involve a spectral transform method. One reason for using this method is that models frequently experience numerical problems for computations in polar latitudes. The ECMWF in 1999 was using a model with grid points spaced every 60 kilometers around the globe at thirty-one levels in the vertical. Its initial conditions used observations over the previous twenty-four hours and added in several early model runs forecasting about twenty minutes ahead to augment the observations. Wind, temperature, and humidity were then forecast at 4,154,868 points throughout the atmosphere. Although these models do very well compared with earlier, more primitive models, there is considerable room for improvement in forecasting ability, especially in the tropical latitudes. Medium-range forecast models continue to be improved periodically.

FUTURE OF NUMERICAL WEATHER PREDICTION

The greatest advancements in numerical weather prediction may arise through continued improvement in quality and quantity of weather data used as input in the models, through better methods of data interpolation, through improved algorithms of atmospheric behavior, and in the use of increasingly sophisticated supercomputers capable of more and faster iterations. Some modelers believe that departing from a latitude-longitude grid would prevent clustering of grid points

near the poles, which now induces some problems in numerical computations.

Nowcasting, especially for rapidly developing severe weather conditions including severe thunderstorms, flash floods, and tornadoes, has received a great deal of attention by scientists and governments. Within the United States, it is hoped that the inclusion of the Nexrad Doppler radar observations into operational models will lead to better nowcasting.

Forecasts have been improving in a slow, steady fashion since the introduction of supercomputers. However, even for short-range time periods, there is still room for improvement. One method for improving any model's forecasting ability is to make ensemble forecasts by running the same model many times, each time with slightly different initial conditions, and then pooling the results statistically. This technique yields better results than any single model run, based on comparing verifications of the individual and ensemble results for a large number of forecasts. A variation of this technique is the superensemble forecast in which forecasts from several different computer models runs or even ensemble runs are pooled. Superensemble forecasts have been found to show greater skill than ensemble forecasts. One way in which superensemble forecasts are thought to achieve greater skill is in the minimization of "forecast bias" in which any one model develops a spurious trend as the increasing numbers of iterations magnify small errors.

Worldwide, better weather satellite imagery for input to more detailed hemispheric models could lead to enhanced quality of numerical weather predictions. Greater availability of weather observations for the world's oceans would also be helpful. While hemispheric models do a good job forecasting at midlatitudes, less skill is shown in tropical forecasting, as determined by model verification. Improved models incorporating more detailed initial data could provide better forecasts in the tropics.

One of the greatest challenges confronting numerical weather prediction lies in the development of reliable long-range predictive models. Long-range predictions often fail because of nondeterministic or random events. El Niño is an example. El Niños are now viewed as virtually nondeterministic events; when a strong El Niño is occurring, the output from some long-range models may be subjectively altered by the meteorologists to reflect past historical events. The goal of numerical weather prediction is to provide objective forecasts using a model that verifies more frequently than do subjective forecasts, even those formulated by highly trained meteorologists. To forecast El Niño better, it would be advantageous to develop a predictive model that could successfully incorporate Pacific Ocean surface temperatures as initial conditions. With longer and better data records, and further study on why existing long-range predictions have failed, better models may be developed.

Dennis G. Baker and Anita Baker-Blocker

CROSS-REFERENCES

Atmosphere's Evolution, 1816; Atmosphere's Global Circulation, 1823; Atmospheric and Oceanic Oscillations, 1897; Atmospheric Inversion Layers, 1836; Climate, 1902; Clouds, 1909; Drought, 1916; El Niño and La Niña, 1922; Floods, 2022; Greenhouse Effect, 1867; Hurricanes, 1928; Ice Ages, 1111; Lightning and Thunder, 1935; Monsoons, 1940; Ocean-Atmosphere Interactions, 2123; Precipitation, 2050; Satellite Meteorology, 1945; Seasons, 1951; Storms, 1956; Tornadoes, 1964; Van Allen Radiation Belts, 1970; Volcanoes: Climatic Effects, 1976; Weather Forecasting, 1982; Weather Modification, 1992; Wind, 1996.

BIBLIOGRAPHY

Bader, M. J., G. S. Forbes, J. R. Grant, and R. Lilley. *Images in Weather Forecasting: A Practical Guide for Interpreting Satellite and Radar Imagery*. New York: Cambridge University Press, 1997. Discusses basic techniques for interpreting mid-latitude satellite images of weather systems in North America and Europe. Radar images are explained in terms of airflow and weather patterns.

Bluestein, Howard B. *Synoptic-Dynamic Meteorology in Midlatitudes*. New York: Oxford University Press, 1992. This two-volume work, designed for upper-level college students, requires a strong background in physics and calculus. For readers with this background, this is a fine reference work.

Browning, K. A., and R. J. Gurney. *Global Energy and Water Cycles*. New York: Cambridge University Press, 1999. This excellent book features reviews by Europeans and Americans on how weather forecasting models are developed and used.

Burroughs, William J. *Watching the World's Weather*. New York: Cambridge University Press, 1991. This book discusses the techniques for the collection and handling of meteorological data from radar and satellites. Data from these observations are used as initial conditions in numerical weather prediction models.

Djuric, Dusan. *Weather Analysis*. Englewood Cliffs,

N.J.: Prentice-Hall, 1994. Written as a text for a college course in weather analysis, this illustrated text, which uses examples from real weather events, covers the basic mathematical approach to numerical weather analysis. Background knowledge of college physics and calculus is essential for understanding this work.

Friedman, Robert Marc. *Appropriating the Weather*. Ithaca, N.Y.: Cornell University Press, 1989. This biography of Vilhelm Bjerknes traces the quest for techniques to accurately predict the weather and how the quest for numerical weather prediction lost out to the use of qualitative weather maps during the first half of the twentieth century.

Hodgson, Michael, and Devin Wick. *Basic Essentials: Weather Forecasting*. 2d ed. Chester, Conn.: Globe Pequot, 1999. This book for nonmeteorologists describes how people are trained to make forecasts using human observations of the weather around them. Before attempting to use computer forecasts, people should have a basic understanding of the physical principles operating in the atmosphere.

Katz, Richard W., and Allan H. Murphy. *Economic Value of Weather and Climate Forecasts*. New York: Cambridge University Press, 1997. An overview of numerical weather prediction and methods used to verify forecasts, this book offers a high-level appraisal of the relative value of objective weather forecasts.

Krishnamurti, T. N., et al. "Improved Weather and Seasonal Climate Forecasts from Multimodel Superensemble." *Science* 258 (September 3, 1999): 1548-1550. Moving to superensemble forecasting is shown to outperform all models and to have higher skills than forecasts based solely on ensemble forecasting. The authors urge establishment of a multimodel superensemble center for weather and seasonal climate forecasting.

Monmonier, Mark S. *Air Apparent: How Meteorologists Learned to Map, Predict, and Dramatize Weather.* Chicago: University of Chicago Press, 1999. Using weather maps, this book shows the growth of technology used in weather forecasting.

Newton, Chester, and E. O. Holopainen. *Extratropical Cyclones.* Boston: American Meteorological Society, 1990. Containing twelve review papers, this book provides many insights into modern meteorology and understanding the weather of the Northern Hemisphere. While a good background in physics is helpful to the reader, only two chapters require a grasp of calculus.

WEATHER MODIFICATION

Human activities can cause intentional or accidental changes in local weather situations. Many intentional weather modification experiments have focused on creating conditions favorable for agriculture.

PRINCIPAL TERMS

CLOUD SEEDING: the injection of cloud-nucleating particles into clouds likely to enhance precipitation

DYNAMIC MODE THEORY: a theory proposing that enhancement of vertical movement in clouds increases precipitation

FOG DISSIPATION: removal of fog by reducing the number of droplets, decreasing the radius of droplets, or both

HAIL SUPPRESSION: a technique aimed at lessening crop damage from hailstorms by converting water droplets to snow to prevent hail formation or, alternatively, by reducing hailstone size

HYGROSCOPIC PARTICULATES: minute particles that readily take up and retain moisture

STATIC MODE THEORY: a theory assuming that natural clouds are deficient in ice nuclei; therefore, a "window of opportunity" exists for cloud seeding whereby clouds must be within a particular temperature range and contain a certain amount of supercooled water

SUPERCOOLED: cooled below the freezing point without crystallizing or becoming solid

DELIBERATE WEATHER MODIFICATION

Inadvertent weather modification, including fog formation and increases or decreases in precipitation downwind from large industrial sites, creates problems in some locales. Scientific attempts to modify weather deliberately have been pursued since World War II. The most popular techniques involve cloud seeding, the injection of cloud-nucleating particles into likely clouds to alter the physics and chemistry of condensation. Proponents of this technique claim that it may enhance precipitation amounts by 5 to 20 percent. However, some scientists believe that deliberate efforts to enhance precipitation often yield questionable results, even in favorable situations. In 1977 the United Nations passed a resolution prohibiting the use of weather modification for hostile purposes because of the threat to civilians. The United States signed the resolution but has continued defense research on operational weather modification in battlefield situations, as summarized in the U.S. Air Force position paper "Weather as a Force Multiplier: Owning the Weather in 2025."

Studies have field-tested various methods of weather modification; results have varied widely. Weather modification has been attempted in many countries around the world, by government agencies, agricultural cooperatives, private companies, and research consortiums. In agricultural areas farmers are convinced that hail suppression and precipitation augmentation have been achieved by weather modification. In some of these same locales, meteorologists have been unable to determine if weather modification has produced any change from what would have occurred without intervention. Attempts to duplicate weather modification efforts that have apparently been successful in one locale have often been met with questionable results. Meteorologists occasionally disagree among themselves as to whether a specific attempt at weather modification has succeeded. Reexamination of data from American studies undertaken in the past has led many scientists to conclude that the efficacy of cloud seeding has been overstated.

It should be understood that it is impossible to change the climate of an entire region for a desired outcome through weather modification. It is also impossible to end a drought by seeding clouds. Cloud seeding for agricultural purposes assumes that some enhancement of regional rainfall amounts over the course of the growing season will increase crop yields. Weather modification for hail suppression assumes that reduction in regional crop losses over the growing season is an attainable goal.

INADVERTENT WEATHER MODIFICATION

Pulp and paper mills produce huge quantities of large-and giant-diameter cloud condensation nuclei (CCN); downwind of these mills, precipitation appears to be enhanced about 30 percent above what was observed prior to construction of the mills. It is also thought that the heat and moisture emitted by these mills may play an active role in precipitation enhancement. One specific study of a kraft paper mill near Nelspruit in the eastern Transvaal region of South Africa has indicated that storms modified by the mill emissions lasted longer, grew taller, and rained harder than other nearby storms occurring on the same day. Radar measurements supported the theory that hygroscopic particulates released by this mill accelerated or amplified growth of unusually large-diameter raindrops.

An egregious example of inadvertent weather modification is the formation of ice fog over Arctic cities in Siberia, Alaska, and Canada. During winter, cities such as Irkutsk, Russia, and Fairbanks, Alaska, experience drastic reductions in visibility as particles released by combustion act as nuclei for the formation of minute ice crystals. No techniques are available to modify ice fogs.

During an investigation of the meteorological effects of urban St. Louis, Missouri, conducted during the 1970's, it was found that urban summer precipitation was enhanced by 25 percent relative to the surrounding area. Most of the increased precipitation occurred in the late afternoon and evening as a result of convective activity. The frequency of summer thunderstorms was enhanced by 45 percent, and the frequency of summer hailstorms was higher by 31 percent over the city and adjacent eastern and northeastern suburbs. During the late 1960's, studies demonstrated that widespread burning of sugar cane fields in tropical areas released large numbers of cloud condensation nuclei. Downwind, rainfall decreases of about 25 percent were noted.

CLOUD SEEDING

For millennia, people attempted to influence the weather by using prayers and incantations. Sometimes rain followed, and sometimes no rain fell for extended periods. Scientists began attempting various techniques to modify weather during World War II. In 1946 Vincent Schaefer of the General Electric Research Laboratory observed that dry ice put into a freezer with supercooled water droplets caused ice crystals to form. On November 13, 1946, Schaefer demonstrated that dry ice pellets dropped from an aircraft into stratus clouds caused liquid water droplets to change to ice crystals and fall as snow. Bernard Vonnegut, a coworker, determined that silver iodide (AgI) particles also caused ice crystals to form. Project Cirrus involved apparently successful scientific attempts to seed clouds with ground-based AgI generators in New Mexico. These researchers then tried seeding a hurricane on October 10, 1947. The hurricane changed direction, making landfall in Georgia, resulting in a number of lawsuits against General Electric.

Early cloud-seeding experiments were empirical. AgI was dropped from aircraft, shot into clouds by rockets, or dispersed from ground-based generators. Researchers could selectively seed a pattern such as an "L" into a supercooled stratus cloud and see a visible "L" appear, thus "proving" that they could achieve results. When any rain occurred near a seeded area, it was attributed to the intervention. The apparent success of cloud seeding using AgI caused the technique to be modified and adopted in France, Canada, Argentina, Israel, and the Soviet Union. Wine-growing regions such as the south of France and Mendoza, Argentina, installed ground-based AgI generators. The Soviet Union opted for rocket-borne AgI, which was launched in agricultural areas during thunderstorms in an effort to suppress hail.

In 1962 the U.S. Navy and Weather Bureau began an ambitious cooperative plan to modify hurricanes called Project Stormfury. Only a few hurricanes were seeded in attempts to reduce the intensity of the storms. Proponents of Stormfury suggested that seeding of Hurricane Debbie in 1969 caused a reduction of 30 percent in wind speed on one day. The following day, no seeding was done, followed by another seeding attempt. The second seeding was thought to have caused a 15 percent reduction in wind speeds. Proponents believed that 10 to 15 percent reductions in wind speeds might result in a 20 to 60 percent reduction in storm damage if similar results could be achieved by seeding other hurricanes. Stormfury was terminated in the late 1970's, with no definitive results.

During winters between 1960 and 1970, the Climax I and Climax II randomized cloud seeding studies were conducted in the Colorado Rockies. Although it was initially thought that precipitation enhancements on the order of 10 percent may have resulted, more recent examination of the results appears to indicate that cloud seeding had no statistically discernible effect on precipitation. During the Vietnam War, the U.S. military attempted to increase precipitation along the Ho Chi Minh Trail in an effort to impede enemy forces. In the United States during the 1970's, some entrepreneurs deployed ground-based AgI generators in selected agricultural regions, billing farmers for their services. Aircraft delivery of AgI became increasingly popular. By the late 1990's, a number of private companies were delivering airborne cloud seeding services in various areas worldwide.

Cloud physicists have explored why cloud seeding might be effective. The evidence suggests that seeding increases the size of droplets or ice crystals, allowing them to fall as precipitation. Two concepts have emerged: a static mode theory, which assumes that natural clouds were deficient in ice nuclei, and a dynamic mode theory, which assumes that enhancement of vertical movement in clouds increases precipitation. The static mode assumes that a "window of opportunity" exists for seeding cold continental clouds during which clouds must be within a particular temperature range and contain a certain amount of supercooled water.

Fog Dissipation and Hail Suppression

During World War II, when improvements in visibility were crucial for military operations, efforts were made to dissipate fog. Fog can be dissipated by reducing the number of droplets, decreasing the radius of droplets, or both. Decreasing droplet radius by a factor of three through evaporation can provide a ninefold increase in visibility. Possible methods of fog re-

moval include using dry ice pellets or hygroscopic materials, heating the air, and mixing the foggy air with drier air. Airports that are plagued by supercooled fog in winter, such as Denver and Salt Lake City, can dissipate the fog by dropping dry ice pellets. Dry ice causes some liquid water droplets to freeze and grow, evaporating the remaining liquid droplets and allowing the larger frozen ice crystals to fall. One way of clearing fog at military airports when there is a shallow radiation fog close to the ground is to use helicopters to provide mixing. Jet engines can be used as heaters, an expensive technique that has been used operationally in France.

Farmers and vintners worldwide fear damaging hailstorms that can devastate crops. There are three approaches to suppressing hail damage: converting all liquid water droplets to snow to prevent hail formation, seeding to promote growth of many small hailstones instead of larger damaging hail, and introducing large condensation nuclei to reduce the average hailstone size. Most weather modification proponents believe that seeding with lead iodide or AgI to cause many small hailstones to form can substantially reduce hailstone size. Because small hailstones are less damaging than large ones, this technique could potentially lessen (but not eliminate) crop losses. It has been claimed that rocket-borne lead iodide seeding in Bulgaria reduced crop losses from hail by 50 to 60 percent. Similar seeding operations in the former Soviet Union were said to have reduced crop damage by 50 to 95 percent. A randomized study in North Dakota over four summers claimed that seeding helped reduce hail severity.

Anita Baker-Blocker

Cross-References

Atmospheric and Oceanic Oscillations, 1897; Climate, 1902; Clouds, 1909; Drought, 1916; El Niño and La Niña, 1922; Hurricanes, 1928; Lightning and Thunder, 1935; Monsoons, 1940; Satellite Meteorology, 1945; Seasons, 1951; Storms, 1956; Tornadoes, 1964; Van Allen Radiation Belts, 1970; Volcanoes: Climatic Effects, 1976; Weather Forecasting, 1982; Weather Forecasting: Numerical Weather Prediction, 1987; Wind, 1996.

BIBLIOGRAPHY

Baer, F., N. L. Canfield, and J. M. Mitchell, eds. *Climate in Human Perspective: A Tribute to Helmut E. Landsberg.* Boston: Kluwer, 1991. Several essays deal with aspects of weather modification, such as "Five Themes on Our Changing Climate," by J. Murray Mitchell, and "Climate of Cities," by Timothy R. Oke.

Cotton, William R., and Roger A. Pielke. *Human Impacts on Weather and Climate.* New York: Cambridge University Press, 1995. A comprehensive overview of weather modification.

Hartmann, Dennis L. *Global Physical Climatology.* San Diego, Calif.: Academic Press, 1994. A thorough, fairly technical textbook on physical climatology. Requires some background in climatology but includes a very good discussion of anthropogenic climate change in chapter 12.

Hess, Wilmot N. *Weather and Climate Modification.* New York: Wiley, 1974. Provides a look at the "glory days" when scientists and governments enthusiastically embraced weather modification.

Hopler, Paul. *Atmosphere: Weather, Climate, and Pollution.* New York: Cambridge University Press, 1994. This brief book offers a wonderful introduction to the study of the Earth's atmosphere and its components. Chapters explain the causes and effects of global warming, ozone depletion, acid rain, and climatic change. Illustrations, color maps, and index.

Horel, John, and Jack Geisler. *Global Environmental Change: An Atmospheric Perspective.* New York: John Wiley & Sons, 1997. Global warming and stratospheric ozone depletion form the major topics of this readable book geared for beginning-level undergraduate students. An interesting feature is its discussion of the many Internet sites where graphics and information on global environmental change can be obtained.

House, Tanzy J., James B. Near, William B. Shields, Ronald J. Celentano, Ann E. Mercer, and James E. Pugh. *Weather as a Force Multiplier: Owning the Weather in 2025.* Maxwell Air Force Base, Ala.: Air University, 1996. Discusses military weather modification.

Lutgens, Frederick K., and Edward J. Tarbuck. *The Atmosphere: An Introduction to Meteorology.* 7th ed. Upper Saddle River, N.J.: Prentice Hall, 1998. A beginning college textbook. The authors have other introductory Earth science books that are very understandable for the beginner. This profusely illustrated book has twelve pages devoted to tornadoes. Contains a glossary and seven separate appendices to explain everything from metric conversions to the reading of daily weather charts.

Mather, Graeme K. "Coalescence Enhancement in Large Multicell Storms Caused by the Emissions from a Kraft Paper Mill." *Journal of Applied Meteorology* 91 (1991). A detailed study of the effects of kraft paper mills on precipitation.

Rangno, A. L., and P. V. Hobbs. "Further Analyses of the Climax Cloud Seeding Experiments." *Journal of Applied Meteorology* 93 (1993). Discusses difficulties in assessing the effects of cloud seeding.

Semonin, Richard G., and Stanley A. Changnon. "METROMEX: Lessons for Precipitation Enhancement in the Midwest." *Journal of Weather Modification* 7 (1975). Presents information gleaned from a study of urban effects on local weather.

Stevens, William Kenneth. *The Change in the Weather: People, Weather, and the Science of Climate.* New York: Delacorte Press, 1999. Stevens describes various natural and human-induced causes of changes in the climate. Includes a twenty-page bibliography and an index.

Strahler, Alan, and Arthur Strahler. *Physical Geography: Science and Systems of the Human Environment.* New York: John Wiley & Sons, 1997. A thorough, well-illustrated book containing considerable information about atmospheric processes and issues. Suitable for college students.

WIND

Wind is the horizontal movement of air; it blows as a result of differences in atmospheric pressure. Pressure differences may develop on a local or global scale in response to differences in heating, which affect the density of air and, therefore, its pressure.

PRINCIPAL TERMS

CONSTANT PRESSURE CHART: a chart that shows the altitude of a constant pressure, such as 500 millibars

CONVERGENCE: air flowing in toward a central point

DIVERGENCE: a net outflow of air from a specified region

GEOSTROPHIC WIND: an upper-level wind that flows in response to a balance of pressure gradient and Coriolis acceleration

HURRICANE-FORCE WIND: a wind with a speed of 64 knots (118 kilometers per hour) or higher

ISOBAR: a line of equal pressure

LOCAL WINDS: winds that, over a small area, differ from the general pressure pattern owing to local thermal or orographic effects

PRESSURE GRADIENT: the rate of change of pressure with distance at a given time

RAWINSONDE: a radiosonde tracked by radar in order to collect wind data in addition to temperature, pressure, and humidity

FACTORS AFFECTING WIND FLOW

Wind, as defined by the meteorologist, is the horizontal movement of air. Differences in heating and internal motion in the atmosphere create differences in atmospheric pressure; when a change in pressure over distance is established, air accelerates down this pressure gradient from higher to lower pressure. The acceleration of this air depends on the amount of pressure change over a given distance.

Moving air associated with pressure change over a distance will either spread out over the surface (diverge) or will flow inward (converge). High-pressure areas are regions of divergence, and low-pressure areas are regions of convergence. The force associated with the air moving from high to low pressure is called the pressure gradient force. Pressure gradient force sets the wind into motion. If it were the only force affecting the wind, then winds would blow directly from high to low pressure. However, other forces affect wind direction and velocity.

A second major factor affecting wind flow is Coriolis acceleration, which results from the Earth turning on its axis. An object moving over the surface of the Earth, except at the equator, moves in a curved path when observed from the rotating

Earth. In the Northern Hemisphere, there is an acceleration to the right of the path of motion; in the Southern Hemisphere, the acceleration is to the left. Thus, in the Northern Hemisphere, a wind blowing from north to south becomes a northeast wind, and a wind blowing from south to north becomes southwesterly. The reverse occurs in the Southern Hemisphere.

A third force affecting wind flow is centrifugal acceleration. Air currents seldom move on a straight path for long but rather tend to develop a curved pattern. When this type of flow pattern evolves, centrifugal acceleration is directed away from the center of the cell or curve. This acceleration is directed outward from both high- and low-pressure cells. Therefore, airflow around a high pressure, centrifugal acceleration is in the same direction as pressure gradient and in a direction opposite to pressure gradient around a low pressure. Centrifugal acceleration is more important in smaller circulations such as hurricanes and tornadoes than in larger, midlatitude cyclones.

A fourth factor affecting wind velocity and direction is frictional drag, which works in a direction opposite to wind motion; therefore, friction tends to slow wind velocity. A decrease in wind velocity, however, is accompanied by a decrease in

Coriolis acceleration, which causes a slight change in wind direction back toward the direction of the pressure gradient. Frictional drag is at a maximum over land where an uneven surface consisting of trees, buildings, and hills provides barriers to the even flow of wind. Also, friction affects the flow of wind only in the first kilometer of the atmosphere. Wind direction and velocity in the lowest kilometer of the atmosphere is based on the sum of pressure gradient acceleration, Coriolis acceleration, centrifugal acceleration, and frictional drag.

Above 1 kilometer, winds blow in response to pressure gradient, Coriolis, and centrifugal acceleration. Frictional deceleration is negligible or completely absent. Consider a situation in which pressure is distributed in a linear fashion so that lines connecting points of equal pressure are straight. In this situation, pressure gradient acceleration and Coriolis acceleration are the only forces working on the wind. Here, pressure gradient force is balanced by Coriolis acceleration so that the wind flows in a direction parallel to the isobars. These winds are called geostrophic winds. Around circular highs and lows above the friction level, pressure gradient acceleration is balanced by both Coriolis and centrifugal acceleration. Thus, winds blow parallel to isobars in a clockwise direction around highs and in a counterclockwise direction around lows. In the Southern Hemisphere, the reverse is true. These winds thus described are called gradient winds.

Monsoon and Pressure Changes

A seasonal wind system, called the monsoon, that changes direction from winter to summer exists over eastern Asia and the adjacent oceans. In winter, over the large landmass of Asia, air is cooled, and a cold, dense high-pressure center forms with a clockwise circulation of winds about it. Generally, these winds flow from land to sea during winter. In summer, a thermal low forms over India, and the airflow pattern reverses itself with cyclonic flow bringing air on shore from the ocean. Reinforcing the thermal low is the migration of the Intertropical Convergence Zone northward over India. Moreover, the jet stream breaks down during summer, which reinforces the monsoon flow. With this annual wind-flow reversal, the climate of Asia is greatly affected. The offshore winds in winter bring dry weather to much of eastern Asia. Conversely, the onshore winds of summer bring copious amounts of precipitation to India and adjacent areas of southeast Asia.

Daily changes in temperature at many places around the world result in daily pressure changes, which cause distinctive wind patterns. One such system is that of the land and sea breezes. This system develops along coastal areas and along shorelines of large lakes. During the day, as the land heats rapidly, the air above heats, expands, and becomes less dense, forming a thermal low. The warm, buoyant air rises, and cooler air from the water surface flows in to replace the rising air. In this fashion, a sea breeze develops during the day and usually reaches a peak in mid-afternoon, when the temperature high is attained. At night, conditions are reversed. The land cools more rapidly than water. In this way, the pressure relationship between land and water is reversed day to night. At night, pressure is higher over land and lower over water, so air flows from land to water, producing a land breeze. The land breeze is usually not as well developed as is the sea breeze because the temperature contrast between land and water is not as great at night as it is during the day. Another wind system that has a day-to-night change in wind direction is that of the mountain and valley breezes. During the day, the mountain slopes warm the air, and it expands. The warm, less dense air rises and is called a valley breeze after its place of origin. At night, the slopes cool, and the air's density increases. The cool, dense air flows downslope in response to gravity and is called a mountain breeze.

Local Winds

Several local winds occur in response to topographic peculiarities and are difficult to explain on the basis of pressure patterns as they might appear on a weather map. The chinook in the Rockies and the foehn in the Alps of Europe result from a combination of topographic effects and large-scale atmospheric systems. In response to these systems, winds flow down the lee side and are heated by compression. The warming brought by these winds is often rapid.

The sirocco (khamsin in Egypt and sharov in Israel) and the haboob are hot, dusty winds that occur on flat terrain. The sirocco precedes a low-pressure system moving across the Sahara Desert.

As it crosses the Mediterranean Sea, it picks up moisture and becomes a hot, humid wind by the time it reaches the coast of Europe. The haboob is created by air spilling out of the base of a thunderstorm and attains high speeds and picks up small soil particles, creating a sand storm extending upward as high as 1 kilometer or higher.

A katabatic wind is a cold wind flowing downslope from an ice field or glacier. Wind velocities range from as little as 10 knots up to hurricane speeds. One such wind is the bora, which originates in the Soviet Union and blows out across the Adriatic coast of Yugoslavia. In France, a wind known as the mistral blows out of the French Alps and through the Rhone Valley to chill the Riviera along the Mediterranean Sea.

STUDY OF WIND

A number of instruments are used to collect data about wind direction and velocity at the surface or in the upper troposphere. The wind vane is commonly used for determining surface wind direction. Most wind vanes are long arrows with a tail. The arrow is attached to a vertical pole about which it can move freely. The arrow always points into the wind, giving the wind direction.

The anemometer is an instrument used to record wind velocity. It consists of three hemispherical cups attached to tri-crossbars, which are in turn attached to a vertical shaft. The cups catch the wind, causing the shaft to turn, and the wind speed is recorded by an electrical counting device at the base of the shaft. An instrument used for recording both wind direction and velocity is the aerovane. It consists of a three-bladed propeller mounted on the end of a streamlined rod, with a vertical fin at the opposite end. The propeller rotates at a rate proportional to the wind speed. The fin and aerodynamic shape keep the propeller blades facing into the wind, so direction is easily determined. When a recorder, often remote, is connected, a continuous record of both wind velocity and direction can be kept.

A series of instruments also have been developed to determine wind directions and velocities at higher levels. One is the pilot balloon, a small balloon released at the surface that rises at a known rate. The balloon is tracked using a small telescope called a theodolite, and periodic measurements of the balloon's horizontal and vertical angles are taken, giving speed and direction of the winds carrying it. The pilot balloon principle can also be applied to a radiosonde accent. Measurements of the radiosonde's vertical and horizontal angles, taken periodically along with its distance from the observing station, can supply information on wind direction and speed.

A rawinsonde can be tracked using radar, so wind speed and direction can be obtained. Radar can also be used in conjunction with rockets to collect wind data at a distance above 30 kilometers. One type of rocket ejects a parachute carrying an instrument that can be tracked by radar. Another type of rocket ejects metallic strips at predetermined levels that can also be tracked by radar. Doppler radar can be used to determine direction and speed of wind. Doppler radar measures speeds of objects moving toward or away from the antenna. When a signal is sent out and reflected from a raindrop or ice crystal, the returning signal will have a high frequency if the particle is moving toward the radar and a low frequency if it is moving away. One drawback of Doppler radar is that velocities of objects at right angles to the unit cannot be determined, so to achieve a three-dimensional effect, two or more units must be used.

Wind directions and speeds are plotted on charts using a symbol called the wind arrow. The shaft of an arrow shows wind direction, while barbs on the end of the arrow indicate speed. A barb represents 10 knots, one-half barb 5 knots, and a flag (a triangle-shaped symbol) represents 50 knots. These symbols may be used singularly or in combination to show any wind speed. On a surface chart, the wind arrows point out from a station in the direction from which the wind is coming. In the upper air above the friction level, the barbed end points into the wind.

Winds above the friction level are plotted on constant pressure charts. A constant pressure chart is drawn using contour lines to show the elevation above the Earth's surface of a constant pressure level, such as the 500-millibar level. When pressure is particularly high in an area in relation to surrounding areas, the height of a constant pressure surface is higher than surrounding areas, and heights of a low pressure region are lower than surrounding regions. The average elevation of the 500-millibar level is 5.5 kilometers but can vary from less than 5 to more than 6 kilometers.

Various constant pressure charts are used, ranging from just above the surface, such as the 850-millibar level, up to the tropopause at roughly the 200-millibar level. With the use of these constant pressure charts, meteorologists can gain a sense of the three-dimensional wind-flow profile from the Earth's surface up to the tropopause.

SIGNIFICANCE

Winds, in conjunction with temperature and humidity, can greatly affect human comfort. The effects of wind influence the exchange of heat between the human body and the atmosphere. The body, particularly the surface of the skin, is continually exchanging heat with the environment. On a cold day when a wind is blowing, air molecules impact the skin, then move away, taking body heat with them. Clothing provides insulation, creating a shallow layer of warm air molecules, which form a shield that protects the skin from heat loss.

Humankind's use of wind power may stem from the use of winds to propel sailboats or ships. Sails have been used as power sources on ships and boats for thousands of years; wind was the chief source of power for water transportation until the use of steam in the latter part of the nineteenth century. The next step in the use of wind for power was through the use of windmills. The first known windmill appeared in Europe in 1105, and by the following century, thousands of windmills were in use in Europe. Then, as the burning of coal as a source of power became less expensive, it and other energy sources replaced windmills. In the early part of the twentieth century, windmills became popular as an inexpensive means of pumping water for agricultural uses on farms and ranches. Today, wind power is again being considered as a partial solution to growing energy needs. The assets of using wind power are that windmills are nonpolluting, and they are not limited to daylight hours as are solar cells. The liabilities of using wind power

are several in number: Windmills can only be used in windy areas where wind flow is steady and neither too weak nor too strong; a weak wind will not turn the blades, and a strong wind might damage the machine. Windmills detract from the aesthetics of the landscape, and their cost factor can be quite high. Finally, the amount of modern energy needs that could be satisfied by wind power is low, at best probably no more than 5 percent.

Ralph D. Cross

CROSS-REFERENCES

Atmosphere's Global Circulation, 1823; Atmospheric and Oceanic Oscillations, 1897; Climate, 1902; Clouds, 1909; Drought, 1916; El Niño and La Niña, 1922; Floods, 2022; Hurricanes, 1928; Lightning and Thunder, 1935; Meteorological Satellites, 2786; Monsoons, 1940; Precipitation, 2050; Satellite Meteorology, 1945; Seasons, 1951; Storms, 1956; Tornadoes, 1964; Van Allen Radiation Belts, 1970; Volcanoes: Climatic Effects, 1976; Weather Forecasting, 1982; Weather Forecasting: Numerical Weather Prediction, 1987; Weather Modification, 1992.

A granite outcrop eroded by windblown sand, Llano de Caldera, Atacama Province, Chile. (U.S. Geological Survey)

BIBLIOGRAPHY

Ahrens, C. Donald. *Meteorology Today: An Introduction to Weather, Climate, and the Environment.* 3d ed. New York: West, 1988. Three chapters in the book—11, 12, and 13—specifically treat atmospheric motion and wind systems. Chapter 11 covers causes of wind flow and the ways in which wind data are collected and portrayed. The next chapter portrays small-scale and local wind systems. The final chapter concerns global wind systems.

Blumenstock, David I. *The Ocean of Air.* New Brunswick, N.J.: Rutgers University Press, 1959. The treatment of winds in this book is rather unique, based on the way the discussion is organized. In chapter 3, the author presents the nature of the wind, including mechanics. The subsequent chapters concern winds of the sea and the land-sea borders and the winds of the land.

Chang, Jen-hu. *Atmospheric Circulation Systems and Climates.* Honolulu, Hawaii: Oriental Publishing, 1972. An excellent nontechnical book, it covers wind systems of all types, from the primary or global system to wind systems of a more regional nature. Of particular note are chapters describing polar fronts and cyclones and the circulation system of the Arctic and Antarctic, subjects not normally discussed extensively in wind analyses.

Gedzelman, Stanley David. *The Science and Wonders of the Atmosphere.* New York: John Wiley & Sons, 1980. In chapters 15 and 20 is presented a thorough coverage of subglobal wind systems and their causes and characteristics. Chapter 15 treats factors causing air to flow as winds, and chapter 20 covers local wind systems of the world. Winds are also mentioned in other parts of the book as they are related to other climatic variables.

Gombosi, Tamaas I. *Physics of the Space Environment.* New York: Cambridge University Press, 1998. This college-level text explores the field of atmospheric physics to determine the make-up of the upper atmosphere and the space environment and how these forces affect the Earth's atmosphere and weather cycles. A focus is placed on solar winds and the magnetosphere. Intended for the advanced reader. Illustrations, index, and bibliography.

Hidy, George M. *The Winds: The Origins and Behavior of Atmospheric Motion.* Princeton, N.J.: Van Nostrand, 1967. A very thorough, although somewhat technical, analysis of how and why winds flow. A cursory knowledge of physics and mathematics is recommended as a prerequisite to reading this book, which covers air flow observations, the static atmosphere, forces and motion in large-scale winds, and eddies, vortices, and turbulence.

Hopkins, J. S. *The Accuracy of Wind and Wave Forecasts.* Sudbury, England: HSE, 1997. In this brief pamphlet, Hopkins examines and critiques the techniques and protocol associated with the science of wind and weather forecasting. Illustrations and bibliography.

Lutgens, Frederick K., and Edward J. Tarbuck. *The Atmosphere: An Introduction to Meteorology.* 7th ed. Upper Saddle River, N.J.: Prentice-Hall, 1998. A beginning college textbook. The authors have other introductory Earth science books that are very understandable for the beginner. This profusely illustrated book has twelve pages devoted to tornadoes. Contains a glossary and seven separate appendices to explain everything from metric conversions to the reading of daily weather charts.

Lydolph, Paul E. *Weather and Climate.* Totowa, N.J.: Rowman & Allanheld, 1985. Chapters 5 and 12 cover the various aspects of wind flow—the former, the forces that affect wind velocity and direction, and the latter, the different types of local winds experienced in various parts of the world. The discussions are brief but to the point. This source provides a general, basic definition of wind systems.

Spiegel, Herbert J., and Arnold Gruber. *From Weather Vanes to Satellites: An Introduction to Meteorology.* New York: John Wiley & Sons, 1983. Chapter 8 is a general coverage of winds of all types. Also presented are the factors governing wind flow, and described are global, secondary, and tertiary wind systems around the world. Included is a section on how wind observations are made and some of the instruments used. The narrative is well written and presented in a style that makes it easy to understand.

Stevens, William Kenneth. *The Change in the Weather: People, Weather, and the Science of Climate.* New York: Delacorte Press, 1999. Stevens describes various natural and human-induced causes of changes in the climate. Includes a twenty-page bibliography and an index.

3
HYDROLOGY

AQUIFERS

Aquifers are the source of water for approximately 40 percent of the U.S. population. The identification and protection of aquifers are important to the future of drinking water supplies in the United States.

PRINCIPAL TERMS

AQUIFER: any saturated geologic material that yields significant quantities of water to a well

CONE OF DEPRESSION: the depression, in the shape of an inverted cone, of the groundwater surface that forms near a pumping well

CONFINED AQUIFER: an aquifer that is completely filled with water and whose upper boundary is a confining bed; it is also called an artesian aquifer

CONFINING BED: an impermeable layer in the Earth that inhibits vertical water movement

GROUNDWATER: water found in the zone of saturation

PERMEABILITY: the ability of rock, soil, or sediment to transmit a fluid

POROSITY: the ratio of the volume of void space in a given geologic material to the total volume of that material

UNCONFINED AQUIFER: an aquifer whose upper boundary is the water table; it is also called a water table aquifer

WATER TABLE: the upper surface of the zone of saturation

ZONE OF SATURATION: a subsurface zone in which all void spaces are filled with water

WATER TABLE

To understand aquifers, one must understand how water occurs beneath the Earth's surface. The world's water supply is constantly circulating through the environment in a never-ending process known as the hydrologic cycle. Natural reservoirs within this cycle include the oceans, the polar ice caps, underground water, surface water, and the atmosphere. Water on the land surface that is able to infiltrate the ground becomes underground water. "Groundwater recharge" occurs when the infiltration reaches the water table.

Underground water exists in three different subsurface zones: the soil moisture zone, the intermediate vadose zone, and the zone of saturation. The soil moisture zone is found directly beneath the land surface and contains water that is available to plant roots. This zone is generally not saturated unless a prolonged period of rainfall or snowmelt has occurred. Water in this zone is held under tension by the attractive forces between soil particles and water molecules, or surface tension forces. The depth of this zone corresponds to the depth to which plant roots can grow.

Water able to infiltrate through the soil moisture zone may pass into the intermediate vadose zone before reaching the water table. This zone is always unsaturated, and the water it contains is held under tension. The thickness of a vadose zone depends on how close the water table is to the surface.

The water table forms the uppermost surface of the zone of saturation and is characterized by a water pressure equal to atmospheric pressure. It may be only a few meters below the land surface in humid regions and hundreds of meters below the surface in desert environments. In general, the water table mimics the land surface topography. If the water table intersects the land surface, the result is a lake, swamp, river, or spring. Below the water table, in the zone of saturation, geologic materials are completely saturated, and the water pressure increases with depth. Water contained within the zone of saturation is known as groundwater. When groundwater occurs in a particular type of geologic formation known as an aquifer, it can feed a well.

POROSITY

An aquifer can be defined as any Earth material—rock, soil, or sediment—that yields a significant quantity of groundwater to a well or spring.

The definition of "a significant quantity" varies according to the intended use; what constitutes an aquifer for an individual homeowner may be quite different from what constitutes an aquifer for a municipal supply. For a geologic formation to be useful as an aquifer, it must be able both to hold and to transmit water.

The ability of geologic materials to hold, or store, water is known as porosity. Simply stated, porosity is the volume of void space present divided by the total volume of a given rock or sediment. This proportion is usually expressed as a percentage. The higher this ratio is, the more void space there is to hold water. There are various types of porosity. Unconsolidated materials (soil and sediment) have pore spaces between adjacent grains referred to as intergranular porosity. The ratio of pore space to total volume depends on several factors, including particle shape, sorting, and packing. Loosely packed sediments composed of well-sorted, spherical grains are the most porous. Porosity decreases as the angularity of the grains increases because such particles are able to pack more closely together. Similarly, as the degree of sorting decreases, the pore spaces between larger grains become filled with smaller grains, and porosity decreases. Values of porosity for unconsolidated materials range from 10 percent for unsorted mixtures of sand, silt, and gravel to about 60 percent for some clay deposits. Typical porosity values for uniform sands are between 30 and 40 percent.

Rocks have two main types of porosity: pore spaces between adjacent mineral grains and voids caused by fractures. Rocks formed from sedimentary deposits, such as shale and sandstone, may have significant intergranular porosity, but it is usually less than the porosity of the sediments from which they were derived because of the compaction and cementation that take place during the process of transforming sediments into rock. Therefore, although sandstone porosities may be as high as 40 percent, they are commonly closer to 20 percent because of the presence of natural cements that partially fill available pore spaces. Igneous and metamorphic rocks are composed of tightly interlocked mineral grains and therefore have little intergranular porosity. Virtually all void space in such rocks is a result of fractures (joints and faults). For example, granite, a dense igneous rock, usually has a porosity of less than 1 percent, but it may reach 10 percent if the rock is fractured.

There are additional types of porosity that occur only in certain kinds of rocks. Limestone, a rock that is soluble in water, can develop solution conduits along fractures and bedding planes. In the extreme case, solution weathering may lead to the development of a cave, which has 100 percent porosity. The overall porosity of solution-weathered limestone sometimes reaches 50 percent. Rocks created by volcanic eruptions may contain void space in the form of trapped air bubbles (called vesicles), shrinkage cracks developed during cooling (called columnar joints), and tunnels created by flowing lava (called lava tubes). In extreme cases, the porosity of volcanic rocks may exceed 80 percent.

PERMEABILITY

The presence of void space alone does not constitute a good aquifer. It is also necessary for groundwater to be able to move through the geologic material in question. The ability of porous formations to transmit fluids, a property known as permeability, depends on the degree to which the void spaces are interconnected. Some high-porosity materials, such as clay and pumice, do not make good aquifers because the void spaces are largely isolated from one another. Materials that have high permeability include sand, gravel, sandstone, and solution-weathered limestone. Rocks with low porosities, such as shale, quartzite, granite, and other dense, crystalline rocks have low permeabilities, unless they are significantly fractured.

Groundwater moves along tortuous paths through the available void space in a given porous formation. Regardless of the material's permeability, groundwater flows much more slowly than surface water in a river. The velocity of stream flow may be measured in meters per second, whereas groundwater velocities commonly range between 1 meter per day and less than 1 meter per year, averaging about 17 meters per year in rocks. Underground rivers are uncommon, occurring only in cavernous limestone or lava tubes in volcanic terrane.

The geologic materials that make good aquifers are those that have both high porosity and high permeability. The response of any given aquifer to a pumping well will also depend, however, on its position beneath the surface and its relationship to the water table. Aquifers near the Earth's surface usually have the water table as their upper boundary. The thickness of these aquifers therefore changes as the water table rises or falls. An aquifer under these conditions is an unconfined aquifer, or water table aquifer. These types of aquifer are the easiest to exploit for a water supply, but they are also the easiest to contaminate. It is therefore important to delineate the extent of unconfined aquifers and to take measures to protect them from various forms of pollution.

Because the water table is free to fluctuate in unconfined aquifers, the amount of water supplied to a well reflects the gravity drainage of water from void spaces. The volume of water available from aquifer storage, or the "specific yield," therefore approaches the upper limit set by porosity. Some groundwater is unable to drain from void spaces under the influence of gravity, because it is tightly held by surface tension forces; this retained water, known as "specific retention," forms a thin film around individual grains. The highest values of specific yield occur in coarse-grained,

permeable aquifers, such as sand, gravel, and sandstone.

CONFINED AQUIFERS

Some aquifers, usually found at depth or those known as "inclined" aquifers, are completely filled with groundwater and bounded at the top by an impermeable layer called a confining bed, or aquiclude. The water in these confined, or artesian, aquifers is under pressure because of the weight of overlying formations and the fact that the confining bed does not allow groundwater to escape. If a well is placed in such an aquifer, the water level will usually rise above the base of the confining bed, creating an artesian well. In some cases, water may rise above the land surface at the point where the well is placed. This condition is known as a flowing artesian well. Water will flow freely out of such wells as long as the aquifer remains under pressure. Many of the Great Plains states (Kansas, Nebraska, and the Dakotas) are underlain by important shale layers. The original pressure in these aquifers was quite high because the sandstone beds are upwarped along the eastern front of the Rocky Mountains and Black Hills, where they receive groundwater recharge.

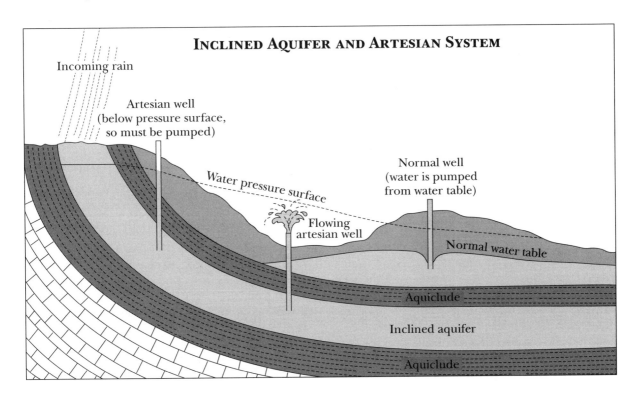

INCLINED AQUIFER AND ARTESIAN SYSTEM

Incoming rain

Artesian well (below pressure surface, so must be pumped)

Water pressure surface

Normal well (water is pumped from water table)

Flowing artesian well

Normal water table

Aquiclude

Inclined aquifer

Aquiclude

Confined aquifers supply water to a well not through the gravity drainage of void spaces but through compression of the aquifer as water pressure is reduced during pumping. The volume of water available from storage in a confined aquifer, or the "specific storage," is only a small fraction of the total volume and is therefore always much less than porosity. When confined aquifers are pumped to the extent that they become dewatered, the accompanying pressure reduction can lead to extensive aquifer compression and land surface subsidence. This problem is most serious in cases where the confining bed is composed of clay because the loss of fluid pressure beneath the clay causes water to be squeezed out of the clay layer by the weight of overlying formations. Once a clay layer is compressed in this way, it will not be able to reabsorb water, even if the surrounding materials become saturated again.

STUDY OF AQUIFERS

The first step in studying aquifers is to utilize data that have already been collected, such as geologic maps. These maps show the distribution of various geologic formations on the land surface and are therefore valuable tools for delineating the outcrop patterns of potential aquifers. If cross sections are also available, they can aid in the estimation of potential aquifer thicknesses and the identification of possible confined aquifers. Geologic maps, however, serve only as a preliminary tool in aquifer study. Any interpretations made from maps need to be verified by field descriptions and, if possible, pumping tests.

Topographic maps can be used to make generalizations about the groundwater flow system. Springs, lakes, streams, and swamps may indicate areas of groundwater discharge. Because the water table is usually a subdued version of the land surface, it may be possible to infer groundwater flow directions from the local topography. That can provide a clue as to where recharge areas occur. Other indications of recharge areas are topographic high points and a general lack of surface-water features. For groundwater recharge to occur, however, permeable materials must be exposed at the land surface.

Reliable estimates of aquifer properties require fieldwork. Often, samples are taken from the field for the purpose of determining aquifer properties, such as grain-size distribution or permeability, in the laboratory. Such tests, however, are performed only in small samples and may not be representative of the overall aquifer unit. That is particularly true in fractured rocks, where the movement of groundwater may be very difficult to predict. In such situations, the injection and monitoring of tracer dyes has proved helpful in understanding groundwater flow.

The most direct way to study aquifers is by boring holes and installing wells. By drilling, geologists are able to discover the exact nature of the subsurface materials. Detailed drilling logs are kept of the different layers and the depths at which they were encountered. Using their knowledge of geologic materials, properly trained geologists can predict which formations will constitute the best aquifers.

Once wells have been installed, additional information about the aquifer can be learned by conducting pumping tests (pumping the wells at known rates for extended periods). Because water moves relatively slowly through an aquifer, the pumping of a well removes groundwater faster than it can be replaced. The resulting water-level drawdown is called a cone of depression; it is a zone of dewatering near the pumping well that resembles an inverted cone. The exact shape of this cone is a function of the pumping rate and the aquifer properties. Therefore, certain aquifer characteristics, such as permeability, are discovered by studying the cone of depression created by a known pumping rate. To identify the shape of the cone, monitoring wells must be placed near the pumped well to detect drawdown at various distances.

The prediction of long-term well yields requires not only an understanding of aquifer properties but also a knowledge of the groundwater recharge rates. To determine the amount of water in a particular drainage basin that is available for groundwater recharge, it is necessary to develop a hydrologic budget for that basin. Hydrologic budgets attempt to account for all water inputs and losses from the basin in question. Inputs include precipitation, surface-water inflow, groundwater inflow, and water imported by humans (as for irrigation). Losses include surface-water runoff, groundwater outflow, evaporation, transpiration by plants, and water exported from the basin by humans. The difference between these inputs and losses equals the amount of water gained or lost by the groundwater reservoir.

SIGNIFICANCE

The term "aquifer" is not precise because its definition depends somewhat on the intended use. For an individual homeowner who requires only 7 to 20 liters per minute from a well, fractured bedrock may serve as a suitable source of water. On the other hand, only a few geologic materials are capable of delivering the larger water supply demanded by a municipality or industrial plant—often more than 1,500 liters per minute. In the case of high-capacity well requirements, the term "aquifer" is restricted to highly permeable geologic materials, such as unconsolidated sands and gravels, sandstone, and solution-weathered limestone.

Most private wells draw water from unconfined aquifers, which have the water table as an upper boundary. The saturated thicknesses of these aquifers can fluctuate as the water table rises or falls. Therefore, shallow wells completed in unconfined aquifers can be pumped dry as the water table declines during periods of prolonged drought. Because of their proximity to the land surface, these aquifers are also susceptible to groundwater contamination. Confined aquifers do not have a water table because they are fully saturated and capped with an impermeable confining bed. The natural groundwater quality in confined aquifers is not necessarily superior to that found in unconfined aquifers, but the presence of a low-permeability confining bed may help to protect these aquifers from surface contamination.

An understanding of aquifer characteristics is important to the proper use of groundwater. The determination of maximum sustainable well yields requires a knowledge of how the aquifer stores and transmits water. Such information is also needed to estimate how close adjacent wells can be to one another without causing interference (overlapping cones of depression). Pumping water out of an aquifer at rates exceeding the rate of groundwater recharge will eventually cause the depletion of a valuable water supply. Many arid regions in the western United States are facing this problem because of years of groundwater mismanagement. The overpumping of confined aquifers is especially troublesome because if these aquifers become dewatered, they compress, which may lead to subsidence of the land surface.

Delineating the extent of aquifers and particularly of their recharge zones is critical to the development of policies that protect groundwater supplies from pollution. In the case of confined aquifers, recharge zones may be restricted to areas where the confining bed is absent, so contaminants entering the ground through a relatively small zone could affect a large number of downgradient wells. Moreover, because groundwater moves very slowly through the subsurface, it takes a long time to be renovated once contaminated.

David L. Ozsvath

CROSS-REFERENCES

BIBLIOGRAPHY

Batu, Vedat. *Acquifer Hydraulics: A Comprehensive Guide to Hydrogeology Data Analysis.* New York: Wiley, 1998. Batu offers a detailed look at the role aquifers play in groundwater flow. This well-illustrated and indexed text is ac- companied by a CD-ROM.

Davis, S. N., and R. J. M. De Wiest. *Hydrogeology.* New York: John Wiley & Sons, 1966. A well-il- lustrated introduction to the geologic aspects of groundwater occurrence. The hydrologic cycle and its relationship to groundwater are discussed in chapter 2. Four chapters are dedicated to a discussion of the occurrence of groundwater in different aquifer materials and geologic settings. Suitable for both high school and college-level readers.

Fetter, Charles W. *Applied Hydrogeology.* 2d ed. Columbus, Ohio: Merrill, 1988. This textbook emphasizes the practical aspects of understanding groundwater occurrence and movement. Chapter 8 contains a detailed discussion of the influence that geologic conditions have on groundwater occurrence, with special emphasis on groundwater regions in the United States. Chapter 7 addresses regional groundwater movement within aquifers and contains helpful illustrations of the principles discussed. A glossary of important terms is included. Suitable for college-level readers.

Freeze, R. Allan, and John A. Cherry. *Groundwater.* Englewood Cliffs, N.J.: Prentice-Hall, 1979. A more advanced book stressing the mathematical derivations of groundwater flow equations and their applications. Chapter 4 presents a discussion of the geologic controls on groundwater occurrence. Chapter 6 contains a number of figures illustrating the principles of regional groundwater flow.

Gilbert, Janine, Dan L. Danielopol, Jack A. Stanford, et al., eds. *Groundwater Ecology.* San Diego, Calif.: Academic Press, 1994. This collection of ecological essays written by experts examines the problem of groundwater contamination and pollution, as well as possible solutions. Color illustrations and maps.

Gorelick, Steven M., et al., eds. *Groundwater Contamination: Optimal Capture and Contamination.* Boca Raton, Fla.: Lewis, 1993. This compilation looks at pollution levels in groundwater and surface water. It provides a thorough description of water quality measures, standards, and procedures. Illustrations, index, and bibliography.

Hamblin, Kenneth W., and Eric H. Christiansen. *Earth's Dynamic Systems.* 8th ed. Upper Saddle River, N.J.: Prentice Hall, 1998. Although not limited to the topic of groundwater, this widely used introductory textbook has a good discussion of groundwater occurrence and movement in chapter 13. The color figures are especially helpful to one unfamiliar with subsurface geology. Suitable for both high school and college-level readers.

Montgomery, Carla W. *Physical Geology.* Dubuque, Iowa: Wm. C. Brown, 1987. An introductory geology textbook which contains a good discussion of groundwater and aquifers. The full-color figures are helpful, as is the author's emphasis on the environmental aspects of water supply and pollution.

Todd, David K. *Ground Water Hydrology.* 2d ed. New York: John Wiley & Sons, 1982. A textbook emphasizing the practical aspects of groundwater occurrence and movement. Chapter 1 introduces the reader to groundwater utilization and its relationship to the hydrologic cycle. Chapter 2 discusses aquifer types and the occurrence of groundwater within the United States. Suitable for college-level readers.

Walton, W. C. *Groundwater Resource Evaluation.* New York: McGraw-Hill, 1970. Although much of this book deals with advanced techniques for analyzing aquifer properties from well pump tests, chapter 2 contains a valuable discussion of the terms pertinent to aquifers and an analysis of groundwater regions in the United States. Chapter 1 provides background on the role of groundwater within the hydrologic cycle.

ARTIFICIAL RECHARGE

Artificial recharge is the technique of capturing water that might otherwise go to waste, such as flood runoff or treated sewage effluent, and using it to replenish groundwater supplies by allowing it to infiltrate into the soil or forcing it underground with recharge wells. Not only can this technique help to conserve drinking water resources, but it can also correct problems caused by overpumping, such as seawater invasion or land subsidence.

PRINCIPAL TERMS

GROUNDWATER: the water contained in soil and rock pores or fractures below the water table

GROUNDWATER RECHARGE: the water that infiltrates from the surface of the Earth downward through soil and rock pores to the water table, causing its level to rise

LOSING STREAM: a stream that is located above the elevation of the water table and that loses water to the ground via infiltration through the stream bed; the opposite of a gaining stream

RECHARGE WELL: a well designed to pump surface water underground in order to recharge the groundwater; sometimes called an injection well

UNSATURATED ZONE: the area of the soil or rock between the land surface and the water table; in this zone, the voids between the soil and rock particles contain both air and moisture

WATER SPREADING: an artificial recharge technique in which floodwaters are diverted from the stream channel and spread in a thin sheet over a flat land surface, allowing the water to infiltrate the ground

WATER TABLE: the upper surface of the saturated zone or groundwater; below this level all soil and rock pores are filled with water

NATURAL RECHARGE

Groundwater is one of the United States' most vital resources. More than 20 percent of all fresh water used in the nation comes from the ground, and use of it has been steadily increasing over the years. Groundwater comes from rainfall, which infiltrates the soil and percolates slowly downward through a region of soil and rock known as the unsaturated zone. At some level below the surface, the small spaces in the soil or rocks become completely filled with water; this area is the saturated zone, or groundwater. The upper surface of the saturated zone is called the water table and is equivalent to the level of water in a well that might intersect it. When a water well is pumped, water is withdrawn from the soil and rock pores, and the level of the water table drops.

Recharge to the groundwater occurs when more rain falls and infiltrates through the unsaturated zone; when this infiltrating water reaches the water table, its level will rise again. Natural recharge can also occur when a stream flows over permeable sands and gravels and water leaks downward through the stream bed. Streams that leak water into the ground in this way are called losing streams, and, for the most part, are found in arid climates. (Many streams in humid climates actually gain water from the ground.)

In many regions of the United States, the amount of rainfall is small enough that natural groundwater recharge occurs very slowly, even though groundwater may be abundant. In addition, humans often modify the land in such a way as to reduce natural recharge. For example, impervious coverings of the soil surface such as parking lots, streets, roof tops, and airports all prevent rainfall from infiltrating the ground. Small losing streams that might also have previously recharged the groundwater are frequently diverted into sewer systems and hastened out of an area. For a combination of these reasons, many cities that pump groundwater have experienced a severe and continuous drop in the level of the water table, which has not only threatened the water supply but also resulted in other undesirable side effects, such as subsidence (sinking) of the land surface or invasion of salt water from the ocean into the fresh groundwater supply.

ADVANTAGES OF ARTIFICIAL RECHARGE

Artificial recharge is a technique used to increase the amount of surface water moving into the ground. Surface water is manipulated by some method of construction such as building pits or basins, spreading it on the surface, or injecting it directly into the ground via recharge wells. In this manner, natural recharge is supplemented, and water that might otherwise go to waste (such as floodwater) can be stored underground for later use. The advantages are numerous. Storing water inside the Earth rather than in traditional surface reservoirs means that little or no water will be wasted as a result of evaporation. There is also no wastage of land because of the flooding of a reservoir and no expensive dam to build that later may threaten to break catastrophically. Underground reservoirs do not eventually fill with sediment as do surface reservoirs, and they are less vulnerable to contamination.

Artificially recharged groundwater can be used not only to replenish water supplies and raise the water table but also to increase the pressure of the underground water enough to prevent seawater from migrating farther into the ground or even to flush the intruded seawater out of an area where pumping wells are located. In some areas, severe lowering of the water table (or the underground water pressure) because of overpumping has caused the soil structure to compress, resulting in a lowering of the actual ground surface. This subsidence is usually gradual, but in some areas with limestone bedrock, sinkholes can collapse quite suddenly. In either case, ground subsidence can result in substantial damage to buildings or even flooding of the area. Artificially recharged water cannot restore the ground surface to its former levels, but it can substantially reduce or halt further subsidence. Another use of artificial recharge is for storage of energy. The demand for heating or air-conditioning is seasonal. During the summer, surplus hot water can be stored underground through recharge wells. This water has a different viscosity (flow behavior) from that of the cold regional groundwater and thus mixes with it surpris-

ingly little. Even after a storage period of up to three months underground, as much as three-fourths of the stored heat can be recovered.

SEWAGE EFFLUENT

The water that is used to recharge groundwater artificially normally comes from excess storm runoff collected by various means. One of the most attractive aspects of artificial recharge, however, is that treated sewage effluent can be used as the water source. The effluent can be disposed of in this way, and, if the operation is carefully designed and monitored, the waste water becomes cleansed and purified during the recharge process—a result of the very nature of groundwater movement: The slow percolation of water through the tiny soil and rock pores allows the Earth itself to act as a filter for the water. Movement of sewage effluent through the unsaturated zone can remove all of the bacteria and viruses, along with a large portion of the solid matter and undesirable chemicals. Artificial recharge of sewage effluent must be carefully planned and carried out so that the water does not percolate too rapidly or reach the water table before most of the contaminants have filtered out—otherwise, the groundwater may become polluted.

Artificial recharge of groundwater has been widely practiced for more than two hundred years throughout the world for a variety of purposes, but there are some difficulties with the method. One of the biggest problems in the United States is that there are often separate laws dealing with surface water and groundwater, as well as separate governing bodies or "owners." That makes the legal and economic aspects of the conjunctive use of surface water and groundwater extremely complicated. Artificial recharge may not be possible in some areas, or it may be too expensive in others. If it is not carried out properly, it can lead to groundwater contamination or other problems. In spite of these drawbacks, artificial recharge has been highly successful in restoring groundwater levels and reducing seawater intrusion and subsidence.

RECHARGE PITS

The most common technique used for artificial recharge involves the excavation of pits or basins to collect local storm runoff or diverted stream flow (to save money, old gravel pits can also be used). The pits or basins must intersect soils or layers of rock that have a high permeability, such as sands and gravels, and be located well above the water table. When a storm occurs, floodwater is diverted to a basin or series of basins paralleling the (naturally losing) stream channel. This water then fills the basin and, over a few days or weeks, gradually infiltrates through the permeable sand or gravel at the bottom of the basin and works its way downward to the water table. The water table then rises directly below the basin, forming a mound shape, which grows upward and spreads outward as recharge of the groundwater occurs.

Recharge pits and basins require continual maintenance, as the infiltration rate decreases sharply with time. Fine-grained material (clay and silt) suspended in the floodwater settles to the bottom and works down into the upper few centimeters of soil pores, clogging them, as do subsequent algae and bacteria growth. To reopen the soil pores, it is necessary to allow the basin to dry out from time to time. This causes the clays to dry and crack and allows the organic matter to decompose. Sometimes it may even be necessary to till and scrape the basin floor. If it is possible to construct the pit so that it is steep-sided and deep, however, clogging of the pit floor may not matter as long as the walls are permeable enough.

Since 1935, Long Island, New York, has used recharge basins to divert storm water, which would otherwise run out to sea via sewers, to the groundwater. There are now more than three thousand recharge basins that dispose of approximately 230,000 cubic meters of water per day. These basins are unlined open pits about 3 meters deep and open to the underlying gravels. Their infiltration rate is high enough so that almost all basins are dry within five days after a 2-centimeter rainfall.

WATER SPREADING

Another common artificial recharge technique is known as water spreading. Small losing tributary streams in a relatively flat area are modified in such a way that floodwaters, instead of racing down the stream bed, are diverted and spread as a thin sheet over the land surface. The floodwaters then infiltrate the permeable soils and recharge the groundwater over an extensive area. The changes to the stream bed are inexpensive and

generally involve the construction of low check dams bulldozed from river-bottom material and perhaps reinforced by vegetation, wire, or rocks. An added benefit of this technique is that by trapping floodwaters harmlessly in the upstream areas, urban areas that are located downstream are more protected from flood damages. Some drawbacks to the technique are the large amounts of land required, ice buildup in winter, and the fact that the small dams are easily washed out by larger floods and must constantly be rebuilt. If the check dams are constructed of permanent materials, they may create a flood hazard upstream. Water spreading recharge techniques can be adapted to steeper terrains by constructing a series of shallow, flat-bottomed, closely spaced ditches or furrows near the losing stream channel. The exact configuration of the ditches (contoured, tree pattern, or trellis) can be adjusted to the local terrain, but the lowest ditches should flow back into the main stream channel to avoid any overflow flood hazard. The ditches are somewhat more costly to construct than are check dams, also require a lot of land area, and, similar to recharge basins, are subject to clogging.

Sometimes it is possible to avoid the construction costs of ditches and check dams entirely by using preexisting irrigation canals for water spreading. In this case, cropland is irrigated by floodwaters during the dormant or winter season as well as the growing season. Care must be taken with this method to ensure that the constant leaching (removal by downward percolation) of salts and nutrients from the topsoil does not adversely affect the crops during the growing season. If treated sewage effluent is used as the recharge water in this technique (although it is usually spray-irrigated), nutrients are actually added to the soil. This not only benefits the crops but also cleanses the recharged effluent. The problem with this technique is that over-irrigation can overwhelm the natural filtering characteristics of the unsaturated zone and allow the groundwater to become contaminated.

RECHARGE WELLS

The artificial recharge techniques already discussed are practical only in areas where permeable soils on rocks allow direct infiltration from the surface to the water table. In many areas, the layers of the Earth bearing the groundwater (aquifers) are deeply buried by impermeable materials, and there is no direct access to the surface. It is still possible, however, to recharge these layers artificially using recharge wells (sometimes called injection wells). Unlike pumping wells, in which water is pumped from the groundwater to the surface, recharge wells allow surface water to be forced under pressure into the ground and underlying permeable rocks. Large volumes of water can be stored in this way and later pumped out as the need arises. The recharged fresh water can also be used to force seawater, which may have invaded the ground and contaminated drinking water supplies, back out to sea. An additional benefit of this particular technique is that it does not require much land (unlike recharge basins or water spreading), so it is practical to use in urban areas.

Recharge wells are not without their drawbacks. Compared with other artificial recharge techniques, they are expensive to construct and require continual maintenance. Their main problem, similar to recharge basins, is well clogging. Any fine-grained sediment in the recharge water will enter the rock pores and clog them. Dissolved air, also very abundant in the recharge water, will bind up the pore spaces, too. Bacteria growing in the water will coat the well sides and rock pores with clogging growths, and chemical reactions between the recharge and groundwaters will also cause clogging substances (such as rust) to grow. For these reasons, the recharge water must be carefully treated before it is injected, and the clogged wells must be pumped frequently to restore some of their original permeability. Finally, recharge wells allow direct access of the surface water to the ground, bypassing the filtering characteristics of the unsaturated zone. That means that if any contaminant finds its way into the recharge water, it may rapidly enter the ground and contaminate drinking water supplies.

Recharge wells have been used successfully in the Los Angeles, California, area to prevent contamination of drinking water supplies by invading seawater. More than ninety wells have been placed in a line about 15 kilometers long and parallel to the coast. Filtered, chlorinated wastewater is injected into four underlying waterbearing layers at an average rate of 1,500 cubic meters per day per well. The wells create a ridge of pressurized water along the coast that separates the invading seawater from pumping wells farther inland. Water levels in the pumping wells can even be drawn below sea level with no worry of saltwater contamination.

Sara A. Heller

CROSS-REFERENCES

BIBLIOGRAPHY

Asano, Takashi, ed. *Artificial Recharge of Groundwater.* Boston, Mass.: Butterworth, 1985. A comprehensive and somewhat technical compendium of articles dealing with artificial recharge, grouped by topic. Emphasis is placed on recharge with reclaimed wastewater and case histories. A short section on the legal and economic concerns of groundwater recharge is also included. Bibliography, index. Suitable for the college-level reader.

Bowen, Robert. *Groundwater.* 2d ed. New York: Elsevier, 1986. This general textbook on groundwater contains a short chapter (chapter 11) on artificial recharge. Short discussions are presented on water spreading methods, recharge wells, wastewater recharge, and incidental recharge. Illustrations are sparse; text is somewhat awkward. Bibliography, index. Suitable for the college-level reader.

D'Angelo, Salvatore, Thomas G. Richardson, Richard P. Arber, et al., eds. *Using Reclaimed Water to Augment Potable Water Resources.* Denver, Colo.: American Water Works Association, 1998. This collection of lectures and essays, compiled by the Water Environment Federation and the American Water Works

Association, focuses on the practice of recharging and recycling contaminated groundwater and other water sources.

O'Hare, Margaret P., et al. *Artificial Recharge of Ground Water.* Chelsea, Mich.: Lewis, 1986. A short and concise text on the basic artificial recharge techniques, including chapters on how to evaluate an area for its recharge potential and case histories of several artificial recharge projects. Also contains a lengthy annotated bibliography on the subject. Included are appendices with data tables on the recharge potential and needs of various geographic areas of the United States. Index. Suitable for the college-level reader.

Saether, Ola M., and Patrice de Caritat, eds. *Geochemical Processes, Weathering, and Groundwater Recharge in Catchments.* Rotterdam, Netherlands: A. A. Balkema, 1997. This slightly technical compilation looks at the geochemical cycles and processes associated with catchment basins. Several essays deal with the artificial recharge of groundwater in such systems.

Sopper, William E., and Louis T. Kardos, eds. *Recycling Treated Municipal Wastewater and Sludge Through Forest and Cropland.* University Park: The Pennsylvania State University Press, 1973. This book is a compendium of articles resulting from a symposium on wastewater and sludge disposal. Includes articles on artificial recharge with treated sewage effluent. Major sections of the book cover how the soil acts as a natural filter and explain specific chemical processes that go on when effluent is filtered through it via spray irrigation or recharge basins. Also includes case studies. No index or summary. Suitable for the college-level reader.

Sopper, William E., and Sonja N. Kerr, eds. *Utilization of Municipal Sewage Effluent and Sludge on Forest and Disturbed Land.* University Park: The Pennsylvania State University Press, 1979. Another compendium of articles resulting from a symposium on wastewater and sludge disposal. The first half of the book covers artificial recharge with treated sewage effluent using spray irrigation or recharge basins. Topics include the effects on water quality, vegetation, wildlife, mosquito populations, and public health. No index or summary. Suitable for the college-level reader.

Thompson, Stephen A. *Hydrology for Water Management.* Rotterdam, Netherlands: A. A. Balkema, 1999. This thorough account of hydrology, groundwater flow, and stream flow focuses on the management of water supplies in efforts to keep them free of pollutants and available to the largest number of people possible. Illustrations, maps, index, and bibliography.

Todd, David Keith. *Ground Water Hydrology.* 2d ed. New York: John Wiley & Sons, 1982. In this classic textbook on groundwater, chapter 13 covers artificial recharge. Detailed descriptions of water spreading and recharge well techniques are provided, in addition to some case studies. Nontechnical, easy to follow, with excellent illustrations and extensive references and index. Suitable for the college-level reader.

Walton, William C. *Groundwater Resource Evaluation.* New York: McGraw-Hill, 1970. A comprehensive if somewhat technical text on the basic concepts of groundwater hydrology, with emphasis on fundamental equations of groundwater flow. Chapter 6 covers groundwater recharge and runoff, with a mathematical treatment of the water table mounds that result from artificial recharge. The text of this chapter is readable but brief and somewhat dated. Bibliography, index. Suitable for the college-level reader.

DAMS AND FLOOD CONTROL

Dams have provided a means of substantially reducing the risk of catastrophic floods, as well as saving lives and dollars. As added benefits—which help to offset environmental costs—dams generate pollution-free hydroelectric power and provide a reliable water supply for drinking, irrigation, industrial use, and recreation.

PRINCIPAL TERMS

CHANNELIZATION: the practice of deliberately rerouting a stream or artificially modifying its channel by straightening, deepening, widening, clearing, or lining it

FLOODPLAIN: a wide, flat, low-lying area, adjacent to a river, that is generally inundated by floodwaters

FLOOD ZONING: passing laws that restrict the development and land use of flood-prone areas

LEVEE: a dike-like structure, usually made of compacted Earth and reinforced with other materials, that is designed to contain the stream flow in its natural channel

OUTLET WORKS: gates or conduits in a dam that are generally kept open so as to discharge the normal stream flow at low water

SPILLWAY: generally, a broad reinforced channel near the top of the dam, designed to allow rising waters to escape the reservoir without overtopping the dam

FLOODPLAIN RISKS

Humans and rivers have competed for the use of floodplains for centuries. The floodplain, a wide, flat, low-lying area adjacent to the river, is built by natural sedimentation as the river carries its load of sand, silt, and clay to the sea. During times of excess flow, the river spills over its banks and covers the floodplain with muddy water. When the water retreats, a new layer of fertile soil is left behind. Humans have long exploited floodplains for these rich soils in order to grow crops. Floodplains are also attractive places to settle because the nearby river provides an accessible source of water, transport, and sewage disposal. In some rugged terrains, the development of communities in any area other than floodplains is almost impossible.

Unfortunately, floodplain development has often proceeded without an awareness of the risks involved, and the result has been disastrous. In the United States, a few flood-related deaths occur almost every year; sometimes one flood kills many. For every death that occurs, many more are left homeless or experience property damage, hardship, or suffering. Heavy rains in the winter of 1926-1927 caused disastrous floods on the Mississippi River that killed several hundred people and left up to 650,000 homeless.

DAM CONSTRUCTION

Human attempts to control floods by the use of dams goes at least as far back as ancient Egypt, where about 2700 B.C.E., a dam was built at Sadd-el-Karfara. The basic principle of dam flood control is to store floodwaters in the reservoir instead of allowing them to spread over the natural floodplain, which contains valuable human-made structures. After the smaller tributary streams below the dam have passed their floodwaters safely, the main reservoir is drained in a controlled fashion. The overall effect is to lengthen the time of passage of the flood, while drastically reducing the peak flow.

Dams are constructed of three basic materials: earth, rocks, and concrete. Eighty percent of all dams in the United States and Canada are earth dams. In these, sand and soil are compacted into a broad triangular embankment surrounding a watertight clay core. The upstream face must be reinforced with rock or concrete to prevent wave erosion. Earth dams are the most practical in broad valleys and are relatively inexpensive to construct. Rockfill dams are similar to earth dams, but the heavy weight of the rock requires a more solid natural foundation. The upstream side must be covered with a watertight material to prevent water from leaking through the dam.

Concrete dams require narrow valleys with hard bedrock floors to anchor and support them. A concrete gravity dam uses its great bulk and weight to resist the water pressure. These dams can generally remain stable when floodwaters overtop them, but they are costly (Washington State's Grand Coulee Dam, which is of this type, required 8.1 million cubic meters of concrete). A concrete buttress dam relies both on its weight and structural elements to support it: The watertight upstream face slopes underneath the reservoir, which helps to distribute the water pressure to the foundation. Buttresses on the downstream face of the dam both counteract the force of the water and help the dam to withstand minor foundation movements—a distinct advantage in earthquake-prone areas. Concrete arch dams have a convex upstream face that spans the steep valley walls. The water pressure transmits the force along the arch to the side abutments and foundation, bonding the dam to the canyon. These dams are less expensive to construct than are gravity dams, but also are more likely to fail in the event of a small rupture.

Outlet Works and Spillways

All dams must contain properly designed outlet works, which are gates or conduits near the base of the dam kept open to discharge the normal low-water flow of the stream. These gates can be operated manually or automatically, but they must be carefully regulated to control the reservoir flood storage in an optimum manner and to prevent overwhelming of the spillway or overtopping of the dam. This means that the engineers of the outlet works must have an intimate knowledge of the design flood (the statistical probability and approximate return period of the maximum flood for which the dam was designed), the reservoir capacity, characteristics of past flood behavior, downstream flood hazards, accurate meteorological forecasts, and a good dose of intuition.

Finally, all dams must be constructed with a spillway. A spillway is generally a broad reinforced channel near the top or around the side of the dam that acts as a safety valve because it allows rising waters, which might otherwise overtop the dam and cause its collapse, to escape harmlessly. Outlet works cannot be depended on to relieve the rising waters because they may become either blocked with debris or inoperative during a flood. The spillway must be large enough to convey the maximum probable flood.

Multiple-purpose Reservoirs

An additional benefit of dams is that they not only provide flood control, but also the reservoir can be used as a water supply for irrigational, industrial, and municipal purposes. When the water falls from the top of the reservoir to the dam base through turbines, it generates inexpensive and pollution-free electricity. The reservoir also can be used for fishing, boating, and swimming and can become a haven for certain kinds of wildlife.

Unfortunately, a multiple-purpose reservoir contains built-in conflicts of interest. Because a reservoir used to control floods must have storage space for the floodwaters, ideally the water level should be kept low, or the reservoir kept nearly empty. Conserving water for irrigation or domestic use, however, requires holding floodwaters in storage, sometimes for years. For hydropower generation, the reservoir must be kept as full as possible and certainly never emptied. The area's fish and wildlife are best served by maintaining a stable reservoir level, as are recreational uses. Thus, the management of a multipurpose reservoir for flood control is a very complicated enterprise. The goal is to derive the maximum dollar value from the water while keeping the threat of flood damage to a minimum. All forecasts of possible and probable floods must be carefully weighed against the need for keeping the reservoir full for other purposes.

Integrated Flood-Control Programs

Flood protection is rarely implemented by the construction of a single dam. An integrated flood-control program also involves the construction of levees, floodways, and channel modifications. Levees are dikes or structures that attempt to confine the stream flow to its natural channel and prevent it from spreading over the floodplain. They have the advantage of increasing the flow velocity in the channel, which diminishes the deposition of sediment in it. The increased velocity, however, also tends to undercut and erode the levee. Levees block off the floodplain, but the increased volume of water in the channel raises the level to which the waters will flood. When levees are breached, the floodwaters spill out over the floodplain suddenly, catching residents by surprise. A breached levee may also trap floodwaters downstream by preventing their return to the channel, thereby increasing the damage. A way to prevent the breaching of levees is to install emergency outlets (like dam spillways) to specially constructed floodways, or flood-diversion channels. These are a means for safely returning the river to its natural floodplain.

Channelization, or modifications to the stream channel, is also used in conjunction with flood-control dams. This generally involves straightening, deepening, widening, clearing, or lining the channel in such a way that the stream flows faster. In this way, potential floodwaters are removed from the area more quickly. The lower Mississippi River was shortened by 13 percent between 1933 and 1936 (by short-cutting meander bends), which reduced the flood levels 61-366 centimeters for equal rates of flow.

DISADVANTAGES OF FLOOD-CONTROL SYSTEMS

Channelization may have deleterious effects. Erosion of the channel from faster flows may drain adjacent wildlife habitats and may undermine levees and bridges. The rapid passage of floodwaters also increases the hazard to places downstream of the channelized area.

Flood protection by means of dams and their attendant structures has some other disadvantages. The term "flood control" is often misinterpreted by the general public to mean absolute and permanent protection from all floods under all conditions. This leads to a false sense of security and promotes further economic development of the floodplain. Every levee and every dam has a limit to its effectiveness. To compound the danger, there is always the possibility that the dam could fail entirely. The main causes of dam failure are overflowing because of inadequate spillway capacity, internal structural failure of earth dams, and failure of the dam foundation material. During the twentieth century, more than 8,000 people perished in more than 200 dam breaks. An earth dam across the Little Conemaugh River above Johnstown, Pennsylvania, was overtopped and failed on May 31, 1889, killing 2,209. To prevent such failures, geologists and engineers must cooperate closely to design the safest possible structure

matching the geology of the dam's foundation.

Dams can also cause undesirable sedimentation on and erosion of a riverbed. As the river enters the reservoir, it is forced to slow and drop its sediment load, which decreases the water-holding capacity of the reservoir and limits its usable lifetime. The resulting delta can grow upstream and engulf adjacent properties and structures. The opposite effect results from the discharge of the reservoir water, now deprived of its sediment load, back into the natural channel below the dam. Here, the water can rapidly erode the bed. At Yuma, Arizona, for example, 560 kilometers downstream of the Hoover Dam, the riverbed has been lowered by 2.7 meters. If a river formerly supplied sediments to beaches, the deprivation of this material could result in coastal erosion. The loss of topsoil deposition as a result of flood control on the Nile (the river is protected by the Aswan Dam) has required farmers to add expensive fertilizers to their crops.

Other harmful effects of the construction of dams and flood-control projects include an increase in the water's temperature, salinity, and nutrient content (from the strong solar heating and evaporation in reservoirs and the return flow of irrigation waters). This decrease in water quality can result in undesirable weed growth in reservoirs and in fish kills. Fish that migrate up rivers to spawn, such as trout and salmon, are physically prevented from doing so by dams. Fish ladders around dams are expensive and have proven to be only partially effective. The loss of wet floodplain habitat to reservoir inundation has been detrimental to many water birds, some of which are already endangered species. Unfortunately, the reservoir itself does not usually substitute for this loss because its elevation must fluctuate. Finally, dams are expensive to build and maintain, and they cause the loss of valuable farmland because of reservoir inundation and sometimes force the relocation of entire communities.

PREVENTION OF FLOODPLAIN MISUSE

One of the best methods to avoid the economic, social, and environmental disadvantages of dams is to prevent floodplain misuse. This is often effectively done through the use of flood zoning laws, which restrict or prohibit certain types of development in flood hazard areas. (Appropriate

The 726-foot-high Hoover Dam harnesses the energy of the Colorado River. (© William E. Ferguson)

uses of flood hazard zones include parkland, pastureland, forest, or farmland.) Flood insurance laws, building codes, and tax incentives can have the same effect.

If necessary, existing buildings can be flood-proofed (raised higher, reinforced, or both) or re-located. Land-treatment procedures can improve the ability of the natural ground surface to retain water and release it slowly to streams. These techniques include reforestation, terracing, and building contour ditches and small check dams. In urban areas, rooftop or underground floodwater retention tanks and porous pavements can be installed to reduce the risk of flooding.

MISSISSIPPI RIVER PROJECT

The Mississippi River is an example of a massive flood-control project. Levees were first constructed on it in 1727 to protect New Orleans, and they continued to be constructed (privately and haphazardly) into the 1830's. Major funds were granted in 1849 and 1850 by Congress to the U.S. Army Corps of Engineers for a flood-control study. After the Civil War, the entire 1,130-kilometer lower Mississippi Valley was organized into levee districts. In spite of the levees, major floods struck the Mississippi River in 1881, 1882, 1883, 1884, 1886, 1890, 1903, 1912, 1913, and 1927. The worst of these came in the spring of 1927 after a long winter of heavy rains, when the levees failed in more than 120 places. Many people were killed or left homeless. This inspired Congress to pass the Flood Control Act of May 15, 1928, which provided for levee expansion and improvement, dams and reservoirs, bank stabilization coverings (revetments), floodway diversion channels, artificial meander cutoffs, and emplacement and operation of river gauging stations (to monitor the water level continuously). The first test of the system came in 1937. Although the potential for another 1927 flood was just as great, the damages caused in 1937 were far less severe.

Work has continued on the Mississippi. A major contribution was the 1944 project on the Missouri

River, a tributary of the Mississippi. On the Missouri, six huge dams and a 1,500-kilometer chain of reservoirs and levees have been built. From 1950 to 1973, the combined effect of natural events and human-made structures produced a lull in the floods. The calm was broken when a massive flood in 1973 moved down the Mississippi, a flood that was a significant test of the control projects. The maximum flow was approximately 56,800 cubic meters per second, enough to supply New Orleans with its daily water needs in less than 10 seconds.

Despite the flood-control facilities, losses were great. Thirty-nine levees were breached or overtopped. Property losses were estimated at $1 billion. In spite of this, the river crested 21 centimeters lower than it had in 1927 in Cairo, Illinois, and 207 centimeters lower than it had the same year in Vicksburg, Mississippi. Engineers estimated that flood-control works had reduced damages from a possible $15 billion. Still, the flood left 23 dead and another 69,000 homeless.

IMPORTANCE OF WISE RIVER-MANAGEMENT POLICIES

Where humans have been short-sighted in settling and building cities on floodplains, wisely built and well-managed dams and their attendant structures are an excellent way to reduce (but not eliminate) the flood hazard. The lives and dollars saved have been great. The environmental losses, which are harder to measure, also have been great. Wild, natural rivers are rapidly becoming one of the nation's rarest possessions. As the demands for electric power generation and water resources continue to increase, managing the rivers wisely is in the nation's best long-term interest.

Sara A. Heller

CROSS-REFERENCES

BIBLIOGRAPHY

Barrows, Harold Kilbrith. *Floods: Their Hydrology and Control.* New York: McGraw-Hill, 1948. A somewhat technical yet practical discussion of flood control projects in the United States, including their costs and problems. Also included are sections on basic flood hydrology and some notable great floods. Suitable for the college-level reader. Appendices, index.

Berkman, Richard L., and W. Kip Viscusi. *Damming the West: Ralph Nader's Study Group Report on the Bureau of Reclamation.* New York: Grossman, 1973. A thorough scrutiny of the politics of water-resource projects in the western United States involving dams financed by the Bureau of Reclamation. Cost-benefit studies are reevaluated in terms of the environmental impact of dams, who really benefits from the projects, and Indian water rights. Suitable for the college-level reader. Index.

Bowker, Michael, and Valerie Holcomb. *Layperson's Guide to Flood Management.* Sacramento, Calif.: The Foundation, 1995. This brief book serves as an excellent handbook and introduction to flood control, floodplain management, and water drainage for the person without prior knowledge in the field. Illustrations.

Chadwick, Wallace L., ed. *Environmental Effects of Large Dams.* New York: American Society of Civil Engineers, 1978. A collection of short articles concerning various environmental effects of large dams and reservoirs. Topics include harmful temperature effects, fish, algae, and aquatic weed problems, loss of wildlife habitat, erosion, and seismic activity. This latter article contains an extensive list of dams in the United States that have experienced some earthquake difficulties. Suitable for the college-level reader. Index.

Clark, Champ. *Flood.* Planet Earth Series. Alexandria, Va.: Time-Life Books, 1982. One of a popular series of books that are nontechnical and abundantly illustrated with full-scale color photography. Chapter 3 describes the history of flood control on the Mississippi River. Chapter 5 contains a short discussion of dam construction and several examples of floods caused by dam failure. Bibliography and index.

Committee on the Safety of Existing Dams, Water Science and Technology Board, Commission on Engineering and Technical Systems, and the National Research Council. *Safety of Existing Dams: Evaluation and Improvement*. Washington, D.C.: National Academy Press, 1983. A practical and not overly technical consideration of dam safety problems intended for the college-level reader. Considers hydrologic, geologic, and seismologic factors, foundation, outlet works, reservoir problems, and available instrumentation. For each problem additional technical references are included. Index.

Ghosh, Some Nath. *Flood Control and Drainage Engineering*. 2d ed. Rotterdam, Netherlands: A. A. Balkema, 1997. This somewhat technical college-level textbook looks at the engineering methods and practices used in the construction and maintenance of dams and other flood-control structures.

Hoyt, William G., and Walter B. Langbein. *Floods*. Princeton, N.J.: Princeton University Press, 1955. A thorough and comprehensive treatment of all aspects of floods that is so well-written as to be a classic reference. Chapters 5 and 6 cover flood damage and flood-control measures. Numerous examples of flood-control projects are described. Also contains a comprehensive list and description of historic floods from 1543-1952 by date and stream. Suitable for the college reader. Bibliography and index.

Leopold, Luna B., and Thomas Maddock, Jr. *The Flood Control Policy: Big Dams, Little Dams, and Land Management*. New York: Ronald Press, 1954. A close examination of the conflict between the advantages of the construction of one large dam and having many small tributary dams to control floods. Also contains chapters on the technical and political difficulties of flood control. Intended for those professionally concerned with water resources as well as for the general public. Bibliography and index.

Leuchtenburg, William Edward. *Flood Control Politics*. Cambridge, Mass.: Harvard University Press, 1953. A detailed history and assessment of the Connecticut River valley flood-control problems from 1927 to 1950, with emphasis on the political and economic issues. Well written and suitable for college-level readers. Extensive bibliography, index.

Smith, Norman. *A History of Dams*. London: P. Davies, 1971. Well written and interesting historical discussion of dam-building through the ages beginning with the Egyptians and continuing to the end of the nineteenth century, but not including India or the Far East. Suitable for the high school-level reader. Photographic plates, glossary, and index.

Tennessee Valley Authority. *Floods and Flood Control*. Technical Report 26. Knoxville, Tenn.: Author, 1961. A comprehensive discussion of the development, costs, operation, and benefits of the Tennessee River system based on the integrated multiple-purpose system of reservoirs. No discussion of environmental impacts. Suitable for a college reader with some technical background. Appendices, index.

Walters, R. C. S. *Dam Geology*. London: Butterworth, 1962. A short general discussion of the influence of geologic structures on dam construction and failure is followed by a lengthy compendium of dams from all over the world that have experienced typical geological difficulties. Suitable for the college-level reader. Well-illustrated, with a bibliography, an index, and appendices with civil engineering data.

FLOODS

Floods are conditions of extreme flow of water. They generally occur because of extreme amounts of rainfall, but various kinds of dam failures also induce them. Floods exert a major role in the shaping of some river systems, and their occurrence is critical to the human use of riverine lands.

PRINCIPAL TERMS

DISCHARGE: the volume of water moving through a given flow cross section in a given unit of time

FLASH FLOODS: rises in water level that occur unusually rapidly; they generally occur because of especially intense rainfall

FLOOD: a rising body of water that overtops its usual confines and inundates land not usually covered by water

HYDROLOGY: the branch of science dealing with water

JÖKULHLAUP: a flood related to water impound-

ment by a glacier; most such glacier outburst floods involve the failure of some type of glacial dam or subglacial volcanic activity

MONSOON: a seasonal pattern of wind at boundaries between warm ocean bodies and landmasses

RECURRENCE INTERVAL: the average time interval in years between occurrences of a flood of a given or greater magnitude than others in a measured series of floods

RUNOFF: that part of the precipitation that eventually appears in surface streams

CAUSES OF FLOODS

Floods involve extremely large flows of water in rivers and streams. Some technical hydrological definitions of floods involve stages, or heights, of water above some reference level, such as the banks of a river channel. In practice, however, floods can be thought of as any extreme flow of water that exceeds usual experience or that damages or threatens life and property.

Floods may be caused by a variety of physical factors, including dam failures and the subsidence of land. The most common kind of flood occurs when extreme precipitation and physical factors of the land combine to produce optimum runoff of water. The precipitation can yield runoff directly, or melting snow can produce the flow. Very intense flash floods are usually associated with heavy, short-duration rainfall from thunderstorms. Such rainfall easily overrides the infiltration capacity of the ground, and water rapidly runs off hillsides into adjacent stream channels. If enough water concentrates in the channel, it will constitute a flood.

Physical factors on the land surface also determine the rate of concentration of floodwater in stream channels. For example, less permeable soils will allow water to run off faster, as will a lack of vegetation. Artificial enhancement of runoff occurs when slopes are covered by impermeable materials. This situation commonly occurs in construction, when the natural ground surface is replaced by buildings and pavement. The result of such construction is extreme enhancement of runoff. In cities, the yield of floodwater off paved surfaces may be ten times greater than the same storm's yield under natural conditions. In this way, construction tends to exacerbate flooding problems in urban areas.

SEDIMENT AND BEDROCK

As water rises in a stream channel, it is usually associated with considerable sediment. If this sediment also composes extensive deposits adjacent to the stream channel, the river is termed "alluvial." An alluvial river commonly has a floodplain of this deposited sediment adjacent to the channel. The floodplain is really an intimate part of the river system, since sediment is added to it every time the river rises above its banks. As the water rises, the river's depth rapidly increases, causing an increased ability to transport sediment that is eroded from the bed and banks. If the stream is appropriately loaded with sediment, it will be deposited when the banks are overtopped and the width of the flow greatly increases. If not enough sediment is supplied by erosion, however, the increasingly energetic flood flows will be erosive. They will attack the banks, widening the channel and restoring the appropriate sediment load to the stream.

When the bed and banks of a river are composed of bedrock, these adjustments of sediment load to flow energy cannot occur. Since the bedrock can be very resistant to erosion, the energy level in the flow can rise spectacularly without being damped by sediment on which work is expended. The excess energy of such sediment-impoverished floods goes into the development of turbulence. Turbulence at high energy levels takes on an organized structure of great vortices that produce immense pressure changes. These pressure effects may be sufficient to erode the bedrock boundary by a "plucking" action.

A famous geological controversy once surrounded the problem of bedrock erosion by great floods. In the 1920's, University of Chicago geologist J. Harlen Bretz proposed that immense tracts of eroded basalt bedrock in Washington state had been created by a catastrophic glacial flood. Bretz subsequently showed that the fascinating landforms of that region, known as the Channeled Scablands, were created when a great lake impounded by glacial ice burst. The lake, glacial Lake Missoula, had been more than 600 meters deep at its ice dam. It took Bretz nearly fifty years to convince his many critics that catastrophic flooding could explain all the bizarre features of the Channeled Scabland. Geologists now know that the physics of catastrophic flooding is completely consistent with Bretz's observations. Missoula flood flows moved at depths of 100 to 200 meters and velocities of 20 to 30 meters per second. The power (rate of energy expenditure) per unit area for such flows is thirty thousand times that for a normal river, such as the Mississippi, in flood. The reason for such immense flow power is that the Missoula flooding occurred on very steep slopes and that the potential energy from the dam burst was great.

The Missoula floods last occurred during the Ice Age, more than twelve thousand years ago. There are, however, modern examples of glacial

floods called jökulhlaups. In Iceland, jökulhlaups occur where glaciers overlie active volcanoes. Volcanic heat melts water that is stored in subsurface reservoirs. Lakes may also form adjacent to the ice masses. Because ice is less dense than water, such juxtapositions are inherently unstable. When the pressure is high enough, the water may lift the ice dam and burst out from beneath the glacier. These jökulhlaups move house-sized boulders and transport immense quantities of sediment.

DISCHARGE AND RECURRENCE INTERVAL

The volume of water released by a flood per unit of time is its discharge. This quantity is the magnitude of the flow that will potentially inundate an area. The chances of experiencing floods of different magnitudes are expressed in terms of frequency. Large, catastrophic floods have a low frequency, or probability of occurrence; smaller floods occur more often. The probability of occurrence for a flood of a given magnitude can be expressed as the odds, or percent chance, of the recurrence of one or more similar or bigger floods in a certain number of years. Analyses of flood magnitude and frequency are achieved by measuring floods and statistically analyzing this experience. Results are expressed in terms of the probability of a given discharge being equaled or exceeded in any one year. The reciprocal of this probability is the return period, or recurrence interval, of the flooding, expressed as a number of years.

The concept of a recurrence interval is sometimes confusing; some examples will illustrate the meaning. Suppose analysis of the flood frequency of a river flow indicates certain discharges for ten- and one-hundred-year recurrence intervals. The smaller, ten-year flood will have a probability of occurrence in one year of 0.1, or 10 percent. The larger one-hundred-year flood's probability of occurrence in one year is 0.01, or 1 percent. Note that these numbers do not preclude several such floods occurring in a given period; for example, two or more one-hundred-year floods theoretically could occur in the same year. Such an event, while possible, is extremely improbable.

Flood magnitude-frequency relationships vary immensely with climatic regions. In the humid-temperate regions of the globe, such as the northern and eastern United States, stream flow is relatively continuous. Even rare floods are not appreciably larger than more common floods. A result of this relationship is that stream channels are adjusted in size to convey, with maximum efficiency, the relatively common, moderate-sized floods. In contrast, the more arid regions of the southwestern United States have immensely variable flood responses. Stream channels may be dry most of the time, filling with water only after rare thunderstorms. The flash floods that characterize these streams may also be highly charged with sediment. Indeed, the sediment may so dominate the flow that the phenomenon is of debris flowage rather than stream flow. Because extreme events dominate in these environments, the stream channels are adjusted in size to the rare, great floods.

A TIME LINE OF HISTORIC FLOODS

2400 B.C.E.	GREAT DELUGE (KNOWN WORLD): According to the Bible, all human and land-based animal life drowned, except for those on Noah's ark.
1228	FRIESLAND, THE NETHERLANDS: More than 100,000 drowned.
Apr. 17, 1421	DORT, THE NETHERLANDS: More than 100,000 dead, 20 villages never found again.
Nov. 1, 1570	FRIESLAND, THE NETHERLANDS: More than 50,000 dead.
Aug. 3, 1574	LEIDEN, THE NETHERLANDS: More than 10,000 Spanish troops dead, siege of Leiden.
Nov. 19, 1824	ST. PETERSBURG, RUSSIA: More than 10,000 dead, millions of dollars in property.
July 10, 1887	ZUG, SWITZERLAND: 70 dead, more than 600 left homeless.
Sept. 28, 1887	ZHENGZHOU, CHINA: Approx. 900,000-7 million dead.
May 31, 1889	JOHNSTOWN, PENNSYLVANIA: Approx. 2,209 dead, 1,600 homes lost, 280 businesses.
Feb.-Apr., 1890	UPPER MISSISSIPPI RIVER: Almost 100 dead, 50,000 homeless.
May 26-June, 1903	KANSAS AND MISSOURI RIVERS: 200 dead, 8,000 left homeless.
June 14, 1903	HEPPNER, OREGON: 325 dead, $250 million in damage.
Aug. 7, 1904	EDEN, COLORADO: 96 dead.
Sept., 1911	CHINESE PROVINCES OF HUBEI AND HUNAN: More than 100,000 dead.
Apr., 1912	MISSISSIPPI RIVER FROM CAIRO, ILLINOIS, TO BOLIVAR: 250 dead, 30,000 homeless, $45 million in damage.
July, 1915	GUANGZHOU, CHINA: More than 100,000 dead.
Jan., 1916	NORTH SEA COAST OF THE NETHERLANDS: 10,000 dead.
Jan., 1916	OTAY VALLEY, CALIFORNIA: 22 dead, $10 million in damage.
June 2-3, 1921	PUEBLO, COLORADO: 120 dead, $25 million in damage.
Sept. 7-11, 1921	SAN ANTONIO, TEXAS: 51 dead, $5 million in damage.
Apr., 1927	SEVEN STATES ALONG THE MISSISSIPPI RIVER: 313 dead, 700,000 homeless, $300 million in damage.
May 29-30, 1927	NORTHEASTERN KENTUCKY: 89 dead, 12,000 homeless, $7 million in damage.
Nov. 3-4, 1927	VERMONT, NEW YORK: 200 dead, $28 million in damage.
Mar. 12, 1928	NEAR SAUGUS, CALIFORNIA: About 450 dead; 1,200 homes and other buildings severely damaged after St. Francis Dam collapses.
Jan., 1937	OHIO RIVER BASIN: 300 dead, more than 900,000 evacuated in Ohio Valley.

MONSOONS AND ANCIENT FLOODS

In tropical areas, some of the greatest known rainfalls are produced by tropical storms and monsoons. Monsoons are seasonal wet-to-dry weather patterns driven by atmospheric pressure changes over the oceans. They result in alternations between dry periods, which inhibit vegetation growth, and immensely wet periods, which facilitate runoff.

Tropical rivers thus have a pronounced seasonal cycle of flood, and some floods may be immense. These rivers also show channel size adjustment to rare, great flows. Some of the most populous places on Earth are situated on seasonal tropical rivers, such as the Ganges and Brahmaputra rivers in India and Bangladesh. Immense tragedies may occur when a particularly severe monsoon or tropi-

June 19, 1938	NEAR TERRY, MONTANA: 46 dead, about 60 injured.
Aug.-Nov., 1939	CHINESE PROVINCES OF HEBEI AND SHANDONG: Estimated 200,000 dead, 25,000,000 homeless.
May-June, 1948	NORTHWEST U.S., BRITISH COLUMBIA: 51 dead, $100 million in damage.
Mar.-Aug., 1950	ANHUI PROVINCE, CHINA: 489 dead, 10 million homeless.
May-July, 1950	EASTERN NEBRASKA, ESP. LINCOLN AND BEATRICE: 23 dead, 60,000 acres inundated, $60 million in damage.
July, 1951	KANSAS AND MISSOURI RIVERS: 19 dead, 500,000 homeless, $870 million in damage.
Nov. 7-29, 1951	PO RIVER VALLEY, ITALY: At least 150 and perhaps 1,000 dead.
Feb. 1, 1953	NETHERLANDS, GREAT BRITAIN, BELGIUM: 1,853 dead.
June 27-July 3, 1954	RIO GRANDE (SOUTHWESTERN TEXAS, NORTHERN MEXICO): 16 dead, 728 homes destroyed.
July-Aug., 1954	SOUTHERN TIBET, ESP. XIGAZÊ AND GYANGTSE: 500-1,000 dead.
Aug., 1954	FARAZHAD, IRAN: 2,000 dead.
Dec., 1955-Jan., 1956	YUBA CITY, CALIFORNIA: 74 dead (61 in California, 13 in Oregon), 50,000 homeless.
Dec. 2-3, 1959	ABOVE FRÉJUS, FRANCE: Between 412 and 421 dead after Malpasset Dam collapses.
Feb. 16-25, 1962	GERMAN COAST, PARTICULARLY HAMBURG AND BREMERHAVEN: 309 dead, tens of thousands homeless, $250 million in damage.
Sept. 26, 1962	BARCELONA AND ENVIRONS, SPAIN: 700 estimated dead, $50 million or more in damage.
June 17-28, 1965	GREAT PLAINS (COLORADO AND KANSAS): 23 dead, at least $100 million in damage.
Jan. 11-13, 1966	RIO DE JANEIRO AND ENVIRONS, BRAZIL: 239-400 estimated dead, 45,000 evacuated.
Nov. 3-4, 1966	FLORENCE, ITALY: 24 dead; 50,000 homeless; 1,300 artworks and 700,000 books lost.
Jan. 24-Mar. 21, 1967	RIO DE JANEIRO AND SÃO PAULO, BRAZIL: 1,250 dead.
Nov. 25-26, 1967	LISBON AREA, PORTUGAL: 457 or more dead, thousands homeless.
July 21-Aug. 15, 1968	GUJARAT, INDIA: 1,000 people and 80,000 cattle dead.
Sept. 29-Oct. 28, 1969	TUNISIA: 542 people and 1 million livestock dead, 100,000 homeless.
Feb. 26, 1972	LOGAN COUNTY, WEST VIRGINIA: 125 dead, 4,000 homeless, $10 million in damage.
June 9, 1972	RAPID CITY, SOUTH DAKOTA: 237 dead, $160 million in damage.
July 18-Aug. 12, 1972	MANILA, PHILIPPINES: 536 dead, 1.2 million refugees, 6 million affected.
Mar.-May, 1973	MISSISSIPPI RIVER, FROM SOUTHERN LOUISIANA NORTH: 33 dead, $1.155 billion in damage.

(continued)

cal storm produces especially great floods.

Studies of ancient floods (paleofloods) through geological reconstruction of past discharges show that monsoons and other flood-generating systems have varied in the past. Between about ten thousand and five thousand years ago, floods were much more intense in many world areas on the boundaries between the tropics and the midlatitude deserts. These intense floods may have been related to long-term glacial-to-interglacial cycles of planet Earth. Because of modern increases in carbon dioxide and other greenhouse gases, it appears quite likely that tropical floods may again become more intense. Such a situation could have grave consequences for flood-prone tropical countries with large populations.

Jan. 26-30, 1974	QUEENSLAND, AUSTRALIA: 12 dead, 8,000 homeless, $160-$320 million in damage.
Mar. 24, 1974	TUBARÃO, BRAZIL: 500-1,000 dead, city of Tubarão 70 percent destroyed.
July to mid-Aug., 1974	BANGLADESH: 2,000 or more dead, two-thirds of the nation flooded.
June 5, 1976	TETON DAM COLLAPSE (NEAR REXBURG, IDAHO): 11 dead, 1,000 injured, 30,000 homeless.
July 31-Aug. 1, 1976	BIG THOMPSON CANYON, LOVELAND, COLORADO: 139 dead, $35 million damage.
Feb., 1978	SOUTHERN CALIFORNIA: More than 20 dead; more than $80 million in damage and additional dead during successive mudslides.
Sept. 4-29, 1978	NORTHERN INDIA: 1,291 dead, 1.5 million dwellings destroyed or damaged, 43 million people displaced.
Aug. 14-Sept. 7, 1981	SHANXI PROVINCE, CHINA: 764 dead, 5,000 injured, 200,000 homeless.
May 26-28, 1982	NICARAGUA AND HONDURAS: 225 dead, 70,000 homeless.
Aug. 29-Sept. 13, 1982	INDIA: 700 dead, thousands of villages damaged or destroyed.
May 1-4, 1983	BANGLADESH: 75 dead, 50,000 homeless.
July 19, 1985	STAVA, ITALY: 224 dead.
July-Sept., 1988	BANGLADESH, ESP. DHAKA: About 2,400 dead, 30 million homeless, $1 billion in damage.
July, 1991	SOUTHEAST CHINA: 1,781 dead, estimated $7.5 billion in economic losses.
June-Aug., 1993	MINNESOTA, WISCONSIN, IOWA, ILLINOIS, MISSOURI: 52 dead, 74,000 homeless, $18 billion in damage during the Great Mississippi River Flood of 1993.
Jan., 1995	NORTHERN CALIFORNIA: 11 dead, $300 million in damage.
Jan.-Feb., 1995	NORTHERN EUROPE: 30 dead, $2 billion in damage.
Aug. 7-8, 1996	NORTHERN SPAIN: 70 campers dead in the Pyrenees.
Apr., 1997	NORTH DAKOTA AND MINNESOTA: Most of Grand Forks under water, nearly 1 million acres of farmland flooded by Red River.
July-Aug., 1997	EUROPE: 104 dead, $6 billion in damage, esp. in Poland and the Czech Republic.
Aug. 12, 1997	ANTELOPE CANYON, ARIZONA: 11 dead.
July 27-Aug. 31, 1998	CHINA: 3,000 dead, millions homeless, $20 billion in economic losses.
Oct. 17-21, 1998	SOUTH TEXAS: 31 dead, at least 1,500 evacuated in 60 counties.
Aug. 27-Oct. 26, 1999	EASTERN AND SOUTHERN MEXICO: 404 dead, approx. 350,000 homeless.
Jan.-Feb., 2000	MOZAMBIQUE, AFRICA: Thousands dead, approx. 1 million homeless.

STUDY OF FLOODS

Hydrologists study floods by measuring flow in streams. Such measurements are taken at stream gauges, where mechanical devices are used to record the water level, or stage. To transform these stage measurements to discharge values, the hydrologist must perform a rating of the stream gauge. That is accomplished by measuring velocities in the stream channel during various flow events. A velocity meter with a rotor blade calibrated to the flow rate is used for this purpose. When the average measured velocity of the stream channel is multiplied by the cross-sectional area of the channel, the result is the discharge for that flow event. When several flows at different stages are measured, the data are used to generate a rating curve for the gauge. This curve will show discharges corresponding to any stage measured at the gauge.

The discharge values obtained at the gauge are collected over many years, constituting a record that can then be used in flood-frequency analysis. Several statistical procedures can be employed to plot the flood experience and to extrapolate to ideal values of the ten-year flood, the one-hundred-dred-year flood, and so on. The discharges associated with these recurrence intervals are then used as design values for hazard assessment, dam construction, and other flood controls. "Flood control" is probably a misnomer, however, because flows larger than the design floods are always possible. Flood control really involves providing various degrees of imperfect flood protection.

Another approach to evaluating extreme flood magnitudes involves careful study of the precipitation values that generate the greatest known floods. By transposing the patterns of known extreme storms to other areas, scientists can, in theory, calculate what the runoff would be from hypothetical great storms. Such calculations involve the use of a rainfall-runoff model. These models are prevalent in hydrology because they can be easily programmed. The models give idealized predictions of how water from a given storm would concentrate. The flood discharge modeled from the assumed maximum rainfall is called a "probable maximum flood."

Unfortunately, there are problems in both these traditional hydrological approaches to flood studies. Both make assumptions in calculating potential flood flows. Another procedure is to study the natural records of ancient floods, or paleofloods, that are preserved in geological deposits or in erosional features on the land. It has been found for some sections of bedrock that nonalluvial rivers act as natural recorders of extremely large flood events. These natural flood "gauges" can be interpreted only by detailed studies that combine geological analysis with hydraulic calculations of the ancient discharges.

Paleoflood hydrology generates real data on the largest floods to occur in various drainages over several millennia. The data include the floods' ages, or the periods in which they occurred, and the discharges. The information can be used directly in a flood-frequency analysis, or it can be used to assess the reasonableness of extrapolations from conventional data on smaller floods. Paleoflood data can also be compared with probable maximum flood estimates. In this way, expensive overdesign and dangerous underdesign can be avoided in flood-related engineering projects.

SIGNIFICANCE

People have lived with floods since the beginning of civilization. The first great nations of the Earth developed along the fertile but flood-prone valleys of rivers such as the Nile, the Tigris and the Euphrates, the Indus, and the Yangtze and the Huang He. Various means of coping with floods have been documented since the biblical accounts of Noah. Early societies merely avoided zones that tradition told them were hazardous. It has only been in the modern era that large cities have systematically developed on immense tracts of flood-prone lands. Thus the natural process of flooding has become an unnatural hazard to humans.

There are immense consequences for the human insistence on occupying those areas near rivers that infrequently receive floodwater. A single tropical storm system, Hurricane Agnes, generated more than 3 billion dollars in flood damage to the northeastern United States in 1972. Floods in Bangladesh have killed millions of people. In the 1930's, a national U.S. program began to respond to such problems by the construction of large dams. Despite (or perhaps because of) this expensive effort, flood damage to life and property is much greater today than it was in the 1930's.

One trend has been to manage flood-hazard zones with multiple approaches that respond to the nature of the flood risk. The river is treated as a whole integrated system, rather than as individual segments, for engineering design. Management alternatives for this system are not limited solely to structural controls, such as dams and levees; instead, options are considered for land-use adjustment. Flood-prone lands can be used for parks, greenbelts, and bikeways instead of industrial warehouses, stores, and housing. Even when construction must be done on floodplains, it may be possible to make provision for flood risks. Warehouses, for example, can be organized for rapid transfer of materials to second stories or to temporary, safe storage sites. Such adjustments require accurate warning systems that involve measuring rainfall in headwater areas and rapidly predicting the flood consequences to downstream sites at risk.

There is a general need to educate the public that floodplains are a natural part of rivers. Living on a floodplain is really choosing to play a game of "floodplain roulette." For this reason, the most accurate and reliable methods of evaluating flood magnitudes and frequencies are necessary. The choices made for land use on floodplains cannot be based merely on idealized theories of how floods behave.

Victor R. Baker

CROSS-REFERENCES

Aquifers, 2005; Artificial Recharge, 2011; Dams and Flood Control, 2016; Earth Resources, 1741; Groundwater Movement, 2030; Groundwater Pollution and Remediation, 2037; Hydrologic Cycle, 2045; Precipitation, 2050; Salinity and Desalination, 2055; Saltwater Intrusion, 2061; Surface Water, 2066; Water Quality, 2072; Water Table, 2078; Water Wells, 2082; Waterfalls, 2087; Watersheds, 2093.

BIBLIOGRAPHY

Baker, Victor R., ed. *Catastrophic Flooding: The Origin of the Channeled Scabland.* Stroudsburg, Pa.: Dowden, Hutchinson & Ross, 1981. This book uses reprints of the original papers to recount the history of the scientific controversy surrounding the origin of the Channeled Scabland. The catastrophic flood hypothesis of J. Harlen Bretz is featured in considerable detail. The book follows the controversy through its resolution, and it develops many modern concepts of erosion and deposition by examining the immense glacial floods that coursed the Scabland region. Well illustrated.

Baker, Victor R., R. C. Kochel, and P. C. Patton, eds. *Flood Geomorphology.* New York: John Wiley & Sons, 1988. This is a modern treatment of broad, interdisciplinary issues involving floods. Floods are analyzed in terms of landscape, climate, and other geological factors. Their causes, effects, and dynamics are documented, as are methods of management of flood-prone areas. The book is a start at defining a science centered on the study of floods in relationship to the landscapes on which they occur.

Comerio, Mary C. *Disaster Hits Home: New Policy for Urban Housing Recovery.* Berkeley: University of California Press, 1998. Comerio describes the destructive power of floods, hurricanes, and earthquakes, then explores the disaster relief measures and housing policies that have been implemented to handle natural disasters and their aftermaths. Appropriate for the layperson.

Costa, John E., and V. R. Baker. *Surficial Geology: Building with the Earth.* New York: John Wiley & Sons, 1981. This textbook treats the broad range of earth-surface geological processes as they affect humankind. Chapter 12 reviews flooding as a geological hazard. It provides the hydrological background on floods, analyzes methods of establishing flood frequency, outlines techniques for delineating flood-prone areas, and discusses various alternatives for flood-plain management. Other chapters deal with various geological hazards and management problems of the natural environment.

Hoyt, W. G., and W. B. Langbein. *Floods.* Princeton, N.J.: Princeton University Press, 1955. This is the only general hydrological treatment of the entire flooding phenomenon. The book is outdated in methodology, but it provides fascinating case studies of famous floods, particularly in the United States. Illustrated.

Knighton, David. *Fluvial Forms and Processes.* London: Edward Arnold, 1984. This book provides a succinct overview of water-related processes on landscapes. Floods are treated as a part of that spectrum of processes. Aspects of drainage basins, flow mechanics, sediment transport, channel adjustments, and changes in river channels are reviewed. Illustrated with technical diagrams.

Leopold, L. B., M. G. Wolman, and J. P. Miller. *Fluvial Processes in Geomorphology.* San Francisco: W. H. Freeman, 1964. A classic treatment of river processes in relation to landscapes. Although modern work has moved ahead, the book still contains many valuable insights, especially on processes in alluvial river channels. Includes some photographs, but mainly uses scientific diagrams to illustrate relevant issues.

Philippi, Nancy S. *Floodplain Management: Ecology and Economic Perspectives.* San Diego, Calif.: Academic Press, 1996. Philippi presents a thorough account of the destructive power of floods, emergency management policies, and floodplain management tactics. This well-written book can be easily understood by the reader without prior knowledge of the field.

Singh, Vijay P., ed. *Hydrology of Disasters.* Boston: Kluwer Academic Publishers, 1996. Part of the Water Science and Technology Library series, this volume looks at the hydrological aspects of floods and other natural disasters. The book also explores preventive measures and disaster relief programs that have been implemented to deal with floods. Illustrations, index, and bibliography.

Ward, Roy C. *Floods: A Geographical Perspective.* New York: John Wiley & Sons, 1978. A brief introduction to floods from a broad, human-environmental perspective. Basic flood hydrology is reviewed, and floods are related to landscape changes. The main issue, however, is the human use of flood-prone lands. Many important issues for land management are raised.

GROUNDWATER MOVEMENT

The flow of water through the subsurface, known as groundwater movement, obeys well-established principles that allow hydrologists to predict flow directions and rates.

PRINCIPAL TERMS

ELEVATION HEAD: the elevation of a given water particle above a certain datum, usually mean sea level

EQUIPOTENTIAL LINE: a contour line connecting points of equal hydraulic head

GROUNDWATER: water found in the zone of saturation

HYDRAULIC HEAD: the sum of the elevation head and the pressure head at any given point in the subsurface

HYDROSTATIC PRESSURE: the pressure at any given point in a body of water at rest from the weight of the overlying water column

PERMEABILITY: the ability of rock, soil, or sediment to transmit a fluid (commonly water)

POROSITY: the ratio (usually expressed as a percentage) of the total volume of void (empty) space in a given geologic material to the total volume of that material

PRESSURE HEAD: the height of a column of water that can be supported by the hydrostatic pressure at any given point in the subsurface

VELOCITY HEAD: the height to which the kinetic energy of fluid motion is capable of lifting that fluid

WATER TABLE: the upper surface of the zone of saturation

ZONE OF SATURATION: a subsurface zone in which all void spaces are filled with water

POROSITY

The movement of water through the Earth's subsurface is only one part of a larger circulation system known as the hydrologic cycle. This cycle involves the continuous transfer of water between natural reservoirs within the physical environment, such as the oceans, polar ice caps, groundwater, surface water, and the atmosphere. The main processes in the hydrologic cycle are precipitation (for example, rain and snow), evaporation, transpiration by plants, surface-water runoff, and subsurface groundwater flow. When precipitation falls to the land surface, some of this water runs off into streams, some evaporates back into the atmosphere, and the remainder soaks into the ground. The water that infiltrates the land surface either is transpired by plants or percolates deeper to become groundwater. For water to percolate into the subsurface, void spaces (openings not occupied by solid matter) must be available in the underlying geologic materials. The ratio of void space to the total rock or soil volume is known as porosity. This proportion is usually expressed as a percentage. The higher this ratio, the more void space there is to hold water.

There are various types of porosity. Unconsolidated materials (soil and sediment) have pore spaces between adjacent grains, referred to as intergranular porosity. The ratio of pore space to total volume depends on several factors, including particle shape, sorting, and packing. Loosely packed sediments composed of well-sorted, spherical grains are the most porous. Porosity decreases as the angularity of the grains increases because the particles pack more closely together. Similarly, as the degree of sorting decreases, the pore spaces between larger grains become filled with smaller grains, and porosity decreases. Values of porosity for unconsolidated materials range from 10 percent for unsorted mixtures of sand, silt, and gravel to about 60 percent for some clay deposits. Typical porosity values for uniform sands are between 30 percent and 40 percent.

Rocks have two main types of porosity: pore spaces between adjacent mineral grains and voids that are a result of fractures. Rocks formed from sedimentary deposits (such as shale and sandstone) may have significant intergranular porosity, but it is usually less than the porosity of the sediments from which they were derived. This dichotomy is a result of the compaction and cementation that takes place during the process of transforming sediments into rock. Therefore, although sandstone porosities may be as high as 40 percent, they are commonly closer to 20 percent because of the presence of natural cements that partially fill available pore spaces. Igneous and metamorphic rocks are composed of tightly interlocked mineral grains and therefore have little intergranular porosity. Virtually all void space in such rocks is a result of fractures (joints and faults). For example, granite (a dense, igneous rock) usually has a porosity of less than 1 percent, but porosity may reach 10 percent if the rock is fractured.

There are additional types of porosity that occur only in certain kinds of rocks. Limestone, a rock that is soluble in water, can develop solution conduits, or channels, along fractures and bedding planes. Given enough time, solution weathering may lead to the development of a cave, which has 100 percent porosity. The overall porosity of solution-weathered limestone sometimes reaches 50 percent. Rocks created by volcanic eruptions may contain void space in the forms of vesicles (cavities left by gases escaping from lava), vertical shrinkage cracks developed during cooling (known as columnar joints), and tunnels created by flowing lava (called lava tubes). In extreme cases, the porosity of volcanic rocks may exceed 80 percent.

Underground water includes all water that exists below the land surface, but the subject of groundwater movement is mainly concerned with the water that occurs in the zone of saturation, where all empty spaces are completely filled with water. Between the zone of saturation and the land surface, void spaces contain mostly air, unless a heavy rainfall or a period of snowmelt has just occurred. Water in this upper zone is held under tension by the attractive forces between soil particles and water molecules (surface tension forces).

The water table forms the uppermost surface of the zone of saturation and is characterized by having a water pressure equal to atmospheric pressure. It may be only a few meters below the land surface in humid regions or hundreds of meters below the surface in desert environments. In general, the water table mimics the surface topography but with more subdued slopes. If the water table intersects the land surface, the result is a lake,

swamp, river, or spring. Below the water table, in the zone of saturation, geologic materials are completely saturated, and water pressure (called hydrostatic pressure) increases with depth. Water contained within the zone of saturation is generally called groundwater. It is this zone that supplies water to a well when it occurs in a particular type of geologic formation known as an aquifer.

Groundwater Energy

To understand groundwater flow, it is necessary to examine the forms of energy contained in groundwater. The total energy in any water mass consists of three components: elevation head (potential energy from a water particle's elevation above mean sea level), pressure head (potential energy because of the hydrostatic pressure of surrounding fluids), and velocity head (kinetic energy resulting from motion). Because groundwater moves relatively slowly, however, velocity head can usually be neglected, leaving the total energy equal to the sum of the elevation head and the pressure head. This quantity is known as the hydraulic head. Thus the hydraulic head of a given water particle varies directly with its elevation (usually expressed in meters above mean sea level) and its hydrostatic pressure.

The water table can be thought of as a surface with variable hydraulic head. Because water pressure at the water table is always the same (equal to atmospheric pressure), the change in hydraulic head across the water table is dependent only upon the variation in elevation. Below the water table, hydraulic head depends on both the elevation and the water pressure, which increases with depth. Therefore, the variation in hydraulic head with depth below the water table reflects the relationship between decreasing elevation head and increasing pressure head. If the increase in hydrostatic pressure exactly offsets the decrease in elevation head, then hydraulic head will not change with depth.

Groundwater moves in response to differences in hydraulic head between two locations. The direction of movement is always from areas of highest hydraulic head toward areas of lowest hydraulic head. The change in hydraulic head over a specified distance is known as the hydraulic gradient. Both horizontal and vertical hydraulic gradients can exist.

Horizontal gradients are usually defined by the

change in water table elevation between any two locations. As water moves through the subsurface, it flows along the steepest hydraulic gradient. Therefore, it is possible to determine the compass direction of groundwater movement from a knowledge of how the water table elevation varies over distance. Because the water table often mimics the land surface, general groundwater flow directions can sometimes be predicted on the basis of surface topography.

Vertical hydraulic gradients describe changes in hydraulic head with depth. As groundwater flows in any given horizontal direction, it may also be rising or sinking, depending on the vertical hydraulic gradient. Where hydraulic head decreases with depth below the water table, groundwater flow has a downward component, resulting in recharge areas. Where hydraulic head increases with depth, groundwater flow has an upward component, creating a discharge area. Recharge areas commonly occur in the higher elevations of a particular landscape, and discharge areas usually occur in the valleys near lakes, streams, and swamps. This year-round flow of groundwater from higher to lower elevations permits streams to flow in the dry summer months when there is little surface

runoff. In certain situations, the water pressure conditions can cause groundwater to move "uphill" with respect to the surface topography. Therefore, the land surface is not always a good indicator of groundwater flow directions.

Groundwater movement can be divided into local and regional flow. In areas of rugged topography, most groundwater flow is local, meaning that it moves from the hilltops to the nearest stream or lake. In more gentle terrains, however, or in areas where the zone of groundwater movement is very thick, some flow escapes the local system into a deeper, regional system. Thus, water may enter the subsurface at a local zone of recharge and move long distances before surfacing at a regional discharge area. Identifying the boundaries of local and regional flow systems requires detailed information about the horizontal and vertical distributions of hydraulic head over a large area.

RATE OF FLOW AND PERMEABILITY

The rate at which groundwater moves through the subsurface can also be determined on the basis of scientific principles. Groundwater flow velocities depend on two factors: the hydraulic gradient and the permeability of the geologic materials involved.

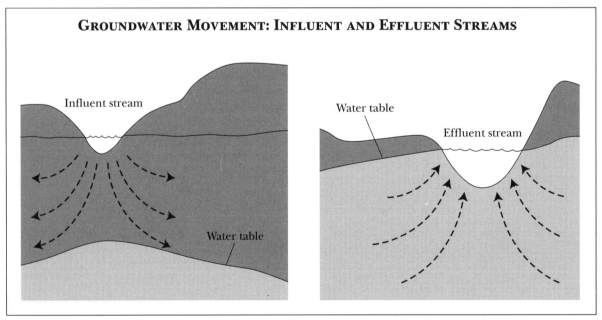

GROUNDWATER MOVEMENT: INFLUENT AND EFFLUENT STREAMS

Two examples of groundwater movement: Influent streams, found in dry climates, cause an upward bulge in the water table as water from the stream migrates downward. Effluent streams, found in wet climates, are produced by a high water table that migrates water into channels in the ground's surface.

Considering that differences in hydraulic head are the driving force behind groundwater movement, it is not surprising that the hydraulic gradient is related to the rate of flow. All other things being equal, the steeper the gradient, the faster groundwater will move. Hydraulic gradients may change during the year, reflecting the influence of recharge. In the spring, when recharge is high, the water table will rise fastest beneath recharge areas, producing a steeper hydraulic gradient and higher flow velocities. As the water table drops throughout the summer, the hydraulic gradient also declines, leading to slower groundwater flow.

Permeability, the ability of porous formations to transmit fluids, is a property of the geologic material in question. This property depends on the size of void spaces and the degree to which they are interconnected. Thus, some high-porosity materials such as clay (up to 60 percent porosity) and pumice (a vesicular volcanic rock with up to 87 percent porosity) are not very permeable because the void spaces are largely isolated from one another. Materials that have high permeability include sand, gravel, sandstone, and solution-weathered limestone. Rocks with low porosities, such as shale, quartzite, granite, and other dense, crystalline rocks have low permeabilities, unless they are significantly fractured.

Groundwater is forced to move along tortuous paths through geologic materials as it follows the connecting spaces between voids. Therefore, even in highly permeable materials, groundwater flows much more slowly than does surface water in a river. Whereas the velocity of stream flow may be measured in meters per second, groundwater velocities commonly range from 1 meter per day to less than 1 meter per year (averaging about 17 meters per year in rocks). The highest velocities occur in rocks that are heavily fractured. In the extreme case, groundwater can actually move as a subsurface stream through cavernous limestone or volcanic rock. This situation is not normal, however, and underground streams are much less common than the average person might suspect.

STUDY OF GROUNDWATER MOVEMENT

The first step in studying groundwater movement in a particular area is to determine the hydraulic gradients that exist. Horizontal gradients are defined by measuring groundwater elevations in wells that intersect the water table. To define vertical gradients that may be present, piezometers are used. Piezometers are essentially vertical pipes that are open at the bottom to allow water to enter. These devices enable the hydraulic head at a particular depth below the water table to be determined. The elevation head is the elevation at the bottom of the piezometer, and the pressure head is the height to which water rises in the piezometer above the intake point.

Contour maps of the variation in hydraulic head are prepared from data collected in a number of wells and piezometers. These maps can show either horizontal variations (such as a map of the water table) or vertical changes in hydraulic head (as shown in a cross-sectional view). The contours used to connect points of equal hydraulic head are called equipotential lines. Groundwater flow directions can be determined once the distribution of known hydraulic head values has been contoured. Flow lines, depicting the idealized paths taken by water particles, are drawn to intersect equipotential lines at right angles, indicating that groundwater moves along the steepest hydraulic gradient. The resulting gridlike pattern of equipotential and flow lines (called a flow net) is a two-dimensional representation of the groundwater flow system.

Flow nets are much better indicators of groundwater flow directions than is the land surface topography. Also, from a study of flow nets, it may be possible to determine accurately recharge and discharge areas. In the map view, flow lines will diverge from areas of recharge and converge in discharge areas. In the cross-sectional view, flow lines will have downward components where recharge is occurring and upward components in discharge areas. If hydraulic head does not change with depth, flow lines will be horizontal, indicating that neither recharge nor discharge conditions exist. Flow nets do not give estimations of flow velocities unless permeability values are known for the materials involved. If an estimation of permeability can be obtained, groundwater flow velocity may be calculated by multiplying the permeability value and the hydraulic gradient and dividing this product by the porosity of the geologic formation. This calculation yields an average linear velocity of groundwater flow through the open area provided by void space.

Groundwater flow directions and velocities can also be determined by introducing a "tracer" into the groundwater and monitoring its migration with observation wells. This technique is especially helpful in areas of fractured rock, where the flow patterns are difficult to predict. Tracers are substances that will dissolve readily and move with the groundwater without reacting with the geologic materials. Ideally, tracers are safe to use, inexpensive, and easy to detect in low concentrations. Examples of tracers used in groundwater studies include salts (sodium or potassium chloride), fluorescent dyes, and the radioactive isotopes of certain elements (helium, hydrogen, and iodine). The choice of tracer usually depends on the subsurface materials involved.

In areas where the hydrogeologic conditions are well defined through wells, piezometers, borings, and permeability tests, the groundwater flow system can be studied with the use of computer models. Computers are used to calculate hydraulic head values for the area modeled given the rates of recharge (infiltration) and discharge (evapotranspiration, discharge to a stream, lake, or well). These computed head values can then be contoured to create a flow net. Although computers are very powerful tools, the accuracy of their predictions cannot surpass the accuracy of the information provided for the model.

SIGNIFICANCE

The understanding of groundwater movement is important to the utilization of aquifers for water supply. From a quantity point of view, flow rates determine which geologic materials will serve as suitable sources of groundwater. The rates of flow must equal or exceed the desired pumping rate for a well to be successful. In this regard, permeability is the limiting factor, because the drawdown (depletion) in a well from pumping will always create a sufficient hydraulic gradient to favor flow toward the well. If the permeability is too low, or the pumping rate is too high, the geologic formation will become dewatered. For domestic wells requiring only 7 to 19 liters per minute, even moderately impermeable materials may supply sufficient water to be considered as an aquifer. High-yield wells (exceeding 190 liters per minute), however, can be sustained only in very permeable materials, such as sand and gravel, sand-

stone, and solution-weathered limestone.

Understanding groundwater flow is also important to water quality. As groundwaters move through the subsurface, they dissolve minerals from the geologic materials with which they have contact. Therefore, groundwater that has traveled great distances in a regional flow system will tend to be the most mineralized and the least desirable for a drinking water source. With an increasing number of possible contamination sources (for example, landfills, leaky underground tanks, and accidental spills), the protection of groundwater supplies is crucial. A knowledge of groundwater flow directions and rates aids in the prediction of contaminant migration. It is particularly important to identify areas that recharge regional flow systems because these zones have the greatest potential impact if they become polluted. Municipalities concerned with the preservation of their well fields are now undertaking wellhead protection studies. The purpose of such studies is to delineate the well field recharge areas that need to be protected from any type of land use that could lead to groundwater contamination.

Contaminant migration is difficult to predict in fractured rock formations, especially in solution-weathered limestone. Because groundwater movement follows a sometimes random network of discontinuous openings in these settings, water levels may not be related from one well to the next. Therefore, maps of the water table are often impossible to construct. Flow through fractured rock is one of the concerns surrounding the choice of a depository for high-level nuclear wastes, which need to be isolated from the environment for at least ten thousand years.

David L. Ozsvath

CROSS-REFERENCES

BIBLIOGRAPHY

Davis, S. N., and R. J. M. DeWiest. *Hydrogeology.* New York: John Wiley & Sons, 1966. A well-illustrated introduction to the geologic aspects of groundwater occurrence. The hydrologic cycle and its relationship to groundwater are discussed in chapter 2. Four chapters (chapters 9-12) are dedicated to a discussion of the occurrence and movement of groundwater in different types of geologic materials. Suitable for both high school and college-level readers.

Fetter, C. W. *Applied Hydrogeology.* 2d ed. Columbus, Ohio: Merrill, 1988. This textbook emphasizes the practical aspects of understanding groundwater occurrence and movement. Chapter 8 contains a detailed discussion of the influence that geologic conditions have on groundwater occurrence, with special emphasis on groundwater regions in the United States. Chapter 7 addresses the regional groundwater movement within aquifers and contains helpful illustrations of the principles discussed. A glossary of important terms is included. Suitable for college-level readers.

_____. *Contaminant Hydrogeology.* 2d ed. Upper Saddle River, N.J.: Prentice Hall, 1999. This textbook emphasizes the conditions that lead to the contamination of groundwater, the problems this contamination can cause, and remediation techniques used in attempts to correct these problems. Includes a detailed discussion of the influence that geologic conditions have on groundwater occurrence and contamination. The book also addresses the regional groundwater movement within aquifers and contains helpful illustrations of the principles discussed. A glossary of important terms is included.

Freeze, R. A., and J. A. Cherry. *Groundwater.* Englewood Cliffs, N.J.: Prentice-Hall, 1979. A more advanced book stressing the mathematical derivations of groundwater flow equations and their applications. Chapter 4 presents a discussion of the geologic controls on groundwater occurrence. Chapter 6 contains a number of figures illustrating the principles of regional groundwater flow. Suitable for college-level readers.

Hamblin, Kenneth W., and Eric H. Christiansen. *Earth's Dynamic Systems.* 8th ed. Upper Saddle River, N.J.: Prentice Hall, 1998. Although not limited to the topic of groundwater, this widely used introductory textbook has a good discussion of groundwater movement in chapter 13. The color figures are especially helpful to one unfamiliar with subsurface geology. Suitable for both high school and college-level readers.

Montgomery, C. W. *Physical Geology.* Dubuque, Iowa: Wm. C. Brown, 1987. Another introductory geology textbook with a good discussion of groundwater movement (chapter 14). The full-color figures are helpful, as is the author's emphasis on the environmental aspects of water supply and pollution. Suitable for high school and college-level readers.

Saether, Ola M., and Patrice de Caritat, eds. *Geochemical Processes, Weathering, and Groundwater Recharge in Catchments.* Rotterdam, Netherlands: A. A. Balkema, 1997. This slightly technical compilation looks at the geochemical cycles and processes associated with catchment basins. Several essays deal with the artificial recharge of groundwater in such systems.

Thompson, Stephen A. *Hydrology for Water Management.* Rotterdam, Netherlands: A. A. Balkema, 1999. This thorough account of hydrology, groundwater flow, and stream flow focuses on the management of water supplies in efforts to keep them free of pollutants and available to the largest number of people possible. Illustrations, maps, index, and bibliography.

Todd, D. K. *Ground Water Hydrology.* 2d ed. New York: John Wiley & Sons, 1982. A textbook emphasizing the practical aspects of groundwater occurrence and movement. Chapter 1 introduces the reader to groundwater utilization and its relationship to the hydrologic cycle. Chapter 2 discusses aquifer types and the occurrence of groundwater within the United States. Suitable for college-level readers.

Walton, W. C. *Groundwater Resource Evaluation.*

New York: McGraw-Hill, 1970. Although much of the book deals with advanced techniques for analyzing aquifer properties from well pump tests, chapter 2 contains a valuable discussion of the terms and principles pertinent to groundwater movement. Chapter 1 provides background on the role of groundwater within the hydrologic cycle. Suitable for college-level readers.

GROUNDWATER POLLUTION AND REMEDIATION

Groundwater pollution is any artificially induced change in the natural quality of the water. Groundwater pollution remediation is concerned with the preservation of a resource's beneficial use, which, with respect to groundwater, is frequently potable supply. Groundwater contaminants are usually removed through the use of "pump-and-treat" technologies. These technologies involve initial groundwater extraction, followed by treatment at the surface prior to reinjection or disposal.

PRINCIPAL TERMS

AIR STRIPPING: the process of passing contaminated water through an aeration chamber, causing the organic contaminants to volatize into the gaseous waste stream

AQUIFER: a rock or sediment from which water may be withdrawn for use

CARBON ADSORPTION: the process of pumping contaminated water directly through carbon filters, to capture contaminants by their binding onto the surface of the carbon adsorbent

CONTAMINANT: any ion or chemical that is introduced into the environment, especially in concentrations greater than those normally present

FLOCCULATION: a slow stirring process that causes coagulated particles to gather together to form larger settleable particles

ION EXCHANGE: involves the reversible switching of ions between the water being treated and the ion exchange resin; undesirable ions in the water are exchanged with acceptable ions on the resin

LEACHATE: fluid produced by the decay of garbage in a landfill

NONPOINT SOURCE: a large, diffuse source of contamination

OSMOSIS: the passage of a liquid from a weak solution to a strong solution across a semipermeable membrane

POINT SOURCE: a single, defined source of contamination

PRECIPITATION: the chemical conversion of a soluble material into an insoluble one

GROUNDWATER CONTAMINATION

Groundwater contamination may be the result of either natural or human causes. When human activities cause degradation of natural water quality, however, the term "pollution" is used. There are at least sixteen U.S. federal statutes related to groundwater quality. One of the most important is the Safe Drinking Water Act of 1974. As a result of this legislation the U.S. Environmental Protection Agency (EPA) issued primary drinking water regulations in 1975, which were modified in 1977. These standards list maximum permissible concentrations, based on health criteria, of a number of chemicals, both organic and inorganic. Measured in milligrams per liter, the largest concentrations were allowed for nitrate (10.0) and barium (1.0), the least for endrin (0.0002) and heptachlor (0.0001). The regulations apply to all public water systems; however, individual ground-

water supplies would ideally be of this quality as well. Although the concentration limits for these chemicals are very low, theoretically, synthetic organic chemicals should not be present at all. Other chemicals, such as chloride, copper, iron, zinc, and manganese, have EPA-recommended levels based on aesthetic or taste criteria.

Much groundwater pollution is the result of microbiological contamination. The desirable limits for the untreated water supply are 100 colonies of total coliforms (bacteria) per 100 milliliters and 20 colonies of fecal coliforms per 100 milliliters. Disease-causing organisms typically come from animal wastes. Contaminated water may cause typhoid fever, cholera, bacillary dysentery, and paratyphoid fever, and even polio and infectious hepatitis. Between 1971 and 1977, there were 192 outbreaks of water-transmitted diseases in the United States and vastly more in Third World countries.

POINT SOURCES

Groundwater pollution is often divided into two major sources—point and nonpoint. These are only potential sources; that is, if designed and operated properly they should not cause pollution. Potential point sources of groundwater contamination include confined animal operations, land application of wastewater and sewage, solid waste landfill, and septic tanks. Industrial point sources can be, for example, injection wells, hazardous and radioactive waste sites, mining activity sites, oil and gas production fields, chemical spills, and leaking underground storage tanks. In addition, saltwater intrusion can cause groundwater contamination, as can urban storm waters. Storm runoff from streets and parking areas contains petroleum products and metals from automobile traffic, which may contaminate the groundwater system if the storm drains have leaks or if the runoff is not collected. Similarly, sewer lines may also cause groundwater contamination. In cold climate areas, salts used for de-icing roads are potential pollutants.

Confined animal operations include milking barns and feedlots for beef production. With hundreds or thousands of animals confined in an area, the soil will not be able to assimilate all the animal wastes. Storm runoff can pollute both ground and surface waters. Animal wastes may supply nitrogen compounds, phosphates, chloride, metals, organic chemicals, and bacteria to the groundwater. Nitrate-nitrogen is perhaps the most important contaminant that may reach the groundwater. Similar contamination is possible from improper land application of sewage and wastewater and from improper operation of septic tanks. Usually bacteria, viruses, and phosphates are removed by the soil. Approximately 20 million septic tank and cesspool systems are in use in the United States, which results in about 2.8 billion gallons of partially treated sewage entering the groundwater system annually—a tremendous source of potential contaminants if the system does not operate properly. Septic system failures result from inadequate flow through the soil or, more commonly, from simply overloading the system.

HAZARDOUS WASTES

Hazardous injection wells and radioactive and hazardous waste sites may contaminate groundwater with a variety of inorganic chemicals (for example, arsenic, lead, chromium, uranium) and organic compounds. The Love Canal incident has received wide publicity but is only one example of groundwater contamination by a hazardous waste site. Between 1947 and 1953, this abandoned canal in New York was used as a hazardous waste dump. Thousands of drums of chemicals were disposed of in the water and in the banks of the canal. After the land was sold in 1953, the canal was filled with dirt, and schools and houses were built near the site. In 1970, chemicals (chloroform, benzene, toluene, and many other dangerous and carcinogenic compounds) were detected in the air in basements of the houses. Chemicals from leaking and disintegrating drums contaminated the groundwater and seeped through basement walls.

Hazardous industrial wastes have sometimes been disposed of by means of deep injection wells that inject the fluids into salty water zones below freshwater aquifers. Although the reported cases of groundwater contamination from such sources is low, the potential for contamination exists if the injection well leaks at the freshwater level. Accidental spills of inorganic or organic chemicals at the Earth's surface may soak through the soil into the groundwater. Chemical spills may occur at the site of production, site of use, or during transportation.

LEACHATE AND PETROLEUM PRODUCTS

There are about 100,000 industrial and municipal landfills in North America. Approximately 16,500 active municipal solid waste landfills in the United States dispose of 160 million tons of waste per year. A city of 1 million people generates enough refuse annually to cover approximately 145 football fields 15 meters deep with garbage. Leachate produced by the decomposition of the large volumes of buried municipal waste poses a possible threat to groundwater quality. Although the composition of leachate is variable, it is enriched in dissolved solids (especially chloride and metals) and will consume a large amount of oxygen (measured as chemical oxygen demand), primarily because of its high concentration of undecomposed organic material. Methods for groundwater protection at landfills include surface-water control, covering the waste with soil (clays) that does not transmit significant amounts of precipitation into the landfill, and liners (natural clay or synthetic) that keep the leachate from leaving the burial site.

Petroleum-product storage tanks constitute the major type of underground tanks. Faulty tanks, a result of defective construction or emplacement or of rust, leak these organic chemicals into the groundwater. Leaky oil wells and distribution pipelines may also contaminate the groundwater with petroleum. Spills of gasoline at gas stations and by tanker trucks ruptured in traffic accidents also have contaminated groundwater. Because oils and gasoline are immiscible with (will not mix with) and are lighter than water, most of the pollution from these sources floats on top of the groundwater. Small amounts of the petroleum products, however, can dissolve in the water and contaminate the entire thickness of the aquifer.

ACID MINE WATERS AND SALTWATER

Acid mine waters, rich in metals (for example, lead, zinc, and cadmium), are common in many ore and coal mining operations. Pyrite (commonly known as fool's gold) is composed of iron and sulfide. When the pyrite comes in contact with air and water, sulfuric acid is produced, which dissolves metals from the rock. Protecting groundwater from contamination produced by mining operations is difficult, especially for deep mines.

Saltwater contamination of groundwater may result from oil and gas wells, as well as from freshwater wells. Most deep oil and gas wells encounter brines (salty water), which are returned to their sources with injection wells. If either the petroleum wells or the injection wells have leaks, salt water may enter more shallow freshwater aquifers. In many areas of the country, especially in coastal areas, salty water underlies fresh water. If freshwater wells (for example, irrigation wells) pump too much water, the level of the fresh water can be lowered enough so that the underlying salt water mixes with the fresh water.

NONPOINT SOURCES

Nonpoint (diffuse) sources of contamination are also important. As much as 50 percent of the total water pollution problem has been attributed to nonpoint sources. Although these two sources of pollution are used for discussion purposes, there is often a gradation between point and nonpoint sources. For example, if only a few septic tanks in an area are causing pollution, each of the septic tanks would be considered a point source of pollution. If many septic tanks are contaminating the groundwater, however, so that there are no distinct points of pollution, the contamination would be considered nonpoint source pollution. The contamination of the groundwater in Long Island, New York, is an example of nonpoint source pollution of groundwater from septic tank discharges. The type of soil and the concentration of a large number of septic tanks in a small area have resulted in contamination of the groundwater by nitrate, ammonia, and detergents. Nitrate-nitrogen and ammonia-nitrogen concentrations in this area both exceed 20 milligrams per liter. The drinking water criterion for nitrate-nitrogen is 10 milligrams per liter; therefore, the water must be treated before use. Atmospheric deposition (both wet and dry) of chemicals downwind of smokestacks in urban areas can cause nonpoint pollution of groundwater over a large area.

Agricultural practices cause a significant amount of nonpoint source pollution. Contamination by fertilizers (nitrate) and pesticides are most significant because they are spread over large areas and may seep through the soil into the groundwater. Irrigation waters can also cause contamination in arid areas. When water evaporates, dissolved chemicals are concentrated in the remaining water, which soaks into the groundwater system. This water, enriched in dissolved chemicals, is then pumped from the ground and used for irrigation again. This cycle can be repeated several times, eventually producing groundwater with very high concentrations of chemicals.

TREATMENT TECHNOLOGIES

Selection of a groundwater treatment system for groundwater pollution remediation is normally a three-step process: an initial groundwater investigation to determine type and concentration of contaminant and extent of pollution, establishment of cleanup goals, and selection of treatment technology. This latter step almost invariably comprises groundwater extraction, surface treatment, and disposal or reinjection, otherwise termed pump-and-treat remediation.

Treatment technologies may be categorized as biological, chemical, or physical. In biological treatment methods, the contaminants are metabolized by aerobic or anaerobic microorganisms. Chemical treatment methods are based on the use of a chemical reactant to immobilize or break down a contaminant and include adsorption and precipitation. Physical treatment technologies utilize a physical property, such as a contaminant's molecular weight or solubility, as the basis for separating the contaminant from the polluted groundwater. These technologies include air stripping, reverse osmosis, and electrodialysis.

AIR STRIPPING

The basic concept behind air stripping is mass transfer, whereby the contaminant in water is transferred to a solution in air. Contaminated water is brought into intimate contact with air; the contaminants become vapors and are removed from the water as the vapors are carried off in the airstream. The transfer of a contaminant is related to its vapor pressure in air relative to its solubility in water. This ratio is formally expressed as Henry's law: $H = C_{ig}/C_{il}$, where C_{ig} is the equilibrium concentration in gas phase (grams per cubic meter), C_{il} is the equilibrium concentration in liquid phase (grams per liter), and H is Henry's law constant. Henry's law constant can be used to predict a contaminant's strippability: The higher the constant, the greater the ease of strippability.

Water is brought into intimate contact with air either by putting air through water or by putting water through air. Two air-through-water systems are diffused air aeration and mechanical surface aeration. Diffused air aeration is a procedure in which compressed air is injected into a tank of water through a porous base plate or through perforated pipes. In mechanical surface aeration, an impeller creates turbulent mixing of air and water.

In each of the air-through-water systems, the mass transfer occurs at the bubble surface. In water-through-air aeration systems, the mass transfer is facilitated by the creation of thin water films or small water droplets. Water-through-air system configurations include crossflow towers and tray aerators, in which the air flows crosscurrent or countercurrent, respectively, to water flowing downward over trays or slats; spray basins, in which water is sprayed from a network of nozzles within a basin into the air in fine droplets to be subsequently collected as they fall back into the basin; and packed towers. Packed towers utilize a countercurrent flow scheme. Air is blown up the tower while water trickles down over an inert (chemically inactive) packing, commonly polypropylene moldings.

CARBON ADSORPTION

In carbon adsorption, contaminants are removed from water by their attraction and binding to the surface of the activated carbon adsorbent. Adsorption performance is estimated through the use of a liquid adsorption isotherm test. When contaminated water is mixed with activated carbon, the contaminant concentration decreases to an equilibrium concentration, at which point the number of molecules leaving the surface of the adsorbent is equal to the number of molecules being adsorbed. The relationship between adsorption capacity and equilibrium concentration, known as an adsorption isotherm, is described by the Freundlich equation: $X/M = KC_e^{1/n}$, where X/M is the adsorption capacity, or milligram volatile organic carbon (VOC) adsorbed per gram of activated carbon; C_e is the contaminant concentration at equilibrium VOC milligram per liter; and K and $1/n$ are empirical constants. From the isotherm, the adsorption capacity of the carbon for a contaminant can be estimated using the X/M value that corresponds to the incoming water contaminant concentration.

Activated carbon is produced by high-temperature pyrolysis (the chemical change of a substance by heat) of an organic material to produce a carbon char, followed by partial oxidation at high temperature in an oxygen-poor atmosphere. During partial oxidation, the oxidation occurs along planes within the carbon, creating macro- and micropores and thereby greatly increasing the surface area of the carbon: The resulting surface area can be up to 1,400 square meters per gram.

During activated carbon treatment, the contaminated water is contacted with the carbon for between fifteen and sixty minutes at a surface loading rate of 2-17 gallons per minute per square foot. In contact systems using granular carbon, the contaminated water flows either through a fixed bed of carbon or a moving bed, in which the carbon moves down the column under gravity countercurrent to the flow of water. The main advantage of the moving bed is reduced suspended solids removal, although this advantage is rarely a factor in groundwater treatment. The flow of water in fixed-bed systems may be either up or down through the bed. System configurations may be single column or multiple columns arranged in parallel or in series. In powdered carbon systems, the carbon is mixed with the contaminated water, typically in the clarifier of an activated sludge system and allowed to settle before disposal. Upon exhaustion of granular carbon's adsorptive capacity, the carbon is removed and regenerated in a furnace, where the high temperature destroys the adsorbed organic matter.

Final sedimentation tanks at a water sewage treatment plant. (© William E. Ferguson)

REVERSE OSMOSIS AND ION-EXCHANGE TREATMENT

The term "osmosis" describes the phenomenon in which certain types of membrane will permit the passage of a solvent but restrict the movement of solutes such that water molecules will pass through a semipermeable membrane from a weak solution to a strong solution, eventually equalizing the solute concentrations on both sides of the membrane. In reverse osmosis, pressure (typically 200-400 pounds per square inch) is applied to force the contaminated groundwater through a membrane, contaminant movement is retarded by the membrane, and purified water is obtained. Cellulose acetate or polyamide membranes in tubular, spiral-wound, or hollow-fin fiber configurations are used.

The electrodialysis process uses electric potential rather than pressure to remove ions from a solution. Two ion-selective membranes, one cation (positively charged) permeable and one anion (negatively charged) permeable, partition the contaminated water from the brine solution and the electrodes. When an electric current is passed across the cell, the cations migrate through the cation-permeable membrane toward the cathode. The anions, however, are prevented from migrating by the membrane. At the anion-permeable membrane, the converse occurs, and, consequently, both cations and anions are removed from the contaminated water. In operational electrodialysis systems, several hundred alternate anion and cation permeable membranes, spaced at 1 millimeter intervals, are placed between a single set of electrodes. The passage of water between the membranes usually takes 10-20 seconds, during which time 25-40 percent of the ions are removed. Depending upon treatment goals, the water may pass between 1 and 6 membrane stacks.

Precipitation is the process whereby soluble inorganic contaminants are converted, by the addition of reagents, to insoluble precipitates, which are then removed by flocculation and sedimentation. With respect to dissolved metals, precipitation is achieved by increasing the concentration of the anion of a slightly soluble metal-anion salt, usually carbonate, hydroxide, or sulfide.

ON-SITE TREATMENT

In addition to contaminant pump-out, reverse osmosis, and ion-exchange treatment of inorganic contaminants, and air stripping-based treatment of inorganic contaminants, two on-site treatments have been under evaluation. On-site nitrate removal involves the use of autotrophic denitrifying bacteria, which are able to catalyze enzymatically the reduction of nitrate to nitrogen gas. Degradation of hydrocarbons by naturally occurring groundwater microorganisms is expected to treat gasoline-related contaminants. Fully implemented pump-and-treat systems, meanwhile, are estimated to yield a total of 70,700 acre-feet per year of remediated groundwater.

STUDYING GROUNDWATER POLLUTION

Although it is possible to determine if a well is polluted, it is difficult to determine the extent of aquifer pollution. A large number of wells must be drilled and designed to collect water from various levels within the aquifer to evaluate truly the extent of aquifer contamination. This approach is obviously expensive and is only used in areas thought to be polluted. Monitoring wells are located near waste disposal sites in order to monitor the quality of the water in the vicinity. The locations and design of the monitoring wells are based on detailed studies of the geology (including soil) and hydrology of the site. Water samples are collected from these wells at regular intervals in order to monitor any degradation of the water. The parameters selected for monitoring are those associated with the wastes that are soluble in water. For example, chromium, lead, and copper might be used for monitoring a hazardous waste site that accepts waste from a metal-processing plant. Early detection of pollution allows the source of contamination and a relatively small portion of the aquifer to be cleaned up before the entire aquifer is damaged beyond repair.

Computer modeling of the transport and deposition of contaminants has become a major method in studying groundwater pollution. These types of studies require not only computer and mathematical expertise but also considerable knowledge of hydrology, geology, and chemistry. The direction and rate of groundwater flow are important. The type of rock and the presence of fractures and faults, which affect the movement of groundwater, are important factors also. Equally important is the solubility of chemicals and their reaction with the soil and aquifer material. These models are very helpful in designing monitoring networks for sites. In addition, these models can be useful in designing a remediation program for polluted portions of aquifers.

IMPORTANCE OF GROUNDWATER QUALITY PROTECTION

High-density populations of organisms, including humans, usually encounter waste disposal problems. These problems are especially critical for industrialized societies that produce large quantities of municipal and hazardous wastes. Disposal of these large volumes of wastes have often resulted in major contamination of groundwater. Approximately one-half the people in the United States utilize groundwater for drinking water. Approximately 60 percent of these people receive their water from a community system, and the remaining 40 percent have private wells. Groundwater is also important as a source of irrigation water, especially in the western part of the United States. The quality of the groundwater is important for its agricultural use. If the water becomes too salty, the plants cannot grow. If the water contains high concentrations of trace metals (for example, selenium), then plants may concentrate the metal and pose a health problem.

Pollution of groundwater is more critical than that of surface water for two reasons. First, it is more difficult to gain access to groundwater; therefore, it is more difficult to determine groundwater pollution and to clean up the contamination. Groundwater pollution cannot be seen and is only detected when a well or a spring becomes noticeably polluted. Second, groundwater movement through aquifers is usually very slow, so remediation of the groundwater will also be slow. Because it is difficult and expensive to clean up an aquifer once it has been polluted, considerable effort should be spent in careful design of installations and monitoring systems and in land-use planning.

Land-use management is crucial to protecting groundwater quality. For example, if soils in an area are too thin or of the wrong type to allow natural treatment of polluted surface waters from feedlots, these operations should be banned in this area. In some cases, agricultural contamination of aquifers has occurred because too much pesticide or fertilizer was used for the soil and vegetation conditions in an area. Mapping of faults and fractures, which often are zones of increased groundwater movement, can be useful in land-use planning. Limestone areas are often very susceptible to groundwater contamination because the rocks are usually fractured and these fractures may be enlarged by the groundwater dissolving the rock. Caves, sinkholes, and disappearing streams also are often associated with limestone aquifers. These features do not allow natural filtration of recharge water; therefore, the groundwater may be quickly polluted by surface waters.

CONTEXT

Groundwater supplies approximately half of the United States' drinking water, providing for the domestic needs of about 117 million people and accounting for 35 percent of municipal drinking water supply and 95 percent of rural drinking water supply. Data on the extent and nature of groundwater contamination are limited. Information compiled by the EPA indicates that 20 percent of all drinking water systems and 30 percent of the systems in municipal areas, using groundwater as their source, show at least trace levels of volatile organic carbons. The true extent of groundwater contamination is, however, probably much greater than that indicated by the EPA data. Potential point sources of pollutants alone include

twenty-nine thousand hazardous waste sites, ninety-three thousand landfills, and twenty-two million septic systems. Given possible widespread groundwater contamination, increasing public awareness of the importance of this resource and the health implications of its contamination, and the increasing prominence of environmental issues on the political agenda, it is clear that the comparatively new technology of groundwater pollution remediation will become an increasingly important component in environmental management.

Kenneth F. Steele and Richard J. Boon

CROSS-REFERENCES

Aquifers, 2005; Artificial Recharge, 2011; Dams and Flood Control, 2016; Floods, 2022; Groundwater Movement, 2030; Hazardous Wastes, 1769; Hydrologic Cycle, 2045; Landfills, 1774; Nuclear Waste Disposal, 1791; Precipitation, 2050; Salinity and Desalination, 2055; Saltwater Intrusion, 2061; Surface Water, 2066; Water Quality, 2072; Water Table, 2078; Water Wells, 2082; Waterfalls, 2087; Watersheds, 2093.

BIBLIOGRAPHY

Canter, L. W., R. C. Knox, and D. M. Fairchild. *Ground Water Quality Protection*. Chelsea, Mich.: Lewis, 1987. This book is an excellent summary of all aspects of groundwater pollution. Although some chapters are at the college level, most of those dealing with pollution are suitable for all readers.

Fetter, C. W. *Contaminant Hydrogeology*. 2d ed. Upper Saddle River, N.J.: Prentice Hall, 1999. This textbook emphasizes the conditions that lead to the contamination of groundwater, the problems this contamination can cause, and remediation techniques used in attempts to correct these problems. Includes a detailed discussion of the influence that geologic conditions have on groundwater occurrence and contamination. The book also addresses the regional groundwater movement within aquifers and contains helpful illustrations of the principles discussed. A glossary of important terms is included.

Heath, R. C. *Basic Ground-Water Hydrology*. Wash-

ington, D.C.: Government Printing Office, 1983. An excellent, concise overview of groundwater hydrology with sections on quality of groundwater, pollution of groundwater, saltwater encroachment, and protection of supplies. Although many sections utilize mathematical equations, those listed above have little mathematical discussion. Suitable for the reader with an advanced high school background.

Huang, P. M., and Iskandar Karam, eds. *Soils and Groundwater Pollution and Remediation: Asia, Africa, and Oceania*. Boca Baton, Fla.: Lewis, 2000. This collection of essays covers a variety of hydrology topics, including groundwater pollution, soil remediation, groundwater purification, and soil pollution. Although this book gets technical at times, on the whole it provides a good overview of hydrology and groundwater movement.

Nyer, Evan K. *Groundwater and Soil Remediation: Practical Methods and Strategies*. Chelsea, Mich.:

Ann Arbor Press, 1998. Nyer aims to provide a broad understanding of contaminated groundwater treatment technologies and their application. The text focuses on treatments, particularly air stripping and carbon adsorption, for organic contaminants. Suitable for college-level readers.

Pettyjohn, Wayne A. *Protection of Public Water Supplies from Groundwater Contamination.* Park Ridge, N.J.: Noyes Data, 1987. Written to provide information for use in groundwater management, this book contains useful chapters on basic groundwater hydrology (chapter 2), groundwater contamination (chapter 5), and treatment of organic compounds contaminating groundwater used for drinking (chapter 7). Suitable for college-level readers.

Thompson, Stephen A. *Hydrology for Water Management.* Rotterdam, Netherlands: A. A. Balkema, 1999. This thorough account of hydrology, groundwater flow, and stream flow focuses on the management of water supplies in efforts to keep them free of pollutants and available to the largest number of people possible. Illustrations, maps, index, and bibliography.

HYDROLOGIC CYCLE

Water circulates through a system called the hydrologic cycle. This cycle operates through vegetation, in the atmosphere and in the ground, and on land, lakes, rivers, and oceans. The Sun and the force of gravity provide the energy to drive the cycle that provides clean, pure water at the Earth's surface.

PRINCIPAL TERMS

BASE FLOW: that part of a stream's discharge derived from groundwater seeping into the stream

CAPILLARY FORCE: a form of water surface tension, forcing water to move through tiny pores in rock or soil, caused by molecular attraction between the water and Earth materials

EVAPORATION: the process by which water is changed from a liquid or solid into vapor

INFILTRATION: the movement of water into and through the soil

INTERCEPTION: the process by which precipitation is captured on the surfaces of vegetation before it reaches the land surface

OVERLAND FLOW: the flow of water over the land surface caused by direct precipitation

PRECIPITATION: atmospheric water in the form of hail, mist, rain, sleet, or snow that falls to the Earth's surface

RUNOFF: the total amount of water flowing in a stream, including overland flow, return flow, interflow, and base flow

SOIL MOISTURE: the water contained in the unsaturated zone above the water table

TRANSPIRATION: the process by which plants give off water vapor through their leaves

EVAPORATION, CONDENSATION, AND PRECIPITATION

The unending circulation of water on Earth is called the hydrologic cycle. This system is driven by the heat energy produced by the Sun. Gravity pulls water that falls on the Earth back to the oceans to be recycled once again. The total amount of water on Earth is an estimated 1.36 billion cubic kilometers. Most of this vast amount of water—97.2 percent—is found in the Earth's oceans. The Greenland and Antarctic ice caps and glaciers contain 2.15 percent of the Earth's water. The remainder—0.65 percent—is divided among rivers (0.0001 percent), freshwater and saline lakes (0.017 percent), groundwater (0.61 percent), soil moisture (0.005 percent), the atmosphere (0.001 percent), and the biosphere and groundwater below 4,000 meters (0.0169 percent). While the percentage of water appears small for each of these water reservoirs, the total volume of water contained in each is immense.

Description of the hydrologic cycle must begin with the oceans, as most of the Earth's water is located there. Each year, about 320,000 cubic kilometers of water evaporate from the world's oceans. Evaporation is the process whereby a liquid or solid is changed to a gas. Adding heat to the water causes the water molecules to become increasingly energized and to move more rapidly, weakening the chemical force that binds them together. Eventually, as the temperature increases, water molecules tend to move from the ocean's surface into the overlying air. Factors that influence the rate of evaporation from free water surfaces are radiation, temperature, humidity, and wind velocity. It is estimated that an additional 60,000 cubic kilometers of water evaporate either from rivers, streams, and lakes or are transpired by plants every year. A total of about 380,000 cubic kilometers of water is evapotranspired from the Earth's surface every year.

Wind may transport the moisture-laden air long distances. The amount of water vapor the air can hold depends upon the temperature. The higher the temperature, the more vapor the air can hold. As the vapor-laden air is lifted and cooled at higher altitudes, the vapor condenses to form droplets of water. Condensation is aided by the ever-present small dust and salt particles or nuclei in the atmosphere. As droplets collide and coalesce, raindrops begin to form and precipitation begins. Most precipitation events are the result of three causal factors: frontal precipitation, or the lifting of an air mass over a moving weather front; convectional precipitation related to the uneven heating of the Earth's surface, causing warm air masses to rise and cool; and orographic precipitation, resulting from a moving air mass being forced to move upward over a mountain range, cooling the air as it rises. Each year, about 284,000 cubic kilometers of precipitation fall on the world's oceans. This water has completed its cycle and is ready to begin a new cycle. Approximately 96,000 cubic kilometers of precipitation fall upon the land surface each year. This precipitation follows a number of different pathways in the hydrologic cycle. It is estimated that 60,000 cubic kilometers evaporate from the surface of lakes or streams or transpire directly back into the atmosphere. The remainder—about 36,000 cubic kilometers—is intercepted by human structures or vegetation, is infiltrated into the soil or bedrock, or becomes surface runoff.

INTERCEPTION, RUNOFF, AND INFILTRATION

Although the amount of water intercepted by and evaporated from human structures—the surfaces of buildings and other artificial surfaces—may approach 100 percent, much urban water is collected in storm sewers or drains that lead to a surface drainage system or is spread over the land surface to infiltrate the subsoil. Interception loss from vegetation is dependent upon interception capacity (the ability of the vegetation to collect and retain falling precipitation), wind speed (the higher the wind speed, the greater the rate of evaporation), and rainfall duration (the interception loss will decrease with the duration of rainfall, as the vegetative canopy will become saturated with water after a period of time). Broad leaf forests may intercept 15 to 25 percent of annual precipitation, and a bluegrass lawn may intercept 15 to 20 percent of precipitation during a growing season.

When the duration and intensity of the rainfall is greater than the Earth's ability to absorb it, the excess water begins to run off, a process termed overland flow. Overland flow will begin only if the precipitation rate exceeds the infiltration capacity of the soil. Infiltration is the process whereby water sinks into the soil surface or into fractures of rocks. It is dependent upon the characteristics of the soil or rock type and upon the nature of the vegetative cover. Sandy soils have infiltration rates of 3.6 to 3.8 centimeters per hour, and clay rock soils average 2.0 to 2.3 centimeters per hour. Nonporous rock would have an infiltration rate of zero, and all precipitation would become runoff. The presence of vegetation impedes surface runoff and increases the potential for infiltration to occur.

Water infiltrating into the soil or bedrock encounters two forces: capillary force and gravitational force. A capillary force is the tendency of the water in the subsurface to adhere to the surface of soil or sediment particles. This tendency may draw the water upward against the downward pull of gravity. Capillary forces are responsible for the soil moisture found a few inches below the land surface.

COMPLETION OF THE CYCLE

Growing plants are continuously extracting soil moisture and passing it into the atmosphere through a process called transpiration. Soil moisture is drawn into the plant rootlet because of osmotic pressure. The water moves through the plant to the leaves, where it is passed into the atmosphere through the leaf openings, or stomata. The plant uses less than 1 percent of the soil moisture in its metabolism; thus, transpiration is responsible for most water vapor loss from the land in the hydrologic cycle. For example, an oak tree may transpire 151,200 liters per year.

The water that continues to move downward under the force of gravity through the pores, cracks, and fissures of rocks or sediments will eventually enter a zone of water saturation. This source of underground water is called an aquifer—a rock or soil layer that is porous and permeable enough to hold and transport water. The top of this aquifer, or saturated zone, is the water table. This water is slowly moving toward a point where it is discharged to a lake, spring, or stream. Groundwater that augments the flow of a stream is called base flow. Base flow enables streams to continue to flow during droughts and winter months. Groundwater may flow directly into the oceans along coastlines.

When the infiltration capacity of the Earth's surface is exceeded, overland flow begins as broad, thin sheets of water a few millimeters thick called sheet flow. After flowing a few meters, the sheets break up into threads of current that flow in tiny channels called rills. The rills coalesce into gullies and, finally, into streams and rivers. While evaporation losses occur from the stream surface, much of the water is returned to the world's oceans, thus completing the hydrologic cycle.

Scientists are interested in how long it takes water to move through the hydrologic cycle. The term "residence time" refers to how long a molecule of water would remain in the various components of the hydrologic cycle. The average length of time that a water molecule would stay in the atmosphere is about one week. Two weeks is the average residence time for a water molecule in a river and ten years in a lake. It would take four thousand years for all the water molecules in the oceans to be recycled. Groundwater may require anywhere from a few weeks to thousands of years

to move through the cycle. This time period may appear extremely long to humans, yet it suggests that every water molecule has been recycled millions of times.

STUDY OF THE HYDROLOGIC CYCLE

Scientists have developed a vast array of mathematical equations and instruments to collect data to quantify the complexities of the hydrologic cycle. The geographic, secular, and seasonal variations in temperature, precipitation, evapotranspiration, solar radiation, vegetative cover, and soil and bedrock type, among other factors, must be evaluated to understand the local, regional, or global hydrologic cycle.

Precipitation, an extremely variable phenomenon, must be accurately measured to determine its input into the hydrologic cycle. The United States has some thirteen thousand precipitation stations equipped with rain gauges placed strategically to compensate for wind and splash losses. Techniques have been developed to determine

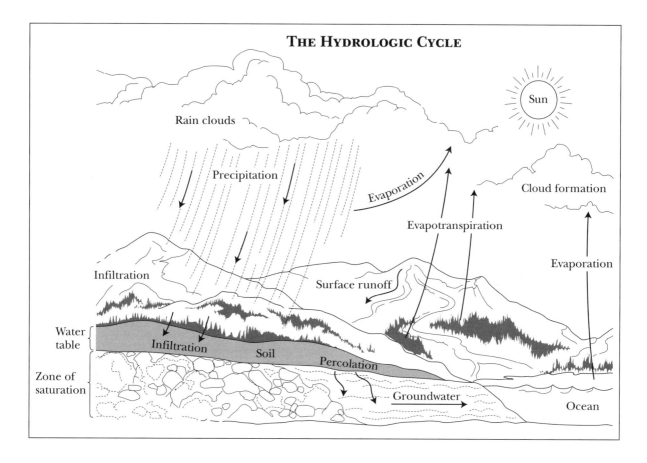

THE HYDROLOGIC CYCLE

the average depth of precipitation falling on a given area or drainage basin. The effective uniform depth method utilizes a rain-gauge network of uniform density to determine the arithmetic mean for rainfall in the area. The isohyetal and polygonal methods are used to determine the arithmetic mean for an area or basin with a nonuniform distribution of rain gauges. The amount of water in a snowpack is estimated by snow surveys. The depth and water content of the snowpack are measured and the extent of the snow cover mapped using satellite photography.

The amount of precipitation lost by interception can be measured and evaluated. Interception is equal to the type of vegetation, the amount of evaporation that occurs during the storm, and the length of the storm. Most often, interception is determined by measuring the amount above the vegetative canopy and at the Earth's surface. The difference would be lost to interception.

The volume of water flowing by a given point at a given time in an open stream channel, measured in cubic meters per second (CMS), is called discharge. Discharge is determined by measuring the velocity of water in the stream channel with a current meter. The Price meter and the pygmy-Price meter meet the specifications of the U.S. Geological Survey. The cross-sectional area of the stream channel is determined at a specific point and multiplied by the stream velocity to determine discharge. Automated stream-gauging stations are located on most streams to supply data for various hydrologic investigations.

The U.S. National Weather Service maintains about five hundred stations using Class A land pans to measure free-water evaporation. These pans are 122 centimeters in diameter and 25.4 centimeters deep, and they are made of unpainted galvanized metal. Water depths of 17 to 20 centimeters are maintained. The wind velocity is also determined. Errors may result from splashing by raindrops or from birds. Because the metal pan will also heat and cool more rapidly than will a natural reservoir, a pan coefficient must be employed to compensate for this phenomenon. A lake evaporation nomograph is employed to determine daily lake evaporation. The mean daily temperature, wind velocity in kilometers per day, solar radiation, and mean daily dew point are the variables required to determine daily lake evaporation.

The amount of evapotranspiration can be measured using a lysimeter—a large container holding soil and living plants. The lysimeter is set outside, and the initial soil moisture is determined. All precipitation or irrigation is measured accurately. Changes in the soil moisture storage determine the amount of evapotraspiration.

Those techniques are utilized to determine the water budget for different geographic areas. Collectively, these data enable scientists to estimate the total water budget of the Earth's hydrologic cycle.

Samuel F. Huffman

CROSS-REFERENCES

Aquifers, 2005; Artificial Recharge, 2011; Dams and Flood Control, 2016; Environmental Health, 1759; Floods, 2022; Groundwater Movement, 2030; Groundwater Pollution and Remediation, 2037; Hazardous Wastes, 1769; Precipitation, 2050; Salinity and Desalination, 2055; Saltwater Intrusion, 2061; Surface Water, 2066; Water Quality, 2072; Water Table, 2078; Water Wells, 2082; Waterfalls, 2087; Watersheds, 2093.

BIBLIOGRAPHY

Cunningham, Floyd F. *1001 Questions Answered About Water Resources.* New York: Dodd, Mead, 1967. Chapter 3, "The Water Cycle," provides forty-nine questions and answers characterizing the hydrologic cycle. There is a short introduction describing the cycle. The answers to each question are well conceived, short, and easily understood by the general public.

Fetter, C. W. *Contaminant Hydrogeology.* 2d ed. Upper Saddle River, N.J.: Prentice Hall, 1999. This textbook emphasizes the conditions that lead to the contamination of groundwater, the problems this contamination can cause, and remediation techniques used in attempts to correct these problems. Includes a detailed discussion of the influence that geologic conditions have on groundwater occurrence and contamination. The book also addresses the regional groundwater movement within aquifers and contains helpful illustrations of the principles discussed. A glossary of important terms is included.

Hamblin, Kenneth W., and Eric H. Christiansen. *Earth's Dynamic Systems.* 8th ed. Upper Saddle River, N.J.: Prentice Hall, 1998. This geology textbook offers an integrated view of the Earth's interior not common in books of this type. The illustrations, diagrams, and charts are superb. Includes a glossary and laboratory guide. Suitable for high school readers.

Leopold, Luna B. *Water: A Primer.* San Francisco: W. H. Freeman, 1974. An excellent introduction to water science for the general public. The book discusses the general principles of hydrology for the reader who desires an overview of the discipline. Suitable for high school readers.

Lutgens, Frederick K., and Edward J. Tarbuck. *Essentials of Geology.* 3d ed. Westerville, Ohio: Charles E. Merrill, 1989. Chapter 6, "Mass Wasting and the Work of Running Water," provides an excellent discussion of the hydrologic cycle and the Earth's water balance. Designed for the college student.

Smith, David G., ed. *The Cambridge Encyclopedia of Earth Sciences.* New York: Crown, 1981. Chapter 17, "Atmosphere, Water and Weather," presents a brief overview of the hydrologic cycle and water budget. The text is suited to the reader with a technical background. An excellent and complete reference volume.

Tarbuck, Edward J., and Frederick K. Lutgens. *Earth: An Introduction to Physical Geology.* 6th ed. Upper Saddle River, N.J.: Prentice Hall, 1999. This introductory college-level textbook provides an excellent discussion of the hydrologic cycle and the Earth's water balance. Illustrations, index, and bibliographical references.

Thompson, Stephen A. *Hydrology for Water Management.* Rotterdam, Netherlands: A. A. Balkema, 1999. This thorough account of hydrology, groundwater flow, and stream flow focuses on the management of water supplies in efforts to keep them free of pollutants and available to the largest number of people possible. Illustrations, maps, index, and bibliography.

PRECIPITATION

Precipitation consists of particles of liquid or frozen water that fall from clouds to the Earth's surface. Thus, precipitation links the atmosphere with the other reservoirs of the global hydrologic cycle, replenishing oceanic and terrestrial reservoirs. In addition, precipitation is the ultimate source of fresh water for irrigation, industrial consumption, and supplies of drinking water.

PRINCIPAL TERMS

ACID PRECIPITATION: rain or snow that is more acidic than normal, usually because of the presence of sulfuric and nitric acid

BERGERON PROCESS: precipitation formation in cold clouds whereby ice crystals grow at the expense of supercooled water droplets

COLD CLOUD: a visible suspension of tiny ice crystals, supercooled water droplets, or both at sub-freezing temperatures

COLLISION-COALESCENCE PROCESS: precipitation

formation in warm clouds whereby larger droplets grow by merging with smaller droplets

RAIN GAUGE: an instrument for measuring rainfall, usually consisting of a cylindrical container open to the sky

SUPERCOOLED WATER DROPLETS: liquid droplets at subfreezing temperatures

WARM CLOUD: a visible suspension of tiny water droplets at temperatures above freezing

CLOUD PARTICLES

Precipitation consists of liquid or frozen particles of water that fall from clouds and reach the Earth's surface. The most familiar types of precipitation are raindrops and snowflakes. Perhaps surprisingly, most clouds, even those associated with large storm systems, do not produce precipitation. A special set of circumstances is required for the extremely small water droplets or ice crystals that compose a cloud to grow into raindrops or snowflakes. A typical cloud particle is about one-millionth the size of a raindrop.

Cloud particle diameters are typically in the range of 2 to 50 micrometers (a micrometer is one-millionth of a meter). They are so small that they remain suspended within the atmosphere unless they vaporize or somehow undergo considerable growth. Upward-directed air currents (updrafts) are usually strong enough to prevent cloud particles from leaving the base of a cloud. Even if cloud droplets or ice crystals descend from a cloud, their fall rates are so slow that they quickly vaporize in the relatively dry air under the cloud. In order to precipitate, therefore, cloud particles must grow sufficiently massive that they counter updrafts and survive thousands of meters of descent to the Earth's surface. Cloud physicists have

identified two processes whereby cloud particles grow large enough to precipitate: the Bergeron process and the collision-coalescence process.

BERGERON AND COLLISION-COALESCENCE PROCESSES

Most precipitation originates via the Bergeron process, named for the Scandinavian meteorologist Tor Bergeron, who, in about 1930, first described the process. It occurs within cold clouds—that is, clouds at a temperature below freezing (0 degrees Celsius). Cold clouds are composed of ice crystals or supercooled water droplets or a mixture of the two. Supercooled water droplets are tiny drops that remain liquid even at subfreezing temperatures. Bergeron discovered that precipitation is most likely to fall from cold clouds composed of a mixture in which supercooled water droplets at least initially greatly outnumber ice crystals. In such a circumstance, ice crystals grow rapidly while supercooled water droplets vaporize. As ice crystals grow, their fall rates within the cloud increase. They collide and merge with smaller ice crystals and supercooled water droplets in their paths and thereby grow still larger. Eventually, ice crystals become so heavy that they fall out of the cloud base. If air temperature is subfreezing during at least most of the descent, crystals reach the Earth's surface as snowflakes. If, however, the air below the cloud is above freezing, the snowflakes melt and fall as raindrops.

Growth of ice crystals at the expense of super-cooled water droplets in the Bergeron process is linked to the difference in the rate of escape of water molecules from an ice crystal versus a water droplet. Water molecules are considerably more active in the liquid phase than in the solid phase; hence, water molecules escape water droplets more readily than they do ice crystals. Within a cold cloud, air that is saturated for water droplets is actually supersaturated for ice crystals. Consequently, water molecules diffuse from the water droplets and deposit on the ice crystals; that is, water droplets vaporize and ice crystals grow.

The collision-coalescence process occurs in warm clouds (clouds at temperatures above 0 degrees Celsius). Such clouds are composed entirely of liquid water droplets. Precipitation may develop if the range of cloud droplet sizes is broad. Larger cloud droplets have greater fall velocities than do smaller droplets; as a result, larger droplets collide and coalesce with smaller droplets in their paths. Collision and coalescence are repeated a multitude of times until the droplets become so large and heavy that they fall from the base of the cloud as raindrops.

Once a raindrop or snowflake leaves a cloud, it enters drier air—a hostile environment in which some of the precipitation vaporizes. In general, the longer the journey to the Earth's surface and the drier the air beneath the cloud, the greater the amount of rain or snow that returns to the atmosphere as vapor. It is understandable, then, why highlands receive more precipitation than do lowlands, which are farther from the base of clouds.

TYPES OF PRECIPITATION

Precipitation occurs in a variety of liquid and frozen forms. Besides the familiar rain and snow, precipitation also occurs as drizzle, freezing rain, ice pellets, and hail. Drizzle consists of small water drops less than 0.5 millimeter in diameter that drift very slowly to the Earth's surface. The relatively small size of drizzle drops stems from their origin in low stratus clouds or fog. Such clouds are so shallow that droplets originating within them have a limited opportunity to grow by coalescence.

Rain falls mostly from thick nimbostratus and cumulonimbus (thunderstorm) clouds. The bulk of rain originates as snowflakes or hailstones, which melt on the way down as they enter air that is warmer than 0 degrees Celsius. Because rain originates in thicker clouds, raindrops travel farther than does drizzle, and they undergo more growth by coalescence. Most commonly, raindrop diameters range from 0.5 to 5 millimeters; beyond this range, drops are unstable and break up into many smaller drops. Freezing rain (or freezing drizzle) develops when rain falls from a relatively mild air layer into a shallow layer of subfreezing air at ground level. Drops become supercooled, then freeze immediately on contact with subfreezing surfaces. Freezing rain forms a layer of ice that sometimes grows thick and heavy enough to bring down tree limbs, snap power lines, disrupt traffic, and make walking hazardous.

Snow is an assemblage of ice crystals in the form of flakes. Although it is said that no two snowflakes are identical, all snowflakes have hexagonal (six-sided) symmetry. Snowflake form varies with air temperature and water vapor concentration and may consist of flat plates, stars, columns, or needles. Snowflake size also depends in part on the availability of water vapor during the crystal growth process. At very low temperatures, the water vapor concentration is low so that snowflakes are relatively small. Snowflake size also depends on collision efficiency as the flakes drift toward the ground. At temperatures near freezing, snowflakes are wet and readily stick together after colliding, so flake diameters may eventually exceed 5 centimeters. Snow grains and snow pellets are closely related to snowflakes. Snow grains originate in much the same way as drizzle, except that they are frozen; diameters are generally less than 1 millimeter. Snow pellets are soft conical or spherical white particles of ice with diameters of 1 to 5 millimeters. They are formed when supercooled cloud droplets collide and freeze together, and they may accompany a fall of snow.

Ice pellets, often called sleet, are frozen raindrops. They develop in much the same way as does freezing rain except that the surface layer of subfreezing air is so deep that raindrops freeze before striking the ground. Sleet can be distinguished readily from freezing rain because sleet bounces when striking a hard surface, whereas freezing rain does not.

Hail consists of rounded or irregular masses of ice, often characterized by concentric layering resembling the interior of an onion. Hail develops within severe thunderstorms as vigorous updrafts transport ice pellets upward into the upper reaches of the cloud. (Often, severe thunderstorms reach altitudes of more than 10 kilometers.) Along the way, ice pellets grow via coalescence with supercooled water droplets and eventually become too heavy to be supported by updrafts. Ice pellets then descend through the cloud, exit the cloud base, and enter air that is typically above freezing. As ice pellets begin to melt, those that are large enough may survive the journey to the ground as hailstones. Most hail consists of harmless granules of ice less than 1 centimeter in diameter, but violent thunderstorms may spawn destructive hailstones the size of golf balls or larger. Hail is usually a spring and summer phenomenon that is particularly devastating to crops.

Changes in Precipitation Chemistry

Over the past few decades, considerable concern has been directed at the environmental impact of changes in the chemistry of precipitation. The global hydrologic cycle purifies water through distillation, but as raindrops and snowflakes fall from clouds to the ground, they wash pollutants from the air. In this way, the chemistry of precipitation is altered. Rain is normally slightly acidic because it dissolves atmospheric carbon dioxide, producing weak carbonic acid. Where air is polluted with oxides of sulfur and oxides of nitrogen, these gases interact with moisture in the atmosphere to produce droplets of sulfuric acid and nitric acid. These acid droplets dissolve in precipitation and increase its acidity. Precipitation that falls through such polluted air may become two hundred times more acidic than normal.

Field studies have confirmed a trend toward increasingly acidic rains and snows over the eastern one-third of the United States. Much of this upswing in acidity can be attributed to acid rain precursors emitted during fuel combustion. Coal-burning for electric power generation is the principal source of sulfur oxides, while high-temperature industrial processes and motor vehicle engines produce nitrogen oxides. Where acid rains fall on soils or bedrock that cannot neutralize the acidity, lakes and streams become more acidic. Excessively acidic lake or stream water disrupts the reproductive cycles of fish. Acid rains leach metals (such as aluminum) from the soil, washing them into lakes and streams, where they may harm fish and aquatic plants.

Study of Precipitation

Precipitation is collected and measured with essentially the same device that has been used since the fifteenth century: a container open to the sky. The standard U.S. National Weather Service rain gauge consists of a cone-shaped funnel that directs rainwater into a long, narrow cylinder that sits inside a larger cylinder. The narrow cylinder magnifies the scale of accumulating rainwater so that rainfall can be resolved into increments of 0.01 inch. (Rainfall of less than 0.005 inch is recorded as a "trace.") Rainwater that accumulates in the inner cylinder is measured by a stick, which is graduated in centimeters or inches. Rainfall is measured at some fixed time once every twenty-four hours, and the gauge is then emptied.

With regard to snow, scientists are interested in measuring snowfall during each twenty-four-hour period between observations, the meltwater equivalent of that snowfall, and the depth of snow on the ground at each observation time. New snowfall is usually collected on a simple board that is placed on top of the old snow cover. When new snow falls, the depth is measured to the board; the board is then swept clean and moved to a new location. The meltwater equivalent of new snowfall can be determined by melting the snow collected in a rain gauge (from which the funnel has been removed). Snow depth is usually measured with a special yardstick or meterstick. In mountainous terrain where snowfall is substantial, it may be necessary to use a coring device to determine snow depth (and meltwater equivalent). Snow depth is determined at several representative locations and then averaged.

The average density of fresh-fallen snow is 0.1 gram per cubic centimeter. As a general rule, 10 centimeters of fresh snow melts down to 1 centimeter of rainwater. This ratio varies considerably depending on the temperature at which the snow falls. "Wet" snow falling at surface air temperatures at or above 0 degrees Celsius has a much greater water content than does "dry" snow falling at surface air temperatures well below freezing.

The ratio of snowfall to meltwater may vary from 3:1 for very wet snow to 30:1 for dry, fluffy snow.

Monitoring the timing and rate of rainfall is often desirable, especially in areas prone to flooding. Hence, some rain gauges provide a cumulative record of rainfall. In a weighing-bucket rain gauge, the weight of accumulating rainwater (determined by a spring balance) is calibrated as water depth. Cumulative rainfall is recorded continuously by a device that either marks a chart on a clock-driven drum or sends an electrical pulse to a computer or magnetic tape. During subfreezing weather, antifreeze in the collection bucket melts snow as it falls into the gauge so that a cumulative meltwater record is produced.

Both rainfall and snowfall are notoriously variable from one place to another, especially when produced by showers or thunderstorms. The siting of a precipitation gauge is particularly important in order to ensure accurate and representative readings. A level site must be selected that is sheltered from strong winds and is well away from buildings and vegetation that might shield the instrument. In general, obstacles should be no closer than about four times their height.

SIGNIFICANCE

Without precipitation, the Earth would have no fresh water and thus no life. When water vaporizes from oceans, lakes, and other reservoirs on the Earth's surface, all dissolved and suspended substances are left behind. Hence, water is purified (distilled) as it cycles into the atmosphere and eventually returns to the Earth's surface as freshwater precipitation. In this way, the global hydrologic cycle supplies the planet with a fixed quantity of fresh water.

As human population continues its rapid growth, however, demands on the globe's fixed supply of fresh water are also increasing. In some areas, such as the semiarid American Southwest, water demand for agriculture and municipalities has spurred attempts to enhance precipitation lo-

cally through cloud seeding. Usually, cold clouds that contain too few ice crystals are seeded by aircraft with either silver iodide crystals (a substance with properties similar to ice) or dry-ice pellets (solid carbon dioxide at a temperature of about −80 degrees Celsius) in an effort to stimulate the Bergeron precipitation process.

Cloud seeding, although founded on an understanding of how precipitation forms, is not always successful and at best may enhance precipitation by perhaps 20 percent. The question remains as to whether the rain or snow that follows cloud seeding would have fallen anyway. Even if successful, cloud seeding may merely bring about a geographical redistribution of precipitation so that an increase in precipitation in one area is accompanied by a compensating reduction in a neighboring area. Cloud seeding that benefits agriculture in eastern Colorado, for example, might also deprive farmers of rain in the downwind states of Kansas and Nebraska. The uncertainties of cloud seeding underscore the need for conservation of the planet's freshwater resource. Conservation should entail not only strategies directed at wise use of fresh water but also measures to manage water quality. Abatement of water pollution not only reduces hazards to human health and aquatic systems but also increases the supply of fresh water.

Joseph M. Moran

CROSS-REFERENCES

BIBLIOGRAPHY

Colbeck, Samuel C. "What Becomes of a Winter Snowflake?" *Weatherwise* 38 (1985): 312-315. This article describes the processes taking place within a snowbank. Well illustrated.

Desbois, Michel, and Françoise Désalmand, eds. *Global Precipitations and Climate Change*. New

York: Springer-Verlag, 1994. A look at meteorology, paleoclimatology, precipitation, and the methodology used in determining the factors involved with climatic changes. Certain essays look specifically at the atmospheric changes that create precipitation. Illustrations, bibliography, and index.

Leopold, Luna B. *Water: A Primer.* New York: W. H. Freeman, 1974. A concise and well-illustrated treatment of the global hydrologic cycle. Provides the context within which precipitation processes take place.

Martin, John Wilson. *Precipitation Hardening.* 2d ed. Boston: Butterworth-Heinemann, 1998. Martin's book looks at the atmospheric factors necessary to evoke precipitation, particularly snow, sleet, and hail. A good introduction to the study of precipitation for the nonscientist.

Moran, Joseph M., and Michael D. Morgan. *Meteorology: The Atmosphere and the Science of Weather.* 2d ed. New York: Macmillan, 1989. A well-illustrated survey of atmospheric science. Includes chapters on the hydrologic cycle, cloud development, and precipitation processes.

Oliver, John E., and Rhodes W. Fairbridge, eds. *The Encyclopedia of Climatology.* New York: Van Nostrand Reinhold, 1987. A comprehensive treatise on the basics of climatology. Includes a detailed discussion of the global and seasonal distribution of precipitation.

Oliver, John E., and John J. Hidore. *Climatology: An Introduction.* Westerville, Ohio: Charles E. Merrill, 1984. A well-written survey of the principles of climate and the climates of the globe. Provides a background on the factors that control the spatial and temporal distribution of precipitation.

Pruppacher, Hans R., James D. Klett, et al., eds. *Microphysics of Clouds and Precipitation.* 2d ed. Boston: Kluwer Academic Publishers, 1997. This lengthy volume, part of the Atmospheric and Oceanography Sciences Library series, gives the reader an overview of precipitation by examining the structure of clouds, the chemical makeup and surface properties of water, cloud chemistry and electricity, and the formation, growth, and diffusion of water drops and snow crystals. Suitable for the nonscientist.

Schaefer, Vincent J., and John A. Day. *A Field Guide to the Atmosphere.* Boston: Houghton Mifflin, 1981. An exceptionally well-illustrated survey of cloud and precipitation processes. Includes suggested simple experiments and demonstrations.

Schindler, D. W. "Effects of Acid Rain on Freshwater Ecosystems." *Science* 239 (1988): 149-156. An excellent summary of what is currently understood about the impact of acidic precipitation.

Snow, J. T., and S. B. Harley. "Basic Meteorological Observations for Schools: Rainfall." *Bulletin of the American Meteorological Society* 69 (1988): 497-507. This article discusses rainfall measurement and evaluates inexpensive rain gauges suitable for classroom use.

SALINITY AND DESALINATION

Salinity is the total amount of solid materials (salts) dissolved per unit of water; the average salinity of seawater, for example, is 35 grams of salts per kilogram of water. Utilization of saline water for domestic purposes requires removal of the dissolved materials. The desalinization of saline or brackish water involves costly operations and is typically performed only in coastal or arid regions where alternative water resources are limited.

PRINCIPAL TERMS

DISSOLVED MATTER: the amount of solid materials that are completely dissolved in water

EVAPORATION: the physical process occurring at the water-air interface where water changes its phase from liquid to vapor

FILTRATION: the removal of particulate matter from the water by passing it through a porous medium

POTABLE WATER: fresh water that is being used for domestic consumption

REVERSE OSMOSIS: the forced passage of seawater through a semipermeable membrane against the natural osmotic pressure

SODIUM CHLORIDE: the main chemical compound found as dissolved material in seawater

SUSPENDED SOLIDS: the solid particles that can be found dispersed in the water column

WATER RESOURCES: all the surface water and groundwater that can be effectively harvested by humans for domestic, industrial, or agricultural uses

WATER SUPPLY: the amount of water that is actually delivered to the various consumer groups

WATER-TREATMENT PLANT: a facility where water is treated by physical and chemical processes until its quality is improved to that of potable water

DISSOLVED SOLIDS

Salinity is one of the most important physical properties of seawater. It is defined as the total amount of dissolved solids found in a unit of water; it is measured in grams of dissolved matter per kilogram of seawater. The average salinity of seawater is approximately 35 grams per kilogram, corresponding to a concentration of 35 parts of salt per 1,000 parts of seawater (35 ppt). The main ionic components of dissolved matter in seawater and the percentages in which they are found in a typical sample are as follows: chlorine ions (55.0 percent), sodium ions (30.6 percent), sulphate ions (7.7 percent), magnesium ions (3.7 percent), calcium ions (1.2 percent), and potassium ions (1.1 percent). One cubic mile of typical seawater contains 120 million tons of sodium chloride, 18 million tons of magnesium chloride, 8 million tons of magnesium sulphate, 6 million tons of calcium sulphate, 4 million tons of potassium sulphate, and 550,000 tons of calcium carbonate. Depending on the prevailing physicochemical conditions, salinity can vary significantly, from brackish water (2 to 15 ppt) in estuaries and riverine deltas to highly saline water such as in the Mediterranean Sea (38.5 ppt) or Red Sea (42.5 ppt). Atmospheric precipitation or freshwater discharges from inland waters (surface water or groundwater) reduce salinity, while evaporation increases salinity content. In spite of the fact that salinity levels vary from one water body to another, the relative abundance of the main dissolved components remains almost unchanged.

The presence of dissolved salts has an effect on the density of seawater. A salinity of 35 ppt amounts to a density difference between fresh and saline water of approximately 2 percent. Thus, whenever there is confluence of fresh and saline water, some interesting phenomena occur. Depending on the degree of energy input by the wind, currents, and tidal action, the fresh and saline waters can be thoroughly or partially mixed or remain stratified. Under quiescent conditions, if all other parameters (such as temperature and pressure) are the same for both the fresh and saline waters, then the heavier saline water will sink.

In many estuaries with high riverine discharge and low tidal range (such as the Mississippi River), seawater can penetrate many kilometers in the upstream direction, moving near the bottom and under the fresh water. This phenomenon is known as "saline wedge"; the separating interface between fresh and saline water is called the "halocline." Mixing conditions between fresh and saline waters are important from an ecological point of view, since salinity levels control the diversity and population of aquatic flora and fauna.

Sea salt is also a major contributor to atmospheric aerosol particles. During wave breaking (the phenomenon of "white caps"), small droplets are carried upward by air currents into the atmosphere. When the droplets evaporate, sea-salt particles of very small sizes (0.5 to 20 millimeters) are transported by the wind over the continents. About 10 percent of the total annual amount of salt that is generated in the oceans (1.8 billion tons) is deposited as airborne sea-salt particles on the continents.

The total volume of the oceanic waters is about 1.37 billion cubic kilometers. This volume constitutes approximately 96.5 percent of the total water on Earth. The other major water resources are groundwater, of which 10.53 million cubic kilometers (0.76 percent) is fresh water and 12.87 million cubic kilometers (0.93 percent) is saline water, and the polar ice, which contains 24.02 million cubic kilometers (1.7 percent) of water. The remaining amount of water is in lakes, rivers, marshes, soil moisture, atmosphere, and biota. Fresh water thus makes up only 2.5 percent of the total amount of Earth's water. Most freshwater supplies are in the polar ice (68.6 percent) and in groundwater (30.1 percent). The percentage of fresh water found in lakes is only 0.26 percent, while the amount found in rivers is even smaller (0.006 percent).

INLAND WATER AND SALINITY MEASUREMENT

Not all inland water is fresh. The amount of the dissolved salts occurring in inland waters, surface or underground, depends on the composition of the soils through which the water passes. Streams and rivers flowing over rocks containing chloride and sodium compounds contribute significantly to the generation of salt. Therefore, some inland waters have a high salt content and are not suitable for any use unless properly pretreated. The most extreme examples of brine inland waters are the Great Salt Lake in Utah (salt concentration of about 120 ppt) and the Dead Sea in the Middle East (salt concentration of about 270 ppt). Salt is also found as surface crust or layers in swamps and dry lake bottoms, particularly in arid climate areas. In coastal regions, excessive pumping of groundwater can lead to seawater intrusion into the aquifer. This can have a long-term negative effect on the water resources of a region, with subsequent devastating impact on the regional economy.

The average amount of fresh water that falls daily on the United States is about 4,200 billion gallons. This water feeds the various surface water bodies (lakes and rivers), recharges the aquifers, evaporates back to the atmosphere, or flows into the oceans. Therefore, only 675 billion gallons of precipitated water can be used for beneficial purposes. From this amount, 336 billion gallons are used by industry, 141 billion gallons by agriculture, and 25 billion gallons by domestic and rural consumers. The daily per-capita water use in the United States is about 1,900 gallons. The average domestic water consumption ranges from 20 to 80 gallons per day. Thus, the average amount of water used by a person during a seventy-year life span is estimated at 1.5 million gallons.

Direct measurement of seawater salinity by evaporation or chemical analysis is too complicated to be used on a routine basis. In the past, salinity (known as "absolute salinity") was estimated in terms of chlorinity. Chlorinity is the amount of chlorine ions plus the chlorine equivalent of bromine and iodine ions. From chlorinity, salinity was estimated by multiplying the value of chlorinity by a factor of 1.80655.

Salinity (known as "practical salinity") is estimated indirectly by measuring the electrical conductivity of the seawater. However, since electrical conductivity is strongly affected by both salinity and temperature, the conductivity readings are properly corrected to compensate for temperature effects.

WATER TREATMENT

Fresh water is a precious commodity that is used for a variety of domestic, rural, industrial, and agricultural purposes. Water is not distributed evenly throughout the world and is subject to temporal short-term and long-term variations. In many regions, the available freshwater resources cannot meet the water demands. After the Industrial Revolution, accelerated anthropogenic pollution added to the water resource problem. In situations of limited water resources, the only alternative solutions are either cleanup and reuse of domestic and agricultural wastewater or, for islands and coastal regions, desalination of seawater. Both of these operations are costly and require construction of appropriate water-treatment facilities. Several countries in the Middle East, including Israel and Saudi Arabia, depend heavily on water desalination for their drinking-water supplies. Under emergency conditions, fresh water is obtained via the use of portable water-purification equipment.

Fresh water contains a number of impurities that have to be removed or treated before it is available for any useful purpose. These impurities include calcium, magnesium, iron, lead, copper, chloride, sulfate, nitrates, fluorides, sodium, different organic compounds, and suspended solids. Impurities can be hazardous to human health or can give water a disagreeable taste, smell, or appearance. In addition, they can create scaling or corrosive problems in pipes or in machinery that uses water. Because saline water contains a large amount of dissolved salts, before use it is always subjected to desalination. Generally, the acceptable quality standards for drinking water are 0.5 micrograms per kilogram of total dissolved solids and 0.2 micrograms per kilogram of chloride.

Water treatment for production of domestic water involves a number of operations such as filtration, softening, distillation, deionization, chemical disinfection, exposure to ultraviolet radiation, and reverse osmosis. The number and the type of operations required depend on the quality and properties of the water supplies. The problem of high salt content can be treated with a variety of methods. Desalination is the process by which dissolved salts are removed from the water; there are a number of desalination methods. Since no one method is applicable in all situations, selection of the most appropriate desalination method is based on such variables as the amount and type of dissolved salts in water, the degree of purification of the water to be produced, and the associated costs. More than one hundred cities worldwide use desalination plants to provide fresh water for their needs.

THERMAL PROCESSING

There are two general methods of desalination: thermal processing (distillation) and membrane separation. The main principle behind thermal processing is as follows: Saline water is heated until it boils, then the released steam condenses as it cools, forming pure water. Membrane separation is achieved by using reverse osmotic pressure, so that the water passes through a membrane while the salt ions are retained by the membrane.

Distillation, the earliest desalination method, was used in steamships as early as 1884. There are several different thermal processing methods for desalination, such as the Thin-Film Multiple Effect Distillation (TFMED), Multi-Stage Flash Desalination (MSFD), Mechanical Vapor Compression Desalination (MVCD), and Thermal Vapor Compression Desalination (TVCD). At standard atmospheric pressure (measured at sea level), water boils at 100 degrees Celsius. However, boiling temperature decreases with decreasing pressure. Also, if the water is heated under high pressure at 100 degrees Celsius and is suddenly released into a vacuum chamber, it flashes into vapor. This technique is used in both the TFMED method and the MSFD method, in which the pressure is continuously reduced in sequential stages. The number of sequential stages can range from fifteen to twenty-five. The MVCD and the TVCD methods use compression to increase the pressure and thus increase the temperature of a constant volume of steam. Heating of the steam stream facilitates the desalination process. Water obtained through thermal processing can easily have a purity of less than 0.1 microgram per kilogram of dissolved matter. The brine wastewater resulting from desalination of seawater has a salinity of approximately 70 grams per kilogram.

MEMBRANE SEPARATION

Membrane separation includes two major desalination methods: Pressure Membrane Processes (PMP) and Electro-Dialysis Reversal (EDR). In PMP, only pure water is able to pass through the membrane, while the majority of the dissolved material is detained by the membrane. The EDR method utilizes ion-specific membranes placed between anodic (positively charged) and cathodic (negatively charged) electrodes. Dissolved material is then collected as the salt ions move under the electric current. Although the Electro-Dialysis Reversal method is relatively common, the most widespread desalination methods involve PMP.

There are four pressure-membrane desalination processes: Microfiltration (MF), Ultrafiltration (UF), Nanofiltration (NF), and Reverse Osmosis (RO). The common principle of these methods is the forced passing (under high pressure) of saline water through a membrane. The main difference among the PMP methods is the size of the particles that are removed from the saline water. For example, MF removes particles larger than 10 microns (1 micron equals 0.000001 meter), UF removes particles of sizes from 0.001 to 10 microns, NF removes particles greater than 0.001 micron, and RO removes particles ranging from 0.0001 to 0.001 micron. Another difference among the PMP methods is the pressure needed for forcing the water through the membrane. MF operates at a pressure of less than 10 pounds per square inch (psi), UF at a range of 15 to 75 psi, NF at a range of 75 to 250 psi, and RO at a range of 200 to 1,200 psi.

Reverse osmosis is an energy-consuming method that can be used very effectively, particularly for production of domestic use water, whenever good-quality saline water is available. Reverse osmosis can be accomplished using different design modules such as the tubular, the plate-and-frame, the spiral-wound, or the hollow fiber. In all these designs, the water that is allowed to pass through the membrane, the "permeate," is collected as the product water, while the water retained by the membrane forms the so-called concentrate or reject.

There are various type of membranes used for reverse osmosis, such as the cellulose acetate group, which includes cellulose acetate (CA), cellulose acetate butyrate (CAB), and cellulose triacetate (CTA), or polyamide (PA). Production of CA membrane involves four stages: casting, evaporation, gelation, and shrinkage. During the first stage, a solution of cellulose acetate in acetone containing certain additives is casted into flat or tubular thin surfaces. Then the acetone evaporates, leaving a porous surface. During the gelation stage, the cast is immersed in cold water, forming a gel, while at the same time the additives dissolve in the water. In the last stage, the film shrinks, forcing reduction of the pore sizes. High temperatures result in smaller pore openings.

Electrodialysis is also an effective desalination process whereby ions are separated from the water by being forced through selective ion-permeable membranes under the action of electric current. The ion-permeable membranes alternate between those allowing only the passage of cations (such as potassium, Na^+) and those allowing only the passage of anions (such as chlorine, Cl^-).

Another potential methodology for desalination involves freezing of the water. Since ice is theoretically free of any dissolved material, various techniques have been proposed for application of freezing for desalination. However, this method does not have any widespread applicability.

Both the thermal and membrane methods require some form of energy to accomplish desalination. The energy efficiency of the desalination methods is expressed either as the Gained Output Ratio (GOR) or the Performance Ratio (PR). The GOR is defined as the ratio between the mass of distillate over the mass of the steam. The PR, also known as economy, is estimated as pounds of distillate per 1,000 British thermal units (BTUs) or as kilograms of distillate per 2,326 kilojoules (KJ). Criteria for the selection of a desalination method include energy consumption, process efficiency, operational and maintenance costs, auxiliary services, and growth of demand. Energy is provided mostly by electrically driven pumps, but diesel engines can also be utilized.

SIGNIFICANCE

Domestic (potable) water is a valuable commodity. Population growth and high demand for water for industrial and agricultural practices have created severe water shortages in many parts of the world. In addition, increased water contamination and extreme hydrologic conditions, such as prolonged drought, can adversely affect the socioeconomic conditions and human health of a region.

In addition, there are other applications that require high-quality water. For example, high-purity water is required by pharmaceutical companies for processing drugs and medications, by hospitals for kidney dialysis, by power and other energy-intensive plants for low-scaling water, and by semiconductor manufacturing for production of high-performance chips.

Since most of the conventional water resources are already under stress, the only alternative solution to the water-resource problem is desalination of the vast oceanic water masses or reuse of wastewater. The various existing thermal or membrane methods can effectively provide fresh high-quality water. However, the high costs associated with these methodologies limit their applicability only to communities that can afford the financial burden. There is ongoing research into improving the efficiency and reducing the cost of existing desalination methods and developing new methodologies. What began as a small distillation process used to provide fresh water to a handful of ships'

crew members has evolved into a complex water-treatment operation that supplies fresh water to large populations.

Panagiotis D. Scarlatos

CROSS-REFERENCES

Acid Rain and Acid Deposition, 1802; Aquifers, 2005; Artificial Recharge, 2011; Climate, 1902; Clouds, 1909; Dams and Flood Control, 2016; Drainage Basins, 2325; Floods, 2022; Freshwater Chemistry, 405; Groundwater Movement, 2030; Groundwater Pollution and Remediation, 2037; Hydrologic Cycle, 2045; Precipitation, 2050; Saltwater Intrusion, 2061; Storms, 1956; Surface Water, 2066; Water Quality, 2072; Water Table, 2078; Water Wells, 2082; Waterfalls, 2087; Watersheds, 2093; Weathering and Erosion, 2380.

BIBLIOGRAPHY

Henry, J. G., and G. W. Heinke. *Environmental Science and Engineering.* 2d ed. Upper Saddle River, N.J.: Prentice Hall, 1996. Lengthy presentation of water quality and wastewater treatment technological issues, including desalination. Contains a discussion of case studies.

Ko, A., and D. B. Guy. "Brackish and Seawater Desalting." In *Reverse Osmosis Technology: Application for High-Purity-Water Production.* New York: Marcel Dekker, 1988. Detailed presentation of technological methods for desalination of brackish and saline waters.

Metcalf & Eddy, Inc. *Wastewater Engineering: Treatment, Disposal, Reuse.* 2d ed. New York: McGraw-Hill, 1979. A comprehensive discussion of wastewater properties, waste-treatment processes, and associated methodologies. Emphasis on engineering design.

Pickard, G. L., and W. J. Emery. *Descriptive Physical Oceanography: An Introduction.* 5th ed. Oxford, England: Pergamon Press, 1995. Thorough presentation of seawater properties and salinity distribution in the various seas and oceans.

Porteous, A. *Saline Water Distillation Processes.* London: Longman, 1975. Detailed presentation of the distillation methods used for desalination of saline water.

Segar, Douglas. *An Introduction to Ocean Sciences.* New York: Wadsworth, 1997. Comprehensive coverage of all aspects of the oceans and salinity. Readable and well illustrated. Suitable for high school students and above.

Shafer, L. H., and M. S. Mintz. "Electrodialysis." In *Principles of Desalination.* 2d ed. New York: Academic Press, 1980. Detailed description of the electrodialysis method for removal of dissolved matter.

Sincero, A. P., and G. A. Sincero. *Environmental Engineering: A Design Approach.* Upper Saddle River, N.J.: Prentice-Hall, 1996. Detailed presentation of water-quality principles and associated water-purification and cleanup technologies and designs.

Speece, R. E. "Water and Wastewater." In *Environmental Health.* 2d ed. Orlando, Fla.: Academic Press, 1980. Provides a brief but comprehensive description of water-quality characteristics and the methodologies used for treatment of both domestic and industrial waters.

Tarbuck, Edward J., and Frederick K. Lutgens. *Earth: An Introduction to Physical Geology.* 6th ed. Upper Saddle River, N.J.: Prentice Hall, 1999. This college text provides a clear picture of the Earth's systems and processes that is suitable for the high school or college reader. In addition to its illustrations and graphics, it has an accompanying computer disc that is compatible with either Macintosh or Windows. Bibliography and index.

Tebbutt, T. H. Y. *Principles of Water Quality Control.* Oxford, England: Pergamon Press, 1971. An introductory textbook on water properties and applied methodologies for water-quality control.

Twort, A. C., R. C. Hoather, and F. M. Law. *Water Supply.* 2d ed. London: Edward Arnold, 1974. A discussion of water-supply systems, including distribution and treatment methodologies.

Viessman, W., Jr., and M. J. Hammer. *Water Supply and Pollution Control.* 5th ed. New York: HarperCollins, 1993. A thorough quantitative coverage of water-supply issues, water-quality treatment methodologies, and engineering designs.

SALTWATER INTRUSION

Saltwater intrusion is the contamination and destruction of freshwater resources by salt water, usually in coastal areas or marine islands. Most saltwater intrusion is caused by human activities, and once started, it can be very difficult to reverse.

PRINCIPAL TERMS

AQUIFER: rock or sediment that is saturated with groundwater and is capable of delivering that water to wells and springs

BRACKISH WATER: water with a salt content between that of salt water and fresh water; it is common in arid areas on the surface, in coastal marshes, and in salt-contaminated groundwater

CONE OF DEPRESSION: a cone-shaped depression produced in the water table by pumping from a well

FRESH WATER: water with less than 0.2 percent dissolved salts, such as is found in most streams, rivers, and lakes

FRESHWATER LENS: the fresh water in aquifers in coastal areas or marine islands that floats on underlying, denser salt water

GROUNDWATER: water located beneath the ground in interconnected pores

MIXING ZONE: the area of contact between a freshwater lens and the underlying salt water

SALT WATER: water with a salt content of 3.5 percent, such as is found in normal ocean water

UPCONING: the upward flexure of the mixing zone toward the ground surface produced by excessive groundwater withdrawal by wells

WATER TABLE: the upper surface of groundwater in an aquifer, with a direct connection overhead to the atmosphere

COASTAL AND MARINE ISLAND AQUIFERS

Saltwater intrusion commonly occurs in coastal or marine island aquifers. In these situations, the fresh water of the groundwater in an aquifer will rest on underlying salt water in the aquifer. This layering of fresh water on top of salt water is caused by the slightly higher density of the salt water. The salt water is denser because of its greater amount of dissolved salts. In the coastal regions of continents and marine islands, the fresh water thins and tapers as the coastline is reached, producing a classic lens-shaped cross section to the freshwater body as it floats on the underlying salt water. For this reason, the fresh water in aquifers in coastal regions is commonly called the freshwater lens.

Under ideal conditions, pure fresh water has a density of 1 gram per cubic centimeter; salt water has a density of 1.025 grams per cubic centimeter, a density difference of one part in forty. Fresh water floats on salt water much like a piece of wood floats on water. How high something floats on an underlying, denser liquid depends on the density difference, or contrast, between the floating material and the underlying liquid. In the case of the freshwater lens, the density contrast of one part in forty means that for every centimeter that the freshwater lens floats above the salt water, it must displace downward into the salt water at least 40 centimeters. Fresh water is not a solid, like a piece of wood; it is a liquid and will flow unless contained. When a freshwater lens floats on salt water, its water table is above the level of the salt water, so it tends to flow in the direction of the source of the salt water. Often, that direction is seaward. As the fresh water flows seaward, it thins the lens. This thinning will continue until the lens almost disappears, unless the fresh water lost seaward is replaced by new fresh water entering the lens from sources above or inland of the lens. A freshwater lens is an example of dynamic equilibrium, or stability through motion. A freshwater lens remains stable only if it is recharged by new fresh water at a rate equal to the flow loss of the lens to the sea. This principle of a dynamic equilibrium in the freshwater lens is called the Ghyben-Herzberg principle, for two of the first scientists to study this topic.

The boundary between the freshwater lens and the underlying salt water is called the mixing zone. Other terms used to describe this boundary include the halocline, the diffusion zone, and the dispersion zone. The boundary can have a variety of characteristics. It can be sharp, in which case the change from fresh water to salt water can occur over a distance of a few centimeters or less, a situation in which the term "halocline" (salt boundary) is used. The water can also change from fresh to salt over a broad zone of brackish water meters or tens of meters thick with an ever-increasing salt content. Fresh water and salt water are miscible liquids: They mix easily. (Oil and water are examples of immiscible liquids.) The reasons that the mixing zone can be sharp or broad are not well understood and may be controlled in part by the size, shape, and chemistry of the pores in the aquifer.

Saltwater intrusion occurs when the dynamic equilibrium of the freshwater lens is upset. If the loss of fresh water from the lens is increased or if the flow of new fresh water to the lens is decreased, the lens will thin and, in the process, migrate inland. Inland migration of the freshwater lens means that some portion of the coastal aquifer that once contained fresh water now contains salt water, as the salt water replaces or intrudes into the freshwater aquifer. Changes in climate, sea level, or river-flow paths are ways in which this process can occur naturally. Most saltwater intrusion, however, occurs because of human modification of the freshwater flow system.

ACTIVE AND PASSIVE INTRUSION

Active saltwater intrusion occurs when excessive pumping of freshwater wells distorts the freshwater lens by water removal. This distortion can cause two separate problems. The first problem is a decrease in the water available in the freshwater aquifer, causing salt water to intrude from the ocean direction, pushing the freshwater lens inland. The second situation is called upconing, where excessive pumping from a well produces an inverted cone of depression in the mixing zone. This upconing draws salt water up into the freshwater aquifer. When a well is pumped in an aquifer, the water is removed from the vicinity of the well and is replaced by flow from the surrounding aquifer. Depending on how fast the aquifer transmits water and how fast water is pumped, the water table around the well is lowered in a characteristic way called the cone of depression. If a well is pumped too fast, the cone of depression can reach the bottom of the aquifer, and the well will go dry. After a period of no pumping, the aquifer will be able to fill the cone of depression enough so that pumping can again produce water.

In a freshwater lens, this situation is doubly complex. Since the fresh water floats on the salt water, as a cone of depression is produced at the top of the freshwater lens, a similar cone forms at the mixing zone but in an inverted shape that matches the shape of the overlying cone of depression. The Ghyben-Herzberg principle requires the mixing zone to rise as the thickness of the fresh water decreases because of pumping. Eventually the bottom of the well is reached by this upconing of the mixing zone; the well begins to draw brackish and then finally salt water, and it is ruined for human use. In some cases, cessation of pumping will allow the freshwater lens to reestablish itself, but in many cases, the upconing process produces long-term saltwater contamination of the freshwater lens.

Passive saltwater intrusion can occur from a subtler distortion of the freshwater lens. Because of the dynamic equilibrium of the lens, it can be distorted not only by freshwater removal at wells but also by interruption of the mechanism by which it is recharged by surface water. Human activities that alter the surface drainage and infiltration of fresh water into the aquifer will cause the loss of recharge for that aquifer. If recharge is interrupted, the freshwater lens retreats inland as the salt water intrudes into the aquifer. Activities such as the digging of flood-control canals, widespread paving of the land surface, and river diversions rob the underlying lens of needed recharge. It may be a long time before the effects of human activities governing recharge are detected in the aquifer, causing difficulty in reversing the process.

OTHER EFFECTS

Coastal and marine island regions are not the only areas where salt water can intrude into a freshwater aquifer. In areas where arid climates predominate, irrigation is often used to grow crops. It is not uncommon for irrigation water to be transported into the area by pipeline or canal because the local aquifer cannot withstand the demand for water. In this situation, the excess water on the surface can mobilize the salts in the ground and transport them downward into the water table. This movement is not true saltwater intrusion, but it can result in loss of the freshwater aquifer by salt contamination.

Many types of rock and sediment are deposited in the ocean. At some later time, they may be uplifted from the sea and form dry land that receives rainfall and develops a freshwater aquifer. The deeper regions of these aquifers may contain waters that were left behind from the original marine environment, called connate water. Connate water originated as salt water, and it can intrude into the freshwater portion of the aquifer during overpumping of the aquifer. The aquifer can become contaminated by salt and ruined. In some cases, the deeper waters of an aquifer are brines, where the salt content has become concentrated above that found in the oceans. Brines are potent contaminators of freshwater aquifers because they have such a high concentration of salt.

The intrusion of salt water into an aquifer can have a major impact on the aquifer material itself. The chemistry of salt water is very different from that of fresh water. Some aquifer materials can become altered by the intrusion of salt water, and they will stay altered even if the salt water is later forced out by replenishment of the original fresh water. Examples of the changes that can occur are shown by aquifers in islands and coastal regions that are developed in limestone. In limestone aquifers, both the freshwater lens and the underlying salt water are usually saturated with the mineral calcite, which composes limestone. When fresh water and salt water mix, however, they are capable of dissolving calcite that neither one could dissolve alone. The result is the development of areas of dissolved rock in the limestone. The development of caves and the increase in the number of pores in the limestone because of this dissolving process change the characteristics of the aquifer. If the mixing zone moves around because of distortions of the freshwater lens, the aquifer can become fundamentally changed as the area of dissolving reaches more of the aquifer. The changes of sea level during the ice ages caused that to happen in many places. The famous blue holes and underwater caves of the Bahamas owe their origin to this process. Human activity and subsequent saltwater intrusion in limestone coastal areas may be causing this process today.

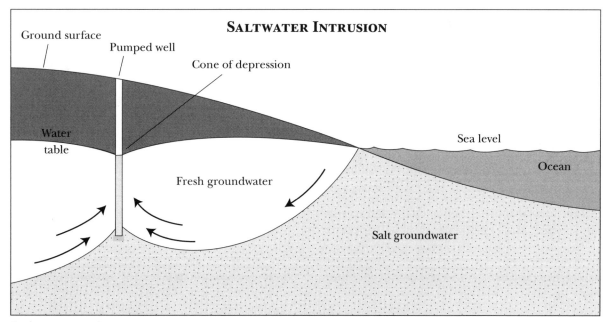

SALTWATER INTRUSION

Fresh water is less dense than salt water and therefore rises to the top of the water table. However, as fresh water is pumped to the surface, salter water "intrudes" to take its place. Eventually the well will yield salt water.

STUDY OF SALTWATER INTRUSION

The study of saltwater intrusion is usually a very straightforward task. The major difficulty lies in the change in water use that the public must accept in order to rectify the problem. The first step in analyzing a freshwater lens is to calculate a water budget for the aquifer. A water budget is much like a bank budget; there is input and output, and when the two are out of balance, a change occurs in the budget. In an aquifer, the input is the recharge of the aquifer by fresh water from the surface. This recharge includes rainfall over the area of the aquifer, of which a certain percentage infiltrates the ground to reach the water table. It also includes water brought into the area from other regions by rivers and streams, a certain percentage of which will also sink to the water table. These input values can be measured by taking into account rainfall, evaporation, the use of water by plants (transpiration), and, finally, the water that runs over the ground surface. The difference between how much water comes in and how much is lost to runoff, evaporation, and transpiration is the amount that infiltrates the ground to recharge the aquifer. If the aquifer is in equilibrium, the amount of recharge must equal the amount of discharge.

To measure what is happening down in the aquifer, observation wells are used. These wells tell where the water table is. If they penetrate deeply enough, they can also tell where the mixing zone and the salt water are. These wells are not used as a water source but as a means of monitoring the aquifer. If the water table is at a constant level (or seasonally fluctuates about a constant level), the aquifer is in equilibrium, and if the input has been calculated, the output can be estimated. As the aquifer is used as a water supply, the observation wells follow any changes produced in the water table.

Saltwater intrusion is easily detected in observation or supply wells because of the change in salinity that occurs. Sometimes it is detected as a change in taste in the water; chemical testing may also be used. Since fresh water is a poor conductor of electricity and salt water is a good conductor, a relatively simple device called a conductivity meter can be used to check, easily and quickly, the presence of salt in a well. In an observation well, a drop in the water table accompanied by a rise in the mixing zone will be directly measurable, and remediation can begin. Once a supply well has been contaminated by salt water, however, it is too late to save that well in the short term. It must be shut down and an alternate water supply located. The observation well can predict when saltwater intrusion will damage supply wells by monitoring the position of the mixing zone and the water table. It is then possible to change water-use patterns at the supply wells, to increase aquifer recharge, or both, thereby preventing a problem. In some cases, the well that is damaged is not the well causing the problem. Imagine a coastal well that is functioning normally. If a well inland of the coastal well pumps too much fresh water, the salt water will intrude into the aquifer and damage the well closer to the sea long before the inland overpumping well is damaged.

Once a saltwater intrusion problem has been detected or predicted through monitoring of the aquifer, a course of action is needed to rectify the situation. Often the necessary action is relatively simple: Water use can be reduced through conservation, allowing the aquifer to keep the salt water out of the freshwater lens. This approach results in a reduction in output for the aquifer. Sometimes, however, conservation will not work because demands are too high. In that case, the situation can be stabilized by increasing the recharge through the development of reservoirs and holding ponds. In extreme cases, water has been brought in by pipeline to an area and recharge accomplished by freshwater injection to the aquifer. When upconing has occurred, it may not be possible to correct the situation by increasing recharge; the fresh water may simply flow around the problem area. It may take many years before the well is again usable.

John E. Mylroie

CROSS-REFERENCES

BIBLIOGRAPHY

Cooper, H. H., F. A. Kohout, H. R. Henry, and R. E. Glover. *Sea Water in Coastal Aquifers.* Washington, D.C.: Government Printing Office, 1964. This government document is a landmark publication on salt water and coastal aquifers. A succession of four papers by the authors, it becomes progressively complex, but the early chapters are very descriptive and useful.

Fetter, Charles W. *Contaminant Hydrogeology.* 2d ed. Upper Saddle River, N.J.: Prentice Hall, 1999. A college-level textbook that is extremely readable, with detailed mathematical treatments handled separately from the main text. It provides a good review of hydrology, with technical specifics on the freshwater lens and a general discussion of saltwater intrusion. Good illustrations.

Griggs, Gary B., and J. A. Gilchrist. *The Earth and Land Use Planning.* North Scituate, Mass.: Duxbury Press, 1977. This textbook goes into the use of the land, planning, and geological interaction at the college level. On pages 319-323, the book provides a detailed discussion of saltwater intrusion with specific examples and case histories.

LaFleur, Robert G., ed. *Groundwater as a Geomorphic Agent.* Boston: Allen & Unwin, 1982. This book is a collection of fifteen papers on groundwater as a landform producer. The twelfth concerns the effect of the mixing zone on coastal aquifers in Yucatan, Mexico, with emphasis on the alteration of the limestone bedrock. Somewhat technical, but very well illustrated.

Leopold, Leo B. *Water: A Primer.* San Francisco: W. H. Freeman, 1974. An introductory text on water and water resources in general. It does not go into the specifics of saltwater intrusion, but it provides an easy-to-read, nonquantitative introduction to the basics of surface and subsurface hydrology.

Manning, John C. *Applied Principles of Hydrology.* Columbus, Ohio: Merrill, 1987. This textbook is more advanced than Leopold's book in scope, detail, and substance, but it is still not very quantitative. It covers a wide variety of water resource topics. Contains many useful examples and sources of data.

Merritt, Michael L. *Assessment of Saltwater Intrusion in Southern Coastal Broward County, Florida.* Tallahassee, Fla.: U.S. Department of the Interior, Geological Survey, 1997. This study, prepared in cooperation with the South Florida Water Management District and Environmental Services, looks at the occurrence of saltwater encroachment and groundwater pollution in southern Florida. Mathematical models are used to describe the state of saltwater intrusion, future developments, and possible tactics to handle the situation.

Palmer, Robert. *The Blue Holes of the Bahamas.* London: Jonathan Cape, 1985. This book is a popular account of cave diving in the Bahamian blue holes. It discusses the science of the freshwater lens, especially the mixing zone, as it relates to the dissolving of limestone and the making of caves. Superbly illustrated.

Saether, Ola M., and Patrice de Caritat, eds. *Geochemical Processes, Weathering, and Groundwater Recharge in Catchments.* Rotterdam, Netherlands: A. A. Balkema, 1997. This slightly technical compilation looks at the geochemical cycles and processes associated with catchment basins. Several essays deal with the artificial recharge of groundwater in such systems.

Strahler, Arthur N., and Alan H. Strahler. *Environmental Geoscience: Interaction Between Natural Systems and Man.* Santa Barbara, Calif.: Hamilton, 1973. A college-level text written for the student with a general interest in the topic. It details numerous human-geology interactions, including the saltwater intrusion problem.

SURFACE WATER

Surface water supplies are among the most basic needs of humankind, but water resources are unevenly distributed across the face of the Earth in both space and time. Outside of the oceans, surface water is most abundant in lakes, both fresh and saline, but streams provide a source of fresh water that is constantly replenished.

PRINCIPAL TERMS

DERANGED DRAINAGE: a landscape whose integrated drainage network has been destroyed by irregular glacial deposition, yielding numerous shallow lake basins

DRAINAGE BASIN: the land area that contributes water to a particular stream; the edge of such a basin is a drainage divide

LAKE BASIN: an enclosed depression on the surface of the land in which surface waters collect; basins are created primarily by glacial activity and tectonic movement

SALINE LAKE: a lake with elevated levels of dissolved solids, primarily resulting from evaporative concentration of salts; saline lakes lack an outlet to the sea

STREAM: a body of flowing water that delivers surplus water from the land to the sea; this term covers all such moving water, including creeks and rivers

THROUGHFLOW: the subsurface movement of surplus water through the soil to a stream; it provides the quickflow of forested watersheds

OVERVIEW

Surface water is a crucial commodity for humans, plants, and animals, but it is available in limited quantities on Earth. Furthermore, it is unevenly distributed geographically and temporally, such that vast areas experience perennial, seasonal, or intermittent shortages, while other regions must cope with excessively abundant water. Replenishment of surface water supplies through the hydrologic cycle assures that water will always be available, though not necessarily at the time and place that it is needed or in the desired quantity.

Surface waters include water in freshwater lakes, streams, and saline lakes such as the Great Salt Lake, the Caspian Sea, and the Dead Sea. Research by the U.S. Geological Survey has revealed that at any given moment, only about 0.001 percent of all the water on Earth is fresh surface water; a similar amount is held in saline lakes. The total volume of water present in freshwater lakes and streams is impressive nevertheless, equaling more than 125,000 cubic kilometers.

Humankind is dependent on surface waters for a wide variety of uses besides direct consumption, including irrigation of crops, generation of electricity, industrial activities, food processing, trans-

portation, and recreation. In moderate climatic zones, humans must ingest only 3 to 4 liters of water per day to survive, but water use in even the least industrialized societies is far higher than that rate. The annual per capita water consumption in underdeveloped nations is approximately 350 cubic meters. In the United States, the annual consumption of water exceeds 2,300 cubic meters per person per year. Only one-tenth of this water is consumed domestically, the remainder going to irrigation, industry, and the generation of electricity. As the Third World nations attempt to industrialize, their demands for water are increasing significantly.

RUNOFF

Surface waters are produced and replenished through precipitation and diminished by evaporation, infiltration to subsurface storage, and return flow to the sea. Precipitation falling on forest or complete grass cover is intercepted by vegetation. Interception is of great importance because it reduces the kinetic energy of raindrops considerably, reducing their erosive potential. This water eventually drops off leaves or runs down the branches of trees, bushes, and stems of grass, where it encounters the ground surface. Studies of steep forested terrain at Coweeta Hydrologic Laboratory in North Carolina revealed that no measurable overland flow was generated in seven years of observation. All runoff reaching the stream channels first infiltrated the soil, later to reemerge in drainageways. It is now believed that all storm flow and base flow of streams draining forested landscapes is generated by infiltrated water, which moves in the shallow unsaturated zone as throughflow, displacement flow, saturated flow in fluctuating belts adjacent to streams, or slow groundwater seepage. In arid areas or on lands disturbed by farming, paving, or compaction, infiltration is greatly reduced, and surface waters are largely generated by direct surface runoff, or overland flow. Runoff generated by overland flow reaches stream channels much more quickly than that which moves by subsurface routes, and it is less substantially depleted by evapotranspiration (the combined loss of water by direct evaporation and transpiration, which is the process by which plants evaporate water in order to sustain their metabolic functions).

All the land area that contributes runoff to a particular stream constitutes the watershed, or drainage basin, of that particular stream. Land use changes within the watershed will affect the runoff timing, volume, and quality of the stream. Activities outside the watershed do not ordinarily influence runoff rates or volumes within the watershed. Thus, the watershed is the basic unit of study in most hydrologic investigations.

The global pattern of seasonal and annual runoff is quite complex; the amount of runoff at any location is determined as much by rates of evapotranspiration as it is by the amount of precipitation that falls. Plants obtain water from soil moisture, the water that adheres to particles in the soil in the same way that a sponge holds water. Runoff, which is the water that flows in streams, is surplus water—water left over after soil moisture is completely replenished. Thus, the tropical and subtropical regions, which receive abundant rainfall, do not necessarily experience the highest rates of runoff, because high temperatures throughout the year cause high rates of evapotranspiration, greatly reducing the water surplus.

The average depth of runoff for all the land area of the Earth is approximately 27 centimeters, but there is considerable variation from that aver-

age. Only a few areas in the world produce more than 100 centimeters of annual runoff. They include tropical areas such as Central America, the lower Amazon basin, equatorial West Africa, Bangladesh and northeast India, Madagascar, and the East Indies, where exceptionally heavy precipitation overcomes the effects of high evapotranspiration. Outside of the tropics, annual runoff in excess of 100 centimeters occurs primarily in coastal alpine settings, where cool temperatures and forced lifting over mountain ranges conspire to produce high rates of runoff. Such areas include coastal Alaska and British Columbia, Norway, Andean Chile and Argentina, Tasmania, and New Zealand. Each of these belts of exceptionally heavy runoff is surrounded by areas receiving 50 to 100 centimeters of runoff per year. The largest areas in this range are the Amazon basin, the Congo basin, southeast Asia, the Appalachians, and Japan.

Areas producing less than 10 centimeters of runoff per year are extensive. The largest such contiguous area covers Africa north of 10 degrees north latitude, the Arabian peninsula, Iran, Afghanistan, Pakistan, and much of interior Asia. The interior of North America west of the hundredth meridian produces little runoff except from the higher mountain ranges, and the Atacama and Patagonian regions of South America also yield less than 10 centimeters of runoff per year. Only the east coast of Australia produces substantial runoff, making it the driest of all continents. Water is a scarce commodity in all these areas, except where unusually abundant groundwater occurs or where exotic streams such as the Nile pass through.

The large remaining continental areas generally produce in the range of 25 to 50 centimeters of runoff per year. Such moderate rates of runoff still produce substantial streams when drainage basins are large. Eastern North America, northern Europe and Asia, East Asia, most of India, east Australia, and East Africa are included in this category. Surface water shortages in these regions occur primarily in watershed divide areas and during seasonal and intermittent droughts.

MOUNTAINS, STREAMS, AND LAKES

In alpine areas, the snowpack generated by winter storms provides the majority of the stream flow. Conversely, areas on the lee side of such mountain ranges are deprived of moisture through the rain-shadow effect. The net result of heavy precipitation in the mountains, accompanied by cool temperatures and reduced evaporation rates, is that the mountain ranges are islands of moisture that generate abundant stream flow during the spring melt. The numerous streams generated by snowmelt flow to the lower, drier adjacent areas and provide them with a source of water. Almost all the streams of the western United States are generated in this manner: the Colorado, the Snake and Columbia, the Missouri, the Arkansas, the Sacramento, and the Rio Grande are but a few examples. The mountain regions of Europe, Africa, Asia, and Australia perform a similar function for adjacent lowlands. When a stream from such locations flows through arid lands, it is especially valued and is known as an exotic stream. Prominent examples include the Nile, Indus, and Colorado Rivers.

The five largest streams of the world account for more than one-third of the world's total stream flow. The Amazon alone accounts for almost 20 percent of the world's stream flow, as it drains the largest watershed on Earth and gathers waters from the world's largest tropical rain forest. The watershed of the Amazon is almost as large as the forty-eight contiguous states of the United States, and, 800 kilometers above its mouth, its channel averages 2.5 kilometers wide and 50 meters deep. The discharge of the next largest stream, the Congo, is only one-quarter as great; the discharge of the mighty Mississippi is only one-tenth that of the Amazon. The combined discharge of all the world's streams is roughly 30 percent of the precipitation that falls on the continents of the Earth; the remainder is returned directly to the atmosphere by evaporation and by transpiration.

Freshwater lakes cover some 1.5 percent of the Earth's land surface and contain the vast majority of the Earth's liquid fresh water, a total of approximately 125,000 cubic kilometers. Saline lakes contain only slightly less water, storing an average of 105,000 cubic kilometers of water. Although the freshwater lakes are generally of greater utility to humans, the saline lakes provide important resources as well, including fish, salt, and magnesium.

Lakes occur where natural basins have developed, such that ponding of surface waters occurs. Because the normal cycle of fluvial (stream-related) erosion of the landscape does not produce such depressions, extensive areas of the continents are without naturally occurring lakes. The vast majority of lake basins were produced by the erratic deposition and scouring action of glaciers, but almost two-thirds of the water held in lakes is contained in six structural basins: the Caspian Sea, the Aral Sea, and Lakes Baikal, Tanganyika, Nyasa, and Victoria. These structural basins were produced by a combination of the downfaulting of basins and uplift of adjacent terrain. Lake Baikal in Siberia contains almost one-fifth of the world's liquid fresh water, because of its large size and depth exceeding 1,700 meters. The rift valleys of east Africa, formed by the downfaulting of linear basins associated with the rending of the continent by plate tectonic movement, contain a series of large, deep freshwater lakes.

Glaciers passing over North America scoured out preexisting river valleys to form deep depressions in the Earth, which were further deepened (relative to their outlets) by postglacial tilting and deposition of recessional moraine. The Great Lakes were thus formed, as were Canada's Lake Winnipeg, Great Slave Lake, and Great Bear Lake. Hundreds of thousands of lakes were formed in the recently glaciated terrains of Europe and North America by erratic deposition of loose materials by glaciers. Such uneven deposition eradicated the well-integrated preglacial drainage networks of the areas affected, leaving vast areas of uneven topography with many shallow depressions, which are now filled with water. This type of hydrologic system is referred to as deranged drainage and is the reason that the state of Minnesota is known as the Land of Ten Thousand Lakes.

Preferential solution of certain areas of limestone and other soluble rock types often results in collapse features known as dolines, or sinkholes. When these solutional depressions are filled with impermeable clays, or where the regional water table is high, the depressions tend to hold water. The Lake District of north-central Florida provides numerous examples of such lakes. Other causes of lake basins include landslides, volcanic flows (Lake Tahoe), volcanic eruption and collapse (Crater Lake), scouring by wind, and meteorite impact. More than 100,000 small, elliptically oriented lake basins, most of them now dry, occur on the coastal plain of the United States from New Jersey to Georgia. The origin of the Carolina Bays has been vigorously debated. Human-made lakes are increasingly common on the landscape, built by damming streams to produce reservoirs for water supply, flood control, power generation, transportation, and recreation.

Lake basins are sediment traps because flowing water entering a lake quickly loses velocity and drops its sediment load. As a result, only the largest and deepest lakes last more than a few thousand years, making them among the most ephemeral geographic features on the landscape over geologic time.

STUDY OF SURFACE WATER

People have been concerned with the origins and volumes of stream flow since the beginnings of civilized culture, as many early civilizations began on the banks of alluvial streams. The ancient Egyptians kept detailed records of the dates and heights of the annual floods of the Nile from as early as 3500 B.C.E., dependent as they were upon its delivery of water and nutrient-rich sediment. Modern hydrologists depend upon similar measurements for much of their research. At more than 16,000 stream sites in the United States alone, gauging stations measure the height of the stream surface every fifteen minutes. These stage recordings are converted to flow rates on the basis of the previously measured cross-sectional area of the stream channel below each gauge height and flow velocity. These measurements are either recorded on paper at the gauging station, to be retrieved later, or are sent directly to the managing agency by telemetry. The stream-flow data are stored in a computer for later tabulation and retrieval. Lake levels are similarly monitored and converted into water volume contained in the lake. In the United States, most stream and lake gauges are operated by the U.S. Geological Survey (USGS), which issues annual reports for all its gauges, organized and distributed by state (available from state offices of the USGS upon request). In addition, the flow data are stored in a computer system that is capable of providing much more detailed flow statistics than do the printed reports. Stream gauges provide valuable warning of ap-

proaching floods in some locales. The cities of Bath, England, and Gatlinburg, Tennessee, have been heavily damaged by floods in the past but now have telemetered gauges upstream that warn of approaching floods. More commonly, such readings are used to allocate limited water resources to downstream claimants.

Hydrologists frequently plot stream-flow rates against time to produce a graphical representation of discharge that is called a hydrograph. Analysis of the hydrograph can provide a wealth of information concerning stream-flow generation and timing mechanisms. By analyzing the form of the hydrograph of a small storm, it is possible to forecast the runoff characteristics of a large storm, such as the peak rate of flow and the time of peak flow.

In order to study how stream flow is generated, hydrologists employ a variety of research methods. On small upland catchments, dye tracers are often used to determine paths and rates of movement. Dye tracers are also used in tracking the paths of disappearing streams in regions of soluble rock. To determine how much water is being added to or lost from a stream channel through bank seepage, stream reach surveys, which involve stream gauging along successive longitudinal segments of a stream between tributaries, are undertaken.

In areas where snowmelt is an important component of stream flow, regular snow surveys are taken. Most commonly, that involves manual coring of the snowpack along predetermined transects to determine snow depth and water content. Depth gauges can be checked from low-flying planes at some locations, and weighing telemetered devices are also used. In the most remote areas, where manual surveys are not possible, aerial photogrammetry and satellite imagery are of great utility in determining snowpack extent and water content. Such information is critical in formulating a basic understanding of snowmelt hydrology as well as for allocating stream flow to irrigators and determining how much flood storage is needed in downstream reservoirs.

SIGNIFICANCE

Water is a critical commodity in even the most primitive society; in warm weather, humans cannot survive much longer than a day without it. In agricultural and industrial societies, immense quantities of water are needed, and delivery of it must be reliable. A hectare of cotton grown in Arizona requires on the order of 15 million liters of water per year; to produce a ton of steel requires roughly 250,000 liters; production of a single Sunday newspaper involves the use of more than 1,000 liters of water. As a result of water-intensive agriculture and industry, the annual per capita use of water in the United States exceeds 2,300 cubic meters. In order to increase standards of living around the world, large quantities of water are needed for a variety of agricultural, industrial, and domestic purposes. In a world with more than 5 billion inhabitants, the acquisition of such volumes of water is increasingly difficult; many experts believe that water supplies will be one of the primary limitations on future population growth. Water shortages have already constrained industrial and agricultural expansion in many areas.

Surface water is unevenly distributed on the Earth. Some locations have such a problem with excess water that vast areas are not arable and remain largely uninhabitable. Such an area is western Siberia, where more than 2 million square kilometers of land are annually flooded for extended periods by the Ob and Yenisey Rivers. Far larger areas are without adequate water supplies, many of them in the regions of the world with the greatest population pressures: Africa and Asia. The largest proportion of liquid surface water is in freshwater lakes, but only ten lakes account for two-thirds of all such water. Thus a few areas (such as the Great Lakes area) are very well endowed with lake waters, while most parts of the world have very little. Most freshwater lakes are very small, contain small volumes of water, are

very easily fouled, and are short-lived. Human-made reservoirs are no exception.

The second most voluminous source of surface water is saline lakes. The Caspian Sea alone contains three-quarters of such water supplies, with lesser bodies such as the Aral Sea, Great Salt Lake, and Dead Sea falling far behind. These bodies of water are not potable, or suitable for drinking, but they greatly influence local weather patterns and provide many important resources.

Although they contain far less water at any given time than do fresh or saline lakes, streams are the most important element of the surface water system on Earth because of the tremendous volumes of water that they convey throughout the year. Although not as limited in extent as is lake water, stream flow evidences great discrepancies in abundance across the globe. The equatorial zone is very well endowed with stream flow, as are the midlatitudes in general, but vast areas of the subtropical arid belts produce very little. Some such areas are blessed with exotic streams, which flow from higher, cooler, wetter climatic regimes.

Michael W. Mayfield

CROSS-REFERENCES

Aquifers, 2005; Artificial Recharge, 2011; Dams and Flood Control, 2016; Environmental Health, 1759; Floods, 2022; Freshwater Chemistry, 405; Groundwater Movement, 2030; Groundwater Pollution and Remediation, 2037; Hydrologic Cycle, 2045; Karst Topography, 929; Precipitation, 2050; Salinity and Desalination, 2055; Saltwater Intrusion, 2061; Seawater Composition, 2166; Water Quality, 2072; Water Table, 2078; Water Wells, 2082; Waterfalls, 2087; Watersheds, 2093.

BIBLIOGRAPHY

Fairbridge, Rhodes W., and Reginald W. Herschy, eds. *The Encyclopedia of Hydrology and Water Resources.* Boston: Kluwer Academic, 1998. This encyclopedia serves as a great reference tool for anyone interested in the study of hydrology. There are sections devoted to surface water and groundwater, as well as water resources. Suitable for college-level readers. Illustrations and maps.

Jones, J. A. A. *Global Hydrology: Processes, Resources, and Environmental Management.* Essex, England: Longman, 1997. Jones looks at the science of hydrology and the environmental issues and aspects relevant to it. The book also focuses on water resource management and development. Suitable for the nonscientist. Illustrations, maps, index, and bibliographical references.

Kirkby, M. J., ed. *Hillslope Hydrology.* New York: John Wiley & Sons, 1978. This collection of works by various authors focuses on the processes by which runoff is generated. Suitable for advanced high school students and above.

Kovar, Karel, et al., eds. *Hydrology, Water Resources, and Ecology in Headwaters.* Wallingford, England: International Association of Hydrological Sciences, 1998. This collection of essays focuses on water supplies, resources, and water resource development. There is a great deal of attention paid to water ecology as well as a nice overview of hydrology in general.

Leopold, Luna B. *Water: A Primer.* San Francisco: W. H. Freeman, 1974. This general treatment of water and the hydrologic cycle was written not for specialists but for the general public. The book is thorough but easy to follow, well illustrated, and contains numerous examples of the phenomena presented. Suitable for high-school-level readers.

Manning, John C. *Applied Principles of Hydrology.* Columbus, Ohio: Merrill, 1987. Designed to be used as the text for a general college course in hydrology, this book is less quantitatively oriented than most engineering hydrology texts, yet it is quite thorough. Suitable for high school students and above.

Pielou, E. C. *Fresh Water.* Chicago: University of Chicago Press, 1998. Pielou examines and evaluates all sources of fresh water, including glaciers, lakes, and rivers. Much attention is given to the measures being taken to preserve and protect freshwater resources.

Scott, Ralph C. *Physical Geography.* St. Paul, Minn.: West, 1989. A very thorough, well-illustrated text in general physical geography. The chapters on streams and stream-flow generation are especially well written, and the full-color illustrations are very helpful. Suitable for high school students and above.

Smith, David Ingle, and Peter Stopp. *The River Basin: An Introduction to the Study of Hydrology.* New York: Cambridge University Press, 1979. This book was written as a text for upper-level high school and introductory college courses in physical geography and environmental studies. As such, it addresses the main topics in surface water hydrology in a clear, concise manner.

United States Department of Agriculture. *Water.* Washington, D.C.: Government Printing Office, 1955. This classic volume on water contains straightforward explanations of the hydrologic cycle, stream flow, soil moisture, groundwater, and numerous applied topics, such as irrigation, gardening, and rural domestic water supply. Suitable for the general public.

WATER QUALITY

The term "water quality" refers to the fitness of water for a defined use, such as for human consumption, and is measured in terms of chemical, physical, and biological parameters. Standards against which the quality of water is compared are neither universal nor constant and depend upon who has established the standards and for what purpose.

PRINCIPAL TERMS

CONCENTRATION: amount of substance present in a given volume of sample water; commonly used units include milligrams per liter or micrograms per liter

CONTAMINANTS: solutes introduced into the hydrologic environment as a result of human activity, without regard to degree of degradation

DETECTION LIMIT: the lowest concentration of a constituent that can be detected reliably

MAXIMUM CONTAMINANT LEVEL GOALS (MCLGs): nonenforceable health goals

MAXIMUM CONTAMINANT LEVELS (MCLs): enforceable standards for drinking water established by the U.S. Environmental Protection Agency under the Safe Drinking Water Act

NATIONAL PRIMARY DRINKING WATER STANDARDS: the list of MCLs and MCLGs for organic and inorganic constituents; the list also includes various standards for asbestos fibers, turbidity, bacteria, viruses, and radioactive emitting constituents, established using less common concentration units

POLLUTION: contaminant levels at objectionable concentrations

CHEMICAL ANALYSIS OF WATER

Pure water composed of hydrogen and oxygen and nothing else does not exist in nature. Even rainwater contains measurable dissolved constituents in the range of 10 milligrams per liter. Water is of acceptable quality for a defined use or purpose if, on analysis, constituents in it do not exceed prescribed concentrations. Water quality involves fitness or suitability of water for a specific use, based upon chemical, physical, and biological parameters. Significant differences have arisen among scientific and engineering disciplines and government agencies as to which indicators are appropriate to evaluate water quality. The results of an analysis of water are commonly presented as indicators of water quality without the provision of a context of use or standards.

Standards against which the quality of water is compared are neither universal nor constant and depend upon who has established the standards and for what purpose. Legal standards are legally binding, prescribe conditions not to be exceeded for a water resource, and allow penalties for violations. Standards are changed through time as more becomes known about health effects, as analysts are able to detect ever smaller quantities of contaminants, and as governments apply standards to more situations.

This methodology of listing constituents and concentrations that must not be exceeded has undergone constant evolution. Prior to the establishment of formal criteria, odor, color, taste, turbidity, and temperature were measures used by humans in assessing the desirability of a drinking-water source. By 1784, less direct chemical criteria involving the ability of a sample to dissolve soap without forming lumps or a residue were applied to establish whether water was drinkable. Chemical criteria alone, however, are insufficient for establishing a water supply as both drinkable and safe. The relationship between disease and pollution by human and animal wastes had been recognized for centuries. An early method of measuring the bacterial quality of water relied upon observing for how long a stored water sample remained free of visible growths. This was replaced by techniques to count bacteria and to compare measurements to a standard.

U.S. GEOLOGICAL SURVEY AND U.S. PUBLIC HEALTH SERVICE

The U.S. Geological Survey (USGS) first reported chemical analyses of natural waters in 1879 and by 1901 had published more than twenty-five reports on the geologic control and chemical and physical properties of natural water. At that time, many water sources were pristine, unaffected by human activity. The composition (quality) of waters from lakes, rivers, and wells resulted from reactions with gases in the air and minerals in soil and rock. The major dissolved constituents, those generally with concentrations of more than 1 milligram per liter, are calcium, magnesium, sodium, potassium, bicarbonate, carbonate, chloride, sulfate, nitrate, and oxygen; these constitute up to 99 percent by weight of the dissolved matter in pristine waters. Under less common natural situations, other constituents such as iron or fluoride may exceed 1 milligram per liter concentration. Minor constituents are detected in pristine natural waters if sought by analysis. Of the major constituents, only sulfate and chloride are included in either the current U.S. primary or secondary drinking-water standards. Other major constituents are listed in standards of other organizations, such as the World Health Organization or the European Community.

After 1901, the USGS expanded its activities to include pollution studies, recognizing that sewage and industrial wastes were degrading the quality of water and adversely affecting municipal water supplies. In 1905, the USGS began the first monitoring program to assess the quality of streams and lakes, efforts that led to estimates of the amounts of dissolved and suspended matter carried to the oceans by rivers.

The U.S. Public Health Service (USPHS) was established in 1912 and was directed to study sanitary water quality. The U.S. Geological Survey continued studying water supplies for public use and for agricultural and industrial purposes and maintained limited networks to evaluate trends in water quality. In the United States, formal chemical water-quality standards originated by action of the USPHS in 1914. Dissolved constituents believed to be harmful to humans were identified, and maximum allowable concentrations were established. Water exceeding the limits could not be used for food preparation or for drinking water on passenger trains (interstate carriers) crossing state lines. The USPHS standards were widely, although unofficially, regarded as the basis for acceptability of a water supply for human consumption. The standards were expanded both in number of constituents covered and in lowered permissible concentrations in 1925, 1942, 1946, and 1962. The final set of standards promulgated by the USPHS, issued in 1962, considered bacterial quality, physical characteristics of turbidity, color, and odor, chemical characteristics, some mandatory and others recommended, and radioactivity. In earlier times of pristine sources in protected watersheds, water that required no extensive treatment was available, and one set of standards was sufficient for both raw water and drinking water. By 1962, modern civilization had affected most water sources, and raw water from degraded sources had to be treated so that the standards for drinking water were met at the point of delivery.

ENVIRONMENTAL PROTECTION AGENCY

By the 1960's, industrial manufacturing of new chemical products and by-products, as well as generation of wastes, had resulted in contamination and pollution, some materials being persistent in the environment and affecting surface and groundwater supplies. Efforts to restrict degradation of water sources were underway in specific geographic areas ranging from small drainage basins to interstate commissions for river basins. The Environmental Protection Agency (EPA) was established in 1970 and given the task of setting goals and standards and identifying sources of polluting effluents. Federal legislation, including the Water Quality Act and Clean Water Act, led to broadened activities in documenting quality, mandating treatment and recovery, and requiring the establishment of standards for discharged waters. The National Pollutant Discharge Elimination System (NPDES) was established for all point source discharges into U.S. waters, involving permits, reporting, and effluent limitation. For situations where the limitations were not stringent enough to improve quality of receiving waters, treatment was required. The EPA was charged with developing a wide array of water-quality standards. An understanding of the degree to which the nation's waters are contaminated and what the water quality trends are was seen as essential to management decisions. The USGS was thus brought back into the process of monitoring surface and groundwater supplies, looking at contamination to a greater extent than before. The agency established the National Water Quality Assessment Program (NAWQA) to systematically study U.S. water, stream basin by stream basin.

The first set of drinking-water standards to be established by the EPA became effective in 1977, under mandates of the 1974 Safe Drinking Water Act. Among the inorganic contaminants, fluoride and mercury were newly added to the earlier USPHS standards, nitrate became a mandatory rather than recommended standard, and radioactive sources were redefined. This came about in large part because of better information about risks and health effects. The standards have undergone changes and extensions since 1977, including modifications authorized by the Safe Drinking Water Act Amendments of 1996. A prior mandate to the EPA to regulate twenty-five new contaminants every three years has been set aside, replaced by a mandate to review at least five contaminants for possible regulation every five years, looking at the following criteria: whether the contaminant adversely affects human health; whether it is known or substantially likely to occur in public water systems with a frequency and at levels of public health concern; and whether regulation of the contaminant presents a meaningful opportunity for health-risk reduction. In addition, there are provisions for monitoring unregulated contaminants, creating an accessible database, and disclosing violations to the public served by water systems as rapidly as within twenty-four hours.

Domestic use, including drinking water, is considered by many to be the most essential use of water, and standards for drinking water receive the greatest attention. Water-quality requirements for other purposes may be more restrictive (for example, for high-pressure boiler water) or less restrictive (for example, for hydraulic cement manufacture) than standards for domestic use.

APPLICATION OF WATER-QUALITY STUDIES

In the Earth sciences, water-quality studies generally have a focus beyond determining whether a water source is suitable for a given purpose. Larger questions of regional water quality, such as determining how and why water of certain characteristics is associated with a specific rock type or recognizing how and explaining why groundwater evolves chemically as it slowly moves through an aquifer, are examples of applications that center on understanding natural processes. The contamination and pollution of water resources has drawn considerable attention, leading to studies of the transport by water of pollutants. One example of applications of the methodology from the Earth sciences should illustrate the disciplinary perspective.

Of the major constituents in natural waters, only sulfate is listed in the national primary drinking water standards. Sulfate in drinking water produces laxative effects and unpleasant taste. It is listed in the unenforceable national secondary drinking water standards for taste and laxative effects at a concentration of 250 milligrams per liter and is prohibited by the Primary Standards at 400 to 500 milligrams per liter. The sources of the sulfate are listed as "natural deposits."

As a result of the Safe Drinking Water Act and using data from the U.S. Geological Survey, a series of state maps was compiled to show the regional variation in concentration of major dissolved constituents in well water. A zone of high-sulfate water extends from Lake Erie across northwest Ohio and into east-central Indiana. Water from wells in this area is of generally unacceptable quality. From the perspective of the Earth sciences, it was important to build upon the recognition of this zone in order to determine probable cause. Obviously, such a study cannot exclude areas from which groundwater just passes the standards. Results from this study provide an understanding of the geologic controls on the system. High-sulfate water is recovered from shallow wells in glacial deposits and also from wells in a deeper bedrock aquifer. The situation is explained in part by the presence of gypsum and anhydrite, calcium sulfate minerals, in a sedimentary terrain characterized by shale and dolomite units. Glacial activity eroded formerly exposed bedrock of shale and anhydrite, moved it, and redeposited it as unconsolidated material over undisturbed bedrock, also containing anhydrite and dolomite. High values of sulfate in the water, at some wells exceeding 1,000 milligrams per liter, would ordinarily not be possible because of the solubility limits of calcium sulfate minerals. In such waters, magnesium values were elevated, and calcium values were lower than expected. The conclusion is that in order for such high values of sulfate to exist in the water, a process known as dedolomitization is taking place, in which dolomite, a calcium magnesium carbonate mineral, is dissolved, and calcite, a calcium carbonate mineral, is deposited, in effect adding magnesium and removing calcium from the water. Without this process, such high values of sulfate would not be possible.

CONTAMINATED WATERS

Contaminated waters may or may not be safe to drink. In addition to the constituents derived from natural sources, these waters contain (by definition) constituents derived from a variety of human activities, such as improperly treated sewage, storage tanks that have leaked, improper disposal of hazardous wastes, and agricultural practices. Many of the chemicals involved are manufactured and do not occur in nature. Standards for safe drinking water for most of these constituents range generally from tenths to several hundred micrograms per liter.

Maximum contaminant levels (MCLs) are derived from studies of risk to humans consuming two liters of water a day over a lifetime of seventy years. Based upon extrapolations of animal studies for a given contaminant, an increased cancer risk to humans of one in one million over a lifetime leads to the standard. By statute, no cancer risk is to be tolerated, but technology is not available to eliminate totally the very small concentrations in water resources of many contaminants. The maximum contaminant level goal is generally zero for MCLGs and a small concentration for MCLs.

Other water-quality standards, including national effluent standards for limiting discharge of pollutants into surface waters, have been established or enforced as a result of a number of federal laws and resultant regulations, including the Clean Water Act, which addresses water pollution from point and nonpoint sources;

the Comprehensive Environmental Response, Compensation and Recovery Act (the "Superfund" Act), which mandates cleanup of hazardous-waste sites and leaking tanks; the Resource Conservation and Recovery Act, which defines and requires tracking and proper disposal of hazardous wastes; the National Environmental Policy Act, which establishes the need for environmental impact statements for federally controlled or subsidized actions; the Endangered Species Act, which protects some habitats; and the Wild and Scenic Rivers Act, which limits development along some rivers. States in some instances have established water-quality standards similar to or more restrictive than the ones put forth by federal agencies.

Management of water quality for effluents takes place through regulations applied to discharges from or into three distinct sites. End-of-pipe standards specify levels that must be met at the point from which effluent is discharged from an industrial facility. Technology-based standards impose regulations on a discharging facility. Assimilative capacity standards define the water-quality conditions that a receiving body of water (stream or lake) must not exceed. Dischargers to the stream or lake are identified, and those causing the stream or lake to exceed standards are required to manage the quality of their discharges.

Robert G. Corbett

Contaminated water at the Argo Tunnel, caused by mining in Idaho Springs, Colorado, 1983. (U.S. Geological Survey, photo by N. Gaggiani)

CROSS-REFERENCES

Aquifers, 2005; Artificial Recharge, 2011; Atmosphere's Global Circulation, 1823; Climate, 1902; Dams and Flood Control, 2016; Floods, 2022; Glacial Landforms, 915; Groundwater Movement, 2030; Groundwater Pollution and Remediation, 2037; Hydrologic Cycle, 2045; Landsat, 2780; Ocean-Atmosphere Interactions, 2123; Precipitation, 2050; River Valleys, 943; Salinity and Desalination, 2055; Saltwater Intrusion, 2061; Surface Water, 2066; Water Table, 2078; Water Wells, 2082; Waterfalls, 2087; Watersheds, 2093.

BIBLIOGRAPHY

Berner, E. K., and R. A. Berner. *The Global Water Cycle.* Englewood Cliffs, N.J.: Prentice-Hall, 1987. An interdisciplinary text that provides chemical characterization of water as rain, soil and groundwater, river water, lakes, estuaries, and oceans, and reports the relative importance of natural and human processes in the cycling of elements through the hydrologic cycle.

Fetter, C. W. *Contaminant Hydrogeology.* 2d ed. Upper Saddle River, N.J.: Prentice Hall, 1999. An applied text for hydrogeologists and engineers that presents technical information on the collection and interpretation of data from contaminated groundwater sites and options for remediation of those sites.

Hem, J. D. *Study and Interpretation of the Chemical Characteristics of Natural Water.* 3d ed. United States Geological Survey Water-Supply Paper 2254. Washington, D.C.: U.S. Government Printing Office, 1992. The term "natural water" refers to water in a real-world environment such as a lake. This introduction to water chemistry was written to serve geologists and hydrogeologists who need to evaluate water quality. The author provides practical information about low-temperature aqueous geochemistry using real-world examples. Offers an extensive list of references.

McKee, J. E., and H. W. Wolf. *Water Quality Criteria.* 2d ed. State Water Resources Control Board, Publication 3A. Sacramento: State of California, 1971. The authoritative compendium of information about water quality through the 1960's, including criteria by state and interstate agencies, judicial expression, quality criteria for beneficial uses, and descriptions of potential pollutants. Provides nearly four thousand references to articles in the field.

Nash, H., and G. J. H. McCall, eds. *Groundwater Quality.* London: Chapman and Hall, 1965. Twenty papers report groundwater quality studies worldwide, excluding North America. Unacceptable groundwater quality results most commonly from anthropogenic pollution of aquifers, saline intrusion, and naturally occurring problems.

Perry, J., and E. Vanderklein. *Water Quality: Management of a Natural Resource.* Cambridge, Mass.: Blackwell Science, 1996. A text claiming that water quality is basically a social and political concern. The authors present a nonmathematical but comprehensive treatment of the subject, addressing broad issues such as ecological responses to stress, cultural dimensions of water-quality policy, and global implications of change. The writers review the evolution of the concepts of water quality and management and provide a perspective that could serve as a framework for further changes. Extensive bibliography.

Tchobanoglous, G., and E. D. Schroeder. *Water Quality.* Reading, Mass.: Addison-Wesley, 1985. A text for civil engineers and physical scientists that is mathematically rigorous and comprehensive in presentation of water quality and quantity, analytical methods for water-quality management, modeling of water quality in the environment, and modification of water quality.

Thompson, Stephen A. *Hydrology for Water Management.* Rotterdam, Netherlands: A. A. Balkema, 1999. This thorough account of hydrology, groundwater flow, and stream flow focuses on the management of water supplies in efforts to keep them free of pollutants and available to the largest number of people possible. Illustrations, maps, index, and bibliography.

U.S. Environmental Protection Agency. Office of Water. *The Quality of Our Nation's Water.* Washington, D.C.: U.S. Government Printing Office, 1992. This government report summarizes the extent of water pollution as compiled from reports from the states in 1990 and 1991 generated under the requirements of the Clean Water Act. The report defines water-quality concepts and uses and summarizes pollutants and sources.

WATER TABLE

The water table is the upper portion of the saturated or groundwater zone beneath the Earth's surface. Depth to the water table is an important consideration in drilling wells for water supply, building and roadway construction, and septic-system and landfill-disposal design.

PRINCIPAL TERMS

CAPILLARY FRINGE: the lowest portion of the unsaturated zone just above the water table

PERCHED WATER TABLE: groundwater that occupies an area above the main or regional water table

PORE: a small opening in a rock or soil

SATURATED ZONE: a subsurface zone in which all of the pore spaces are filled with water

UNSATURATED ZONE: a subsurface zone in which the pore spaces are filled with both air and water

SUBSURFACE HYDROLOGY

The water table forms the boundary between the unsaturated and saturated zones that lie below the Earth's surface. The unsaturated zone, which is also known as the zone of aeration and the vadose zone, forms the uppermost layer in the ground. The pore spaces in the subsurface material—such as sand, silt, clay, gravel, or consolidated rock—contain both air and water in varying proportions. Water in the unsaturated zone, which includes capillary water, is contained under pressure that is lower than that of the atmosphere, whereas air and gases are generally under atmospheric pressure. Water in this zone may move up and down or sideways. Soil water can also be evaporated back into the atmosphere.

Starting at the top, the unsaturated zone is divided into the belt of soil water (or root zone), the intermediate belt, and the capillary fringe. The top and bottom of the belt of soil water is the land surface and the intermediate zone, respectively. It contains plant roots and soil water that is available for plant growth. The depth of the belt varies over the landscape, but it is generally a few meters.

The intermediate belt in the unsaturated zone lies between the belt of soil water and the capillary fringe. The depth of this belt varies substantially, being much deeper in dry climates and much shallower in wet areas. The capillary fringe is the lowermost subdivision of the unsaturated zone. It is located immediately above the water table. The interstices are filled with water under pressure that is lower than that of the atmosphere. The water in the capillary fringe is continuous with the water in the saturated zone below the water table, but it is held above it by surface tension. The upper boundary with the intermediate belt is somewhat indistinct, although it is sometimes arbitrarily defined as the level at which 50 percent of the pore spaces or interstices are filled with water. The vertical extent of the capillary fringe depends upon the soil texture since capillary rise is greater when the openings are smaller. Thus, the thickness in silty material may be as large as 1 meter, where the openings are very small, to as little as 1 centimeter in coarse sand or fine gravel, where the openings are much larger.

The saturated zone's pore spaces are completely filled with water or other fluids. The water table is at the top of the saturated zone and will therefore rise or fall as the saturated zone increases or decreases because of variations in precipitation or pumping. In some locations, a perched water table can exist above the regional water table. The separation is caused by a relatively impermeable layer, such as a clay lens, which impedes infiltration. The perched water table is generally shallow and of limited areal extent, although in irrigated areas it can become much larger.

WATER-TABLE VARIATIONS

The word "table" suggests a surface that is flat and static. In the case of the water table, neither term is appropriate. Rather, the water table tends to be a subdued replica of the topography. It is highest under the highest portions of the landscape (hilltops and divides) and lowest in the valleys, where it approaches the surface close to streams, lakes, or swamps. This topographic configuration is caused by water percolating through the unsaturated zone, which raises the water table, in contrast to streams, lakes, and swamps that receive base flow or seepage from groundwater in the saturated zone, which lowers the water table.

The depth to the water table from the ground surface varies enormously. It is at or close to the surface in swamps and marshes, which means that the unsaturated zone is nonexistent. The depth in arid regions can be measured in tens of meters below the surface. Under natural conditions, the water table will rise after recharge during a period of precipitation and fall during a period of drought. The greater the amount of precipitation and resultant recharge, the larger the rise; conversely, the greater the intensity and length of the dry period, the larger the fall.

The rise and fall of the water table can be greatly magnified by anthropogenic intervention. One obvious example is the decline in the water table that is caused by groundwater extraction. Pumping for water supplies or for dewatering operations at construction sites can radically lower the water table. The resulting decline in the water table forms a cone of depression, which can spread out tens and hundreds of meters from the well. In homogeneous subsurface materials, the shape of the cone of depression forms a smooth parabola. If the subsurface materials are heterogeneous, which is more common, the cone of depression would be irregular in shape but would still spread out a large distance. The vertical distance between the original water table (or static water level) and the new lowered water table (or pumping water level) is known as the "drawdown." The greater the rate and duration of pumping, the larger and deeper the cone of depression and the greater the drawdown. Water-bearing geologic formations called aquifers vary enormously in their permeability or ease of movement of fluids. Thus, water-table fluctuations would be less in aquifers that have high transmission rates (high permeability) and much more in poorer aquifers that have low permeability and consequently less capability for groundwater movement.

ENVIRONMENTAL ISSUES

Water-table fluctuations can present many environmental problems. For example, at the residential level, septic system disposal fields require a minimum depth to the water table of at least 1.2 meters for the effluent to be properly absorbed into the soil. If the water table is too shallow, microbial decomposition of the effluent would not fully occur, which could then lead to groundwater contamination. Other problems occur with overpumping of an aquifer when the declining water table can fall below the depth of the well, which results in a dry well. Depending upon the magnitude of the decline, either the well would have to be deepened or a new well in another location would have to be drilled. The normal hydraulic gradient generally follows the topography from high locations to lower ones, but it can be affected by pumping so that the gradient is reversed. This means that leachate from landfills or septic-system disposal fields can be induced to flow into a nearby well.

Overpumping, in which the pumping rate is in excess of the natural recharge from precipitation, can create a cone of depression large enough to interfere with other wells and cause them to go dry. This is a particular problem with large-diameter public wells that supply water to many users in a community. As a result, many governmental units at the local, county, or state level require pumping tests at specified rates and duration so that the decline and recovery of the water table can be measured to minimize interference with neighboring wells. These pumping tests are often seventy-two hours in duration. In areas where the groundwater yield is marginal and the water-table decline can be substantial, some municipalities even require four-hour pumping tests for domestic wells that serve only one residential dwelling unit.

One interesting illustration of the environmental impact of pumping and water-level change occurred on a large scale in Brooklyn, New York. There were many wells in Brooklyn in the early part of the twentieth century that pumped large amounts of groundwater. The resulting decline in the water table facilitated the construction of subways. When water-quality problems developed as urbanization spread in Brooklyn, the pumping stopped, and the rise of the water table started flooding the subway tracks. In order to keep the subways working, a certain amount of pumping had to be resumed even if the water was simply fed into storm sewers that drained into the ocean.

Irrigated lands often have problems with perched water tables that develop over shallow impermeable beds such as clay lens. The water that is applied to the irrigated soils causes the perched water table to rise toward the surface. As a result, the capillary fringe will finally reach the surface and allow groundwater to be continuously discharged to the atmosphere by evapotranspiration (evaporation plus plant transpiration). The evaporation will produce a buildup of salts in the surface soil since groundwater contains dissolved mineral matter. In addition, the rising water table can drown out plant roots that are near the surface.

Taken together, these two effects have created numerous problems for farmers in irrigated areas, from the days of the ancient farmers in Mesopotamia (the Tigris and Euphrates Valleys in what is now Iraq) to the vast irrigated fields in the Indus River Valley in Pakistan, the San Joaquin and Imperial Valleys in California, the Nile Delta in Egypt, and the wheat belt of Western Australia. For example, a water table that rises at an average annual rate of 0.3 meter and consequent salinization and water logging of the soil in the lower Indus River in Pakistan have cause 36,000 hectares of land to go out of production each year. The solution to this problem requires an elaborate system of drains that remove the salty water from the affected area.

Overpumping along a coast can cause a decline in the water table, which could result in saltwater or saline intrusion. Proximity to the sea means that salt water can be drawn into the well and make the water unfit for human consumption. Saline intrusion problems have been well documented in Miami, Florida, which not only has large wells close to the coast but also is located above the extremely permeable Biscayne Aquifer. The solution to this problem is either to move the wells inland or to install recharge wells between the coast and the contaminated wells.

On an even larger scale, one of the major factors in the selection of Yucca Mountain in Nevada as a high-level radioactive waste disposal site was the great depth to the water table. This area is about 145 kilometers northwest of Las Vegas and has an average annual precipitation of under 200 millimeters. The deep water table and the great extent of the unsaturated zone meant that waste disposal canisters would presumably not be affected by the movement of groundwater, which could become radioactive and flow down-gradient into larger, regional groundwater flow systems. The disposal site is supposed to isolate the waste materials for 10,000 years. Whether the water table would move upward and flood the disposal site during this period of time as a consequence of climatic change is a matter of some controversy.

SIGNIFICANCE

The water table is a critical part of the subsurface portion of the hydrologic cycle. The depth to the table governs the viability of subsurface septic disposal systems and landfills, which require specified depths of the unsaturated zone in order to protect the underlying saturated zone from contamination. Hazardous and radioactive waste disposal also require placement in an unsaturated zone that is sufficiently large and distant from groundwater movement in the saturated zone. Water-table depths are an important factor in wetland determinations and in agriculture where most plants require soils that have pore spaces that contain both air and water.

The water table marks the undulating top of the saturated zone. The importance of groundwater is best illustrated by the fact that it comprises an estimated 25 percent of the total volume of fresh water on the Earth, as compared to 0.3 percent for lakes and 0.03 percent for streams. The U.S. Geological Survey estimates that groundwater accounts for 38 percent of the total amount of water withdrawn for public supply purposes in the nation. Wells must go below the water table in order for groundwater to be extracted; declining water levels resulting from overpumping or extended drought present immediate water supply problems. The decline of the water table in the Ogallala Aquifer in Nebraska, Kansas, Oklahoma, and Texas and the potential demise of farming in this region is but one instance, albeit a large regional one, of the impact of water-table fluctuations.

Robert M. Hordon

CROSS-REFERENCES

Aquifers, 2005; Artificial Recharge, 2011; Dams and Flood Control, 2016; Drought, 1916; Environmental Health, 1759; Floods, 2022; Freshwater

BIBLIOGRAPHY

Dingman, S. Lawrence. *Physical Hydrology*. Upper Saddle River, N.J.: Prentice Hall, 1994. This textbook is aimed at the upper-level undergraduate to graduate level for students in Earth sciences and natural resources. Some physics and calculus background is needed. The water table is discussed in several chapters.

Fetter, C. W. *Applied Hydrogeology*. 3d ed. Englewood Cliffs, N.J.: Prentice Hall, 1994. A standard textbook for advanced undergraduate or graduate courses. There are numerous case studies in the book, many of which involve the application of mathematics to the particular problem discussed. This is a lengthy text with many references and illustrations.

Hornberger, George M., Jeffrey P. Raffensperger, Patricia P. Wiberg, and Keith N. Eshleman. *Elements of Physical Hydrology*. Baltimore: The Johns Hopkins University Press, 1998. The book is designed for upper-level students in environmental or Earth sciences. Some familiarity with calculus and physics is needed. The orientation of the material is on the quantitative aspects of hydrology.

Jones, J. A. A. *Global Hydrology*. Essex, England: Addison Wesley Longman, 1997. A very good introduction to the science of hydrology, with a focus on environmental issues. Although the treatment of topics is technical, no particular background in mathematics is required. It is well indexed and contains many references.

Manning, John C. *Applied Principles of Hydrology*. 3d ed. Upper Saddle River, N.J.: Prentice Hall, 1997. A very readable, nonmathematical discussion of the various aspects of hydrology. The simple line drawings and photographs are useful. There are numerous case studies of problems involving water-table fluctuations.

Price, Michael. *Introducing Groundwater*. 2d ed. New York: Chapman and Hall, 1996. An excellent, very readable introduction to the study of groundwater. The book is aimed for the nonspecialist reader with technical terms and mathematical formulas held to a minimum. It is well illustrated and indexed.

Strahler, Alan H., and Arthur N. Strahler. *Modern Physical Geography*. 4th ed. New York: John Wiley, 1992. One of the best textbooks available in physical geography. It has excellent illustrations and examples. The writing is descriptive and nonmathematical. The water table is well covered in one chapter.

WATER WELLS

More than one-half of the world's population depends upon groundwater for daily water supply requirements. Virtually all groundwater is obtained from wells, more specifically termed production water wells.

PRINCIPAL TERMS

AQUIFER: any geologic material capable of yielding groundwater for domestic, public, or industrial requirements

CASING: conventionally, a tubular material, usually a metal, that is inserted into a raw borehole for the purpose of preventing the collapse of geologic material outside the borehole

CEMENT GROUTING: the injection of a sand-cement mix by means of pressure, resulting in an envelope of concrete lining between the outside of the casing and the undisturbed geologic material of the well bore

FULLY CONFINED AQUIFERS: those that lie sandwiched between impermeable formations and are therefore termed artesian aquifers

PRODUCTION WATER WELLS: human-made subsurface structures designed and constructed to make available a water supply from an aquifer

PUMP PIPE: a tube that extends downward from the wellhead to a level below the top of the water in the well and that serves as the conductor of water being pumped to the surface

RAT HOLE: that portion of a well that extends below the base of the aquifer, deliberately so constructed

SEMICONFINED AQUIFERS: those that perform as if fully confined but receive replenishment from overlying or underlying water-bearing formations of lesser permeability than the aquifer itself

UNCONFINED AQUIFERS: those that have no covering material, whereby the depth to water in the well is the water table; also known as water-table aquifers

WELLHEAD: that portion of the casing that extends upward above ground surface; its strength and stability permit the wellhead to be used as an anchor for pumping devices

WELL SCREEN: a tubular device containing openings that permit groundwater to flow into the well from the aquifer; some aquifers need no screen

OVERVIEW

A water well is constructed for the sole purpose of obtaining water from an aquifer. Other types of well structures, not intended as sources of water supply, appropriately have other names, such as observation well, monitor well, and piezometer (for pressure measurement). Some aquifers are capable of yielding groundwater in only minimum quantities per day, perhaps meeting the requirements of a single household. Other aquifers are capable of supporting a cluster of production wells: for example, ten generously spaced wells each with a safe capacity of 16,000 cubic meters per day (about 15.9 million liters, or about 4.2 million U.S. gallons per day). Such a cluster, or well field, of ten wells would supply the needs of a city of at least 200,000 residents.

The safe yield of water wells, however, is not solely a function of the capability of the aquifers that they tap. The suitability of well design and the quality of well construction are also critical. In the science of geohydrology, the excellence of a production water well, without respect to the character of its aquifer, is known as well efficiency. Some wells have an efficiency of 100 percent, and some have an efficiency of only 40 percent. A well at 40 percent efficiency would have a safe yield two and one-half times greater if both design and construction were executed as a perfect match with the geologic character of the aquifer being tapped.

AQUIFER PERFORMANCE

No perfection in well design and construction will make a poor aquifer into a good one. The performance of an aquifer is totally dependent upon its hydrogeologic setting and its physical character. An aquifer's physical character is judged scientifically by its hydraulic conductivity, thickness, and areal dimensions. For example, a water-bearing sandstone with a consistent thickness of 50 meters throughout a county-sized area might support twice the quantity that it would if its thickness were 25 meters. Other aspects of an aquifer's physical character are also involved.

How a good production well operates in its relation to an aquifer is quite straightforward. Take the case of a well that penetrates a fully confined aquifer, say, 1,000 meters deep. At the time the construction is completed and before any pumping has commenced, the well's static (nonpumping) water level is entirely determined by the natural water pressure within the aquifer. Last year's rainfall has no bearing on that confined fluid pressure. This scenario is an artesian situation. The well might actually flow without any pump; however, it is assumed that the water in this well stands steady at a depth of 30 meters. This assumption means that the fluid pressure in the aquifer must be sufficient to boost the water up the well to within 30 meters of ground surface. In this example, the water in the aquifer must have a confined fluid pressure of nearly 100 kilograms per square centimeter.

At the moment pumping commences, whatever the rate of withdrawal, the well enters a nonequilibrium condition, whereas, prior to pumping, the condition was under equilibrium (also called steady state). The pumping water level starts to decline, rather quickly at first and then ever more slowly. At the 1,000-meter depth, groundwater in the aquifer immediately reacts to loss of pressure and begins to migrate within the aquifer toward the well's intake area. That new water replaces the water that has been withdrawn at ground surface. The groundwater migration toward the well is radially inward, like spokes on a wheel. The pumping water level in the well continues to fall so long as nonequilibrium conditions prevail, perhaps for hours or even weeks. Eventually, the pumping water level stabilizes, and at that point in time hydraulic equilibrium has been reached. The moment that pumping of this hypothetical well is stopped, nonequilibrium conditions again prevail. Now, however, the situation reverses, and the well's water level rises, rapidly at first and then ever more slowly, until the well has fully recovered to its original state. Hydraulic equilibrium has then been reached again.

The scenario is very different in the case of wells that tap unconfined aquifers, such as alluvial sands of river valleys or in deep basins of the American West. In unconfined aquifers, the loss of water level in a well upon start-up of pumping is usually at a slower pace. So, too, is the rate of recovery when pumping ceases. In unconfined aquifers, withdrawals of groundwater result in a draining of a cone of the aquifer itself; that is, the water table actually declines in a cone-shaped area around the production well. It is this draining that causes slower responses because the act of dewatering requires time. Confined aquifers, on the other hand, react to pumping with pressure changes only. Such aquifers remain full of water continuously.

WELL DRILLING

The initial major task in well construction is the drilling operation. The drilling plan often depends upon the area's geological situation. For example, where there is no prior record of drilling in an area, there may be no known aquifer at any depth. In that case, it would be folly to drill a full-diameter borehole initially. A slim hole, perhaps 10 centimeters in diameter, is sufficient to make a determination of the geology and of the presence or absence of an aquifer. The lower-cost slim hole, then, can be reamed to full well diameter once an aquifer has been identified, sometimes even evaluated as to hydrologic performance. If the initial drilling reveals no aquifer, the site is wisely abandoned with manageable loss from the cost of slim-hole drilling. This case is analogous to a wildcat oil test drilled far from any existing production.

Say that a specific aquifer is either known or has been discovered by means of the slim-hole drilling. The diameter of the drill hole is then largely determined by the anticipated pump size, because the hole must accommodate the casing and the casing must accommodate the pump of whatever variety. High-capacity wells, especially where the pumping water level is deep, require large-diameter pump turbines. Domestic wells for household water need pumps of only minimum diameters. Such wells can be designed and constructed with casings of only 15 centimeters or less, whereas wells for large industrial or public water supplies may need to have inside diameters of 30 or even 40 centimeters for high-performance pumps.

It is ordinarily unwise simply to increase a well's diameter in the expectation of greatly increasing the maximum yield of the well. Assume a well 10 centimeters in diameter is tested and found to have a safe capacity of 2 liters per second. The water supply requirement, however, may be 4 liters per second. For that same well to have a safe capacity of 4 liters per second, solely by enlarging the diameter, the well would need a size of about 1 meter. Therefore, it is more economic to have multiples of wells of modest diameter than to attempt to achieve the same total water supply with one huge well. The multiple-well plan is also advantageous for geohydrologic reasons, so long as the wells have spacing between them. Optimum spacing is a calculation performed professionally by geohydrologists.

In some geologic situations, production water wells can be designed and constructed with neither casing nor well screen in the wells' lower portion. For this less costly design to be feasible, the rock left with open drill holes must be physically competent, such as limestone or granite. In all situations, however, the upper portion of the borehole requires a casing, primarily to protect the well against surface contaminants and partly to provide for a stable wellhead. Surface casing needs positive sealing between the outside of the casing and the drilled borehole, as by cement grout.

In geologic situations where the borehole penetrates any soft sediment or rock lacking the strength to sustain itself with open-hole construction, a full-length casing is essential. A well screen is set in that portion of the borehole that penetrates the aquifer. Commercial well screens are fabricated in a variety of designs and metals—for example, bronze and stainless steel. Below a well screen is a short section of blank casing known as the rat hole, which is set in the geologic material under the aquifer. The rat hole serves as a collecting zone for fine-grained material that may enter the well during its lifetime.

Well screens are not filters and cannot serve as filters. Such a function would cause the well to cease its normal yield of water after a few weeks of service, and the well would require expensive rehabilitation at frequent intervals. Screens are for the sole purpose of giving the well physical integrity at aquifer depth, preventing weak aquifer material, such as sand, from collapsing into the well bore. The too-common practice of using casing with sawed slots or bored holes invariably results in poor well efficiency and usually dooms the well to a short life.

WELL DEVELOPMENT

The final stage of construction for most wells with screens is known as well development. Its purpose is to enhance the well's safe yield by increasing the permeability of the aquifer immediately outside the well. Fine-grained aquifer material is stimulated to enter and pass through the screen, thereupon to be removed from the well. The reasoning behind the maneuver is easily understood. During the time a pump is extracting water from the well, groundwater in the aquifer is moving radially inward to replace the water that is being withdrawn. As the specific quantity of groundwater moves closer and closer to the intake area of the well, the space it occupies within the aquifer becomes ever more confined. Therefore, the water's velocity must increase. (An analogy would be water flowing in a pipe of fixed diameter. Assume the pipe has a dent in it that reduces its diameter. When the water passes into the restricted place, its velocity increases until the restriction has been passed.) The removal of silt and fine sand from the aquifer just outside the well has the effect of locally increasing the permeability of the aquifer, which, in turn, helps allow the increased flow velocity. Performance of wells can sometimes be improved 100 percent by the standard practice of applying well-development techniques, used by all good well contractors where the geologic conditions favor that treatment.

Because wells that flow from natural pressure are generally rare, most wells are useless until fitted with an appropriate pump. Where the standing water level in a well is no deeper than about 7 meters, the well can be pumped by suction for supplying small domestic needs. Suction pumps remove a portion of the prevailing atmospheric pressure within the pump pipe, thus causing the water in the pump pipe to rise to ground surface.

They are sometimes referred to as shallow-well pumps, even though the depth of the well has nothing to do with the feasibility or infeasibility of using suction pumps.

By far, most well pumps are of the type where the suction principle cannot be applied. These are crudely referred to as deep-well pumps, also a misnomer. Pumps of a dozen varieties and a score of ratings as to pumping quantities of water are on the market. A few of the most common varieties need mention, including the shaft turbine, submersible, and deep-well cylinder pumps. The shaft turbine pump is commonly used in high-capacity wells for industrial and public water supplies. The energy used to run it is usually supplied by electricity or diesel fuel. A rotating stainless steel shaft extends down the pump pipe to the turbine bowls set at some carefully predetermined depth below the well's pumping water level. These pumps are capable of lifting water 100 meters or more.

Submersible pumps are in worldwide use for pumping both small domestic wells and high-capacity industrial and public water supply wells. They are always powered by electricity. The submersible pump operates essentially like a shaft turbine, except that there is no shaft. The electric motor drive is arranged in vertical orientation under the pump bowls and, therefore, submersed in the well water. All turbine pumps are sensitive to any fine sand or silt that some wells tend to produce along with the water. Turbine blades have close tolerances and suffer loss of efficiency when worn by grit. Deep-well cylinder pumps are widely used for small to modest pumping needs, often powered by wind energy, especially for livestock water supplies in remote areas. These pumps are hung deep in the well, suspended by the pump pipe. A stiff rod extends from ground surface inside the pump pipe and performs by reciprocating action. The cylinder contains a valved piston that is open on the downstroke and closed on the upstroke. Operation is virtually the same as many oil well pumps, driven by a working-head jack and run with an electric motor.

John W. Foster

CROSS-REFERENCES

Aquifers, 2005; Artificial Recharge, 2011; Dams and Flood Control, 2016; Drought, 1916; Earth Science and the Environment, 1746; Floods, 2022; Groundwater Movement, 2030; Groundwater Pollution and Remediation, 2037; Hydrologic Cycle, 2045; Landfills, 1774; Nuclear Waste Disposal, 1791; Precipitation, 2050; Salinity and Desalination, 2055; Saltwater Intrusion, 2061; Surface Water, 2066; Water Quality, 2072; Water Table, 2078; Waterfalls, 2087; Watersheds, 2093; Well Logging, 1733.

BIBLIOGRAPHY

Ahearns, T. P. "Basic Considerations of Well Design." *Water Well Journal*, April-June, 1970. This series of articles offers a comprehensible explanation of how well design must be adapted to the prevailing geologic conditions of the well site. Covers the making of decisions in regard to well diameters, primarily controlled by the well's intended use, the owner's volumetric water requirements, and the capability of the aquifer.

Brassington, Rick. *Finding Water: A Guide to the Construction and Maintenance of Private Water Supplies*. 2d ed. New York: John Wiley, 1995. This college-level text can serve as a handbook about well construction and upkeep for the layperson. Brassington explains the testing and planning that must be done in preparation for the construction of a private well. Illustrations, maps, index, and bibliography.

Calhoun, Donald E., III. "Water Well Design Considerations." *Water Well Journal* 33, no. 3 (1989). This article provides useful suggestions about optimum sizes of well casings, the selection of casing materials, and well screens. Explanation is provided for the avoidance of wells that produce silt-sand particles, which shorten the life of pumps and cause other problems for the owner. Logical argument is set forth for elimination of an old technique used by some well contractors: the cutting of slots in casing as a substitute for the commercial well screen.

Detay, Michel. *Water Wells: Implementation, Maintenance, and Restoration*. Translated by M. N.

S. Carpenter. New York: Wiley, 1997. This is a detailed look into the fields of hydroengineering and water-resource management. Detay examines water supplies and reserves, as well as the management policies surrounding these resources. Illustrations, index, and bibliographical references.

Erickson, Claud R. "Submersible Water Well Pumps." *Journal of the American Water Works Association* 52 (September, 1960): 1145-1158. This major article constitutes a quality analysis of the submersible-type pump, which has become the most popular pumping system for domestic wells and for many industrial and public water supply wells. Manufacturers originally had to overcome the suspicion that the placing of an electric motor under water in the well would be unreasonably hazardous. It has long since been shown that any problems with short circuits were readily overcome and that the submersible motor-driven turbine pump has high efficiency and very low maintenance cost.

Fetter, C. W. *Contaminant Hydrogeology.* 2d ed. Upper Saddle River, N.J.: Prentice Hall, 1999. While written primarily as a textbook on groundwater geology, those portions that address the nature and varieties of aquifers are comprehensible to the lay reader. Admirably serves the understanding of the nature of groundwater, which is the geologic resource tapped by all successful water wells. Water-well technology, however, is not included.

Kasenow, Michael. *Production Well Analysis: New Methods and Computer Programs in Well Hydraulics.* Highlands Ranch, Colo.: Water Resources Publications, LLC, 1996. This book looks at the factors involved in well construction, including groundwater flow and measurement, aquifers, and fluid dynamics. As well as illustrations, the text is accompanied by a CD-ROM. Intended for the reader with some background in the subject.

Plumb, C. E., and J. L. Welsh. *Abstracts of Laws and Recommendations Concerning Water Well Construction and Sealing in the U.S.* Water Quality Investigations Report 9. Sacramento: California Division of Water Resources, 1955. This book addresses by abstracts the all-important technology by which water wells are made safe from pollution and contamination caused by surface leakage. Also serves as an excellent reference to prior publications on the subject, many of which are entirely suitable for the lay reader.

State of California. Department of Water Resources. *Water Well Standards.* Bulletin 74-81. Sacramento: Author, 1982. A book written for the water-well industry and layperson alike, providing cautionary explanations of well designs aimed at assuring the owner of freedom from surface contamination and pollution. Recommendations on materials suitable for use in well construction, and materials that are not suitable, are emphasized. While published by the state of California, the work is generally applicable to all geographic areas.

Thomas, Harold E. *Conservation of Ground Water.* 2d ed. Westport, Conn.: Greenwood Press, 1970. This book was written as a broad-view reference for legislators and readers seeking an understanding of a complex geological engineering field. Its most useful section, specifically for comprehension of the geologic control on where groundwater is and where it is not, addresses the geographic distribution of freshwater aquifers in the United States. This section clearly shows by map reference the regions of the United States where aquifers are known and available for well construction—for example, the entire Gulf coastal plain from south Texas to Florida and the Atlantic coastal plain north to New Jersey. Also reveals in a practical way those regions where the geology is basically less encouraging, where dry holes are commonplace, for example, in the uplands of northern New England and the mountains of the American West.

WATERFALLS

Waterfalls are created by the steep descent of a stream over an escarpment or cliff. The fall may be perpendicular and free from the rock surface or may run across the rock, frequently in a series of falls. Waterfalls attract viewers for their scenic and aesthetic qualities, but they also possess scientific value as an aid in the interpretation of Earth history and economic importance as possible sites for the generation of hydroelectric power.

PRINCIPAL TERMS

CASCADE: a small waterfall or series of small falls

CATARACT: most frequently, an overwhelming flood or a great volume of flow over a cliff; sometimes, however, the term is used interchangeably with "cascade"

GRADE: a hypothetical profile of a stream seeking to achieve quasi-equilibrium

KNICKPOINT: a break in the stream profile, generally caused by a resistant rock layer that retards the rate of erosion

RAPIDS: a turbulent flow in a stream caused by obstructions in the channel or by resistant rock layers

RIFFLES: a smaller version of rapids, a turbulent flow between calm pools, found in nearly all streams for hydraulic reasons

TECTONICS (TECTONIC ACTIVITY): vertical or horizontal movements in the Earth's crust, displacing rocks, landforms, and stream gradients

CHARACTERISTICS

Waterfalls are breaks in the relatively smooth profiles of streams. The break in slope may be minor and result in only a riffle, or slight roughening of the water surface. All streams possess riffles, which are part of the hydraulic adjustment that permits streams to carry sediment. Rapids are a more profound interruption of the profile, sufficient to block the passage of most boats; rapids serve notice that the streambed has encountered resistance in the form of bedrock, boulders, or other obstructions. Some purists maintain that true waterfalls exist only where water falls free of its bed, plunging downward through the atmosphere. Except for the largest and most spectacular falls, however, free fall is generally a function of the amount of discharge in the stream channel, and discharge varies with weather and climate. Most of the world's streams have dry spells every year. During those droughty periods, a free-falling waterfall may become a cascade, a cataract, or even rapids. The Great Falls of the Potomac River dried up to very weak rapids during a late 1960's drought, but they were still an obstruction to navigation and a potential site for waterpower development, possessed scenic appeal, and presented evidence to geologists of the underlying rock

structure and its place in deciphering Earth's history. These are the major reasons why people are interested in waterfalls, whether they are 1,000 meters high or less than 10, and whether they are free-falling or are better described as cascades, cataracts, or rapids.

FORMATION

Streams, on their route to the sea, lake, or some other termination, cut downward through loose sediments and bedrock, eroding and transporting some of this material as their sedimentary load. If the Earth materials through which they are downcutting are relatively soft or weak, the stream ultimately will achieve grade, denoted by a smooth, upwardly concave profile. Waterfalls are an indication that something has prevented the stream from achieving an equilibrium between its discharge, or quantity of flow, and the sediment load it is carrying. The cause may lie in the structure of the underlying rocks, with particularly resistant layers being encountered in the course of the downcutting. Frequently, the stream cannot achieve a smooth profile because of some aberration of geologic history, such as rapid tectonic uplift or glacial deepening of a major valley. Most of Earth's relief is a consequence of tectonic uplift in the past few million years. It comes as no surprise that this is insufficient time for streams to achieve even quasi-equilibrium or grade in the highlands of the world, where complex geology often exposes rocks of varying resistance to weathering and stream downcutting. Resistant rock layers may be sandstones, as in much of the Appalachian Mountains and Colorado Plateaus. Dolomite forms the Niagara Escarpment, over which the Niagara River falls.

The Niagara story, however, is more complicated than merely the encountering of a resistant dolomite rock layer by the downcutting river. Continental glaciation of the eastern and central United States and Canada obliterated the pattern of preglacial stream drainage, and the roughly twelve thousand years that Niagara Falls has been in existence is insufficient time for the stream to achieve grade. Similarly, the head of navigation on the Mississippi River is at St. Anthony Falls in Minneapolis, Minnesota; upstream of the falls, the Mississippi flows on a new, postglacial course. The head of navigation of the Ohio River is the falls at Louisville, Kentucky, only a short distance upstream from the mouth of the Ohio, where it joins the Mississippi. This peculiar situation led geologists to search for the preglacial Ohio River, which was discovered buried by glacial sediments in northern Indiana and Ohio. Niagara Falls, St. Anthony Falls, and the falls of the Ohio at Louisville may be considered as large and dramatic knickpoints, or breaks in the smooth stream profile. Knickpoints migrate upstream as erosion proceeds on its path toward grade or quasi-equilibrium, eventually becoming smaller and smaller until they are eliminated as significant breaks in the stream profile.

Waterfalls therefore represent unusual and temporary geologic circumstances and contribute to scientific knowledge both in the quest for explanation of each individual waterfall and in the contribution to understanding of the geologic history of a region. The largest number of waterfalls worldwide undoubtedly occurs in younger, recently uplifted mountains that have been subjected to alpine glaciation. Just as major rivers have greater eroding capability than their tributaries, so larger valleys are occupied by larger glaciers capable of cutting deeper into the bedrock. The glaciers of tributary valleys are much smaller and have less erosion potential than those in the main valleys. After the glaciers melt, the streams occupying the tributary valleys are left hanging, and they descend to the main stream as waterfalls. Hanging tributary valleys also occur in unglaciated uplands, but they are far more common in the spectacular scenery of alpine mountains, especially in the world's major fjord regions: Alaska and British Columbia, Norway, Chile, and New Zealand's South Island.

SCENIC APPEAL

The scenic appeal of waterfalls is the reason most people find them of interest. The sheer grandeur of the falling water inspires artists, photographers, and writers, whose products enhance the waterfalls' fame and encourage multitudes to experience the view personally. The attraction of waterfalls is a significant factor in decisions to visit state and national parks and even roadside

Although, at about 160 feet, it is not one of the tallest waterfalls in the world, Niagara Falls, straddling the border between New York State and Canada, is famous for its breadth and scenic views. (© William E. Ferguson)

waysides. As a consequence, a sizable fraction of the expenditures of tourists can be attributed to the existence of waterfalls. Tourist counts are notoriously unreliable, but it is safe to say that millions have visited Niagara Falls, owing in part to their location near the large population centers of the United States and Canada. By contrast, Victoria Falls on the Zambezi River in southern Africa, Angel Falls in Venezuela (the world's highest at 979 meters), the Iguazu Falls on the border of Argentina and Brazil, and Guaira Falls on the upper Parana River between Brazil and Paraguay all possess characteristics that encourage many writers to proclaim they are even greater, more attractive, or more spectacular than Niagara; however, they are located too far from large centers of population to be visited easily.

There is also disagreement as to which waterfall has the greatest discharge. Unlike the height of a waterfall, which is static unless a catastrophic rockfall occurs, stream discharge is variable throughout the year, and from dry years to wet years. Nearly all waterfalls occur in places with seasonal precipitation or in environments where the winter precipitation is in the form of snow, leading to spectacular flows during the snowmelt season but disappointing conditions when the stream is chiefly ice. Niagara retains an appeal when it is frozen, but the weather outside is generally uncomfortable at that time of year, and most tourists visit during the warm season. Thousands of people annually register disappointment at the appearance of Yosemite Falls and the other falls of Yosemite National Park during the long, dry summers. This is a characteristic of waterfalls in regions of alpine glaciation; they are frozen in winter and are most spectacular during the short snowmelt season of spring and early summer.

Tropical waterfalls, such as those of Africa, South America, and India may have tremendous discharges during the rainy season but much lower flows during the dry season. One of Niagara's great advantages as a tourist attraction is that it has huge natural reservoirs upstream in the form of the Great Lakes, which ensure an even and high discharge year after year, wet season and dry. The difference between average discharge and the peak discharges of rainy seasons and wet years probably accounts for much of the uncertainty as to which waterfall is the biggest.

RATING SYSTEMS

Arguments concerning which waterfall among the giants is greatest frequently involve rating systems employing the height of the falls, its width, its discharge, and other quantifiable factors. Curiously, vista, or the opportunity to view the falls from a particular point, is often overlooked. Whereas Niagara Falls can be observed from several viewpoints, all of which overwhelm the observer with the perception of a tremendous amount of water flowing over the edge of the ledge, the wider Victoria Falls actually descends into a complicated narrow chasm or canyon; little of Victoria can actually be seen from one point on the surface. This is also true of Iguazu Falls, which is up to several kilometers wide and consists of as many as twenty cataracts. Aircraft flights are the best way to observe these giants and are, in fact,

Bridal Vail Falls, in Yosemite National Park, California, drops from a U-shaped hanging valley 620 feet to the valley floor. (© William E. Ferguson)

WORLD'S TALLEST WATERFALLS

Waterfall	River	Location	Height in Meters (Feet)
Angel Falls	Tributary of Caroni River	Venezuela	977 m (3,212 ft)
Tugela Falls	Tugela River	Natal, South Africa	948 m (3110 ft)
Cuquenan Falls	Cuquenan River	Venezuela	610 m (2000 ft)
Sutherland Falls	Arthur River	South Island, New Zealand	580 m (1904 ft)
Maradalsfossen	Eikesdalsvatnet	Mardola District, Norway	517 m (1696 ft)
Ribbon Falls	Ribbon Creek	California (Yosemite)	491 m (1612 ft)
Della Falls	Drinkwater Creek	British Columbia, Canada	440 m (1445 ft)
Upper Yosemite Falls	Tributary of Merced River	California (Yosemite)	436 m (1430 ft)
Grande Cascade	Gave de Pau River	Southwest France	425 m (1400 ft)
Krimmler Waterfall	Krimmler River	Salzburg, Austria	380 m (1247 ft)
Takakkaw Falls	Yoho River	British Columbia	373 m (1223 ft)
Silver Strand Falls	Tributary of Merced River	California (Yosemite)	357 m (1170 ft)
Basaseachic Falls	Piedra Volada Creek	Mexico	311 m (1020 ft)
Staubbach Falls	Staubbach Falls	Switzerland	300 m (984 ft)
Yosemite Falls	Yosemite Creek	California	277 m (909 ft)
Vettisfossen	Morkedola River	Jotunheim Mountains, Norway	275 m (902 ft)
King George VI Falls	Courantyne River	Guyana	259 m (850 ft)
Jog Falls	Sharavati River	Karnataka, India	253 m (830 ft)
Skykje Falls	Hardanger Fjord	Norway	250 m (820 ft)
Kalambo Falls	Kalambo River	Zambia/Tanzania	215 m (704 ft)
Fairy Falls	Stevens Creek	Washington, USA	213 m (700 ft)
Trummelbach Falls	Trummelbach River	Switzerland	213 m (700 ft)
Aniene Falls	Tiber River	Italy	207 m (680 ft)
Marmore Falls	Tibutary of Nera River	Italy	198 m (650 ft)
Feather Falls	Fall River	California	195 m (640 ft)
Maletsunyane Falls	Maletsunyane River	Lesotho	192 m (630 ft)
Bridal Vail Falls	Yosemite Creek	California (Yosemite)	189 m (620 ft)
Multnomah Falls	Multnomah Creek	Oregon	189 m (620 ft)
Nevada Falls	Merced River	California (Yosemite)	181 m (594 ft)
Voringfossen	Bjoreia River	Hardanger Fjord, Norway	170 m (558 ft)

SOURCE: Data compiled by Paul Brainard 2000 for http://web3.foxinternet.net/xplatypusx/cause.html.

the only way to experience more than a small fraction of the falls.

In a hypothetical rating system, should the amount of time a waterfall is in full flow each year be a factor? Should availability to the general public be a factor? Egalitarian considerations would force a vote in Yosemite's favor, as opposed to Angel Falls in Venezuela or several other competitors in remote regions. In addition, aesthetic considerations come into play. Which waterfalls are the most beautiful, or inspiring, or dramatic? These and other considerations can be important if a governmental decision must be made regarding the protection of a particular waterfall. Such scenarios confront many water projects, which must

procure a favorable ratio of benefits to costs in order to survive. The allocation of limited financial resources increasingly involves such decisions. If waterfalls were forced to pay for themselves in terms of tourist revenues in order to be protected from destruction, few waterfalls other than Niagara and Yosemite would likely survive. However, the increasingly popular sport of canyoning (also called canyoneering), in which people rappel down waterfalls, is a potential source of revenue for many areas with waterfalls.

HYDROELECTRIC POWER

The height of waterfalls creates another area of interest in this phenomenon: the potential for hydroelectric power generation. "Head" is the term used to describe the difference in elevation between the water level at the top of a fall and the bottom. The higher the head, the greater the potential energy of position, which can be converted into electrical energy as the water descends. It must be emphasized that the waterfall itself is not harnessed. Water flows through tubes to the bottom elevation of the falls, where turbines generate electricity. Niagara Falls was for years the world's largest single hydroelectric generating facility, and it is still among the giants. Most of the waterfalls of the tropics have a potential for power generation, but this has not yet been developed extensively. Europe, Japan, the United States, and Canada have the most fully developed hydroelectrical generating capacities. Regions of alpine mountain glaciation have particularly high potential. The economic benefits of this aspect of waterfalls can be enormously significant locally, and frequently nationwide, although highly industrialized nations with significant numbers of automobiles demand far more energy than waterpower can generate. Most hydroelectric power generation today is actually not from natural waterfall sites but from what can be called "artificial waterfalls": the creation of head by building a high dam in a narrow valley or gorge.

At the beginning of the Industrial Revolution, before the development of electricity and its distribution through transmission lines, water-power sites furnished the energy for countless small factories and mills. Many of these were the sites of small waterfalls and rapids, such as along the Fall Line of the Piedmont of the southeastern United States. The Great Falls of the Potomac are an example of such a site. The development of electrical transmission lines freed industry from these locations near falls and rapids and allowed an expansion throughout the countryside. The smaller water-power sites fell into disuse, and only a few have been preserved for historic reasons. The larger sites, such as Niagara Falls, remain an important part of the economy.

Neil E. Salisbury

CROSS-REFERENCES

BIBLIOGRAPHY

Benn, Douglas I., and David J. A. Evans. *Glaciers and Glaciation*. New York: Wiley, 1998. A comprehensive account of the formation and evolution of glaciers. This easily understandable book describes the natural processes of glaciation and the creation of waterfalls. Suitable for the high school reader and above. Color illustrations, index, and a large bibliography.

Berton, Pierre. *Niagara: A History of the Falls*. New York: Kodansha International, 1997. Berton provides the reader with a complete and entertaining history of Niagara Falls. The book is filled with illustrations and maps that complement the text and make it easily understood and enjoyed by readers of all levels. Index and bibliography.

Easterbrook, Donald J. *Surface Processes and Landforms*. New York: Macmillan, 1993. A not-overly-technical discussion of how waterfalls and rapids relate to attempts by streams to achieve equilibrium and a graded profile. Suitable for college-level readers.

Fairbridge, Rhodes W., and Reginald W. Herschy, eds. *The Encyclopedia of Hydrology and Water Resources*. Boston: Kluwer Academic, 1998. The entry on waterfalls provides a succinct but illuminating classification of waterfalls and a brief description of rates of recession of falls, centering on Niagara with a comparison to the Nile. Suitable for college-level readers.

Forrester, Glenn C. *Niagara Falls and the Glacier*. Hicksville, N.Y.: Exposition Press, 1976. A highly readable account of the geological context of Niagara Falls, together with a discussion of the role of glaciation in the origin and development of the falls. Includes considerable information on hydropower developments and measures taken to preserve the falls. Well illustrated, with photographs and diagrams. Suitable for high school readers.

Snead, Rodman E. *World Atlas of Geomorphic Features.* Huntington, N.Y.: Robert E. Krieger, 1980. Includes a readable discussion of the nature of major waterfalls of the world, as well as a listing of more than seventy major worldwide waterfalls and more than thirty major U.S. waterfalls. Accompanying maps show the general location of each falls and, in addition, regions where smaller waterfalls are common. Suitable for high school readers.

World Almanac. *The World Almanac and Book of Facts, 1997.* Mahwah, N.J.: World Almanac Books, 1997. Contains a succinct listing of the world's most famous waterfalls, including their heights and certain other characteristics.

WATERSHEDS

A watershed, or drainage basin, is a region that is drained by a stream, lake, or other type of watercourse. The Earth is divided into millions of watersheds of varying sizes and shapes, all of which act as collectors of runoff that flow from higher to lower elevations. Most watersheds join other larger watersheds and eventually flow into the oceans.

PRINCIPAL TERMS

BASE FLOW: that portion of stream flow that is derived from groundwater

DISCHARGE: the volume of water per unit time that flows past a given point on a stream

DRAINAGE DIVIDE: the ridge of land that marks the boundary between adjacent watersheds

INTERIOR DRAINAGE: watersheds in arid areas where the runoff does not flow into the oceans

PERENNIAL STREAM: a stream that has water flowing in it throughout the year

RELIEF: the difference in elevation between the highest and lowest points of land in a particular region

RUNOFF: that portion of precipitation in a watershed that appears in surface streams

STREAM: water flowing in a narrow but clearly defined channel from higher to lower elevations under the influence of gravity

CHARACTERISTICS

The term "watershed" has several meanings. The term is derived from the German term *Wasserscheide*, which means "water parting," or the line or ridge of higher ground that separates two adjoining drainage basins. This definition is also used in Great Britain, where "watershed" refers to a drainage divide between adjacent drainage basins. The usage of the term in the United States and by several international agencies has been modified to refer to the land area that is drained by, or that contributes water to, a stream, lake, or other water body. For this discussion, the term will refer to the region that serves as the collecting system for all the water that is moving downslope from higher to lower elevations on its eventual path to the ocean. In this physical context, which is governed by topography, the terms "watershed" and "drainage basin" are synonymous.

The relationship of watersheds to topography was recognized years ago. For example, Philippe Buache in 1752 presented a memoir to the French Academy of Sciences in which the concept of the topographic unity of a watershed was outlined. This concept was followed by European cartographers of the late eighteenth and early nineteenth centuries who prepared maps that showed the major drainage basins of each country. Although these early cartographers would often exaggerate the height of the divides between watersheds, the basic concept was to show how the land was divided into a variety of drainage basins that acted as efficient collection systems for runoff that resulted from precipitation.

Watersheds transport water from upland areas to lower elevations in a variety of pathways. The most obvious path is by perennial streams that flow in channels. This form of surface runoff includes not only overland flow, which is the water moving over the ground surface, but also base flow, which comes from that portion of the precipitation that has infiltrated through the soil into the underlying groundwater that enters the stream at some downgradient point. Thus, surface runoff from a watershed is a mix of "stormflow" or "quick flow," which occurs right after a precipitation event, and base flow from groundwater, which takes a much longer time to join the surface water. The rates at which surface water and groundwater move through a watershed depend upon many factors, such as precipitation amounts and intensity, geology, soils, topography, and vegetation.

STREAM CHANNELS

Stream channels vary enormously in length and width, from a channel that can easily be jumped across to a river such as the Mississippi, which is as wide as 1.5 kilometers before it empties into the Gulf of Mexico. Watersheds also have an enormous range in length, area, and discharge. The largest watershed in the world by far is the Amazon, with a drainage area of 6,150,000 square kilometers, approximately one-third of the entire area of South America. The second- and third-largest watersheds in the world are the Zaire in Africa (3,820,000 square kilometers) and the Mississippi in the United States (3,270,000 square kilometers). Thus, the Mississippi River watershed, which includes the Missouri River, drains an astonishing 40.5 percent of the entire area of the conterminous United States.

The Amazon is also the largest river in the world by far in terms of discharge, averaging 6,300 cubic kilometers per year. The second- and third-largest dischargers in the world are the Zaire and the Orinoco in Venezuela (1,250 and 1,100 cubic kilometers per year, respectively). The longest rivers in the world are the Nile in Africa (6,671 kilometers), the Amazon (6,300 kilometers), the Yangtze in China (6,276 kilometers), and the Mississippi (6,019 kilometers). At the other end of the spectrum are innumerable small streams that are in the headwaters of their watersheds near the divides with lengths of only a few meters.

Although there is obvious variation in shape from watershed to watershed, most tend to be pear-shaped. This shape is the most probable one to occur, as ground slopes and branching stream networks become adjusted to dispose of the runoff and the sediment load in the water efficiently. Departures from the usual pear shape are attributed to structural control by bedrock formations. For example, some basins are elongated in shape when they occupy long, narrow valleys; such valleys are often found in regions such as the Appalachians in the eastern United States, where long, resistant ridges of sandstone and quartzite run approximately parallel with less resistant valleys underlain by shale and limestone.

Most of the runoff in the humid land areas of the world eventually flows into the oceans via a series of hydrologically connected watersheds. Thus the waters and sediment load of the Missouri and Ohio Rivers join the Mississippi River at St. Louis, Missouri, and Cairo, Illinois, respectively, and eventually flow into the Gulf of Mexico below New

Orleans. Another large-scale example is that of the watersheds for the Great Lakes, which furnish the water for the St. Lawrence River, which flows into the Gulf of St. Lawrence and the Atlantic Ocean. However, there are areas of the world where runoff flows into interior basins surrounded by high mountains that do not allow the stream to get to the ocean. This type of drainage system, called interior drainage, is common in semiarid and arid climates. Major examples of watersheds with interior drainage include the Caspian Sea in Asia (3,626,000 square kilometers), the Aral Sea in Kazakhstan and Uzbekistan in Asia (1,618,750 square kilometers), Lake Eyre in Australia (1,424,500 square kilometers), and the Great Basin in Utah, Nevada, and eastern California (500,000 square kilometers).

WATERSHED DIVIDES

The divides that separate watersheds vary from sharply defined ridges in mountainous terrain to poorly defined boundaries in glaciated landscapes, regions of low relief, and areas of limited topographic expression. For example, the highest land in the Everglades (12,950 square kilometers) in Florida is only 2.1 meters above sea level, which means that natural runoff (excluding canals) flows in directions sometimes governed more by wind than by topography. Another prominent instance of a poorly defined divide occurs in southern Wyoming, where the Continental Divide in the United States, which separates the waters that flow into the Pacific Ocean from those that flow into the Gulf of Mexico, splits into two divides that surround the Great Basin Divide. This unusual situation means that anyone who drives along Interstate 80 in Wyoming, for example, will be able to cross the Continental Divide twice in an east-west direction.

Watershed divides, especially in mountainous areas, often have been used as political boundaries. Examples include the Andes between Argentina and Chile, the Pyrenees between France and Spain, and the Bitterroot Range between Idaho and Montana. Watershed divides also often serve as starting points for major cities. For example, Atlanta developed as a rail center in the nineteenth century because it was on the divide between the streams that flowed into the Gulf of Mexico (the Chattahoochee and Flint Rivers) and those that flowed into the Atlantic Ocean along the east coast of Georgia (the Ocmulgee River).

The drainage pattern or network of stream channels that develop within a watershed are related to local geologic and geomorphic factors. The most common drainage pattern that develops on horizontal and homogeneous bedrock or on crystalline rock that offers uniform resistance to erosion is called "dendritic," since it resembles the branching pattern of trees. All other types of drainage patterns reflect some form of structural control, such as the trellis pattern that is associated with the

A large watershed of snowmelt in the Sierra Nevada that enters the Merced River. (© William E. Ferguson)

elongated watersheds in the "ridge and valley" regions in the Appalachians of the eastern United States. Rectangular patterns can develop in faulted areas where the drainage paths follow the lines of least resistance that develop along the fault lines.

Watershed size and flow can change either naturally or by anthropogenic means. Major natural examples include the deflection by continental glaciation of the upper Missouri River from Hudson Bay in eastern Canada to its present-day confluence with the Mississippi River, and the geologic subsidence and tilting that diverted the drainage of the Nyanza area in East Africa from the Zaire River (formerly the Congo), which flows into the Atlantic Ocean to Lake Victoria, which drains into the Nile and the Mediterranean Sea. The flow in the Florida Everglades has been substantially altered by drainage activities and canal building for agricultural purposes that started in the late nineteenth century and continued into the twentieth century. Water that used to flow into the Everglades and Florida Bay from Lake Okeechobee was diverted to the canalized Caloosahatchee River, which empties into the Gulf of Mexico, and the Miami, North New River, Hillsboro, West Palm Beach, and St. Lucie Canals,

which are connected with the Atlantic Ocean on the east coast of Florida. Another instance of anthropogenic intervention with watershed flow is illustrated by the diversion of water from Lake Michigan, which is part of the Great Lakes and St. Lawrence River system, to the Chicago Ship and Sanitary Canal, which is connected with the Illinois River, which flows into the Mississippi River. The purpose of the canal was to transport sewage from the Chicago metropolitan area away from Lake Michigan, which is used as a water source.

Robert M. Hordon

CROSS-REFERENCES

Alpine Glaciers, 865; Aquifers, 2005; Artificial Recharge, 2011; Cascades, 830; Continental Glaciers, 875; Dams and Flood Control, 2016; Floods, 2022; Great Lakes, 2270; Groundwater Movement, 2030; Groundwater Pollution and Remediation, 2037; Hydroelectric Power, 1657; Hydrologic Cycle, 2045; Mississippi River, 2280; Nile River, 2285; Precipitation, 2050; River Bed Forms, 2353; River Valleys, 943; Salinity and Desalination, 2055; Saltwater Intrusion, 2061; Surface Water, 2066; Water Quality, 2072; Water Table, 2078; Water Wells, 2082; Waterfalls, 2087.

BIBLIOGRAPHY

Albert, R. C. *Damming the Delaware*. University Park: Pennsylvania State University Press, 1987. An interesting history of two hundred years of water management for the Delaware River, which serves as a water source for New York City, Trenton, and Philadelphia. This is a good case study of the institutional factors that have resulted in a model interstate compact and the historical absence of dams on the main stem of the river.

Black, P. E. *Watershed Hydrology*. 2d ed. Chelsea, Mich.: Ann Arbor Press, 1996. A fine introductory book that presents in an integrated fashion the various hydrologic processes that occur in a watershed. The complex of interactions between climate, soils, vegetation, atmosphere, streamflow and channels, humans, and water-resource infrastructure are thoroughly discussed. The treatment is qualitative, with only minimal uses of quantitative methodology.

Dunne, T., and L. B. Leopold. *Water in Environmental Planning*. New York: W. H. Freeman, 1978. A comprehensive book about watershed hydrology (818 pages). In addition to the useful chapters on hydrology, the book has an extensive section (220 pages) on drainage-basin analysis, hillslope processes, stream channels and changes, and sediment production and transport. Many of the chapters conclude with solved sample problems involving basic hydrologic equations and methodology.

Fairbridge, Rhodes W., and Reginald W. Herschy, eds. *The Encyclopedia of Hydrology and Water Resources*. Boston: Kluwer Academic, 1998. This encyclopedia serves as a great reference tool for anyone interested in the study of hydrology. There are sections devoted to surface water and groundwater, as well as water resources. Suitable for college-level readers. Illustrations and maps.

Hillel, D. *Rivers of Eden*. New York: Oxford University Press, 1994. An engrossing book about water shortages in the Middle East, set against the backdrop of competing territorial claims in a highly politicized and arid environment. The major watersheds selected for study include the Tigris and Euphrates, with headwaters in Turkey and downstream areas in Iraq and Syria; the Nile, with a source region in east-central Africa; and the Jordan River in Israel, Jordan, Lebanon, and Syria. This is a well-written and well-documented study of historical watershed mismanagement and the compelling need for cooperation and water-sharing arrangements if peace is ever to come to the region.

Leeden, F. V. D., F. L. Troise, and D. K. Todd, eds. *The Water Encyclopedia*. 2d ed. Chelsea, Mich.: Lewis, 1990. A huge compendium of information about virtually all aspects of water. Includes hundreds of tables and many maps covering surface and groundwater, water use and quality, environmental problems, water resources management, and water law and treaties. An excellent reference work that contains a bewildering array of facts about water.

McDonald, A. T., and D. Kay. *Water Resources: Issues and Strategies*. New York: John Wiley & Sons, 1988. Chapter 8, "River Basin Management," provides short but useful examples of the Tennessee Valley Authority, the Senegal and Volta River basins in West Africa, and regional water authorities in England and Wales. The institutional factors pertaining to watershed management are stressed.

Marsh, W. M. *Landscape Planning: Environmental Applications*. 2d ed. New York: John Wiley & Sons, 1991. An interesting blend of the principles and processes in physical geography, planning, and landscape architecture as they pertain to environmental issues in landscape planning. Contains some very useful chapters on watersheds, land use, stormwater discharge and landscape change, soil erosion and stream sedimentation, and soils and development suitability.

Newsom, M. *Hydrology and the River Environment*. Oxford, England: Clarendon Press, 1994. Focuses on watersheds as the fundamental unit for water management. Although many of the examples discussed are from Great Britain, there is an inherent universality to the watershed analysis methodology. A very good set of pertinent references for each chapter is included at the end of the book.

Strahler, Arthur N., and Alan H. Strahler. *Modern Physical Geography.* 4th ed. New York: John Wiley & Sons, 1992. One of the best college-level texts in the field. Provides an excellent introduction to the geologic and hydrologic processes that govern watershed development and change. Notable for its numerous and exceptionally lucid diagrams.

Thompson, Stephen A. *Hydrology for Water Management.* Rotterdam, Netherlands: A. A. Balkema, 1999. This thorough account of hydrology, groundwater flow, and stream flow focuses on the management of water supplies in efforts to keep them free of pollutants and available to the largest number of people possible. Illustrations, maps, index, and bibliography.

4
OCEANOGRAPHY

CARBONATE COMPENSATION DEPTHS

Carbonate compensation depths are the levels within ocean basins that separate calcium carbonate-containing seafloor sediments from carbonate-free sediments. The carbonate compensation depth (CCD) may be different in different ocean basins, and it may rise or fall at different times in the same ocean basin as a result of the balance between surface production of carbonate and deep-water carbonate dissolution in the basin. Carbonate solubility increases with increasing pressure and decreasing temperature.

PRINCIPAL TERMS

DEPOSITION: the process by which loose sediment grains fall out of seawater to accumulate as layers of sediment on the seafloor

PALEODEPTH: an estimate of the water depth at which ancient seafloor sediments were originally deposited

PLANKTON: microscopic marine plants and animals that live in the surface waters of the oceans; these floating organisms precipitate the particles that sink to form biogenic marine sediments

PRECIPITATION: the formation of solid mineral crystals from chemicals dissolved in water

PRODUCTIVITY: the rate at which plankton reproduces in surface waters, which in turn controls the rate of precipitation of calcareous or siliceous shells or tests by these organisms

RED CLAYS: fine-grained, carbonate-free sediments that accumulate at depths below the CCD in all ocean basins; their red color is caused by the presence of oxidized fine-grained iron particles

TEST: an internal skeleton or shell precipitated by a one-celled planktonic plant or animal

MARINE SEDIMENTS

Marine sediments are composed of a variety of materials of biological or terrestrial origin. Individual sediment grains in oceanic deposits may be either clastic or biogenic particles. Clastic sediments are materials derived from the weathering, erosion, and transportation of exposed continental rocks, and these grains are classified by particle diameter into gravels, sands, silts, and clays. Clastic sediment particles become less common with increasing distance from the continental landmasses and are nearly absent from deep-water sediments on the abyssal plains. Biogenic sediment particles are composed of skeletons and tests precipitated by planktonic plants and animals living in the shallowest waters of the ocean. Biogenic particles composed of calcium carbonate or of opaline silica make up the majority of oceanic sediments, which are deposited at great distances from land. Deposition of biogenic sediments is controlled by two factors: the biological productivity of surface waters and the dissolution of biogenic particles by corrosive bottom waters. Increased levels of atmo-

spheric carbon dioxide lead to an increase in the corrosiveness of bottom waters.

Biogenic sediment particles are produced in shallow, well-lighted surface waters as a result of the biochemical activity of microscopic plants and animals, which precipitate solid shells and tests from minerals dissolved in seawater. Biological productivity is a measure of the number of the organisms present and their rate of reproduction. The productivity of planktonic organisms is directly related to chemical and physical conditions in the surface waters. High-productivity waters have abundant supplies of oxygen, with dissolved nutrients and chemicals needed for precipitation of shells and tests and enough light for photosynthesis by plants. Generally, high productivity is found in warm-water areas with abundant dissolved oxygen and nutrients.

When planktonic plants and animals die, their shells fall through the water column to be deposited on the bottom of the ocean. This "planktonic rain" causes deposition of seafloor sediments by the sinking of biogenic particles produced in the surface waters, and it moves chemicals from surface waters to deep waters. The higher the productivity values in surface waters, the greater will be the supply of biogenic sediment particles to the seafloor. As biogenic particles sink through the water column to the seafloor, they may be dissolved by corrosive seawater. Surface waters are saturated with dissolved calcium carbonate, so most dissolution takes place in deep waters, which are undersaturated with carbonate.

Calcareous sediments are produced by the accumulation of biogenic particles of calcium carbonate that survive the fall through the water and are deposited and buried on the seafloor. Accumulation rates of biogenic sediments are controlled by the balance between surface productivity and deep-water dissolution: The higher the biological productivity in the surface, the greater the number of shells that will sink to the ocean bottom. In certain high-productivity areas associated with upwelling of cold, nutrient-loaded water masses to the surface, seafloor sediment accumulation rates may be as high as 3 to 5 centimeters per thousand years. In areas with higher dissolution rates, fewer calcareous particles will survive to be deposited on the ocean floor. In these low-productivity areas, all the carbonate produced at the surface may be dissolved, and sediment accumulation rates may be as low as 1 millimeter per million years.

LEVEL OF CARBONATE COMPENSATION DEPTH

The carbonate compensation depth (CCD) demarcates the boundary between carbonate-rich sediments (calcareous oozes and chalks) and carbonate-free sediments (red clays) in the oceans; it is the result of deep-water dissolution rates exceeding the rate of supply of calcium carbonate to the deep sea by surface productivity. The depth of the CCD marks the point at which the supply of carbonate sinking in the planktonic rain from surface waters is exactly balanced by the rate of removal of carbonate dissolution in deep waters. Calcareous sediments will be deposited on the seafloor in water depths shallower than the CCD, because in these areas, the rate of supply of calcium carbonate is higher than the dissolution rate. In these areas, individual particles of calcium carbonate will survive the trip through the water column and are deposited as biogenic sediments on the seafloor.

Below the compensation depth, dissolution exceeds the rate of supply, so all carbonate particles supplied from the surface are dissolved, and seafloor sediments are carbonate-free. Surface sediments in water depths below the CCD tend to be red clays, or combinations of fine-grained materials derived from continental sources and carried to the deep sea by wind, mixed with micrometeorites and other particles from extraterrestrial sources. Red clays generally lack fossils, as a result of complete dissolution of carbonate and opaline silica, so only a few solution-resistant fossils, such as phosphatic fish teeth and whale ear bones, are found in these sediments. The reddish-brown color of deep-sea clay deposits is a result of the presence of iron particles, which have reacted with oxygen in seawater to form the rust-brown color.

Much information on the past history of the compensation depth in different ocean basins has been provided by ocean-floor drilling programs. The level of the CCD has changed dramatically throughout geologic history, with fluctuations of up to 2,000 meters being recorded in deep-sea sediments. Changes in the CCD are believed to be caused by changes in either the rate of supply of carbonate to the oceans or the rate of carbonate dissolution in the deep sea, which may be caused by changes in the shape of ocean basins, by changes in the location of carbonate deposition within different ocean basins, or by changes in the concentration of carbon dioxide in the atmosphere. During the last 100,000 years, carbonate sediments formed at greater depths during glacial intervals than during interglacials, indicating a less corrosive deep-water environment resulting from lower concentrations of atmospheric carbon dioxide.

CALCIUM CARBONATE DEPOSITION

Calcium carbonate is delivered to the oceans by rivers draining eroded continental rocks. Ocean sediments are the primary geochemical reservoir for calcium carbonate, so most of the calcium carbonate on the Earth remains dissolved in ocean water or in the form of calcareous sediments on the seafloor. The amount of carbonate deposition on the seafloor depends on the input of calcium carbonate derived from continental weathering and delivered to the oceans by rivers. Because oceanic plankton can precipitate solid calcium carbonate at a much faster rate than the rate of input of dissolved carbonate from rivers, most of the calcium carbonate that is deposited in the oceans must dissolve in order to maintain the chemical balance between dissolved carbonate and solid carbonate in oceanic sediments.

Any changes in the locations of carbonate deposition may cause corresponding changes in the level of the CCD as the compensation depths in different oceans change as a result of bathymetric fractionation or basin-basin fractionation. In bathymetric fractionation, a balance is established between the rates of carbonate deposition in shallow-water and in deep-water sedimentary basins. Greater deposition of calcium carbonate in shallow waters atop the continental shelves will cause a shallowing of the CCD as more deep-water carbonate deposits are dissolved so as to balance the shallow-water deposition. Similarly, the level of the compensation depth may vary by basin-basin fractionation of carbonate, which establishes a balance between the compensation depths in different ocean basins. For example, greater deposition of calcium carbonate in the Pacific Ocean will cause the Pacific CCD to become deeper, while at the same time, the Atlantic compensation depth must become shallower, because greater deposition of carbonate in Pacific sediments will leave less dissolved carbonate for precipitation in the Atlantic Ocean.

Even within an ocean basin, the level of the CCD may vary, depending on the balance between carbonate productivity and dissolution in a local area. For example, in the equatorial Pacific Ocean, the compensation depth is 500 to 800 meters deeper than in areas immediately to the north and south of this high-productivity area as a result of the greater supply of carbonate in the planktonic rain below these high-productivity surface waters. Also, the compensation depth tends to shoal near the edges of ocean basins, because higher biological productivity in shallow water near the continents causes rapid sinking of large amounts of organic carbon produced by planktonic plants and animals. Breakdown of this organic carbon by seafloor bacteria produces increased amounts of dissolved carbon dioxide gas, which reacts with water molecules to form carbonic acid; carbonic acid is corrosive to solid cal-

cium carbonate. Greater carbonic acid concentrations lead to increased carbonate dissolution in bottom waters and cause upward migration of the CCD into shallower waters.

STUDY OF CARBONATE COMPENSATION DEPTHS

Carbonate compensation depths may be studied by obtaining a series of deep-sea sediment samples from different depths within an ocean basin to determine the relationship between sediment type and water depth. In 1891, a study was published describing the global patterns of seafloor sediment type in each ocean basin, based on sediment samples obtained on the HMS *Challenger* oceanographic expedition. It was discovered that virtually no calcium carbonate was present below depths of 4,500 meters as a result of dissolution of carbonate.

The CCD in the modern ocean was first described in detail in 1935 based on sediment core transects taken across the South Atlantic Ocean by the 1925-1927 German *Meteor* expedition. Similar studies of sediment cores from different depths in the Pacific Ocean revealed that calcareous oozes are common seafloor sediments in water depths to 4,400 meters, with noncalcareous red clays being found in surface sediments deeper than 4,400 meters. Cores with calcareous oozes underlying red clays were obtained, however, in depths well below the present CCD, indicating that this geochemical boundary has migrated vertically throughout geologic history.

One innovative experiment to measure the rate of calcium carbonate dissolution with increasing water depth involved the placement of a stationary mooring for a period of months in deep water in the Pacific Ocean. Calcite spheres and calcareous microfossils were hung in permeable nylon bags at different water depths on the mooring; the nylon bags allowed seawater to come in contact with the calcium carbonate and thus permitted carbonate dissolution to occur. By measuring the weight loss of spheres suspended for a few weeks to months on the mooring, the rate of dissolution was determined for different water depths. In this experiment, little carbonate dissolution was observed at water depths shallower than 3,700 meters, while a rapid transition from minimal dissolution to extreme dissolution was seen. Rapid loss of carbonate by dissolution occurred in carbonate spheres suspended between the lysocline (the depth at which carbonate dissolution first begins to occur) and the compensation depth. Below 4,500 meters, the carbonate compensation depth, all carbonate was removed within a matter of weeks, demonstrating the ability of bottom waters to dissolve calcium carbonate.

Information on compensation depths may also be provided from microfossils preserved in seafloor sediments. Calcite dissolution can be measured by enrichment of solution-resistant forms of planktonic organisms, by benthic-planktonic foraminiferal ratios (foraminifera are one-celled animals that secrete a calcium carbonate internal test), by fragmentation indices (the percentage of broken planktonic tests compared to whole tests), or by the coarse-fraction ratios of seafloor sediments. Different microfossils will have differing susceptibilities to dissolution, depending on the thickness of the walls of the microfossil tests. Thin-walled plankton will be more solution-susceptible, while thicker-walled tests will resist dissolution. Thin-walled planktonic tests in seafloor sediments deposited in water depths between the lysocline and the CCD will be removed by dissolution more rapidly than thicker-walled tests, leading to greater enrichment of solution-resistant fossils with greater carbonate dissolution. Relative dissolution rates may be measured by the ratio between solution-susceptible and solution-resistant planktonic shells in seafloor sediments.

Similarly, the deeper-water bottom-dwelling benthic foraminifera tend to have thicker walls than the tests of planktonic foraminifera, which live floating in the shallow surface waters. Seafloor sediments may contain both benthic and planktonic varieties of foraminifera. Dissolution of calcium carbonate will preferentially remove the thinner-walled planktonic foraminifera, thus leading to enrichment in the proportion of benthic forms remaining in the sediment. The relative amount of carbonate dissolution in sediments may be determined by measuring the proportion between benthic and planktonic foraminifera in sediment samples.

Finally, fragmentation of planktonic tests may provide an indication of dissolution of carbonate from marine sediments. The percentage of broken tests to whole tests will increase with greater dissolution of carbonate. Also, coarse-fraction per-

centages of sediments provide information on carbonate dissolution, because unbroken foraminiferal tests are sand-sized, with particle diameters greater than 63 microns in size (1 millimeter is equal to 1,000 microns). As foraminifera are dissolved, they tend to break into smaller fragments, so dissolution tends to break down sand-sized particles into smaller silt-sized fragments. By measuring the proportion between sand-sized and silt-sized particles (the coarse-fraction percentage) in calcareous sediments, it is possible to obtain an indicator of the relative amount of dissolution that has affected those sediment deposits. The more dissolution that has occurred, the smaller will be the percentage of coarse (sand-sized) sediment particles.

DEPTH ESTIMATES

All these methods provide similar depth estimates for the CCD at approximately 4,500 meters below the sea surface, about halfway between the crests of the mid-ocean ridges and the abyssal plains. Individual compensation depths may vary in the different ocean basins of the world, however, as a result of basin-basin fractionation of carbonate. For example, in the Pacific Ocean, the CCD is typically between 4,200 and 4,500 meters, while in the Atlantic and Indian Oceans the CCD is deeper, being found near a depth of 5,000 meters. Even within an ocean basin, the level of the CCD may vary, depending on the balance between carbonate productivity and dissolution in a local area. In the centers of the North and South Pacific Oceans, the average compensation depth is between 4,200 and 4,500 meters, while near the equator it is found near 5,000 meters because of the higher biological productivity of equatorial surface waters.

Analyses of deep-sea cores drilled by the *Glomar Challenger* have revealed that the level of the carbonate compensation depth has changed by up to 2,000 meters in the South Atlantic, Indian, and Pacific Ocean basins. For example, one of the results of Deep Sea Drilling Project Leg 2 was the discovery of significant vertical excursions in the compensation depth of the Atlantic Ocean. In seafloor boreholes drilled in the North Atlantic, 2- to 5-million-year-old calcareous ooze sediments were found atop older (5- to 23-million-year-old) red clays, which in turn were deposited atop carbonate deposits older than 23 million years. In order for these sediments to have accumulated in that order, large vertical changes must have occurred in the compensation depth, starting when the seafloor was shallower than the CCD more than 23 million years ago. Red clays were deposited between 23 and 5 million years ago, when the CCD became shallower. After these red clays accumulated, deepening of the compensation depth allowed the deposition of younger calcareous sediments atop the carbonate-free red clays.

In order to study the past history of the compensation depth within an ocean basin, it is necessary to obtain a series of cores that were deposited at different paleodepths at the same time in the past. (The paleodepth is the depth at which ancient seafloor sediments were deposited.) Paleodepth estimates for seafloor sediments are calculated by studying the cooling history of seafloor basement rocks. After new seafloor is produced by volcanic activity at the mid-ocean ridge system, these rocks cool and contract, thus sinking to greater water depths as they move away from the ridge system. The older the seafloor, the greater its water depth will be, so sediments deposited atop volcanic seafloor will accumulate in progres-

sively deeper water. Once a series of sediment cores deposited at the same time in the past has been obtained, it is possible for the oceanographer to determine the paleodepth of the compensation depth by finding the paleodepth below which no calcium carbonate is present in ancient sediment deposits.

Dean A. Dunn

CROSS-REFERENCES

Dams and Flood Control, 2016; Deep Ocean Currents, 2107; Drainage Basins, 2325; Floodplains, 2335; Floods, 2022; Geologic and Topographic Maps, 474; Gulf Stream, 2112; Hydrothermal Vents, 2117; Ocean-Atmosphere Interactions, 2123; Ocean Pollution and Oil Spills, 2127; Ocean Tides, 2133; Ocean Waves, 2139; Oceans' Origin, 2145; Oceans' Structure, 2151; River Flow, 2358; River Valleys, 943; Sea Level, 2156; Seamounts, 2161; Seawater Composition, 2166; Sediment Transport and Deposition, 2374; Soil Erosion, 1513; Surface Ocean Currents, 2171; Surface Water, 2066; Tsunamis, 2176; Turbidity Currents and Submarine Fans, 2182; Water Quality, 2072; World Ocean Circulation Experiment, 2186.

BIBLIOGRAPHY

Allen, Philip A. *Earth Surface Processes*. Oxford, England: Blackwell Science, 1997. This book serves as a clear introduction to oceanography and the Earth sciences. Allen details the processes, properties, and composition of the ocean. Color illustrations, maps, index, and bibliography.

Berger, Wolfgang H. "Sedimentation of Deep-Sea Carbonate: Maps and Models of Variations and Fluctuations." *Journal of Foraminiferal Research* 8 (October, 1978): 286-302. A summary of the oceanographic influences on carbonate dissolution and carbonate sediment deposition patterns, with abundant illustrations.

Berger, Wolfgang H., and E. L. Winterer. "Plate Stratigraphy and the Fluctuating Carbonate Line." In *Pelagic Sediments: On Land and Under the Sea*, edited by Kenneth J. Hsü and Hugh C. Jenkyns. Oxford, England: Black-

well Scientific, 1974. A thorough review of the factors that may cause vertical migrations of the compensation depth through time. The text is suitable for college-level readers and contains many explanatory figures.

Duxbury, A. C., and A. B. Duxbury. *An Introduction to the World's Oceans*. 6th ed. Boston: McGraw-Hill, 2000. A freshman-level review of the marine environment. Sections are devoted to the physical, chemical, biological, and meteorological structure of the continental margins. Color plates review satellite and research submarine technology.

Hay, William A. "Paleoceanography: A Review for the GSA Centennial." *Geological Society of America Bulletin* 100 (December, 1988): 1934-1956. This review of all aspects of the study of ancient oceans, from the work of the first geologists to current research, covers the history of oceanographic examination of seafloor sediments, the development of deep-sea sediment sampling, and the history of research on the lysocline and the CCD.

Kennett, James P. *Marine Geology.* Englewood Cliffs, N.J.: Prentice-Hall, 1982. A college-level textbook on all aspects of marine geology and geological oceanography. Chapter 14, "Biogenic and Authigenic Oceanic Sediments," describes the deposition of sea-floor sediments and the factors influencing the positions of the lysocline and the CCD, as well as changes in these depths through time.

Segar, Douglas. *An Introduction to Ocean Sciences.* New York: Wadsworth, 1997. Comprehensive coverage of all aspects of the oceans and their chemical makeup. Readable and well illustrated. Suitable for high school students and above.

Sliter, William V., Allan W. H. Be, and Wolfgang H. Berger, eds. *Dissolution of Deep-Sea Carbonates.* Washington, D.C.: Cushman Foundation for Foraminiferal Research, 1975. This book contains a number of research papers analyzing the dissolution of carbonates in the water column and on the seafloor, along with studies of the factors influencing the position of the lysocline and the compensation depth in ocean basins. Suitable for college-level readers.

DEEP OCEAN CURRENTS

Deep ocean currents involve significant vertical and horizontal movements of seawater. They distribute oxygen- and nutrient-rich waters throughout the world's oceans, thereby enhancing biological productivity.

PRINCIPAL TERMS

BATHYMETRIC CONTOUR: a line on a map of the ocean floor that connects points of equal depth

BOTTOM CURRENT: a deep-sea current that flows parallel to bathymetric contours

BOTTOM-WATER MASS: a body of water at the deepest part of the ocean identified by similar patterns of salinity and temperature

CONTINENTAL MARGIN: that part of the Earth's surface the separates the emergent continents from the deep-sea floor

CORIOLIS EFFECT: an apparent force, acting on a body in motion, caused by the rotation of the Earth

SALINITY: a measure of the quantity of dissolved solids in ocean water

SURFACE WATER: relatively warm seawater between the ocean suface and that depth marked by a rapid reduction in temperature

THERMOHALINE CIRCULATION: vertical circulation of seawater caused by density variations related to changes in salinity and temperature

TURBIDITY CURRENT: a turbid, relatively dense mixture of seawater and sediment that flows downslope under the influence of gravity through less dense water

UPWELLING: the process by which bottom water rich in nutrients rises to the surface of the ocean

EVIDENCE FOR EXISTENCE

Deep-sea currents, ocean currents that involve vertical as well as horizontal movements of seawater, are generated by density differences in water masses that result in the sinking of cold, dense water to the bottom of the ocean. For many years, however, most oceanographers refused to accept the presence of these currents. Even when the Deep Sea Drilling Project, an international effort to drill numerous holes into the ocean floor, was initiated, most researchers envisioned the deep sea as a tranquil environment characterized by sluggish, even stationary, water. More recently, however, oceanographers and marine geologists have accumulated abundant evidence to suggest the opposite: that the deep sea can be a very active area in which currents sweep parts of the ocean floor to the extent that they affect the indigenous marine life and even physically modify the seafloor.

In the 1930's, Georg Wust argued for the likelihood that the ocean floor is swept by currents. Furthermore, he suggested that these currents play an important role in the transport of deep-sea

sediment. Wust's ideas were not widely accepted; in the 1960's, however, strong evidence for the existence of deep-sea currents began to accumulate. In 1961, for example, oceanographers detected deep-sea currents moving from 5 to 10 centimeters per second in the western North Atlantic Ocean. These researchers also determined that the currents changed direction over a period of one month.

In 1962, Charles Hollister, while examining cores of deep-sea sediment drilled from the continental margin off Greenland and Labrador, noted numerous sand beds that showed evidence of transport by currents. The nature of these deposits suggested to Hollister that they did not accumulate from turbidity currents, dense sediment-water clouds that periodically flow downslope from nearshore areas. Moreover, it appeared to Hollister that the sand was transported parallel to the continental margin rather than perpendicular to it, as might be expected of sediment transported by a turbidity current. He argued that the sand beds in the cores were transported by, and deposited from, deep-sea currents moving along the bottom of the ocean parallel to the continental margin. Since then, extensive photography of the ocean floor has provided direct evidence for the existence of deep-sea currents. Such evidence includes smoothing of the seafloor; gentle deflection, or bending, of marine organisms attached to the seafloor, as though they were standing in the wind; sediment piled into small ripples; and local scouring of the seafloor.

THERMOHALINE CIRCULATION

Essentially all Earth scientists now agree that the deep-sea floor is swept by rather slow-moving (less than 2-centimeter-per-second) currents. The driving force behind these currents, and all oceanic currents for that matter, is energy derived from the Sun. Differential heating of the air drives global wind circulation, which ultimately induces surface ocean currents. The vertical circulation of seawater, and thus the generation of deep-sea currents, is controlled by the amount of solar radiation received at a point on the Earth's surface. This value is greatest in equatorial regions; there, the radiation heats the surface water, the seawater that lies within the upper 300 to 1,000 meters of the ocean. As this water is heated, it begins to move toward the poles along paths of wind-generated surface circulation, such as the Gulf Stream current of the northwestern Atlantic Ocean.

The cold waters that compose the deep-sea currents originate in polar regions. There, minimal solar radiation levels produce cold, dense surface waters. The density of this water may also be increased by the seasonal formation of sea ice, ice formed by the freezing of surface water in polar regions. When sea ice forms, only about 30 percent of the salt in the freezing water becomes incorporated into the ice. The salinity and density of the nearly freezing water beneath the ice are therefore elevated. This cold, saline seawater eventually sinks under the influence of gravity to the bottom of the ocean, where it moves slowly toward the equator. Deep-sea circulation driven by temperature and salinity variations in seawater is termed "thermohaline circulation" and is much slower than surface circulation; the cold, dense water generated at the poles moves only a few kilometers per year. After moving along the bottom of the ocean for anywhere from 750 to 1,500 years, the cold seawater rises to the surface in low-latitude regions to replace the warm surface water, which, as noted above, moves as part of the global surface circulation system back to the polar regions.

Thermohaline circulation and related deep-sea currents are commonly affected by the shape of the ocean floor. Although sinking cold seawater seeks the deepest route along the seafloor, deep-sea currents may be blocked by barriers. The Mid-Atlantic Ridge, the large volcanic ridge running down the middle of the Atlantic Ocean, may prevent the movement of water from the bottom of the western Atlantic to the eastern Atlantic. On the other hand, the funneling of deep-sea currents through narrow passages or gaps in seafloor barriers will lead to an increase in the velocity of the current. Once beyond the passage, however, the current spreads and velocity is reduced.

CORIOLIS EFFECT

The circulation pattern of deep-sea currents is controlled to a large extent by the Earth's rotation. The Coriolis effect, the frictional force achieved by the Earth's rotation that causes particles in motion to be deflected to the right in the Northern Hemisphere and to the left in the Southern Hemisphere, induces deep-sea currents to trend along the western margins of the major oceans. Thus, water sinking from sources in the North Atlantic Ocean and moving south toward the equator will be deflected to the right, causing it to run along the western side of the North Atlantic. Similarly, north-directed deep-sea currents generated by the sinking of cold water from the Antarctic region will also be deflected to the western margin of the Atlantic.

The Coriolis effect guides deep-sea currents along bathymetric contour lines, lines on a map of the ocean floor that connect points of equal depth. Deep-sea currents that have a tendency to move parallel to the bathymetric contours are known as bottom currents. Barriers to flow may locally deflect deep-sea currents from the bathymetric contours; nevertheless, bottom currents are most conspicuous along the western margins of the major oceans.

SHORT- AND LONG-TERM CONTROLS

The formation of the cold seawater required to set deep-sea currents in motion can itself be considered in terms of short- and long-term controls. Seasonal sea ice formation is probably the most important process in the production of the north-flowing water generated at the south polar region, or the Antarctic bottom water (AABW). Velocities of the AABW are highest in March and April, that period of the year when sea ice production in the ocean surrounding Antarctica is greatest. During Southern Hemisphere summers, however, the sea ice melts and there is an increase in the freshwater flux to the ocean from the continent, both of which actions reduce the salinity and therefore the density of the seawater, thereby decreasing AABW production.

Many oceanographers and marine geologists have argued that long-term variations in the production of the cold, dense bottom water required to generate deep-sea currents may be related to global climatic changes. More specifically, deep-sea currents appear to be most vigorous during glacial periods, when sea ice production is enhanced and the sea ice remains on the ocean surface for a greater proportion of the year. Nevertheless, there is also evidence to suggest that the velocities of deep-sea currents in the North Atlantic Ocean were much lower during the most recent glacial periods than they were during the times between glacial phases. Much more work is required to gain a more complete understanding of long-term controls on deep-sea currents.

MEASUREMENT TOOLS AND TECHNIQUES

The most common methods of study of deep-sea currents include direct measurement of current velocities, bottom photography, echo sounding, and the sampling of ocean-floor sediment. The speed and direction of deep-sea currents have been determined by the use of free-fall instruments, such as the free-instrument Savonics rotor current meter. This device, dropped unattached into the ocean, is capable of recording current velocities and directions over a period of several days. It returns automatically to the surface of the ocean, at which time a radio transmitter directs a ship to its position. Other current-measuring devices can be suspended at various depths in the ocean from fixed objects, such as buoys or light ships, to monitor currents for long periods. One such anchored meter measures the flow of water past a fixed point. Flowing water causes impeller blades, similar to the blades of a fan, to rotate at a rate proportional to the current's speed. In addition, the blades cause the meter to align with the current's direction. Electrical signals indicating the direction and speed of the current are transmitted by radio or cable to a recording vessel. Current velocities of less than 1 centimeter per second can be detected by this meter.

To get the most complete picture of the variability of the ocean, a combination of various measurement techniques with remote sensing may be employed. Such a multidimensional approach may involve the measurement of current velocity, pressure (a measure of depth), water temperature, and water conductivity (a measure of salinity). These data can be transmitted via satellite to a land station or even directly to a computer.

ADDITIONAL STUDY METHODS

Perhaps the most persuasive evidence for the existence of deep-sea currents and their influence on the ocean bottom has been gained through bottom photography. Sediment waves, or ripples, apparently formed from sediment carried by deep-sea currents, along with evidence of current-induced scour of the ocean floor, were first photographed in the Atlantic Ocean in the late 1940's. Since then, the technology of bottom photography has advanced greatly. This advancement became most apparent with the exploration, in the late 1980's, of the wreck of the SS *Titanic* in the northwestern Atlantic. Bottom photography permits detailed study of some of the smaller features on the ocean floor apparently formed by deep-sea currents. Benthonic, or benthic, organisms, marine organisms that live attached to the ocean floor, bending in the flow of the current, are a particularly intriguing example of the phenomena recorded by this technique.

Echo-sounding studies of the seafloor have yielded abundant information on ocean-floor features that are either formed or modified by deep-sea currents. Notable among these are very long ridges in the North Atlantic evidently constructed from sediment carried by deep-sea currents. In echo sounding, a narrow sound beam is directed from a ship vertically to the sea bottom, where it is reflected back to a recorder on the ship. The depth to the seafloor is determined by multiplying the velocity of the sound pulse by one-half the amount of time it takes for the sound to return to the ship. The depths to the ocean floor are recorded on a chart by a precision depth recorder, which produces a continuous profile of the shape of the seafloor as the ship moves across the ocean.

Sediment transported by and deposited from deep-sea currents can be studied directly by actually sampling the ocean floor. Sampling of these deposits is best accomplished by the use of various coring devices capable of recovering long vertical sections, or cores, of seafloor sediment. Sediment recovery is achieved by forcing the corer, a long pipe usually with an inner plastic liner, vertically into the sediment. The simplest coring device, the gravity corer, consists of a pipe with a heavy weight at one end. This type of corer will penetrate only 2-3 meters into the seafloor. The piston corer, used to obtain longer cores, is fitted with a piston inside the core tube that reduces friction during coring, thereby permitting the recovery of 18-meter or longer cores. Analysis of the sediment recovered from the ocean floor by these and other coring devices reveals much information about small-scale features formed by deep-sea currents.

IMPORTANCE TO LIFE ON EARTH

Because cold bottom-water masses often are nutrient-rich and contain elevated abundances of dissolved oxygen, deep-sea currents are extremely important to biological productivity. There are areas of the Earth's surface where nutrient-rich cold bottom waters rise to the ocean surface. These locations, known as areas of upwelling, are generally biologically productive and therefore are important food sources. Especially pronounced upwelling occurs around Antarctica. Bottom waters from the North Atlantic upwell near Antarctica and replace the cold, dense, sinking waters of the Antarctic.

The great amount of time required for seawater to circulate from the surface of the ocean to the bottom and back again to the surface has become an important practical matter. If pollutants are introduced into high-latitude surface waters, they will not resurface in the low latitudes for hundreds of years. This delay is particularly important if the material is rapidly decaying radioactive waste that may lose much of its dangerous radiation by the time it resurfaces with the current. The introduction of toxic pollutants into a system as sluggish as the deep-sea circulation system, however, means that they will remain in that system for prolonged periods. Nations must, therefore, be concerned with the rate at which material is added to this system relative to that at which it might be redistributed at the surface of the ocean by wind-induced surface circulation. The multinational Geochemical Ocean Sections (GEOSECS) program, introduced as part of the International Decade of Ocean Exploration, attempted to assess better the problem of how natural and synthetic chemical substances are distributed throughout the world's oceans. The GEOSECS program, carried out from 1970 to 1980, yielded abundant information regarding the movement of various water masses and, among other things, the distribution of radioactive material in the oceans. For example, GEOSECS demonstrated that tritium produced in the late 1950's and early 1960's by atmospheric testing of nuclear weapons had been carried to depths approaching 5 kilometers in the North Atlantic Ocean by 1973.

Gary G. Lash

CROSS-REFERENCES

BIBLIOGRAPHY

Baker, D. J. "Models of Oceanic Circulation." *Scientific American* 222 (January, 1970): 114. A somewhat complex discussion of surface circulation in the world's oceans. Generally suitable for college-level readers.

Hollister, C. D., A. Nowell, and P. A. Jumar. "The Dynamic Abyss." *Scientific American* 250 (March, 1984): 42. This excellent article addresses the formation of bottom waters that flow away from the polar regions toward the equator. Suitable for high school students.

Ittekko, Venugopalan, et al., eds. *Particle Flux in the Ocean*. New York: John Wiley and Sons, 1996. This volume contains descriptions of the chemical and geobiochemical cycles of the ocean, as well as the ocean currents and movement. Suitable for the high school reader and beyond. Illustrations, index, bibliography.

Jeleff, Sophie, ed. *Oceans*. Strasburg: Council of Europe, 1999. This collection of debates and lectures is a detailed account of oceanography and ocean ecology. Among the topics discussed are ocean circulation, marine chemistry, and the ocean's structure. Illustrations and bibliography.

Kennett, James P. *Marine Geology*. Englewood Cliffs, N.J.: Prentice-Hall, 1982. This book contains an excellent discussion of deep-sea currents and thermohaline circulation. The major methods of study of deep-sea currents, including bottom photography (there are two pages of black-and-white bottom photographs), are discussed in detail. Best suited to the college student.

Ross, David A. *Introduction to Oceanography*. Englewood Cliffs, N.J.: Prentice-Hall, 1988. A fine introductory oceanography textbook with an informative discussion of deep-sea currents and their mechanisms of generation. There is also a section on oceanographic instrumentation. Suitable for high school students.

Shepard, Francis P. *Submarine Geology*. 3d ed. New York: Harper & Row, 1973. An excellent, if somewhat dated, text on marine geology. The section on methods and instrumentation employed in the study of the oceans is particularly good. Suitable for general audiences.

Smith, F. G. "Measuring Ocean Currents." *Sea Frontiers* 18 (May, 1972): 166. This article discusses methods used to determine the speed and direction of ocean currents.

Steward, R. W. "The Atmosphere and the Oceans." *Scientific American* 221 (September, 1969): 76. A good discussion of the energy exchange between the atmosphere and the ocean and the resulting phenomena. For all readers.

Teramoto, Toshihiko. *Deep Ocean Circulation: Physical and Chemical Aspects*. New York: Elsevier, 1993. This college-level text provides a detailed look at ocean circulation and currents. Much information is provided on the chemical processes that occur in the deep ocean. Illustrations, maps, and bibliographical references.

GULF STREAM

The Gulf Stream is a geostrophic surface current that constitutes the northwestern part of the North Atlantic Gyre. It moves huge quantities of water at remarkably fast velocities through vast distances with many geological, physical, and biological repercussions. By itself, however, it is not responsible for the mild climate of Western Europe.

PRINCIPAL TERMS

CORIOLIS EFFECT: an apparent force acting on a rotating coordinate system; on Earth this causes things moving in the Northern Hemisphere to be deflected toward the right and things moving in the Southern Hemisphere to be deflected toward the left

GEOSTROPHIC CURRENT: a current resulting from the balance between a pressure gradient force and the Coriolis effect; the current moves horizontally and is perpendicular to the pressure gradient force and the Coriolis effect

GYRE: the major rotating current system at the surface of an ocean, generally produced by a combination of wind-generated currents and geostrophic currents

PRESSURE GRADIENT: a difference in pressure that causes fluids (both liquids and gases) to move from regions of high pressure to regions of low pressure

THERMOHALINE CIRCULATION: a mode of oceanic circulation that is driven by the sinking of denser water and its replacement at the surface with less dense water

WIND-DRIVEN CIRCULATION: the surface currents on the ocean that result from winds and geostrophic currents

WATER CIRCULATION

It is convenient to consider any systematic movement of water at sea as being part of either a wind-driven circulation system or a thermohaline circulation system. In the former, the linkage with atmospheric movement is direct, the currents are usually at or near the sea surface, and the velocities of the flows are often in the range of several centimeters per second (or several knots). In thermohaline circulation, the driving force is gravity, which causes denser water to sink and flow to the deepest parts of the sea, and the currents are usually much slower. In the North Atlantic Ocean, thermohaline circulation occurs on a vast scale as saline surface water gives up its heat, sinks, and eventually flows across the bottom of both the North and South Atlantic basins.

Wind-driven circulation develops a gigantic clockwise circular motion called the North Atlantic Gyre. The most intense flow of this gyre is along its northwestern boundary and is called the Gulf Stream. Because the flow is circular, it does not actually have a beginning or end. As it is usually geographically defined, however, the Gulf Stream begins in the straits of Florida, where water leaving the Gulf of Mexico and the Caribbean joins with water continuing to go around the gyre. There the Gulf Stream moves about 30 million cubic meters per second past any point. The volume of water entrained in this flow continues to increase, and by the time it reaches Cape Hatteras, there are about 85 million cubic meters moving by any point every second. When the Gulf Stream reaches longitude 65 degrees west, off the Grand Banks, it moves 150 million cubic meters per second.

To put this in perspective, consider that 1 cubic meter of water has a mass of 1,000 kilograms. A 150-pound person has a mass of 68 kilograms. Therefore, 1 cubic meter of water has the mass of about fifteen people. The flow off Cape Hatteras would be equivalent to 1.2 billion people, roughly the population of China, streaming by every second. By the time it gets to the Grand Banks, the flow would be larger by 65 million cubic meters, or 975 million more people. The Gulf Stream truly dwarfs most of nature's other wonders. The total hydrological cycle, which includes all the rain, snow, sleet, and hail that falls on the Earth (oceans plus continents), moves an average of only 10 million cubic meters of water per second.

CORIOLIS EFFECT

The secret to maintaining such huge flows of water lies in their circular nature. A hill of warm, low-density water sits over the core of the North Atlantic Gyre. Water on or within this hill is driven by gravity or a pressure gradient to move down the slope or away from the center of this hill. This moving water is deflected by the Coriolis effect, which is a consequence of the Earth's rotation. In the Northern Hemisphere, the Coriolis effect causes things moving with a horizontal velocity to move to their right. Therefore, the parcels of water trying to move out from the center of this hill are deflected to move around the gyre instead. Eventually a balance is achieved between the Coriolis effect and the pressure gradient forces. As a result, the parcels of water do not move away from the center of the hill but instead circle it in a clockwise fashion. This kind of current is called a "geostrophic" current.

The Earth's spherical shape means that the effects of the Coriolis effect vary with latitude. One result is that the flow within the North Atlantic Gyre varies with location. The hill is not symmetrical but has its steepest slopes on its northwest edge. Because currents must balance slopes, the gyre is most intense there. The center of the gyre is in the western Atlantic not far from Bermuda. The hill also slopes away to the east, but very gradually, so that the southern flows of the gyre are slow and spread out over a very large region.

Driving the gyre and maintaining the Bermuda High are the winds over the Atlantic Ocean. This subtropical atmospheric feature is a region of descending, dry air near the center of the North Atlantic Ocean. After descending, this air pushes out across the ocean. It, too, is deflected by the Coriolis effect, developing into somewhat circular, clockwise winds. Near the center of this system, winds are weak, and precipitation is uncommon. Early sailors, faced with long, dry periods of calm, sometimes made their thirsty horses walk the plank. This is the origin of the term "horse latitudes," sometimes used to describe this region.

With little cloud cover and a subtropical latitude, this region receives intense solar radiation that warms the surface water and causes intense evaporation. This causes the water of this area, the Sargasso Sea, to become saline and very warm. Just as oil floats on vinegar in a salad dressing, this warm, saline water floats on cooler water below.

It is this warm water that forms the hill driving the Gulf Stream. The warm water tries to spread out and flow over cooler surface waters far from the center of the gyre. The Coriolis effect makes it move around the hill, not down it, and the boundary between this warm, rotating mass of water and the cooler water it is trying to flow over is where the currents are most obvious. This is the Gulf Stream, a distinct boundary between the productive coastal waters, which are green and teeming with life, and the dark blue, nearly lifeless Sargasso Sea waters.

GULF STREAM AS A BOUNDARY CURRENT

For decades, schoolchildren have been taught that the mild climate of the British Isles and Western Europe gets its heat from the Gulf Stream. The Gulf Stream is often presented as a river of warm water moving north and then east to deliver its heat to the European continent. The significance of the Gulf Stream as a source of heat changes when it is recognized for what it is, just a boundary current. It separates the huge hill of warm, saline, Sargasso Sea water from surrounding waters.

The Gulf Stream is a very active boundary region with motions driven by the wind and geostrophic currents that surround the huge quantity of surface water, heated by the Sun in subtropical high-pressure zones and made especially salty by accompanying evaporation. The Gulf Stream gets all the press, but it is this huge quantity of water that conveys heat to Europe.

In the North Atlantic, beyond the northern extent of the circulating gyre, cold winds remove heat from the surface waters. In the process, these winds are warmed and bathe Europe in the pleasant temperatures to which the continent is accustomed. However, by removing heat, these winds cool the saline surface waters until they become dense enough to sink to the bottom of the sea. This is the thermohaline circulation. The sinking waters are replaced by a gradual northward flow of surface water. It is likely that most of this surface water spent time in the North Atlantic Gyre, but its transport northward was independent of that circular motion. If the gyre were to stop tomorrow, the thermohaline circulation would continue, and Europe would stay just as warm as it is today. In fact, some researchers have suggested that the strength of the Gulf Stream actually reduces the warming effects of this thermohaline circulation. If they are correct, were the Gulf Stream to stop, Europe might grow warmer.

Although its role as a heat-delivery system may have been overstated, the Gulf Stream is still an incredibly powerful element of oceanic circulation and, as such, greatly influences the biology and chemistry of the surface ocean. This significance is easily seen where meanders develop on the Gulf Stream, typically beyond Cape Hatteras. Just as meanders can grow and develop on a slow-moving river, they also grow on the Gulf Stream. Whereas a stream meander may form an oxbow lake if the course of the stream closes in upon itself, a meander in the Gulf Stream produces a circulating eddy separated from the rest of the stream when one closes upon itself. These eddies will have a core of warmer, less fertile water and rotate in a clockwise manner if they close off on the southeastern side of a meander. They will have a core of cooler, more fertile water and rotate in a counterclockwise manner if they close off on the northwestern side of a meander. These rings persist for months to a year or more and may establish their own ecosystems during their lifetimes.

The Gulf Stream disperses eggs, seeds, and juvenile and adult organisms. As a chemical agent, it stirs up the surface waters, keeping its warm waters well mixed. As a physical agent, it moves enormous quantities of water. It is clearly a remarkable current and a very important part of the global ecosystem.

STUDY OF THE GULF STREAM

The Gulf Stream is studied by directly measuring the strength of its currents at different depths and locations and by examining the effects of its currents through monitoring the position of floats released within it and designed to stay at particular depths.

Floating current meters can be moored to anchors at the bottom of the sea. Their depth is controlled by the length of the tether keeping them attached to the anchor. They can record data electronically, storing it in computer memory. When a vessel is at the surface, ready to retrieve the meter and its data, it transmits a special coded sound pulse. This instructs the meter to release itself

from the tether and rise to the surface, where it transmits a radio signal allowing the vessel to home in on it and recover it. The data is incorporated in complex computer models that tie together the results obtained from hundreds of current meters deployed during overlapping time periods. Snapshots of the current system can then be obtained, and sequences of these snapshots reveal the behavior of the currents over time.

Floating objects can be released at sea and tracked by satellite. To ensure that these objects are being moved by ocean currents rather than surface winds, they usually have a large parachute or sail deployed in the water beneath them. Because the density of seawater increases with depth, floats can be designed to be neutrally buoyant at some particular depth (neither sinking below, nor floating above that depth). A layer of the ocean (the SOFAR channel) acts as a wave guide for sound waves. Floats in this layer can transmit sounds over tremendous distances, permitting them to be tracked efficiently by a small number of surface ships with sonar receivers suspended into this layer.

Because geostrophic currents are driven by the slopes of the ocean's surface, any technique that can measure those slopes can provide valuable insight into the driving forces behind the Gulf Stream. These slopes are very gradual—total dynamic relief over all the world's oceans is about 2 meters—and, consequently, direct measurement is difficult. Satellite techniques, coupled with computer models to filter out waves, tides, and dozens of other confounding effects, are approaching the point where they will be able to measure this topography. Yet the dynamic topography is generally determined indirectly by measuring temperature and salinity as a function of depth and position. These data are used to determine the density of the seawater as a function of depth and position. By assuming that at some depth the horizontal pressure gradients have disappeared, it is possible to reconstruct the differences in the height of the water column needed to accommodate these variations in density. Then the velocities and directions of the resulting currents can be calculated. When the theoretical models are compared with currents measured by moored meters or revealed by the paths of floating objects, the results are in agreement. This gives strong support to the theoretical concepts underlying the study of ocean currents.

Many of the approaches to study of the Gulf Stream are exercises in applied mathematics. That computed and measured results agree so well is a triumph of geophysical fluid dynamics.

SIGNIFICANCE

The North Atlantic Gyre, which has the Gulf Stream as its northwest boundary current, dominates surface flow in the Atlantic Ocean. Voyages of discovery, exploration, conquest, and exploitation all were affected to some extent by this system and the winds that accompany it.

The Gulf Stream, studied since the eighteenth century, has provided the basis for much of what scientists know of ocean currents. Elaborate mathematical constructs, including the entire concept of geostrophic currents, have been developed to describe, analyze, and comprehend this mighty system.

As scientists have learned more about the Gulf Stream, they have discovered many new areas of study, including branches in the stream, countercurrents at the surface and at depth, and fluctuations in flow and velocity with time and place. Some researchers devote their entire careers to studying just the rings of the Gulf Stream, which contain entire ecosystems. Others investigate the location and strength of the Gulf Stream in the distant past, during and even before the ice ages.

There is evidence that at times the current has taken different paths across the continental shelf, perhaps scouring out valleys in the ocean floor in the process.

Scientists' understanding of the dynamics of the planet relies on comprehending the transfer of energy from the equator to the poles. The Gulf Stream is an important component of this transfer. As people have become more aware of the fragility of the environment and become more concerned about issues of global climate change, the role of the Gulf Stream and thermohaline circulation in influencing temperatures in Europe and elsewhere on the planet has taken on a new importance.

Otto H. Muller

CROSS-REFERENCES

Abyssal Seafloor, 651; Carbonate Compensation Depths, 2101; Continental Shelf and Slope, 584; Deep Ocean Currents, 2107; Hydrothermal Vents, 2117; Ocean-Atmosphere Interactions, 2123; Ocean Drilling Program, 359; Ocean-Floor Drilling Programs, 365; Ocean Pollution and Oil Spills, 2127; Ocean Tides, 2133; Ocean Waves, 2139; Oceans' Origin, 2145; Oceans' Structure, 2151; Sea Level, 2156; Seamounts, 2161; Seawater Composition, 2166; Surface Ocean Currents, 2171; Tsunamis, 2176; Turbidity Currents and Submarine Fans, 2182; World Ocean Circulation Experiment, 2186.

BIBLIOGRAPHY

Allen, Philip A. *Earth Surface Processes.* Oxford, England: Blackwell Science, 1997. This book serves as a clear introduction to oceanography and the Earth sciences. Allen details the processes, properties, and composition of the ocean. Color illustrations, maps, index, and bibliography.

Briggs, Peter. *Rivers in the Sea.* New York: Weybright and Talley, 1969. A very easy-to-read book, suitable for a high school student. The author includes a fair amount of history and lore, with anecdotes and some sketches related to early scientific expeditions.

Chapin, Henry, and F. G. Walton Smith. *The Ocean River.* New York: Charles Scribner's Sons,

1952. A popular source of both information and misinformation. In attempting to review all of Earth's history before taking on the Gulf Stream, it describes many theories now considered groundless. Although the book is loaded with many discussions of atmospheric and oceanic dynamics, it is not clear that the authors ever grasped the nature of the North Atlantic Gyre or the importance of thermohaline circulation. Still, the book contains considerable information of historical interest and is important because of the influence that it has had. Suitable for high school readers.

Gaskell, Thomas Frohock. *The Gulf Stream.* New York: John Day, 1972. An interesting overview from a slightly technical perspective. The book emphasizes history and anecdotes but includes some well-chosen excerpts of original reports from some early expeditions. Good descriptions of many oceanographic techniques and excellent summaries of some of the biological consequences of the Gulf Stream.

MacLeish, William H. *The Gulf Stream.* Boston: Houghton Mifflin, 1989. Compelling reading, this book offers a personal approach to the Gulf Stream and many of its repercussions. Organized as a narrative rather than a scientific tome, it is easy to read, suitable for a high school reader. It has no equations and few technical discussions but represents an obviously thorough research effort.

Open University Oceanography Course Team. *Ocean Circulation.* Oxford, England: Pergamon Press, 1989. This carefully written textbook provides an introduction to the analysis and understanding of all sorts of circulation in the world's oceans. Intended for an undergraduate college-level reader, it provides the necessary mathematical background to understand many aspects of geophysical fluid dynamics.

Segar, Douglas. *An Introduction to Ocean Sciences.* New York: Wadsworth, 1997. Comprehensive coverage of all aspects of the oceans, their chemical makeup, and circulation. Readable and well illustrated. Suitable for high school students and above.

Stommel, Henry. *The Gulf Stream.* Berkeley: University of California Press, 1958. The author of this book is one of the leading pioneers in the study of the Gulf Stream. Based on geophysical fluid dynamics, this book is likely to challenge most undergraduate college students. Contains much qualitative information, however, that is accessible to even a high school reader. Although much has been learned since this book was written, the essentials are all here.

Teramoto, Toshihiko. *Deep Ocean Circulation: Physical and Chemical Aspects.* New York: Elsevier, 1993. This college-level text provides a detailed look at ocean circulation and currents. Much information is provided on the chemical processes that occur in the deep ocean. Illustrations, maps, and bibliographical references.

HYDROTHERMAL VENTS

Hydrothermal vents are openings on the seafloor where hot water is released. Hydrothermal water forms when cold seawater percolates down into and is heated by hot rocks. Both rocks and water are changed by this interaction. Deposits rich in metals form around these vents, and many ore deposits originated in similar environments. Numerous kinds of organisms live around hydrothermal vents, supported by bacteria that derive their energy from reduction of sulfide to sulphate.

PRINCIPAL TERMS

BASALT: a typical volcanic rock of the oceanic environment

CONVERGENT PLATE MARGIN: an area where the Earth's lithosphere is returned to the mantle at a subduction zone, forming volcanic "island arcs" and associated hydrothermal activity

DIVERGENT MARGIN: an area where the Earth's crust and lithosphere form by seafloor spreading

ENDOSYMBIONT: an animal hosting autotrophic bacteria, with both host and bacteria enjoying the benefits of symbiosis

HYDROSTATIC PRESSURE: the pressure resulting from an overlying, continuous column of water, approximately 1 bar for every 10 meters of water

MAGMA: molten rock generated by melting of the Earth's mantle

MID-OCEAN RIDGE: a region of the seafloor where new oceanic crust is created by seafloor spreading

OPHIOLITE: a section of oceanic crust and upper mantle that has been thrust out of the ocean floor and up onto the continental crust

CHARACTERISTICS

The late 1970's discovery of hydrothermal vents and the unique communities that live around them astounded scientists. Studies of these features have provided important and exciting research opportunities for biologists, geologists, and marine chemists. Almost all of the deep seafloor is a cold and quiet place, but this is not true near hydrothermal vents. In contrast to the homogeneous and stable environment of the deep sea, hydrothermal vents are constantly changing, and scientists have seen many changes in hydrothermal vents and their associated communities over the period that they have been studied. Hydrothermal vents are easily the most spectacular features, both geologically and biologically, on the floor of the abyss. There are almost certainly many more hydrothermal vents remaining to be discovered along mid-ocean ridges, where hydrothermal vents are concentrated.

Hydrothermal vents are hot springs on the ocean bottom, the "exhaust pipes" of seafloor hydrothermal systems. These vents are the most accessible parts of seafloor hydrothermal systems, which are driven by heat sources that lie several

hundred meters or even kilometers beneath the ocean bottom. Hydrothermal systems form where near-freezing seawater from the bottom of the ocean penetrates deep into the crust along fissures until it comes into contact with hot rocks. The water may be heated to as high as 400 degrees Celsius before it rises back up through the crust and jets back into the deep sea at the vents. Development of hydrothermal systems requires two things: hot rocks and a way for seawater to percolate along cracks in the hot rocks and return to the surface. How vigorous the resulting hydrothermal system is depends on both these factors. The hottest rocks are found at mid-ocean ridges and other submarine volcanic centers such as the Loihi seamount off the coast of Hawaii, where frequent eruption of basalt lavas maintains conditions that are most favorable for developing and maintaining hydrothermal systems.

Yet it is not enough for hot rocks to lie beneath the surface; there must also be a way for seawater to be brought into contact with the hot rocks and kept there until the water is sufficiently heated. This requires just the right configuration of fractured hot rocks: If the cracks are too wide or the hot rocks are not deep enough below the seafloor, the water will not be heated enough before it returns to the surface. Similarly, if the cracks are too narrow, not enough water will flow through the system to develop a robust hydrothermal system, or the channels may be easily blocked by minor Earth movements or mineral deposits. The sensitivity of deep-sea hydrothermal systems to the sub-seafloor circulation geometry may be why hydrothermal events experience rapid changes and are generally short-lived features.

Seafloor hydrothermal systems are similar to hot springs or geysers found on the continents in that both form as hot rocks heat and chemically modify cold water. Seafloor hydrothermal systems differ, however, in their much higher temperatures. This is because the boiling temperature of water increases as pressure increases, and pressures on the seafloor are several hundred times greater than they are on the surface. This effect is well known to most cooks, who are familiar with the fact that water boils at a lower temperature at high altitude, where the pressure is lower, than it does at sea level. Hydrothermal systems on land, such as those of Yellowstone National Park, have a maximum water temperature of not much more than about 100 degrees Celsius; at greater temperatures, the water "flashes," or turns to steam. This is what makes a geyser such as Old Faithful erupt, as the water in contact with the hot rock turns to steam and violently forces out the colder, overlying water. The higher pressure of the seafloor prohibits flashing and allows water to be heated to much higher temperatures.

Although most hydrothermal systems are found along the mid-ocean ridges, where magma lies not more than a few kilometers beneath the ocean floor, hydrothermal water is not heated by magma; water heated by magma would be much closer to magmatic temperatures of 1,200 degrees Celsius. Instead, the water is heated as it passes through hot but solid rocks, although the hot rocks are kept hot by underlying magma. The upper limit of about 400 degrees Celsius for seafloor hydrothermal systems may reflect the maximum rock temperature at which fractures can form and remain open, or it may reflect separation of the fluid into immiscible fluids. It may be that at temperatures much higher than 400 degrees Celsius, rocks begin to slowly flow, closing any fractures

that do form. Another possibility is that, at pressures corresponding to typical depths of mid-ocean ridges (about 300 bars), seawater separates into two fluids at about 400 degrees Celsius. One phase is enriched in salt relative to seawater, while the other fluid contains a lower concentration of salt than seawater. The more concentrated phase will be the denser phase, and this may be why it is not found among waters issuing from hydrothermal vents. About one-third of the total heat lost from the Earth's interior through the seafloor is lost as a result of seafloor hydrothermal systems.

METAL DEPOSITS

The seawater moving through a seafloor hydrothermal system is chemically changed as it is heated, and it alters the rocks through which it passes as well. All of the magnesium and sulphate in the seawater is absorbed by the hot rocks, and large amounts of metals such as manganese, cobalt, copper, zinc, and iron are lost from the hot rocks to the circulating water. In fact, magnesium is so completely removed from seawater that the depletion of this element in seawater can only be explained by seafloor hydrothermal activity. Oxygen-rich seawater is transformed into oxygen-poor hydrothermal water, and a high concentration of metals can be dissolved in such water. Seawater, for example, contains negligible magnesium and iron, whereas hydrothermal waters contain up to 1 millimole of magnesium and 6 millimoles of iron per liter.

When hot, chemically transformed water returns to the ocean at hydrothermal vents, the result is often spectacular. Near the vent, the hot, oxygen-poor, hydrogen-sulfide-rich hydrothermal water mixes with cold, oxygen-rich, hydrogen-sulfide-poor seawater. This may cause the metals in the hydrothermal fluid to precipitate. Some of it may precipitate in fractures just beneath the seafloor. These "stockwork" deposits are most likely to be preserved and many ore deposits represent ancient stockwork deposits. Some metals precipitate around the vent itself, forming metal-rich chimneys. These chimneys are typically composed of an outer layer of anhydrite and inner deposits of copper-iron sulfides. These chimneys can have fantastic shapes, resembling spires, columns, cones, and beehives. Scientists visiting these chimneys give them fanciful names: "nail," "fir tree," and "moose" are names given to vent features from one hydrothermal field. Chimneys have been found that are up to 30 meters in diameter and up to 45 meters tall. These growth rates can be on the order of 1 meter per year, and vent chimneys are thus among the fastest-changing of all geologic phenomena. Fast-growing vent chimneys rapidly become unstable and collapse, then rise again. Cycles of growth and collapse continue as long as the hydrothermal fluids continue to issue from the vent.

Chimneys of iron oxide on Loihi volcano, Hawaii. (National Oceanic and Atmospheric Administration)

Metals that do not precipitate as stockwork or vent chimneys will rise with the hydrothermal water issuing from the vent. Be-

A black smoker at a mid-ocean-ridge hydrothermal vent. (National Oceanic and Atmospheric Administration)

cause this water is so hot, it is less dense than seawater and jets out of the vents. The vent is the base of a hydrothermal plume, and the plume becomes larger as it rises. The plume becomes larger as it rises because cold seawater mixes turbulently with the hydrothermal water, and these waters quickly become well mixed. The hottest hydrothermal fluids (those heated to 300 degrees Celsius or more) are sufficiently metal-rich and acidic that they precipitate large quantities of microscopic grains of iron sulfides, zinc sulfides, and copper sulfides upon mixing with seawater. Much of this precipitates as the plume rises, producing a sulphide "cloud." The rapid precipitation of sulphides darkens the hydrothermal plume, giving it the appearance of smoke. Those vents releasing the hottest fluids have the greatest density of sulfide precipitation and are therefore often referred to as "black smokers." In contrast, vents releasing cooler water (100 to 300 degrees Celsius) do not contain sufficient sulfide or metals in solution to cause this effect. Instead, mixing of these hydrothermal waters with seawater causes white particles of silica, anhydrite, and barite to form, forming a white cloud in the mixing plume and giving rise to the name "white smoker." White smokers may reflect mixing just below the surface, with the result that most of the metals associated with white smoker vents may be deposited just below the seafloor. Because the original hydrothermal vent waters are cooler and thus denser, plumes from white smokers rise less far than those from black smokers before obtaining neutral buoyancy and spreading laterally.

Mixing progressively dilutes the hydrothermal water until the mixed water cools to the point that it has the same density as ambient seawater and no longer rises. A tremendous amount of seawater must be mixed with the hydrothermal fluids before the plume attains neutral buoyancy. This may ultimately involve 10,000 or even 100,000 times as much seawater as hydrothermal water. During dilution, the mixture rises tens to hundreds of meters to a level of neutral buoyancy, eventually spreading laterally as a distinct hydrographic and chemical layer, recognizable hundreds or even thousands of kilometers away from the hydrothermal vent. Continued settling of hydrothermal iron and magnesium from these layers is the source of most of the metals in manganese nodules of the abyssal seafloor, far away from the mid-ocean ridges.

Vast quantities of metals are deposited as sulfides around hydrothermal vents, especially as stockwork and in sediments around chimneys. These are particularly rich in iron, copper, zinc, and lead, and the exact composition of these deposits reflects several controls, including temperature of the hydrothermal fluid, water depth, composition of the source rock, and flow regime in the portion of the hydrothermal system that lies beneath the seafloor. Many of the world's great ore deposits seem to be fossil seafloor hydrothermal systems, with the possible exception that these may have mostly formed at convergent plate margins instead of divergent plate margins, where most modern hydrothermal systems are known. This may be part of the reason why ancient massive sulfide deposits are typically much larger than modern ones.

STUDY OF HYDROTHERMAL VENTS

To a marine biologist, hydrothermal vents are the oases of the deep seafloor. In contrast to most of the ocean bottom, which supports few animals, hydrothermal vents teem with life. In fact, the analogy to a desert oasis falls short because there is much more life around a hydrothermal vent. Life in most of the deep sea is scarce because food is scarce. There is no sunlight, so plants, the basis of most food chains, cannot grow. In contrast, food is abundant around hydrothermal vents; of-

ten, life is so crowded that the animals obscure the seafloor.

Vent fields can be divided into three biotic zones reflecting distance from the hydrothermal vents: the vent opening, the near-field, and the periphery. The bulk of the biomass is at the vent openings, where the density of life is so great that it appears to be limited by space, not food. Around hydrothermal vents on the East Pacific Rise and Galápagos Rift, vent openings are dominated by endosymbionts such as tube worms, clams, and mussels. These endosymbionts can grow to great size; some tube worms are 1 meter long with 3-meter tubes, while clams up to 30 centimeters long are common. Other animals live with these, including limpets, polychaete worms, bresiliid shrimp, crabs, and fish. Vent chimneys at several Mid-Atlantic Ridge vents are almost entirely populated by bresiliid shrimp.

Autotrophic bacteria are the primary producers around hydrothermal vents. These use sulfur to convert carbon dioxide, water, and nitrate into essential organic substances in a fashion similar to the way in which plants use sunlight during photosynthesis. This process is called "chemosynthesis." Autotrophic bacteria live within the subsurface plumbing system of the vents, on the seafloor, and suspended in and around the hydrothermal plume itself, sometimes in such abundance that they color the water a milky blue or carpet the seafloor in white or bright yellow mats. Some of these bacteria can tolerate incredibly high temperatures, up to 110 degrees Celsius or more. All the other life around a hydrothermal vent ultimately feeds off the autotrophic bacteria. Some of the autotrophic bacteria are symbiotic, living within larger vent animals in a mutually beneficial relationship. The bacteria provide food, and the animals provide essential inorganic nutrients.

The flow of water around hydrothermal vents controls the distribution of life. Because cold water mixes with the hydrothermal plume, cold seawater flows in from all directions to converge on the rising plume. This means that the sulfur on which the autotrophic bacteria depend is not distributed around the vent. For a similar reason, the animals that feed or depend on the bacteria cannot survive away from the vent opening. The near-field is mostly populated by suspension feeders, animals that capture bacteria and other organisms that drift away from the vent opening. It is presumed that these animals live as close as possible to the vent but are forced to maintain a certain distance because of toxic effects resulting from very high concentrations of heavy metals. Animals living on the vent periphery include scavengers and other types that sustain themselves from bacteria that settle out of the hydrothermal plume.

Most of the animals that live around hydrothermal vents live nowhere else except in other sulfur-rich, reducing environments, such as in the rotting carcasses of whales or in "cold seeps," where cold, chemically altered seawater percolates up through the seafloor. Many vent animals may be ancient, originating in Mesozoic or earlier times. These animals may have been insulated from the effects of surface catastrophes, such as the meteor collision hypothesized to have killed the dinosaurs about 65 million years ago.

There are a number of fascinating features about the life around hydrothermal vents, but one of the most intriguing is the suggestion that this environment is similar to what existed when life was just developing about 4 billion years ago. It is very likely that the development of life involved synthesis of organic chemicals in vents and that the first autotrophic life-forms were chemosynthetic, not photosynthetic.

Robert J. Stern

CROSS-REFERENCES

BIBLIOGRAPHY

Allen, Philip A. *Earth Surface Processes*. Oxford, England: Blackwell Science, 1997. This book serves as a clear introduction to oceanography and the Earth sciences. Allen details the processes, properties, and composition of the ocean. Color illustrations, maps, index, and bibliography.

Anderson, R. N. *Marine Geology: A Planet Earth Perspective*. New York: John Wiley & Sons, 1986. A textbook intended for college undergraduates, with an excellent chapter on metal deposits and marine geology. This well-written chapter (chapter 7) and other parts of the book are accessible to a general, scientifically interested audience.

Francheteau, J. "The Oceanic Crust." *Scientific American* 249 (1983): 114-129. A good overview of the nature, composition, and mode of formation of the oceanic crust.

Humphris, Susan E., et al., eds. *Seafloor Hydrothermal Systems*. Washington, D.C.: American Geophysical Union, 1995. The most technically comprehensive overview of physical, chemical, biological, and geochemical interactions. A wealth of pictures, figures, and references are included in this volume. Technical but useful.

Karl, David M., ed. *The Microbiology of Deep-Sea Hydrothermal Vents*. Boca Raton, Fla.: CRC Press, 1995. This book looks into the microbiology of extreme and unusual environments such as hydrothermal vents on the deep ocean floor. There is special attention given to what causes the vents, their benefits, and the environments they create. Suitable for the nonscientist.

Macdonald, K. C., and B. P. Luyendyk. "The Crest of the East Pacific Rise." *Scientific American* 244 (1983): 100-118. Provides a good look at an outstanding example of a mid-ocean ridge.

Parson, L. M., C. L. Walker, and D. R. Dixon, eds. *Hydrothermal Vents and Processes*. London: Geological Society, 1995. A thorough look into deep-sea and hydrothermal vent ecology. There is much discussion about the seafloor and plate techtonics as well. Suitable for the high school reader or someone without a background in oceanography. Color illustrations and maps.

Rona, P. A., and S. D. Scott. "A Special Issue on Sea-Floor Hydrothermal Mineralizations: New Perspectives." *Economic Geology* 88, no. 8 (1993): 1933-2249. Deals with how mineral deposits form around hydrothermal vents, with many examples of individual vent fields from around the world. The preface includes a complete listing and summary of all known seafloor hydrothermal systems. Intended for the scientist, but abundant pictures, figures, and references make it useful to all who are interested in the topic.

OCEAN-ATMOSPHERE INTERACTIONS

The interactions between the oceans and the atmosphere are basic to the understanding of oceanography and meteorology. The liquid and gaseous envelopes of Earth have a powerful effect on weather and climate on a global scale.

PRINCIPAL TERMS

ATMOSPHERE: five clearly defined regions above the Earth's surface composed of layers of gases and mixtures of gases, water vapor, and liquid particles

CORIOLIS EFFECT: the deflection of any moving body on Earth to the west or east, depending on whether the latitude is north or south, respectively; this effect is the result of the Earth's rotation

EL NIÑO: the phase of a gigantic meteorological system called the Southern Oscillation that links the ocean and atmosphere in the Pacific; it is characterized by a weakening of the trade winds

GYRES: the major wind systems responsible for broadly symmetrical patterns of surface-water transport

KUROSHIO: the current, also known as the Japan Current, where cold continental air flows over warm ocean currents moving toward the poles

LA NIÑA: the phase of the Southern Oscillation that brings cold water to the South American coasts, which makes easterly trade winds stronger, the waters of the Pacific off South America colder, and ocean temperatures in the western equatorial Pacific warmer than normal

ATMOSPHERE AND OCEAN

The largest fraction of the heat energy the atmosphere receives for maintaining its circulation is derived from the condensation of water vapor originating mainly from marine evaporation. Therefore, fundamental to the understanding of atmospheric behavior and oceanic behavior is an understanding of the processes occurring at the air-sea boundary. The interactions between the marine and atmospheric environments involve constant interchanges of moisture, heat, momentum, and gases.

The study of oceanic and atmospheric interactions is the examination of a huge gaseous body and a massive liquid body, neither of which is ever homogeneous in content. As the makeup of the atmosphere varies, dependent on the areas over which it flows, so the content of the oceans varies in density, temperature, salinity, rate of movement by regular currents, and surface movement under the influence of winds. As a consequence, this interaction is very complex and not entirely understood. Certain conditions, however, that are regularly met can act as a general guide to understanding ocean-atmosphere interactions.

Atmospheric circulations depend on heat rising from the Earth's surface. Because of the large area of the Earth covered by oceans, the main source of heat to the atmosphere is the sea surface. In the oceans, the heat supply is primarily from the atmosphere and the Sun. Heat-supply processes are important to the development of convection currents in the surface layer of the ocean, to the local exchange of energy with the atmosphere, and to slower, deep-water circulation. The heat exchange between the ocean and atmosphere has a pattern similar to evaporation. Wherever the surface of the ocean is warmer than the atmosphere, heat is transferred from the ocean to the air via latent heat and is moved to great heights by eddies and convection currents in the air.

The range of weather extremes is smaller over the ocean than over land. Because of the enormous heat-storage capacity of the ocean, it tends to stabilize atmospheric conditions and qualities. The upper layer of the oceans (approximately 70 meters in depth) can store approximately thirty times more heat than the atmosphere. Ocean climates, therefore, are mostly determined by the atmospheric circulation and latitude. Ocean climate and atmospheric circulation are both affected by the solar distribution over the Earth's surface, which is a function of the latitude and the season of the year. Northern latitudes receive proportionately less solar radiation than do latitudes at the equator; in winter, polar latitudes receive no solar radiation. Because of the greater capacity of the oceans to store heat, for a given change in heat content, the temperature change in the atmosphere will be around thirty times greater than in the ocean. Therefore, the ocean will lose its heat by radiation much more slowly than will the air. Land, which is intermediate with regard to heat storage and heat loss, can be modified by the effect of heat storage and heat loss of the ocean.

PROCESSES AT THE AIR-SEA BOUNDARY

The atmosphere adjacent to the oceans is constantly interacting with the oceans. Air does not simply flow along the surface of the sea but has a frictional effect or wind stress, which causes the surface water to be carried along with the wind. Wind stresses on the surface of the sea produce ocean waves, storm surges, and shallow ocean currents. Pure wind-driven currents are the result of frictional wind surface stresses and the Earth's rotational motion. This rotational motion can be seen in the Coriolis effect, which causes a deflection of air currents and ocean currents to the east or west, depending on the hemisphere. Because the sea is continually in contact with the atmosphere, the gases that occur in the atmosphere are also found in seawater. The concentration of the gases depends on their solubilities and on the chemical reactions in which they are involved. These concentrations are affected by temperature, which is determined by many factors, and by wind and wave actions.

The sea also has a large storage and regulating capacity with respect to processes involving carbon dioxide in the atmosphere and in the sea, including those processes relevant to plant life and photosynthesis. This whole group of reactions concerning carbon dioxide is extremely complex. Carbon dioxide, a powerful reflector of heat, forms a heat-conserving umbrella in the atmosphere that reflects heat back to the Earth's surface. This "greenhouse effect" acts much like a horticultural greenhouse, which traps heat and raises the temperature of the air inside the greenhouse.

The largest fraction of radiant energy absorbed by the oceans is used in evaporation. The maximum evaporation and heat exchange between the sea and the atmosphere occur where cold continental air flows over warm ocean currents moving toward the poles. Examples of this phenomenon are the Kuroshio (Japan Current) and the Gulf Stream. The radiant energy that is absorbed and stored by the oceans at tropical latitudes may be given off to the atmosphere elsewhere. This process is important to understand in terms of the Southern Oscillation and the effect of El Niño and La Niña.

The major wind systems are responsible for broadly symmetrical patterns of surface-water transport known as gyres, which rotate clockwise around the North Pacific and North Atlantic and counterclockwise around the South Pacific and South Atlantic oceans. Their tropical segments are the North and South Equatorial currents, which are driven westward by trade winds. The Equatorial Countercurrent, a compensating flow, travels west to east in the Pacific between the North and South Equatorial currents along a course that averages a few degrees north of the equator.

EL NIÑO AND LA NIÑA

Marked by warm water and high winds from the western Pacific, El Niño typically brings heavy winter rains to Peruvian deserts and warm weather to the West Coast of the United States. El Niño arises through interaction between the oceanic and atmospheric systems. During El Niño, the southeast trade winds over the equatorial Pacific collapse, allowing warm water from the western Pacific to flow eastward along the equator. This warm-water flow suppresses the normal upwelling of cold, nutrient-rich water and leads to the northward displacement of fish normally feeding on the nutrient-laden cold water.

El Niño is part of a gigantic meteorological system called the Southern Oscillation that links the ocean and atmosphere in the Pacific. This system normally functions as a kind of huge heat pump, distributing energy from the tropics at the equator to the higher latitudes through storms that develop over the warm western Pacific. Another part of the Southern Oscillation has been dubbed La Niña, which brings cold water to the central Pacific. La Niña exaggerates the normal conditions of the system. During this activity, easterly trade winds are stronger, the waters of the eastern Pacific off South America are colder, and ocean temperatures in the western equatorial Pacific are warmer. Atmospheric and oceanic conditions in the equatorial Pacific region can generate powerful effects on global weather. Therefore, the study of this interaction is essential.

STUDY OF OCEAN-ATMOSPHERE INTERACTIONS

Because the study of the ocean-atmosphere interactions is concerned with the boundary between marine and air masses, it is of necessity an interdisciplinary type of study. The data used are gathered by oceanographers and meteorologists who have made this interaction the area of their research. The same instruments are used, drawing heavily upon the type of data collected, that are employed in oceanographic and meteorologic research, with a shift in emphasis. Because gathering the data is expensive and utilizes costly, specialized kinds of equipment, nearly all studies are conducted by government scientists or are sponsored by government grants.

Scientists who study ocean-atmosphere interactions are interested in seeing how these huge bodies of matter affect each other and how these effects influence the weather and climate in the rest of the world. They are also interested in the ability to predict weather and climate changes more accurately. Each element researched, such as salinity, is compared with some other element, such as temperature, to see what relationship may exist. Air temperature and air movement are compared with wave movement, changing ocean currents, and water temperature. Each of a vast number of data is examined for possible interrelations and interactions. When interrelations are found, the data are fed into a computer model for correlation with other data. Computer models have been effective to some extent in predicting the results of interactions between marine and atmospheric environments. Separate computer models are used with the output aimed at the interactions between these forces.

Scientists do not fully understand all the mechanisms that link certain phenomena of the ocean and atmosphere interactions, such as El Niño and La Niña. It is their hope that in studying and understanding the mechanisms involved, they may be able to make accurate long-range predictions of the amount and area of precipitation in specific regions.

SIGNIFICANCE

The interaction between the oceans and the atmosphere can cause immense problems on a global scale. An example of this interaction and its consequences is seen in the phenomenon known as El Niño. Every three to five years, the surface waters of the central and eastern Pacific Ocean become unusually warm at the equator. Warm currents and torrential rains are brought to the normally dry desert area of central Peru, and nutrient supplies for marine life along the west coast of South America are disrupted. Hardship can occur as a result, such as widespread flooding in Peru.

La Niña brings an effect opposite to that of El Niño; easterly trade winds are stronger, the waters of the eastern Pacific off South America are colder, and ocean temperatures in the western equatorial Pacific are warmer than normal. As a result, the deserts in Peru and Chile become drier than normal, and the Indian subcontinent is inundated by heavier-than-usual rainfall and flooding. In Bangladesh in late 1988, heavy rains and flooding killed more than 1,000 people, destroyed the homes of 25 million people, inundated 5 million acres of rice land, and damaged 70,000 kilometers of roads. While the storms were attributed to the La Niña phenomenon, much of the flooding was the result of massive deforestation in the Himalaya and foothills, which allowed water to rush down from the barren, eroded hills onto Bangladesh near sea level.

George K. Attwood

CROSS-REFERENCES

Abyssal Seafloor, 651; Basaltic Rocks, 1274; Carbonate Compensation Depths, 2101; Deep Ocean Currents, 2107; Geysers and Hot Springs, 694; Gulf Stream, 2112; Heat Sources and Heat Flow, 49; Hydrothermal Mineralization, 1205; Hydrothermal Vents, 2117; Life's Origins, 1032; Magmas, 1326; Manganese Nodules, 1608; Ocean Basins, 661; Ocean Drilling Program, 359; Ocean-Floor Drilling Programs, 365; Ocean-Floor Exploration, 666; Ocean Pollution and Oil Spills, 2127; Ocean Tides, 2133; Ocean Waves, 2139; Oceans' Origin, 2145; Oceans' Structure, 2151; Ophiolites,

BIBLIOGRAPHY

Curry, Judith A., and Peter Webster. *Thermodynamics of Atmospheres and Oceans*. San Diego, Calif.: Academic Press, 1999. This book offers a look at the effects of the interaction between oceans and the atmosphere on weather patterns and climatic changes. Provides good insight into the role that atmospheric thermodynamics play in meteorology. Illustrations, maps, and index.

Glantz, Michael H., and J. Dana Thompson. *Resource Management and Environmental Uncertainty: Lessons from Coastal Upwelling Fisheries*. New York: John Wiley & Sons, 1981. Excellent information on and discussions of El Nino. Considers social, economic, and political values in terms of the effects from the ocean-atmosphere interactions and the weather and climate changes they produce. Not a difficult reading level.

Ittekko, Venugopalan, et al., eds. *Particle Flux in the Ocean*. New York: John Wiley and Sons, 1996. This volume contains descriptions of the chemical and geobiochemical cycles of the ocean, as well as the ocean currents and movement. Suitable for the high school reader and beyond. Illustrations, index, bibliography.

Kerr, R. A. "La Niña's Big Chill Replaces El Niño." *Science* 241 (August 26, 1988): 1077-1078. An excellent source, although *Science* magazine is not available in all libraries. Illustrated. For advanced high school science majors and college students.

Knox, C. E. "Hot and Cold Pacific Fed Midwest Drought." *Science News* (October 15, 1988): 247. Describes the influence of El Niño on the mid-1980's drought suffered by the midwestern United States. Includes map. *Science News* gives accurate summaries of articles appearing in less accessible publications.

Linden, E. "Big Chill for the Greenhouse." *Time* 132 (October 31, 1988). Similar to the Kerr article (above) but at an easier reading level. *Time* magazine frequently offers brief but accurate coverage of new scientific research and discoveries.

Majumdar, Shyamal K., et al., eds. *The Oceans: Physical-Chemical Dynamics and Human Impact*. Easton: Pennsylvania Academy of Science, 1994. This compilation of essays provides an overview of oceanography while focusing on the relationship between the atmosphere and oceans. Many of the sections deal with both the effects of this relationship on humans and the effects of humans on the ocean environment. Suitable for the college-level reader.

Schmitt, Raymond W. *The Ocean Freshwater Cycle*. College Station: Texas A&M University, 1994. This brief book examines the importance of the interaction between the atmosphere and the oceans in the study of meterology and climatology. Suitable for the careful high school reader. Color illustrations and bibliographical references.

Thurman, Harold V. *Introductory Oceanography*. 8th ed. Columbus, Ohio: Merrill, 1997. An introductory college-level text. Chapter 7, "Air-Sea Interactions," describes the Coriolis effect, the "heat budget" of the world ocean, and weather and climate. Chapter 8, "Ocean Circulation," describes currents. Includes glossary and index. Well illustrated.

OCEAN POLLUTION AND OIL SPILLS

Oil spills resulting from human error often affect marine and coastal areas. Past oil spills in different areas of the world demonstrate that environmental damage depends on the toxicity and the persistence of the oil; both vary widely depending on a variety of factors.

PRINCIPAL TERMS

BOOM: a floating oil fence made of a weighted flexible sheet hanging vertically above and below the sea surface; used to contain or move floating oil during a spill, they are most effective during calm seas

ENVIRONMENTAL PERSISTENCE: the relative length of time that oil remains in the environment with the possibility of causing negative environmental effects

MECHANICAL TOXICITY: the process by which most organisms are impacted or killed by spilled oil; ingesting or being coated by oil may lead to death by suffocation or exposure

MOUSSE: a gelatinous oil-water emulsion resembling chocolate pudding that is created when crude oil is spilled in churning seawater

SKIMMER: a specialized oil-spill response vessel that picks up floating oil with an absorbent conveyor belt; skimmers are most effective during calm seas

TOXICITY: a measure of the dose required to produce a negative health effect

WHY OIL SPILLS OCCUR

Oil is the lifeblood of the American lifestyle. With only about 6 percent of the world's population, the United States uses more than 25 percent of world's petroleum, the majority of which is imported from other nations. The percentage of imported oil that Americans use has grown steadily since 1982, largely negating the lessons learned about cartel politics and energy conservation in the late 1970's. In fact, the amount of oil imported into the United States continued to grow even during periods when its merchandise trade deficit was in decline. Americans have also demonstrated that they are willing to go to war for oil. In the Gulf War with Iraq, the United States paid a huge cost to protect world oil markets in spite of the fact that very little of the oil produced in the Arabian Gulf is imported to the United States. Most of the nation's imported oil comes from Africa, Central and South America, and the Indo-Pacific region. Domestic production continues to be an important but dwindling source, with most coming from the north slope of Alaska via the Alaska Pipeline and Port Valdez in Prince William Sound.

Almost all imported and Alaskan oil is transported to U.S. refineries and consumers by ocean-going tankers. Most oil spills result from marine transportation accidents, with human error usually playing a major role. Navigation errors, equipment malfunctions, bad judgment, even the inability of all crew members to speak a common language have all been major contributing factors in the largest and most environmentally damaging oil spills. Not all oil spills are environmental disasters, but spilled oil can decimate plant and animal populations by a combination of mechanical toxicity and chemical toxic effects resulting from an organism's physiological reaction to the chemicals present in oil.

ENVIRONMENTAL DAMAGE

The most common sight during an oil spill is dark, gelatinous masses of "mousse"—an oil and water emulsion that floats on the water, sticking to everything with which it comes into contact. Mousse usually causes the majority of the environmental damage during an oil spill by the process of mechanical toxicity, as it suffocates and smothers organisms that ingest it or are covered by it. Seabirds and furry marine mammals are highly susceptible to this process, succumbing to exposure, dehydration, or starvation.

Crude oil is a complex mixture of thousands of different chemicals called hydrocarbons, named

HISTORIC OIL SPILLS

Mar. 24, 1989   EXXON VALDEZ OIL SPILL: Shortly after midnight, the oil tanker *Exxon Valdez* struck Bligh Reef in Prince William Sound, Alaska, spilling more than 11 million gallons of crude oil. The spill was the largest in U.S. history. The spill posed threats to the food chain that supports Prince William Sound's commercial fishing industry, as well as ten million migratory shore birds and waterfowl, hundreds of sea otters, porpoises, sea lions, and several varieties of whales. Through direct contact with oil or because of a loss of food resources, many birds and mammals died. In the aftermath of the *Exxon Valdez* incident, Congress passed the Oil Pollution Act of 1990, which required the Coast Guard to strengthen its regulations on oil tank vessels and oil tank owners and operators.

Mar. 28, 1993   FAIRFAX COUNTY OIL SPILL: A rupture occurred in an oil pipeline, sending a 100-foot plume of fuel oil into the air. The high-pressure pipeline, owned by the Colonial Pipeline Company, released more than 400,000 gallons of oil to the environment before it could be shut down and fully drained. The rupture resulted in one of the largest inland oil spills in recent history. The oil affected nine miles of the nearby Sugarland Run Creek as well as the Potomac River.

Aug. 10, 1993   TAMPA BAY SPILL: Three vessels collided at the entrance to Tampa Bay, Florida, resulting in the release of an estimated 328,000 gallons of oil. Two barges, one containing number six fuel oil and one containing jet fuel, gasoline, and diesel fuel, and a Philippine freighter carrying phosphate rock were involved. A damaged fuel tank aboard the barge *Bouchard 155* was the source of most of the oil released, forming a 17-mile oil slick on the water surrounding the vessel. The second barge, the *Ocean 255*, exploded and caught fire upon impact. Two miles of sandy beach were affected, including birds, turtles, and other wildlife in a park along the Gulf of Mexico.

Jan., 1998   ASHLAND OIL SPILL: A four-million-gallon oil storage tank owned by Ashland Oil Company, Inc., split apart and collapsed at an Ashland oil storage facility, releasing diesel oil over the tank's containment dikes, across a parking lot on an adjacent property, and into an uncapped storm drain that emptied directly into the river. Within minutes the oil slick moved miles down river, washing over two dam locks and dispersing throughout the width and depth of the river. The oil was carried by the Monongahela River into the Ohio River, temporarily contaminating drinking water sources for an estimated one million people in Pennsylvania, West Virginia, and Ohio, contaminating river ecosystems, killing wildlife, damaging private property, and adversely affecting businesses in the area.

SOURCE: Environmental Protection Agency, http://www.epa.gov/oilspill.

after a molecular structure based on hydrogen and carbon atoms. Different hydrocarbons vary in their chemical properties, toxicity, and behavior during an oil spill. The major groups are classified by molecular geometry and weight. The low-molecular-weight molecules (aliphatics) are single-bonded, chain-shaped molecules, such as gasoline. They are the most chemically reactive and volatile, and they are acutely toxic. These compounds tend to evaporate or burn easily during an oil spill and therefore do not persist in the environment for long periods. Intermediate-molecular-weight hydrocarbons, or aromatics, are ring-shaped molecules, such as benzene. They are also highly reactive and cause biological impacts be-

cause of both acute and chronic toxicity. Aromatic hydrocarbon compounds are more environmentally persistent than aliphatics. Since many are carcinogens, they can cause different forms of biological damage, disease, and death even after a long time period and in low doses. The high-molecular-weight oil compounds are mostly polycyclic aromatic hydrocarbons structured of ring shapes bonded together to form molecules. Although they are not very chemically reactive and do not dissolve well in water, many are carcinogenic. They tend to be very environmentally persistent.

For hundreds of millions of years before humans evolved, oil was "spilled" naturally into the world's oceans by natural oil seeps—fractures in

Earth's crust that tap deep, oil-bearing rocks. A variety of natural processes act to reduce the environmental impacts of this oil, and these same processes also take place during a human-caused oil spill. Oil is dispersed from the oil slick and into the larger environment by five basic processes. Evaporation of the low-molecular-weight hydrocarbon compounds removes most of the oil relatively quickly. Sunlight can degrade additional oil in a process called photodegradation if the oil is exposed for enough time. Because oil is an organic substance, additional oil is removed by natural biodegradation thanks to "oil-eating" microorganisms. Most of the rest of the oil either washes up onto a coastal area or breaks up into heavy "tar balls" rich in high-molecular-weight hydrocarbons that eventually sink.

Some oil spills put so much oil into the environment that these processes cannot respond quickly enough to prevent environmental damage. Other factors can also enhance environmental damage from oil spills. Some types of oil or refined petroleum products are more toxic than others. Oil spills in cold climates generally cause more damage because cold temperatures retard evaporation and the microbial metabolic rates necessary for rapid oil removal. Furthermore, sunlight is often of low intensity, which retards photodegradation. Wave conditions and tidal currents can affect how much oil washes up onto a coastal area and how rapidly it is moved elsewhere or removed. Finally, the amount of environmental damage from an oil spill is highly dependent on the type of coastal environment oiled in the spill, as coastal environments vary in the density (or biomass) and varieties of wildlife. Coastlines also vary in the degree to which they are sheltered from natural oil-removal processes. In general, rocky headlands, wave-cut rock platforms, and reefs exposed to high wave activity suffer far less damage during an oil spill than do sheltered marshes, tidal flats, and mangrove forests. The damage on beaches is related to the grain size of the beach sediment. Fine-sand beaches are relatively flat and hard-packed, and oil does not soak into the sediment or persist for long. Oil will soak deeply into coarse sand, gravel, and shell beaches, causing more damage over a longer period.

Most of what has been learned about oil spill behavior, environmental damage, and oil spill cleanup techniques comes from studying past spills. In most cases, spill prevention is far cheaper and more effective than spill response, and cleanup efforts usually capture very little of the spilled oil.

Ixtoc I

The Ixtoc I spill of June 3, 1979, was the result of an explosion, or "blowout," of an offshore oil well that was drilling into a subsurface oil reservoir. Although human error was definitely a factor, the cause of the blowout remains unresolved. It has been blamed on the use of drilling mud that was not dense enough to counteract the pressure of the oil and gas at depth, as well as on the improper installation of the blowout preventer—a fail-safe device used on drilling rigs to prevent just this type of disaster. The result was a continuous 290-day oil spill, during which an estimated 475,000 metric tons of crude oil (one metric ton equals approximately five barrels) were released into the environment. In addition to doing considerable environmental damage on the coast of Mexico, oil fouled much of the barrier island coast of Texas. However, most of the oil did not make it to shore, and the final accounting for this spill gives a good indication of the long-term fate of spilled oil in offshore areas: 1 percent burned at the spill site, 50 percent evaporated, 13 percent photodegraded or biodegraded, 7 percent washed up on the coast (6 percent in Mexico, 1 percent in Texas), 5 percent was mechanically removed by skimmers and booms, and 24 percent sank to the seafloor (assumed by mass balance).

The *Exxon Valdez*

The *Exxon Valdez* oil spill—which occurred in Prince William Sound, Alaska, on March 24, 1989—is a good example of how environmental damage follows human error and inadequate response. After departing Port Valdez with a full cargo, the *Exxon Valdez* oil tanker struck a well-charted submerged rock reef located 1.6 kilometers outside the shipping lane. The ship was under the command of an unlicensed third mate in calm seas and left the shipping lane with permission from the Coast Guard to avoid ice. However, it strayed too close to the reef before evasive action was attempted. The captain, who had a history of drunk driving convictions, was in his cabin under the influence of alcohol during events leading to accident. His blood alcohol level nine hours after the grounding was measured at 0.06 percent; the estimate at the time of the accident was 0.19 percent—almost twice the legal level for drivers in California. Convicted of negligence and stripped of his commander's license, he was subsequently employed as an instructor to teach others to operate supertankers.

Leaking oil was observed immediately. Oil-spill response crews funded by Exxon and the Alyeska Pipeline Consortium—oil companies that used the Port Valdez terminal—were poorly prepared and reacted too slowly and with inadequate equipment. The first response arrived ten hours after the accident with insufficient booms and skimmers. Chemical dispersants applied to break up the oil slick were ineffective in the calm seas and caused the oil slick to thin and spread more rapidly. Four days later, the weather changed: 114-kilometer-per-hour winds mixed the oil with seawater, creating a frothy mousse. More than 65,000 metric tons of oil spilled over several weeks. About 15,600 square kilometers of ocean and 1,300 kilometers of shoreline were affected. Federal estimates of wildlife mortality include 3,500 to 5,500 otters; 580,000 seabirds; and 300 deer poisoned by eating oiled kelp. Economic damages totaled more than $5 billion. The long-term effects on commercial marine organisms, larval organisms, and bottom-dwelling life are not known.

Exxon promised to clean nearly 500 kilometers of shoreline by September, 1989, but only cleaned 2 kilometers during the first month after the spill. Exxon and its contractors used a variety of cleanup techniques, including placing booms and skimmers, sopping up oil with absorbent materials, scraping oil by hand from rocks, stimulating the growth of oil-eating bacteria cultures, and washing coastal areas with cold water, hot water, and steam. The use of hot water and steam was effective at cosmetically removing surface oil, but it did not remove oil that had soaked into the sediment; the technique killed most organisms that had escaped the oil. The oil washed from the beach was to be collected by booms and skimmers offshore, but this process was so inefficient that much of the oil migrated to tide pools that had not been affected by the spill directly. Ironically, only eighteen months after the spill, life had re-

turned to oiled coasts that had received little or no cleanup, while beaches cleaned with hot water were still relatively sterile and required several years to repopulate. Exxon announced that it would not return to clean more shoreline in 1990 but relented under threat of a court order from the Coast Guard to enforce federal cleanup requirements. During the summer of 1990, shoreline cleanup resumed, including application of fertilizer to stimulate growth of naturally occurring oil-eating bacteria, a technique that is not very efficient in the cold waters of southern Alaska.

The tale of the *Exxon Valdez* is not complete without mentioning that the Port Valdez Coast Guard did not have state-of-the-art radar equipment for monitoring ship movement in this heavily used and environmentally sensitive area. In the early 1980's, federal and state funds for monitoring the Port Valdez oil companies' compliance with oil-spill preparedness legislation had been cut by more than 50 percent. The original environmental impact statement for oil-handling activity in Prince William Sound included an agreement that defines cleanup responsibility for oil spills. Exxon, as the company responsible for the spill, was to pay the first $14 million of cleanup costs, with $86 million in additional cleanup funds from the Alyeska contingency fund. Thus, the maximum financial responsibility to oil companies from a spill was $100 million unless the spill was judged to be caused by negligence. Cleanup activities ceased eighteen months after the spill with total expenditure of $2.2 billion; most of this at taxpayer expense. In 1994, a federal court unanimously awarded $5.3 billion in punitive and compensatory damages, the largest-ever jury award, to some thirty-five thousand people impacted by the spill. By June, 1999, Exxon had yet to pay a single dollar as the case continued through the legal process. Finally, it is interesting to note that Exxon's estimate of cleanup costs in late 1989 were $500 million, and it carried $400 million of oil spill liability insurance. Exxon saved $22 million by not building the *Exxon Valdez* with a double hull; its 1988 annual profits were $5,300 million.

According to National Oceanic and Atmospheric Administration (NOAA) estimates, less than 1 percent of *Exxon Valdez*'s oil burned at the site, 20 percent evaporated, 8 percent was mechanically removed, and nearly 72 percent was deposited on the seafloor. According to Exxon's estimates, 7 percent of the oil burned at the site, 32 percent evaporated, 9 percent photodegraded or biodegraded, 15 percent was mechanically removed, and 37 percent was assumed deposited on the seafloor.

OPERATION DESERT STORM

The February, 1991, Operation Desert Storm oil spill—the largest oil spill in history—occurred when the Iraqi military opened valves and pumps at Sea Island Terminal, a tanker loading dock located 16 kilometers off the coast of Kuwait. This facility has a production capacity of 100,000 barrels per day, about three *Exxon Valdez* loads each week. The Iraqis also opened plugs on five Kuwaiti tankers, spilling an additional 60,000 barrels. The estimate for the entire spill is 6 million barrels, or roughly 30 times the volume of the *Exxon Valdez*. About 650 square kilometers of coast were heavily contaminated.

Three days after starting the spill, the Iraqis ignited the oil leaking from the terminal. This was the best thing to happen from an environmental perspective. During most spills, more oil is removed by natural evaporation than by any cleanup technique; igniting the oil merely speeds up this process. Burning can be an important mechanism for removing oil from the sea and avoiding environmental damage, and tests have shown purposeful ignition in open water away from the coast to be an excellent oil-slick fighting strategy. However, this must be done within the first few hours of the spill—in order to maintain the fire, the slick must be more than 1 millimeter thick and must contain relatively little emulsified water. To maintain thickness, the slick is best surrounded with fireproof booms. However, at the time of the Operation Desert Storm spill, almost all the fireproof boom in the world was in Prince William Sound. Saudi Arabia also used dispersants on portions of the slick, but this effort was too late to be effective before a thick mousse had formed.

The prime objectives of causing the spill were to hamper an amphibious military landing by oiling the beaches and to disrupt desalinization of drinking water at Khafji and Jubail, the two primary sources of potable water for Saudi Arabia. The Saudis used booms to protect the plant intakes with great success. The retreating Iraqis also ignited more than seven hundred of about one thousand inland wells, resulting in an additional 6 million barrels per day burned. This volume eventually made the marine spill insignificant, and the burning created 3 percent of total global carbon emissions during the time period of the event.

The Arabian Gulf is an unusual body of water. It is very shallow (average 33 meters) and is nearly enclosed as a marine basin. Because it is also microtidal (the tidal range is less than 0.6 meter), it flushes out slowly (once every two hundred years, compared with once every few days for Prince William Sound). It is also important to remember that this is not a pristine marine environment. Natural oil seeps are very common, there is a general lack of environmental standards and poor cooperation among Persian Gulf nations, and virtually no oil spill preparations or equipment were present in this part of the world. Earlier spills had occurred in the region, but they typically were associated with ongoing wars; the hostile environment made it difficult to utilize spill abatement specialists and equipment. For example, during the Iran-Iraq War, Iraq attacked an Iranian offshore platform (Nowruz) in 1982, spilling more than 2 million barrels of oil, a volume nearly half as large as the Operation Desert Storm spill. Losses of marine mammals and birds were great, and populations had not rebounded by the time of Operation Desert Storm.

During the Operation Desert Storm oil spill, about 180 kilometers of Saudi Arabian coastline were oiled (65 kilometers were severely damaged), and oil reached south as far as the United Arab Emirates and Bahrain. Much of the southern Kuwaiti coast (sea-grass beds, marsh, and mangroves) was severely damaged, and about 25 percent of the Saudi shrimp industry was lost. Although some twenty thousand wading birds were killed, no deaths of dolphins or dugongs were reported. However, these animals suffered greatly during the Iran-Iraq War. Estimates of the time required for ecological renewal of the Persian Gulf following the Operation Desert Storm spill (one to four years) were relatively short for two reasons: The high water temperature results in high microbial activity and biodegradation of the oil, and much of the oil was burned.

James L. Sadd

BIBLIOGRAPHY

Alaska Wilderness League. *Preventing the Next Valdez: Ten Years After Exxon's Spill New Disasters Threaten Alaska's Environment.* Washington, D.C.: Alaska Wilderness League, 1999. This short book, published by the Alaska Wilderness League in conjunction with the Sierra Club, reports on the status of the environment in Prince William Sound ten years after the *Exxon Valdez* oil spill.

Etkin, Dagmar Schmidt. *Financial Costs of Oil Spills in the United States.* Arlington, Mass.: Cutter Information, 1998. Etkin discusses the economic aspects of water pollution, particularly oil spills, in the United States. Includes information on liability for oil pollution damages. Bibliography, illustrations.

_____. *Marine Spills Worldwide.* Arlington, Mass.: Cutter Information, 1999. Etkin's report includes statistics concerning oil spills, the offshore oil industry, and oil pollution in the sea worldwide between 1960 and 1999. Illustrations and charts.

Hall, M. J. *Crisis on the Coast.* Portland, Or.: USCG Marine Safety Office, 1999. This report, published by the U.S. Coast Guard Safety Office, focuses on the grounding of the *New Carissa* oil tanker off the coast of Coos Bay, Oregon, in 1999.

Smith, Roland. *The Sea Otter Rescue: The Aftermath of an Oil Spill.* New York: Puffin Books, 1999. This book, intended for juveniles, recounts the rescue of sea otters following the 1989 *Exxon Valdez* oil spill in Prince William Sound off the coast of Alaska.

OCEAN TIDES

Tides are the displacements of particles on Earth caused by the differential attraction of the Moon and the Sun. There are atmospheric tides, land or crustal tides, and ocean tides. Of these, ocean tides are the most apparent because the ocean, as a fluid, is more easily stretched out of shape by the pull of the Moon.

PRINCIPAL TERMS

BASINS: container-like places on the ocean floor, usually elliptical, circular, or oval in shape, varying in depth and size

BORE: a wall of incoming tidal waters

DIURNAL TIDE: having only one high tide and one low tide each lunar day; tides on some parts of the Gulf of Mexico are diurnal

MIXED TIDE: having the characteristics of diurnal and semidiurnal tidal oscillations; these tides are found on the Pacific coast of the United States

NEAP TIDE: a tide with the minimum range, or when the level of the high tide is at its lowest

RANGE: the difference between the high-tide water level and the low-tide water level

SEMIDIURNAL: having two high tides and two low tides each lunar day

SPRING TIDE: a tide with the maximum range, occurring when lunar and solar tides reinforce each other a few days after the full and new moons

CAUSES OF TIDES

Each particle in the ocean moves in response to the force of gravitational attraction exerted on it by both Sun and Moon. Although the Sun is 27 million times the size of the Moon, the Moon is the primary factor in the ebb and flow of ocean waters. In fact, the Moon's power is more than double the periodic tidal-stretching force exerted by the Sun because it is much closer to Earth. This is explained by Sir Isaac Newton's universal law of gravity, which states that a force is proportionate to the mass of the attracting body, but its power of attraction diminishes in proportion to the square of the distance. Thus, proximity counts for more than distant mass in solar and lunar relations with Earth.

In order to explain how tides are caused, tidal scientists use the concept of a theoretical "equilibrium tide." This concept is based on an ideal in which the ocean waters are always in static equilibrium and in which no continents obstruct the flow of water on Earth's surface. When the Moon is directly above a particular location, its force of attraction causes water bulges to pile up directly under the Moon and also on the opposite side of Earth. Meanwhile, in the opposite quadrants of the globe, two low-water troughs result from the

pull of the water away from these areas. Thus, Earth rotates beneath tidal bulges and troughs, which result in high and low tides.

Tides generally follow the lunar day, which is twenty-four hours and fifty minutes, or the time it takes the Moon to orbit Earth. Some complex tidal cycles, however, result from the combined influences of Sun and Moon. If there were no Moon, the Sun's influence alone would cause tides to occur at the same time each day; however, because the plane of the Moon's orbit around Earth is in a different plane from that of Earth's orbit around the Sun, mixtures of full diurnal and semidiurnal tides result. In most areas, semidiurnal tides are the rule, with high and low tides occurring twice each lunar day, averaging twelve hours and twenty-five minutes apart. Diurnal tides occur when only one high tide and one low tide take place in a lunar day. They are found in parts of the China Sea and the Gulf of Mexico. Mixed tides result from a combination of both diurnal and semidiurnal tidal oscillations. Such tides are found along the United States' Pacific coast and in parts of Australia.

OTHER TIDE VARIABILITY FACTORS

Other factors that contribute to the variability of ocean tides are the phases of the Moon, the po-

sition of Sun and Moon relative to Earth, and the latitude
and topography of the tide's
location on Earth. When the
Sun, Moon, and Earth are
aligned (in "conjunction" or
"opposition"), then the combined gravitational effects of
these bodies will exert additional pull on Earth, resulting
in increased tidal amplitude.
This phenomenon, when lunar
and solar tides reinforce each
other, is called spring tide and
occurs around the full and new
Moons. In between the spring
tides, a neap (from the word
"nip") tide takes place when
the Sun, Moon, and Earth are
positioned at the apexes of a
triangle. At this time, during the first and third
quarters of the Moon's phase, solar high tides are
superimposed on lunar low tides, so resulting tides
are the lowest in the month. In the open ocean,
spring tides may be more than 1 meter high, while
neap tides may be less than 1 meter. Tidal amplitude will also vary, depending on latitude and the
declination of the Moon.

High tide at Bolinas Lagoon, Marin County, California, 1906. (U.S. Geological
Survey)

Despite the complexity of the many variables
that determine tidal behavior, tidal scientists can
now predict the time and height of a tide anywhere, on any past or future date, given one condition: that they have sufficient information on
how the local topography of the site modifies the
tide. Local geographical conditions such as the
width of a bay's mouth, the uneven slope of the
bottom, and the depth of the body of water are
the types of features that determine the range,
amplitude, and time of the local tide. Why does
the island of Nantucket experience a difference of
no more than 0.3 meter between high and low water, while only a few hundred kilometers away, the
Bay of Fundy has the highest tides in the world,
with a rise of 15 meters during spring tides? Scientists have developed a theory of tidal oscillation,
which holds that the ocean is divided up among a
great many basins of water, each with its own depth,
length, and resulting period of oscillation. The
boundaries of each basin are determined by the
surrounding land above and below the ocean, and
the influences of gravitational attraction in each
are always changing, as are the currents that flow
in. Ordinarily, when water rocks up and down in a
basin, the water at the rim is most active, while the
least amount of motion occurs in the center of the
basin around a tideless node. Thus the physical dimensions of these basins determine the period of
oscillation of the waters throughout the basin.

OPPOSING TIDAL BULGES

When the pull of the Moon creates a high tide
on the side of Earth closest to it, why should a
high tide occur simultaneously on the opposite
side? Logically, one would not expect this to be
the case. To answer that, it is necessary to know
that the Moon and Earth revolve not only around
each other but also around a common center of
gravity located 1,600 kilometers below the surface
of Earth. As the Earth-Moon system revolves, the
centrifugal force stretches the oceans out into
space, but Earth's gravity keeps the oceans from
actually flying off. At its center, Earth is not still
but is moving in a circle that is a small fraction of
the size of the Moon's orbit. This invisible revolution of Earth around the Moon produces a centrifugal force throughout Earth, which varies as the
Moon revolves and pulls Earth's surface out of
shape. It is the resulting "prolate," or lemon-
shaped, elongations of Earth that are observed as
the tides. The tides nearest the Moon are caused

Low tide at Bolinas Lagoon, Marin County, California, 1906. (U.S. Geological Survey)

by gravitational attraction. On the side of Earth farthest from the Moon, however, the centrifugal force is greater than the pull of the Moon. To compound the complexity of the situation, the Moon's gravitational force is also pulling the ocean floor of that area away from the waters there. Thus high tides are produced on both sides of Earth in line with the Moon.

Because tides are actually long waves, an observer on the Moon might expect to see the two tidal bulges move around Earth at a speed consistent with the pace of the Moon. Instead, tidal waves move out of step with the Moon. Keeping pace with the Moon would be possible on two conditions: if the oceans of the world were 22 kilometers deep (they average a bit more than 3 kilometers deep) and if there were no continents obstructing the movement of tidal waves. Thus, the speed of the movement of the tides is only 1,100 to 1,300 kilometers per hour, and the tides do not keep up with the Moon as it travels westward around the terrestrial globe.

EARTH'S ANGULAR MOMENTUM

The angular momentum of Earth is slowed by ocean tides. Because the momentum of the Earth-Moon system is conserved, the slowing of Earth results in a speeding up of the Moon's rotation around Earth. For unknown geophysical reasons, several abrupt increases in the length of Earth's day, some approaching 1 millisecond, have been observed over the past several decades. Ocean tides are not responsible for these abrupt changes but rather for small, steady increases over millennia. However, the U.S. National Bureau of Standards must sometimes add a "leap second" to the Earth year to take into account the observed slowing of Earth. Assuming that the position of Earth's orbital distance with respect to the Sun has not changed, the number of days in a year has changed from about 400 during the Devonian period to about 365 at the current time. Thus, the current length of a day is now quite a bit longer than it was during the Devonian, about 400 million years ago. Over this same time period, the Moon's angular momentum is estimated to have increased by about 1.6 percent. As Earth's rotation slows and the Moon revolves around Earth more quickly, the Moon moves farther away from Earth. The Moon has moved about 1.6 kilometers away from Earth in the last 100,000 years.

STUDY OF OCEAN TIDES

Tides are not simple to predict. Qualitative prediction of the tides has been going on in harbors around the world for centuries, but quantitative prediction began in the past century, when people first designed tide-predicting machines to help forecast the tides. The first such machine was invented in 1872 by Lord Kelvin, who is often referred to as the first electrical engineer. Kelvin's machine was capable of drawing a line picture of the curve of the tide, and for this achievement he was knighted. Soon after this breakthrough, an employee of the United States Coast and Geodetic Survey invented a tide-predicting machine that showed the times and heights of the tides. The survey later designed a simpler machine that combines the capabilities of both previous inventions: It gives the curve of the tides as well as the times and the heights of the tides. Unfortunately, these machines are not completely reliable because other factors, such as heavy storms, winds, or accumulation of sand as a result of wave action, can have dramatic impact on the water levels. Tide tables can only give the approximate high and low tides.

Today, tide-predicting machines have been replaced by faster digital computers. Tidal analysis still involves many complex computations, and predictions can be made only for places where a long series of observations are available. For each given spot on Earth, and for given time intervals, observations must be made which provide the value of the gravitational acceleration, the deflection of the vertical, and the measurements for the elevation of the water level. With this set of numbers in hand, the matter of prediction becomes one of extrapolating from the past into the future. Thus, around the coasts of the world, in inlets, in tidal rivers, and on islands, the water levels caused by tidal forces are carefully recorded. These measurements are analyzed locally. Once these measurements of the water level of a given place are taken within specified time intervals, the phases and amplitudes of the tide can be determined by a number of mathematical methods. Then, knowing the phases and amplitudes, scientists can reproduce the measurements according to a harmonic series. This harmonic method is a fairly reliable means of tidal prediction for deep-water ports, but in shallow-water areas, nonharmonic methods may need to be used. Finally, a determination of harmonic constants is made by national authorities, and the resulting data are sent to the International Hydrographic Bureau in Monaco. As a result, hydrographic offices in countries around the globe publish tide tables forecasting the high- and low-water times and water heights for the world's ports.

The problem of the measurement of tidal displacements in the open ocean has yet to be solved. This situation results from the fact that tidal sea-level records pertain primarily to the coastal locations, and there are few or no measurements from the open ocean.

SIGNIFICANCE

Knowledge of ocean tides serves crucial purposes in the fields of navigation, coastal engineering, and tidal power generation. In addition, it holds a key position in relation to geophysics, marine geodesy, and astronomy.

Tides are of vital importance in navigation. Although the tides have become tamer, they still both help and hinder all mariners. The *Coast Pilots and Sailing Directions* for different parts of the world reveals the menacing possibilities which tides in various places are known to cause. Tidal currents often move violently when opposed by winds or confined in narrow channels. Men have been swept off boats by the onslaught of giant waves when sailing in a flood tide through narrow straits. At certain stages of the tide, the waters can have dangerous eddies, whirlpools, or bores. A bore is created when a large portion of the flood tide enters a channel at once as one wave. Bores are walls of water which can be highly destructive and dangerous. Wherever they occur, they control the schedule of all shipping, as well as the rhythm of

Ebb tide near Windsor, Nova Scotia. (Geological Survey of Canada)

harbor life in the area. Even where there are no bores, the largest oceangoing liner must wait for slack water before entering a harbor where rushing tidal currents can fling it against piers. Since it is to the ship's advantage to sail in the direction of the tidal current flow, knowledge of tides is invaluable. All navigators approaching a coast rely on tide tables to supplement the information on depths in their nautical charts.

Coastal-engineering work is dependent on knowledge of the tides for such undertakings as the management of tidal estuaries, construction of harbors, and damming of tidal rivers. Another practical aspect of tide information concerns the handling of problems that arise from the pollution of coastal waters and the ocean.

Tidal power generation is a new field which has gained increasing attention because of the shortage of available energy sources. Humans have long dreamed of harnessing the tidal forces for their energy needs. In 1966, the first tidal power station ever built was completed in the La Rance estuary in France. Construction of the project took place just after the Suez crisis, when France felt uncertain about the future of its oil supply. The half-mile Rance Dam was built to harness energy from the very large tides of the area—with a mean range of nearly 8.5 meters and rising to more than 13 meters at equinoctial spring tides. This power installation transmits electricity to Paris and the surrounding area, producing more than 580 billion watt-hours of energy a year but costing slightly more than the cost of operation of hydroelectric plants. In 1968, the Soviet Union finished construction on a 400-kilowatt tidal plant north of Murmansk, at a site where the maximum tide is less than 4 meters in height. In China, more than forty small tidal plants have been built, according to dated reports. Plans for an additional eighty-eight more include one which will be China's largest and will be sited where a famous bore occurs. In the United States, at Passamaquoddy Bay in Maine, a major tidal power plant project was abandoned, because of the expense of maintaining the pipes and machinery in salt water, and of transmitting the electricity generated to the nearest big users.

Additional areas of practical concern involving tides include the correlation of tides with earthquakes, volcanic eruptions, and geyser activity. Some scientists believe that tides trigger earthquakes.

Nan White

CROSS-REFERENCES

BIBLIOGRAPHY

Clemons, Elizabeth. *Waves, Tides, and Currents.* New York: Alfred A. Knopf, 1967. Written for young adults, this book will appeal to interested nonscientists of all ages. Complexities of tidal phenomena are lucidly explained and read like a story. Includes photographs and easy-to-understand diagrams, a bibliography, and a glossary.

Curry, Judith A., and Peter Webster. *Thermodynamics of Atmospheres and Oceans.* San Diego, Calif.: Academic Press, 1999. This book offers a look at the effects of the interaction between oceans and the atmosphere on weather patterns and climatic changes. Provides good insight into the role that atmospheric thermodynamics play in meteorology. Illustrations, maps, and index.

Freuchen, Peter. *Peter Freuchen's Book of the Seven Seas.* New York: Julian Messner, 1957. A well-written general-interest book which gives a very clear explanation of the tides. It includes scientific explanations for laypersons of all ages, as well as folklore and history as the author imagined it. An entertaining book full of photographs.

Godin, Gabriel. *The Analysis of Tides.* Toronto: University of Toronto Press, 1972. This book is primarily a study of the mathematical principles underlying the analysis of the tides, but it covers the origin of tidal phenomena and the measurement of time, which is intimately related to celestial motion. The book contains an excellent chapter on Newtonian mechanics and celestial motion. Suitable for college-level students who want to gain a more technical knowledge of the tides.

Gregory, R. L., ed. *Tidal Power and Estuary Management.* Dorchester, England: Henry Ling, 1978. A collection of papers presented at the Symposium on Tidal Energy and Estuary Management, held under the auspices of the Colston Research Society at the University of Bristol in 1978. The papers were written by eminent authorities in the field of estuary management, many of them associated with tidal power production. This text presents a holistic picture of current research and thinking in two of the fields most intensively concerned with the tides. The viewpoints of engineers, botanists, zoologists, mathematicians, and economists make interesting reading.

Hamblin, Kenneth W., and Eric H. Christiansen. *Earth's Dynamic Systems.* 8th ed. Upper Saddle River, N.J.: Prentice Hall, 1998. This geology textbook offers an integrated view of the Earth's interior not common in books of this type. The illustrations, diagrams, and charts are superb. Includes a glossary and laboratory guide. Suitable for high school readers.

Komar, Paul D. *Beach Process and Sedimentation.* Upper Saddle River, N.J.: Prentice Hall, 1998. Extensive treatment of waves, longshore currents, and sand transport on beaches. Equations and mathematical relationships are presented and elaborated upon. College-level material. This book is for those interested in the specifics of coastal processes.

Melchoir, Paul. *The Tides of the Planet Earth.* Elmsford, N.Y.: Pergamon Press, 1978. Written by the foremost authority on tides, this text is suited for college-level readers who are not intimidated by technical language and who understand some mathematics or are willing to skip through it. The 146-page bibliography covers all papers to 1978 published on the subject of Earth tides and related topics. The introduction gives a brief summary of the relation of tidal research to the fields of astronomy, geodesy, geophysics, oceanography, hydrology, and tectonics, as well as a brief history of discoveries made about tides.

Wilhelm, Helmut, Walter Zuern, Hans-Georg Wenzel, et al., eds. *Tidal Phenomena.* Berlin: Springer, 1997. A collection of lectures from leaders in the fields of Earth sciences and oceanography, this book examines the Earth's tides and atomospheric circulation. Complete with illustrations and bibliographical references, this book can be understood by someone without a strong knowledge of the Earth sciences.

Wylie, Francis E. *Tides and the Pull of the Moon.* Brattleboro, Vt.: Stephen Greene Press, 1979. A lucid account of lunar and tidal phenomena and their influences on daily life. This well-written book includes information from science, history, and marine lore. It contains extensive bibliographical notes at the end of each chapter to guide readers to excellent sources on each topic covered. A complete introduction for anyone interested in the subject.

OCEAN WAVES

Waves shape beaches, and wave energy can be harnessed to generate power. Storm waves have inflicted great damage on human-made structures and have killed thousands of people.

PRINCIPAL TERMS

DEEP-WATER WAVE: a wave traveling in water with a depth greater than one-half of its wave length

FETCH: the area or length of the sea surface over which waves are generated by a wind having a constant direction and speed

STORM SURGE: a general rise above normal water level resulting from a hurricane or other severe coastal storm

SWELL: ocean waves that have traveled out of their wind-generating area

TSUNAMI: a long-period sea wave produced by a submarine earthquake or volcanic eruption

WAVE HEIGHT: the vertical distance between a wave crest and the adjacent wave trough

WAVE LENGTH: the horizontal distance between two successive wave crests or wave troughs

WAVE ORBIT: the path followed by a water particle affected by wave motion; in deep water, the orbit is nearly circular

WAVE PERIOD: the time (usually measured in seconds) required for two adjacent wave crests to pass a point

WAVE REFRACTION: the process by which a wave crest is bent as it moves toward shore

CAUSES OF WAVES

The waves that agitate a lake or ocean are rhythmic, vertical disturbances of the water's surface. Their appearance may vary from a confused seascape of individual hillocks of water, each with a rounded or peaked top, to long, orderly swell waves with parallel, rounded crests. Waves involve a transfer of energy from place to place on the ocean's surface. The earthquake that jolts the Japanese coast one evening may generate a tsunami that races across the Pacific and destroys a pier in Hawaii the next morning. The water itself, however, does not move; it is the wave form, or the energy impulse, that travels. The water stays where it is but oscillates as the wave form goes past.

Waves can originate in many ways. Tsunamis are shock waves resulting from a sudden disturbance of the water's surface by a submarine earthquake or volcanic eruption. Shock waves can also be generated when a pebble is tossed into a pond or a ship creates a wake. A second type of wave is the type produced by the gravitational pull of the Sun and the Moon—the tides that raise and lower the ocean's surface. Tide waves are the largest ocean waves of all, stretching halfway around the

world as they race along the equator at speeds of up to 1,600 kilometers per hour.

Waves can also originate through the action of the wind. Ordinary waves on the ocean or a lake form in this way. If there is no wind, the water surface is calm. If a slight breeze arises, the water surface is instantly roughened by patches of tiny capillary waves, the smallest waves of all. As the wind continues to blow steadily and in the same direction, ripples will appear because the surface roughness created by the capillary waves has given the wind something to push against. Soon the crests of adjacent ripples are being pushed together to create larger and larger crests. This process continues as the intensity of the wind increases, with small waves steadily giving way to larger and larger ones.

Three factors determine the size of the waves ultimately produced: the wind speed; the duration, or the length of time that the wind blows in a constant direction; and the fetch, or the extent of open water over which the wind blows. A 37-kilometer-per-hour wind blowing for ten hours along a fetch of 120 kilometers will generate waves 3 meters high, but a 92-kilometer-per-hour wind

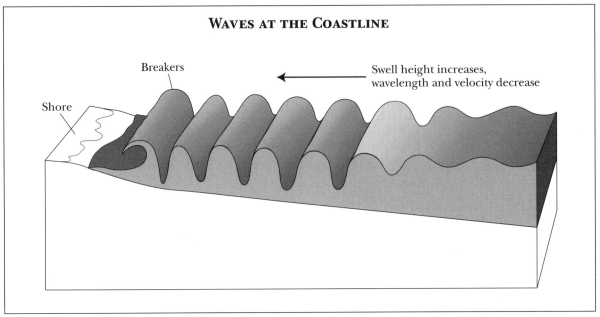

WAVES AT THE COASTLINE

Breakers

Shore

Swell height increases, wavelength and velocity decrease

Waves are vertical disturbances in water caused by wind, earthquakes, or the gravitational pull of the Sun (tides). As they approach shallower water, such as a coastline, the ocean bottom interrupts their flow, causing both wavelength and velocity to decrease and eventually disrupting the top of the wave, resulting in "breakers."

blowing for three days along a 2,400-kilometer fetch will generate waves 30 meters high. Fortunately, such waves are very rare.

One way to understand the motion of water particles within a wave is to analyze the direction of water movement at various places in the wave. One can do so by sitting in a boat beyond the breakers. As the forward slope of a wave crest approaches, a lifting motion is experienced, followed by a forward push as the crest passes beneath the boat. This forward push is seen when waves break at the beach and their crests are thrown forward in a violent rush of water. Once the crest has passed, the boat is on the back slope of the wave, and now a downward motion is experienced. Next comes a backward motion, as the trough passes beneath the boat. This backward motion is also in a beach's breaker zone; after the crest has crashed forward on the beach, there follows a strong outward surge of water. This outward surge represents the backward water motion in the wave's trough. When all the preceding observations are combined, it can be seen that when a wave passes, the water particles move first up, then forward, then down, and finally back. This circular path is known as the wave orbit.

OTHER WAVE TYPES

The term "heavy sea" is often encountered in descriptions of the ocean. A heavy sea results from the prolonged action of strong winds over the open ocean. The waves are large, peaked, and confused, totally lacking in orderly arrangement by size. Frequently, there is much spray in the air as a result of the tops being blown off the waves. A heavy sea is what one would expect to encounter in a hurricane or a violent storm. The term "swell," on the other hand, refers to waves that have moved out of the wind-generating area. As these waves approach the beach, they appear as long rows of smoothly rounded wave crests, evenly spaced at wide intervals and of uniform height. These swells have been produced by a distant storm at sea and have then moved out of the wind-generating area. As they travel outward, their original irregularities are diminished. Very little energy is lost, however, because a wave traveling at the ocean's surface encounters very little friction.

Groups of larger swell waves will be interspersed with groups of smaller ones. Oceanographers believe that such variation is caused by two or more wave patterns traveling together across the ocean's surface. When the crest of the larger wave pattern is superimposed on the crests of the smaller wave pattern, larger swell waves will result. When the trough of the larger pattern is superimposed on the crests of the smaller pattern, the swell waves will be smaller.

Oceanographers also recognize two major categories of wind-generated waves: deep-water and shallow-water waves. Deep-water waves travel in water depths greater than one-half of their wave length, and wind waves in the open ocean are generally in this category. Shallow-water waves, on the other hand, travel in water so shallow that their wave orbits are affected by friction with the bottom. The shallower the water becomes, the slower they go. This reduction in speed as waves approach the shoreline results in a process known as wave refraction, in which apparently straight wave crests approaching a shoreline from an angle are seen to be bent when viewed from above.

Close to the beach is the surf zone. There, the forward speed of waves is slowed and their crests are bunched together. The shape of the crests changes from nearly flat to broadly arched, and there is a conspicuous increase in the height of the wave. In addition, the water in the crest of the wave begins moving faster than the water in the trough because of the friction created by the bottom, and this friction soon causes the crest to collapse in a torrent of water. The wave has "broken." Oceanographers recognize two types of breaker. The first is a plunging breaker, in which the wave crest curls smoothly forward, trapping a tube-shaped pocket of air below. The other type is known as a spilling breaker, in which foaming water spills down the forward slope of the crest as the breaker advances. This type of breaker has no air-filled tube.

The final zone, found between the breakers and the beach, is a narrow strip characterized by the rhythmic alteration of water rushing shoreward on the beach and water sliding back out to sea. The inward rush of water is known as the swash; it is a miniature wall of foaming water filled with air bubbles. The backward flow is a thin, glistening film of water known as the backwash.

TSUNAMIS AND STORM SURGES

Two additional wave types require special mention. The first is a giant ocean wave known as a tsunami, which can be caused by submarine earthquakes, volcanic eruptions, or a landslide dumping massive amounts of debris into a bay, lake, or reservoir. In the open ocean, tsunamis behave just as any other ocean wave. They have crests and troughs that vary in height by 1 meter or so while the wave is still at sea. The tsunami wave length is enormously long, however, averaging perhaps 240 kilometers between crests. Tsunamis also have astonishingly high speeds—sometimes 650 kilometers per hour or more. A tsunami does not come ashore as a plunging breaker; rather, the crest rushes in as a surge of foaming water.

The second special wave type is known as a storm surge. Storm surges are drastic rises in sea level accompanying hurricanes or other severe coastal storms. Several factors combine to create such a surge. One factor is the reduced atmospheric pressure that occurs in the eye of a hurricane. This reduction may allow the ocean's surface to rise 1 meter or more. If, in addition, the Sun and Moon are aligned in such a way as to pro-

Storm-driven ocean waves viewed through the demolished front of a beachfront home in Aptos, California, during the January, 1983, El Niño event. (© William E. Ferguson)

duce unusually high tides, the storm surge may rise 1 meter higher. A third contributing factor is the presence of strong onshore winds. In a major hurricane, these winds will push massive waves against the coast, increasing the impact of the storm surge. Finally, the nature of the offshore bottom plays a role. Shallow offshore bottoms permit wind to get a better "grip" on the water, raising its level higher. As a result, a hurricane that creates a 4-meter storm surge along Florida's east coast, with its deep offshore waters, would be able to raise a 10-meter storm surge on Florida's west coast, where the bottom is flat and shallow for a distance of 160 kilometers offshore.

STUDY OF OCEAN WAVES

Until the early 1940's, the principal method for studying waves was to observe the sea's surface and to record the length, height, speed, and period of individual waves. Based on an analysis of wave period, it was determined that the various types of ocean wave could be arranged in an increasing spectrum of size. Capillary waves were found to be the smallest ocean waves, with periods of less than 0.1 second. Ripples came next, with periods of 0.1 to 1 second. Ordinary wind waves followed, with periods ranging from 1 second to 1.4 minutes. Larger still were tsunamis, with periods averaging 17 minutes, and finally the tides, with periods of 12 or 24 hours.

One basic wave-measuring instrument is the tide gauge, which is used to study tsunamis. The tide gauge is usually mounted on a pier in the quiet waters of a harbor, where it will not be exposed to damaging surf. It consists of a float inside a vertical, hollow pipe. The float is free to rise and fall with the water level, and a continuous record is made of the float's movement. The pipe is sealed at the bottom in such a way that only the long-period waves associated with a tsunami can force the float to rise. In this way, the tide gauge can record the preliminary waves of a tsunami and serve as a warning of the larger waves that follow. After the destruction of Hilo, Hawaii, by a tsunami on April 1, 1946, a seismic sea wave warning system was set up for the Pacific Ocean utilizing seismograph records and the type of tide gauge just described.

Other instruments measure the impact of storm waves against pilings, piers, and deep-water structures. The measurements obtained from these instruments have enabled engineers to design structures that can better withstand the impact of storm waves. Before the 1950's, lighthouses and breakwaters were the structures most vulnerable to wave attack, but since that time, large oil drilling and production platforms have been built in the open ocean many kilometers from shore. During severe storms, several of these platforms have capsized, causing tragic losses of life.

In laboratory experiments, ocean waves can be simulated in wave tanks. These range from tabletop models with glass sides that look like aquariums to outdoor tanks that can hold large boats and generate breakers several meters in height. The mathematical treatment of waves is facilitated by the regularity of their pattern, and computers are able to predict wave heights and other wave characteristics with a high degree of accuracy.

Beginning with the energy crisis in the late 1970's, intensive consideration has been given to the possibility of harnessing wave energy to generate power. Many systems have been designed, and several have been tested. The fundamental principle on which most wave generators work is that the motion of a passing wave is similar to the rise and fall of a piston.

SIGNIFICANCE

Beaches owe their origin to wave action. The waves bring in the needed sand to build the beach and then smooth it daily, erasing imprints with their in-and-out motion. Storm waves, however, are capable of great damage. The height of such waves in the open ocean can be dramatic; the USS *Ramapo* measured waves 34 meters high in 1933, for example. When these waves finally reach shore, the destruction can be enormous. Concrete blocks weighing 65 tons or more have been torn loose from breakwaters by such waves. Even sandy coastlines are not immune from attack. A longshore current is set up when the waves reach the shoreline at an angle, and this current can transport vast quantities of sand along a coast. Where this transported sand is trapped by an obstacle, such as a harbor jetty, excessive deposition will take place, perhaps requiring expensive dredging. At other points along the coastline, beach erosion may occur, also causing problems.

One of the most feared wave types is the tsunami, which is most frequently encountered on the shores of the Pacific. Although tsunamis are almost imperceptible in the open ocean, they can engulf boats, buildings, and people when they reach shore. There are records of tsunamis rushing up mountain sides to elevations of 30 meters or more. An added danger is that tsunamis have a series of crests that arrive twenty minutes or so apart, and often the third or fourth to arrive is the largest.

Storm surges are another dangerous wave type. They are commonly encountered along the Atlantic and Gulf coasts of the United States and have been known to carry oceangoing ships 1 kilometer or more inland. Six thousand people lost their lives in one such storm surge in Galveston, Texas, in 1900, and hundreds of thousands have drowned in a single storm surge on the shores of India's Bay of Bengal. Expensive storm surge barriers now guard Providence, Rhode Island; London; and the Netherlands.

Two other wave types that may present hazards for humans are rogue waves and seiches. Rogue waves are huge, solitary waves occasionally encountered in the ocean. They are particularly associated with the southward-flowing Agulhas current off South Africa and have been credited with sinking or severely damaging several large cargo ships. Seiche waves are oscillations in an enclosed water body such as a lake or bay. One such seiche, created by a hurricane, overflowed the dike surrounding Florida's Lake Okeechobee in 1928 and drowned two thousand people.

Donald W. Lovejoy

CROSS-REFERENCES

Beaches and Coastal Processes, 2302; Carbonate Compensation Depths, 2101; Dams and Flood Control, 2016; Deep Ocean Currents, 2107; Earth's Rotation, 106; Earthquake Prediction, 309; Forecasting Eruptions, 746; Geysers and Hot Springs, 694; Gulf Stream, 2112; Hydrothermal Vents, 2117; Ocean-Atmosphere Interactions, 2123; Ocean Basins, 661; Ocean Pollution and Oil Spills, 2127; Ocean Tides, 2133; Oceanic Crust, 675; Oceans' Origin, 2145; Oceans' Structure, 2151; Sea Level, 2156; Seamounts, 2161; Seawater Composition, 2166; Surface Ocean Currents, 2171; Tsunamis, 2176; Turbidity Currents and Submarine Fans, 2182; World Ocean Circulation Experiment, 2186.

BIBLIOGRAPHY

Bascom, Willard. *Waves and Beaches: The Dynamics of the Ocean Surface.* Rev. ed. New York: Anchor Books, 1980. A thorough treatment of the nature of waves and beaches and the principles that govern them. A good introduction for the nonscientist and an excellent source for the professional oceanographer. Contains many helpful diagrams and tables and numerous black-and-white photographs. Suitable for college-level readers.

Bird, Eric C. *Coastline Changes: A Global Review.* New York: John Wiley & Sons, 1985. This book is based on the results of a project conducted by the International Geographical Union's Working Group on the Dynamics of Coastline Erosion. It presents a detailed picture of the effects of wave erosion on the coastlines of 127 countries. Illustrated with numerous photographs and maps. Suitable for college-level readers.

Fairbridge, Rhodes W., ed. *The Encyclopedia of Oceanography.* New York: Reinhold, 1966. An oceanography source book for the student and the professional. It has excellent sections on waves and the related topics of fetch, tsunamis, wave energy, wave refraction, and wave theory. A well-illustrated and carefully cross-referenced volume. The text is aimed at readers with some technical background.

Gross, M. Grant. *Oceanography: A View of the Earth.* 7th ed. Englewood Cliffs, N.J.: Prentice-Hall, 1996. A well-written and well-illustrated text. Chapter 9 provides a comprehensive overview of all aspects of ocean waves, with diagrams, tables, and photographs. The appendix contains a table of conversion factors, and there is an excellent glossary. Suitable for general audiences.

Haykin, Simon S., and Sadasivan Puthusserypady. *Chaotic Dynamics of Sea Clutter.* New York: John

Wiley, 1999. This technical book offers the reader mathematical models designed to simulate the conditions necessary for wave generation. There is much attention given to the equipment to monitor ocean movement, including radar and remote sensing systems.

Ingmanson, Dale E., and W. J. Wallace. *Oceanography: An Introduction.* 5th ed. Belmont, Calif.: Wadsworth, 1995. Chapter 11 of this oceanography text provides a comprehensive treatment of all aspects of ocean waves. It includes many fine photographs, particularly of tsunami damage. The appendix contains a table of conversion factors, a complete set of charts of the U.S. coastline, and a glossary. Suitable for college-level readers or the interested layperson.

Komar, Paul D. *Beach Process and Sedimentation.* Upper Saddle River, N.J.: Prentice Hall, 1998. Extensive treatment of waves, longshore currents, and sand transport on beaches. Equations and mathematical relationships are presented and elaborated upon. College-level material. This book is for those interested in the specifics of coastal processes.

Murty, T. S. *Storm Surges: Meteorological Ocean Tides.* Ottawa: Department of Fisheries and Oceans, 1984. This book covers storm surges throughout the world. The introductory sections pertaining to the origin of storm surges are too technical for the lay reader, but chapter 7, "Case Studies of Storm Surges on the Globe," gives detailed storm surge data for every coastal zone that is subject to them.

Myles, Douglas. *The Great Waves.* New York: McGraw-Hill, 1985. A book on the subject of tsunamis designed for the lay reader and scientist alike. It gives background information pertaining to the origin of tsunamis and describes major destructive tsunamis throughout world history. Eyewitness accounts are included whenever available. There are no photographs, but the endpapers reproduce a fine etching of the tsunami accompanying the Lisbon earthquake of 1775. Suitable for high school readers.

Tricker, R. A. R. *Bores, Breakers, Waves, and Wakes: An Introduction to the Study of Waves on Water.* New York: Elsevier, 1964. A thorough treatment of water-wave behavior, written with the aim of simplifying the mathematics so as to make the discussion understandable to the general reader. Chapter 17, "Ships' Wakes," provides fine coverage of that topic. There are many black-and-white and color plates illustrating various wave types. Suitable for college-level readers.

Young, Ian R. *Wind Generated Ocean Waves.* New York: Elsevier, 1999. Young's book looks at the environmental factors involved in producing waves and the role waves play in coastal processes. Suitable for the high school reader or above. Illustrations, index, and bibliographical references.

OCEANS' ORIGIN

Oceanic waters are derived from the outgassing of hydrated minerals bound up during the formation of the Earth. Subsequent evolution of the waters involves primarily ions dissolving in the fluid medium by interactions with the continental and oceanic bottom sediments to give the basic saltiness characteristic of the Earth's oceans.

PRINCIPAL TERMS

CARBONACEOUS CHONDRITES: a class of meteoritic bodies found to contain large amounts of carbon in conjunction with other elements; used to date the solar system and provide chemical composition of original solar nebula

GEOCHEMICAL SINKS: the means by which elements and compounds are removed from the crustal area and oceans to be recycled in active chemical cycles

OUTGASSING: the process by which trapped volatiles in the Earth leak out gradually over time to form the terrestrial atmosphere and oceans

PRIMORDIAL SOLAR NEBULA: the original collection of dust and gases that comprised the basic cloud from which the solar system was formed

SMOKERS: undersea vents on the active rift areas that emit large amounts of superheated water and dissolved minerals from deep inside the Earth

SOLAR WIND: particles ejected from the Sun's surface as a result of the action of solar flares of great energy

VOLATILES: chemical elements and compounds that become gaseous at fairly low temperatures, allowing them to be released from solids easily

WATER OF HYDRATION: water that is bound up loosely to minerals present in the rocks, much of which went to form the oceans

EARTH'S OCEANS

Of all the planets in the solar system, the Earth stands out as basically a watery world, distinguished from its peers by excessive quantities of liquid water. Covering more than 1.35 billion cubic kilometers, the Earth's seas contain enough salt to cover all Europe to a depth of 5 kilometers. This salty solution is composed primarily of sodium chloride, some 86 percent of the ions by weight, in association with other ions of magnesium, calcium, potassium, sulfate, and carbonate groups, providing a salinity of 35,000 parts per million with pH 8 for the hydrogen-ion concentration, or slightly basic in nature. Of all the water on Earth, 97 percent of it is salty, the remainder being proportioned among ice (77 percent of total fresh water) and continental and atmospheric waters. The ice itself, principally in the Arctic-Greenland area (1.72 million square kilometers, 3,200 meters thick) and the Antarctic area (12 million square kilometers, 4,000 meters thick), provides effects ranging from climatic control to habitats for living organisms and sources of new seawater that, in the past, have caused sea levels to rise more than 100 meters.

The problem of the oceans' origin is twofold: the primordial origin of the water itself and the origin and rate of addition of the salt-inducing ions present in the past and contemporary waters. The database for solving the problems includes chemistry of water, the amounts and types of runoffs delivered by rivers into the sea, and the composition of volcanic gases, geysers, and other vents opening to the surface. In addition, most researchers find the oceans and atmosphere to be linked in origin, providing additional data for analysis.

Numerous sources for the Earth's water have been proposed, although the problem has not been resolved completely. Sources include the primordial solar nebula, the solar wind acting over time, bodies colliding with the Earth, impact degassing, and outgassing from the planetary interior. A final solution involves the investigation of factors controlling water on Earth, particularly rates, amounts, and types of outgassing, modes of planetary formation, possible chemical reactions providing water, loss rates of gases to space, and, finally, internal feedback mechanisms such as changes in Earth's albedo (reflective power), temperature, alteration of mass, and other factors not clearly understood.

SOLAR WIND, COLLIDING BODIES, AND IMPACT DEGASSING

The solar wind as a primary source can be eliminated for several reasons. The basic constituents, charged protons, may help formulate water in the atmosphere by reactions with oxygen, but all evidence points to no free oxygen in the primordial atmosphere. The geologic record shows the presence of liquid water at least 4 billion years ago that was substantially devoid of free oxygen. Astrophysical evidence suggests that the solar flux of energy, associated with the solar wind, was such that, early in the Earth's history, water on Earth should have been frozen, not liquid, if that was the primary source. The presence of liquid through at least 80 percent of Earth's history, however, has been established.

Colliding bodies would be from two primary sources: meteorites and comets. The presence of cometary bodies striking the Earth has yet to be documented; however, the basic chemical makeup, consisting of water, various ions, metals, organic molecules, and dust grains, would supply enough water, providing that gigantic numbers of cometary objects struck the Earth during the first half-billion years of history. No evidence for such happenings is available at present, although a theory of the Earth still being bombarded incessantly by small comets containing large quantities of water has been fiercely debated.

Meteoritic impact, particularly during the earlier stages after final planetary accretion, would definitely add water to the crust via two mechanisms. Through the study of carbonaceous chondrites (the oldest and most primitive meteorites), abundant volatiles, such as water, are found to be bound chemically to minerals such as serpentine. Additional waters, trapped in crustal and mantle rocks, would have been released during impact, quite justifiably in terms of large meteoritic rocks. It has been calculated that such impact degassing should have released 10^{22} kilograms of volatiles, quite close to the currently estimated value of 4×10^{21} kilograms for the Earth as a whole. Remnants of such ancient astroblemes are lacking, however, because of subsequent erosion, filling in by molten magma, or shifting of the continental masses over 4 billion years.

OUTGASSING

The most widely accepted origin for the oceans and atmosphere combines the features of the primordial solar nebula and slow outgassing from within the solidifying Earth. Original water would have been combined, under gravitational collapse, with silicates and metallic materials during the planetary accretion process, the hydration of minerals assisted by the heating of the Earth by infalling bodies and radioactive elemental decay. Such wet silicates appear to hold large quantities of bound water for indefinitely long periods of time. The primordial Earth, believed to have accreted cold, trapped the water molecules; if it had started too hot, all the minerals would have been dehydrated, and if too cold, no water would have been released. A delicate balance of temperature must have been achieved. Further, the volatiles forming the atmosphere must have outgassed first, since water must be insulated from solar radiation in order to form a liquid phase.

A secondary problem deals with how swiftly the fluids outgassed, either all at once, as individual events, or in a continuous fashion. Most data suggest the continuous mode of emission, with greatest reliance on data from still active sources, mainly volcanoes, undersea vents, and associated structures. Fumaroles, at temperatures of 500 to 600 degrees Celsius, emit copious quantities of water, sulfur gases, and other molecules. These bodies grade gradually into hot spots and geysers, areas where water is moved crustward from great depths. Magmatic melts rising in volcanoes release water and other gases directly to the surface.

In Hawaii, for example, the Halemaumau Pit, the volcano Kilauea's most active vent, emits, in terms of material, 68 percent water, 13 percent carbon dioxide, and 8 percent nitrogen, with the rest mostly sulfurous gases. Similar types of values are found in ridge-axis black and white smokers, where hydrothermal accretions result in spectacular deposits of minerals falling out of solution from the emerging hot mantle waters. Detailed studies show water trapped in the structures of altered minerals within basaltic crust of the oceanic plates, 5 percent of the rocks, by weight, in the upper 2 to 3 kilometers being water and hydroxide ions. Free water is known to be extremely buoyant, rising in the crust along shallow dipping faults. Bound water, subducted to great depths, would be expected to cook, moving upward as the rock density lessens, then acting as a further catalyst for melting the surrounding rocks.

In the Earth's earliest stages, the primordial atmosphere was released, to vanish from the Earth's gravitational pull because of overheating. In the second phase, gases are released from molten rocks, with a surface temperature of 300 degrees Celsius, providing 70 percent water and large quantities of carbon dioxide and nitrogen. In stage three, the atmosphere and oceans gradually change, with gases and liquid water from volcanoes and weathering action, more and more water deposited as liquid as the temperature falls. The rest of the atmosphere added oxygen either by thermal dissociation of water molecules, photochemical breakdown of high-altitude water, or photosynthetic alteration of carbon dioxide to oxygen in plants.

OCEAN SALT

The saltiness of the oceans can be accounted for by the extreme dielectric constant of water, essentially ensuring that it does not remain chemically pure. Geologic evidence shows the general composition to be similar over time, the stability of water content attributable to the continuous seawater-sediment interface. John Verhoogen has shown that only 0.7 percent of the ocean has been added since the Paleozoic era, primarily from lava materials. The salty quality is a product of acidic gases from the volcanoes (hydrochloric, sulfuric, and carbonic acids) acting to leach out the common silicate rocks. Paleontological studies indicate that the change in ions must have been extremely slow, as demonstrated by the narrow tolerance of organisms then alive, such as corals, echinoderms, brachiopods, and radiolarians. Present river ion concentrations differ drastically from the ocean's values, however, indicating a different atmospheric environment in the past. Robert M. Garrels and Fred T. Mackenzie have divided the oceans into three historical periods: earliest, with

water and volcanic acidic emissions actively attacking the crust, leaching out ions, leaving residues of alumina and silicates; the next period, from 1.5 to 3.5 billion years ago, slow continuous chemical action would go on attacking the sedimentary rocks, adding silica and ferrous ions; and period three, from 1.5 billion years onward, adding ions to the modern concentrations, until the composition is in apparent equilibrium with a mixture of calcite, potassium-feldspars, illite-montmorillonite clays, and chlorite.

Because it is known that output equals input of ions, a new problem, that of geochemical "sinks," has been identified. Calcium carbonate (limestone) is removed by living organisms to form skeletons, as is silica for opaline skeletons. Metals are dropped from seawater as newly formed mineral clays, oxides, sulfides, and zeolites, and as alteration products at the hot-water basaltic ridges. Sulfur is removed as heavy-metal sulfides precipitating in anaerobic environments, while salts are moved in pore waters trapped in sediments. Residence times for many of the ions have been determined: for example, sodium cycles in 210 million years, magnesium in 22 million, calcium in 1 million, and silicon in 40,000. With such effective removal systems, it is truly a measure of the geochemical resistivity of the Earth's oceans to change that allows the composition to remain so stable for 4 billion years.

STUDY OF THE OCEAN

Numerous avenues of approach have been used to investigate the ocean and its ions, including geological, chemical, and physical means. Geology has supplied basic data on the types and makeup of rocks from the earliest solidified materials to present depositional formations. Use of the petrographic microscope, involving thin sections of rocks seen under polarized light, allows the identification of minerals, providing quantity measurements of water attached to the minerals themselves. Paleontological studies of fossil organisms and paleosoils indicate the range of ions in the sea at diverse geologic periods, both by the ions themselves left in the deposited soils and rocks and through studies of the tolerance ranges for similar, twentieth century organisms. Such studies, along with sedimentology investigations of rates and types of river depositions, dissolved ion concentrations, and runoff rates for falling rain, provide determinants for comparing ion concentrations with those in the past for continentally derived materials.

Chemical analysis reveals the various ions present in seawater and rocks, via two principal methods. Use of the mass spectrometer identifies types and quantities of ions present by use of a magnetic field to accelerate the charged ions along a curved path, the size of the path based on the weight and charge of the ions. Collection at the end of the path provides a pure sample of the different ions present. For solid samples, electron beam probe studies provide analysis from an area only one micron in diameter. The electrons, fired at the sample, cause characteristic X rays to arise from the point. Each type of X ray, energywise, is affixed to a specific element or compound. By use of various optics to focus the X rays, identification of even minute variations in concentration is possible.

Solubility studies provide residence times as a means of geochemical analysis for cyclical research. Similar laboratory projects, testing the ability of water to dissolve and hold ions in solution, argue for a primordial Earth atmosphere being essentially neutral, or mildly reducing in nature. Such reduction characteristics are based on the study of planetary composition for Earth and, presently, for other planets as supplemented by the various "lander missions," such as Vikings 1 and 2 and the Soviet Venera series. Chemical analysis from such missions in interplanetary space has also determined compositions for meteoritic gases, cometary tails and nuclei, and the mixing ratios for noble gases, important for determining the origin of the solar system. The latter study, involving physics in analyzing radioactive isotopes such as helium 3, an isotope of mantle origin, has allowed geophysicists to consider the Earth's mantle as a major elemental source and sink for the various geochemical cycles.

Laboratory analysis reaches two other areas. Petrographic studies of returned lunar rocks reveal that the Moon is empty of water, lacking even hydroxyl ions. This discovery helps eliminate the solar wind and meteoritic impact as major factors in forming oceans. Furthermore, high-temperature/high-pressure metallurgical and chemical studies indicate molten granite, at temperatures of 900 degrees Celsius and under 1,000 atmospheres of pressure, will hold 6 percent water by weight, while basalt holds 4 percent. Based on geochemical calculations of the amounts of magma in the planet and lavas extruded over the first billion years, all the ocean's waters can be accounted for, particularly if parts of the fluid, as steam under pressure, are a result of oxidation of deep-seated hydrogen deposits trapped within or combined with mantle rocks. This supposition is considered an excellent likelihood from evidence gathered on radioactive decay inside the Earth.

Arthur L. Alt

CROSS-REFERENCES

Beaches and Coastal Processes, 2302; Carbonate Compensation Depths, 2101; Deep Ocean Currents, 2107; Gulf Stream, 2112; Hurricanes, 1928; Hydrothermal Vents, 2117; Lakes, 2341; Landslides and Slope Stability, 1501; Monsoons, 1940; Ocean-Atmosphere Interactions, 2123; Ocean Pollution and Oil Spills, 2127; Ocean Power, 1669; Ocean Tides, 2133; Ocean Waves, 2139; Oceans' Structure, 2151; Sand, 2363; Sea Level, 2156; Seamounts, 2161; Seawater Composition, 2166; Sediment Transport and Deposition, 2374; Sedimentary Mineral Deposits, 1637; Storms, 1956; Surface Ocean Currents, 2171; Tsunamis, 2176; Tsunamis and Earthquakes, 340; Turbidity Currents and Submarine Fans, 2182; Weathering and Erosion, 2380; Wind , 1996; World Ocean Circulation Experiment, 2186.

BIBLIOGRAPHY

Brancazio, Peter J., ed. *The Origin and Evolution of Atmospheres and Oceans*. New York: John Wiley & Sons, 1964. This work is a collection of papers dealing with the chemical problems relevant to the early formation of the fluid parts of the Earth. Tracing all the basic arguments, the criteria for water formation is clearly explained and its relationship to minerals and rocks elucidated. Some heavy reading, charts, extra references.

Chamberlain, Joseph W. *Theory of Planetary Atmospheres: An Introduction to Their Physics and Chemistry*. New York: Academic Press, 1978. A detailed analysis of the characteristics of diverse atmospheres in the solar system, including water contents. By comparisons of chemical compositions and meteorological observations, criteria are established for examining the possible origins for atmospheric gases and oceans. Some mathematics, heavy reading, numerous charts, comprehensive references.

Frakes, L. A. *Climates Throughout Geologic Time*. New York: Elsevier, 1980. A well-written explanation of how the interaction of the Earth's atmosphere and oceans has caused the climate of the Earth to change over the history of the planet. Beginning with the possible origin of ocean and atmosphere, changes are traced as revealed through the geological and paleontological records. Numerous graphs and charts and extensive references.

Hamblin, Kenneth W., and Eric H. Christiansen. *Earth's Dynamic Systems*. 8th ed. Upper Saddle River, N.J.: Prentice Hall, 1998. This geology textbook offers an integrated view of the Earth's interior not common in books of this type. The illustrations, diagrams, and charts are superb. Includes a glossary and laboratory guide. Suitable for high school readers.

Henderson-Sellers, A. *The Origin and Evolution of Planetary Atmospheres*. Bristol, England: Adam Hilger, 1983. This work details the theories of where the volatiles for the Earth came from and how the oceans and other planetary atmospheres came into existence from the creation of the solar system. Water as a direct result of outgassing of planetary interiors is considered, as are the effects from lack of water molecules. Harder reading, but an advanced layperson should find it comprehensible. Includes an extensive bibliography.

Holland, Heinrich D. *The Chemistry of the Atmosphere and Oceans*. New York: John Wiley & Sons, 1978. A very detailed reference on the basic chemical elements present in the two media and the wide variety of reactions occurring in each area. The interactions of the two areas are stressed, as are their common origin from materials outgassed from within the Earth. The action of their chemicals on the terrestrial areas is described in detail. Contains references and numerous charts of data but is difficult reading.

Imberger, Jeorg, ed. *Physical Processes in Lakes and Oceans*. Washington, D.C.: American Geophysical Union, 1998. This extensive volume details the origins, processes, and phases of oceans, lakes, and water resources, as well as the ecology and environments surrounding them. Illustrations and maps.

Ittekko, Venugopalan, et al., eds. *Particle Flux in the Ocean*. New York: John Wiley and Sons, 1996. This volume contains descriptions of the chemical and geobiochemical cycles of the ocean, as well as the ocean currents and movement. Suitable for the high school reader and beyond. Illustrations, index, bibliography.

McElhinny, M. W., ed. *The Earth: Its Origin, Structure, and Evolution*. New York: Academic Press, 1979. A more readable work dealing with all the basic elements of Earth science. Starting with the theory of planetary formation, the work covers the origin of oceans, atmosphere, land, and life-forms. Excellent description of the changes occurring on the planet throughout geologic time. Well written, it has good pictures and extra references.

Majumdar, Shyamal K., et al., eds. *The Oceans: Physical-Chemical Dynamics and Human Impact*. Easton: Pennsylvania Academy of Science, 1994. This compilation of essays provides an overview of oceanography while focusing on the relationship between the atmosphere and oceans. Many of the sections deal with both the effects of theis relationship on humans and the effects of humans on the ocean environment. Suitable for the college-level reader.

Plummer, Charles C., Diane H. Carlson, and David McGeary. *Physical Geology*. 5th ed. Boston: McGraw-Hill, 1999. A general introduction to physical geology and oceanography. Intended for use at the college-freshman level. Contains many tables, illustrations, and photographs; also includes a glossary and an index.

Ponnamperuma, C., ed. *Cosmochemistry and the Origins of Life.* Dordrecht, Netherlands: Reidel, 1982. A collection of works dealing with the distribution of elements in the universe, particularly those necessary for life. Provides information on the formation of the planetary system, showing how the chemicals combined at various temperatures to make the planets as different as they are. Discusses origins of oceans, atmospheres, and life; detailed reading with many charts and an extensive bibliography.

Seibold, E., and W. Berger. *The Sea Floor.* New York: Springer-Verlag, 1982. A delightful book covering the chemistry, geology, and biology of the bottom of the Earth's oceans. That the oceans are a result of outgassing is emphasized. Well written, with a very interesting section on the white and black smokers and their relation to the origin of waters and life. Contains excellent illustrations, along with additional references.

OCEANS' STRUCTURE

The ocean has a complex structure, both at its surface and in the vertical dimension descending to the ocean floor. This internal structure results in layering with respect to temperature, salinity, density, and the way in which the ocean responds to the passage of light and sound waves.

TEMPERATURE LAYERING

One highly significant aspect of ocean structure is a layering of water based on temperature and salinity differences. In order to understand the reasons for the temperature layering of the ocean, one must bear in mind that the primary source of heating for the ocean is sunlight. About 60 percent of this entering radiation is absorbed within the first meter of seawater, and about 80 percent is absorbed within the first 10 meters. As a result, the warmest waters in the ocean are found at its surface.

That does not mean, however, that surface temperatures in the ocean are the same everywhere. Since more heat is received at the equator than at the poles, ocean surface temperatures are closely related to latitude. As a result, they are distributed in bands of equal temperature extending east and west, parallel to the equator. Temperatures are highest along the equator because of the near-vertical angles at which the Sun's rays are received here. Toward the poles, ocean temperatures gradually cool as a result of the decreasing angle of the incoming solar radiation.

Measurements of ocean surface temperature range from a high of 33 degrees Celsius in the Persian Gulf, a partly landlocked, shallow sea in a desert climate, to a low of −2 degrees Celsius in close proximity to ice in polar regions. There, the presence of salt in the water lowers the water's freezing point below the normal 0 degree level. Because salinity of this cold water is so high, it sinks to the ocean floor and travels along it for substantial distances. Ocean surface temperatures may also vary with time of year, with warmer waters moving northward into the Northern Hemisphere in the summertime and southward into the Southern Hemisphere in the wintertime. These differences are most noticeable in midlatitude waters. In equatorial regions, water and air temperatures change little seasonally, and in polar regions, water tends to be cold all year long because of the presence of ice.

Vertically downward from the equator, toward the ocean floor, water temperatures become colder. That results from the facts that solar heating affects the surface waters only and that cold water is denser than warm water, since its molecules are more closely packed. When waters at the surface of the ocean in polar regions are chilled by extremely low winter temperatures, they become denser than the underlying waters and sink to the bottom. They then move slowly toward the equator along the seafloor, lowering the temperature of the entire ocean. As a result, deep ocean waters have much lower temperatures than might be expected by examination of the surface waters alone. Although the average ocean surface temperature is 17.5 degrees Celsius, the average temperature of the entire ocean is a frigid 3.5 degrees.

Oceanographers recognize the following layers within the ocean, based on its temperature stratification: First, there is an upper, wind-mixed layer, consisting of warm surface water up to 500 meters thick; this layer may be lacking in polar regions. Next is an intermediate layer, below the surface layer, where the temperature decreases rapidly with depth; this transitional layer can be 500 to 1,000 meters thick and is known as the main thermocline. Finally, there is a deep layer extending to the ocean floor; in polar regions, this layer may reach the surface, and its water is relatively homogeneous, the temperature slowly decreasing with depth.

Because the upper surface layer is influenced by atmospheric conditions, such as weather and climate, it may contain weak thermoclines as a result of the daily cycle of heating and cooling or because of seasonal variations. These are temporary, however, and may be destroyed by severe storm activity. Nevertheless, the vast bulk of ocean water lies below the main thermocline and is uniformly cold, the only exception being hot springs on the ocean floor that introduce water at temperatures of 300 degrees Celsius or higher. Plumes of warm water emanating from these hot springs have been detected within the ocean.

SALINITY

A second phenomenon responsible for layering within the ocean is variation in the water's salinity. For the ocean as a whole, the salinity is 35 parts per 1,000. Considerable variation in the salinity of the surface waters from place to place results from processes that either add or subtract salt or water. For example, salinities of 40 parts per 1,000 or higher are found in nearly landlocked seas located in desert climates, such as the Red Sea or the Persian Gulf, because high rates of evaporation remove the water but leave the salt behind. High salinity values are also found at the surface of the open ocean at the same latitudes where

there are deserts on land (the so-called horse latitudes). There, salinities of 36 to 37 parts per 1,000 are common.

At the equator, however, much lower salinity values are encountered, despite the high temperatures and nearly vertical rays of the Sun. The reason is that the equatorial zone lies in the so-called doldrums, a region of heavy rainfall. The ocean's surface waters are therefore diluted, which keeps the salinity low. Low salinities are also found in coastal areas, where rivers bring in large quantities of fresh water, and in higher latitudes, where rainfall is abundant because of numerous storms.

Despite the variation in salinity in the ocean's surface water, the deep waters are well mixed, with nearly uniform salinities ranging from 34.6 to 34.9 parts per 1,000. Consequently, in some parts of the ocean, surface layers of low-salinity water overlie the uniformly saline deep waters, and in other parts of the ocean, surface layers of high-salinity water overlie the uniformly saline deep layer. Between these layers are zones of rapidly changing salinity known as haloclines. An important exception to this picture is a few deep pools of dense brine—such as are found at the bottom of the Red Sea, for example—where salinities of 270 parts per 1,000 have been recorded.

Haloclines are very common in coastal areas. Off the mouth of the Amazon River, for example, a plume of low-salinity river water extends out to sea as far as 320 kilometers, separated from the normally saline water below by a prominent halocline. In many tidal rivers and estuaries, a layer of heavier seawater will extend many kilometers inland beneath the freshwater discharge as a conspicuous saltwater wedge.

DENSITY STRATIFICATION

A prominent density stratification within the ocean results from the variation in ocean temperatures and salinity just described. As has been noted, two things make water heavier: increased salinity, which adds more and more dissolved mineral matter, and lowering of temperature, which results in the water molecules being more closely packed together. Therefore, the least-dense surface waters are found in the equatorial and tropical regions, where ocean temperatures are at their highest. Toward the higher latitudes, the density of these surface ocean waters increases because of the falling temperatures. In these areas, low-density surface water is found only where large quantities of fresh water are introduced, by river runoff, by high amounts of precipitation, or by the melting of ice.

Vertical density changes are even more pronounced. As water temperatures decrease with depth, water densities increase accordingly. This increase in density, however, is not uniform throughout the ocean. At the poles, the surface waters are almost as cold as the coldest bottom waters, so there is only a slight increase in density as the ocean floor is approached. By contrast, the warm surface waters in the equatorial and tropical regions are underlain by markedly colder water. As a result, a warm upper layer of low-density water is underlain by an intermediate layer in which the density increases rapidly with depth. (This middle layer is known as the pycnocline.) Below it is a deep zone of nearly uniform high-density water.

Convective overturning takes place when this normal density stratification is upset. In a stable density-stratified system, the less dense surface water floats on top of the heavier, deeper water. Occasionally, however, unstable conditions will arise in which heavier water forms above lighter water. Then, convective overturning takes place as the mass of heavier water sinks to its appropriate place in the density-stratified water column. This overturning may occur gradually or quite abruptly. In lakes and ponds, it occurs annually in regions where winter temperatures are cold enough. In the ocean, convective overturning is primarily associated with the polar regions, where extremely low winter temperatures result in the sinking of vast quantities of cold water. In addition, convective overturning has been observed in the Mediterranean during the wintertime, when the chilled surface waters sink to replenish the deeper water.

LIGHT PENETRATION AND SOUND WAVES

Oceanographers also recognize stratification in the ocean based on the depths to which light penetrates, and they divide the ocean into two zones. The upper zone, which is known as the photic zone, consists of the near-surface waters that have sufficient sunlight for photosynthetic growth. Below this zone is the aphotic zone, where there is insufficient light for photosynthetic growth. The lower limit of the photic zone is generally taken as the depth at which only 1 percent of the surface intensity still penetrates. In the extremely clear waters of the open ocean, this depth may be 200 meters or more.

Stratification in the ocean based on the behavior of sound waves has also been observed. Because sound waves travel nearly five times faster under water than in the air, their transmission in the ocean has been extensively studied, beginning with the development of sonar, the echo sounder. So-called scattering layers have been recognized; they are regions that reflect sound, usually because of the presence of living organisms that migrate vertically, as layers within the water column, depending on light intensity. The sofar (sound fixing and ranging) channels are density layers within the ocean where sound waves can become trapped and can travel for thousands of kilometers with extremely small energy losses. These channels have the potential to be used for long-distance communications. Shadow zones are also caused by density layers within the ocean. These layers trap the sound waves and prevent them from reaching the surface. One advantage of shadow zones is that submarines can travel in them undetected.

STUDY OF THE OCEAN

The measurement of water temperatures at the ocean's surface is quite simple. A thermometer placed in a bucket of water that has been scooped out of the ocean at the bow of a boat will suffice, provided the necessary precautions have been taken to prevent temperature changes caused by conduction and evaporation. For oceanwide or global studies, satellites provide near-simultaneous readings of ocean surface temperatures within an accuracy of 1 degree Celsius. These satellites utilize infrared and other sensors, which are capable of measuring the amount of heat radiation emitted by the ocean's surface to within 0.2 degree.

Measuring the temperature of the deeper subsurface waters has always posed a problem, however, because a standard thermometer lowered over the side of a ship will "forget" a deep reading on its way back to the surface. As a result, the so-called reversing thermometer was developed in 1874. This thermometer has an S-bend in its glass tube. When the thermometer is inverted at the desired depth, the mercury column breaks at the S-bend, thus recording the temperature at that depth. Subsequently, electronic instruments that record subsurface water temperatures continuously were introduced. These devices can be either dropped from a plane or ship or moored to the ocean floor.

The measurement of the salinity of seawater is not as easy as one might think. An obvious way to determine water salinity would be to determine the amount of dried salts remaining after a weighed sample of seawater has been evaporated, but in actual practice, that is a messy and time-consuming procedure, hardly suitable for use on a rolling ship. A variety of other techniques have been used over the years, based on such water characteristics as buoyancy, density, or chloride content. By far the most popular relies on seawater's electrical conductivity. In this method, an electrical current is passed through the seawater sample; the higher the salt content of the water, the faster this current is observed to pass. Using this method, oceanographers have been able to determine the salinity of seawater samples to the nearest 0.003 part per thousand—an important advantage, in view of the fact that the salinity differences between deep seawater masses are very minute.

Measurement of the densities of surface water samples can easily be accomplished by determining the water sample's buoyancy or weight, but the real difficulty comes with attempts to determine a subsurface water sample's density. If this water sample is brought to the surface, its temperature, and therefore its density, will change. Although sophisticated techniques are available for the determination of density at depth, in actual practice the density is not measured at all; instead, it is computed from the sample's known temperature, salinity, and depth. It turns out that water's density is almost wholly dependent on these three factors.

Various methods are available for measuring the depth of light penetration in seawater. A crude estimate can be made using the Secchi disk, which was first introduced in 1865. This circular, white disk is slowly lowered into the water, and the depth at which the disk disappears from sight is noted visually. More sophisticated measurements can be made using photoelectric meters. Another good indicator of the maximum depth of light penetration in the sea is the lowest level at which photosynthetic growth can take place. For sound studies within the ocean, various methods are used to create the initial sound, including the use of explosives in seismic profiling. The returning echo is detected by means of a receiver known as a hydrophone.

Donald W. Lovejoy

CROSS-REFERENCES

Atmosphere's Structure and Thermodynamics, 1828; Carbonate Compensation Depths, 2101; Deep Ocean Currents, 2107; Elemental Distribution, 379; Evolution of Earth's Composition, 386; Fluid Inclusions, 394; Geochemical Cycle, 412; Gulf Stream, 2112; Hydrologic Cycle, 2045; Hydrothermal Vents, 2117; Life's Origins, 1032; Magmas, 1326; Ocean-Atmosphere Interactions, 2123; Ocean Pollution and Oil Spills, 2127; Ocean Tides, 2133; Ocean Waves, 2139; Oceans' Origin, 2145; Precipitation, 2050; Sea Level, 2156; Seamounts, 2161; Seawater Composition, 2166; Solar System's Elemental Distribution, 2600; Solar System's Origin, 2607; Surface Ocean Currents, 2171; Tsunamis, 2176; Turbidity Currents and Submarine Fans, 2182; World Ocean Circulation Experiment, 2186.

BIBLIOGRAPHY

Fairbridge, Rhodes W., ed. *The Encyclopedia of Oceanography.* New York: Reinhold, 1966. An outstanding oceanography source book for the student or the professional. It has sections on the temperature structure of the ocean, salinity, density, underwater light properties, and underwater sound channels. The text is suitable for college-level readers who have some technical background. A well-illustrated and carefully cross-referenced volume.

Gross, M. Grant. *Oceanography: A View of the Earth.* 7th ed. Englewood Cliffs, N.J.: Prentice-Hall, 1996. A well-written and well-illustrated oceanography text. Chapter 7, "Ocean Structure," provides a thorough treatment of the topic. There are many useful diagrams and charts, including color plates showing temperature distribution in the ocean based on satellite imagery. Suitable for college-level readers or the interested layperson.

Imberger, Jeorg, ed. *Physical Processes in Lakes and Oceans.* Washington, D.C.: American Geophysical Union, 1998. This extensive volume details the origins, processes, and phases of oceans, lakes, and water resources, as well as the ecology and environments surrounding them. Illustrations and maps.

Ingmanson, Dale E., and William J. Wallace. *Oceanography: An Introduction.* 5th ed. Belmont, Calif.: Wadsworth, 1995. Chapter 7 has useful information on ocean-surface salinities and gives typical salinity profiles for the ocean. Chapter 8 gives similar information regarding temperature, density, light, and sound. The text is well illustrated throughout and is suitable for college-level readers.

McLellan, H. J. *Elements of Physical Oceanography.* Elmsford, N.Y.: Pergamon Press, 1965. A thorough treatment of temperature, salinity, and density distribution within the ocean and the methods used for their measurement. Data are given for selected locations, with helpful tables and charts. There are photographs that help to explain the workings of the reversing thermometer and the bathythermograph.

Pickard, G. L., and W. J. Emery. *Descriptive Physical Oceanography: An Introduction.* 5th ed. Boston: Butterworth-Heinemann, 1995. A text designed to introduce oceanography majors to the field of physical oceanography. Chapter 3 has useful sections on the temperature, salinity, density, light, and sound structure of the ocean. Chapter 6 has a discussion of the various instruments and methods used for measuring the above properties.

Segar, Douglas. *An Introduction to Ocean Sciences.* New York: Wadsworth, 1997. Comprehensive coverage of all aspects of the oceans, their chemical makeup, and circulation. Readable and well illustrated. Suitable for high school students and above.

Sverdrup, H. W., M. W. Johnson, and R. H. Fleming. *The Oceans: Their Physics, Chemistry, and General Biology.* Englewood Cliffs, N.J.: Prentice-Hall, 1942. This text is one of the classics in oceanography. It provides much valuable, detailed information relating to the distribution of temperature, salinity, and density in the oceans. The charts showing the worldwide distribution of temperature and salinity in the ocean have been reproduced in many other oceanography texts. Suitable for college students and the interested nonspecialist.

Teramoto, Toshihiko. *Deep Ocean Circulation: Physical and Chemical Aspects.* New York: Elsevier, 1993. This college-level text provides a detailed look at ocean circulation and currents. Much information is provided on the chemical processes that occur in the deep ocean. Illustrations, maps, and bibliographical references.

Von Arx, W. S. *An Introduction to Physical Oceanography.* Reading, Mass.: Addison-Wesley, 1962. Written for readers lacking a strong background in physics and applied mathematics, this book has helpful sections on temperature, salinity, and density distributions in the sea. Also contains an important section on stratification that discusses the concepts of the thermocline, halocline, and pycnocline. For college-level readers.

SEA LEVEL

Sea level is the average position of the surface of the Earth's ocean relative to the land, providing the frame of reference for land elevations and ocean depths. Sea levels change through time because of a variety of factors, and these changes have a major influence on Earth geology.

PRINCIPAL TERMS

CONTINENTAL SHELF: the extension of the continent beneath the sea; a flat or gently sloping platform usually 10 to 100 kilometers wide, extending to a depth of 100 to 150 meters

EUSTATIC SEA-LEVEL CHANGE: a change in sea level worldwide, observed on all coastlines on all continents

GLACIATION: commonly known as an "ice age," the cyclic widespread growth and advance of ice sheets over the polar and high- to midlatitude regions of the continents

GREENHOUSE EFFECT: the warming of the Earth's atmosphere, caused by accumulation of excess carbon dioxide and other gases, primarily through human activity

ISOSTASY: the passive, vertical rise or fall of the Earth's crust caused, respectively, by the removal or addition of a load to the crust

LOCAL SEA-LEVEL CHANGE: a change in sea level only in one area of the world, usually by land rising or sinking in that specific area

MEAN SEA LEVEL: the average height of the sea surface over a multiyear time span, taking into account storms, tides, and seasons

REGRESSION: the retreat of the sea from the land; it allows land erosion to occur on material formerly below the sea surface

TECTONICS: the process of origin, movement, and deformation of large-scale structures of the Earth's crust

TRANSGRESSION: the advance of the sea over the land; it allows marine sediments to be deposited where they could not form before

CHANGING SEA LEVEL

Sea level is a major aspect of the modern world. The distinction between the land and the sea depends on the level of the sea. A rise in sea level means the flooding of low-lying land; a drop in sea level means exposure of some of the seafloor. Much of the world's commerce, food supply, and recreation is associated with shallow coastal waters. The position and rate of change of sea level is important to all these activities. Sea level is most accurately defined as mean (average) sea level, which is the average height of the sea surface measured over an extended period of time for all conditions of tides, seasons, and storms. Land elevations are measured with sea level as the frame of reference. Nautical charts, however, use mean low water, the average position of low tide, to measure ocean depths; mean high water, the average position of high tide, is used for marking the exposure of adjacent land areas. The reason for use of low- and high-tide values, for depth and land, respec-

tively, is purely practical: Successful navigation requires accurate knowledge of water depths and land exposures. By referring depths to mean low tide, ship captains are assured that, regardless of tide position, they know the minimum water depth available for their vessel. The same principle applies to land areas: Using a high-tide value to show land allows land users to be sure that the low-lying areas will not flood with the high-tide cycle.

The most fascinating aspect of sea level is that it is not a constant. Through time, it can change. The time of change can be hours, days, years, or centuries. When sea level rises and floods the land, it is called a marine transgression. The flooded land becomes part of the ocean environment, and marine sediments are deposited on what was once dry land. Marine regression, by contrast, occurs when sea level drops and the shallow seafloor is exposed. The exposed seafloor becomes part of the land environment and is subject to erosion. Almost every continental coastal area contains a continental shelf, or a flat-lying or gently sloping portion of the continent that extends under the sea to depths up to 150 meters over a width of several tens of kilometers to, in some cases, well over 100 kilometers. The gentle slope of the continental shelf means that a minor rise in sea level results in a major amount of the shelf being drowned; a minor drop in sea level produces a major exposure of the continental shelf seafloor. The rocks of the continents show that over the Earth's history, sea level has transgressed onto, and regressed from, the continents many times to many different levels.

Sea level can be changed in six basic ways: by oscillating the ocean surface; by changing the force of gravity; by moving the land up or down; by changing the characteristics of the ocean water; by changing the amount of water in the oceans; and by changing the volume of the ocean basins. The change in sea level brought about by these six ways results in two major types of sea-level change: local and eustatic. Local sea-level change means that only a specific area of coastline is involved and that coastlines relatively far away or on other continents are not changed. Oscillation of the water surface, changing gravity, and land moving up and down produce local sea-level change. Eustatic sea-level change means that coastlines around the planet all experience a sea-level change of the same magnitude at the same time. Changing water characteristics, changing the amount of water on the planet, or changing the volume of the ocean basins produces eustatic sea-level change.

LOCAL SEA-LEVEL CHANGE

Sea-level change produced by oscillation of the ocean surface refers to waves, storm surges, and tides. A wave breaking on the beach runs up the beach face, then slides back down. Each wave event can be considered a short-term, low-magnitude sea-level change. In extreme cases of large waves, erosion and coastline modification can occur during each sea-level "micro-event." Storm surges are changes in sea level brought about by the movement of surface waters under strong winds and low atmospheric pressures, such as hurricanes. Low atmospheric pressure literally sucks sea level up a small amount. High onshore winds can pile water up on coastal areas by 3 or 4 meters;

offshore winds can lower sea level by lesser amounts.

Tides are produced by the force of lunar gravity, and to some extent by solar gravity, on the oceans of the Earth. Coastline configuration, latitude, time of the lunar month, and many other factors control the timing and magnitude of ocean tides. Tide characteristics are extremely variable from place to place on the Earth, but in a given area, the changes in sea level are extremely predictable, with magnitudes from a fraction of a meter to more than 10 meters possible. Because ocean surface oscillations are common, are low in magnitude, and occur regularly, they are often not recognized as sea-level changes. They are different for every coastline, producing local sea-level change.

The force of gravity is not exactly uniform over the Earth's surface. It depends on the amount of mass beneath the Earth's surface and thus, because of tectonics and other activities, may vary by a small amount from location to location across the Earth. The ocean, as a fluid, responds to the force of gravity: If gravity is slightly weaker in one area than in another, the sea will rise slightly higher; if it is slightly stronger, the sea will sink slightly lower. This sea-level variation occurs only on time frames of millions of years, but it is used to explain local sea levels that are different in certain areas from predicted values.

In many areas, the land is moving up or down. This movement has three sources: tectonics, isostasy, and subsidence. Tectonics is the movement and distortion of the Earth's crust by forces generated within the Earth; it forms volcanoes, mountains, and earthquakes. In an area where tectonics is active, the land may be forced up or down, causing it to rise or sink with respect to sea level. In tectonically active areas such as Southern California or in the Mediterranean basin, this type of land movement is common, and sea-level changes of many meters up and down have been historically documented. Isostasy is the vertical movement of the Earth's crust downward when a load is applied or its rebound upward when a load is removed. When isostasy occurs in a coastal area, the land will rise or sink, with a subsequent change in sea level. Glaciers are examples of how a load can be applied to the crust, causing isostatic subsidence. When the glacier melts, the load is removed, and the crust can isostatically rebound. Coral islands and volcanic islands are other examples of a load being placed on the crust, which then isostatically sinks. Land can sink beneath sea level by a process called subsidence. Subsidence is often caused by compaction of the land material, which commonly happens when oil or water is withdrawn from the ground. The loss of the fluid allows the rock to compact, and the land sinks. All three of these methods of moving land up or down occur in localized areas and so result in local sea-level change.

EUSTATIC SEA-LEVEL CHANGE

Changing sea level eustatically, or worldwide, requires changes that affect the oceans as opposed to the land. The ocean basin has a fixed volume; if the nature of the water in it is changed, if the amount of water is changed, or if the shape of the container (ocean basin) is changed, sea level will change. The two characteristics of water that control sea level are its temperature and its salinity.

For each degree Celsius that the oceans warm, thermal expansion will raise sea level 2 meters. Increasing the salinity makes the water denser and causes sea level to fall. Sea-level variations caused by changes in salinity measure approximately 1 to 2 meters.

The amount of water available to the oceans is changed in three ways: by the growth and melting of glaciers, by steam released from volcanoes, and by water lost to sediments. Glaciation is one of the most important sea-level controls. As ice sheets grow, they are fed by evaporated seawater falling as snow; this seawater is then "trapped" as ice on the continents, and sea level falls. If the ice melts and the water flows back to the ocean, sea level rises. During the last 2 million years, glaciations have come and gone at least four times, and sea level has risen and fallen over a range of approximately 125 meters. This range is enough to almost totally expose continental shelves during maximum ice advance on the continents. Only eighteen thousand years ago, sea level was 125 meters

A crack in the Weddell Sea, Antarctica, in February, 1997. Satellite images reveal that a huge chunk of the Larsen B ice shelf has separated from the Antarctic Peninsula, an event that many scientists see as evidence of global warming. (AP/Wide World Photos)

below where it is today, yet by three thousand years ago, it was essentially at today's level. If the remaining ice on Antarctica were to melt entirely, sea level would rise an additional 60 or more meters, drowning coastal cities worldwide.

The oceans are thought to have originated by the release of water as steam from early volcanic activity. Volcanoes are still active presently, adding water to the oceans. When sediments are deposited in the ocean, they contain some ocean water. Tectonic activity can return these water-bearing sediments back into the Earth's crust. There appears to be a rough balance between water escaping from the Earth's crust from volcanoes and water returning to the crust through sediment deposition, with no overall sea-level change caused by these processes in the modern world.

Sea level is also affected by changes in the volume of the ocean basins. Continents are continually dumping sediments into the sea. This addition of sediment is filling in the ocean basins, pushing sea level up. At the same time, tectonic activity is distorting the edges of continents, often folding them, which increases the volume of the ocean basins. Tectonics on the ocean floor can force the seafloor upward, limiting ocean basin volume. Tectonics on the Earth tends to occur episodically, in fits and jumps, as opposed to a smooth, continuous process. When tectonics are active, they may increase or decrease ocean volume, causing a drop or rise in sea level. Sediments deposited on the ocean floor may be plastered back on the continents by tectonic activity, first decreasing and then increasing ocean basin volume. Sea-level changes, in response to sediment balance and tectonic location and magnitude, occur slowly over millions of years.

STUDY OF SEA LEVEL

When sea level changes, it leaves an imprint on the land. Transgressing ocean waters erode the land surface with waves and then, as waters deepen, deposit marine sediments. When regression occurs, the shallowing water leads to wave erosion of the seafloor. Further regression exposes the seafloor to the air and to erosion by wind, rain, and running water. Casual examination of the continents reveals clear evidence of past marine transgression and regression. Limestones, marine shales, and marine sandstones are com-

mon around the world on the dry land of the continents. They range in age from billions of years old to only a few hundred years old. The sea has transgressed and regressed many times in the Earth's history. The timing and exact nature of the sea-level change may be more difficult to determine. In the Bahama Islands, fossil corals exist several meters above sea level. By sampling the corals and dating them by means of trace amounts of radioactive elements in the samples, scientists can determine their age. In this case, the age would be about 125,000 years old. More than 100,000 years ago, the land to which the fossil coral is attached was under water. Did the land rise, or did the water drop? Measurement of tectonic activity shows that the Bahama Islands are not rising or sinking. Examination of the history of glaciation shows that 125,000 years ago, the ice sheets had melted back a little more than they are presently. From this information, scientists conclude that the coral grew when sea level was eustatically higher than it is today.

Further evidence of sea-level change can be obtained by examining the tectonic, deposition, and erosion history of the Earth. Examination of major episodes of tectonic activity, sediment deposition, and erosion can allow the construction of graphs showing how sea level rose and fell, in general terms, over much of the Earth's history. Most of the sea-level changes preserved in the rock record were tectonically generated. Glaciation has occurred at only a few times in the past, and the sea-level changes caused by glaciation have left a unique and distinctive record in the Earth's rocks.

If sea level never changed, the geology of the surface rocks of the Earth would be very simple. There would be no exposed marine rocks at all. If there were no tectonics to help drive sea-level change, the continents would erode to sea level. The Earth would be a relatively featureless place.

John E. Mylroie

CROSS-REFERENCES

BIBLIOGRAPHY

Kennet, James P. *Marine Geology*. Englewood Cliffs, N.J.: Prentice-Hall, 1982. A college-level text on the geology of the oceans, this book assumes some science background on the part of the reader. Contains an excellent, in-depth overview of sea-level change starting on page 265.

Komar, Paul D. *Beach Process and Sedimentation*. Upper Saddle River, N.J.: Prentice Hall, 1998. Extensive treatment of waves, longshore currents, and sand transport on beaches. Equations and mathematical relationships are presented and elaborated upon. College-level material. This book is for those interested in the specifics of coastal processes.

McCormick, J. Michael, and John V. Thiruvathukal. *Elements of Oceanography*. 2d ed. New York: Saunders College Publishing, 1981. This introductory-level textbook on oceanography contains a superb definition of sea level and related topics on page 212.

Montgomery, C. W. *Environmental Geology*. 4th ed. Dubuque, Iowa: Wm. C. Brown, 1995. An introductory college text, this book discusses the causes and ramifications of the greenhouse effect on pages 200-202.

Neshyba, Steve. *Oceanography Perspectives on a Fluid Earth*. New York: John Wiley & Sons, 1987. This introductory text on oceanography describes sea-level change with respect to planetary geology, with specifics on pages 68-70.

Patrusky, Ben. "Dirtying the Infrared Window." *Mosaic*, Fall/Winter, 1988: 19. *Mosaic* is a journal put out by the National Science Foundation and is available through the Government Printing Office. The author provides a detailed review of the greenhouse effect easily readable by the layperson.

Perry, A. H., and J. M. Walker. *The Ocean-Atmosphere System*. New York: Longman, 1977. This introductory textbook provides an excellent review of ocean surface oscillations, especially surges, on pages 68-71.

Sinha, P. C., ed. *Sea Level Rise*. New Delhi, India: Anmol Publications, 1998. This collection of essays reviews the continuing debate over changes in sea level, taking into account changes in the atmosphere and climate caused by global warming and the greenhouse effect. The volume also looks at the effects that sea-level change may have on the environment.

Stowe, Keith. *Exploring Ocean Science*. 2d ed. New York: John Wiley & Sons, 1996. The text is introductory and provides the layperson with a good discussion of ice volumes and sea level on pages 339-342.

Weinhaupt, John G. *Exploration of the Oceans: An Introduction to Oceanography*. New York: Macmillan, 1979. This college-level text is very easy to read, and chapter 5 is devoted to a discussion of sea-level change, including the mechanisms involved and the results over the span of Earth history.

SEAMOUNTS

Seamounts, or undersea volcanoes, are far more numerous and much larger than are their volcanic counterparts on land. Some may become large enough to build above sea level as islands for a brief period, but almost all are eventually doomed to resubmerge as guyots or atolls. Radiometric dating of seamounts has helped to verify the theory of continental drift.

PRINCIPAL TERMS

ATOLL: a tropical island on which a massive coral reef, often ringlike, generally rests on a volcanic base

BASALT: a rock that results when lava rich in iron and magnesium and low in silica is cooled rapidly, resulting in a fine-grained, dark-colored appearance

GUYOT: a drowned volcanic island with a flat top caused by wave erosion or coral growth

HOT SPOT: a column or plume of molten rock that rises from deep within the Earth and can cause volcanic eruptions if it penetrates the lithosphere

HYALOCLASTITE: the rock type that results when lava is chilled rapidly and explosively beneath

the sea at shallow depths, resulting in a fragmental, glassy texture

LAVA: magma that has erupted from a volcano

LITHOSPHERE: the rigid outermost layer of the Earth, which floats on the softer layer (the asthenosphere) beneath; it is thinner under the oceans than on the continents

MAGMA: a general term for molten rock

MID-OCEAN RIDGE: a roughly linear, submarine mountain range where new seafloor lithosphere is created by seafloor spreading

PILLOW LAVAS: lavas that have been rapidly cooled by water as they erupt and consequently develop crusted, rounded, baglike structures

ORIGIN AND GROWTH

Seamounts are volcanoes that arise from the seafloor but that lack sufficient height to be classified as islands. Like their volcanic counterparts on land, seamounts are roughly conical or domelike in shape, and many have craters at the top. Compared to volcanoes on the continents, seamounts are much larger and higher; they may rise to as much as 10,000 meters above the ocean floor. A seamount that has a flat top is called a guyot. Almost all guyots were islands at one time, and they acquired their flat tops from wave erosion, the growth of corals, or a combination of both processes. The islands underwent resubmergence caused by subsidence or sinking of the seafloor or by eustatic rise in sea level. The tops of some guyots can be as much as 3,000 meters below present sea level, but most are about 1,000 meters deep. A guyot that is less than 200 meters deep is called a bank. Seamounts and guyots are a prominent feature of the ocean floor, particularly in the southwestern Pacific Ocean. Some are associated

with the vast mid-ocean ridges, but many rise from the deep, flat sections of the ocean floor known as the abyssal plains. Some seamounts are isolated, others occur in clusters and form a large volcanic plateau, and still others form a long chain.

Little is known regarding the origins of seamounts. It is known, however, that magma (molten rock) rising through the seafloor squeezes its way beneath the saturated but relatively buoyant bottom sediments and crystallizes. After many of these intrusions, a solid base is built up, and cone development can begin. The magma that makes up seamounts is not of the explosive variety, so seafloor eruptions are relatively quiet affairs. In fact, the water pressure in the deep ocean is so great that the gas and steam normally associated with volcanoes on the continents cannot escape. Instead, the lava (magma that has escaped the Earth's interior) oozes out quietly in submarine flows. Because deep seawater is quite cold, the lava is chilled very rapidly; often, the lava cools into wrinkled, glassy crusts that enclose round, sacklike

Nauru Island, a seamount, or guyot, in the Pacific that has risen above sea level and is covered with bird guano used for fertilizer. (©William E. Ferguson)

blobs. These structures, common on the seafloor, are known as pillows. The rock that makes up the pillows is a dark, fine-grained rock known as basalt, which is rich in iron and magnesium-bearing minerals.

As the seamount grows, its increasing size begins to bring it nearer the sea's surface and consequently into areas of lower water pressure. At a depth of about 2,000 meters, the gases associated with the lava and the heated seawater itself can begin to expand and explode violently as the lava cools, resulting in a glassy, fragmental structure known as hyaloclastite. It is also at these or shallower depths that the submarine eruption may be noticed by passing ships as an area of dark, turbulent water, floating fragmental and bubbly rocks, and gas and ash clouds at the surface. During both of these stages, magma may continue to be injected along cracks and layers inside the growing seamount and crystallize internally. If the magma finds a place where it can escape along the sides or base of the mountain, a flank eruption may build a parasitic cone.

The next stage in the growth of a seamount, when the eruption is close to sea level, is critical. The effects of the exploding gases, combined with the effects of wave erosion on the loose, fragmental debris, tend to prevent the seamount from emerging above the water to become an island. If the volume of lava erupted is enough to keep pace

with these destructive forces, however, subaerial lava flows may begin to cover the hyaloclastites with an armor plating that is resistant to wave attack, and a volcanic island is born.

EROSION AND SUBSIDENCE

During its life as an island (which is likely to be only a few million years, a relatively brief period compared to its life as a seamount and guyot), the mountain is attacked by wind, running water, and glaciers (depending on its latitude). All these factors tend to reduce its elevation even as subsequent eruptions may build it up. If the conditions are tropical, coral reef growth may encrust the wave-battered shores. Every island eventually succumbs, however, to the combined effects of erosion and subsidence (sinking) of the seafloor, which begin to submerge it once again. Coral growth may be able to keep pace with the effects of subsidence for a time. Sometimes, this extensive coral growth results in a completely coral-capped volcano known as an atoll. Many atolls in the South Pacific have more than 1,000 meters of massive coral growth above their volcanic bases.

If coral growth cannot keep pace with the rate of seafloor subsidence, the island or atoll becomes submerged once again, this time in the form of a bank and then finally as a guyot once it passes a depth of 200 meters. The mountain weathers gradually on the seafloor: The sides are subject to erosion by currents, crumbling, and slumping, and the rocks undergo chemical reactions with the seawater. The entire mountain is also gradually buried in the ubiquitous "rain" of dust and the tiny remains of microscopic plants and animals, which cap the top and drape the sides in thick layers of fine sediments.

CONTINENTAL DRIFT

No guyot lasts a very long portion of geologic time, however, because all are eventually carried by continental drift to subduction zones, where the seafloor is recycled. The study of seamounts and guyots has helped to verify the idea of continental drift (plate tectonics) and to show how fast plate movement occurs. The seafloor consists of one or more slabs (plates) of quasi-rigid rock 50 to 80 kilometers thick. The oceanic lithosphere, as it is called, is created at the great mid-ocean ridges, where two plates are being pulled directly apart from each other. Magma wells up to fill the gap thus created. As the magma cools and crystallizes, it is added to the slab and becomes part of the plate. Thus, new seafloor lithosphere is added to both of the diverging plates at mid-ocean ridges. Sometimes, a seamount will grow at the mid-ocean ridge (Iceland is an example of a volcanic island in this location) and actually become split in half as the two plates pull apart.

The slabs of oceanic lithosphere are able to slide about on a slippery zone deeper inside the Earth known as the asthenosphere, just as slabs of ice on a winter pond can drift on the water. Where stresses push two plates together, one is forced to override the other. The plate that is forced underneath (subducted) plunges deep into the Earth and eventually melts and creates another type of submarine volcanic chain called an island arc. Because oceanic lithosphere is thin and dense compared to continental lithosphere, it is the seafloor that gets subducted, or recycled, most often. Thus, the seafloor, carrying its seamounts, is geologically much younger than are the continents.

HOT SPOTS

Although some seamounts erupt near mid-ocean ridges, many that erupt far from plate boundaries are associated with hot spots. Hot spots, or plumes, are believed to be long, narrow fountains of magma that rise from deep within the Earth, well below the asthenosphere. These magmas have a chemical composition that is quite different from that of the magma of the mid-ocean ridges. Seamounts are generally made up of alkaline basalts (basalts enriched in the elements sodium and potassium compared to mid-ocean-ridge basalts).

The upwelling plume of magma at the hot spot has a dramatic effect on the seafloor lithosphere passing over it. The rising plume pushes the lithosphere upward as much as 1,000 to 1,500 meters into a broad arch or swell and heats the lithosphere from below, making it thin and stretched. Finally, if magma can break through the weakened lithosphere, a seamount is born. Eventually, however, the oceanic lithosphere is carried by plate motion away from the hot spot, which, unlike the lithosphere, has a relatively fixed location. The seamount or volcanic island begins to move, along with the lithosphere in which it is embedded, down the side of the bulge caused by the upwelling magma. In addition, the heavy burden of a large volcano on the thin oceanic lithosphere causes it to sag downward into the soft asthenosphere below. Both of these effects combine to pull a seamount downward or to submerge an island into a guyot.

At the same time that the old volcano is moving off the hot spot and subsiding, magma may continue to erupt through the new lithosphere over the hot spot, and a new volcano will emerge next to the old one. In this way, over millions of years, a whole chain of seamounts, some of which may have been islands at one time, extend from the location of the hot spot in the direction of plate motion for thousands of kilometers. It was in this fashion that the Hawaiian Islands and the closely related Emperor Seamounts were formed as the Pacific plate was pulled over a hot spot in a northwesterly direction toward the subduction zone near the Aleutian Islands of Alaska.

Sara A. Heller

CROSS-REFERENCES

BIBLIOGRAPHY

Heezen, Bruce C., and Charles D. Hollister. *The Face of the Deep.* New York: Oxford University Press, 1971. An exposition of the deep seafloor in close-up black-and-white photographs. The explanatory text is informative and nontechnical. Chapter 12, "Pedestals and Plateaus," contains diagrams and seismic profiles of seamounts and guyots, plus close-up photographs of the volcanic rocks and sediments associated with them. Appendices, index. Suitable for high school readers.

Hekinian, R. *Petrology of the Ocean Floor.* New York: Elsevier, 1982. A comprehensive, detailed, and rather technical textbook covering the chemical, mineral, textural, and genetic aspects of the rocks of the seafloor. Chapter 5, "The Ocean Basins and Seamounts," contains chemical analyses from a number of seamounts, as well as a few photographs of rock samples. The final chapter discusses the magma sources for the seafloor. Appendices, index. Suitable for the college-level reader.

Keating, Barbara H., Patricia Fryer, Rodney Batiza, and George W. Boehlert, eds. *Seamounts, Islands, and Atolls.* 2d ed. Washington, D.C.: American Geophysical Union, 1988. This rather lengthy volume contains a collection of twenty-six technical articles concerning seamounts and related features. These articles generally focus on a particular seamount or group of seamounts and demonstrate the use of a particular method of study such as sonar, geophysics, sediment analysis, or geochemistry. A final section contains articles about the effect of seamounts on ocean circulation and biology. No index or general conclusions. Suitable for the college-level reader.

King, Cuchaine A. M. *Introduction to Marine Geology and Geomorphology.* London: Edward Arnold, 1974. An introductory text on the geologic origins of submarine landforms and their methods of study. Contains chapters on plate tectonics, continental margins, ocean sediments, sea-level changes, and the landforms of the open oceans such as seamounts, guyots, and atolls. Extensive bibliography and index. Suitable for the college-level reader.

Menard, H. W. *Islands.* New York: W. H. Freeman, 1986. An up-to-date, nontechnical, and well-illustrated volume describing the geological influences upon islands, seamounts, and guyots, with emphasis on continental drift. Chapter 5, "Growth of Isolated Volcanic Islands," discusses the birth and development of submarine volcanoes and their tectonic effects on the oceanic crust. Index. Suitable for the college-level reader.

_____. *Marine Geology of the Pacific.* New York: McGraw-Hill, 1964. This book discusses the geological aspects and history of the Pacific Ocean. Chapter 4, "Vulcanism," contains an extensive discussion of the distribution, rock types, geography, development, and origin of seamounts, guyots, archipelagic aprons, and islands. A useful text for the physical attributes of submarine volcanoes, but has been supplanted by the author's later volume (listed above) because of the subsequent development of plate tectonic theory. Bibliography, index. Suitable for the college-level reader.

Nairn, Alan E. M., Francis G. Stehli, and Seiya Uyeda, eds. *The Ocean Basins and Margins.* Vol. 7A, The Pacific Ocean. New York: Plenum Press, 1985. This volume and its companion (7B) summarize the geology of various regions of the Pacific Ocean. Chapter 3, "Pacific Plate Motion Recorded by Linear Volcanic Chains," by Robert A. Duncan and David A. Clague, summarizes knowledge about several chains of seamounts in the Pacific and what they reveal about hot spots and the motion of the Pacific plate. Extensive references. Suitable for the college-level reader.

Seibold, E., and W. H. Berger. *The Sea Floor.* 3d ed. New York: Springer-Verlag, 1996. A well-illustrated text about the landforms of the sea-floor and continental margins, seafloor sediments and biota, and how sediments can be used as indicators of past climates. Contains a brief section on seamounts. Appendices, index. Suitable for the college-level reader.

Williams, Howel, and Alexander R. McBirney. *Volcanology.* San Francisco: Freeman, Cooper, 1979. A comprehensive if somewhat technical treatment of volcanoes, volcanic landforms, magma, volcanic rocks, and types of eruptions. Chapter 12, "Oceanic Volcanism," discusses the intraplate volcanism that produces seamounts and guyots and compares it to volcanism at mid-ocean ridges. Includes examples and methods of study. Bibliography, index. Suitable for the college-level reader.

SEAWATER COMPOSITION

The properties of seawater are determined largely by the properties of pure water. Because water has great capacity as a solvent, the seawater is well mixed and is salty because of the ions of elements that have been dissolved from the rocks that make up the Earth's crust. Seawater is a source of mineral wealth for humankind.

PRINCIPAL TERMS

ELEMENT: one of a number of substances composed entirely of like particles; atoms that cannot be broken into smaller particles by chemical means

FREE OXYGEN: the element oxygen by itself, not combined chemically with a different element

HYDROSPHERE: the region of the Earth that is covered by water, including the oceans, seas, lakes, and rivers

MINERAL: an inorganic substance occurring naturally in the Earth and having definite physi-

cal properties and a characteristic chemical composition that can be expressed by a chemical formula

NODULE: a lump of mineral on the ocean floor

PRIMARY CRYSTALLINE ROCK: the original or first solidified molten rock of Earth

SALINITY: a measure of the quantity of dissolved solids in ocean water; this measurement is given in parts per thousand by weight

WEATHERING: the breaking down of rocks by chemical, physical, or biological means

FORMATION OF THE EARTH

The hydrosphere consists of the water area of the Earth, which is made up of ponds, lakes, rivers, groundwater, and the oceans. The oceans form the largest portions of the hydrosphere—71 percent of the Earth's surface. The composition of these waters derives from the circumstances surrounding the formation of the solar system and the planet Earth.

In the beginning, gas and space dust came together, and eddies started forming separate small clouds, which later consolidated into planets. The planets started to cool and become solid, although the crust, being thin, cracked, giving off heat, molten material, and gaseous material. This gaseous material formed the first atmosphere. It was probably made up of hydrogen and helium molecules. These molecules, being fairly light in weight and highly energized, drifted off into space. Eventually, these gases were replaced with other gases by the ongoing volcanic activity as the Earth's surface continued to cool, filling the atmosphere with carbon dioxide and water vapor.

Clouds formed from condensing water vapor, covering the Earth's surface and allowing less than 60 percent of the Sun's energy to penetrate. As

the surface continued to cool, the water vapor condensed into liquid and began to fall and accumulate in depressions. With widespread volcanic activity continuing, water vapor was continuously being released, along with smaller amounts of carbon dioxide, chlorine, nitrogen, and hydrogen gas. The atmosphere now contained water vapor, carbon dioxide, methane, and ammonia. As the surface continued to cool, vast amounts of condensation formed and, over time, oceans formed. Geologic evidence shows that the oceans have existed for at least 3 billion years. This evidence comes in the form of algal fossils assumed to have grown in a marine environment.

The crust of the Earth was formed by primary crystalline rocks—that is, the first molten material that solidified into rock as the Earth cooled. These rocks were weathered and eroded into the particles that became dissolved in water, carried into the ocean basins, and accumulated. Also carried into the ocean were great loads of particulate material weathered from the primary crystalline rocks and deposited as sediments. Thus the components of the primary rocks were freed by chemical weathering and dissolved in ocean water or chemically bonded with sediments and carried into the ocean by rivers. The chemical weathering took place because of the different compounds in the atmosphere, some of which combined with the water vapor to form mild acidic rainwater. There is also the possibility that nitric acid (nitrates and water) and sulfuric acid (sulfates and water) formed, also aiding in the chemical weathering of the rocks. Seawater gained its characteristic saltiness through these processes.

Some components are more abundant than can be accounted for. The substances found in excessive amounts compared to the products of chemical weathering are volatile or gaseous substances at slightly above-average temperature at the Earth's surface. They include water vapor, carbon dioxide, chlorine, nitrogen, sulfur, hydrogen, and fluorine gases. One explanation for this excess is that gases are released from molten interior by volcanic activity. A percentage of these gases may be recycled gases that have been dissolved in groundwater and carried down from the Earth's surface, and a small percentage of the gases are actually new gases. Comparing the discharge of gases by hot springs in the United States to the average rate throughout the 3 billion years of the oceans' existence, one finds that enough water vapor is produced to fill the oceans to one hundred times their present volume. Water is recycled and not all new; it does not represent new water release from the crystallization of magma. Only 1 percent of this water requires such an origin to account for the present volume of the Earth's oceans.

Water on land comes daily from the sea. The seas hold about 4.4 billion cubic meters of salt water. Of this amount, about 12 million cubic meters are sucked up each year by evaporation and returned by rainfall and flow of rivers, and about 3 million cubic meters descend each year over continents, replenishing ponds, lakes, and rivers.

GASES AND SOLIDS IN SEAWATER

Ocean water is a mixture of gases and solids dissolved in pure water (96 percent pure water and 4 percent of dissolved elements). Nearly every natural element has been found or is expected to be found in seawater, some in very small percentages. The most abundant mineral found in ocean water is sodium chloride (table salt), which makes up 85 percent of the dissolved minerals. The composition of human blood is similar to that of seawater;

in it are all the elements of the sea, dispensed in different proportions.

The seven most abundant minerals in seawater include sodium chloride, 27.2 parts per 1,000; magnesium chloride, 3.8 parts per 1,000; magnesium sulfate, 1.7 parts per 1,000; calcium sulfate, 1.3 parts per 1,000; potassium sulfate, 0.9 part per 1,000; calcium carbonate, 0.1 part per 1,000; and magnesium bromide, 0.1 part per 1,000. Six elements actually make up 99.3 percent of the total mass of the dissolved material: chlorine, 55.2 percent; sodium, 30.4 percent; sulfate, 7.7 percent; magnesium, 3.7 percent; calcium, 1.2 percent; potassium, 1.1 percent; and others, 0.7 percent.

The salinity of seawater is fairly uniform across the oceans at different latitudes and at various depths, mainly because of winds, waves, and currents. Variations in mineral content are attributed to positions of freshwater streams entering oceans, glacial melt, and human activity, but the variations are small. The salinity averages about 35 parts per 1,000. This has been verified by the *Glomar Challenger* expedition, using Nansen bottles to take samples of seawater at different depths around the world and salinometers to measure salinity, as well as many other types of tests. There is more variation of salinity at the surface than at depths because of fresh water (rivers, glacial melt) entering the ocean at a given location, plus the biological activity and climate at different latitudes.

In areas where fresh water is entering the ocean, the salinity will decrease. Biological activity changes the salinity according to which species and how many marine plants and animals reside in the area. In climates that are hot and dry, the evaporation is high, and because rainfall is low, the ratio of dissolved salts to water is higher, making the water more saline (saltier). Similarly, in the polar regions in winter, the water freezes but the minerals do not, which changes the water-to-mineral ratio and thus increases salinity. In most climates, rainfall is greater than evaporation, thus diluting seawater and decreasing salinity.

The most abundant dissolved gases in the ocean are nitrogen, carbon dioxide, and oxygen. The amount varies with depth, and as oxygen and carbon dioxide are vital to life, most living plants and animals are found in the top 100 meters of the ocean, or the sunlight penetration layer. The amount of dissolved gases will also vary with temperature. Warm water holds less dissolved gas than does cold water because cold water is heavier and sinks, carrying oxygen-rich water to the ocean depths, which, in turn, allows some fish and marine life to live in the deepest parts of the ocean. At the surface, gases (oxygen and carbon dioxide) are exchanged between the ocean and the atmosphere as well as between the plants and animals that live in the top layers of the ocean, where the most abundant life is found. Marine plants take in carbon dioxide along with water and then, with the aid of sunlight, produce sugar and oxygen in the process called photosynthesis. The marine animals take in the oxygen and exhale carbon dioxide in the process called respiration.

Most precipitation that falls finds its way back to the ocean largely by rivers, bearing salts from soils and rock in solution. Many things affect the salinity of the ocean, including the exchange of water between the ocean and the atmosphere, which is determined by climate, and the absorption of salts by plants and animals. Considering the history of ocean salinity, one might ask if the oceans have possessed a relatively uniform salinity throughout their history or if they are becoming more saline. By far the most important component of salinity is the chlorine ion, which is produced in the same manner as is water vapor. The ratio of chlorine ions to water vapor has not fluctuated throughout geologic time, so scientists conclude, on the basis of present evidence, that ocean salinity has been relatively constant.

It is now evident that the oceans became salty early in their history because of weathering by slightly acidic rainfall and the eroding of the primary crystalline rock. Also, through the continuous volcanic activity, water vapor and gases were amply supplied to the atmosphere to aid in this weathering and eroding process. These particles were carried to the ocean and dissolved in seawater, making the water salty. This process has been going on since the formation of the oceans, so it is safe to assume that the ocean has been salty since its birth approximately 3 billion years ago.

Comparing the salinity of oceans both past and present by studying vents and hot springs, the salinity appears to have a certain balance. Salinity ranges from 33 to about 38 parts per 1,000. This range is caused by the different latitudes, the fresh water entering into the oceans from rivers and gla-

cial melt, and by the biological activity in the ocean itself; the plants and animals remove minerals needed for their growth and development (photosynthesis, respiration, shell building, and so on) from the seawater and put into it other minerals (waste products, gases, and so on). In local areas, such as bays, coves, and estuaries, salinity of seawater is further controlled by human activities such as industry, agriculture, and the pollution from these activities.

STUDY OF SEAWATER COMPOSITION

The investigation of seawater composition is accomplished primarily by collecting samples of seawater at different depths around the world and then measuring the salinity of the sample. A variety of instruments are used to collect samples. Nansen bottles are special metal cylinders fastened at a measuring point on a strong wire, which is then lowered into the sea to the desired depth. A messenger weight is dropped down the wire; when it strikes the bottle, it releases a catch. The bottle then turns upside down, and its valves close, trapping the water at that depth inside. The Nansen bottle commonly used does not seal completely, however; a better apparatus is the Fjorlie sampler or Niskin bottle. The Fjorlie sampler or Niskin bottle is attached to a line at both ends with spring-closing hinged ends. A messenger closes the bottle with a good seal.

Corrosion of metal-lined samplers may cause changes in the water composition in an hour or so. Copper, zinc, lead, and iron in metal linings often contaminate seawater samples. Plastics have solved this problem for both collection and storage of seawater. It is also necessary to filter out any organic matter that could alter the seawater composition. For maximum accuracy of testing, samples should be tested as soon as possible and not stored.

A salinometer is used to measure salinity. It is easy to use and gives immediate readings. Because the minerals dissolved in seawater make it a conductor of electricity, the more mineral matter in the sample, the better it conducts. The results are then compared to a table of standard measurements to obtain the sample's salinity.

Joyce Gawell

CROSS-REFERENCES

Abyssal Seafloor, 651; Carbonate Compensation Depths, 2101; Deep Ocean Currents, 2107;

BIBLIOGRAPHY

Attenborough, David. *The Living Planet.* Boston: Little, Brown, 1984. Based on the television series *The Living Planet,* produced by WGBH-TV in Boston in association with Mobil Oil Company, this book tells the story of the Earth, how it came to be and how it has been shaped and continuously altered by volcanism, mountain building, weathering, erosion, and the drifting of the continents. Also explores the interactions of the living species with their environments. Each environmental niche is explored, from the Arctic to the dry deserts and the freshwater ecosphere to the open ocean, as well as the artificial environment created by humans to make life fit their needs and desires.

Christy, F. T., Jr., et al. *The Law of the Sea: Caracas and Beyond.* Cambridge, Mass.: Ballinger, 1975. An examination of the results of the Caracas Session of the Third United Nations Conference on the law of the sea. It deals with the problems of seabed mining, and the impact of technological changes on future use and law of the sea is emphasized.

Coble, Charles R. *Earth Science.* Englewood Cliffs, N.J.: Prentice-Hall, 1988. This grade-nine textbook covers the Earth sciences, from astronomy, space science, and physical and historical geology to weather, climate, and the Earth's oceans. Well written and organized logically, so the younger student and layperson can search out single facts or examine the complex interrelationships of this planet's physical and biological environments.

Davis, Richard A., Jr. *The Ocean.* San Francisco: W. H. Freeman, 1972. A book-format reprint of the September 1969 issue of *Scientific American,* which was devoted to the oceans. Included are articles covering a wide range of subjects and oceanic research, from the origin of the oceans and deep-ocean floor to seawater chemistry and marine resources. Reading level is senior high and above.

_____. *Oceanography: An Introduction to the Marine Environment.* 2d ed. Dubuque, Iowa: W.C. Brown Publishers, 1991. This text is written to acquaint students with the principles of physical, chemical, biological, and geological oceanography. Included are sections covering a wide range of subjects and oceanic research, from the origin of the oceans and deep-ocean floor to seawater chemistry and marine resources. Broad in scope and shallow in depth, this book is best suited to a nonscience major or a person entering teaching where a large knowledge base is important.

_____. *Principles of Oceanography.* 2d ed. Reading, Mass.: Addison-Wesley, 1977. Designed for use in a one-semester course in general oceanography, this text is written to acquaint students with the principles of physical, chemical, biological, and geological oceanography. Broad in scope and shallow in depth, this book is best suited to a nonscience major or a person entering teaching where a large knowledge base is important.

Grasshoff, Klaus, Klaus Kremling, and Manfred Ehrhardt, eds. *Methods of Seawater Analysis.* 3d ed. New York: Wiley-VCH, 1999. This third edition of a widely used college text is completely revised and extended, devoting more attention to the techniques and instrumentation used in determining the chemical makeup of seawater. Illustrations, bibliography, and index.

Horsfield, Brenda, et al. *The Great Ocean Business.* New York: Coward, McCann and Geoghegan, 1972. In this book, developments in oceanography are viewed against the political and economic considerations governing the United States' ocean policy. The authors focus on developments in marine geology, including seafloor spreading and continental formation, but mineral and biological resources of the sea and research on waves, tides, and currents are also discussed. Reading level is senior high and above.

Luce, Harry R., et al. *The World We Live In.* New York: Time Incorporated, 1955. This book is a classic study of the Earth and its position in the grand scheme of cosmic events. The table of contents reads like a master's work in the physical and biological sciences. Although the material is somewhat dated, many of the ideas that have been replaced still find their roots in the research that went into this book—it is probably safe to conjecture that this volume provided the inspirations for many of today's scientists. Good reading for the layperson, with excellent pictures to interest all levels.

Namowitz, Samuel N., and Donald B. Stone. *Earth Science.* New York: American Book, 1981. A textbook used in the ninth through twelfth grades in the public school system. As do most Earth science textbooks, it covers the complete spectrum of the physical and biological relationships of the Earth, plus the position and relationship of the Earth in space. The text contains an excellent section on the composition and environmental structure of the world's oceans. An excellent reference text for the layperson, readily available in most public school libraries.

Sverdrup, Harold U., et al. *The Oceans.* Englewood Cliffs, N.J.: Prentice-Hall, 1957. A graduate-level university textbook that presents a now-classic and very scientifically detailed analysis of the entire ocean environment, including the chemistry of the sea, distribution of oceanic variables, physical properties of seawater, and the relationship of plants and animals to the physical-chemical properties of the ocean environment. Also includes a very complete treatment of statistics

and ocean kinematics. Not for the layperson.

Thurman, Harold V. *Essentials of Oceanography.* 6th ed. Upper Saddle River, N.J.: Prentice-Hall, 1999. An introductory college text designed to give the student a general overview of the oceans in the first few chapters. This overview is followed by a well-designed, in-depth study of the ocean—involving chemistry of the ocean, currents, air-sea interactions, the water cycle, and marine biology—and a well-developed section on the practical problems resulting from human interaction with the ocean, such as pollution and economic exploitation.

Usdowski, Eberhard, and Martin Dietzel. *Atlas and Data of Solid-Solution Equilibria of Marine Evaporites.* New York: Springer, 1998. This college-level textbook offers the reader an illustrated guide to seawater composition, including phase diagrams of seawater processes. The book is accompanied by a CD-ROM that reinforces the concepts discussed in the chapters.

Weiner, Jonathan. *Planet Earth.* New York: Bantam Books, 1986. Based on the television series *Planet Earth*, produced by WQED-TV Pittsburgh, in association with the National Academy of Sciences. Covers the full range of science as it applies to the Earth, from its cosmic beginnings to its expected final demise. Main topics include the Sun, climate, oceans, mineral and economic wealth, geology, and associations with other planets in the solar system.

Yasso, Warren E. *Oceanography.* New York: Holt, Rinehart and Winston, 1965. A basic textbook designed to introduce the science of oceanography to the beginning student. The contents are restrictive in that they concentrate on the potential and progress of the deep sea or open ocean; nearshore processes are not included, nor is the continental drift theory presented in detail. The major areas or topics covered are the chemistry of the sea, ocean circulation, topography of the seafloor, ocean sediments, and marine biology. A good book for the layperson.

SURFACE OCEAN CURRENTS

Ocean currents represent a dynamic system that, along with atmospheric circulation, helps to distribute heat evenly over the Earth. Responding to the seasons, the currents play important roles in Earth climates, marine life, and ocean transportation.

PRINCIPAL TERMS

CORE RING OR CORE EDDY: a mass of water that is spun off of an ocean current by the current's meandering motion

CORIOLIS EFFECT: the apparent deflection of any moving body or object from its usual course, caused by the rotation of the Earth

CURRENT: a sustained movement of seawater in the horizontal plane, usually wind-driven

DRIFT: a movement similar to a current but more widespread, less distinct, slower, more shallow, and less easily delineated

EARTH'S HEAT BUDGET: the balance between the incoming solar radiation and the outgoing terrestrial reradiation

GYRE OR GYRAL: the very large, semiclosed circulation patterns of ocean currents in each of the major ocean basins

PLANETARY WINDS: the large, relatively constant prevailing wind systems that result from the Earth's absorption of solar energy and that are affected by the Earth's rotation

THERMOHALINE CIRCULATION: any circulation of ocean waters that is caused by variations in the density of seawater resulting from differences in the temperature or salinity of the water

OVERVIEW

The "heat budget" of the Earth results in temperatures that make life on the planet possible. The ocean currents play a vital role in the heat budget. The currents are major determinants of climates and strongly influence the distribution of marine life. Ocean currents must be studied in relation to other aspects of the environment with which they are interwoven. The currents are, for example, closely associated with atmospheric circulation. The planetary winds are the prime movers of the currents. The friction of the wind blowing over the ocean surface establishes the slow movements in the shallow surface waters that become a global circulation of immense volumes of seawater. There are deeper ocean currents that are much slower moving and difficult to monitor, whose significance, therefore, is less well understood. The deep currents are primarily caused by thermohaline circulation. They are driven by slowly responding to slight differences in the densities of the water resulting from differences in temperature and salinities. This study concerns the shallow, wind-driven currents, the ones affecting the surface waters, although they may be hundreds of meters deep.

Most significant about ocean currents are their geographic locations and their directions of flow. It is helpful to recognize overall patterns. There are large-circulation gyres in each of the major ocean basins. These gyres, or gyrals, move clockwise in the Northern Hemisphere and counterclockwise in the Southern Hemisphere. The North Central Atlantic Gyre, for example, located east of the United States, is one of the best known and most studied. The Florida Current (part of the Gulf Stream system) is on the west side of the gyre and is a warm current flowing poleward (northward). The Canaries Current, on the east side of the gyre, is a cold current that flows equatorward (southward). The North Atlantic Drift and the North Equatorial Current form the eastward and westward components of the gyre, respectively. The result of the circulation in the gyre is that warm water from the equatorial region is transported poleward to the heat-deficient areas. Simultaneously, the Canaries Current transports colder water back toward the equator. The ocean currents thus help to distribute heat more evenly over the Earth's surface. In the Atlantic Ocean south of the equator, the large gyre is counterclockwise. The warm Brazil Current on the west side of the gyre flows poleward (southward), transporting heat away from the equator. The Benguela Current on the east side of the gyre moves colder water toward the equator.

In the Pacific Ocean, similar patterns of clockwise and counterclockwise gyres are apparent. North of the equator, the Japan Current (also known as Kuroshio) transports the warm water poleward, and the California Current moves the colder water equatorward. In the Pacific Ocean south of the equator, the cold, equatorward-flowing, nutrient-rich Humboldt Current lies off the west coast of South America and is renowned as one of the most fertile commercial marine fishery areas on Earth. The Indian Ocean possesses similar gyres, although the attenuated portion north of the equator presents some special features.

WIND AND SOLAR ENERGY

The forces that drive the oceanic current circulations are the planetary winds. The planetary winds are in turn driven by solar energy. The ocean currents, therefore, are Sun-driven. It is sunshine that energizes the Gulf Stream and the other currents. Some general principles about Earth's heat budget can be stated. The Sun heats Earth—its atmosphere, oceans, and land—but each portion heats differently. The atmosphere, the most fluid and most responsive of the three, has developed large planetary bands of alternating pressure belts and wind belts, such as the Northeast Trade Winds and the Prevailing Westerlies. The Northeast Trade Winds lie between 5 and 25 degrees north latitude. The winds are predominantly from the northeast and form one of the most constant of the wind belts. The friction of the wind moving over the ocean surface causes the surface waters to move with the wind, but because of the Coriolis effect, caused by Earth's rotation, the movement of the water current in the Northern Hemisphere tends to be about 45 degrees to the right of the winds that cause the current. The resultant current, the North Equatorial Current, is fragmented into different oceans because of the intervening continents. Largely as a result of the Coriolis effect, the current deflects to its right and, in the Atlantic Ocean, eventually becomes the Gulf Stream. In the Pacific Ocean, the comparable current is the Japan Current. One can thus see the origins of the clockwise gyrals in the Northern Hemisphere. Another wind belt in the Northern Hemisphere, the Prevailing Westerlies, is located between 35 and 55 degrees north latitude. The winds are not as constant as the Northeast Trade Winds, and they prevail from the west. The correlation of the latitudes of the Westerlies and the west-to-east-moving currents of the gyres is apparent. The currents slow down and become more widespread, shallower, and less distinguishable but are urged on toward the east by the Westerlies. The North Atlantic Drift and the North Pacific Current are the results.

NORTHERN HEMISPHERE

Again analyzing the North Central Atlantic Gyre, the blocking position of the Iberian Peninsula causes the North Atlantic Drift to split, part moving southward toward the equator as the cold Canaries Current and part moving poleward into the Arctic Ocean as the warm Norwegian Current. The Canaries Current merges into the North Equatorial Current to complete the gyre. The temperature characterizations of currents and drifts as "warm" or "cold" are relative. There are no absolute temperature divisions. Some warm currents are actually lower in temperature than some cold currents. For example, the Norwegian Current is considered a warm current only because it is warmer than the Arctic water into which it is entering. Itself only a few degrees above freezing in winter, it nevertheless transfers significant amounts of heat into these high latitudes and moderates the winter temperatures in Western and Northern Europe. A compensating movement of cold water out of the Arctic is accomplished by the southward flowing Labrador Current between Greenland and North America.

The Gulf Stream is the world's greatest ocean current; however, there is some confusion about what comprises the Gulf Stream. The Gulf Stream is generally taken to include the entire warm-water transport system from Florida to the point at which the warm water is lost by diffusion into the Arctic Ocean. It would include the North Atlantic Drift and the Norwegian Current. To the professional oceanographer, the Gulf Stream is a smaller segment of that transport system—that portion off the northeast coast of the United States. The Gulf Stream system thus includes the Florida Current, the Gulf Stream, the North Atlantic Drift, and the Norwegian Current.

SOUTHERN HEMISPHERE

Ocean current patterns in the Southern Hemisphere are almost a mirror image of those in the Northern Hemisphere, adjusted for differences in continent configurations. The Southeast Trade Winds drive the South Equatorial Current. The Coriolis effect causes deflections to the left. Resultant gyres are counterclockwise, but again, the poleward-moving currents transfer the heat away from the equator, and the equatorward-moving currents return colder water. In general, cold currents are richer in nutrients, have a higher oxygen content, and support a greater amount of life than warm currents. Most products of the world's commercial fisheries are yielded by these cold waters. On the other hand, cold currents offshore are associated with desert climates onshore. The atmospheric circulations that drive the ocean currents also create conditions that are not conducive to precipitation in the latitudes of these cold currents. Examples are the Sahara Desert adjacent to the Canaries Current, the Peru-Chile Desert adjacent to the Humboldt Current, the Sonoran Desert adjacent to the California Current, and the Kalahari Desert adjacent to the Benguela Current.

There are other currents that are sporadic in occurrence, such as the warm El Niño current that periodically develops off the west coast of northwestern South America for reasons that are not well understood. Local currents are caused by tides, storms, and local weather conditions.

STUDY OF OCEAN CURRENTS

The study of ocean currents has acquired new significance as scientists have learned of the role of currents in Earth climates and marine life. Information comes from many sources. One of the earliest attempts to identify and chart an ocean current was by Benjamin Franklin when he was postmaster general of colonial America. His map of the Gulf Stream was published in 1770 and has proven to be remarkably accurate when one considers his sources of information. Franklin had

noted that vessels sailing westward from England to America in the midlatitudes of the Atlantic Ocean were taking longer than ships moving eastward and longer than ships moving westward but in lower latitudes. He correctly concluded that the vessels were moving against a slow, eastward-moving current.

Since that time, vast amounts of data have been acquired to detect, measure, and chart the currents. One of the early methods still employed is the use of drift bottles. Sealed bottles are introduced into the sea at various locations and dates and are allowed to float with the currents. Finders are requested to note the date and location of the bottle-find and to return the data to the address in the bottle. Various types of current meters are also used. Some are moored to the bottom and can transmit results by radio. It is difficult for a ship at sea to measure currents because the ship itself is drifting with the current. Currents are generally very slow and difficult to measure. A few currents may be measured at 6 to 8 kilometers per hour, but much more common are those less than 1 kilometer per hour. The average surface velocity of the North Atlantic Drift is about 1.3 kilometers per hour. The currents also vary in width and depth. The Florida Current off Miami is about 32 kilometers wide, 300 meters deep, and moving at about 5 to 8 kilometers per hour. It transports more than 4 billion tons of water per minute. The volume of flow is more than one hundred times that of the Mississippi River. As the flow proceeds north and then east as the North Atlantic Drift, it spreads, thins, slows, and splits into individual meandering flows that are difficult to follow. Spin-off eddies, or "core rings," occur that can persist for months.

One useful method of tracking ocean currents is to be able to identify water of slight temperature variations and salinity differences. When the flow movement is so slow as to be practically undetectable with current meters, the slight temperature and salinity differences can be used as "tracers" to identify current movements. This method is also used in identifying the even slower-moving deep ocean water currents. Another technique is the use of satellite imagery and high-altitude aerial photography. Using sensors that detect radiation at selected bands of the electromagnetic spectrum, satellites collect data on broad patterns of seawater temperatures and thus help scientists to understand the movements and extent of the currents. This type of sea monitoring may be useful in detecting any changes that might occur in the oceans in the future.

SIGNIFICANCE

Ocean currents play a vital role in the environment. Energy from the Sun is absorbed by Earth and especially by the oceans. Along with atmospheric circulation, the ocean currents serve to distribute the absorbed heat more evenly over Earth's surface. Immense volumes of warm seawater are slowly moving poleward, transporting heat from the heat-surplus equatorial regions of Earth to the heat-deficient regions nearer the poles. Cold-water currents in turn move chilled water back toward the equator. Although neither solar energy nor rainfall is evenly distributed over Earth, the mixing actions of the ocean currents keep the Earth environment in a steady state. These moderating effects of the ocean currents are the climates of the west coasts of the continents in the middle and high latitudes, especially in Europe. The densely populated nations of northwestern Europe experience much milder winters than would otherwise be expected for such high latitudes. Northwestern North America is similarly benefited.

Life in the sea is also aided by these ocean-water mixings. In addition to heat, ocean currents distribute oxygen and nutrients, the result being certain areas in the oceans where very favorable life-supporting conditions occur. These fertile areas of mixing are concentrated sources of commercial marine fishery products. Where mixing is limited, nutrient-poor regions arise in the ocean such as the Sargasso Sea, located in the center of the North Central Atlantic Gyre. Global warming caused by the greenhouse effect may alter ocean currents, creating a further need to study these currents and their effects on climate and marine life.

John H. Corbet

CROSS-REFERENCES

Carbonate Compensation Depths, 2101; Deep Ocean Currents, 2107; Earth Resources, 1741; Gulf Stream, 2112; Hydrologic Cycle, 2045; Hydrothermal Vents, 2117; Ocean-Atmosphere Interactions, 2123; Ocean Pollution and Oil Spills, 2127; Ocean Tides, 2133; Ocean Waves, 2139; Oceans' Origin, 2145; Oceans' Structure, 2151; Saltwater Intrusion, 2061; Sea Level, 2156; Seamounts, 2161; Seawater Composition, 2166; Tsunamis, 2176; Turbidity Currents and Submarine Fans, 2182; World Ocean Circulation Experiment, 2186.

BIBLIOGRAPHY

Duxbury, Alyn C., and Alison B. Duxbury. *An Introduction to the World's Oceans.* 6th ed. Boston: McGraw-Hill, 2000. This book includes a clearly written section on the Earth's planetary winds and their effects on the ocean currents. The general patterns of ocean-current circulation are mapped, and diagrams are used to explain the Coriolis effect. Suggested readings are listed.

Gaskell, T. F. *The Gulf Stream.* New York: New American Library, 1972. Following a general discussion of ocean currents, a more detailed description of the Gulf Stream is given. Life in the current, its meanderings, and its effects on climate are discussed. This book is informative and easy to read. A brief bibliography is included.

Gross, M. Grant. *Oceanography: A View of the Earth.* 7th ed. Englewood Cliffs, N.J.: Prentice-Hall, 1996. This is a general introductory text for oceanography. It is well illustrated in all aspects of the study of the oceans. Ocean currents are explained and mapped. Both horizontal and vertical aspects of oceanic circulation are related to winds, temperatures, and salinities. Comprehensive and easy to read. Contains a glossary and an index.

Ingmanson, Dale E., and William J. Wallace. *Oceanography: An Introduction.* 5th ed. Belmont, Calif.: Wadsworth, 1995. A general introduction to oceanography. Written as an introductory college text but can be read by the high school student who is interested. The text is well illustrated, and important terms are in bold print. Includes a glossary, and each chapter contains a list of further readings.

Pickard, George L., and William J. Emery. *Descriptive Physical Oceanography: An Introduction.* 5th ed. Boston: Butterworth-Heinemann, 1995. This book covers only the physical aspects of oceanography, in a less comprehensive and more technical manner than general-introduction oceanography texts. It presents a detailed description of ocean circulation, both surface currents and deep currents. Includes details on the types of current meters and the methods of current measurements. An extensive bibliography is provided.

Stowe, Keith. *Exploring Ocean Science.* 2d ed. New York: John Wiley & Sons, 1996. An introductory text for college students who have little background in the sciences. Three chapters discuss oceanic and atmospheric circulation; one deals specifically with surface ocean currents. All chapters include review questions and suggestions for further reading. A glossary is included.

Thurman, Harold V. *Introductory Oceanography.* 8th ed. Upper Saddle River, N.J.: Prentice-Hall, 1997. This is an introductory text to oceanography. It is comprehensive but not too technical for the general reader. It is very well illustrated and includes some high-quality color maps and diagrams. Each chapter includes questions and exercises and also lists references and suggested readings. The circulations of the ocean currents are addressed according to the major ocean basins, with maps and diagrams for each basin.

TSUNAMIS

A tsunami is a series of potentially catastrophic ocean waves most commonly caused by movement of the seafloor associated with large earthquakes. Once known as tidal waves, tsunamis have resulted in thousands of deaths, primarily around the Pacific Ocean.

PRINCIPAL TERMS

BORE: a nearly vertical advancing wall of water that may be produced by tides, a tsunami, or a seiche

EARTHQUAKE: vibrations in the Earth produced by the sudden movement of large masses of rock

RICHTER SCALE: a scale devised by C. F. Richter used for measuring the magnitude of earthquakes

SEICHE: an oscillation in a partially enclosed body of water, such as a bay or estuary

TSUNAMI WARNING: the second phase of a tsunami alert; it is issued after the generation of a tsunami has been confirmed

TSUNAMI WATCH: the first phase of a tsunami alert; it is issued after a large earthquake has occurred at the seafloor

CAUSES OF TSUNAMIS

Giant ocean waves that were once called tidal waves or seismic sea waves have resulted in thousands of deaths in countries around the Pacific Ocean. These destructive waves are now generally known as tsunamis. In the simplest terms, a tsunami is a series of waves usually caused by violent movement of the seafloor.

The movement at the seafloor that causes the tsunami can be produced by three different types of violent geologic activity. By far the most important of these is submarine faulting, which occurs when a block of the ocean floor is thrust upward or suddenly drops. Such fault movements are accompanied by earthquakes. Probably the second most common cause of tsunamis is landslides; a tsunami may be generated by a landslide starting above sea level and then plunging into the sea or by a submarine landslide. The highest tsunami waves ever reported were produced by a landslide at Lituya Bay, a confined fjord in Alaska, on July 9, 1958. A massive rock slide at the head of the bay produced a tsunami wave that attained a high-water mark more than 500 meters above the shoreline.

The third cause of tsunamis is nearshore or submarine volcanic activity. In most cases, the flank of a volcano is suddenly uplifted or depressed, producing a tsunami in much the same way as faulting activity does; however, tsunamis have also been produced by the actual explosion of submarine or shoreline volcanoes. In 1883, the violent explosion of the famous island volcano Krakatau sent tsunami waves as high as 40 meters crashing ashore in Java and Sumatra. More than thirty-six thousand people were killed.

Although tsunamis caused by landslides or volcanic activity may be very large near their sources and may cause great damage there, they have relatively little energy. They decrease rapidly in size, becoming small or even unnoticeable at any great distance. The giant tsunami waves that cross entire oceans are almost all caused by submarine faulting associated with large earthquakes. Most tsunamis occur in the Pacific Ocean because the Pacific Ocean basin is surrounded by a zone of very active features in the Earth's crust: deep ocean trenches, explosive volcanic islands, and dynamic mountain ranges.

CHARACTERISTICS OF TSUNAMIS

Tsunami waves are very different from ordinary ocean waves, most of which are caused by the wind blowing over the water. These wind-generated waves rarely have a wavelength (distance from crest to crest) greater than 300 meters. Tsunami waves may have a wavelength of as much as 160 kilometers. Wind-generated ocean surface waves never travel at more than about 100 kilometers per hour and are usually much slower. In the deep water of an ocean basin, tsunami waves may travel at 800 kilometers per hour; they may, however, be only 50 centimeters high and pass by ships at sea unnoticed.

A tsunami is not a single giant wave but rather consists of several waves, perhaps ten or more, that form what is called a tsunami wave train. These individual waves follow one behind the other, between five and ninety minutes apart. When tsunami waves move into shallower water and approach shore, they start to change. The shape of the nearshore seafloor has an effect on how the tsunami waves behave. The waves tend to be smaller near small, isolated islands, where the bottom drops away quickly into deep water. Near large islands, such as the main Hawaiian Islands, the waves are strongly influenced by the bottom; they bend around the land and may be reflected from the shoreline. The reflected waves may augment other waves and create extremely large wave heights in unexpected places.

Though it is not yet possible to predict the size or effects of tsunami waves as they arrive in coastal areas, prediction of the velocity of tsunamis is made possible by their great wavelength. To understand how the arrival time of a tsunami is determined, it is necessary to look at ocean waves in general. Oceanographers divide waves into various categories based on the relationship between wavelength and the depth of water through which they are passing. When the water depth is less

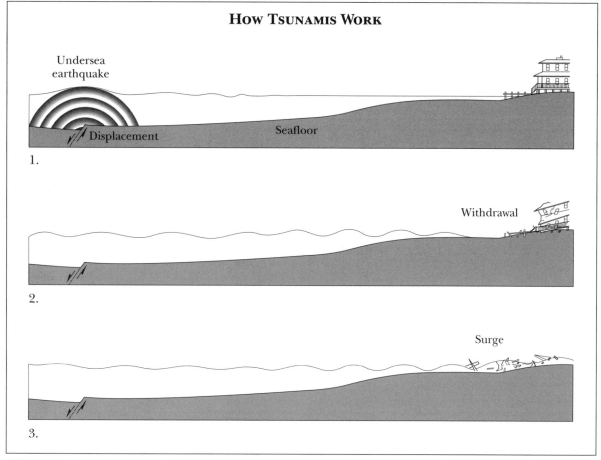

How Tsunamis Work

Although tsunamis can result from other causes, most tsunamis are caused by earthquakes in the ocean floor, which result in deep-water waves with nearly undetectable wave height (panel 1). Upon approaching a shoreline, however, the rising seafloor causes these very long waves to pile up, initially causing the water at the shoreline to withdraw suddenly (panel 2), but ultimately resulting in a huge wave surge (panel 3) that can reach miles inland, destroying everything in its path.

than one-twentieth of the wavelength, the waves are known as shallow-water waves, and their speed is determined solely by the depth of water. Knowing the water depth, one can calculate the velocity of any shallow-water wave. Tsunami waves may have a wavelength of more than 160 kilometers, and if the water depth is less than one-twentieth of the 160-kilometer wavelength, or 8 kilometers, then the tsunami waves would be shallow-water waves. Most of the deep Pacific Ocean basin is less than 5 kilometers deep, so most tsunamis there are shallow-water waves. To determine the speed of such tsunamis, all that is necessary to know is the depth of the water through which they pass. In most of the deep Pacific Ocean, this speed works out to be around 725 kilometers per hour, but it will vary depending on the exact water depth along the path of the tsunami.

EFFECTS OF TSUNAMIS

As the waves continue to approach the shore, they travel progressively more slowly, with the forward speed dropping to around 65 kilometers per hour. At this point, the wave height usually begins to increase dramatically. A tsunami wave that was 50 centimeters high at sea may become 10 meters high at the shoreline.

Like storm waves, tsunami waves are often more severe on headlands, where the wave energy is concentrated. The presence of a well-developed coral reef off a shoreline also appears to have a strong effect on tsunami waves; a reef may serve to absorb a significant amount of the wave's energy, reducing its height and the intensity of its impact on the shoreline. Unlike ordinary waves, however, tsunamis are often quite large

in bays. In this way, tsunami waves resemble tides.

Another wave phenomenon may also be produced in bays when a tsunami strikes. The water in any basin, be it a small bay or a large sea, will tend to oscillate back and forth in a fixed period determined by the size and shape of the basin. This oscillation is known as a seiche. A tsunami wave may set off a seiche, and if the following tsunami wave arrives with the next natural oscillation of the seiche, the water may reach even greater heights than

TSUNAMIS WITH MAGNITUDES OF 4 OR GREATER

Date	Magnitude	Intensity	Location
Nov. 29, 684	4.0	3.0	Nankaido, Japan
July 13, 869	4.0	4.0	Sanriku, Japan
Aug. 3, 1361	4.0	3.0	Nankaido, Japan
Sept. 20, 1498	4.0	3.0	Nankaido, Japan
Oct. 28, 1562	4.0	3.5	Chile
July 9, 1586	4.0	3.5	Peru
Nov. 24, 1604	4.0	3.5	Peru
Dec. 2, 1611	5.0	4.0	Sanriku, Japan
Aug. 1, 1629	4.0	3.0	Banda Sea, Indonesia
Dec. 31, 1703	4.0	3.0	Tokaido-kashima, Japan
July 8, 1730	4.0	3.5	Chile
Oct. 17, 1737	4.5	4.0	Kamchatka Peninsula
Oct. 29, 1746	4.6	3.5	Lima, Peru
Apr. 24, 1771	4.0	3.0	Ryukyu Trench
Dec. 29, 1820	4.0	3.5	Flores Sea, Indonesia
Feb. 20, 1835	4.0	3.0	Chile
Dec. 23, 1854	4.0	3.0	Tokaido, Japan
Dec. 24, 1854	4.0	3.0	Nankaido, Japan
Apr. 3, 1868	4.1	3.4	Hawaii
Aug. 13, 1868	4.3	3.5	Chile
Mar. 2, 1871	4.0	3.5	Sulawesi, Molucca islands
May 10, 1877	4.0	3.5	Chile
June 15, 1896	4.0	3.8	Sanriku, Japan
Feb. 3, 1923	4.0	0.0	Kamchatka Peninsula
Apr. 13, 1923	4.0	3.5	Kamchatka Peninsula
Apr. 1, 1946	5.0	4.0	Aleutian Islands
Nov. 4, 1952	4.0	4.0	Kamchatka Peninsula
July 9, 1956	5.0	6.0	Amorgos
May 22, 1960	4.5	4.0	Chile
Mar. 28, 1964	4.5	5.0	Gulf of Alaska, Alaska Peninsula
Jan. 24, 1965	4.0	1.5	Ceram Island, Indonesia

NOTES: The tsunami magnitude and tsunami intensity are measurements of the size of the tsunami based on the logarithm of the runup. *Tsunami magnitude* is defined by Iida and others (1967) as M = log2h, where "h" is the maximum runup height of the wave. *Tsunami intensity* is defined by Soloviev and Go (1974) as I = log2/2h, where "h" is the maximum runup height of the wave.
SOURCE: National Oceanic and Atmospheric Administration, National Environmental Satellite, Data, and Information Service, National Geophysical Data Center, World Data Center-A, Colorado. URL: http://www.ngdc.noaa.gov

from the tsunami waves alone. Much of the great height of tsunami waves in bays may be explained by this constructive combination of a seiche wave and a tsunami wave arriving at the same time.

The popular image of a tsunami wave approaching shore is a nearly vertical wall of water, similar to the front of a breaking wave in the surf. Actually, most tsunamis probably do not form such wave fronts; the water surface instead is very close to horizontal, and the surface itself moves up and down. The waves arrive like a very rapidly rising tide. Under certain circumstances, however, an arriving tsunami wave can develop an abrupt, steep front that will move inland at high speeds. This phenomenon, generally encountered only as a tidal phenomenon, is known as a bore.

Bores produced by tides occasionally occur in the mouths of rivers; well-known examples occur on the Solway Firth and the Severn River in Great Britain, on the Petitcodiac River in Maine, near the mouth of the Amazon River in Brazil, and, most strikingly, on the Chientang River in China, where the bore may attain a height of nearly 5 meters. In place of the usually gradual rise of the tide, the onset of high tide is delayed; when it does occur, it takes place very quickly, and the rapidly moving wall of water is followed by a less steep, but still quite dramatic, rise in the water level, accompanied by swift upstream currents.

Because the height of tsunami waves is strongly influenced by the submarine topography and shape of the shoreline and by reflected waves, and because they may be further modified by seiches, tides, and wind waves, the actual inundation and flooding produced by a tsunami may vary greatly from place to place over only a short distance. Though the image of a bore is the most dramatic—such a wall of water can raze nearly everything in its path—it is the flooding effect of a tsunami that causes the most damage. Two different terms are often used to describe the extent of tsunami flooding: "inundation" and "run-up." Inundation is the depth of water above the normal level and is usually measured from sea level at average low tide. Inundation may be measured at any location reached by the tsunami waves. Run-up is the inundation at the maximum distance inland from the shoreline reached by the tsunami waters.

Even the withdrawal of the tsunami waves can cause significant damage. As the water is rapidly drawn back toward the sea, it may scour out bottom sediments, undermining the foundations of buildings. Entire beaches have been known to disappear as the sand is carried out to sea by the withdrawing tsunami waves.

STUDY OF TSUNAMIS

The best way to learn about tsunamis is to study the tsunami waves themselves. Much of this work is being carried out in the Hawaiian Islands because of their susceptibility to tsunamis from all parts of the Pacific Ocean basin and because Hawaii is the headquarters of the Pacific Tsunami Warning System. Because tsunamis are, fortunately, not everyday events, it is imperative to collect the most information possible from each occurrence. When a tsunami is en route to the Hawaiian Islands, trained observers head toward preselected shoreline vantage points, where time-lapse surveillance cameras are set up to film the waves. Portable tsunami gauges are deployed from piers and in designated shoreline areas. These gauges sense the change in water pressure as the waves pass over them and record their measurements.

U.S. Navy patrol planes take to the air before the arrival of the first waves. The aircraft fly at an altitude of 300 meters in a racetrack pattern over critical shoreline areas, covering the same spot every fifteen minutes. Special cameras mounted in the belly of each plane take 480 exposures on 70-millimeter film, documenting the arrival of each tsunami wave on the shores of the islands.

After a tsunami, a ground survey and damage assessment are conducted by a team from the U.S. Army Corps of Engineers, which is joined by volunteers from the American Society of Civil Engineers. A post-tsunami aerial photographic survey is undertaken jointly by the Navy, the Coast Guard, and the Civil Air Patrol; National Weather Service U-2 surveillance aircraft are used.

Laboratory techniques have also been employed to study tsunami waves. Much of the damage caused by tsunamis results from the run-up of the giant waves on shore. Various modeling techniques have been used to try to simulate the run-up phase. One such technique is called hydraulic modeling; it uses a physical scale model. Hilo Bay on the island of Hawaii is particularly sensitive to tsunami waves. After the disastrous tsunami of 1960, a hydraulic model of the bay was constructed. The model measured 25 by 19 meters and represented the triangular shape of Hilo Bay. Model tsunami waves were produced by releasing water from large tanks according to a program. Experiments with this hydraulic model have shown that almost any wave that enters the bay either hits downtown Hilo directly or is bounced off the northeast coast into the town. Sometimes the direct and reflected waves interact constructively to produce especially large waves in the center of the bay.

Another type of model being increasingly used to study tsunamis is the numerical model, which uses a high-speed digital computer to calculate mathematical simulations of tsunami waves. The models, however, are only as good as the data on which they are based, and these data can only come from measurements and observations of real tsunamis. A recent improvement in data collection has been the use of satellites to transmit data from remote tide stations.

Satellite-telemetered tide stations have been installed at some twenty-five sites across the Pacific. These stations operate on their own independent power sources, secure from electricity outages resulting from earthquakes. Sea-level measurements are made every two seconds and averaged over a three- or four-minute interval, and the data are routinely transmitted by satellite to the Pacific Tsunami Warning Center every three to four hours. In the event of a tsunami wave, however, an "event detector" almost instantaneously sends a message to the warning center over a special emergency satellite channel.

Even more sophisticated measuring devices are being tested, including devices that do not have to be installed on land. A warning system has been proposed that would be based on the detection of tsunamis by highly sensitive bottom-pressure gauges located on the seafloor in mid-ocean. According to specialists, a tsunami wave as small as 1 centimeter could be detected with such instruments. In fact, a small tsunami resulting from a Mexican earthquake in March of 1979 was successfully measured using a device on the seafloor 3,000 meters deep, off Baja California. To study how tsunami waves change upon entering shallow water, an observational program has been conducted off the Galápagos Islands, with instruments at depths of 3,000 meters, 10 meters, and 1 meter.

In addition to seafloor tsunami detectors, seismographs have been deployed on the ocean bottom. The Japanese have successfully operated a permanent ocean-bottom seismograph system off the southern coast of central Honshu. Attached to their seismographs is a tsunami gauge. In the future, the system will ultimately use both shore-based and ocean-bottom earthquake and tsunami sensors that transmit real-time data via satellite to the center, where computers will provide scientists with the information they need to issue confirmed tsunami alerts and will increase their knowledge of tsunami wave generation, propagation, and run-up.

Walter C. Dudley, Jr.

CROSS-REFERENCES

BIBLIOGRAPHY

Adams, W. M., and N. Nakashizuka. "A Working Vocabulary for Tsunami Study." *Tsunami Hazards* 3 (1985): 45-51. This article provides an elementary scientific discussion of the different terms used to describe tsunami waves and their effects.

Dudley, W. C., and M. Lee. *Tsunami!* 2d ed. Honolulu: University of Hawaii Press, 1998. This well-written book includes eyewitness accounts of tsunamis striking the Hawaiian Islands. It provides a thorough, nontechnical explanation of the generation, propagation, and run-up of tsunami waves and an account of the origin and operation of the Pacific Tsunami Warning System. Includes maps, charts, and numerous photographs of tsunamis and their devastating effects.

Finkl, Charles W., ed. *Coastal Hazards: Prediction, Susceptibility, and Mitigation.* Charlottesville, Va.: Coastal Education and Research Foundation, 1994. This volume from the Coastal Research series focuses on techniques used to prepare for such coastal hazards as tsunamis and hurricane landfall. Illustrations, maps, index, and bibliography.

Jaggar, T. A. "The Great Tidal Wave of 1946." *Natural History* 55 (1946): 263-268. A good historical account of the 1946 tsunami that did great damage in the Hawaiian Islands. Jaggar was the founder of the Hawaiian Volcanoes Observatory and an early expert on tsunamis.

Myles, Douglas. *The Great Waves.* New York: McGraw-Hill, 1985. This book relates the history of tsunamis from the beginning of re-corded history to the mid-1980's. It discusses the nature and origins of various tsunamis and provides personal accounts of people affected by the tsunami disasters.

Tsuchiya, Yoshito, and Nobuo Shuto, eds. *Tsunami: Progress in Prediction, Disaster Prevention, and Warning.* Boston: Kluwer Academic Publishers, 1995. This thorough book provides information on advancements made in the study of tsunamis. Much attention is paid to the technologies and procedures used to predict tsunamis, warn populations of their approach, and prevent damage. Color illustrations, maps, bibliographical references.

Wylie, F. E. *Tides and the Pull of the Moon.* Brattleboro, Vt.: Stephen Greene Press, 1979. Though this book discusses primarily the phenomenon of tides, it contains a chapter on tsunamis. This chapter mentions tsunamis ranging from those striking Hawaii to the one produced by the explosion of Krakatau. It discusses the possibility that tsunamis are the basis for the biblical story of the Flood and the legend of Atlantis.

TURBIDITY CURRENTS AND SUBMARINE FANS

Turbidity currents are a major mechanism whereby sediment from nearshore areas is transported to the deep parts of the oceans. This sediment often accumulates as part of large depositional systems called submarine fans, which will most likely be exploited as a major source of petroleum in the future.

PRINCIPAL TERMS

ABYSSAL PLAIN: the flat, sediment-covered area of the seafloor that merges with the base of the continental rise

CONTINENTAL MARGIN: that part of the Earth's surface that separates the emergent continents from the deep-sea floor; it generally consists of the continental shelf, continental slope, and continental rise

CONTINENTAL RISE: the broad and gently sloping ramp that rises from the abyssal plain to the base of the continental slope; submarine fans are found here

CONTINENTAL SHELF: the gently seaward-sloping submerged edge of a continent that commonly extends to a depth of about 200 me-

ters or to the landward edge of the continental slope

CONTINENTAL SLOPE: the relatively steep region of the continental margin between the continental shelf and the continental rise

SUBMARINE CANYON: a submerged, V-shaped canyon cut into the continental shelf and continental slope, through which turbidity currents funnel into the deeper parts of the oceans

TURBULENT FLOW: a high-velocity sediment flow in which individual sediment particles move in very chaotic or nonstreamline directions above the seafloor

DEVELOPMENT OF THE CONCEPT

A turbidity current is a dense mass of water and sediment that flows downhill along the bottom of an ocean or any other standing body of water such as a lake. Turbidity currents may reach high speeds during their movement and are capable of carrying large quantities of sand and of eroding or scouring the ocean floor as they move.

The concept of the turbidity current has a long history. In the late nineteenth century, geologists observing the waters of the Rhone River entering Lake Geneva noted that instead of mixing with the lake water, the river water moved along the bottom of the lake in a channel. This behavior was interpreted to reflect the elevated density of the cold, sediment-laden Rhone River water. These dense, bottom-hugging currents were called "density currents." It was suggested in the late 1930's that dense currents composed of sediment and water could be produced by wave activity on the continental shelves, the flat or gently sloping submerged edges of the continents, during periods when global sea level is lower than it is today. Fur-

thermore, it was postulated that these dense, turbulent mixtures of suspended sediment and water would flow across the continental shelves and down large canyonlike features into deeper parts of the ocean.

From the 1930's to the 1950's, researchers conducted numerous laboratory studies to test the hypothesis that turbidity currents could actually erode the ocean floor and produce such features as submarine canyons on the continental shelf. This work was done by pouring muddy water into the end of an inclined flume (a long, straight trough in which the hydrodynamic properties of moving fluids can be studied) in order to produce artificial turbidity currents. Results of these investigations indicated that sediment in a turbidity current moves in a very chaotic or random pattern. Additionally, geologists recognized that turbidity currents can move rapidly and that their velocity is dependent upon the slope of the sea bottom along which they move as well as the density of the sediment-water mixture.

These experimental studies also yielded insight

into the nature of the sedimentary deposits that accumulate from turbidity currents. These deposits, referred to as turbidites, form as graded beds, sedimentary layers in which the largest or most coarse-grained sediment particles are concentrated at the bottom of the bed and grade gradually upward to the smallest or most fine-grained sediment at the top. The graded nature of the bedding was inferred correctly to be a consequence of a reduction in the flow velocity of the turbidity current upon reaching the very gentle slopes that are typical of the bottom of the ocean, so that the largest sediment particles settle out of the current to the bottom of the ocean first, followed by progressively finer particles. Thus, each turbidity current produces a single-graded bed or turbidite.

Geologists working in the field began to realize that the graded beds such as those produced in the laboratory studies could be observed in sequences of sedimentary rocks exposed on land in various mountain belts throughout the world (the Appalachians, Apennines, and Carpathians). Moreover, oceanographic research demonstrated the presence of turbidite sands in some of the deepest areas of the modern oceans. It is now apparent that turbidity currents are and were a dominant, if not the dominant, mechanism of transport of sediment from shallow-water environments on the continental shelf to the deeper parts of the ocean. Evidence of this includes the recovery of shallow-water organic remains from the very deep parts of the oceans far removed from land. The laboratory studies and subsequent oceanographic research indicated that turbidites differ from most other sedimentary deposits in that they are nearly instantaneous deposits that may accumulate over a few hours to a few days. Moreover, turbidity currents are quite capable of eroding the great submarine canyons that cut into many continental shelves.

CAUSES OF TURBIDITY CURRENTS

Downhill movement of turbidity currents can be triggered by various causes. Commonly, earthquakes affecting the continental shelves will cause sediment on the continental slope, that part of the continental margin characterized by an increase in gradient immediately seaward of the continental shelf, to slide downslope, thereby mixing with

seawater to form a turbidity current. A rapid sequence of successive breaks in transatlantic cables on the continental margin bordering southern Newfoundland was apparently caused by generation and downslope movement of a dense, turbulent mixture of seawater and sediment generated from the continental shelf and slope in response to the Grand Banks earthquake of November 8, 1929. Those cables closest to the point on the ocean floor directly above the earthquake, the epicenter, broke first, whereas cables farther from this point broke later. The breaking of the cables indicated the erosive ability of turbidity currents. Subsequent drilling of that part of the ocean floor traversed by the turbulent flow recovered a 1-meter-thick graded bed containing shallow-water organic remains; this discovery strengthened the argument that the flow was indeed a turbidity current. The earthquake-induced turbidity current averaged 27 kilometers per hour, although it reached velocities in excess of 70 kilometers per hour in steeper parts of the continental margin, and it covered an area of more than 195,000 square kilometers.

Turbidity currents can also be generated by wave activity on continental shelves, an idea, as already noted, that was postulated in the 1930's. More specifically, large waves produced during great storms such as hurricanes are capable of creating the turbulence required to mix sediment and seawater, thereby creating a turbidity current. An equally plausible mechanism of turbidity-current generation involves oversteepening of the continental slope by the sudden addition of sediment. This is especially common where large rivers deposit their sediment load near the head of a submarine canyon.

The rate at which turbidity currents are generated was greatest during periods when sea level was much lower than it is today. This was especially typical of glacial periods, when more seawater was locked up in the enlarged polar ice caps, resulting in the lowering of the global level of the oceans. During these times of lowered sea level, rivers were capable of transporting their sediment loads directly across the previously submerged continental shelf into the head of a submarine canyon. There, the sediment mixed with seawater and was fed, via the submarine canyon, directly into the deeper parts of the ocean as turbidity currents. When sea level rose and the shoreline retreated landward as the exposed continental shelf was once again submerged, the river was cut off from the canyon head, thereby precluding the infusion of sediment directly to the canyon and reducing the likelihood of turbidity-current generation.

CHARACTERISTICS OF SUBMARINE FANS

Turbidity currents triggered on continental shelves or continental slopes move downhill until they reach such a point at which the reduced gradient of the ocean floor causes a reduction in the velocity of the sediment flow. This leads to deposition of a graded bed or turbidite. Many turbidite deposits that are funneled through submarine canyons ultimately accumulate as part of large fan-shaped or cone-shaped sediment bodies called submarine fans. The submarine fans, which spread outward from the mouths of the submarine canyons, merge with the bottom of the continental slope and comprise the continental rise, the broad, gently sloping feature that rises from the abyssal plain of the ocean floor and merges with the base of the continental slope. Where submarine canyons are close together along the continental shelf and continental slope, the attached submarine fans may coalesce to form a wide, laterally extensive continental rise.

In general, submarine fans are subdivided into three major morphologic elements or parts: the upper fan, midfan, and lower fan. The upper fan, also referred to as the inner fan, is typically characterized by a single submarine channel that is connected to the submarine canyon. Upper fan channels range from 2 to 18 kilometers wide and may be as deep as 900 meters. The single channel is commonly flanked on both sides by levees, low ridges that run along the length of the channel. Most of the sediment transported through the upper fan channel via the submarine canyon is deposited into the midfan, that area of the submarine fan composed of raised, lobelike sequences of turbidites called depositional lobes. The depositional lobes are fed by numerous shallow and unstable distributive channels that branch off the main upper fan channel. Because these channels are generally relatively shallow (several tens of meters deep), they are more likely to be filled in during passage of extraordinarily dense turbidity currents. When that happens, the channels are abandoned and subsequent turbidity currents erode or scour out new channels in adjacent areas of the midfan part of the submarine fan. The lower fan is characterized by a smooth ocean floor that passes imperceptibly seaward into the abyssal plain. The low gradient of the lower fan relative to that of the midfan leads to a much-reduced abundance of turbidites in the lower fan. Turbidite deposits are not common on the abyssal plain because ocean floor gradients in this area are typically too low to sustain movement of the dense sediment cloud.

Submarine fans display great variations in size. The Bengal Fan in the Bay of Bengal (northeast Indian Ocean) is the largest submarine fan on the Earth's surface. Its total length exceeds 3,000 kilometers. The main or upper fan channel ranges from 13 to 18 kilometers wide and from 150 to 900 meters deep. Most submarine fans, however, are much smaller than the Bengal Fan.

The growth of submarine fans, because they are major sites of turbidite sedimentation, is controlled by variations in sea level. As noted previously, turbidity currents are generated more frequently during periods of low sea level. Accordingly, submarine fans are likely to grow fastest, as manifested by increased rates of turbidite sedimentation, during periods of global lowstands of sea level. Indeed, geologists have demonstrated that the Mississippi submarine fan, which is fed by the Mississippi River, was the site of abundant turbidite sedimentation at the end of the most recent ice age, approximately 15,000 to 20,000 years ago, when sea level was as much as 120 to 130 meters lower than it is today. Since then, the rate of turbidite sedimentation has dropped off greatly as sea level gradually rose, and the shoreline retreated landward to its present position.

Gary G. Lash

CROSS-REFERENCES

BIBLIOGRAPHY

Black, J. A. *Oceans and Coasts*. Dubuque, Iowa: Wm. C. Brown, 1986. This high school or beginning-level college text provides a discussion of continental margin environments as well as an overview of turbidity currents and submarines.

Ross, D. A. *Introduction to Oceanography*. 5th ed. New York: Harper-Collins College Publishers, 1995. This oceanography textbook offers a good discussion of continental margin environments, including the shelf, slope, and rise as well as the abyssal plain. Turbidity currents and submarine fans are also discussed. Suitable for high school or first-year college readers.

Scholle, P. A., and D. Spearing, eds. *Sandstone Depositional Environments*. Tulsa, Okla.: American Association of Petroleum Geologists, 1982. A well-illustrated overview of the various depositional environments of sandstone. The section on submarine fans is excellent and easy to follow. The book includes useful photographs of turbidites. Suitable for high school students.

Shepard, F. P. *Submarine Geology*. 3d ed. New York: Harper & Row, 1973. A good discussion of the methods used by marine geologists to study sediments of the ocean floor. Suitable for high-school-level readers.

Thurman, H. V. *Introductory Oceanography*. 8th ed. Upper Saddle River, N.J.: Prentice-Hall, 1997. This introductory textbook offers a good overview of the science of oceanography. A discussion and diagram of turbidity currents and the submarine canyons they create is provided in chapter 3. This text is well illustrated and contains a helpful glossary.

Walker, Roger G. *Facies Models: Response to Sea Level Change*. 2d ed. Tulsa, Okla.: Society of Economic Paleontologists and Mineralogists, 1994. This publication contains an excellent discussion on turbidites and their environments of deposition. Perhaps the best feature of this book is the reading lists, which contain references on the history and philosophy of the turbidity current concept, hydrodynamic properties of turbidites, and submarine fans. This publication is best suited for the first-year or second-year college student.

WORLD OCEAN CIRCULATION EXPERIMENT

The World Ocean Circulation Experiment is an ambitious international project designed to increase knowledge about the movement of water, heat, and various substances in the sea. This data is expected to be of major importance in predicting future long-term changes in climate.

PRINCIPAL TERMS

CORIOLIS EFFECT: the phenomenon, caused by the Earth's rotation, that causes the path of a moving object to curve away from a straight line

CURRENT: the horizontal movement of ocean water

DOWNWELLING: the sinking of ocean water

EKMAN LAYER: the region of the sea, from the surface to about 100 meters down, in which the wind directly effects water movement

OCEAN CIRCULATION: the worldwide movement of water in the sea

THERMOHALINE CIRCULATION: movement of ocean water caused by differences in temperature and salt concentration

UPWELLING: the rising of ocean water

WIND-DRIVEN CIRCULATION: movement of ocean water caused by the motion of air

WIND STRESS: the friction between moving air and the surface of the ocean

MOVEMENT OF SEAWATER

The World Ocean Circulation Experiment (WOCE), started in 1990, brings together scientists from around the globe to study the way in which water moves in all parts of the sea. The WOCE is also intended to obtain information on the movement of heat in the ocean, as well as the movement of substances such as salt and oxygen.

The movement of seawater, known as ocean circulation, consists of horizontal and vertical motion. Horizontal movements are known as currents. Vertical motions are known as upwellings and downwellings. Currents vary widely in speed, ranging from a few centimeters per second to as much as 4 meters per second. Surface currents typically move between 5 and 50 centimeters per second, with deeper currents generally moving more slowly. Vertical movement of seawater is much slower, with a typical speed of only a few meters per month.

Ocean circulation is primarily caused by two major factors. Wind-driven circulation is caused by air moving across the surface of the sea. This causes friction, known as wind stress, between the water and the air, setting the water in motion. Thermohaline circulation is caused by differences in temperature and salt concentration. These differences cause variations in the density of seawater, leading to differences in water pressure, result-

ing in motion. Surface currents are mostly caused by wind-driven circulation. Deeper currents and vertical movements are mostly caused by thermohaline circulation.

Several factors are involved in determining the size, shape, and speed of ocean circulation patterns. An important influence is the Coriolis effect, named for the French scientist Gustave-Gaspard de Coriolis. This effect, caused by the rotation of the Earth, causes the path of a moving object to be curved away from a straight line. The Coriolis effect causes ocean currents to bend to the right in the Northern Hemisphere and to bend to the left in the Southern Hemisphere.

Friction also influences the nature of ocean circulation. Layers of water moving at different speeds produce friction where they meet. This causes the faster layer to move more slowly and the slower layer to move more quickly. Friction also occurs between moving water and the continents, and between currents at the bottom of the sea and the ocean floor. This friction tends to slow the motion of seawater.

Important effects on ocean circulation are seen in the region of the sea known as the Ekman layer, named for the Swedish scientist Vagn Walfrid Ekman. The Ekman layer extends from the surface of the ocean to a depth of about 100 meters.

In this layer, the wind has a direct effect on water movement. Wind stress, the Coriolis effect, and friction between layers of water combine to move the Ekman layer in complex ways.

Water at the surface of the ocean tends to move at an angle of about 45 degrees to the direction of the wind because of the Coriolis effect. This angle is bent to the right in the Northern Hemisphere and to the left in the Southern Hemisphere. With increasing depth, the water moves more slowly and the angle increases. At a depth where the speed of the water is about 4.3 percent of the surface speed, the water moves in the opposite direction to the wind. The overall effect is that the average movement of the water is at about 90 degrees to the wind.

This movement in the Ekman layer, combined with differences in wind stress, creates areas on the surface of the ocean where water converges or diverges. Where it converges, water sinks in downwellings. Where it diverges, water rises in upwellings. Downwellings and upwellings also occur where the wind blows parallel to the coast of a landmass. They also occur because of differences in temperature and salt concentration. Cold water and salty water tend to sink, while warm water and less salty water tend to rise.

WORLDWIDE PATTERNS OF OCEAN CIRCULATION

The numerous factors involved in ocean circulation, combined with the irregular shapes of the continents, result in complex patterns of water movement. Although major surface currents have been known since the earliest days of ocean travel, much less is known about deeper currents. Less is also known about the Southern Hemisphere than the Northern Hemisphere. The WOCE was designed to fill these gaps in scientific knowledge.

Before the WOCE, the basic pattern of surface currents was fairly well understood. In the Northern Hemisphere, strong currents tend to move northward along the eastern coasts of the continents. These include the Gulf Stream-North Atlantic-Norway Current, along North America, and the Kuroshio-North Pacific Current, along Asia. In the Southern Hemisphere, strong currents tend to move northward along the western coasts of the continents. These include the Peru Current, along South America; the Benguela Current, along Africa; and the Western Australia Current, along Australia.

In the regions north and south of the equator, major surface currents move westward. These are the Pacific North Equatorial Current, between Central America and Asia; the Pacific South Equatorial Current, between South America and Asia; the Atlantic North Equatorial Current, between Africa and Central America; the Atlantic South Equatorial Current, between Africa and South America; and the Indian South Equatorial Current, between Australia and Africa.

At the equator, narrow eastward currents are found between the wider westward currents. These are the Pacific Equatorial Countercurrent, between Asia and South America; the Atlantic Equatorial Countercurrent, between South America and Africa; and the Indian Equatorial Countercurrent, between Africa and Asia. Another major surface current is the Antarctic Current, moving eastward around Antarctica.

Although less is known about deep currents, certain broad patterns of movement are understood. Cold water in the northern part of the Atlantic Ocean sinks in downwellings. This deep, cold water tends to move southward along the eastern coasts of North and South America to join the deep, cold water that sinks in downwellings near Antarctica. This water then tends to flow eastward in a deep current around Antarctica. Some of this water then moves northward along the coasts of Asia and Africa, rising in upwellings as it warms.

STUDYING OCEAN CIRCULATION

Scientific studies of surface currents began in the eighteenth century in order to aid navigation. Later studies concentrated on the effect of changes in ocean circulation on the weather. The need for a major effort to increase the amount of information known about the movement of seawater became clear in the 1980's. The best models of ocean circulation, based on the available data, failed to describe the conditions actually observed in the sea with complete accuracy.

A major factor in the decision to create the WOCE was the development of new techniques for studying ocean circulation. Temperature measurements could be made from a ship without stopping the movement of the vessel using an instrument known as a bathythermograph. Devices designed to drift in the ocean, both on the surface and at specific depths, were developed that could be tracked for months or years. Advanced methods of accurately measuring the concentration of substances present in seawater in very low concentrations were also developed. In addition, computers able to handle the enormous amount of data that the WOCE would generate began to appear.

Perhaps the most important new instruments available for the WOCE were satellites capable of obtaining data on ocean circulation. In 1979, the Seasat satellite mission, lasting one hundred days, demonstrated that detection of radar echoes and microwave radiation from the sea could be used to produce detailed information. After years of planning, the WOCE project was ready to begin collecting data in 1990.

Scientists from more than thirty nations participate in the many studies involved in the WOCE. In the United States, the headquarters of the WOCE is located at the Department of Oceanography at Texas Agricultural and Mechanical University. Data collection ended in 1998, but analysis of the information was expected to last until at least 2002. Many scientists expected processing of the data to last until at least 2005.

The first WOCE study began with the launching of the German research ships *Polarstern* and *Meteor* in 1990. These ships collected data in the southern part of the Atlantic Ocean between Antarctica and South Africa. Other early WOCE studies also concentrated on the Southern Hemisphere because this area had been studied in less detail prior to the WOCE than the Northern Hemisphere. Later WOCE studies moved into the Indian Ocean, the North Atlantic Ocean, and the Pacific Ocean.

Satellites used by the WOCE included the ERS series, launched by the European Space Agency; the TOPEX/POSEIDON, a joint project of France and the United States; and the Japanese ADEOS. More than one thousand drifting instruments, designed to remain at specific depths far below the surface of the sea, were also used. The movement of these instruments was measured by satellites or by sonic equipment. Tens of thousands of measurements were made at the surface of the ocean as well.

SIGNIFICANCE

The WOCE project is likely to be one of the most important sources of oceanographic data in the early twenty-first century. The official goals of

the WOCE include providing a complete description of the general circulation of the ocean; creating a numerical model of ocean circulation for use in advanced computers; accounting for seasonal changes in ocean circulation; obtaining data on the exchange of substances between layers of water in the ocean; providing detailed information on the interaction between the ocean and the atmosphere; and obtaining data on the movement of heat within the ocean.

The most important application of WOCE data will be in the study of the effect of ocean circulation on climate changes. This information is expected to aid scientists in predicting the effect on long-term weather patterns of various human activities, such as the increase in carbon dioxide in the atmosphere. Such data will also be useful in predicting natural changes in climate that take place over years or decades.

Several examples of the interaction between ocean conditions and changes in weather are well documented. The Sahel, a region of Africa along the southern fringe of the Sahara Desert, experienced severe droughts in the 1970's and 1980's after having experienced much wetter conditions in the 1950's. These droughts were associated with higher-than-normal surface temperatures in the South Atlantic Ocean, Indian Ocean, and southeast Pacific Ocean, and with lower-than-normal temperatures in the North Atlantic Ocean and most of the Pacific Ocean. Ocean temperature is also an important factor in the formation of tropical cyclones and other powerful storms.

Currents have a powerful effect on weather patterns. The Gulf Stream-North Atlantic-Norway Current brings warm tropical water northward, warming the climates of eastern North America, Ireland, the British Isles, and the coast of Norway. The Kuroshio-North Pacific Current does the same for Japan and western North America. These warm currents also increase water evaporation, increasing rainfall in these areas. The Peru Current brings cold polar water northward along the western coast of South America, decreasing water evaporation and creating deserts in Peru and Chile. The Benguela Current, running northward along the western coast of Africa, has the same effect in Namibia.

Perhaps the best-known example of the effect of changes in ocean circulation on weather is the

El Niño phenomenon. This situation occurs at irregular intervals in the eastern Pacific Ocean. Increased water temperatures, typically 2 to 8 degrees Celsius higher than normal, are associated with changes in climate. Typical effects seen during an El Niño condition are droughts in Australia, northeastern Brazil, and southern Peru; excessive summer rainfall in Ecuador and northern Peru; severe winter storms in Chile; and warm winter conditions in North America.

The El Niño effect is also associated with large reductions in fish populations along the western coast of South America. During normal conditions, the water near the coast consists of a thin layer of warm, nutrient-poor water above a thick layer of cool, nutrient-rich water. The top layer is thin enough to allow coastal upwellings to bring nutrients to the surface, supporting marine life. During El Niño conditions, the top layer of warm water is much thicker, preventing nutrients from reaching the surface. Data from the WOCE project is expected to aid in the prediction of climate changes such as El Niño, with the possibility of having a major impact on human activities.

Rose Secrest

CROSS-REFERENCES
Abyssal Seafloor, 651; Carbonate Compensation Depths, 2101; Continental Shelf and Slope, 584; Deep Ocean Currents, 2107; Gulf Stream, 2112; Hydrothermal Vents, 2117; Ocean-Atmosphere Interactions, 2123; Ocean Drilling Program, 359; Ocean-Floor Drilling Programs, 365; Ocean Pollution and Oil Spills, 2127; Ocean Tides, 2133; Ocean Waves, 2139; Oceans' Origin, 2145; Oceans' Structure, 2151; Sea Level, 2156; Seamounts, 2161; Seawater Composition, 2166; Surface Ocean Currents, 2171; Tsunamis, 2176; Turbidity Currents and Submarine Fans, 2182.

BIBLIOGRAPHY

National Research Council Ocean Studies Board. *The Ocean's Role in Global Change: Progress of Major Research Programs.* Washington, D. C.: National Academy Press, 1994. An official report sponsored by the Office of Global Programs of the National Oceanic and Atmospheric Administration. Discusses the WOCE and compares it to other international projects studying the ocean.

Needler, George T. "WOCE: The World Ocean Circulation Experiment." *Oceanus* 35 (Summer, 1992): 74-77. A clear introduction to the goals of WOCE and the methods used to obtain data. Written by the first scientific director of the WOCE, who was instrumental in planning the project.

Pedlosky, Joseph. *Ocean Circulation Theory.* New York: Springer, 1996. A detailed description of modern models of ocean circulation and the effect it has on climate. An excellent resource for advanced students, this volume demonstrates the importance of the data provided by the WOCE.

U.S. WOCE Office. *U.S. WOCE Implementation Plan: The U.S. Contribution to the World Ocean Circulation Experiment.* College Station, Tex.: U.S. WOCE Office, 1993. Discusses the goals of WOCE and the role of the United States in the project.

Wefer, Gerold, et al. *The South Atlantic: Present and Past Circulation.* New York: Springer, 1996. An example of modern scientific analysis of data obtained on ocean circulation. Deals with the region of the ocean where the earliest WOCE studies were performed.

Woods, J. D. "The World Ocean Circulation Experiment." *Nature* 314 (April 11, 1985): 501-510. An outstanding introduction to the WOCE project. Discusses the importance of ocean circulation to changes in climate, the failure of models prior to the WOCE to accurately predict ocean conditions, the goals of the WOCE, and the technology that allowed the project to take place.

5
OCEANS AND SEAS

ARAL SEA

The Aral Sea was once the fourth largest body of fresh water on the Earth. Located in the desert of the former Soviet Union's central Asian republics of Kazakhstan and Uzbekistan, its waters once supported productive fisheries. As the result of unwise water management practices, the Aral Sea lost about 40 percent of its surface area, is saltier than the world ocean, and markedly changed the local climate and ecology.

PRINCIPAL TERMS

CLIMATE MODERATION: a change in the climate or average weather of a region that reduces the extremes of heat and cold

ECOLOGICAL BACKLASH: the unanticipated ecological effect of what appears, at first, to be a harmless activity

IRRIGATION: the application of water by ditches or sprays to help crop production

SALINITY: a measure of the quantity of dissolved salts in water

WATER EQUILIBRATION: a condition in a lake or other body of water in which the water lost by evaporation is equal to the water added by rainfall or runoff

OVERVIEW

Great tectonic movements of the crust of the Earth during the Tertiary period (from about 65 million years to about 3 million years ago) formed large basins in parts of central Asia. In time, these basins filled with water from rains that poured in by way of several ancient rivers. As the climate and rainfall varied over millions of years, the number and sizes of these water-filled basins also varied. The Caspian Sea and Aral Sea (actually large freshwater lakes) are remnants of much larger, ancient bodies of water. The Aral Sea formed in a depression in the Earth's surface at the beginning of the Pleistocene epoch (the last Ice Age) about 1.6 million years ago.

Aquatic vegetation eventually established a presence in the water-filled basins. The nearby wetlands supported lush growths of plants to form extensive, highly productive marshes. Farther back from the marshes, shrubs and trees developed into riparian forests. Beyond the influence of the waters, shrubby grasslands and sparse vegetation gave way to near-desert conditions.

Flocks of migratory waterfowl—including ducks, geese, and egrets—fed among the lush vegetation, nested, and raised their young. Deer, wild pigs, muskrats, and a host of small mammals also occupied the extensive wetlands and meadows. Fertilized by the abundant aquatic vegetation, shoals of sturgeon, carp, roach, pike, perch, bream, and other fish provided a ready supply of food for humans from the time they first ventured into the area.

HUMAN DEVELOPMENT

During the early decades of the twentieth century, human settlement increased around the shores of the Aral Sea, and the harvest of the rich fisheries increased. In addition to the catch consumed by local human populations, large volumes of fish were packed and shipped to other parts of what had, by then, become the Soviet Union. To satisfy the demand for Aral Sea fish as food, the waters yielded an annual harvest of fifty thousand tons of fish. Trawlers, some as long as 15 meters, plied the lake waters, gathering up Lake Aral's fish.

In addition, cattle grazed on the lush grasslands surrounding the Aral Sea, and crops were grown and harvested to feed the local human population. Although the region was surrounded by the sparse vegetation and terrain of desert, the availability of water from the Aral Sea and its riverine systems made agriculture possible.

The area surrounding the Aral Sea enjoyed what has been described as a desert-continental climate with cold winters, hot summers, and sparse rainfall. The climate was moderated somewhat by the heat capacity of the volume of water in the Aral Sea. During the summer, the waters absorbed heat from the air and slowly released it in the winter. This had the effect of reducing the extremes of seasonal temperatures. Although the Aral Sea never became a popular resort area, its waters were used by villagers around its shores for swimming and boating during the warm season.

In addition, the waters of the Aral Sea and the rivers that supplied it—the Amu Darya and the Syr Darya—were an important source of potable water and water for domestic and farming uses. Both humans and livestock, as well the abundant wildlife of the region, depended on these waters for life.

COTTON AT A PRICE

In the early days of the Soviet Union following World War I, the communist authorities developed a plan designed to make the new nation self-sufficient in cotton. The program—which involved irrigating vast, near-desert regions in central Asia and planting and growing cotton—began in the mid-1920's and appeared to be successful. By the 1950's, vast volumes of water were being diverted from the two rivers that fed the Aral Sea. The waters flowed through hundreds of kilometers of canals that extended from the rivers to the surrounding desert to irrigate the fields of new cotton. Poor or hasty planning, however, had led to the construction of canals that were open to the searing heat of the Sun, resulting in great losses of water to evaporation. In addition, the canals were unlined, and much water was lost by percolation into the sandy desert soils.

Despite these problems, the irrigated cotton project yielded rich dividends, and the Soviet Union joined the United States and China as a world leader in cotton export. However, the ecological backlash of the project became apparent during the early 1960's. The first sign of trouble was the shrinking of the Aral Sea as billions of liters of water were diverted from the feeder rivers to irrigate the cotton crops. Between 1950 and 1988, irrigated cotton fields were expanded to cover more than 7 million hectares of land. The fishermen and cattle workers in the Aral Sea region were forced to become cotton pickers.

LOSS OF WATER VOLUME

The impact of the huge withdrawal of water from the feeder rivers of the Aral Sea, the Amu Darya and Syr Darya, soon became apparent as an increasing amount of sea bottom was exposed. Details about the dismal state of the Aral Sea were largely unknown outside the Soviet Union. The communist government maintained a wall of silence about what many eventually called one of the world's worst environmental disasters, perhaps second only to the nuclear power plant accident at Chernobyl in 1986. However, in the mid-1980's, under the principle of *glasnost* (openness) practiced by Soviet president Mikhail Gorbachev, the world soon learned the truth about the Aral Sea.

What had been the world's fourth largest lake—slightly bigger than Lake Huron—had, by the 1980's, lost 80 percent of its volume. This loss was largely caused by evaporation. With its riverine input diverted to irrigate cotton, there was little water to resupply the lake basin. More than 3 million hectares of seabed, now exposed to the drying sunlight and wind, became an expanse of white salt. The mineral content of the remaining Aral Sea waters increased dramatically until it was three times the salinity of the world ocean. The resulting brine killed off all twenty-four of the Aral Sea's native species of fish. With the collapse of the once-thriving fisheries and related economic adversities, 100,000 workers were displaced.

The loss of water volume caused both immediate and long-term problems for the Aral Sea, the surrounding region, and the local people. Not only were the fish gone, but also the environment of the sea became inhospitable to birds and other wildlife. Deer and small mammals deserted the sea edges where the lush grasses and shrubs had been replaced by growing layers of crusting salts. Even the cotton harvest slowly declined, with the plants producing fewer and smaller bolls. The shrinking of the Aral Sea, which formerly had moderated the local climate, resulted in a harsher climate, a shorter growing season, and frequent dust storms. The salts and chemical residues blowing off the dry seabed reduced the soil's fertility.

The long-term impact on human populations in the Aral Sea region has extended well beyond the loss of the fisheries and jobs. Many of the people in the region have said that they fear they are slowly being poisoned. Runoff from the irrigation system supplies the local drinking water, which has become a brew of pesticides, defoliants, and fertilizers. Worse, domestic sewage only partially treated in the old, Soviet-era equipment is also part of the drinking-water supply. The frequent wind storms blow clouds of salt, dust, and agricultural chemicals from the exposed, dry seabed, resulting in a high incidence of respiratory problems. The windborne pollutants, coupled with the contaminated drinking water, have been blamed for many digestive upsets, typhoid, dysentery, and birth defects. Populations close to the Aral Sea report infant mortality rates as high as one hundred deaths per one thousand births, the highest in the former Soviet Union. Further, life expectancy at birth tends to be lower than the average in more economically developed countries.

SIGNIFICANCE

The story of the Aral Sea is not unfamiliar. A number of environmental disasters have been induced by those who elevate the need for economic gain above sound, broad environmental planning. Some of the disasters have come about simply because the needed information was not available, while others have occurred because available information was ignored by planners.

The Aswan Dam across the Nile River in Egypt was built in the early 1960's to provide electricity to the city of Cairo and to furnish irrigation water to the lower Nile basin. The completion of the dam caused a number of ecological backlashes. Nutrient-rich silt, which provided annual enrichment of the local farms, was suddenly trapped behind the dam and could no longer flow downriver. The silt had also enriched the nearby waters of the Mediterranean Sea, thereby boosting its productive sardine fishery. Deprived of this natural fertilizer, farmers were forced to import chemical fertilizers at great expense. The sardine fisheries collapsed. Furthermore, Lake Nasser, the reservoir that formed behind the Aswan Dam, provided

habitat for snails that carry the organism that causes schistosomiasis, a disease that is sometimes fatal in humans.

Other examples of environmental backlash include the unregulated dumping of organic household waste in many communities, which often results in the pollution of drinking water; the introduction of nonnative species to new areas, which sometimes causes the extinction of native plants and animals; and the widespread use of the agricultural insecticides, which may accumulate in food chains and lead to the deaths of larger animals that prey on the target insects.

The restoration of the Aral Sea may prove to be beyond human technology and capabilities. In the long run, the most feasible course of action may be one of inaction to allow the natural processes of ecological succession to eventually restore the area to some semblance of its natural state. Certainly the greatest benefit to be gained from the example of the Aral Sea is that of a lesson learned. While it is difficult, if not impossible, to foresee all or most of the consequences of human action in the natural environment, at least as much attention should be paid to a soundly produced environmental impact plan as is paid to engineering and economic plans.

Albert C. Jensen

CROSS-REFERENCES

Arctic Ocean, 2197; Atlantic Ocean, 2203; Black Sea, 2209; Caspian Sea, 2213; Deep Ocean Currents, 2107; El Niño and La Niña, 1922; Gulf of California, 2217; Gulf of Mexico, 2221; Hudson Bay, 2225; Indian Ocean, 2229; Mediterranean Sea, 2234; North Sea, 2239; Pacific Ocean, 2243; Persian Gulf, 2249; Red Sea, 2254; Surface Ocean Currents, 2171.

BIBLIOGRAPHY

Ellis, William S. "The Aral: A Soviet Sea Lies Dying." *National Geographic* (February, 1990): 70-93. This well-written, accurate text—coupled with detailed color photographs—clearly describes the destruction of the once-productive Aral Sea. The impact on the local people

is grimly obvious. Suitable for high school readers and upper-elementary grades.

Kotlyakov, V. M. "The Aral Sea Basin: A Critical Environmental Zone." *Environment* 23 (1): (January, 1991). Kotlyakov describes the role of Soviet scientists in covering up possible adverse environmental impacts of the diversion of water from the Aral Sea. Suitable for high school readers.

Morgan, Michael D., Joseph M. Moran, and James H. Wiersma. "The Vanishing Aral Sea: Can It Be Saved?" *Environmental Science: Managing Biological and Physical Resources.* Dubuque, Iowa: Wm. C. Brown, 1993. This article places the Aral Sea disaster in the context of managing water and aquatic food resources. Suitable for high school students.

Pearce, Fred. "Poisoned Waters." *New Scientist* 2000 (October 21, 1995): 29-33. Pearce examines the impact on the human population in the area around the Aral Sea, especially the human health aspects. Grimly perceptive photographs. Suitable for high school and college students.

Schneider, David. "On the Level: Central Asia's Inland Seas Curiously Rise and Fall." *Scientific American* (July, 1995): 14. Schneider discusses the historic changes in water volume of the Aral Sea and the nearby Caspian Sea. Suitable for high school students.

Stone, Richard. "Coming to Grips with the Aral Sea's Grim Legacy." *Science* 284 (April 2, 1999): 30-33. An on-the-spot report of the weather and the ecological environment of the Aral Sea and some possible efforts to ameliorate the conditions. Suitable for high school and college students.

"Tamerlane v. Marx." *Bulletin of the Atomic Scientists* 50 (January/February, 1994): 48-51. The author, who resides in the vicinity of the Aral Sea, requested anonymity for this article, which is highly critical of the Soviet Union and the subsequent government. The writer lays the blame for the social, economic, and environmental disasters surrounding the Aral Sea at the feet of Soviet and Russian government officials. Provides a good background on the topic. Suitable for high school and college students.

ARCTIC OCEAN

The Arctic Ocean, the fourth-largest of the world's oceans with an area of 12,257,000 square kilometers, lies completely within the Arctic Circle. Large segments of it remain frozen throughout the year. It has an average depth of 1,000 meters but in some parts is nearly 5,500 meters deep.

PRINCIPAL TERMS

BRASH: splinters that become detached from ice floes and float in the Arctic Ocean

ICE FLOES: large formations of ice, usually 2.5 to 3.5 meters thick, that float in the waters of the Arctic Ocean

IGLOO: a temporary Inuit structure made from blocks of ice or snow

INUIT: the native dwellers in the northern polar regions; often referred to as Eskimos

PHYTOPLANKTON: tiny floating sea plants that are most plentiful in the presence of sunshine and rich nutrients

PINGO: a large, stable ice intrusion that lurks on the ocean floor much like a coral reef; pingoes must be charted to prevent ships from sailing into them and being badly damaged

POLAR ICE CAP: a large sheet of ice, often hundreds of square kilometers in size, that covers the polar portions of the Arctic Ocean and does not melt seasonally

SALINITY: the salt content of such substances as water and food

UMIAK: a large boat, covered with animal skins, that the Inuit use as they hunt marine mammals such as seals and whales

LOCATION OF THE ARCTIC OCEAN

Lying wholly within the Arctic Circle, the Arctic Ocean was once viewed by geographers as a part of the Atlantic Ocean. It is now viewed as a discrete body of water with definite boundaries. As such, it is the fourth-largest ocean on the Earth, smaller than the Pacific, the Atlantic, and the Indian Oceans, which are, respectively, the largest, second-largest, and third-largest oceans in the world. Waters from all oceans intermingle as tides and currents carry them along and disperse them. The Arctic Ocean is unique in that it is almost landlocked. Essentially circular, it extends from the North Pole south to about 80 degrees north latitude or, if one includes its smaller fringe seas, to about 70 degrees north latitude. The main landmasses that it touches are Canada, Alaska, Siberia, Greenland, Iceland, and Norway.

The Arctic Ocean spills about 60 percent of its water into the Atlantic Ocean between Greenland and Spitsbergen, a group of islands that belong to Norway. It is largely the surface waters that are exchanged because a high range of submerged mountains known as the Faeroe-Icelandic Ridge, which in some places breaks the surface and cre-

ates islands, blocks the exchange of the deepest water. Of all the major oceans, the Arctic, because of its unique pattern of temperatures, currents, and ice conditions, probably has the most independent existence. There is virtually no flow of water from the Arctic Ocean into the Pacific because land barriers and the Earth's rotation prevent such a flow. Some 35 percent of the water that comes to the Arctic Ocean flows in from the Pacific. Most of the other water that enters it comes from the Norwegian Sea.

Crucial to the ecology of the Arctic Ocean is the Greenland Sea. Several large rivers flow from Canada and Siberia into the sea, bringing enormous quantities of water into its somewhat constricted basin. In this part of the world, evaporation is not great, so if the water that comes into the ocean were not expelled into the Greenland Sea, serious problems would ensue.

The water from the Greenland Sea creates a cold current, termed the East Greenland Current, that flows south along Greenland's east coast. The much weaker Labrador Current flows through Smith Sound and Baffin Bay. Yet another weak current flows from the Bering Strait. Water that does not flow out through the Greenland Sea is deflected by Greenland's northern shore. It forms a current that, off the northwest portion of the Arctic Archipelago, runs southwest and west, then turns again while seeking an outlet, which creates a unique circular current in the Arctic Ocean.

This current explains why the part of the Arctic Ocean that is bounded by Siberia has less ice than the same ocean has in Greenland and in parts of the Arctic Archipelago, notably Ellesmere Island. Large ice floes tend to drift southward and westward, many of them melting before they have a chance to drift into the congested shipping lanes of the Atlantic Ocean. Icebergs that reach the Atlantic Ocean are usually brought there by the Labrador Current from western Greenland's fjords. Although some of the ice melts as it moves southward, the polar ice cap that covers part of the Arctic Ocean has not melted in recorded history.

It was once thought that the Arctic Ocean had a considerable effect upon the climate and ecology of all the other oceans. Researchers, however, have questioned this supposition, concluding generally that the Arctic Ocean is more affected by conditions in the world's other oceans than they are affected by conditions in the Arctic Ocean.

DERIVATION OF THE NAME "ARCTIC"

In early times, the Arctic Ocean lay in what the ancients called "terra incognita," or unknown

Sea ice covers the Arctic Ocean near the Northwest Territories, Canada. (© William E. Ferguson)

land. The areas that it touched were thought to be incapable of sustaining life, although it is now recognized that the Arctic area is teeming with life of many varieties, from complex vertebrates such as humans, seals, and polar bears to the simple microorganisms such as phytoplankton, which flourish in both the Arctic and the Antarctic.

The ancient Greeks named the Arctic after the astronomical constellation known as "the great bear." The Greek word for bear is *arktos*, hence "Arctic." The ancients observed that the great bear constellation appeared to revolve around the North Pole. Convinced that there must be another pole at the Earth's other extremity, the Greeks also coined the term "anti-Arctic," which in time evolved into the term "Antarctic."

In early times, the Arctic Ocean was often referred to as the Frozen Ocean because so much of it was permanently covered with the ice of the polar ice cap. With winter temperatures as low as −33 degrees Celsius, it was generally thought that the ocean and the area that surrounds it precluded human habitation, a fallacy that has since been disproved.

PEOPLE OF THE POLAR REGIONS

Despite early conjecture that suggested that the polar regions were uninhabited, it is now clear that they have sustained human existence for thousands of years. The Eskimos, or Inuits, have long been permanent residents of areas within the Arctic Circle and around the Arctic Ocean. These inhabitants are largely hunters who kill the native animals, particularly seals, for food. They also use animal skins for clothing, bones for buttons, and the fattier parts of the animals they kill as a food that helps them survive the long, intensely cold polar night, which, during most of December, lasts for most or all of the day's twenty-four hours.

The Inuit are expert at fishing through the ice and take a great deal of their food from the Arctic Ocean, particularly in winter when they tend to dwell along the ocean's shore. They once hunted whales, which they harpooned from umiaks, large open boats covered with animal skins. Whales provided the Inuit with oil to fuel their fires and lamps and also with blubber, the large, fatty layers the keep the whale warm and buoyant. Eating this blubber helps sustain the Inuit during the long winters. Despite their high-fat diet, these natives generally do not have high cholesterol and are not subject to heart disease as people in the Western world are.

A Navy supply ship sits frozen in the Arctic Ocean near Barter Island, Alaska North Slope, 1948. (National Oceanic and Atmospheric Administration)

During the winter, the Inuit often build igloos, domed shelters made from blocks of ice, that are quite warm inside. Sometimes they cut holes in the ice of the igloo's floor and fish through them, protected from the driving winter winds. In summer, most of the Inuit go inland to the tundra to hunt for game and to fish in the lakes. They gather the wild berries and roots that grow there. They preserve the meat of some of the fish and game they catch by drying or smoking it so that they will have food during the harsh winter.

The people of the polar regions are of Mongoloid stock. It has been suggested that they possess physiological qualities that make it possible for them to adapt to the polar climate. They consist essentially of people from Siberian tribes, North American Eskimos, and European Lapps from the far reaches north of Scandinavia. These people have been touched by technology, usually living in villages that have schools, stores, churches, and medical centers. They live in houses made from imported materials. Such houses usually have an air lock, an area between the door that opens to the outdoors and the second door that opens into the house so that the cold Arctic air does not penetrate the building's interior warmth.

Only sketchy and inaccurate information was available about the Arctic and its people until the nineteenth century, when exploration of the region began to be undertaken. Earlier European sailors had gotten to the region in their search for a short route to Asia. Most, however, found the area so forbidding that they turned back.

EXTENT OF THE ARCTIC OCEAN

The Arctic Ocean occupies an area of more than 12 million square kilometers, about one-seventh the size of the Atlantic Ocean. Many scientists believe that global warming will gradually increase the size of the Arctic Ocean and will cause flooding in the land areas that touch it.

Several seas lie on the fringes of the Arctic Ocean, including the Barents, Beaufort, East Siberian, Greenland, Kara, and Laptev Seas, as well as Baffin Bay. These fringe seas extend 10 degrees further south than the Arctic Ocean proper, reaching as far as 70 degrees north latitude. Although some geographers consider the polar regions to be those that cannot sustain the growth of trees, the polar circles that most geographers accept are generally calculated as being 66.33 degrees north and south of the equator, so that even the fringe seas are wholly within the Arctic Circle.

The whole of the Arctic Ocean is roughly the size of Antarctica, whose ocean is very different from its northern counterpart in that it surrounds a continent, whereas the Arctic Ocean is surrounded by landmasses. That more life thrives in the Arctic than in Antarctica is attributable to the fact that the landmasses around the Arctic Ocean

An iceberg in the Arctic Ocean near Greenland. (© William E. Ferguson)

are warmed by the ocean's currents, with much of the water being quite shallow because of the continental shelves. Nevertheless, much of the Arctic Ocean is frozen all year long, with its water temperature hovering between −1.1 and 1 degree Celsius.

The northern and southern polar regions are generally defined as being those points on the Earth that experience at least one day each year when the Sun does not set. This phenomenon is true at both the North Pole and the South Pole in June and December of each year, respectively.

APPEARANCE OF THE ARCTIC OCEAN

Looking down upon the Arctic landscape from above, one is impressed by how much water and ice constitute the landscape. The Earth below seems solidly frozen, rimmed with ice from the sea, much of which is covered with ice. The view of the frozen sea from the air is dramatic beyond imagining. At times it is not punctuated for hundreds of kilometers by any stretches of water. As one moves out from the land, the ocean waters, if they are visible at all, appear dark, but they gradually moderate into a lighter green, their surfaces studded by huge chunks of ice called brash that have splintered off from the larger ice floes.

If one could look below the surface to the floor of the ocean, it would present the appearance of a warped, rifted surface with great irregularities. In some places, it sinks to depths of 5,500 meters. Freshwater ice formations called pingoes are present in some of the shallower water, each moored soundly to the bottom on which they rest.

It may seem odd that freshwater islands exist in a saltwater ocean. This peculiarity is explained by the fact that when it ages, ice resulting from the freezing of saltwater becomes less and less saline in its constitution. Ice loses one-half of its salinity in its first year of being frozen; eventually, it reaches the point where it has virtually no salinity, so that such ice can be returned to its liquid state and drunk with no ill effects.

Beneath its surface, the Arctic Ocean consists of two basins. They developed separately more than 100 million years ago when the tectonic plates of the Earth's surface drifted apart. The Eurasian Basin resulted when the seafloor spread in a line along the Nansen Cordillera, a range of submerged mountains that constitutes the northernmost part of the Mid-Atlantic Range. In time, this movement pushed a narrow portion of Asia away from the mainland. Known as the Lomonosov Ridge, this sliver lies directly beneath the North Pole. At roughly the same time, the land that is now Alaska moved away from North America and left a basin, known as the Canada Basin, on the other side of the Lomonosov Ridge. This basin is also sometimes referred to as the Amerasian Basin.

CONTINENTAL SHELVES

The Arctic Ocean contains the widest continental shelf in the world, between 490 and 1,780 kilometers in width around the Eurasian basin, stretching north from Siberia toward the North Pole. A similar shelf, the Amerasian Basin, which is from 97 to 200 kilometers wide, extends north from North America. Its exposed portions form the Arctic Archipelago, which consists of Wrangel Island, the Franz Josef Archipelago, the New Siberian and Lyakhov Islands, Severnaya Zemlya, Novaya Zemlya, and Spitsbergen.

The continental shelf that extends north from North America eventually drops off into a deep, oval basin reaching south from the North Pole to the Bering Sea. This and parts of the Greenland Sea east of Greenland are the deepest parts of the Arctic Ocean. It was once thought that no form of life could exist at such depths, but it has now been

established that in some of the deepest parts of the oceans, hot water, sometimes as hot as 367 degrees Celsius, flows out of deeply submerged structures that are comparable to volcanoes. Life exists in many forms around such vents.

Nothing in nature remains the same forever. Cores of ice taken from the Arctic reveal that the area once had a temperate climate. The fossils found in the ice cores suggest that all sorts of vegetation grew where now a wholly different kind of vegetation exists in the harsh climate within the Arctic Circle. Evidence also substantiates the theory that the polar climate is still changing, once again becoming warmer. This warming has implications for the entire world: If the polar ice cap begins to melt at a rapid rate, developed areas of the world may be flooded and may eventually become uninhabitable. Studying the Arctic Ocean and the circumpolar regions will raise awareness that everything in nature is connected and that no alteration in nature, whether natural or human-made, is without consequences.

R. Baird Shuman

CROSS-REFERENCES

BIBLIOGRAPHY

Ballard, Robert D., and Malcolm McConnell. *Explorations: My Quest for Adventure and Discovery Under the Sea.* New York: Hyperion, 1995. In this autobiographical volume, one of America's leading oceanographers and explorers presents fascinating information about the very deep sea, correcting many previously held notions about it.

Broad, William J. *The Universe Below: Discovering the Secrets of the Deep Sea.* New York: Simon and Schuster, 1997. Although Broad does not focus on the Arctic Ocean specifically, the information he gives about the very deep sea has considerable oceanographic significance. In this highly readable and informative book, Broad explodes the notion that there is little life in the very deep sea. Indeed, he demonstrates that all sorts of life that have hitherto been undetected inhabit the abyssal depths, sometimes enduring high temperatures from hot flows that would kill most known forms of life.

Cousteau, Jacques-Yves. *Guide to the Sea.* Vol. 20 of *The Ocean World of Jacques Cousteau.* Danbury, Conn.: Danbury, 1975. Richly illustrated, this oversized volume aimed at general readers presents especially compelling information about animal life in the Arctic Circle and about human adaptation to it.

_____. *Outer and Inner Space.* Vol. 15 of *The Ocean World of Jacques Cousteau.* Danbury, Conn.: Danbury, 1975. Cousteau offers valuable information about life in the Arctic and about how the Arctic Ocean affects existence there. As usual in this series, the volume is well-illustrated, has a useful index, and is written with the general reader in mind.

_____. *The White Caps.* Vol. 16 of *The Ocean World of Jacques Cousteau.* Danbury, Conn.: Danbury, 1975. Of the three Cousteau books mentioned here, this is perhaps the most relevant for those interested in the Arctic Ocean. It presents interesting and vital information that compares the Arctic Ocean with the Antarctic Ocean.

Penny, Malcolm. *Seas and Oceans: The Polar Seas.* Austin, Tex.: Raintree Steck-Vaughn, 1997. Written specifically for the juvenile audience, this book articulates clearly a great deal of information about the Arctic Ocean and the people who live near it. Penny's information is accurate and is presented at a level that most middle school students might easily understand.

ATLANTIC OCEAN

The Atlantic Ocean separates the North and South American continents from Europe and Africa. With an area, excluding its dependent seas, of about 83 million square kilometers, it is second in size only to the Pacific Ocean. The dependent seas add another 23 million square kilometers to the Atlantic's total size.

PRINCIPAL TERMS

CONTINENTAL DRIFT: the gradual movement of one landmass away from another

CONTINENTAL SHELF: the part of the seafloor, generally gently sloping, that extends beneath the ocean from the continents it touches

CONTINENTAL SLOPE: the part of the continental shelf that drops off sharply toward the ocean's floor

EQUATOR: an imaginary line, equidistant from the North and South Poles, around the middle of the Earth where day and night are of equal length

ESTUARY: an area where the mouth of a river broadens as it approaches the sea; estuaries are characterized by the mixing of fresh water and salt water

LAGOON: a long, narrow body of salt water that is separated from the ocean by a bank of sand

TECTONIC PLATES: vast rock surfaces upon which landmasses rest

TIDAL RANGE: the difference in water depth between high and low tides

LOCATION AND ORIGIN

The Atlantic Ocean, with an average depth of 3,300 meters and a maximum depth of 8,380 meters, touches the eastern coastlines of North and South America and the western coastlines of Europe and Africa. The Atlantic Ocean is often divided into the North Atlantic and the South Atlantic, although the entire oceanic mass may properly be considered a single ocean. The North Atlantic extends from the equator north to the Arctic Ocean. The South Atlantic, that part of the ocean south of the equator that extends as far as Antarctica, meets the Pacific Ocean at Cape Horn, the southernmost tip of South America, and the Indian Ocean at the Cape of Good Hope, the southernmost tip of Africa.

Over 200 million years ago, the Earth consisted of a single landmass, presently referred to as Pangaea, surrounded by an enormous sea, presently called Panthalassa. The landmass, like contemporary landmasses, rested on large, solid rock plates, beneath which were vast seas of molten rock. Molten rock in the form of volcanoes can seep through the fissures of tectonic plates and erupt as volcanoes. When this happened in ancient times, eruptions divided the landmass and

formed continents. Looking at a map of the modern world and regarding it as a huge jigsaw puzzle, it is easy to visualize how the western coast of Africa fits into the eastern coast of the Americas and how the coastlines of other geographical areas fit neatly into those facing them.

After the fissure developed that separated the current American continents from Africa and Europe, continental drift continued over millions of years; the ever-widening space between the continents filled with water to form what is now the Atlantic Ocean. Dependent seas such as the Mediterranean, the Aegean, and the North Seas spilled over through narrow passages into the space between such landmasses as southern Europe and northern Africa.

Two major tectonic plates moving apart annually at the rate of about 1.3 centimeters run down the middle of the Atlantic Ocean. Underwater volcanoes have occurred along this fissure and have cooled into a long, underwater, mineral-rich mountain range called the Mid-Atlantic Ridge. Where the volcanic activity has been intense, the molten rock that has cooled has sometimes protruded above the water's surface to create islands; however, in the Atlantic, unlike the Pacific (whose islands are scattered throughout the ocean), more islands are found close to shore than toward the middle of the ocean. Islands are actually the exposed peaks of underwater mountains, some of which, such as Mount Tiede in the Canary Islands, extend more than 3,600 meters above sea level.

Ancient people named the Atlantic Ocean after the mythological giant Atlas, thought to have carried the world on his shoulders. Many ancients, believing that the Earth was flat, feared that anyone sailing far enough into the ocean would eventually fall over the edge of the Earth into an abyss. This belief inhibited early explorers, although even in ancient times, some skeptical adventurers sailed far into the ocean and returned safely to shore. Not all ancients accepted the notion of a flat Earth.

CONTINENTAL SHELF AND COASTLINE

Close to shore, the Atlantic Ocean is generally shallow, seldom exceeding a depth of 545 meters. Waters cover a shelf of rock that, in most of the North Atlantic, extends between 65 and 80 kilometers into the ocean. At the end of this shelf, the ocean drops precipitously to a depth of about 3,940 meters, gradually leveling off into what is designated the abyssal plain. The continental shelf off the coast of Africa is much narrower than the continental shelf off the coast of North America.

The coastlines that border the Atlantic Ocean have been forming for millions of years, shaped largely by the motion and activity of the water that laps the shore. The rockiest areas are found in the northern parts of the North Atlantic and in the South Atlantic. The level of the ocean has risen considerably over the past 10,000 years as glaciers have melted. With the end of the Ice Age, huge quantities of water have poured down rivers into the oceans. The point at which these rivers meet the ocean is called an estuary. Among the deepwater estuaries are the Chesapeake Bay in the United States and the Falmouth Estuary in Great Britain, both of which are sufficiently deep to accommodate large ships.

Less rocky portions of coastline are often characterized by long stretches of sandy beaches that have been formed over the years as the moving waters of the ocean have pulverized rocks and shells, transforming them into sand. Sometimes sandbars form near the coast. These may be long expanses of sand, often permanently covered by water, while others are below the water's surface only at high tide. Some sandbars that are permanently above water may be broad enough to be inhabited, as are the Outer Banks of North Carolina, although at times devastating floods occur in communities built on sandbars. Sandbars endanger ships that venture too close to them and sometimes founder. Such ships may float off a sandbar at high tide, but in many cases this does not happen; the ship, trapped on the sandbar, must be hauled off. Most seaports employ pilots whose duty it is to guide incoming ships through deep channels away from dangerous sandbars or rock formations beneath the surface of the water. Lagoons sometimes form between the coast and sandbars that are not permanently covered with water. These are salty bodies of water in which marsh grasses and other vegetation grow. Many of them are rich in the microorganisms that fish need for their survival. Extensive fish and bird populations cluster around lagoons.

No ocean is fed by more rivers than the Atlantic. These rivers have, over hundreds of thousands of years, carried silt and sand toward the ocean they feed. These deposits have, in many places, built up to form deltas, triangulated areas whose broad base fronts the ocean. Some deltas, particularly those of the Amazon and Niger Rivers, are hundreds of miles wide with channels running through them. Often in tropical areas, mangrove swamps flourish in the deltas, the roots of the mangrove trees being covered with salt water at high tide to provide them with the nutrients they require for their growth. Fish and birds flourish in mangrove swamps.

TIDES AND TRADE WINDS

Ocean tides are predictable. High tide and low tide each occur twice in any twenty-four-hour period. Tidal ranges—depth variations between high tide and low tide—are most pronounced on continental shelves and in deepwater bays. In some parts of the world, the tidal range is dramatic. In the Canadian province of Nova Scotia, the Bay of Fundy has a tidal range of more than 12 meters, one of the largest in the world. The tide rushes out so rapidly that it leaves small boats tilted on their sides beside their moorings, only to be refloated when the high tide rushes in.

Gentler tides are observed as one approaches the equatorial regions, although there can be dramatic tidal activity in such areas during hurricanes and tropical storms, which occur in August through October in the Northern Hemisphere and February through April in the Southern Hemisphere. Hurricanes usually form in tropical areas but can move rapidly to more temperate zones and, if they strike land, can result in injury, death, and substantial destruction. In some cases, such storms erode entire beaches and completely inundate and reclaim waterside property. The Atlantic coastline in many places is shrinking rapidly as the result of water erosion from such storms.

Predictable wind currents, much like jet streams in the upper atmosphere, blow across the Atlantic Ocean. The trade winds near the equatorial areas blow in a westerly direction. Such winds made it possible for early explorers to sail from European ports to the Americas. For their return trips, explorers depended upon winds to the north of the areas where the trade winds blow called the westerlies, which blow from west to east. Along the equator is a narrow area where waters are calm and where there is virtually no wind. In this area, referred to as the doldrums, sailors can drift languidly for long periods of time with no wind to propel them.

ATLANTIC CURRENTS

Ocean water is never still. Much ocean water moves in one direction in predictable flows, in essence forming rivers through the ocean. Such rivers are called ocean currents. The warm tropical currents near the equator push the waters of the Atlantic Ocean westward toward North and South America, moving clockwise in the Northern Hemisphere and counterclockwise in the Southern Hemisphere.

The northern equatorial current runs from just north of the equator along the northeastern coast of South America and on toward the southern coast of North America. There, in the ocean east of the United States, the Gulf Stream runs through the Atlantic, flowing in a northerly direction. The Labrador Current that flows from the Arctic Ocean meets the Gulf Stream off Newfoundland, causing temperature variations that result in high humidity and dense fogs. North of Labrador, the Gulf Stream divides into the North Atlantic Drift, which flows north toward Greenland, and the Canaries Current, which flows southeast through the Atlantic off the west coast of Africa. These clockwise currents flow in what is termed the North Atlantic Gyre.

South of the equator, the southern equatorial current moves counterclockwise off South America's east coast, proceeds south to the Brazilian Current, which then veers southeast to meet the Benguela Current that flows from just north of Antarctica to the Tropic of Capricorn. These counterclockwise currents are referred to as the South Atlantic Gyre.

The oceanic rivers that form the currents of the North and South American Gyres have a profound affect upon climate. The Gulf Stream, which brings warm, tropical waters north at the rate of some 130 kilometers per day, makes far northern areas such as Greenland and Iceland warm enough for human habitation. Its southeasterly branch gives Great Britain a more temperate climate than one would expect at such latitudes. The northern regions of the United States are considerably warmer than they would be if the Gulf Stream did not flow along their coasts. Even areas within the Arctic circle feel the effects of the Gulf Stream, which keeps Russia's Arctic port of Murmansk free of ice all year.

The South Atlantic Ocean's Benguela Current propels cold waters from the Antarctic Ocean north along Africa's southwestern coast, keeping its temperatures much more temperate than would be expected in such latitudes. Cold air, however, cannot hold moisture for long, so when warm, humid winds from Africa's southwestern coast strike the cold air of the Benguela Current, they cause rain to fall over the ocean. The result is that the whole southwestern coast of Africa is arid. The South Atlantic Ocean there is bordered by desert.

FOOD CHAIN

The world's oceans are teeming with life. In a somewhat hierarchical arrangement called the food chain, the sea's larger creatures feed on the smaller ones. At the base of the food chain are microscopic organisms called plankton, which form the basis for life in oceans. Plankton, which exists in both animal and vegetable form, is the basic diet of many aquatic creatures. Plankton cannot live without sunlight, so it is found close to the water's surface. It depends upon nutrients such as minerals for sustenance, and such nutrients are usually found in coastal areas that are not very deep, such as mangrove swamps, where plankton often flourish. Some of the minerals the plankton require are washed into the ocean through estuaries that carry silt from inland areas. Other minerals come from deep in the sea, carried to the surface by ocean currents. In the Atlantic, the most abundant plankton is found in the extreme northern and southern reaches of the ocean, as well as off the west coast of Africa. These areas have a high concentration of the nutrients—abundant in deep, cool waters—on which plankton feed.

The presence of copious plankton in these areas results in a wealth of other aquatic creatures, most of which live near the surface of the ocean where plankton is found, although exploration of the very deep oceans has revealed a remarkable amount of sea life in parts of the ocean so deep that it was previously thought that nothing could survive there.

The Atlantic has an intriguing population of creatures that range from among the smallest on the Earth, such as plankton, to the largest, such as whales. Most of the fish caught by commercial fisherman for human consumption—notably sardines, smelts, cod, halibut, mackerel, hake, sole, and anchovies—swim in large schools that rove the ocean near the surface to harvest the plankton and smaller fish that constitute their diets. The larger fish—sharks, whales, swordfish, and sailfish—are fast-moving and live on a diet of smaller fish. They are often found in the cooler waters of the Atlantic, where plankton is abundant enough to nourish the small fish on which larger fish feed.

Some of the sea's creatures are extremely mobile. Whales move from the polar regions to areas thousands of kilometers away for spawning. Dolphins swim fast enough to cover 245 kilometers in a single day. On the other hand, some shellfish, such as mussels, attach themselves to formations in the sea and are immobile, protected from predators by their thick shells. Some forms of sea life are protected by camouflage that allows them to blend in with the surrounding area. Others are able to emit noxious jets of fluid into the waters around them when they are threatened.

Considerable bird life is associated with the Atlantic Ocean. Most of the birds that depend upon the Atlantic for their food supply can fly, but some of them, such as the cormorant or the puffin, can swim as well. They dive into the water to catch their prey and can swim to catch it. Penguins, found in the Antarctic, cannot fly and can barely walk, but they swim as well as many fish. Cranes and other long-legged birds flourish in shallow waters, where they eat tiny sea creatures that they strain through their beaks.

In the middle of the North Atlantic is a large, relatively calm area of water known as the Sargasso Sea, whose surface is covered with huge fields of nutritious seaweed. The Sargasso Sea has become a spawning ground for eels that migrate to it during mating season from Europe and North America.

SEASIDE SETTLEMENTS

Throughout history, humans have tended to settle beside the sea or along rivers. The transportation opportunities provided by such locations still make them desirable places for settlement. Such areas usually offer a temperate climate as well. The ancient cultures of Rome and Greece grew up on the shores of the Mediterranean and the Aegean, two of the dependent seas of the Atlantic Ocean. The Iberian Peninsula borders on the Atlantic to the west, as do France and Great Britain. Advanced civilizations have flourished on the shores of the Atlantic, particularly in Europe and Africa, since long before recorded history.

The islands of the Atlantic—the products of violent volcanic activity that created large, hilly outcroppings—were slow to develop. The exceptions are the Canary Islands and Madeira, both of which became necessary stopover points for explorers sailing from Europe to the Americas. In recent times, considerable development of the Atlantic islands has taken place in such areas as the Caribbean. The Falkland Islands off the southeastern coast of South America have a stable, permanent population, as do such island enclaves as Nova Scotia and Newfoundland in the north.

Shipping and commerce have been the backbone of the economy in most of the areas that have developed along the Atlantic coastline. Raw materials are brought into Atlantic port cities such as New York, Philadelphia, Rio de Janeiro, and Lis-

The Atlantic Ocean off Cape Cod, Massachusetts. (PhotoDisc)

bon to feed the manufacturing industries of their countries. The fishing industry has flourished for many centuries along the Atlantic coastline. Tourism has also become a major economic factor in the more temperate and scenic regions on the Atlantic coast.

Trade patterns were substantially altered with the opening in 1896 of the Suez Canal, which connects the Mediterranean Sea with the Indian Ocean, and with the opening in 1914 of the Panama Canal, which links the Atlantic and Pacific Oceans. These strategic canals opened up a great deal of trade worldwide and overcame the neces-

sity of plying the turbulent and dangerous waters around Cape Horn and the Cape of Good Hope in order to deliver goods to distant markets.

R. Baird Shuman

CROSS-REFERENCES

Aral Sea, 2193; Arctic Ocean, 2197; Black Sea, 2209; Caspian Sea, 2213; Gulf of California, 2217; Gulf of Mexico, 2221; Hudson Bay, 2225; Indian Ocean, 2229; Mediterranean Sea, 2234; North Sea, 2239; Pacific Ocean, 2243; Persian Gulf, 2249; Red Sea, 2254.

BIBLIOGRAPHY

Ballard, Robert D., and Malcolm McConnell. *Explorations: My Quest for Adventure and Discovery Under the Sea.* New York: Hyperion, 1995. One of America's leading oceanographers relates a great deal about his explorations of the world's oceans, with particularly pertinent information about the Atlantic Ocean, its currents, its animal life, and its general place among the oceans of the world.

Broad, William J. *The Universe Below: Discovering the Secrets of the Deep Sea.* New York: Simon and Schuster, 1997. This is one of the most compelling books about the very deep ocean. Broad explodes many myths about the deep ocean, pointing out that considerable life flourishes in its very deepest parts, even at volcanic underwater temperatures of more than 500 degrees Celsius.

Collins, Elizabeth. *The Living Ocean.* New York: Chelsea House, 1994. This environmentally oriented book provides crucial information about the Atlantic Ocean and about environmental factors that threaten it and its sea life.

Stevenson, R. E., and F. H. Talbot, eds. *Oceans.* New York: Time-Life, 1993. Provides excellent coverage of the Atlantic Ocean. The contributors write in direct, easily understandable prose. Numerous excellent illustrations.

Waterlow, Julia. *The Atlantic Ocean.* Austin, Tex.: Raintree Steck-Vaughn, 1997. This well-illustrated volume is directed toward the young adult market. Although the book is short, its coverage is comprehensive and lucid. The writing style is excellent, and the research is thorough. The glossary should prove useful to those unfamiliar with the subject.

BLACK SEA

The Black Sea is a nearly landlocked body of water that was formerly linked to its eastern neighbor, the Caspian Sea. Geological changes separated it from the Caspian and created two narrow waterways, the Bosphorus Straits and the Dardanelles Straits, connecting it with the Mediterranean. Some of its most interesting physical features make it unique among the world's small to medium-sized seas.

PRINCIPAL TERMS

BOSPHORUS AND DARDANELLES STRAITS: straits leading to and from the Sea of Marmara; they are the only sea lanes linking the countries with shores on the Black Sea to the Mediterranean

CRIMEAN PENINSULA: a large peninsula located on the north central shore of the Black Sea; it is the most significant irregularity on the otherwise regular coastline

DANUBE RIVER: probably the most famous river that enters the Black Sea; the river is a vital means for international maritime access, via the Black Sea, for several European countries

ISTANBUL: a major international city located in Turkey just at the entry point of the Bosphorus into the Sea of Marmara

KERCH STRAITS: the strategic narrows that join the Black Sea to the Aral Sea, located northeast of the Crimean Peninsula and sharing a similar ecology

ODESSA: the major Black Sea city of the Ukraine; its port, Ilichyovsk, is possibly the busiest maritime center on the Black Sea

TETHYS SEA: the much larger geological predecessor of both the Black Sea and the Caspian Sea to its east; named after the Titan and wife of the Greek god of the great outer sea, Oceanus

GENERAL FEATURES

The Black Sea, which encompasses approximately 414,960 square kilometers of saline water, is located on the southern margin of the Eurasian land mass. Its shores include territories in several countries. From the southern and southwestern coast in Turkey, the shoreline proceeds in a clockwise direction through the territory of Bulgaria, Romania, Moldavia, Ukraine, Crimea, and Georgia. The southeastern shore is the border zone between the Republic of Georgia (until the 1990's part of the Soviet Union) and Turkey.

The name Black Sea is a literal translation of the Turkish Kara Deniz. This name, suggesting a somewhat ominous image (particularly in comparison to the Turkish name White Sea, or Ak Deniz, referring to the Mediterranean), dates from the Turkish occupation of the southern shores of the Black Sea between the thirteenth and the fifteenth centuries. The sense of blackness refers not to any impression of the color of its waters, but rather to a presumed inhospitable environment. When the Turks captured the Greek-named Pontus Euxinus (meaning "hospitable sea") from the Byzantine Greeks, they adopted a much earlier Greek denomination for the same body of water, namely Pontus Axeinus (or "inhospitable sea").

The major ports of the Black Sea littoral states are Trabzon, Samsun, Sinop, and Zonguldak (Turkey); Burgas (Bulgaria); Varna (Romania); Odessa (Ukraine); Sevastopol (Crimea); and Batumi (Georgia). In the 1990's, Turkey began developing a new port between Trabzon and Batumi to accommodate major increases in the movement of goods, including petroleum, from the former southern Caspian Sea coastal territories of what had been the Soviet Union.

GEOLOGICAL ORIGINS

Geologically, the Black Sea appears to be a basin left behind by the retreating ancient Tethys Sea, a process that took place over about 200 million years. This long geological evolution reached a significant stage about 50 million years ago when the upward thrust of the Anatolian landmass (Turkey) and Western Iran split the Caspian Sea basin off from the Black Sea. Further mountain-forming activity, including the upheavals that created the Pontic, Caucasus, Carpathian, and Crimean landmasses, also affected the basin that eventually held the waters of the Black Sea. Finally, the geology associated with the famous Black-Sea-to-Mediterranean connection via the Bosphorus and Dardanelles Straits, which lead into and out of the Sea of Marmara, appears to have been the most recent stage contributing to the configuration of the Black Sea, occurring less than 10,000 years ago.

The total area of the Black Sea, including its immediately contiguous arm, the Aral Sea, is about 423,800 square kilometers. At its widest, the sea extends more than 1,134 kilometers from west to east. Its narrowest north-to-south point, running from the tip of the Crimean Peninsula to the Turkish coast at the Kerempe Burmi Cape, is slightly more than 240 kilometers.

The sea is fed by a number of important freshwater rivers. The most famous of these, the Danube, enters the sea on the coast of Romania. The Dniester, which flows parallel to the northern fringes of the Carpathian Mountain range, ends at the delta that feeds the farthest northwest area of the Black Sea. The Dniester Delta forms part of the major port complex of Odessa in the Ukraine. To the east of the Dniester is the Dnieper, the last major river to enter the north coast of the Black Sea proper before its coastline is broken by the Crimean landmass and the waters and rivers associated with the Sea of Azov.

The depth of the sea varies; the deepest area, in the southern part of the basin about midway from east to west, measures more than 2,120 meters. The fact that the deeper zones of the Black Sea receive very low levels of ventilation (circulation of oxygen) means that these areas do not support any significant plant growth or fish populations. In the deep waters, a high concentration of hydrogen sulfide causes whatever oxygen that exists to dissolve. These conditions mean that only certain species of bacteria can survive at lower depths. The beginnings of low oxygen levels, with consequential drops in marine life, occur at depths of around 150 meters. Where waters are sufficiently

ventilated, there are substantial numbers of fish, although only a few species are sought by commercial fisheries.

The most outstanding irregularity at surface level along the coast is the Crimean Peninsula, which juts southward into the Black Sea from the mainland just east of the Dnieper River Delta. The Crimean Peninsula is the site not only of the traditional seaside vacation town of Yalta (where the Western allies met with the Soviet Union's Joseph Stalin near the end of World War II) but also of the strategic port of Sevastopol, founded by the Russian czars as a strategic maritime outpost for Russian access to the Mediterranean.

There are no major islands in the Black Sea. Most geographical accounts overlook the few smaller islands, mentioning only Zmiyini, located east of the Danube River delta, and Berezan, near the mouth of the Dniester.

The Kerch Strait, located to the east of the Crimean Peninsula where a city by that name stands, connects the Black Sea to the Sea of Azov. The latter, which forms part of the same residual basin complex left behind by the receding Tethys Sea, encompasses approximately 67,600 square kilometers. The Sea of Azov receives the waters of two major rivers, the Don and the Kubani.

With the one major exception of the Crimean Peninsula, the coastline of the Black Sea is quite regular, reflecting its geological origins as a residual basin. The shallow coastal shelf is seldom any deeper than 90 meters. It can be very wide but generally extends out from the shore about 11 kilometers. If one takes the average of 11 kilometers and adds areas where the shelf is considerably wider, these shallow waters make up almost one-quarter of the total surface of the Black Sea.

TIDAL PHENOMENA AND CURRENTS

There is very little regularly predictable tidal movement in the Black Sea. When moderate tidal movements do occur, their levels can be affected by a sudden rise or fall of atmospheric pressures in different regions either over the sea or on the landmasses around it. Only rarely do such tides cause dangers for boats or pier structures along the shorelines.

Currents are strongest near the shores of the sea, where the water is not very deep. Such currents run an average of between 40 and 50 centimeters per second. In areas of deeper water, very slow currents are difficult to measure but generally do not run faster than 2.5 to 5 centimeters per second. A particular phenomenon of currents exists in the Bosphorus Straits zone leading from the Black Sea to the Sea of Marmara. Here, a current near the surface carries water from the Black Sea past the city of Istanbul into the Sea of Marmara and eventually into the Mediterranean. At deeper levels, the current moves in the opposite direction, carrying water with higher saline content that originates in the Mediterranean into the Black Sea.

The salinity level of the Black Sea is roughly one-half that of the world's major oceans. Salinity counts (relative proportions of salt to water) near the surface are between seventeen and eighteen parts of salt per thousand parts of water. In the cold band that lies beneath the first layer of Black Sea surface water, the salinity count increases rapidly, reading twenty-one parts per thousand. The proportion of salt to water continues to increase as one descends to deeper zones, but the rate of increase is much more gradual than the apparently sudden increase that occurs in the intermediate-level cold band. The highest salinity count in the deep water is about thirty parts salt to one thousand parts water. Measurements are higher in the Bosphorus Straits, but this phenomenon is linked to the transfer of Mediterranean water into the Black Sea via subsurface currents.

Because the Black Sea is not a very large body of water, its surface temperature varies considerably from season to season. Winter temperatures in the northern zones hover just below 0 degrees Celsius. Where the inflow of fresh water from major rivers is substantial, this means some freezing can occur near the coast. The saline content of most zones is sufficiently high, however, to prevent widespread freezing of Black Sea waters. In the southern half of the sea, winter temperatures are more moderate, reaching about 10 degrees Celsius in the southeast corner near the Turkish border with Georgia. Summer brings surface temperatures up to nearly 21 degrees Celsius. The warming effect, however, is limited to depths down to about 45 meters. Below this level, down to about 90 meters, there is a more or less constant cold band, where temperatures remain at about 7 degrees Celsius throughout the year. In fact, deeper waters may even be a bit warmer than the cold band, but this phenomenon stems from an insulating effect and is not tied to rising temperatures at the surface.

NATURAL RESOURCES AND ECONOMIC POTENTIALS

Until the late twentieth century, perhaps the best-known natural resource exploited on an industrial scale in the Black Sea region was coal, shipped in large quantities from the Turkish port of Zonguldak. Caspian Sea petroleum transit shipments from the Black Sea port of Batumi were always a factor in the economic history of the Black Sea during the period of the Soviet Union, but their importance grew rapidly in the last decade of the twentieth century, as newly opened free international markets for Caspian Sea oil provided incentives for expanded construction of pipelines and port facilities in the southeast coastal region of the Black Sea.

Fishing is the most traditional form of economic activity in the Black Sea. It appears to have a limited capacity for expansion beyond levels reached by the middle of the twentieth century. The most common species of fish in the Black Sea are Mediterranean varieties, some of which actually move into and out of the Black Sea via the Sea of Marmara from season to season. Specialists have noted nearly two hundred different species of fish, but only about forty varieties are exploited commercially. The most important locally processed fish include horse mackerel and another species bearing the local name khamsa.

SIGNIFICANCE

Study of the Black Sea region can be particularly useful for comparison with small to moderately large inland seas that exist around the world. The scale of ecological considerations in such circumstances is drastically different from what would apply in the study of large oceans. In this case, two comparable bodies of water, the Black Sea and the Caspian, are located very much in the same region but demonstrate quite different characteristics. So far, for example, the Black Sea has not been seriously affected by large-scale petroleum development, something that has had major effects on the ecology of the Caspian. On the other hand, the Black Sea has considerably more large urban port cities around its coasts. Examination of pollution levels and marine biology in the Black Sea can lead to a better understanding of its own environmental prospects as well as those of similar bodies of water in other areas of the world.

Byron D. Cannon

CROSS-REFERENCES

Abyssal Seafloor, 651; Aral Sea, 2193; Arctic Ocean, 2197; Atlantic Ocean, 2203; Caspian Sea, 2213; Continental Drift, 567; Continental Shelf and Slope, 584; Deep Ocean Currents, 2107; Earth Tides, 101; Gulf of California, 2217; Gulf of Mexico, 2221; Gulf Stream, 2112; Hudson Bay, 2225; Indian Ocean, 2229; Mediterranean Sea, 2234; Mid-Ocean Ridge Basalts, 657; North Sea, 2239; Ocean Tides, 2133; Ocean Waves, 2139; Oceans' Origin, 2145; Pacific Ocean, 2243; Persian Gulf, 2249; Red Sea, 2254; Sand Dunes, 2368; Storms, 1956; Surface Ocean Currents, 2171; Surface Water, 2066.

BIBLIOGRAPHY

Brewer, Peter G. *Hydrographic and Chemical Data from the Black Sea*. Woods Hole, Mass.: Woods Hole Oceanographic Institution, 1971. Although somewhat technical in nature, this detailed study of the chemical content of Black Sea waters helps explain why, past certain depths, marine flora and fauna are practically absent.

Conservation of Biological Diversity as a Prerequisite for Sustainable Development in the Black Sea Region. Dordrecht, Netherlands: Kluwer, 1998. This book contains the proceedings of an international workshop held in the Georgian port city of Batumi in 1996. Emphasis is on the prevention of further deterioration of plant and animal life in the Black Sea and along its coasts.

Cooper, Bill. *Back Door to Byzantium: To the Black Sea by the Great Rivers of Europe*. London: Adlard Coles Nautical, 1997. This is a skillfully written travel book emphasizing the importance of several major waterways that allow the interior of Europe to connect with the Black Sea and the Mediterranean.

Mavrodiev, Strachimir Chterev. *Applied Ecology of the Black Sea*. Commack, N.Y.: Nova Science, 1999. This is essentially a technical report covering the geological history of the Black Sea as a route to understanding the origins of some of the natural resources of the region. Emphasis is on the need to educate populations living around the Black Sea and to regulate activities that can lead to depletion of productive potential contained in its ecology.

Ukraine. Black Seas Environmental Series 7. New York: United Nations, 1998. This is a collaborative effort by United Nations-sponsored scientists to study local flora and fauna in key zones around the Black Sea. This volume focuses on the coast of Ukraine, with particular attention to the effects of the major urban port area around Odessa.

CASPIAN SEA

The Caspian Sea—the biggest inland body of water on the Earth—holds petroleum and wildlife resources that are crucial for Eurasia's economic and political stability.

PRINCIPAL TERMS

HYDROCARBON: an organic compound consisting of hydrogen and carbon

LITTORAL: adjacent to or related to a sea

PETROCHEMICAL: a chemical substance obtained from natural gas or petroleum

PETROLEUM: a naturally occurring liquid composed of hydrocarbons found in the Earth's strata

CASPIAN SEA GEOGRAPHY

The Caspian Sea, the world's largest landlocked body of water, is located in Eurasia. During the Miocene Era 26 million years ago, the Caspian Sea was part of a larger sea that included the Black and Aral Seas before they became distinct basins. Actually a saltwater lake, the Caspian Sea is bordered by five countries. Kazakhstan lies to the north and east of the sea, while Turkmenistan is also on the sea's east coast. Azerbaijan and Russia (containing the two autonomous republics of Daghestan and Kalmykia) are on the Caspian Sea's western border. Iran is south of the sea. Known to humans for centuries, the Caspian Sea was called by the ancient name of Caspium Mare. Its proximity to the trade route known as the Silk Road assured its importance as a commerce center until ships dominated trade.

Unsymmetrically shaped and resembling a seahorse, the Caspian Sea's size varies because it undergoes cycles of contraction and expansion. On average, the sea stretches 1,210 kilometers from north to south and from 210 to 436 kilometers across varying expanses from east to west, encompassing an approximate area of 371,00 square kilometers. Six rivers feed fresh water into the Caspian Sea, diluting its brine. The Volga and Emba Rivers enter from the north. The Ural River, considered the geographical division line between Europe and Asia, also ends at the sea's northern edge. The Gorgan and Atrek Rivers flow from the east, and the Kura River is the sea's western tributary. The Elburz and Greater Caucasus mountain ranges are located south and southwest of the Caspian Sea.

Deepest in its southern part, the Caspian Sea has an average depth of 170 meters. There are two gulfs along the eastern coast, the Krasnovodsk Gulf and the shallow Kara-Bogaz-Gol, which evaporates and leaves salt deposits. Although the Caspian Sea's level fluctuates every year, it measures an average of 28 meters below sea level. During the 1960's and 1970's, the Caspian Sea's level was substantially lowered because water that usually flowed into the sea from its tributaries was diverted to irrigate agricultural fields. In an attempt to stop water loss, engineers built a dike across the mouth of the Kara-Bogaz-Gol in 1980. Instead of creating a long-lasting lake, the dammed-up water evaporated within three years. Subsequently, an aqueduct was necessary to deliver water to the Kara-Bogaz-Gol. Other river and sea dams block both water and fish from moving naturally while protecting oil refineries on land.

Lacking a natural outlet to other large bodies of water, the Caspian Sea is connected by tributaries to adjacent waterways (such as the Black Sea) that can be navigated to reach distant markets. The Caspian Sea's primary ports are Krasnovodsk in Turkmenistan, Makhachkala in Russia, and Baku on the Apsheron Peninsula in Azerbaijan. The frigid winter climate causes ice to form in the northern parts of the Caspian Sea, hindering travel and fishing. Weather patterns produce dangerous storms, mostly moving southeast across the sea, damaging vessels and oil-drilling platforms. These storms are also hazardous to humans, plants and animals living in the water, and reefs and geological structures.

PETROLEUM AND GEOPOLITICS

Humans have known about the Caspian Sea's valuable natural resources for hundreds of years. Thirteenth century explorer Marco Polo commented that Caspian oil was so plentiful that it gushed like fountains. By the 1870's, speculators equipped with drilling technology became wealthy selling oil, and the Caspian Sea region enjoyed prosperity until the 1917 Russian Revolution, when oil barons were arrested and their wells seized. Caspian Sea oil was considered a strategic raw material during both world wars. Soviet leaders abused Caspian oil production, negligently removing oil with inadequate tools and polluting adjacent areas in the process. Because the Caspian Sea contains some of the world's greatest untapped petroleum deposits, international competition to secure this valuable fuel was the catalyst for an oil rush in the late twentieth century that filled the sea with oil derricks.

The largest known oil deposits in the Caucasus region and Central Asia are located in the Caspian states of Kazakhstan and Azerbaijan. Geologists surveyed three major areas of petroleum lying in deposits that began about 3,600 meters beneath the Caspian Sea's surface. Although the size of these deposits was not determined, their potential seemed to rival the measured deposits of other oil-producing countries. One supply is underneath the Caspian Sea. The others are onshore. The second deposit reaches eastward from Baku to Turkmenistan. The third area spreads west of Kazakhstan underneath the sea's northern tip. The Caspian Sea's oil offered an alternative source of fuel to countries relying on Persian Gulf reserves.

Some experts suggested that almost two hundred billion barrels of oil remained in the Caspian Sea area. On average, 1.1 million barrels were removed daily, in addition to by-products of sulfur. The area was also rich in natural gas, with 7.89 million cubic meters. During the 1990's, European and American oil companies such as Amoco and Chevron secured agreements with governments of several of the countries bordering the Caspian Sea to invest money, equipment, and labor to build pipelines and extract petroleum. The Azerbaijan State Oil Academy prepared workers for the technical demands of oil extraction. Geologists rely on seismic imaging to analyze the Caspian Sea for petroleum. Drillers encountered problems with currents and mud volcanoes on the seafloor, and scientists developed refined technology and geophysical extraction methods to locate and remove oil from deep reservoirs of hydrocarbons existing at high temperatures and pressures.

Investors were aware of tensions in the Caucasus region between the independent Russian states and their neighbors and carefully planned to place pipelines to avoid territorial claims on oil. Competing commercially for the oil, foreign nations strived to achieve strategic alliances and consider diplomatic consequences. They were aware of the vulnerability and instability of the Caspian Sea countries. Several treaties were signed by the littoral Caspian states to legally divide the sea into zones for each state to claim resources. These nations encouraged competition to boost prices because access to Caspian Sea energy resources became an urgent post-Cold War geopolitical objective.

CASPIAN SEA INHABITANTS

The Caspian Sea and its surrounding landmasses provide habitats for an eclectic assortment of wildlife and plants that enhance the area's environment and economy. The sea's primary water source, the Volga River, forms a swampy delta near the confluence with the sea. This marsh is home to waterfowl, including ducks, swans, flamingos, and herons, as well as eagles who rely on the wetlands for shelter and food. Along parts of the uninhabited eastern shore, camels flourish, and wildflowers grow among reeds on the dunes. Fish living in the Caspian Sea include salmon, perch, herring, and carp. The sea is also home to seals, tortoises, and crustaceans unique to the region. Most significantly, three species of sturgeon have flourished in the Caspian's salty water. The Beluga sturgeon stretches as long as 4 meters and weighs 1,000 kilograms, the Asetra sturgeon is about 2 meters long and weighs 200 kilograms, and the Sevruga sturgeon is just over 1 meter in length and weighs approximately 25 kilograms.

Approximately 90 percent of the global sturgeon population lives in the Caspian Sea, producing what gourmets consider the world's tastiest caviar. Eggs compose as much as 15 percent of a sturgeon's weight. Russian czars generously served Caspian caviar, and Soviet and Iranian entrepreneurs became wealthy exporting the processed sturgeon eggs for hundreds of dollars per ounce. In an effort to maintain a monopoly and keep prices high, these countries carefully controlled fishing and cultivated sturgeon in fisheries to replenish the population. After the collapse of the Soviet Union in 1991, sturgeon poaching became a serious problem that threatened the Caspian Sea. Nineteenth century Russians reported yearly catches of 50,000 tons of sturgeon. In the 1980's, the Soviet Union caught as much as 26,000 tons of sturgeon annually according to statistics maintained by the Caspian Fishery Research Institute. By the late 1990's, only 3,000 tons of sturgeon were reported to be harvested by commercial fisherman from Caspian countries each year.

Motivated by potential profits and desperate economic conditions plaguing the independent Russian states, Caspian fishermen poached sturgeon to survive financially by selling fish eggs and meat. One ounce of caviar often sold for more than ten times the amount of an individual's monthly salary. Poachers also saved portions of fish to feed their families. Placing nylon line traps beneath the surface of the Caspian's tributaries, poachers waited for spawning sturgeon to swim upstream. Such uncontrolled snaring reduced the sturgeon population to the point that it was eventually considered an endangered species. Experts stressed that overfishing resulted in the deaths of approximately 90 percent of sturgeons too immature to spawn (most sturgeons require an estimated eighteen years to develop reproductive capabilities), which detrimentally reduced the number of fish returning to the Caspian's rivers every year. Fewer Beluga sturgeon were captured annually, and not many reached their potential to be as heavy as 1 ton and survive for a natural one-hundred-year life span.

ENVIRONMENTAL ISSUES

Pollution also posed substantial dangers for sturgeons. The Caspian Sea was contaminated by oils and petrochemicals from drilling platforms, as well as by pollutants and industrial residues dumped into rivers by factories. Careless drilling methods caused oil films as thick as 6 millimeters to float on the Caspian Sea, creating an unhealthy environment for fish. Raw sewage was also discharged from drains into the sea's tributaries. Sci-

entists explained that the waters near the cities of Baku and Sumqayit were so saturated with poisonous wastes and were so depleted of oxygen and nutrients that they could not harbor aquatic life.

The Caspian Sea benefits from the marshes in the Volga River's delta, which strain out some of the debris and contaminants before the water reaches the sea. Near the river's mouth, which consists of thousands of slender streams moving through the stalks of indigenous water plants, the Caspian Sea's shallow pools of sparkling blue water seem clean. Moving toward the sea's interior, however, the water becomes murky and appears gray in color.

Experts warn that the Caspian Sea is experiencing an environmental crisis that will only worsen as extraction of natural resources increases and industrialization expands in the countries bordering the Caspian Sea. Authorities blame these pollutants for contributing to the growing rates of miscarriages and stillbirths in Caspian nations. Industries are encouraged to reduce pollution and to cease using rivers for effluent disposal. Petroleum companies operating onshore oil fields and platforms in the Caspian Sea have been advised to adapt their operations to rigid standards regarding drilling to prevent tainting the water with unclean fluids and detritus.

Ecologists want extractors of natural resources to become more aware of their impact and accountability to maintain the Caspian Sea's environmental quality. Scientists demand that bans on poaching and illegal caviar processors be enforced more diligently, with police monitoring the sea and arresting offenders. They also seek to prohibit open-sea fishing, which kills young sturgeons, and recommend that Caspian Sea nations agree on caviar quotas so that fair prices are set to discourage black-market profiteering and poaching. Public relations campaigns attempted to promote public knowledge of the Caspian Sea's environmental crises and provoke more responsible treatment of its resources.

RISING OF THE CASPIAN SEA

Biologists monitoring the Caspian Sea noted that after a lengthy period of contraction, its level rose 2 meters from the late 1970's to the 1990's. They optimistically believed that the increased amount of water might weaken the effect of pollutants on the sea and on its inhabitants unless the amount of toxic material and oil spills entering the sea drastically increased.

Many scientists speculated that global warming and the region's climate might have contributed to the rise in sea level. Other scholars hypothesized that the water level increases might have been caused by natural processes, possibly initiated by climatic deviations that affected the atmosphere and sea over periods gauged by single years, decades, or greater time measurements. Erratic changes in the sea's composition and contents that were documented in the past and that periodically recur will crucially affect future development of the Caspian Sea's resources. The quan-

tity and quality of the Caspian Sea's water is vital to ensure Eurasian economic and political stability and maintain international technological and business interests in the region.

Elizabeth D. Schafer

CROSS-REFERENCES

BIBLIOGRAPHY

Amineh, Mehdi Parvizi. *Global Change: Towards the Control of Oil Resources in the Caspian Region.* New York: St. Martin's Press, 1999. A technical text that analyses the political aspects of the Caspian Sea petroleum industry. Provides bibliographical sources and index.

Amirahmadi, Hooshang, ed. *The Caspian Region at a Crossroad: Challenges of a New Frontier of Energy and Development.* New York: St. Martin's Press, 1999. Scholarly essays discussing the geopolitical aspects of future oil extraction from the Caspian Sea. Suggests additional sources to consult. Indexed.

Croissant, Cynthia. *Azerbaijan, Oil, and Geopolitics.* Commack, N.Y.: Nova Science, 1998. Focuses on historic, current, and future Caspian Sea oil developments, outlining political concerns and potential economic benefits. Illustrated with maps, tables, and figures and includes a bibliography and an index.

Cullen, Robert. "The Rise and Fall of the Caspian Sea." *National Geographic* 195 (May, 1999): 2-35. A well-illustrated article offering insight

to the experiences and motivations of people who live in the Caspian Sea region and seek profits from the oil and caviar industries. Useful for high school students to supplement geophysical descriptions.

Marvin, Charles Thomas. *The Region of the Eternal Fire: An Account of a Journey to the Petroleum Region of the Caspian in 1883.* London: W. H. Allen, 1888. This narrative details the nineteenth century Caspian Sea oil boom, providing historical perspective to modern studies. Illustrated and suitable for high school readers.

Rodionov, Sergei N. *Global and Regional Climate Interaction: The Caspian Sea Experience.* Dordrecht, Netherlands: Kluwer, 1994. Rodionov describes how the cyclical contraction and expansion of the Caspian Sea affects the environment and the petroleum industry. Suggests possible scientific reasons for rising levels. Contains maps, bibliographical notes, and index.

GULF OF CALIFORNIA

The Gulf of California, located between Baja, California, and mainland Mexico, is home to an abundant array of land and marine animals. However, overfishing and industrial runoff have led to environmental problems that threaten to destroy the ecological balance of the region.

PRINCIPAL TERMS

chubasco: a type of severe storm that occurs in the Gulf of California and along the west coast of Mexico from time to time

GILL NET: a large, meshed fishing net that allows the head of a fish to pass through, entangling its gill covers and preventing it from backing out; because of their size, gill nets

often catch many marine animals in addition to the target fish

TECTONICS: the study of the processes that formed the structural features of the Earth's crust; it usually addresses the creation and movement of immense crustal plates

GEOGRAPHY AND GEOLOGY

The Gulf of California, also known as the Sea of Cortez, lies north of the Tropic of Cancer. The Mexican states of Sonora and Sinaloa on the country's mainland are located on the gulf's eastern shore. To the north, a narrow strip of Mexico's states of Baja California North and Sonora separate the gulf from the southwestern U.S. states of California and Arizona. The mouth of the Colorado River is located on the northern reaches of the gulf. Mexico's states of Baja California North and Baja California South lie to the west of the gulf. The Pacific Ocean borders its southern reaches.

Geologists estimate that the gulf originated more than 4 million years ago, the result of a clash of tectonic plates beneath the Earth's surface. This action caused the Baja California peninsula to drift westward from Mexico's mainland. The area now containing the gulf subsided, allowing ocean water to flow in from the south.

The Pacific Ocean interacts continuously with the waters of the gulf. As the ocean tide rises, the sea floods into it. As it recedes, the water level in the smaller body also drops. The tide surge is most noticeable in the north, where the incoming tide floods the surrounding lowlands for miles inland and then departs, leaving miles of sandy beaches and mud flats saturated with rich sea life. The most extreme wave action occurs twice per lunar month, with the advent of the new and full Moons.

Over one hundred islands dot the gulf's surface. Most are barren, and few have a supply of potable water. As a result, wildlife is scarce and is limited to only a few of the bigger islands. Only three have any human habitation: San José and Isla del Carmen have colonies of salt mine workers, and San Marcos contains a contingent of employees of a gypsum mining operation.

The weather in the gulf remains fairly constant. The occasional storms that occur are mild compared to those found out in the Pacific Ocean. More tempestuous storms, the *chubascos*, occur from time to time, threatening the small boats out on the gulf's waters. The watercraft are also threatened by local currents. The islands of the Salsepuedes group, two-thirds of the way north from the ocean entrance, present a particularly dangerous area to sailors. In the mid-summer, especially in July and August, temperatures on the gulf can reach an uncomfortable 38 degrees Celsius.

The Colorado River, for millennia, fed its waters into the gulf. At the beginning of the twentieth century, humans interrupted the natural process by erecting a series of dams and canals along the river's course. Today the Colorado delivers only a trickle of surface water to the Gulf of California, although some does flow into the gulf from underground sources. The output is heavily loaded with pesticides, fertilizers, and human waste as well. The dam system has also reduced the amount of sedimentation that flows into the gulf. The loss of sediment has resulted in a decline in the amount of nutrients fed into the gulf's waters.

MARINE LIFE

The Gulf of California's outstanding characteristic lies in the quality and quantity of its sea life. The diversity of its marine occupants varies from great whales to plankton, tiny invertebrates that serve as the basic food source for the gulf's feeders. Its open access to the Pacific promotes a constant interchange of a wide variety of species between the two bodies of water.

Huge finback whales, the second largest of the species, travel well up into the gulf. The gray whale, once thought to be on the verge of extinction, migrates from its home in Alaska to Scammon Lagoon on the Pacific side of the peninsula. It is at Scammon that the females of the species give birth to their offspring. This migration is now a common site along the shores of Washington, Oregon, and California during the annual pilgrimage south. The gulf is also home to many other mammals. Dolphins, porpoises, orcas— the so-called killer whales—and sea lions have taken up residence. There is an ample food supply within the gulf for all species. Huge turtles are found in the north, although their popularity as a dish for human consumption threatens the long-term survival of the species.

There are thirty different types of sharks among the estimated 650 different species of fish. Lower on the food chain, in addition to the plankton, a side variety of clams, mussels, starfish, flatworms, and sea anemones make their home in the mud flats and shallow pools surrounding the gulf. Overhead, huge flights of birds hunt continuously for marine prey. The gulf is a living biological museum.

HISTORY

Cultural anthropologists believe that the first humans settled in the area surrounding the sea as early as ten thousand years ago. Its abundant sea food and small land animals, coupled with the area's temperate climate, served as magnet to attract immigrants. Sites uncovered by archaeologists reveal the gulf's inhabitants to have had a simple social organization. The clement weather obviated the need for either protective clothing or permanent shelter. They used only the most basic types of tools and occupied themselves with gathering food rather than growing it.

In 1535, a Spanish expedition organized by Hernán Cortés, the conqueror of Mexico, landed at La Paz, now a port on Baja California's southeastern shore. A plan by the Spaniards to establish a permanent colony there ended in failure when conflicts with the local natives broke out. As had been the case with other Spanish expeditions, an attempt by the explorers to exploit or enslave the natives caused the fight.

A more sympathetic group of Spaniards arrived at the gulf about 150 years later. Jesuit missionaries landed at Loreto, further up Baja's east coast, to establish Spain's first permanent colony. The Jesuits sought to build a string of missions throughout the largely barren peninsula. They erected some thirty-five churches at one time or another in Baja California. Most failed to last. In most cases, the combination of water shortages and the susceptibility of the natives to European communicable diseases proved to be, in most locations, insurmountable. Even today, because of its aridity, most of the peninsula remains sparsely populated.

ECONOMY

The gulf today depends on fishing and tourism, as well as some cattle raising, farming, and mining, for its economic existence. The principal cities, especially those on the Baja peninsula, reflect this. Only in recent years has the Mexican government taken an active role in the gulf's development.

Numerous ports, both on the Baja peninsula and the Mexican mainland, surround the gulf. Many of the peninsula's economic necessities must be imported from the Mexican mainland or the United States. Much of the needed imports arrive by sea. The gulf's many ports furnish evidence of the areas's economy. There are six key harbors on the gulf's western reaches and five on the opposite mainland Mexico side.

San Felipe, in Baja California North, is the gulf's northernmost facility. The small village has become a popular destination for fishermen of all types, as well as tourists from the United States. The tremendous tidal surge in the area allows boat owners to beach their craft during one of the monthly high tides. When the water recedes, the boats are left high and dry for both minor and major repairs and maintenance. The craft can then be launched easily at the next high tide. Santa Rosalía, in Baja California South, is the terminal for the gulf's mining operations. A gypsum mine exists on nearby San Marcos Island. A French concession operated a copper mine nearby during the nineteenth century until it was taken over by the Mexican government. The government closed it down in 1954 but is considering revitalizing the operation as part of its overall plan for the development of the area.

Mulege, in Baja California South, looks very much like an oasis in the desert. It is surrounded by a virtual jungle of date palms and semi-tropical fruits planted originally by the Jesuit order in the early eighteenth century. The port contains several resort hotels catering to both Mexican and American tourists. Loreto, in Baja California South, is considered the oldest settlement in Baja California. The port also seeks to attract tourists to its resorts. As is the case with Mulege, Loreto has an extensive growth of date palms surrounding it. The town has been destroyed several times in the past by earthquakes, severe *chubascos*, and fire. Loreto is the home of the Museum of the Missions of California.

La Paz is the state capital for Baja South. The port has excellent beaches, a fleet of fishing boats for hire, guides for hunting in the nearby hills, and government-sponsored duty-free stores for visitors. Regular ferry service crossing the gulf itself exists between La Paz and Mazatlán, Sinaloa, on Mexico's mainland. The first-class resort town of Cabo San Lucas is situated on the gulf's entrance to the Pacific. It contains the most luxurious hotels found in Baja California. Campgrounds are also available for tourists traveling with their own accommodations. Game fishing is among the best available throughout the area. Ferry service is available between the port and Puerto Vallarta, Jalisco, on the Mexican mainland.

Puerto Peñasco and Puerto Kino, Sonora, have a modest tourist trade, including many visitors from Arizona. Entrepreneurs have built some small resort hotels near the ports and have laid out campgrounds for travelers with trailers. Sport fishing is the prime attraction at both locations. Guaymas, Sonora, is home to commercial fishermen and acts as the staging area for commercial goods to be shipped inland as well. It has a natural harbor that is spacious and sheltered. Guaymas also exports local products such as cotton and grains from the hinterland. Good hotels and trailer parks are available at the sea front.

A small shipping line operates some small freighters between Topolobampo in Sinaloa and La Paz. Shipping is restricted by shifting sand bars at the entrance to the bay in which the port is situated. Fishing is excellent, and a wide variety of game fish can be found in the surrounding waters. Mazatlán, Sinaloa, is the key port for all of northwestern Mexico. The Mexican navy maintains a base at the port, and a number of small warships are in evidence. The largest amount of sea traffic between Mexico and the Baja peninsula passes between Mazatlán and La Paz. An international airport services the area as well. Both hunting and sport fishing are available to the visitor. Good hotels, restaurants, and night clubs can be found in abundance.

ENVIRONMENTAL PROBLEMS

The marine resources of the Gulf of California includes some of the most outstanding varieties of fish and other denizens of the sea on the face of the globe. The gulf is home to more than nine hundred species of fish and marine animals. The question remains as to how long this unusual concentration of marine riches will continue. Some species of fish have already disappeared from the gulf's waters. Others have reached such low numbers that they are seriously threatened with the same fate. Already lost are cabrillas, black seabass, white seabass, gulf groupers, yellowtails, manta rays, roosterfish, dog snappers, sierras, and vaquetas. The problem lies with overfishing; fish are being taken illegally or through the bribing of local law enforcement officials.

Commercial shrimp boats from Mexico's mainland have raided protected areas on the gulf designed to ensure a continuation of the species. Authorities in Baja California are calling for the banning of commercial fishing boats from the mainland. Commercial fishing permits should be restricted to Baja residents, they feel. Many feel that effective control of overfishing is better left in the hands of local authorities.

Drug smuggling operations have also had a negative impact. In one instance, a cyanide-based dye used to mark drug drop-off spots killed both fish and mammals in the area. Government patrols have not been able to put a stop to the smuggling. Gill net fishing, illegal in the gulf, has resulted in an overkill as well. This indiscriminate method of harvesting wholesale lots of fish also scoops up many fingerlings in addition to mature species. One Baja hotel owner stated that the nets have turned the seas around his property into a graveyard for marine life.

Corruption of government authorities is a major problem. The fisheries on the mainland have millions of dollars at stake. Bribes have reached the highest level of officialdom. The solution, in the eyes of most protectors of the environment, is to transfer the surveillance and law enforcement authority to the local level, where control will be in the hands of those who have the most to lose, the locals who make their living from fishing the gulf. During the 1990's, the central government began the process of creating civilian surveillance committees.

In 1996, Mexico's president Ernesto Zedillo declared the bay at Loreto a national park, banning mainland shrimp trawlers from exploiting the area. As a result, marine life locally is rebounding, and the local fishermen are able to earn a living. The shrimp inside the park are multiplying, and the average size of the shrimp in the catch is growing larger. Loreto could become an example for environmental protection throughout the gulf.

The conflict over environmental protection is not easily solved. Mexican government officials and the Japanese firm Mitsubishi have clashed with environmentalists over a proposed saltworks to be built at the Laguna San Ignacio on the peninsula s Pacific side. A group of dedicated scientists is convinced that the implementation of the plan will have a seriously adverse effect not only on the gray whale migrants who come there to calve but also on a number of other rare marine species making their homes in the area. The government's experts deny that the program will have a negative environmental effect.

Carl Henry Marcoux

CROSS-REFERENCES

Aral Sea, 2193; Arctic Ocean, 2197; Atlantic Ocean, 2203; Black Sea, 2209; Caspian Sea, 2213; Engineering Geophysics, 353; Gulf of Mexico, 2221; Hudson Bay, 2225; Indian Ocean, 2229; Mediterranean Sea, 2234; North Sea, 2239; Ocean Drilling Program, 359; Ocean-Floor Drilling Programs, 365; Ocean Pollution and Oil Spills, 2127; Offshore Wells, 1689; Oil and Gas Distribution, 1694; Oil and Gas Exploration, 1699; Oil and Gas Origins, 1704; Onshore Wells, 1723; Pacific Ocean, 2243; Persian Gulf, 2249; Petroleum Reservoirs, 1728; Red Sea, 2254.

BIBLIOGRAPHY

Cannon, Ray. *The Sea of Cortez.* Menlo Park, Calif.: Lane Magazine and Book Company, 1966. The author has spent more than twenty years travelling throughout the area, mapping and writing about it.

Knudson, Tom. "An Overview of the Destruction of the Sea of Cortez." *Sacramento Bee*, December 10-13, 1995. The writer's investigation of the environmental problems developing in the gulf led to the formation of the Sea Watch Foundation, committed to restoring the ecological balance of the Gulf of California.

_____. "Ray of Hope Glints off Sea of Cortez's Troubled Waters." *Sacramento Bee*, July 4, 1999. The writer's follow-up report on what has been done to restore the gulf's environment.

Los Angeles Times. "An Unacceptable Risk." July 12, 1999. An appeal by scientists to the Mexican government to discontinue the plans for opening more commercial salt factories near Scammon lagoon.

Miller, Tom, and Elmar Baxter. *The Baja Book.* Santa Ana, Calif.: Baja Trail Publications, 1974. A detailed description and mapping of the ports and fishing grounds in the gulf.

Steinbeck, John. *The Log from the Sea of Cortez.* New York: Penguin Books, 1995. The report by Steinbeck and biologist Edward F. Ricketts of their 6,500-kilometer, six-week scientific expedition to study the gulf's invertebrate marine population.

GULF OF MEXICO

The Gulf of Mexico is often called the American Mediterranean since it is almost completely enclosed by landmasses. The waters of the gulf yield millions of pounds of food fish and millions of dollars worth of oil and gas each year. It is also the source of such weather systems as destructive hurricanes.

PRINCIPAL TERMS

ESTUARY: a highly productive, partially enclosed tidal body of saltwater, with freshwater input from a river or from coastal runoff

OCEAN CURRENT: horizontal movement of seawater induced by the wind and affected by rotation of the Earth, nearby landmasses, and temperature and salinity of the water

SALINITY: a measure of the quantity of dissolved salts in water

TIDAL RANGE: the difference in height between high tide and low tide at a given point

TIDE: the periodic, predictable rising and falling of the sea surface as a result of the rotation of the Earth and the gravitational attractions of the Moon and Sun

OVERVIEW

Geographers and oceanographers define the Gulf of Mexico as a Mediterranean sea, one that is partially landlocked. The gulf is bounded on the east, north, and west by the United States, from Key West at the tip of Florida; across Alabama, Mississippi, Louisiana; to Brownsville, Texas. Mexico forms part of the extreme western reaches of the gulf to the Yucatán Peninsula. The long, east-west-oriented island of Cuba forms the southern boundary of this marginal sea.

Water circulation in the Gulf of Mexico is generated by the great volume of water that enters the gulf from the Caribbean Sea by way of the Yucatán Strait. The water circulates in the gulf in a more or less clockwise direction, then exits to the east through the Florida Straits between Cuba and Key West. Loaded with warm, salty water, the current forms the beginning of a great oceanic gyre that includes the Gulf Stream and the North Atlantic Current.

Fresh water flows into the Gulf of Mexico via a number of rivers and streams, the greatest being the Mississippi River. Draining nearly one-half the land area of the United States, this river drops abundant nutrients into the gulf waters. Other rivers around the rim of the gulf basin add their burden of sediment and nutrients, making the Gulf of Mexico a rich—if thin—soup supporting a host of marine animals including fish, shrimp, and some whales. The productive gulf waters supply approximately 20 percent of the commercial fish catch of the United States' fisheries.

Part of the productivity of the gulf's waters comes from the broad continental shelf that forms the rim of the gulf basin. The sunlit waters over the shelf—which is up to 320 kilometers wide in some areas—support a host of marine organisms that form an elaborate ecosystem, ranging from microscopic plants to large fish and giant whales. At greater depths, oil seeps support fish and shellfish whose food supply is based not on sunlight energy but on energy from oil and gas oozing from the seafloor. The oil and gas deposits in the gulf floor also provide energy sources to keep industry, transportation, and homes functioning.

The basin of the Gulf of Mexico is believed by geologists to have formed about 500 million years ago, during a time of crustal movement. Over geologic time, the gulf waters evaporated, leaving vast deposits of salt that formed immense domes. During periods when the gulf basin was flooded by a warm, shallow sea, tiny marine plants and animals called plankton flourished in unimaginable numbers. As the plankton died, their remains settled to the seafloor and were overlain by sediments. Eventually, after millions of years, the plankton remains were changed into petroleum and natural gas.

Over long periods of time, the Earth's crust shifts in vast, slow-moving tectonic movements.

However, geologic researchers offer evidence of a sudden catastrophic event that not only affected the shape of the young Gulf of Mexico but also adversely affected all life on the Earth. About 65 million years ago, a giant meteorite slammed into what is now the northern edge of the Yucatán Peninsula. The impact shattered the continental edge, raised a seismic sea wave (tsunami) of astronomical proportions, caused an underwater landslide that moved at a speed of hundreds of kilometers per hour, and created vast geologic and atmospheric disasters. In the process, nearly one-half of the species of plants and animals then inhabiting the world were wiped out, including the dinosaurs.

OCEANOGRAPHIC OBSERVATIONS

The waters of the gulf fill a basin of more than 1.5 million square kilometers. The basin measures about 1,600 kilometers between Florida and Mexico and 1,280 kilometers from New Orleans, Louisiana, to the Bay of Campeche. The floor of the basin features a variety of topography, including rises and depressions. The deepest depression is the Sigsbee Deep, almost in the center of the basin, which has been plumbed to a depth of 5,203 meters below the sea surface. The average depth of the gulf basin is 1,430 meters. Numerous knolls and flat-topped underwater mountains called guyots break up the monotony of the abyssal plain.

The gulf's productivity results from a combination of the nutrients swept in by the many rivers and the temperature and salinity regime. Sea-surface temperatures in the summer range between 18 degrees Celsius in the northern gulf to 24 degrees Celsius off the Yucatán Peninsula. In some years, summer highs of 32 degrees Celsius have been recorded. Bottom-water temperatures of about 6 degrees Celsius occur in the southern part of the Yucatán Channel. Salinity of Gulf of Mexico waters, measured in parts per thousand, is lowest near the shore, especially off the mouth of the Mississippi River, which sometimes measures less than twenty-four parts per thousand. Elsewhere in the gulf, salinities as high as thirty-seven parts per thousand have been recorded.

The tidal range varies from place to place in the gulf but is, on average, small. The gulf features diurnal tides—one high and one low tide each tidal day. Depending on where in the gulf they are measured, the tidal range varies between 0.8 meters and 1.2 meters.

Climatically, the region over and around the Gulf of Mexico varies from tropical to subtropical. The heat from the Sun and warm water that sweeps in from the Caribbean Sea produce unstable atmospheric conditions over the gulf that help produce more precipitation in the United States than the Atlantic and Pacific Oceans combined. Warm, moist air flowing northward from the gulf, with no topographic boundaries to interfere, often collides with cold, dry continental Canadian air flowing southward. The meteorological disturbances that result are often of monumental proportions. These conflicting air masses may produce deluges and floods, dangerous lightning storms, and, occasionally, fearsome, destructive tornadoes. During the Atlantic hurricane season (from June 1 to November 1), tropical cyclones may form in the gulf or enter from the eastern equatorial Atlantic Ocean, causing massive destruction, flooding, and loss of life.

The climatic regime of the Gulf of Mexico tends to feature monsoonal weather. In the summer, winds blowing onshore from the gulf bring an abundance of warm, moist air and copious rainfall. In the winter, cool, dry air wafts southward on offshore winds.

ORGANIC AND MINERAL RESOURCES

Numerous estuaries ring the Gulf of Mexico, forming a sort of border between the open waters of the gulf and the coastal uplands. The organisms that live in these estuaries exist in ever-changing environments. They are at the mercy of the river runoff and the daily tides, which cause the estuarine waters to fluctuate in salinity, temperature, and nutrient levels. Yet they are among the most productive marine areas on the planet. Approximately 60 to 80 percent of marine organisms valued by humans for food or recreation spend at least part of their life cycle in estuaries. Oysters, scallops, shrimp, flounders, redfish, and mullet—nourished by a food chain that begins with the dead and decaying marsh plants—grow fat and abundant. Shrimp, for example, spawn in the gulf, get swept by tides and currents into estuaries to fatten, then move back into the gulf to be caught by shrimp trawlers. The catch, about 242 million pounds annually, is worth close to $421 million.

In addition to the living resources of finfish and shellfish, the Gulf of Mexico contains valuable mineral resources. These include sulfur, sand, gravel, shells used for construction, petroleum, and natural gas. In 1937, the first successful offshore gulf oil well was drilled in water 4 meters deep. Since then, more than twenty thousand wells have been drilled into the near-coastal waters of the gulf, some as deep as 1,800 meters. Oil reserves in the gulf are believed to be worth an estimated $75 billion.

Oil has leaked naturally into the waters of the Gulf of Mexico for centuries, producing some curious ecosystems. Researchers from the University of Texas and Pennsylvania State University descended more than 500 meters below the surface in a submersible to explore such an ecosystem. At that depth, where no sunlight penetrates, the researchers observed tube worms, clams, and mussels that were thriving on a diet of bacteria that produced organic energy from petroleum. Instead of producing energy-rich food by photosynthesis with sunlight, as surface plants do, the bacteria produced energy-rich food through the process of chemosynthesis from the petroleum. The presence of fish near the oil seeps suggest that a petroleum-based food chain may have developed.

SIGNIFICANCE

The Gulf of Mexico, often referred to as the American Mediterranean Sea, is a productive, biologically rich, highly unique, and economically valuable ecosystem. More than two-thirds of the North American watershed drains into the gulf, mostly through the Mississippi River. Its coastal wetlands provide feeding areas, nursery sites, and breeding grounds for a host of marine and coastal animals. Its waters provide food and recreation for millions of people from Mexico and the United States, as well as visitors from other countries. The submerged lands of the gulf contain mineral resources, particularly petroleum, that are vital to the commerce of nations adjacent to the gulf but also to the world at large.

Albert C. Jensen

CROSS-REFERENCES

Aral Sea, 2193; Arctic Ocean, 2197; Atlantic Ocean, 2203; Beaches and Coastal Processes, 2302; Black Sea, 2209; Caspian Sea, 2213; Colonization of the Land, 966; Earth Tides, 101; Environmental Health, 1759; Future Resources, 1764; Gulf of California, 2217; Hudson Bay, 2225; Indian Ocean, 2229; Mammals, 1038; Mediterranean Sea, 2234; North Sea, 2239; Ocean Tides, 2133; Pacific Ocean, 2243; Persian Gulf, 2249; Red Sea, 2254; Surface Ocean Currents, 2171.

BIBLIOGRAPHY

Appenzeller, Tim. "Travels of America." *Discover* 17 (September, 1996): 80-87. Appenzeller discusses the findings of two geologists concerning the origin of the Gulf of Mexico basin about five hundred million years ago. The event is traced to the breakup of the crust of North America during a period of continental drift. Suitable for high school readers.

Galtsoff, Paul S., ed. "Gulf of Mexico: Its Origin, Waters, and Marine Life." *Fishery Bulletin of the Fish and Wildlife Service* 55 (1954): 1-604. Fifty-five scientists provide authoritative, detailed information about the Gulf of Mexico. As a background source, there is no other like it. Available in most larger libraries and

college and university libraries. Suitable for upper-level high school and college students.

Gore, Robert H. *The Gulf of Mexico: A Treasury of Resources in the American Mediterranean.* Sarasota, Fla.: Pineapple Press, 1992. The author, a professional marine scientist, offers an overview of the history, geology, geography, oceanography, biology, ecology, and economics of this important body of water. Written in a readable style for high school students and lower-division college students.

MacDonald, Ian R., and Charles Fisher. "Life Without Light." *National Geographic* 190 (October, 1996). This article details the findings of two research oceanographers who explored the floor of the Gulf of Mexico in a submersible. They discovered a productive ecosystem of mussels and tube worms subsisting on bacteria that thrive on gas and oil seeping from cracks in the seabed. The authors describe the process of chemosynthesis in a lightless environment. Excellent illustrations. Suitable for high school readers.

Malakoff, David. "Death by Suffocation in the Gulf of Mexico." *Science* 5374 (July 10, 1998): 190-192. Large areas in the Gulf of Mexico feature low dissolved oxygen levels. These "dead zones" drive away shrimp and fish. Clams, starfish, and marine worms that cannot escape the oxygen-depleted waters remain to die on the bottom. Experts suggest that the phenomenon is caused by fertilizers carried more than 1,600 kilometers from farms in the midwestern United States. The fertilizers may also help produce "red tides." Malakoff provides an excellent review of the problem. Suitable for high school readers.

Warrick, Joby. "Death in the Gulf of Mexico." *National Wildlife* 37 (June/July, 1999): 48-52. A readable review of the dead water in the gulf, with some details on the specifics of the adverse water quality. Farm fertilizers and other pollutants carried down by the Mississippi River are labeled as the cause of the low-oxygen water. Suitable for high school readers.

Weber, Michael, Richard T. Townsend, and Rose Bierce. *Environmental Quality in the Gulf of Mexico: A Citizen's Guide.* 2d ed. Washington, D.C.: Center for Marine Coordination. 1992. In brief, readable fashion, this book discusses the physical features of the Gulf of Mexico, including geology, climate, and oceanography. It also describes the ecosystems and principal economic activities in the gulf, especially tourism, fisheries, offshore oil and gas production, and hard-mineral mining. Suggestions for managing the diverse resources are included. Suitable for high school students.

HUDSON BAY

Although considered an extension of the Atlantic Ocean, Hudson Bay—a shallow gulf covering 827,000 square kilometers in east-central Canada—is often treated as a separate entity because of its isolation. The area remains a vast, unspoiled wilderness and preserve for wildlife. Uplift of lands surrounding Hudson Bay provide Earth scientists with clues about processes associated with the retreat of ice sheets, especially crustal rebound.

PRINCIPAL TERMS

CANADIAN SHIELD: the geological core of North America that extends over central Canada and that experienced glaciation during the Pleistocene epoch; underlain by Precambrian rocks, the shield has an undulating surface of moderate relief

CRANTON: an area of the Earth's surface that has been stable for millions of years

ICE SHEET: a broad, flat glacial mass with relatively gentle relief; ice sheets once covered extensive portions of North America

ISOSTATIC REBOUND: a process based on the opposing influences of buoyancy and gravity within the Earth's crust in which the surface adjusts itself vertically until these forces are balanced and isostatic equilibrium has been reached

PLEISTOCENE EPOCH: the most recent ice age period, during which the Earth experienced cycles of continental glaciation

LOCATION AND DISCOVERY

Ranking twelfth in area among the Earth's seas and oceans, the horseshoe-shaped Hudson Bay is approximately 1,370 kilometers long and 1,050 kilometers wide. It is bounded by the Canadian provinces of Quebec to the east and south, Ontario to the south, Manitoba to the west, and the Northwest Territories to the northwest. The bay is linked to the Atlantic Ocean by Hudson Strait and to the Arctic Ocean by the Foxe Channel and Roes Welcome Sound. Hudson Bay and its islands are administered by the district of Keewatin, a region within Canada's Northwest Territories. Its largest islands are Southampton (41,214 square kilometers) and Mansel (3,181 square kilometers), both located in the north. Other large islands include Coats (5,499 square kilometers), located on the southeast side of Fisher Strait, and Akimiski (3,002 square kilometers) within James Bay. Island groups include the elongated Belcher Islands and smaller Nastapoka Islands positioned in the southeastern portion of the bay and the Ottawa Islands to the west of Ungava Peninsula.

Early exploration of the bay was driven by the hope of locating an Arctic shortcut to Asia. Systematic investigations of a route began with John Cabot's visit to Canada's eastern shore in 1497. Hudson Bay was discovered by and named for English navigator Henry Hudson. During his first expedition in 1607, Hudson and the crew of his ship, the *Hopewell*, landed on the shores of Greenland and the Svalbard Islands, later turning north in an unsuccessful attempt to find a route to East Asia by way of the Arctic Ocean. He returned the following year to renew the search, passing the Novaya Zemlya Islands in the Barents Sea. Hudson began his last expedition in 1610, during which he reached Hudson Bay and spent three months investigating its eastern shores and islands. Pack ice prevented a departure from the bay, and, after a winter of extreme deprivation, the crew of his ship mutinied. Placed in a boat and set adrift by the mutineers, Hudson and eight others were never seen again.

Further exploration of the bay was carried out by Sir Thomas Button, who reached the west shore near Churchill in 1612. The following summer, Button departed from the bay, passing Southampton Island on his way to Hudson Strait. Others who explored the bay included William Baffin in 1615 and Luke Fox and Thomas James in 1631. In 1662, Pierre Esprit Radisson and Médard Chouart de Groseillers became the first to reach Hudson Bay using an overland route. Since its first exploration, the bay has served as a graveyard for many mariners. Numerous sunken vessels, along with the only stone military fortification in the Arctic, are part of the bay's historical legacy.

GEOLOGY

Hudson Bay is located within a depression of the vast Canadian Shield and is underlain with ancient Precambrian rocks that are more than 500 million years old. Covering 4.8 million kilometers in Canada and the northern United States, the shield encompasses Labrador, Baffin Island, and portions of Quebec, Ontario, Saskatchewan, the Northwest Territories, Wisconsin, Minnesota, New York, and Michigan. The oldest region within the North American crustal plate, the shield is considered a cranton, or an area of the Earth's crust that has been stable for millions of years. Rocks within the shield contain the fossils of some of the earliest forms of life on the Earth. Hudson Bay is bordered by gently dipping Paleozoic limestones, sandstones, and dolomite rocks.

The bay was formed by glacial activity during the Pleistocene epoch, which began about 2.4 million years ago. Centered over what is now Hudson Bay, the Laurentide ice sheet stripped away soil and deposited glacial drift as it moved across the landscape. The enormous weight of accumulated ice on the continent caused portions of the Earth's crust to sink. As the ice sheet began to melt, numerous remnants were left in the form of till deposits and long, sinuous ridges of sand and gravel called eskers. The retreat of ice sheets of the Wisconsin glacial period began about 13,000 years ago, although large portions of glacial ice remained until about 7,000 years ago. As the ice melted, relatively warm seawater invaded Hudson Bay through Hudson Strait, further shrinking the ice mass. Eventually, the ice sheet was split into the Labrador and Keewatin ice centers, which disappeared completely between 6,500 and 5,000 years ago. As the ice melted, the crust began to slowly rebound. Isostatic adjustment within the Hudson Bay region is not complete, and the entire area surrounding the bay continues to rise at a rate of about 0.6 meters per century. Uplifting of the landscape is especially obvious along portions of

the coast where lines representing former beaches run parallel to the shore. The rapid rate of rebound suggests that Hudson Bay will become much shallower and may disappear completely when isostatic equilibrium has been reached.

The underwater physiography of Hudson Bay demonstrates broad contours that are generally concentric with its periphery. The bay averages 128 meters in depth, with a deepest known depth of 183 meters. Its floor is predominantly smooth but is incised with cuts and banks in some places. There are a few submarine troughs and ridges with trends that are controlled by the geologic structure. The basin forming Hudson Bay has a north-south elongation, which is a reflection of its bedrock structure.

Marine currents are the most significant agents of sediment transport within the bay. Because Hudson Bay is covered by snow and ice during several months of the year, the normal movement of sedimentary discharge from streams is inhibited. In some cases, sediments, coarse gravel, and boulder-sized rocks are carried by moving ice away from shore in a process known as ice rafting. Although concentrated within 30 to 36 kilometers of the shoreline, materials carried by ice rafting are strewn over the entire floor of the bay. In most places within the bay, but particularly in the western half, postglacial sedimentation has become a less important factor than marine erosion in shaping underwater physiography.

COASTAL LANDSCAPE AND WATER

Typical shoreline areas of the northern portion of the bay are low in elevation, rocky, and indented with numerous inlets and small islands. In other areas, the bay is surrounded by a broad, flat plain. A few higher areas (with elevation exceeding 305 meters) are located to the north and northeast on Southampton Island and Quebec's Ungava Peninsula. The size of Hudson Bay's drainage basin is more than 3,800,000 square kilometers. Shoreline areas on the west and southwest coasts form an extensive region of drowned swampland. Deltas and estuaries are common, and, in some locations, tidal flats extend up to 9 kilometers inland. Tundra surrounding much of the bay is, in essence, a cold desert in which moisture is scarce. Plant cover in the tundra includes grasses, lichens, and a scattering of low growing shrubs. Muskeg or bogs with small black spruce trees are found in the south, especially around James Bay. Plants must complete their annual cycles during the brief summer in waterlogged environments, which are the result of poor drainage tied to the underlying permafrost. Soils on land areas surrounding the bay include infertile entisols, histosols, inceptisols, and spodosols. A discontinuous region of permafrost extends southward to 54 degrees north latitude on the western side of Hudson Bay and 61 degrees north latitude on the eastern side.

Arctic in nature, waters found within Hudson Bay demonstrate fairly uniform temperatures that average near freezing. Decreasing from the periphery of the bay to the center, temperatures are slightly warmer over underwater shoals and cooler over deeper areas, reflecting seasonal warming near the surface layer. The general pattern of water circulation is counterclockwise. River outflow inhibits Atlantic waters from entering Hudson Strait, resulting in the low salinity of bay water, especially in spring and summer months. The difference between high and low tide ranges from just less than 1 meter to greater than 4 meters in the western portion of the bay. Pack ice blankets the water from October to June each year. Strong prevailing winds along most of the shoreline help to separate land from pack ice. The bay is navigable for a short period of time extending from early July to October.

Hudson Strait connects Hudson Bay with the Atlantic Ocean to the east. Unlike the bay, water within the strait is predominantly deep, with recorded depths of up to 500 meters. Undersea cliffs and canyons are found near the strait's shoreline. Water from the West Greenland Current mixes with water passing out of Hudson Bay within the strait, creating conditions that support plankton and a diversity of invertebrate and fish species.

CLIMATE

Hudson Bay's climate is influenced by cold, dry, and stable continental polar and Arctic air masses. During much of the year, the bay is covered with a bleak and inhospitable blanket of snow and ice. A diagonal line running from Chester Inlet in the northwest to the Belcher Islands in the southeast divides Hudson Bay into two climatic regions. To the northeast, the bay has a Köppen classification of ET (tundra), while the southwest is classified Dfc (continental taiga climate). Average January temperatures for most of the land area surrounding Hudson Bay range from −30 to −20 degrees Celsius. In contrast, average July temperatures range from 10 to 20 degrees Celsius. The coldest land area adjacent to the bay is located on its northwest side, extending from Manitoba's border with the Northwest Territories to Southampton Island. Precipitation totals throughout the region surrounding Hudson Bay are highest in the summer. The influence of the bay's continental climate can be illustrated in a comparison with Scotland, which is located at about the same latitude. In contrast to conditions around Hudson Bay, Scotland's maritime climate, moderated by the Atlantic Ocean, supports pastures for cattle and sheep.

WILDLIFE AND HUMAN OCCUPATION

During the summer months, the fauna of Hudson Bay is dominated by birds and insects, especially mosquitoes and flies. The rocky coastline and islands, along with tidal flats and inland marshes, provide nesting sites for one of the world's largest concentrations of migrating and moulting waterfowl and shorebirds. Along with almost one-half of the eastern Arctic's population of lesser snow geese are Canada geese, brant, oldsquaws, loons, black guillemots, and common eiders. Also found within the area are gulls, Hudsonian godwits, whimbrels, snowy owls, horned larks, ptarmigan, and the world's largest concentration of peregrine falcons. Elk, arctic hare, and lemmings are also found on its shores. One of the highest densities of polar bear denning in Canada is found adjacent to Churchill, where large number of bears gather during the autumn to await the return of the ice and better feeding. More than forty freshwater, Arctic, and subarctic marine fish species are found within the bay. Examples include capelin, Arctic cod, ogac, and Arctic charr, as well as salmon, cod, halibut, and plaice. Also inhabiting the bay are seals, whales, dolphins, walruses, and beluga whales.

The earliest occupants of the area were nomadic Inuit (Eskimo) Indians who lived in igloos and skin tents. In 1670, the Hudson's Bay Company received a charter from the English crown for exclusive trading rights within the watershed of Hudson Bay. Soon after, trading posts were constructed at the mouths of the Moose and Albany Rivers. Between 1682 and 1713, the French attempted to force the British out of the bay. However, by the terms of the 1713 Treaty of Utrecht, the French handed over all posts in the area to the British. In 1929, the Hudson Bay Railway was completed to the town of Churchill, facilitating shipments of grain produced in Canada's prairie provinces by boat to world markets. During World War II, the United States operated an air base near Churchill on the western side of the bay. The shoreline of Hudson Bay remains sparsely settled, with a few small trading villages located at the mouths of rivers entering the bay in Quebec, Manitoba, and Ontario. Pack ice filling the bay makes villages on its shores inaccessible for much of the year. The shoreline has a population density of fewer than three people per square mile.

Thomas A. Wikle

CROSS-REFERENCES

Abyssal Seafloor, 651; Aral Sea, 2193; Arctic Ocean, 2197; Asteroid Impact, 2633; Atlantic Ocean, 2203; Black Sea, 2209; Caspian Sea, 2213; Gulf of California, 2217; Gulf of Mexico, 2221; Gulf Stream, 2112; Hurricanes, 1928; Indian Ocean, 2229; Mediterranean Sea, 2234; Mississippi River, 2280; Monsoons, 1940; North Sea, 2239; Ocean-Floor Exploration, 666; Ocean Tides, 2133; Pacific Ocean, 2243; Persian Gulf, 2249; Plate Tectonics, 86; Red Sea, 2254; Salt Domes and Salt Diapirs, 1632.

BIBLIOGRAPHY

Bally, A. W., and A. R. Palmer, eds. *The Geology of North America: An Overview.* Boulder, Colo.: Geological Society of America, 1989. A good general reference book describing the geology of Canada. Includes illustrations and an index.

Hood, Peter J., ed. *Earth Science Symposium on Hudson Bay.* National Advisory Committee on Research Geological Survey of Canada Paper 68-53 (1968). Although somewhat dated, this compendium of scientific papers on Hudson Bay provides a good overview of the geomorphology of Hudson Bay and the Hudson Bay Lowlands.

Lewry, J. F., and M. R. Stauffer. *The Early Protero-* *zoic Trans-Hudson Orogen of North America.* Newfoundland: Geological Society of Canada, 1990. Contains a compendium of papers illustrating the structural geology of Hudson Bay and surrounding areas. Maps and illustrations are included.

Ruddiman, W. F., and H. E. Wright. *North America and Adjacent Oceans During the Last Deglaciation.* Boulder, Colo.: Geological Society of America, 1987. This volume examines physical processes and causes associated with ice-sheet erosion in North America during Pleistocene glaciation. The long-term history of North American ice sheets is examined. Includes detailed charts, diagrams, and maps.

INDIAN OCEAN

The Indian Ocean shares the broad ecological and oceanographic features of the other major oceans of the world, but it possesses many unique and interesting characteristics. In general terms, the Indian Ocean's immense size means that both its shores and its depths are so varied that it is difficult to describe it as a single geographical unit. Its westernmost tides arrive at the shores of a continent that bears little resemblance to the shores of its easternmost limits. Most of the truly unique characteristics of the Indian Ocean, however, lie beneath its surface in the form of extensive mountain ridges and, in one case at least, a very deep rift that represents an unparalleled underwater world of its own.

PRINCIPAL TERMS

GONDWANALAND: an ancient continent in the Southern Hemisphere that geologists theorize broke into at least two large segments; one segment became India and pushed northward to collide with the Eurasian landmass, while the other, Africa, moved westward

GYRES: spiral patterns in the movement of surface currents that create nearly self-contained local subsections within the larger pattern of a typical ocean current

JAVA TRENCH: one of the deepest areas of the Indian Ocean, located off the southern coasts of Java in Indonesia; it is a form of geological canyon created by the upward thrust of mountain ridges from the ocean floor

MONSOON: a seasonal movement of winds into the Indian Ocean region caused by vast low atmospheric pressure over the interior land mass of Asia; such winds carry the much-needed monsoon rains into the countries of south Asia and Southeast Asia

SOMALI CURRENT: a seasonally reversing current that moves between the eastern coasts of Africa and the Arabian Peninsula

GEOLOGICAL ORIGINS

The geological origins of the Indian Ocean make it unique among the world's major oceans. Compared with the Pacific and Atlantic, the Indian Ocean is considerably smaller (covering an area of about 73 million square kilometers, in contrast to the Atlantic's nearly 84 million and the Pacific's nearly 166 million square kilometers). It is also of more recent geological origin than the Atlantic, Pacific, or Arctic (the only other major body of water considered to be a world ocean, despite obvious climatic characteristics that make it distinct). Geologists specializing in the evolutionary history of the Earth's crust and plate tectonics estimate that about 150 million years ago, a giant southern continent called Gondwanaland began to break apart. The movement of segments both westward (to what became the African continent) and northeastward (to what became India, a central section of which became known as Gondwana) took at least 100 million years. The collision of the Indian subcontinent with the Eurasian landmass about 50 million years ago contributed

to the violent upheaval of the Himalaya Mountains. One of the effects of this phenomenon was to define new shorelines of the "youngest" of the world's major oceans.

The final product of these major geological upheavals was an oceanic body extending over a vast area between Australia in the east and Africa to the west. Its northernmost point corresponds to the Tropic of Cancer, where the Indian subcontinent joins the Eurasian landmass. From India's western coast to the southeastern tip of Arabia, the waters of the Indian Ocean form the Arabian Sea. On the opposite side of India, the Bay of Bengal and the Andaman Sea extend eastward to the coasts of Southeast Asia (Myanmar, Thailand, the Malay Peninsula, and the Indonesian Archipelago). If one includes two more smaller subsidiary seas, the Persian Gulf and the Red Sea, the Indian Ocean extends even farther north, to 30 degrees north latitude. To the south, the Indian Ocean technically goes as far as Antarctica. Two features of the Indian Ocean's floor define the point at which it separates the Atlantic and Pacific Oceans:

the Atlantic Indian Basin and the South Indian Basin.

ASSOCIATED SEAS AND MAJOR RIVERS

Although formation of the Indian Ocean created several important seas and gulfs as distinct subsections of its total surface, there are fewer such bodies here than in the other oceans of the world. One should contrast general geographical denominations such as the Bay of Bengal (comprising most of the area east of India and touching the coasts of Southeast Asia) or the Arabian Sea (separating western India from the coasts of the Arabian Peninsula) with the geographical uniqueness of the Red Sea and the Persian Gulf. The latter two are, in fact, clearly separated from the main body of the Indian Ocean by narrow straits, the Mandab Straits and the Straits of Hormuz, respectively. Both the Red Sea and the Persian Gulf have ecologies that are very different from that of the main body of the Indian Ocean. This is not the case of the two other semicontained gulfs off the Arabian coast, the Gulf of Oman and the Gulf of Aden. The area known as the Great Australian Bight is simply the slightly curved central southern coast of Australia and is therefore even less circumscribed than the Bay of Bengal west of India. The Andaman Sea lies north of the Indonesian Archipelago and is enclosed geographically from the Bay of Bengal by a line of islands (actually an extension of the Indonesian islands) called the Nicobar and Andaman Islands.

Several major rivers pour vast amounts of fresh water into the Indian Ocean. Such rivers are probably much older than the Indian Ocean itself, even though their pattern of flow was different in earlier geological ages. The Zambezi in East Africa, the Indus in northwest India, and the Ganges in northeast India were all probably flowing from their respective continental freshwater sources toward what eventually became the Indian Ocean's coastline. Each of these has had, over the long period since the formation of the ocean, a notable effect on the configuration of the coast where it empties into the ocean. Freshwater currents have, for example, cut actual canyons into the continental shelf area adjacent to the coast. In the case of the Ganges, an immense zone of sediment has built up, affecting both marine life and local currents in its delta area in the Bay of Bengal.

OCEAN DEPTHS AND SUBMARINE GEOLOGICAL FEATURES

The average depth of the Indian Ocean is in the range of 3,636 to 3,940 meters. Several extremely deep but limited areas, notably the Java Trench to the south of Indonesia, are nearly twice as deep. The continental shelf along the coasts of the Indian Ocean is generally narrower than that of the other oceans, averaging 122 kilometers before deeper waters begin. The area west of the Indian coast, off the major city of Bombay, is an exception. There the continental shelf extends almost 325 kilometers into the ocean.

The floor of the Indian Ocean is crisscrossed by a number of underwater mountain ranges. Although notable ridges exist, its underwater topography is nowhere near as complex or spectacular as the eastern half of the Pacific, where extensive archipelagos with many small islands and some very large island formations (Japan, the Philippines, New Guinea, and New Zealand, for example) predominate. The most concentrated area of subsurface mountains in the Indian Ocean is centered near 30 degrees longitude, which is about halfway between the west coast of India and the Gulf of Aden, south of Arabia. A number of small but historically important islands mark high points along the mountain ridges between 30 and 60 degrees longitude. Mauritius and the Seychelles are examples of these. The huge island of Madagascar (588,000 square kilometers) is the most prominent surface example of the complex north-to-south submarine mountain systems located east of the African coast. The first of these is the Mauritius Ridge, marked on the surface at its southernmost point by Mascarene Island (due east from Madagascar) and by the Seychelles Islands to the north. The next range, the Carlsberg Ridge, is longer than the Mauritius Ridge, but none of its peaks emerge to form islands.

Finally, a long ridge extends due south off the southwest coast of India. This range is marked at the surface by the Laccadive and Maldive Islands. Again, like Mascarene Island and the Seychelles near Madagascar, the Maldives are dwarfed by the single major island just to the southeast of the tip of India, Sri Lanka (formerly Ceylon). Sri Lanka, however, is part of the Indian subcontinental landmass rather than the tip of a submarine mountain range.

Beginning with the 90-degree east longitude

line and moving toward the eastern shores of the Indian Ocean, the topography of the ocean floor is quite different from that of the western half. First, the very name of one subsurface range, the Ninety East Ridge, suggests a very regular pattern extending from north to south. The Ninety East Ridge, discovered as late as the 1960's, has gained the distinction of being the longest and straightest underwater mountain range in the world. Unlike the other ridges of the eastern basin of the Indian Ocean, most notably those south of the Indonesian coast, the Ninety East Ridge appears to be seismically inactive.

Between 90 degrees east longitude and the western shores of Australia one finds the third-deepest point in the Indian Ocean, the Wharton Basin, measuring nearly 6,364 meters deep. Farther south, off the southwestern tip of Australia, is the Diamantina Deep. Neither of these deep points is, however, associated with the rapid fall-off from mountain ridges to deep valleys, which is characteristic of ocean trenches. Trenches are very characteristic of the eastern rim of the Pacific Ocean, but only one such phenomenon occurs in the Indian Ocean. The Java Trench off the southern coast of Indonesia is more than 6,060 meters deep. Pioneer scientists examining flora and fauna in this area of what were then still undiscovered ocean trenches hypothesized that Indonesia was very close to the dividing line between tectonic plates. They observed that plants and animals to the east of Java appear to be biologically isolated in their evolution from species farther west.

The long, curved pattern of the ranges that constitute the Indonesian Archipelago actually extends far beyond the northern tip of Sumatra. Its peaks can be found in the chain of islands known as the Nicobar and Andaman Islands. The presence of these island chains west of the coasts of Thailand and Myanmar helps define the Andaman Sea area east of the main body of the Bay of Bengal.

TIDES AND CURRENTS

The immense size of the Indian Ocean means that tidal phenomena are variable, both in type and in the volume of tidal movement registered. The most common tides are semidiurnal (occurring twice daily). These are characteristic of the subequatorial eastern shores of Africa and, farther north and much further east, in the Bay of Ben-

gal. Australia's southwest coast, which is roughly opposite the subequatorial eastern coast of Africa, has an entirely different tidal pattern. Australia's Indian Ocean shores experience diurnal (once per day) tides that are extremely light by comparison to those of other coasts.

The Indian Ocean is the only ocean in the world with asymmetric reversing surface currents. Asymmetric conditions apply when currents in the northern half of the ocean are moving in a different direction from those of the southern half. The complexity of Indian Ocean currents close to the surface goes well beyond the relatively simple question of north-south asymmetry. One finds, for example, that wind conditions contribute to the creation of gyres, or spiral movements that break the broad pattern of the surface current into localized segments.

It is particularly important to note that broad patterns of currents in the Indian Ocean reverse according to the season. Currents, like so many other factors determining the overall ecology of the Indian Ocean, are affected in large part by the major monsoonal wind and weather conditions that are characteristic of this region of the world.

Probably the most famous current in the Indian Ocean is the so-called Somali Current, which moves, in certain seasons of the year, in a fairly rapid clockwise direction from the northeast coast of the Horn of Africa. In the summer season, this current goes as far as the coast of India. At its farthest point moving east in the summer months, it meets the southwesterly monsoon current off the Indian subcontinent. During the winter, the direction of the Somali Current reverses, a situation that created near-ideal seasonal sailing conditions for centuries for sailing ships sailing between the Arabian peninsular zone—particularly the Persian Gulf—and the Horn of Africa.

CLIMATIC ZONES AND WIND PATTERNS

Because of the great north-to-south distance covered by Indian Ocean waters and its associated seas, there are several quite distinct climatic zones according to geographical location. The most famous, and most important for sustaining seasonal agriculture in the entire Indian Ocean area, is the so-called monsoon zone, which runs from 10 degrees north of the equator to about 10 degrees south. The region between 10 and 30 degrees south of the equator is the zone of what have traditionally been called the trade winds. It is the predictability of steadily blowing southeast winds in this wide region (as distinct from the area of the Somali Current) that made maritime communication, and therefore trade, between the opposite shores of the Indian Ocean possible.

Farther south, near the global band of the Tropic of Capricorn (running through the island of Madagascar on the western shores and Australia on the east) lies the subtropical to temperate zone, between 30 and 45 degrees latitude. Once one is some 20 degrees south of the Tropic of Capricorn, climatic conditions begin to show the temperate cooling influence of the extreme southern extent of Indian Ocean waters leading to the last climate zone. From 45 degrees latitude southward to the Antarctica ice cap, the beginnings of sharp antarctic cold waters mark the end of the gradational transitions separating some of the world's hottest climates in the Indian Ocean proper from the extreme cold of the southern zone of the globe. Here, three of the world's four oceans almost literally fuse in the Antarctic ice cap.

MONSOON WIND CONDITIONS

Generally stated, monsoon conditions are semiannual reversing wind patterns. Extensive areas of high pressure "empty" their air in the direction of equally vast low-pressure zones. When this happens, winds crossing water carry the moisture they pick up, which is precipitated as rain even before they reach their low-pressure destinations. In the case of Indian Ocean monsoons, the widespread heating of the landmass in the Northern Hemisphere during summers creates conditions of low atmospheric pressure over Asia. This low-pressure zone becomes an attractive force for masses of air that are pressed downward toward the Earth's surface by high-pressure conditions over Australia. The resultant winds that move in a northwesterly direction across Southeast Asia and the Indian subcontinent bring with them a much-needed monsoon season of heavy rains that lasts until the particular atmospheric conditions cease to apply. Generally, the monsoon rainy season is predictable, but the gift of torrents of rain to quench the dry agricultural fields of south Asia and Southeast Asia is not guaranteed. When the typical monsoon wind pattern develops but insufficient moisture is collected to bring rains, areas that depend on monsoon waters can face serious drought for at least one year, since there is no chance of humid air movements over the landmass once the directional wind pattern created by Asia's summer heating ends.

NATURAL RESOURCES

The Indian Ocean contains many key minerals that are extracted to supplement the local economies of several of the countries along its shores. Manganese is found in several areas off South Africa and Australia. Other minerals include tin and chromite. Mineral wealth in the water is, however, overshadowed by the vast petroleum reserves concentrated in one of its seas: the Persian Gulf. These are estimated to be the largest oil reserves in the world. Other locations where petroleum wealth is important are on the island of Sumatra and in its offshore waters. Similar intermediate potential for petroleum production is found in the Red Sea and off the shores of western India.

The fishing industries that depend on the Indian Ocean are varied. Many depend on the phenomenon of upwellings, movements of water from lower depths that carry phytoplankton—basic nutrients for many fish species—close to the surface. The most common commercialized fisheries seek large schools of sardines, mackerel, and anchovies. By far the largest single species fished in many areas of the Indian Ocean is the shrimp.

SIGNIFICANCE

The Indian Ocean and its associated seas represent one of the most diversified ecological marine environments in the world. This applies both to its coastlines and to the world beneath its surface. The African landmass along the ocean's western shores reveals a variety of natural characteristics that hardly resemble what one finds along its eastern shores. Dry coastal regions characterize both the southern and northern reaches of the African landmass, with some of the driest deserts in the world extending from the Somalian Horn of Africa into the Red Sea and around the southern shores of Arabia into the Persian Gulf. Although these conditions continue all the way along the northwestern coastline to the Gujerati coast of India, the Indian Ocean world is clearly distinct once one passes the Indian subcontinent. From the Bay of Bengal southward along the eastern shores of the Indian Ocean, the environment becomes increasingly tropical, passing through some of the most extensive tropical rain forests in the world, especially in the Indonesian Archipelago. Indeed, parts of the west coast of Australia—generally thought to be nearly a desert environment—exhibit a tropical ecology that contrasts notably with the dry climates of the African coastline of the Indian Ocean. With this physical ecological diversification comes an enormous variation in plant and animal life along the ocean's shorelines and in the relatively shallow waters of its continental shelf. Studying the diversity of the Indian Ocean's biological ecology and submarine geology is like observing two or more different worlds on one segment of the planet.

Byron D. Cannon

CROSS-REFERENCES
Aral Sea, 2193; Arctic Ocean, 2197; Atlantic Ocean, 2203; Black Sea, 2209; Caspian Sea, 2213; Gulf of California, 2217; Gulf of Mexico, 2221; Hudson Bay, 2225; Mediterranean Sea, 2234; North Sea, 2239; Pacific Ocean, 2243; Persian Gulf, 2249; Red Sea, 2254.

BIBLIOGRAPHY

Bowin, Carl O. *Origin of the Ninety East Ridge from Studies near the Equator.* Woods Hole, Mass.: Woods Hole Oceanographic Institution, 1973. This research represents the result of about ten years of work following the discovery, in the 1960's, of the famous north-to-south submarine ridge running along the 90 degree east longitude.

Hastenrath, Stefan, and Peter J. Lamb. *Climatic Atlas of the Indian Ocean: Surface Climate and Atmospheric Circulation.* Madison: University of Wisconsin Press, 1979. The introduction to this valuable reference guide provides a full explanation of a main body of charts dealing with the thermodynamics of all regions of the Indian Ocean. Information covers typical patterns and variations in climate, including air pressure and winds, for each of the months of the year.

Kerr, Alex, ed. *The Indian Ocean: Resources and Development.* Boulder, Colo.: Westview, 1981. This edited work covers the natural resource potential of the Indian Ocean and human efforts to exploit various aspects of these resources for economic benefit. Although most of the contributions deal with questions of political and economic development policies followed by countries around the Indian Ocean periphery, several are concerned with the relationship between climate and soils and particular types of agricultural potential, notably prospects for sugar production and large-scale plantations.

Neprochnov, Y. P., et al., eds. *Intraplate Deformation in the Central Indian Ocean Basin.* Bangalore, India: Geological Society of India, 1998. Studies of local features of underwater plate tectonics have attempted to uncover signs of change in geological activity in the Earth's crust. Because the Indian Ocean was formed by relatively "recent" movement of very large segments of the plates, some of the scientists associated with this research report cite arguments suggesting that the geological formation of the ocean's origins is not actually over.

Rao, P. V., ed. *The Indian Ocean: An Annotated Bibliography.* Delhi, India: Kalinga, 1998. Although there tends to be major emphasis in this bibliography on literature dealing with the various islands of the Indian Ocean, subjects covered include ecological scientific data gathering, both in the realm of hydrology and marine life, and comparable studies of plant and animal life in various land environments.

Wyrtki, K. *Oceanographic Atlas of the International Indian Ocean Expedition.* Washington, D.C.: National Science Foundation, 1971. Although the data contained in this volume do not reflect up-to-date findings of the separate subfields of oceanic investigation, it is perhaps the most important single-volume collection of scientific findings on the Indian Ocean. The data represent the product of a multinational scientific team's collaborative work.

MEDITERRANEAN SEA

The Mediterranean Sea, located between southern Europe and North Africa, is one of the most historically important and largest seas in the world. However, environmental problems caused by tourism and industrial discharge threaten to destroy the ecological balance of the region.

PRINCIPAL TERMS

CONTINENTAL SHELF: the submerged offshore portion of a continent, ending where water depths increase rapidly from a few hundred to thousands of meters

EDDY: turbulence attributed to water or wind

MEDITERRANEAN CLIMATE: a pattern of weather conditions characterized by a long, hot, dry summer and a short, cool, wet winter

SALINITY: a measure of salt presence in oceanic water controlled by the difference between evaporation and precipitation and by the water discharged by rivers

SEMIDIURNAL: having two high tides and two low tides each lunar day

LOCATION AND HISTORY

The Mediterranean Sea separates Europe geographically from the continents of Africa and Asia. It is surrounded by southern Europe, western Asia, and North Africa. It extends about 3,700 kilometers from east to west and almost 1,600 kilometers from north to south at its widest point. Because of the irregular shoreline, the width varies considerably but averages almost 600 kilometers. The European coast includes Spain, France, Italy, Yugoslavia, Albania, Greece, and the European part of Turkey. The Asian portion continues with the Asian part of Turkey, Syria, Lebanon, and Israel, while the African part includes—from east to west—Egypt, Libya, Tunisia, Algeria, and Morocco. Its total area is about 2.5 million square kilometers.

The Mediterranean Sea is connected to the Atlantic Ocean by the Strait of Gibraltar in the west, to the Red Sea by the Suez Canal in the southeast, and to the Black Sea by the Strait of Bosporus in the northeast. The Strait of Gibraltar is approximately 20 kilometers wide at the boundary, while the Bosporus varies in width from as little as 0.8 kilometer to about 3.2 kilometers at the Black Sea boundary between Rumeli Burnu in European Turkey and Anadolu Burnu in Asiatic Turkey. The Suez Canal is the only human-made channel of the three and was first opened in the late 1860's to speed the naval transportation between Europe and the countries of the Persian Gulf and the In-

dian Ocean. It is almost 150 kilometers long, 82 meters wide, and about 10 meters deep.

Historically, the Mediterranean Sea has been one of the most important as well as one of the largest seas on the Earth. Its name is derived from the Latin words *medius* (middle) and *terra* (land). The Mediterranean area has also been known as "the cradle of Western civilization" because of the advanced civilizations that had an impact on the human race, especially those that flourished in the eastern part of the sea.

All three straits and canals have long been considered critical strategic points for commercial, political, naval, and military reasons, and the great military powers have, historically, been interested in controlling them. Such powers included the Phoenicians, the Greeks, the Romans, the Persians, the Byzantines, the Arabs, the Turks, the British, the French, the Italians, and the Russians. The Romans referred to the Mediterranean as *mare nostrum* (our sea); Catherine the Great of Russia had dreams about controlling the Dardanelles in the late 1790's; British admiral Horatio Nelson understood that blocking Gibraltar was the only way to beat Napoleon's expansion; Winston Churchill realized that controlling the Dardanelles was critical during World War I; Benito Mussolini attempted to control the Balkan Peninsula and North Africa to revive the Ancient Roman Empire; and the British and French declared war against Egyptian President Gamal Abdel Nasser after he nationalized the Suez Canal.

GEOGRAPHY

The present coastline of the Mediterranean is irregular and indented with small gulfs and capes. Historical geological studies suggest that during the Triassic period—200 million years ago—the sea, known as Tethys Sea, extended as far east as Turkestan and the Aral Sea territory and as far north as what is now the Danube valley. The area north of it was called Laurasia, while the part south of it was known as Gondwanaland. That much larger and wider body of water is believed to have been contracted by factors such as faulting, folding, continental drift, and volcano influence. Beginning in the Cretaceous era (65 million years ago) and ending during the Miocene period (13 million years ago), the smaller Aral, Caspian, Black, Marmara, and Mediterranean seas were created out of that much larger sea basin. Moreover, the two natural outlets of Gibraltar and the Dardanelles appear actually to have been completely landlocked by land bridges that were eventually worn off by the tides. There is also an indication that the islands that are off the eastern Greek mainland were once attached to what is now mainland Greece and Turkey.

The Italian peninsula splits the Mediterranean into two basins of nearly equal size, the western and the eastern, which are connected by the Strait of Sicily and the Strait of Messina. The Strait of Sicily—which lies between Cap Bon in Tunisia and the Italian island of Sicily—is almost 130 kilometers wide, while the Strait of Messina, which lies between Punta Sottile in Sicily and the Italian mainland, is no more than 3.3 kilometers across.

Each of the basins is also subdivided into four smaller bodies of water. The western basin is composed of the Alboran Sea near Spain, the Balearic or Iberian Sea between mainland Spain and the Islas Baleares, the Ligurian Sea just north of the French island of Corsica, and the Tyrrhenian Sea, which is surrounded by Corsica, Sardinia, Sicily, and the western part of mainland Italy. The eastern basin includes the Adriatic Sea, which lies between the western parts of some of the former territories of Yugoslavia and the eastern part of mainland Italy all the way down to Capo Santa Maria de Leuca and above the Greek island of Corfu; the Ionian Sea, which is an extension of the Adriatic east of Sicily and on the western part of Greece; the Aegean Sea, which stretches over the eastern Greek shores as high as the Dardanelles; and the Sea of Marmara, which is engulfed by the Dardanelles and Bosporus.

The Mediterranean has many islands, several of which are of volcanic origin. Most of them lie in the eastern basin in the Adriatic Sea and, in particular, the Aegean Sea. The Aegean has more than two thousand islands that form small basins and narrow passages with an irregular coastline and topography. The largest islands are Sicily, with an area of 25,818 square kilometers; Sardinia, with an area of 24,138 square kilometers; independent Cyprus, with an area of 9,282 square kilometers; Corsica, with an area of 8,762 square kilometers, and the Greek Crete, with an area of 8,424 square kilometers. Other significant islands in the western basin include Alboran Island and the Balearic islands (Menorca, Mallorca, Ibiza, and Formentera) near Spain, and Elba, the Pontine islands, Ischia, Capri, Stromboli, and the Lipari islands near Italy. In the eastern basin are Malta off the coast of Tunisia, as well as the Greek Ionian, Cyclades, and Aegean islands. The floor of the Mediterranean is generally made up of carbonate-rich yellow mud, which covers the blue mud floor.

Among the most important rivers that flow into the Mediterranean are the Spanish Ebro, the French Rhone-Saone-Durance, and the Italian Arno, Tiber, and Volturno. Others include the Po and the Tagliamento in Italy, the Vardar in Greece and Yugoslavia, and the Nestos in Bulgaria and Greece. North Africa has only the Nile, which is the only large river that flows toward the north. Over the years, several deltas have slowly formed at the mouths of the rivers that flow into the sea. The most important one is the Nile Delta, which is the most fertile land in Egypt. More than five thousand years ago, Memphis, the Pharaonic capital of Egypt (located close to the current capital, Cairo) was a port.

The coastline is often covered with small hills, while the high-altitude mountains lie on the inner land. The presence of volcanoes such as Vesuvius and Aetna in Italy have been responsible for the loss of thousands of lives over the years but have also contributed to the fertility of the neighboring land.

WATER FLOW

The total water area of the Mediterranean is about 2.5 million square kilometers, with an average depth of about 1,500 meters. Generally, the eastern basin is deeper, with the greatest depth of about 5,450 meters in the Ionian Sea, approximately 57 meters off the Greek mainland. The western basin's deepest spot lies in the Tyrrhenian Sea at about 3,600 meters. The sill depth in both the straits of Sicily and Gibraltar is about 275 meters. The Adriatic Sea is the shallowest of all seas because it is an overextension of the continental shelf. The water of the western basin is a little cooler and fresher than that of the eastern basin. The average surface temperature in the west is as low as 11 degrees Celsius in February and as high as 23 degrees Celsius in August. The corresponding temperatures in the eastern basin are 16 degrees and 26 degrees Celsius. The dissolved salt increases the density of water. Seas that are isolated and have few rivers flowing into them are traditionally saltier. This effect is particularly enhanced by arid climates, where evaporation is greater than precipitation.

The Mediterranean, which fulfills all these requirements, has a sill at the entrance of Gibraltar; as a result, water from the lower-density Atlantic Ocean flows in at the surface. At the same time, the Mediterranean water flows over the sill into the ocean, where it sinks to a much lower depth until it reaches levels of the same density. Although a considerable degree of mixing takes place at the upper level because of the initial flow of salty water to the Atlantic, once the water reaches the high-density region, it spreads horizontally and can be followed far into the Atlantic. In general, however, the deep outflowing current does not equal the intake of surface water. The measurement and comparison of salinities of the Atlantic and the Mediterranean waters in the Strait of Gibraltar provide data that give a good measurement of the outflow of water. Taking into consideration also the evaporation rates, the outflow has been estimated to be in the neighborhood of 35 to 63 million cubic feet per second. Temperature and wind variability have an impact on the actual outflow.

Generally, the salinity of the western Mediterranean basin is less than that of the eastern. The water mass flowing from the Black Sea through the Bosporus and the Dardanelles to the north of the

Aegean Sea is estimated to be close to 200 cubic kilometers per year. The Dardanelles boundary condition results in the spreading of the outflow to the Western Aegean, as well as to the north, along the Greek mainland coast. Black Sea waters appear to be able to reach the central Aegean; their influence is greatest in summer and fall during the time of maximum discharge through the Dardanelles in September and minimum salinity in July. Moreover, a westerly current enters the area between Asia Minor and the Greek island of Rhodes, which reverses in spring and summer. Along the northern coast of Crete, easterly currents appear to move all year long. The tide that is attributed to the Atlantic Ocean appears to lose strength in the Strait of Gibraltar.

The Mediterranean tides are generally semidiurnal, with wave systems that have nodal lines that extend from Barcelona, Spain, to Bougie, Algeria, and from Turkish Korfezi to Libyan Tobruch. The tide range in most other places does not surpass the 1-meter mark. Strong hydraulic currents occur at the ends of the two straits of Sicily and Messina because of the formation of high water in exactly opposite phase. These are the same waves that made the Messina strait such a dangerous place to navigate in the ancient times; this is the area that was given the name of the legendary Scylla and Charybdis in Homer's *Odyssey*. In contrast, the Adriatic has a progressive tide that has the highest range of 1 meter.

Historically, the study of currents was investigated because of the observed continuous in-flow of water through the strait of Gibraltar on which mythology indicated that Atlas was standing holding the Earth sphere on his shoulders. Scientists became aware of the fact that the evaporation of water over the Mediterranean was larger than the precipitation and river water outpouring. As a result, the possibility of underground canals that would bring back the water to the Atlantic was theorized. It is estimated that Gibraltar provides more than 97 percent of the water entering the Mediterranean, while precipitation, the Black Sea, and river flow provide the rest. On the other hand, of the water lost by the Mediterranean, 93 percent leaves through Gibraltar, more than 6 percent is evaporated, and less than 0.5 percent is lost to the Black Sea. There is no doubt that if the Atlantic water in-flow through Gibraltar did not occur, the

Mediterranean Sea would be a salty pond with little life, similar to the Dead Sea.

CLIMATE AND PLANT LIFE

The Mediterranean climate is transitional between the harsh weather of central Europe and the desert-dominated conditions found in North Africa. Because of the decreased humidity, the overall climate is comfortable for its inhabitants. A small degree of sleet and snow precipitation takes place, especially in the mountains, and is supported by occasional polar or Arctic invasions of cold winds. However, the monthly temperature averages never drop below freezing, which is particularly favorable to crops. The southern parts of North Africa and the eastern basin, which are closer to the equator, experience hotter and drier summers since they lie far from the Atlantic Ocean's influence.

Overall, the eastern basin is roughly 6 degrees of latitude nearer to the equator than the western basin. In particular, the Sahara Desert in North Africa makes the greatest parts of Egypt, Libya, and Algeria practically uninhabitable. Often there are three- to four-month spans where no precipitation occurs in the eastern basin, while the western basin territory gets numerous storms and substantial amounts of rains.

In general, the term "Mediterranean climate" refers to a long, hot, dry summer and a short, cool, wet winter. Several spots on the Earth experience the same conditions, such as Los Angeles in California, Santiago in Chile, Capetown in South Africa, and several cities of southwestern Australia. The mild, dry summers and the beautiful scenery of many areas in the Mediterranean prompt a great number of tourists to visit the beaches of the French and Italian Rivieras, southern Italy, the Adriatic coast of Yugoslavia, the coasts of Lebanon, and the islands of Greece. The unique climate provides characteristic crops and vegetation. The dry weather favors the growing of the long-rooted olive trees and oaks, as well as the dominance of the dense scrub 2 to 4 meters in height known as *maquis* in France and *macchia* in Italy. All lose very little water by evaporation and have a thick bark and small, thin leaves. Interestingly, the climate helps the growing of fruit that is characteristic of several other places on the Earth, such as grapes and oranges, as long as there is enough water during the dry season.

SIGNIFICANCE

The Mediterranean Sea's impact on civilization has also added to problems that the human race is encountering. The ecosystem has been substantially eroded by the destruction of forests, which has been taking place for more than thirty centuries. The Phoenicians, Athenians, Romans, Byzantines, and Turks were significant naval powers and used timber for ship construction. The citizens of Athens, Rome, Jerusalem, Alexandria, and Constantinople routinely used trees in the building of their cities.

Tourism and industry are responsible for one of the greatest concerns, pollution. The use of iron, steel, concrete, petrochemicals, and other industrial products has affected the air quality of cities such as Rome, Athens, Marseilles, and Barcelona. The presence of millions of tourists has led to the need for more sanitary facilities and for recycling projects. Often the natives encounter the great problem of raw sewage and other refuse that overwhelm the beaches.

Finally, oil spills have also added to environmental pollution. The 1975 United Nations-sponsored Mediterranean Action Plan (MAP) convinced most Mediterranean nations to sign a pact to eliminate toxic spillage and chemical waste from the Mediterranean. In 1982, the MAP Coordinating Unit was established in Athens, followed in 1995 by the adoption of MAP Phase I, Barcelona Resolution, Priority Fields of Activities for Environment and Development in the Mediterranean Basin (1996-2005) and the Protocol Concerning Specially Protected Areas and Biological Diversity in the Mediterranean.

Soraya Ghayourmanesh

CROSS-REFERENCES

BIBLIOGRAPHY

Chester, R., and Stefano Guerzoni, eds. *Impact of Desert Dust Across the Mediterranean.* Norwell, Mass.: Kluwer Academic Publishers, 1996. This is a compilation of specific topic papers on the effect of the Saharan dust transport to the Mediterranean Sea. Topics include the modeling of the Saharan dust transport, the chemistry and mineralogy of the dusts and their effect on precipitation, the contribution of the dust to marine sedimentation, the aerobiology of the dusts, and the Saharan dust's impact on the different climates.

Fabbri, Paolo. *Coastlines of the Mediterranean.* Reston, Va.: American Society of Civil Engineers, 1993. This proceeding contains the papers presented during the Eighth Symposium on Coastal and Ocean Management (coastal zone 93) held in New Orleans on July 19-23, 1993. Data on coastal management planning, conservation, and development, as well as public information and citizen participation, are included.

Jeftic, L., et al., eds. *Climatic Change and the Mediterranean: Environmental and Societal Impacts of Climatic Change and Sea-Level Rise in the Mediterranean Region.* New York: E. Arnold, 1992. The climate change of the Mediterranean is discussed, with sections on changes in precipitation, hydrological and water resources, and land degradation in such areas as the Nile Delta, the Thermaicos Gulf, and the Lion Gulf. A chapter deals exclusively with the predictions of relative coastal sea-level change in the Mediterranean based on archaeological data.

Kalin Arroyo, Mary T., Paul H. Zedler, and Marilyn D. Fox, eds. *Ecology and Biogeography of Mediterranean Ecosystems in Chile, California, and Australia.* Vol. 108. New York: Springer-Verlag, 1995. The economorphological characteristics of Mediterranean-like vegetation in Chile, California, and Australia are presented and analyzed.

Rendel, Phillip W., F. M. Jaksic, and G. Montenegro, eds. *Landscape Disturbance and Biodiversity in Mediterranean-Type Ecosystems.* New York: Springer-Verlag, 1998. This book compares worldwide ecosystems that are similar to the Mediterranean one. Such ecosystems include those found in South Africa, California, southwestern Australia, and central Chile.

NORTH SEA

The North Sea, located between the United Kingdom and continental Europe, is one of the most economically important bodies of water in the world. The impact of human activity on the North Sea is a major concern of environmentalists.

PRINCIPAL TERMS

BALTIC SEA: the body of water between Scandinavia and Eastern Europe

BANK: an elevated area of land beneath the surface of the ocean

NORWEGIAN SEA: the body of water north of the North Sea

STRAIT: a narrow waterway between two larger bodies of water

TRENCH: a long, narrow, depressed area in the ocean floor

PHYSICAL CHARACTERISTICS

The North Sea is an arm of the Atlantic Ocean located between the islands of Britain and the mainland of northwestern Europe. It is bordered by the island of Great Britain to the southwest and west, by the Orkney Islands and the Shetland Islands to the northwest, by Norway to the northeast, by Denmark to the east, by Germany and the Netherlands to the southeast, and by Belgium and France to the south.

To the north, the North Sea opens to the Norwegian Sea. To the south, it is connected to a narrow waterway known as the Strait of Dover, located between southeastern England and northwestern France. The Strait of Dover connects the North Sea to a wider waterway known as the English Channel, located between southern England and northern France. The English Channel opens to the Atlantic Ocean. To the east, the North Sea is connected to a strait known as the Skagerrak, located between Norway and Denmark. The Skagerrak is connected to a strait between Sweden and Denmark known as the Kattegat, which opens to the Baltic Sea.

The North Sea covers an area of about 570,000 square kilometers. It contains about 50,000 cubic kilometers of water. The North Sea is generally shallow, with an average depth of about 94 meters. By comparison, the Atlantic Ocean has an average depth of about 3,930 meters. The southern part of the North Sea is the most shallow, with the northern part growing deeper as it approaches the much deeper Norwegian Sea.

The floor of the North Sea is rough and irregular. In the southern part, where the water is often less than 40 meters deep, many areas of elevated underwater land known as banks are shifted and reworked by tides and currents. These moving banks often present a hazard to navigation. The Dogger Bank, a large bank located roughly in the center of the North Sea, is only about 15 to 30 meters below sea level.

Several areas of greater-than-average depth, known as trenches, are located in the North Sea. In the otherwise shallow waters of the south, a trench known as Silver Pit reaches a depth of about 97 meters. Not far north of the Dogger Bank, a trench known as Devils Hole reaches a depth of more than 450 meters. The deepest part of the North Sea is the Norwegian Trench, a large trench that runs parallel to the southern coast of Norway. The Norwegian Trench is between 25 and 30 kilometers wide, with depths ranging from 300 to 700 meters.

The coastline of the North Sea varies from rugged highlands in the north to smooth lowlands in the south. The coast of Norway is mountainous and broken, with thousands of rocks and small islands. The coasts of Scotland and northern England are high and rocky, but less broken. The coasts of middle England and the Netherlands are low and marshy. The coasts of southern England, France, and Belgium are low and sandy.

GEOLOGICAL HISTORY, HYDROLOGY, AND CLIMATE

The shape and size of the North Sea have varied greatly over time. At the end of the Pliocene

Epoch, about 1.6 million years ago, the southern half of the North Sea was part of the mainland of Europe. At that time, the Rhine River, which now empties into the North Sea in the southern part of the Netherlands, ran to a point about 400 kilometers north of London. The Thames River, which now empties into the North Sea east of London, continued eastward until it met the Rhine River.

During the Pleistocene Epoch, from about 1.6 million to 10,000 years ago, vast ice sheets advanced and retreated several times and deposited a thick layer of clay on the bottom of the North Sea. At its greatest extent, the ice covered the entire North Sea. About 8,000 years ago, the ice retreated for the last time. A few hundred years later, the rising waters of the North Sea broke through the land bridge connecting England and France, forming the Strait of Dover. The modern coastlines of the North Sea were formed about 3,000 years ago.

The movements of the ice sheets were largely responsible for the rugged floor of the North Sea. The Dogger Bank and smaller banks in the southern part of the North Sea were created when the ice deposited large amounts of earth and stones in a particular area. Some of the trenches in the North Sea are believed to be located in areas where ancient rivers emptied into the North Sea when it was much smaller.

Water enters the North Sea from the Atlantic Ocean by way of the Strait of Dover and the Norwegian Sea. This water is relatively warm and salty. It is heated by the warm North Atlantic Current, which moves north along the western side of the British Isles and enters the Norwegian Sea. Colder, less salty water from the Baltic Sea enters by way of the Skagerrak, creating a counterclockwise current in the North Sea.

Fresh water enters the North Sea from the Thames, the Rhine, and other large rivers. The salt content of the North Sea varies from about thirty-four to about thirty-five parts per thousand. Higher concentrations are found off the coast of Great Britain, and lower concentrations are found off the coast of Norway.

The average temperature of the surface water in the North Sea in January varies from about 2 degrees Celsius east of Denmark to about 8 degrees Celsius between the Shetland Islands and Norway. In July, the average temperature varies

from more than 15 degrees Celsius along the coast from the Strait of Dover to Denmark, to about 12 degrees Celsius between the Shetland Islands and the Orkney Islands.

The average air temperature varies from between 0 and 4 degrees Celsius in January to between 13 and 18 degrees in July. Winters are stormy, and gales are frequent. The average difference between low tide and high tide is between about 4 meters and 6 meters along the coast of the British Isles and the southern coast of the European mainland. Along the northern coast of the European mainland, the difference is usually less than 3 meters.

HUMAN ACTIVITIES

Because of its long coastline and the many rivers that empty into it, the North Sea has long been an area of important human activity. The exchange of people, goods, and ideas made possible by the North Sea had a profound influence on the cultural development of northwestern Europe during the Renaissance.

In modern times, the North Sea is one of the busiest shipping areas in the world. Adding to its economic importance is the fact that it provides the only waterway between the Baltic Sea and the Atlantic Ocean. Because of the North Sea, the Netherlands and the United Kingdom are among the world's leading nations in the volume of cargo carried by sea. The Europoort complex at Rotterdam, in the Netherlands, handles more cargo than any other seaport in the world. Other major ports located on the North Sea include Antwerp, Belgium; Dunkirk, France; London, England; and Hamburg, Bremerhaven, and Wilhelmshaven, all in Germany.

The accessibility of the North Sea also made it an early area of scientific research. The Challenger Expedition, launched by the British in 1872, began a new era in oceanography. In 1902, the International Council for the Exploration of the Seas was founded in Denmark for the purpose of studying the North Sea and the Baltic Sea. It has since compiled the longest record of marine ecological conditions in the world. In more recent years, a large number of marine laboratories, research centers, and scientific vessels have been active in the area.

The North Sea has also long been an area of

land reclamation and flood-control projects, particularly in the Netherlands. For centuries the Dutch have reclaimed land from the North Sea by building dikes. In the 1930's, a dike 30 kilometers long was built in the Netherlands, creating a large freshwater lake. Three-fifths of an area of land formerly under the North Sea was then reclaimed as farmland. After abnormally high tides flooded a large part of the Netherlands in 1953, the Dutch built a large flood-control system on rivers emptying into the North Sea. The British built a similar system on a smaller scale on the Thames River in 1984.

Fishing has long been a major activity in the area. The constant mixing of shallow water in the North Sea provides a rich supply of nutrients to plankton, microscopic organisms that support a wide variety of commercially important fish. The main species caught are cod, haddock, herring, and saithe. Lesser quantities of plaice, sole, and Norway pout are also caught. Sand eel, mackerel, and sprat are caught for the production of fish meal.

The major fishing nations in the region are Norway, Denmark, the United Kingdom, and the Netherlands. In 1983, the nations of Europe created the Common Fisheries Policy. This arrangement establishes the amount of each species that each nation may catch in the open waters of the North Sea. This policy does not apply to fish caught in coastal waters, which are considered to belong to a particular nation.

Natural Gas and Oil

New economic resources were discovered in the North Sea in the second half of the twentieth century. In 1959, the first known source of natural gas in the North Sea was identified. This source was an extension into the sea of a large natural gas field in the northeastern part of the Netherlands. By the end of the 1970's, a large number of natural gas production sites were developed in the North Sea. These sites are located primarily along a line running east and west for about 150 kilometers between the Netherlands and England. A smaller number of natural gas sources are located in the central and northern regions of the North Sea.

The first nation to obtain oil from the North Sea was Norway, which began operating its first offshore oil well in 1971. The United Kingdom began producing oil from the North Sea in 1975.

The oil wells are primarily located in the northern and central regions of the North Sea. By the 1980's, offshore oil wells were in operation from north of the Shetland Islands to an area about 650 kilometers to the south.

Oil is found in the North Sea in a basin of sediments thousands of meters thick. Almost all the world's supply of oil is found in similar basins. Such basins are believed to exist in areas where the outer layer of the Earth has been stretched and thinned. This causes a basin to be formed beneath the outer layer, which then collects sediments for millions of years. If the sediments are subjected to certain temperatures and pressures, organic matter within them is transformed into oil. Seismic measurements using controlled explosions have confirmed that the outer layer of land beneath the central portion of the North Sea is about half as thick as the same layer elsewhere. The stretching and thinning of this area took place millions of years ago, when Europe and the island of Great Britain were drifting apart.

Significance

The most critical issue facing the nations that border the North Sea is the impact of human activity in the area. Pollution from ships and from land-based operations is a major concern. Although international agreements on limiting pollution of the North Sea have been in place since 1969, enforcement remains difficult. Events such as the Ecofisk oil well blowout, which spilled more than 30 million liters of oil into the North Sea in 1977, point out the importance of controlling contamination.

An important problem is the disposal of offshore equipment that is no longer in use. As long ago as 1958, international agreements stated that all such equipment must be removed from the sea. However, because of the extremely high cost of removing large installations, new agreements were made in 1989 that allowed such equipment to be disposed of at sea. The agreement required that the water where the equipment is disposed be at least 100 meters deep and that at least 35 meters of water remain above the disposed debris. This policy met with worldwide controversy in 1995, when the Shell company decided to dispose of an unused oil well known as the Brent Spar in this way. Under intense pressure from environmentalists, Shell reversed this decision and made plans to remove the Brent Spar entirely.

Less dramatic, but possibly even more important, is the problem of oil pollution from muds used to lubricate drills in offshore oil wells. Studies released in 1995 indicated that oil from this and other sources was far more widespread than previously thought. The most significant impact was on an organism known as the burrowing brittle-star. The digging of this animal brings oxygen into sediments, encouraging other organisms to grow there. The burrowing brittle-star is also an important food source for many fish. The number of burrowing brittle-stars per square meter fell from more than one hundred in unpolluted waters to zero within 1 or 2 kilometers of oil wells.

Such losses may have been a major factor in the decrease of the numbers of North Sea fish such as cod, whose population fell by two-thirds between 1980 and 1995.

Rose Secrest

CROSS-REFERENCES

BIBLIOGRAPHY

Harvie, Christopher. *Fool's Gold: The Story of North Sea Oil.* London: Hamish Hamilton, 1994. A well-researched, detailed account of the discovery and production of oil in the North Sea. Deals largely with the environmental, economic, and political consequences of extracting oil from the region.

Ozsoy, Emin, and Alexander Mikaelyan, eds. *Sensitivity to Change: Black Sea, Baltic Sea, and North Sea.* Boston: Kluwer Academic Publishers, 1997. The proceedings of a workshop sponsored by the North Atlantic Treaty Organization (NATO) held in Bulgaria in 1995. Discusses the impact of human activity and climate change on the North Sea and other major arms of the Atlantic Ocean.

Warren, E. A., and P. C. Smalley. *North Sea Formation Waters Atlas.* London: Geological Society, 1994. An extremely detailed description, with numerous maps, of the physical structure of the North Sea. Includes information on the location of natural gas and oil fields.

Ziegler, Karen, Peter Turner, and Stephen R. Daines, eds. *Petroleum Geology of the Southern North Sea: Future Potential.* London: Geological Society, 1997. A scientific account of the possibility of producing oil from regions of the North Sea not usually associated with oil fields. For advanced students.

PACIFIC OCEAN

The world's largest ocean, the Pacific has an overall area of 182 million square kilometers. It is twice the size of the next largest ocean, the Atlantic. The Pacific Ocean covers approximately one-third of the Earth's surface and is larger than the Earth's entire landmass.

PRINCIPAL TERMS

CONTINENTAL DRIFT: the gradual movement of a landmass away from another landmass where a fissure has occurred

CONTINENTAL SHELF: part of the seafloor, usually gently sloping, that extends beneath the ocean from the continent that it touches

CONTINENTAL SLOPE: the part of the continental shelf that drops off sharply toward the ocean's floor

DELTA: a triangular area with its longest side abutting the sea where a river deposits silt, sand, and clay as it flows into the ocean

EQUATOR: an imaginary line, equidistant from the North and South Poles, around the middle of the Earth where day and night are of equal length

LAGOON: a body of salt water separated from the ocean by a bank of sand

TECTONIC PLATE: a vast rock surface upon which a landmass rests

TIDAL RANGE: the difference in water depth between high and low tides

LOCATION OF THE PACIFIC OCEAN

The Pacific Ocean, with a mean depth of 4,255 meters and a maximum depth of 10,970 meters, extends from its southern extreme at Antarctica some 15,550 kilometers north to the Bering Strait, which separates North America from Asia. Its east-west area extends almost 19,440 kilometers from the western coast of Colombia in South America to Asia's Malay Peninsula.

The Pacific and Arctic Oceans meet at the Bering Strait in the north. In the far south at Drake Passage, south of Cape Horn, the Atlantic and Pacific Oceans come together. Exactly where the Pacific meets the Indian Ocean is more difficult to define because the two bodies of water intermingle along a string of islands that extends to the east from Sumatra through Java to Timor and across the Timor Sea to Australia's Cape Londonderry. South of Australia, the Pacific runs across the Bass Strait to the Indian Ocean and continues from Tasmania to Antarctica.

The eastern Pacific generally follows the Cordilleran mountain system, which runs the entire length of North and South America from the Bering Strait to Drake Passage and includes both the Rocky Mountains and the Andes. The east coast of

the Pacific is quite regular save for the Gulf of California and the fjord regions in its northern and southern extremes. The continental shelf of the eastern Pacific is narrow, and the continental slope at times quite steep.

The Pacific's western extreme in Asia is, by contrast, quite irregular. As on the eastern coast of the Pacific, the western coast is bordered by mountain systems that run roughly parallel to the coast. The western Pacific has many dependent seas, notable among them the Bering Sea, the East China Sea, the Sea of Japan, the Sea of Okotsk, the South China Sea, and the Yellow Sea. The eastern extremes of these seas are characterized by peninsulas jutting out toward the south, by island arcs, or by both.

Unlike the Atlantic, much of whose water flows into it from the rivers that feed it, only about one-seventh of the Pacific's water comes from direct river flow. Most of its water comes from the dependent seas that are nourished by such East Asian rivers as the Amur, the Xi, the Mekong, the Yangtze, and the Yellow Rivers.

ORIGINS AND DIVISIONS

The Pacific Ocean evolved some 200 million years ago, when the single landmass on the Earth began to split. Through continental drift that oc-curred over millions of years, the land divided at various fissures into large areas resting on the tectonic plates that float on seas of molten rock that underlie the Earth's entire crust. As fissures, at first quite narrow, became larger, they filled with the water that surrounded the landmass and, through eons, formed oceans, seas, lakes, lagoons, and rivers. The rivers that flowed over the land cut deeply into it, but it took millions of years for canyons and deep channels to form. As huge landmasses drifted, continents took shape, each continent separated from other continents by oceans that flowed into the widening fissures that continental drift created.

The Pacific Ocean is so large that it is difficult to discuss it as a single ocean. Various parts of it are substantially different from other parts. For purposes of discussion, the Pacific is usually thought of as comprising three distinct regions.

The western region begins in Alaska's western Aleutian Trench, extends through the Kuril and Japanese Trenches, runs south to the Tonga and Kermadec Trenches, then continues on to an area northeast of New Zealand's North Island. This region is characterized by large strings of islands, the largest constituting Japan, New Guinea, New Zealand, and the Philippines. Some of these strings—notably Japan, New Guinea, and New Zealand—were, through time, sheered off from the continent. The entire western Pacific is also dotted with volcanic islands, which are the peaks of high underwater mountains. These were formed when molten rock from successive volcanic eruptions occurring over millions of years cooled and solidified.

The ocean floor of the central area of the Pacific is the largest underwater expanse on the Earth. It is the most geologically stable area of the Earth's crust on the Pacific floor. It contains sprawling underwater plains, most at a depth of about 4,500 meters, that are essentially flat, although they contain some irregularities and some geological formations that resemble the mesas found in above-water plains.

The eastern part of the Pacific, which abuts the United States' West Coast states, has a narrow continental shelf and a steep continental slope so that close to shore there are deep areas, such as the Monterey Canyon, which is more than 3 kilome-ters below the water's surface. The two most significant trenches in this part of ocean are the Middle America Trench in the North Pacific and the Peru-Chile Trench in the South Pacific.

Volcanic activity has been greater in the western part of the Pacific than in its eastern portion, although the East Pacific Rise, a chain of underwater mountains that runs from Southern California almost to the tip of South America, is volcanic. In the western Pacific, as, to a lesser extent, in the central area, there is a profusion of islands that are the tops of underwater mountains.

PACIFIC CURRENTS

The Pacific, like the Atlantic, has two large systems of currents called gyres. The Northern Gyre moves in a westerly direction from near the equator and carries warm surface water into the Kuroshio Current and gives Japan moderately warm temperatures. The Kuroshio Current then moves northeast away from the Kamchatka Current, which brings cold water from the area around the Bering Strait. Part of it divides, moving north into the North Pacific Drift. The other part heads southeast.

Off the west coast of North America, part of the North American Drift moves north and becomes the Alaska Current. It brings to the Alaskan coastline waters warm enough to keep its shoreline and harbors from freezing in winter. Another part of the North American Drift veers south and forms the California Current, whose waters are cool as they flow south to join the North Equatorial Current, which again flows west, completing its clockwise motion.

The South Equatorial Current also moves west toward Australia and New Guinea, but it then veers south to begin its counterclockwise course. The East Australian Current skirts Australia's east coast and passes between Australia and New Zealand, south of which it feeds into the West Wind Drift, which moves east toward the coast of South America. Just north of South America's tip, it feeds into the Peru or Humboldt Current, which runs north along South America until it connects with the South Equatorial Current, which moves west to complete the South Pacific Gyre.

In both the northern and southern areas of the Pacific, winds from the west move the water in an easterly direction. Closer to the equator, however, trade winds move the water toward the west. This balance permitted early sailors to head east by riding the westerly winds and to return to the east by riding the trade winds. There are virtually no winds on the equator itself. Sailors can founder there, in the so-called doldrums, for days with none of the propulsion that nature provides both north and south of the Earth's dividing line.

The currents of the Pacific are important both for transportation and for bringing warm and cold waters to parts of the ocean, thereby giving them and the land adjoining them more moderate temperatures than might be expected at their latitudes. Occasionally, nature runs afoul of this balance as the Southern Trade Winds diminish and fail to push the cool Peru Current north. Instead, the Peru Current is replaced by the Pacific Equatorial Countercurrent. The Peruvians first dubbed this phenomenon El Niño, which means "the Christ child," because it usually occurs close to Christmas. In some years, El Niño is followed by La Niña, resulting in two successive years of hardship. As waters reach unaccustomed temperatures, much sea life dies, bringing catastrophe to people dependent upon fish and seaweed for their diets or for income.

FORMATION OF PACIFIC ISLANDS

The Pacific Ocean has more islands than any other ocean. In the west, most of these islands are volcanic, although some—such as Japan, New Guinea, and New Zealand—were formed when large chunks of land broke away from continents in prehistoric times. The volcanic islands in the western Pacific often rise just barely above sea level. Any change in the ocean's level or tidal range can inundate them. Global warming threatens to melt ice in the polar regions, which may cause oceans to rise and, in some cases, islands to disappear. Indeed, some small, politically independent island nations face total inundation. It is estimated that by the year 2050, the oceans of the world might rise as much as 50 centimeters, which could wipe out the habitations of millions of people in the South Pacific.

Another major structure in the South Pacific is the atoll, a tropical island on which a massive coral reef, often ringlike, generally rests on a volcanic base. Kwajalein is the largest atoll in the world, with a circumference of nearly 324 kilometers.

Other coral formations, such as ridges or reefs composed of coral, can grow to enormous lengths, such as the Great Barrier Reef east of Australia. Coral formations are composed of living organisms that require light to survive. They are, therefore, always close to the surface of the ocean and often protrude above the surface. Most of them teem with aquatic life. In some cases, when they form around low-lying islands, the island disappears as the ocean surrounding it rises. The coral formation, however, often remains, rising above the water's surface as an atoll.

Such island chains as the Hawaiian, Pitcairn-Tuamotu, and Tubai Islands are the result of great extrusions of molten rock that have risen through the Earth's mantle over millions of years and burned a hole in the oceanic plate above it. The resulting formation that occurs when the molten rock cools is an island composed largely of basaltic rock. Over great time spans, the island is slowly carried away to the northwest, clearing the way for another such island to be formed above the hole in the oceanic plate. Hundreds of islands, lined up almost like pearls in a necklace, can result from this ongoing volcanic activity.

PACIFIC RIM

The areas around the Pacific have many volcanic mountains, large ranges of which line the coasts and still larger ranges of which are submerged. The highest mountains on Earth are the Himalayas, with Mount Everest towering to 8,795 meters; many of the mountains submerged in the Pacific, however, are higher. From the ocean's floor at the Mariana Trench, mountains rise more than 10,600 meters as they approach the ocean's surface.

Many of the mountains on the Pacific Rim arose when, over eons, great portions of the oceanic crust intruded gradually beneath sections of the continental crust. Intense heat from the Earth's interior eventually caused the crusts above it to melt, permitting molten rock to erupt through the continental crust as volcanoes, which occur from the North Island of New Zealand to the Aleutian Islands and on down the west coast of the United States to Peru and Chile.

The coast along the Pacific Rim is narrow, giving way to the mountain ranges that run close to the shoreline. An exception to this is in eastern China where the Yellow and the Yangtze Rivers

The Pacific coast. (PhotoDisc)

have, for many centuries, carried silt toward the ocean, creating fertile coastal plains.

Geological activity beneath the Pacific results in thousands of small earthquakes every year along the Pacific Rim as oceanic plates run into continental plates. These earthquakes, seldom felt by humans, result in little or no damage. They relieve pressure that builds up when tectonic plates collide. When such pressure builds up without relief over a period of many years, however, a major earthquake, usually with considerable property damage and loss of life, may occur. Earthquakes occur beneath the sea as well as on land. Severe earthquakes that rock the bottom of the sea generate huge waves called tsunamis. These waves, sometimes more than 30 meters high, can engulf large stretches of shoreline, as happened in Alaska in 1964.

FOOD RESOURCES

The feeding hierarchy that exists among the creatures of the sea is usually referred to as the food chain. One can envision this hierarchy as a triangle, at whose broad base are phytoplankton, microscopic plant organisms that require sunlight for their survival and are plentiful near the surfaces of oceans. Zooplankton, microscopic animal organisms, feed on the phytoplankton. They, in turn, constitute the diet of small fish called anchoveta. These fish feed such larger fish as tuna and dolphin, as well as such birds as cormorants and pelicans, which dive into the water to catch fish. Animal waste, decaying plants, and dead fish and birds sink to the ocean floor, where scavengers eat them. Bacteria at those levels cause such droppings to decay, in the process releasing nutrients into the sea water that provide the phytoplankton with nourishment, thereby completing the food chain.

The food chain triangle becomes smaller as it approaches the top. Small fish survive by eating plankton. Larger fish eat the smaller fish, then sharks and whales eat the larger fish. A food chain that begins with millions of plankton may end up producing hundreds of thousands of anchoveta;

thousands of cod, hake, and mackerel; and one-half dozen sharks for every whale that is part of this intricate hierarchy.

Although fish is the major food source harvested from the world's oceans, the Pacific also yields a great deal of kelp and seaweed, which, particularly in Asian countries, constitute a major part of human diets. Seaweed is often harvested and laid in strips beside the ocean, where the Sun dries it, thereby preserving it and making it easy to transport and to store. Even though they are not consumed as readily in the eastern parts of the Pacific Rim as in Asia, kelp and other sea plants are used in many pharmaceutical preparations. They have valuable pharmacological properties that are in great demand by drug manufacturers throughout the world. The iodine contained in saltwater fish and plants is necessary for the proper functioning of the thyroid gland. Inland people, deprived of iodine, may develop goiters; however, with the widespread introduction of iodized salt, the incidence of goiters has been greatly reduced.

MINERAL RESOURCES

The mineral wealth beneath the Pacific Ocean has barely been tapped. Among the minerals available near the shore are chromite, gold, iron, monazite, phosphorus, tin, titanium, and zircon. The greatest exploitation of the ocean's treasure trove has taken place in the petroleum industry. Offshore drilling takes place worldwide for the recovery of crude oil and natural gas. In addition, large quantities of sand and gravel are also harvested every year for use in construction and manufacturing.

The deep sea has remained a mysterious place. The historic descent in 1960 of the bathyscaphe *Trieste* into the Mariana Trench, the deepest area of ocean anywhere in the world, unlocked many mysteries. Subsequent exploration of the deepest areas of the ocean has challenged many long-held beliefs and revolutionized deep-sea research. It was long thought impossible for life to exist under the enormous pressure of the ocean below 760 meters, but the *Trieste* discovered life at the very bottom of the Mariana Trench, much to the surprise of oceanographers.

The deep ocean is richest in its deposits of cobalt, copper, manganese, and nickel. These deposits remain largely unharvested because of the difficulty of getting to them, but such problems will undoubtedly be overcome as deep-sea mining technology is developed.

R. Baird Shuman

CROSS-REFERENCES

Aral Sea, 2193; Arctic Ocean, 2197; Atlantic Ocean, 2203; Black Sea, 2209; Caspian Sea, 2213; Gulf of California, 2217; Gulf of Mexico, 2221; Hudson Bay, 2225; Indian Ocean, 2229; Mediterranean Sea, 2234; North Sea, 2239; Ocean Pollution and Oil Spills, 2127; Offshore Wells, 1689; Persian Gulf, 2249; Red Sea, 2254.

BIBLIOGRAPHY

Allen, Gerald R., and D. Ross Robertson. *Fishes of the Tropical Eastern Pacific.* Honolulu: University of Hawaii Press, 1994. This is one of the best books on the marine life of the eastern part of the Pacific Ocean.

Bottoni, Luciana, Valeria Lucini, and Renato Masso. *The Pacific Ocean.* Milwaukee: Raintree, 1989. This brief book about the Pacific Ocean is profusely illustrated, clearly written, and carefully researched. It is directed toward young adult readers, who will find that it is a good starting point for information.

Broad, William J. *The Universe Below: Discovering the Secrets of the Deep Sea.* New York: Simon and Schuster, 1997. Broad's information about the very deep parts of the ocean is fascinating and cogent. In his well-written discourse, Broad presents compelling information about the Monterey Canyon, which is close to the mainland of central California in the eastern Pacific. He illustrates the pressing need for accelerated exploration of the deep sea, noting that scientists know more about outer space than about the deepest oceans.

Lambert, David. *The Pacific Ocean.* Austin, Tex.: Raintree Steck-Vaughn, 1997. This book, rich with illustrations and maps, is perhaps the best such volume for the young adult audience. Lambert, whose *Seas and Oceans* (1994) is also excellent, understands his subject well and presents it engagingly. His information about the people of the Pacific Rim is particularly useful.

Severin, Tim. *The China Voyage: Across the Pacific by Bamboo Raft*. Reading, Mass.: Addison-Wesley, 1994. Anyone interested in the migrations of ancient inhabitants of the South Pacific will find this book fascinating. It tells of the author's long voyage across the Pacific in a primitive craft similar to the ones used by the ancient Polynesians and Melanesians during their migrations.

Taylor, Brian, and James Natland, eds. *Active Margins and Marginal Basins of the Western Pa-*

cific. Washington, D.C.: American Geophysical Union, 1995. The contributors focus on geological activity in the western Pacific and its implications. Although the contributions are specialized, they provide an excellent background for those interested in the volcanic activity of the area.

Trillmich, Fritz, and Kathryn A. Ono, eds. *Pinnipeds and El Niño: Responses to Environmental Stress*. New York: Springer Verlag, 1991. Although the contributions to this book are mostly specialized, the information it contains about the effects of El Niño is indispensable to those interested in this phenomenon. The selections can be read selectively by those not overly conversant with the field.

PERSIAN GULF

The Persian Gulf, located in the southwest part of Asia, is an extension of the Indian Ocean. The surrounding region is characterized by an arid climate, oil-rich desert land, and lightly populated areas. The Desert Storm operation that ended the Gulf War in 1991 had a dramatic effect on the ecology and environment of the territory.

PRINCIPAL TERMS

LIMESTONE: a common sedimentary rock containing the mineral calcite; the calcite originated from fossil shells of marine plants and animals or by precipitation directly from seawater

PETROLEUM: a natural mixture of hydrocarbon compounds existing in three states: solid (asphalt), liquid (crude oil), and gas (natural gas)

SALINITY: a measure of the degree of salt content in sea water

GEOGRAPHICAL LOCATION

The Persian Gulf is the marginal part of the Indian Ocean that lies in the southwest part of Asia between the Arabian peninsula and southeast Iran. The sea has an area of almost 250,000 square kilometers. Its length is almost 1,000 kilometers, and its width varies from a maximum of 330 kilometers to a minimum of 55 kilometers at the Strait of Hormuz. It is a relatively shallow sea with an average depth of about 25 meters.

The Persian Gulf is bordered on the north, northeast, and east by Iran; on the northwest by Iraq and Kuwait; on the west and southwest by Bahrain, Qatar, and Saudi Arabia; and on the south and southeast by the United Arab Emirates and Oman. The gulf is referred to as the Arabian Gulf in all Arab states. The term Persian Gulf is often used to refer not only to the Persian Gulf itself but also to its outlets, the Strait of Hormuz and the Gulf of Oman, which open into the Arabian Sea. The Gulf of Oman is an extension of the Arabian Sea lying between southeastern Iran and Oman in southeastern Arabia. It is connected with the Persian Gulf via the Strait of Hormuz, and its most important cities are Jask in Iran and Muscat in Oman.

Several islands lie on the Iranian side of the gulf, including Minoo, Kharg, Sheikh Saas, Sheikh Sho'ayb, Hendurabi, Kish, Farur, Sirri, Abu Mussa, the Greater and Lesser Tunb Qeshm, Hengam, Larak, Frsi, Hormuz, and Lavan. Other small islands belong to the United Arab Emirates, Kuwait, and Saudi Arabia, with Bahrain and Abu Dhabi being the largest islands off the Saudi Arabian coast.

The only rivers that flow into the Persian Gulf are the Tigris and Euphrates. The Tigris River has its source in eastern Turkey and flows southward for about 1,800 kilometers. Just south of the Iraqi city of Mosul, it is joined by two smaller rivers, the Great Zab and the Little Zab. The combined river then flows southeast past Baghdad—the Iraqi capital—and joins the Euphrates to form a single waterway, Shatt-al-Arab. About 200 kilometers farther south, the waterway enters the Persian Gulf. The Euphrates River also has its sources in eastern Turkey and travels a much longer path before it meets the Tigris.

Other than the two rivers, there are only a few small streams that discharge into the Persian Gulf, mostly on the Iranian coast south of Bushehr. No fresh water flows into the gulf on its southwest side. The frequently occurring dust storms bring large quantities of fine dust into the sea via the predominant northwest winds from the desert areas of the surrounding lands. As a result, the lands of the Persian Gulf that are adjacent to the Iranian coast, as well as the land around the Tigris-Euphrates Delta, are covered with green-gray soil that is rich in calcium carbonate.

The Zagros Mountains stretch along the Iranian coastline east of the Euphrates and Tigris Rivers. Mountains as high as 1,200 meters rise abruptly in parts of the coastline of the United Arab Emirates, whose coastal plain blends southward into the hot and dry Ru Al-Khali Desert.

Desert lands also occupy the greatest part of Kuwait with occasional oases such as the Al-Jahrah. The lightly populated Hasa Plain lies on the coastline of the eastern coast of the Arabian peninsula, with occasional hills also found to rise above the shore. Elsewhere the coastal plain is covered with beaches and intertidal flats.

Geological studies of the past have suggested that during the Triassic period (200 million years ago), the Tethya Sea extended as far east as Turkestan and the Aral Sea area. This placed the Persian Gulf between the Tethya Sea and the Indian Ocean in a land known as Gondwanaland. Faulting, earthquakes, continental drift, and the river flow slowly created what is now the land around the Persian Gulf.

HISTORY

The huge valley created by the Euphrates and the Tigris has often been mentioned as the biblical Garden of Eden. There is no doubt that the people living in that area had the most significant civilization several thousands of years ago. The many civilizations—the Chaldeans, the Assyrians, the Babylonians, the Medes, the Persians and, later, the Arabs in cities such as Ur, Babylon, and Uruk— were among the leaders in farming, housing, animal domestication, and building construction.

The Persians eventually became the dominant force until 330 C.E., when Alexander the Great defeated them. The area residents later struggled under the Romans, the Byzantines, the Arabs, and the Turks. The British and the Portuguese fought over the Persian Gulf during the seventeenth century, with the British eventually winning it after aligning with the Persian forces and destroying key fortresses at Qishm and Hormuz. The Persian Gulf's importance increased dramatically during the twentieth century, with the British using their military bases to combat the Ottoman Empire in Mesopotamia during World War I. During World War II, the Persian Gulf served as a supply route by the Allies to the Soviet Union through Iran.

The importance of the Persian Gulf further increased after the end of World War II with the growing dependence of the world on oil. Since the Strait of Hormuz is narrow, a naval blockade is relatively easy and has led to several frictions between Iran and the neighboring Arab states, especially on the claim over several small islands. A long, enduring war between Iran and Iraq started in 1980 and ended eight years later after a loss of about 1 million lives. In 1990, Iraqi president Saddam Hussein invaded Kuwait and occupied it for about one year. He retreated only after most of the countries of the world, led by the United States and its allies, attacked Kuwait and defeated the Iraqis decisively.

OIL RESERVES

The Persian Gulf is the single largest source of petroleum in the world. All Persian Gulf states combined hold more than 50 percent of the world's known oil reserves, which gives evidence to the existence of the so-called Arabian Platform under and around the Persian Gulf. Most of the oil is extracted from a limestone or dolomite reservoir, unlike the rest of the world's fields where oil is extracted from sandstone. Approximately one-half of the fields are totally on shore, while the remainder are either partly or totally under the Persian Gulf. Generally, the Persian Gulf fields have a large anticlinal or dome structure with little faulting and structural complications that are usually encountered in most of the fields in the rest of the world. This leads to a larger oil content, as well as an easier and less costly excavation. This is the main reason why the total number of wells in the Persian Gulf area is no more than 4,300 with an average production of 3,500 barrels per day per well.

Saudi Arabia has the largest (Ghawar) and the fourth largest (Sanfaniya) oil fields in the world. The Ghawar field is approximately 240 kilometers long and 16 kilometers wide. Kuwait's largest oil field is the Burgan field, which is relatively shallow and was discovered in 1938. The largest oil fields in Iran (Ahwaz, Marun, Gach Saran, and Agha Jari) were also discovered late (between 1928 and 1963) and have not been extensively depleted.

Oil and gas pipelines have been built to provide oil to other countries. The Trans-Arabian Pipeline, which was built after World War II, is more than 1,600 kilometers long and carries oil from Saudi Arabia to the Lebanese city of Sidon. Two other major crude oil pipelines start from the Iraqi city of Kirkuk. One is 1,000 kilometers long and sends oil to Dortyol and Yumurtalik in Turkey, and the other is 800 kilometers long and ends in the Lebanese city of Tripoli.

CLIMATE

The Persian Gulf is considered to be the hottest part of Asia, with land temperatures reaching 49 degrees Celsius. Temperatures are high, although winters may be rather cool, especially in the north-western regions. Despite an extremely high degree of humidity, the sparse rainfall occurs mainly as sharp, heavy downpours between November and April, and is heavier in the northeast. Kuwait, for example, has as much as 10 centimeters of rain each year. The region has few clouds, and thunderstorms and fog are rare throughout the year. However, dust storms, sand storms, and haze occur frequently, especially in the summer. Because of the lack of fresh water sources, water is valuable and may be either imported or treated in desalination plants.

The pattern of sea-surface temperatures changes significantly with the seasons. In February, the Intertropical Convergence is near 10 degrees south and the heat equator is also in the Southern Hemisphere. As a result, most of the area between the equator and 20 degrees south has temperatures as low as 27 degrees Celsius. The water temperature is even cooler, down to almost 20 degrees Celsius in the northern parts of the Persian Gulf.

The different climatic conditions over various parts of the Indian Ocean cause the formation of characteristic surface water masses. During the Southwestern monsoon, water flows inward toward the Strait of Hormuz, while from January to April, there is a slight outward flow. The Arabian Sea water has a great degree of salinity for three reasons. First, there are very few rivers that discharge their waters into it. Moreover, the arid climate allows more water evaporation than precipitation to take place. Finally, water of even higher salinity originally formed in the Red Sea, and the Persian Gulf leaves these basins and spreads as a layer of high salinity that lies below the surface. The Persian Gulf water was proven to spread at a depth of almost 300 meters, while the Red Sea water spread as deep as 1,000 meters. This subsurface water was traced as far south as the island of Madagascar in southeastern Africa and as far east as Sumatra in Indonesia. Because of all these factors, the Arabian Sea has a rather moderate seasonal temperature variation, which is classified as subtropical.

OPERATION DESERT STORM

In 1991, Iraqi soldiers, retreating from the forces of Operation Desert Storm, set fire to 732 Kuwaiti oil wells. Operation Desert Storm's bombing campaign also accidentally created some fires but spared the desalination plants to the south of Saudi Arabia, as well as the coastal industries. The effort required by firefighting companies to stop the flow of oil from the wells and to extinguish the blazes was enormous. The environment was made hazardous by choking soot, airborne gases, incredibly high heat, and a vast blanket of smoke that turned daylight into darkness for many weeks. This was the largest oil spill in history, estimated to be about forty times greater than the amount of oil that leaked from the *Exxon Valdez* in the 1980's.

The expensive task ended much earlier than expected, primarily because of the advanced technology and the persistent efforts of the hired private companies. Modern techniques of remote sensing were applied and provided critical information that determined the location of wells, as well as fire and plume movement. Ground stations helped monitor the air quality and particulate matter, while modeling techniques were used to study the dispersion characteristics of pollutants and the deposition of soot. Unfortunately, the postwar cleanup efforts were undermanned and underfunded, and a lot of spilled oil was eventually buried under the shifting sand. Moreover, the effect of contaminated air must have been considerable, but little critical health data was accumulated partly because of the insistence of the Kuwaiti government on curbing the reports of the dangers associated with the hazards of breathing contaminated air.

SIGNIFICANCE

Until the discovery of oil in the late nineteenth and early twentieth centuries, the Persian Gulf area was important mainly for fishing, pearling, sailcloth making, camel breeding, and growing date crops, as well as for mining red ocher from the islands of the south. Most of these traditional industries declined dramatically after the economy of the region began to be determined by oil. The first major discovery of Iranian oil in 1908 triggered the rivalry among many European nations for control of the Persian Gulf. Britain eventually succeeded in controlling the fields until the late 1930's, when American companies discovered the oil reserves in Kuwait and Saudi Arabia and won the dominance of the world market. Oil has also served as a primary cause of many of the region's internal conflicts and uprisings, such as the Iranian Revolution, the Iraqi invasion of Kuwait, the Desert Storm operation, and subsequent arms races among the Persian Gulf countries. The strategic importance of the Persian Gulf will therefore be great as long as oil, which is fundamental for transportation, industrial polymers, heating, food distribution, and construction, is available there. However, the threat of embargoes in the 1970's created the need to look for alternative energy sources. Modern scientific technology has provided nuclear energy, solar heating, and electric cars that are not oil dependent.

The pollution in the Persian Gulf is different from that seen in other seas such as the Mediterranean, where tourism is much more significant and where the waste produced is often hard to dispose of properly. Oil spills and the poor air quality attributed to the oil-extracting industry and the haziness of the climate are important contributors to pollution in the Persian Gulf. The Kuwaiti oil fires produced by the retreating Iraqi forces of Saddam Hussein at the end of the 1991 created an environment whose long-range impact on health is yet to be seen.

Since the Euphrates and Tigris are the only major rivers that discharge into the Persian Gulf and since earthquakes are not as prominent as in other Middle Eastern regions, such as northern Iran and Turkey, the Persian Gulf's geographical area is not expected to change dramatically. However, should the greenhouse effect and global warming occur, melting of the Antarctic icebergs may cause a considerable change in the salinity of the Arabian Sea and the Persian Gulf, as well as a change in the formation of the surface water masses.

Soraya Ghayourmanesh

CROSS-REFERENCES

Abyssal Seafloor, 651; Aral Sea, 2193; Arctic Ocean, 2197; Atlantic Ocean, 2203; Black Sea, 2209; Caspian Sea, 2213; Continental Crust, 561; Continental Drift, 567; Continental Shelf and Slope, 584; Continental Structures, 590; Deep Ocean Currents, 2107; Earth's Mantle, 32; El Niño and La Niña, 1922; Global Warming, 1862; Gulf of California, 2217; Gulf of Mexico, 2221; Hawaiian Islands, 701; Hudson Bay, 2225; Indian Ocean, 2229; Mediterranean Sea, 2234; North Sea, 2239; Ocean-Floor Exploration, 666; Ocean Tides, 2133; Oceanic Crust, 675; Pacific Ocean, 2243; Plate Tectonics, 86; Red Sea, 2254; Reefs, 2347; Tsunamis, 2176; Tsunamis and Earthquakes, 340.

BIBLIOGRAPHY

Hawley, T. M. *Against the Fires of Hell: The Environmental Disaster of the Gulf War.* New York: Harcourt, Brace and Jovanovich, 1992. This is a careful assessment of the environmental damage created during the Operation Desert Storm, both globally and locally. It describes the cleanup efforts after the army coalition was dissolved, as well as the attempts of the firefighters to extinguish the fires while preventing the oil from flowing off the Kuwaiti wells. It also analyzes the impact of that disaster by gathering critical health data.

Husain, Tahir. *Kuwaiti Oil Fires: Regional Environmental Perspectives.* New York: Elsevier Science, 1995. This very readable text analyzes the effect that the oil-well fires set by the Iraqi army had on the environment. Topics such as the success of the technology used to extinguish the fires, the smoke plume characterization, the monitoring of the gaseous products and other particulates, and outlines for future emergency response are discussed in different chapters. The book also includes several impressive photographs.

McKinnon, Michael. *Arabia: Sand, Sea, Sky.* BBC/Parkwest, 1992. This book examines the dramatic changes that have shaped Arabia and its wildlife in the last several thousands of years. It also examines the profound and rapid environmental transformations and their impact on the future of the Persian Gulf's ecology, people, and wildlife.

Regional Report of the State of the Marine Environment. Safat, Kuwait: Regional Organization for the Protection of the Marine Environment, 1999. This report, which includes a bibliography, provides detailed information on marine pollution and the coastal ecology in and along the Persian Gulf.

Sadiq, Muhammad, and John C. McCain. *Gulf War Aftermath: An Environmental Tragedy.* Dordrecht, Netherlands: Kluwer Academic Publishers, 1993. This book discusses the impact that the 1991 Desert Storm operation had on the environment. Individual chapters analyze the effect that the military operations and oil fires had on air quality, land resources, human health, and the marine environment.

RED SEA

The Red Sea is one of the most dynamic and interesting geological features on the Earth. Its location and continuing geological activity make it important in the geological history of both Africa and Asia, a major geographic and cultural area in world culture, and an important source of information about geological processes and their impacts.

PRINCIPAL TERMS

ASTHENOSPHERE: the layer of the mantle beneath the lithosphere; the asthenosphere exists in an almost "plastic" state and therefore behaves like a very thick liquid

CONTINENTAL SHELF: the gentle slope that extends from the coast into the ocean to a depth of about 500 meters

GRABEN: a down-thrown block of rock between two steeply angled normal faults

HYDROTHERMAL VENT: a location on the surface of the Earth where superheated liquid and gases are released because of volcanic activity

PANGAEA: the supercontinent that geologists hypothesize existed about 280 million years ago, when all the landmasses of the world were one

RIFT: a graben on a very large scale that results in a massive depression in the Earth's surface with steep sides; the most famous rift is the Great Rift Valley, which extends from Turkey in Asia to Mozambique in Africa

SALINITY: the amount of salts present within water, usually expressed in parts per thousand

SEISMIC ACTIVITY: movements within the Earth's crust that often cause various other geological phenomena to occur; the activity is measured by seismographs

GENERAL DESCRIPTION AND LOCATION

The Red Sea is one of several important seas associated with the Indian Ocean. It is almost totally landlocked by the Sinai Peninsula and Egypt to the north and the narrow (32 kilometers) strait of Bab el-Mandeb—which divides Yemen from the kingdom of Djibouti—to the south. The southern end connects with the Gulf of Aden, which, in turn, connects to the Arabian Sea and thence to the Indian Ocean. The Suez Canal, completed in 1869 after a ten-year effort, directly links the waters of the Red Sea with the Mediterranean Sea through the innovative use of three natural lakes and a set of canals.

The northern end of the Red Sea splits into two small gulfs, the relatively shallow Gulf of Suez as the western branch and the much deeper Gulf of Aqaba as the eastern branch. Eight nations (Egypt, Sudan, Eritrea, Djibouti, Yemen, Saudi Arabia, Jordan, and Israel) border one or more sides of the Red Sea, which creates a major divide between the continents of Africa and Asia. A number of islands (many of them actually exposed coral reefs) lie within its waters, particularly in the southern end.

A body of water with brilliant blue-green hue, the Red Sea likely takes its name from the occasional blooms of the red algae *Trichodesmium erythraeum*, which, upon death, give the waters a reddish-brown tint. Other explanations that have been proposed include the reflection of the sunburnished cliffs in its waters or the reddish skin color of peoples living near it. Records of ancient Egypt, Palestine, and Mesopotamia provide references indicating that the sea was termed "red" by the earliest civilizations within the region, although at that time it referred to an area extending all the way to the northwest coast of India (today's Indian Ocean).

GEOLOGY

The Red Sea, technically a graben, is part of the African rift valley system or Great Rift Valley, a major geological depression and fault zone on the Earth's surface. This is the point where the African and Arabian plates—two of the thirteen semirigid plates of continental crust that cover the Earth—have been slowly tearing apart (rifting) because of pressures exerted from deep within the asthenosphere that push up through the lithosphere. This large rift valley extends south from the Sinai Peninsula 3,500 kilometers to Tanzania and about 450 kilometers north through the Dead Sea-Jordan Rift Valley. The Red Sea valley itself cuts through a mass of Precambrian igneous (basalt) and metamorphic rocks known as the Arabian-Nubian Massif, the upper portions of which form the rugged mountains—technically known as steep fault scarps—that ring the sea. On top of the Precambrian rock strata sit layers of Paleozoic marine sediments laid down some 544 to 245 million years ago, as well as Mesozoic and Cenozoic sediments as much as 57 million years old.

The 30-million-year-old Red Sea was created when the Arab Peninsula began to break away from Africa and started to move north in two distinct phases. The northern part of the Red Sea was created over a process of about 10 million years. Subsequent geological movement that commenced about 3 to 4 million years ago created the much deeper Gulf of Aqaba and the southern half of the Red Sea. Even today the movement continues, adding about 15 millimeters per year to the width of the Red Sea and supporting unique forms of marine life among the hydrothermal vents at the deepest parts of the sea where extensive volcanism and seismic activity occurs. There are a number of active undersea volcanoes at the southern end just south of the Dahlak Archipelago, as well as a recently extinct volcano on the island of Jabal at-Ta'ir.

There are five major types of mineral resources present in the Red Sea region. Oil and natural gas deposits (more than 120 fields or discovery wells) have been found and extensively tapped near the junction of the Gulf of Suez and the main body of the Red Sea. Evaporites such as halite, sylvite, gypsum, and dolomite are mined along the Sinai Peninsula, although not in proportion to what is available. Sulfur deposits have been extensively mined since the beginning of the twentieth century, while the phosphate deposits are of such a low grade that present extraction techniques make them unattractive resources. There are extensive and valuable heavy metal deposits in the deep portions of the Red Sea along the Atlantis II Deep area, but these deposits are not yet commercially mined. Because they are found in the form of fluid oozes rather than solid rocks, it is believed that they may be able to be pumped up to the surface, although attempts to do so in more than a casual manner have so far proven elusive.

PHYSICAL FEATURES AND CLIMATE

The sea extends some 2,000 kilometers from north to south and is about 350 kilometers at its widest point near the southern end. The maximum depth is about 3,000 meters, although there is an extensive continental shelf around the periphery of the entire sea that is only a few hundred meters deep or less.

The Red Sea contains some of the Earth's hottest waters—the temperature averages 25 degrees Celsius—and some of the saltiest at forty parts per thousand. Little precipitation occurs in the entire area, although there is some evidence of greater precipitation during some periods in the past. A remarkable circulatory system maintains the water level in the face of very high evaporation rates of more than 2 meters per year along its length and breadth. The denser, saltier water of the north

The Red Sea at sunset. (National Oceanic and Atmospheric Administration)

sinks and flows south along the lower depths, while winds drive less dense and less salty waters (about thirty-six parts per thousand) above it north from the shallow southern end of the strait of Bab el-Mandeb. This results in a complete replacement of the sea's water about every twenty years. Underneath this exchange of salt water, very deep in the central trough, brine with an average temperature of almost 60 degrees Celsius and a salinity of 257 parts per 1,000 surfaces from underground. This upward movement adds to the general circulation within the Red Sea and measurably increases its overall salinity.

Fierce windstorms can arise suddenly, especially coming off the desert sands to the northwest, which are referred to as the Egyptian winds. Daily air temperature ranges from 8 to 28 degrees Celsius during the fall, winter, and spring, with scorching highs of up to 40 degrees Celsius in the summer season (July and August) accompanied by intense relative humidity.

Scientific and Economic Importance

The Red Sea provides scientists with a glimpse of geologic processes that likely occurred in the early stages of the formation of the Atlantic and Pacific Ocean basins during the breakup of Pangaea hundreds of millions of years ago. The continuing movement of the Arabian plate northward adds to the width of the Red Sea each year in a small but measurable way and provides a living laboratory to observe ongoing volcanic and seismic activity associated with the evolution of crustal margins, along with the action of hydrothermal vents. The Red Sea's complex water chemistry and geology have been the focus of many scientific expeditions in the modern era, including those conducted in such submersibles as the Swedish *Albatross* (1948) and the American *Glomar Challenger* (1972), the latter of which drilled and removed core samples from some of the sea's deepest locations.

The undersea vents and the extreme saltiness of the Red Sea support a unique and varied marine life, including spectacular coral reefs—notably the Protector Reef near Port Sudan, Ras Muhammad at the southern tip of the Sinai (one of the top ten diving spots in the world), and the Deadalus Reef parallel with Aswan, Egypt. Vibrantly colored exotic fish and an abundance of rare and endangered species of marine life found nowhere else in the world attract visitors from all over the globe. More than one thousand species of invertebrates, one thousand species of fish, and more than two hundred coral types have been identified within its teeming waters. Its natural beauty has been a lure for divers since ancient times, a popular subject of nature films, and a playground for underwater photography enthusiasts.

The Red Sea has been important since ancient times as a commercial and cultural waterway that provides access to and from Africa, Arabia, and the Indian subcontinent. This is despite the fact that its many reefs that lie just below the surface have brought more than one ship to the bottom and the need to keep the southern channel open at Bab el-Mandeb Strait by blasting and dredging on a regular basis. The Egyptians—from the time of

The Sinai Peninsula juts into the Red Sea in this satellite image. The Mediterranean Sea lies to the north, extending to the curved horizon. (PhotoDisc)

Ramses II and Seti I—used shallow canals to connect the Red Sea to branches of the Nile River delta and permit the easy passage of goods via sea rather than by an arduous land route through blistering desert sands. The Greco-Romans and early Muslim empires continued this practice until the end of the eighth century C.E. Today the modern Suez Canal has assured the Red Sea an important place in world commerce, although hostilities among Arab nations, Israel, and African nations sometimes led to complete closure of this important sea artery during the twentieth century. Major seaports in the Red Sea include Suez and Al Ausayr in Egypt, Jiddah in Saudi Arabia, Elat in Israel, Massawa in Eritrea, and Port Sudan and Sawakin in Sudan. Fishing, a major regional industry at more than 8,000 metric tons per year, is supported by the extensive reefs that plunge thousands of meters to the ocean floor.

Dennis W. Cheek

CROSS-REFERENCES

Aral Sea, 2193; Arctic Ocean, 2197; Atlantic Ocean, 2203; Black Sea, 2209; Caspian Sea, 2213; Deep Ocean Currents, 2107; Gulf of California, 2217; Gulf of Mexico, 2221; Hudson Bay, 2225; Indian Ocean, 2229; Mediterranean Sea, 2234; North Sea, 2239; Ocean Basins, 661; Ocean Tides, 2133; Pacific Ocean, 2243; Persian Gulf, 2249; Saltwater Intrusion, 2061; Surface Ocean Currents, 2171; Volcanoes: Climatic Effects, 1976.

BIBLIOGRAPHY

Coleman, Robert G. *Geologic Evolution of the Red Sea.* New York: Oxford University Press, 1993. A geological treatment of the Red Sea that describes in considerable detail—via both words and maps—the geological history of the Red Sea. The book is technical in nature.

Cousteau, Jacques-Yves. *World Without Sun.* Edited by James Dugan. New York: Harper and Row, 1965. A classic introductory treatment of the Red Sea including both geology and marine biology perspectives. Illustrated with a wealth of maps, diagrams, and color photographs of both geological features and marine life.

Edwards, Alasdair J., and Stephen M. Head. *Red Sea.* New York: Oxford University Press, 1987. A publication for the general public emphasizing the unique geological features and marine life of the Red Sea. The foreword by the duke of Edinburgh urges that the world concentrate its efforts on preserving this unique environment for future generations.

Lapidoth, Ruth E. *The Red Sea and the Gulf of Aden.* Hingham, Mass.: Kluwer, 1982. A detailed geological discussion of the Red Sea, with some attention to marine life.

Taylor, Leighton R. *The Red Sea.* Woodbridge, Conn.: Blackbirch, 1998. A presentation for adolescent readers that succinctly summarizes the state of knowledge about the geology and biology associated with this unique body of water. Color illustrations, maps, glossary, and an index aid reader comprehension.

Waterlow, Julia. *The Red Sea and the Arabian Gulf.* Austin, Tex.: Raintree Steck-Vaughn, 1997. A text suitable for younger readers that surveys both the biological and geological realms of the Red Sea.

6
LAKES AND RIVERS

AMAZON RIVER BASIN

The Amazon River Basin is the largest in the world, covering an area of more than 4,810,000 square kilometers. It is home to more plant and animal species than any other region in the world and contains thousands of species that have not yet been discovered or named.

PRINCIPAL TERMS

BASIN: the region drained by a river system, including all of its tributaries

DELTA: the area at the mouth of a river that is built up by deposits of soil and silt

FAUNA: the animal population of a region

FLORA: the plant species of a region

TRIBUTARIES: rivers that flow into larger rivers

VÁRZEA: the part of the rain forest that is flooded for up to six months per year

GEOGRAPHY

The Amazon River of South America is the world's largest river, though, at 6,400 kilometers, not quite the longest. Africa's Nile River is 6,650 kilometers long, about 200 kilometers longer than the Amazon. The amount of water flowing down the Amazon is, however, about sixty times greater than that of the Nile. The Amazon carries more water than the Nile, the Mississippi, and China's great Yangtze River combined, a total that adds up to more than one-fifth of all the river water in the world. Its basin, the region drained by a river and its tributaries, includes parts of Brazil, Bolivia, Peru, Ecuador, and Colombia and measures more than 7 million square kilometers, an area about three-fourths the size of the continental United States. More than a thousand tributaries flow into the Amazon, and the basin contains about 22,680 kilometers of navigable water.

There are two different types of tributaries, so-called black-water and white-water rivers. The black-water rivers, such as the Urubu, the Negro, and the Uatama, come from very old geologic areas that are poor in salts, very acidic, and filled with sediments. They are avoided by fishermen because they have small fish populations. The water is actually tea-colored because of heavy concentrations of tannic acid. The black-water rivers have few nutrients and carry much decaying vegetation from river banks. The decaying matter, chiefly leaves and branches, consumes much of the oxygen in the water. Because of low oxygen levels, these rivers support few forms of life; they are so

muddy that little light can penetrate and few plants can live in them, further reducing the food supply for fish. Indians living along these rivers call them "starvation rivers" because little food can be obtained from them. On the other hand, they can also support fewer insects and mosquitoes, making the areas around them more livable for human beings. The white-water, or clear-water, tributaries come from areas of little erosion. The waters from these streams, such as the Xingu and the Tapajos, are clear enough that the bottom is visible in many places.

The Amazon begins high in the Andes Mountains in Peru, less than 160 kilometers from the Pacific Ocean. There, the elevation is about 5,200 meters; for its first 970 kilometers, the Amazon headstream, known as the Apurímac River, drops through the mountains at a rate of about 5 meters per kilometer. It passes through grasslands used for grazing sheep and alpaca and by Cuzco, the old capital of the Inca Empire and the region's largest city. After passing through the Montana, an area of dense forests and steep valleys, the Apurímac flows into the Urubamba, forming a river called the Ucayali. It then heads north for about 325 kilometers, where the stream is joined by several smaller tributaries coming from Peru and Ecuador.

Where the Ucayali joins the Maranon, about 160 kilometers above the city of Iquitos, the middle course of the Amazon begins. There, the elevation is about 105 meters, and the river is more than 3 kilometers wide. It then travels through

405 kilometers of unpopulated territory before it enters Brazil. For the next 1,296 kilometers, the Amazon becomes a slow-moving, meandering stream, forming many huge lakes. There the Río Negro, named for its very dark waters, enters from the north. The Río Negro is 6.5 kilometers wide and carries water from Colombia and Venezuela. A few hundred miles downstream, the Madeira, the Amazon's longest tributary, enters from the south. Its waters come from as far away as La Páz, the capital of Bolivia, to the Paraguay River in far southern Brazil. Below the Madeira, the Amazon narrows to little more than 0.5 kilometer wide, but it is about 130 kilometers deep. It still has about 800 kilometers to go before it reaches the ocean.

The Xingu is the last major river to enter the Amazon, and its junction is considered the place the delta—the triangular part of a river formed by soil deposits just before it enters the sea—begins. The Xingu is about 1,620 kilometers long and rises in the Brazilian highlands. The Amazon Delta stretches along the Atlantic coast for about 325 kilometers to the north and inland for about 485 kilometers. The delta contains the island of Marajó. The city of Belem is in the delta and is the official port of entry for the Amazon basin.

CLIMATE AND FLORA

Most of the basin has a rainy tropical climate. Annual rainfall in the delta averages about 215 centimeters per year, with 180 centimeters in the middle region and 300 to 400 centimeters per year in the upper course. The middle course and the delta have a rainy season that lasts from December through May, during which as much as 45 centimeters of rain can fall every month. During this time, the floodplain, called the *várzea*, is soaked and becomes a giant lake that can be anywhere from 50 to 200 kilometers wide. Above the mouth of the Madeira, the Amazon flows through a different type of floodplain called the *terra firme*. There the soil is generally very poor in nutrients and unsuitable for agriculture. Even in the *várzea*, however, less than 2 percent of the floodplain can be used for farming. Mostly, there is a tropical rain forest of light woods, brush, wild cane, and grasses. The soil is poorly drained and often clayey. Cattle are grazed on Marajó Island, and Japanese farmers have brought in water buffalo, but little else can be grown or raised in the region.

There are more than thirty thousand flowering plant species in the basin. That number represents about one-third of all the flowering plants found in South America and three times the number found in all of Europe. The amount of vegetation in various areas of the river is influenced by the rise and fall of the water. Some plants are drowned during floods, while others thrive. During periods of low water, the floodplain has little vegetation. There are, however, grasses and wild rice that grow at low water. The grasses become floating islands during flood season, sometimes reaching sizes of more than 0.6 kilometer in length and several hundred meters wide. The islands are home to passionfruit, morning glories, and giant water lilies more than 0.6 meter in diameter. Ants, spiders, and grasshoppers also live on the islands. The root zone becomes a home for insects and an important source of food for fish. Many fish also eat the fruit of the plants. At low water, the islands are trapped in bushes and trees and the plants begin to rot, turning the water a deep, murky brown.

In the *várzea*, forest trees and bushes must adapt to survive during the flood season, which reaches its peak from May until July. During this period, plants become waterlogged, and the forests, or *igapo*, constantly are under water from 1 to 2 meters deep. In some cases, the shoreline is swamped all year, and many plants and trees will die. Young plants are often under water from seven to ten months per year. There, growth is also limited by lack of light, as bigger trees block out the light. In many cases, it takes seedlings ten to twenty years to reach a size that will enable them to survive. There are also other problems; for example, *kapok*, or silk-cotton trees, have been among the most successful plants in the *várzea*. This species has been almost entirely eliminated by loggers cutting down trees to be made into boxes.

FAUNA

About nine hundred species of birds live in the Amazon, about 10 percent of the total number of bird species on the planet. About 50 percent of these species are endemic, meaning that they are found in this region and nowhere else. The bird population includes hummingbirds, macaws, owls, parakeets, parrots, toucans, and many others.

Most species favor a particular level of rain forest, either on the ground or in the lower, middle, or upper level of trees. Only a few kinds of birds move between levels. The Amazon is the richest bird region in the world. There is also an enormous population of another kind of flying animal, bats, with more than fifty different species represented, making it the most diverse group of mammals in the region.

There are no crocodiles in the Amazon, but there is a smaller relative, the caiman. The black caiman can reach a length of 5.5 meters. Unfortunately, it is endangered by overhunting. Its shiny black skin, which is used for handbags and expensive shoes, is sought by hunters.

There are believed to be somewhere between 1,300 and 2,000 species of fish in the river; in comparison, the Congo River in Africa has an esti-

The Amazon River Basin in 1943 during a transport of survey crews with the Army Air 311th Air Photo Wing. (National Oceanic and Atmospheric Administration)

mated 560, and the Mississippi River has fewer than 300. There are so many species in the Amazon because there were no mass extinctions in the river caused by glaciers or ice ages. The huge size of the basin also provides for numerous kinds of different environments, each of which can be exploited by a unique species. There are rapids, waterfalls, lakes, and streams, each with different types of water and vegetation. Both freshwater and saltwater species are found. Lungfish, which have survived from the Paleozoic era 150 million years ago, are common.

About 43 percent of the fish are characoids, including piranhas and neons, which are familiar to many aquarium owners. Piranhas are a diverse group of twenty species; they are red-bellied and very dangerous. They hunt in groups of up to one hundred fish, and they are constantly in search of food. Normally, they are found in the floodplain lakes of the *várzea*. Various kinds of catfish and twenty species of stingrays are found in deeper water. One species, the piraiba catfish, is the largest fish in the river, reaching a length of 4.5 meters and a weight of 230 kilograms. The Amazon River is also home to many freshwater dolphins, which can reach lengths of 2.4 meters.

The phylum Arthropoda is well-represented in the Amazon region, with 1 million or more species in the area, a large number of them still unclassified. They are without a doubt the most successful animals in the rain forest. About 90 percent are insects (jointed invertebrates, or animals without backbones), 9 percent are spiders, and 1 percent are other, more obscure kinds of creatures. Insects include flies, beetles (the most diverse group), ants, scorpions, millipedes, centipedes, and symphylans. The latter are centipede-like animals that live on the floodplain floor. The floor-dwelling species generally migrate upward into the trees when the rainy season begins. In the middle and upper tree levels are spiders, moths, butterflies, and various other winged species. The most abundant species at all levels are the ants, many of them adapted to living on one particular kind of tree or plant at one specific level. Millions more live on the ground, where huge numbers of mollusks, segmented worms, and flatworms also thrive.

Several types of lizards live in the basin, iguanas being the best known. Iguanas are typically green in color but can quickly change color to blend in with their background. They have long toes and sharp nails that enable them to climb trees, and most live at the tops of trees. Iguanas are vegetarians and can reach a size of more than 0.3 meter in length. Another kind of lizard, the teiid, can reach lengths of 1.5 meters. There are only a few frog species in the region, primarily because they have difficulty competing with fish for food and are easy prey for piranhas and other flesh-eating fish. There are more toads than frogs in the Amazon, as toads can more easily hide in the leaf litter that covers the ground.

The Amazon is home to the world's largest rodent, the capybara, which looks something like a large guinea pig and lives in groups of a dozen or more along the riverbanks, feeding on grasses and other plants. The capybara is being rapidly reduced in numbers, however, because it is intensively hunted for food by native peoples. Other rodents include the torospiny rat, a common night hunter; tree-dwelling porcupines; anteaters; leaf-eating sloths; five species of opossums; and several species of armadillos. Most of these groups migrated into the basin from the north when Central America became connected to South America via Panama from 3 million to 4 million years ago.

There are some hoofed mammals in the basin, but most species of this type have become extinct. The tapir is the oldest hoofed mammal in the basin and the only one that enters the river, usually for bathing and to eat fish. Monkeys are abundant in the region, with more than forty species found. There are two main groups. Marmosets and tamarins form one group; they live in small groups in trees and are most active during the day. Cebids, the other major group, include capuchins and howlers. They are much larger in size than species in the first group, have large brains, and resemble African apes.

There are 160 species of snakes in the Amazon; most live on the floodplain or on the riverbank. Only the coral snakes and vipers are poisonous. The area's pit vipers are among the most dangerous snakes in the world. They hunt at night, reach lengths of 1 meter, and live on a diet of small animals. The anaconda is the largest snake in the basin, reaching a length of 6 meters. It is not poisonous but kills by squeezing its victims to death. It is most active at night, grabbing birds, rodents, turtles, or caimans by the neck and coiling around and crushing them until they are soft enough to digest. Boa constrictors use the same method of capturing their food and are common in the water and on the shore.

Manatees, the largest animal in the Amazon, weighing 450 kilograms or more, are found in large numbers. These relatives of the elephant eat grasses and plants and live in the water. They can stay under water for more than an hour before coming to the surface to breathe. Other animals include coatis, which resemble raccoons; several kinds of deer; and peccaries, or wild hogs. Human beings were the last species to enter the basin, perhaps about ten thousand years ago.

Leslie V. Tischauser

CROSS-REFERENCES

African Rift Valley System, 611; Earth's Crust, 14; Faults: Normal, 213; Ganges River, 2265; Great Lakes, 2270; Hot Spots, Island Chains, and Intraplate Volcanism, 706; Hydrothermal Mineral Deposits, 1584; Hydrothermal Mineralization, 1205; Hydrothermal Vents, 2117; Lake Baikal, 2275; Marine Terraces, 934; Mid-Ocean Ridge Basalts, 657; Mississippi River, 2280; Nile River, 2285; Plate Margins, 73; Plate Tectonics, 86; Reefs, 2347; Yellow River, 2290.

BIBLIOGRAPHY

Davis, Wade. *One River: Explorations and Discoveries in the Amazon Rain Forest.* New York: Simon & Schuster, 1996. A history of human exploration of the region. Includes much information on local peoples and the impact of modern economic development. Well written, with many photographs and a useful index.

Furneaux, Robin. *The Amazon: The Story of a Great River.* New York: Putnam, 1970. A still-useful resource, with many photos and detailed descriptions of major plant and animal species. Recounts the history of human settlement and exploration in the region.

Malkus, Alida S. *The Amazon: River of Promise.*

New York: McGraw-Hill, 1970. Especially written for younger readers. Contains a wide amount of useful and interesting information about the major plant and animal species in the area.

Moran, Emilio. *Developing the Amazon.* Bloomington: Indiana University Press, 1981. An economist's view of development in the river basin. Pays some attention to the ecological losses caused by development. Written mainly for specialists in development; however, contains some excellent maps and descriptions of the region's geography.

Smith, Nigel J. H. *Man, Fishes, and the Amazon.* New York: Columbia University Press, 1981. A brief but detailed exploration by an anthropologist who lived in the basin for several years. Contains a lengthy discussion of the different kinds of fish found in the river and how they are trapped and caught by local Indians. Ends with a plea for respect for the traditional lives of native peoples. Also has information on climate and the impact of industrialization on the region. Interesting descriptions of traditional ways of life, some photographs, and a detailed index.

GANGES RIVER

The Ganges River flows from the Himalayas through the northern and eastern portions of India, then through Bangladesh to the Bay of Bengal. Although the Ganges is not a long river, it is the second muddiest river in the world, depositing sediment along the fertile floodplains and at the delta. The Ganges system annually floods and periodically causes great loss of life and damage.

PRINCIPAL TERMS

AVULSION: a natural change in a river channel, usually caused by flooding and excess deposition of sediment

BASE FLOW: the natural flow of groundwater into the river, which commonly maintains the flow of perennial rivers during the dry season

CAPACITY: the total amount of sediment transported by a river at a given time

COMPETENCY: a measure of the largest particle that a river can carry at the time of measurement

DELTA: a landform created when a river enters a still body of water and deposits much of the sediment it was transporting; deltas may form in lakes, but the largest deltas form where a silt-laden river flows into the ocean

DISCHARGE: the volume of water flowing past a measurement point in a river during a given time interval; usually the discharge is measured in cubic meters per second

DISTRIBUTARY: a river that diverges from the main river and carries water across a delta

SUSPENDED LOAD: the total amount of sediment carried in suspension (floating) in a river; suspended load may be measured over any time period but commonly is determined on an annual basis

TRIBUTARY: a smaller river that flows into the main river; a tributary may be a significant river in its own right, but it loses its identity when it merges with the main river

HISTORY

The Ganga River is known in the Western world by its anglicized name of the Ganges River. The Ganges River has been one of the most important Asian rivers in recorded human history. Indian civilizations dating back to the kingdom of Asoka in the third century B.C.E. have developed along the Ganges River or in the floodplains of the Gangetic Plain. Today, millions of people in India and Bangladesh rely on the waters of the Ganges River. The population of the Gangetic Plain is now more than 380 million people, making it the most heavily populated river basin in the world. It is estimated that the basin population will exceed 500 million people by 2010.

The Ganges River is the holiest of all rivers for the Hindus, many of whom travel miles to bathe in its waters at Varanasi. Bathing in the river at India's oldest living city is considered a purification act, and for many it is the culmination of one's lifetime. Hindus also believe that dying and hav-ing one's ashes scattered on the Ganges will ensure *moksha*, the release from the constant cycle of death and reincarnation. Because of these beliefs, there are numerous crematoria along the Ganges at Varanasi, and many ashes and bodies are given over to the river.

GEOGRAPHY OF THE GANGES

The Ganges River arises from the runoff from the southern side of the Himalayas, the world's highest mountain range. The Ganges River flows about 2,500 kilometers across the Gangetic Plain and through a coastal delta in Bangladesh to empty into the Bay of Bengal. The river system drains a basin totaling more than 580,000 square kilometers, nearly one-quarter of the area of India. The average discharge of the Ganges is about 11,610 cubic meters per second, but floods can increase the discharge by a factor of five or more.

Many believe that the source of the river is an ice cave at the base of a glacier at Gangatri, from which the Bhagairathi River flows. The true source of the Ganges is, however, considered to be some 21 kilometers to the southeast at Gaumukh. After flowing southward off the southern face of the Himalayas, the Ganges River flows predominantly eastward along the front of the Himalayan system, later turning southward as it makes its final way to the sea. The headwaters of the Ganges consist of five rivers that all arise in the northern Indian state of Uttar Pradesh. The main tributaries are the Bhagirathi, Alaknanda, Mandakini, Dhauliganga, and Pindar Rivers. The Bhagirathi and the Alaknanda Rivers merge at Devaprayag to form the main stream of the Ganges River. The Ganges then cuts its way through the southern outer mountains of the Himalayas and emerges from the mountains at Rishikesh. From Rishikesh, the Ganges flows out onto the Gangetic Plain at Hardiwar and continues eastward, still within the state of Uttar Pradesh. From this area eastward, the Ganges River is joined by the Yarmuna River—which flows past the capital city of New Delhi—as well as the Tons, Ramganga, Gomati, and Ghagara Rivers. The Ganges River then flows eastward into the Indian state of Bihar, where it is joined by the Gandak, Buhri Bandak, Ghugri, and Kosi Rivers, all of which arose in the mountains to the north.

These rivers provide more than 40 percent of the entire flow of the Ganges River. The surface waters of the area south of the Ganges are drained into the main river primarily by the Son River. After flowing northward around the Rajmahal Hills, the Ganges River turns southward and flows into the state of West Bengal. The river here is generally known as the Padma River and enters the delta region at Farakka. Flowing eastward into the delta and joining the main river is the Hooghy (or Hugli) River, near which the city of Calcutta was built. To the south, the Ganges

River and the great Brahmaputra River merge and flow southward into the Bay of Bengal. The name of Padma is given to this part of the river, where many tributaries and many distributaries carry the water across the wide delta in Bangladesh.

HYDROLOGY

The hydrologic regime in the Ganges River system is highly seasonal, with up to 80 percent of the rainfall occurring in the interval from July through October. This heavy rainfall is caused by the southwesterly monsoons that cross the region during this time. Significant quantities of water are also added by the melting of snow in the Himalayas during the spring months. Rainfall, however, varies considerably over the entire river basin. In the western part of the basin, precipitation averages about 750 millimeters per year, but up to three times that much falls in the eastern part of the basin. In addition to the annual rainfall and snowmelt, water is added to the rivers by the base flow of the groundwater systems that are contiguous with the river.

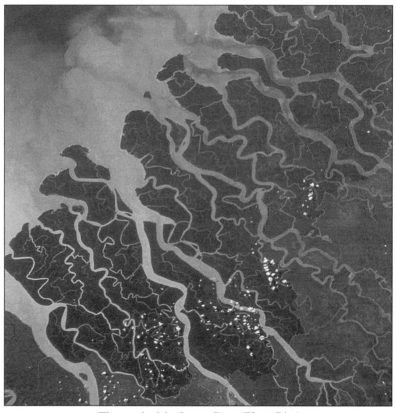

The mouth of the Ganges River. (PhotoDisc)

The strong seasonal nature of precipitation is shown in the variations in the discharge of the Ganges system throughout the year. Discharge at the Hardinge Bridge in Bangladesh averages 39,400 cubic meters per second in August but may rise to a maximum of more than 53,000 cubic meters during flood stage. In contrast, the mean discharge at the same location during the height of the dry season from February through May is slightly more than 21,000 cubic meters. Most of the flow during the dry season is derived from the northern tributaries of the Ganges River.

In addition, the delta region is commonly buffeted by severe cyclonic storms that generally just precede the onset of the monsoon season or occur at the end of the main rainy season. These storms have been responsible for many of the very damaging and deadly floods in the delta region.

SEDIMENTATION

The rivers that flow out of the highlands of the southern Himalaya generally have high velocity and therefore a very high competency. The land in many areas is highly erodible, particularly if recently deforested; as a result, the rivers carry large amounts of sediment in suspension and bed load. Based upon suspended load calculations, the Ganges River is second behind only the Yellow River in China in terms of sediment load. The average suspended load is about 1.4 billion metric tons per year. This leads to an average of approximately 1,500 metric tons of suspended sediment per square kilometer of the river basin. Although this is high, it is not as high as the Yellow, Ching, or Lo Rivers of China. The very gentle gradient of the Ganges River after it leaves the Himalayas results in relatively little erosion and a low production of suspended load. In contrast, the Kosi River, a northern tributary to the Ganges, has an annual suspended load of only 172 million metric tons per year. However, because of its high gradient and relatively small drainage basin, the average sediment load per square kilometer of basin is almost twice that of the Ganges River as a whole. The high sediment load, coupled with large-scale flooding caused by the rapid runoff from the Himalayas, has led to the frequent avulsion of the Kosi River. Its channel shifted more than 100 kilometers in the years between 1736 and 1964.

The gradient of the river across the Gangetic Plain (south of Himalayas to the delta) is very shallow. The 1,600-kilometer stretch of river from the Yarmuna River at New Delhi to the Bay of Bengal drops only about 1,100 meters in elevation. This shallow gradient leads to low water velocity and rapid deposition of sediment in the river channel and in the floodplains during flood stage. The combination of rapidly rising, very high mountains as a sediment source and a slowly moving major river leads to rapid aggradation of the floodplain surfaces along the river. The Gangetic Plain is composed of this rapidly deposited sediment and is, in some places, up to 2,000 meters thick. Although the age of the bottom sediments in the basin is not definitely known, it is probable that the deposition of this thick accumulation of sediments was rapid, perhaps occurring within the last 10,000 years.

Sediment that is transported to the coastal area by the Ganges and Brahmaputra Rivers is deposited in the deltaic area. The delta built by these two great rivers is huge, totaling almost 60,000 square kilometers. Like most deltas, this landmass has been built up by the deposition of silts and sands along the main channel and in the many distributaries that channel water to the ocean. The areas between the channels are often heavily vegetated, which leads to the formation of peat when the plants die and are buried in the swampy areas. In some places, the peat has been mined and used as an energy source. The sediments that compose the delta include sand, silt, peat, and marl. Because of the heavy sediment load and the seasonally variable discharge of the river system, the delta has grown larger and extended further seaward throughout historic times. New land areas, called *khadar*, are generally formed in the areas near the present-day channels where the river water meets the oceanic waters. As sediment accumulates, it blocks water flow and forces the river to change its course. In this way the sediment depositions continually change location, and the delta continues to prograde out into the Bay of Bengal. The current flow pattern of the river is toward the eastern side of the delta in Bangladesh, causing the buildup of land areas and some offshore islands. The western side of the delta has remained essentially unchanged since the eighteenth century. The delta will continue to grow as long as water and sediment are transported by the river to the coast. However, with increasing population

The Ganges River teams with boat traffic. (PhotoDisc)

growth and the resultant increase in agricultural and urban use of water, the total flow of the Ganges and Brahmaputra Rivers may be significantly reduced. This reduction will lead to the eventual erosion of the coastal delta.

The seaward edge of the delta is characterized by a vast area of tidal forests and swamps that contains immense biological diversity. The tidal forests—known as the Sudarbans—have been declared protected areas by both the Indian and Bangladesh governments, which have instituted conservation measures to preserve the biota, including the Bengal tiger.

FLOODS

As in most areas, flooding has had both beneficial and detrimental effects on the land and the people in the floodplain of the Ganges River. The floods bring down sediment and nutrients into the lower reaches of the river and are responsible for building a fertile floodplain. Records from the sixteenth and seventeenth centuries indicate that the Ganges basin was once heavily forested and was home to a wide variety of animals and plants

that are no longer found in the area. Deforestation of the area led to the loss of the animal life; the floodplain today sustains a high level of agricultural productivity. The rapid increase in human population in the region has also led to increasing residential use of the floodplain and therefore a growing flood risk to the region. There are now more than 17 million people who are directly susceptible to flooding along the Ganges.

Even though the area of the state of Bihar near the Ganges is swampy land, the population has grown in the region. Settlement in the floodplain has led to extensive loss of life and property damage during the frequent flooding of the area. Flooding in the upper part of the basin within Bihar is responsible for more than one-half of India's annual flood toll of 1,200 people. There is generally some damage each year from flooding, but major flood events have occurred in 1890, 1898, 1899, 1922, 1924, 1954, 1974, 1978, 1987, and 1988. The 1987 floods resulted in more than 400 deaths, the destruction of 70,000 homes, and damage to another 350,000 homes. The government of Bihar blamed the apparent recent increase in flooding in the region on Nepal, arguing that deforestation of upstream regions led to higher levels of runoff and flooding of the downstream basin. Although it is known that deforestation can lead to downstream flooding, scientific evidence does not clearly indicate that this has been the case for the more recent Bihar floods.

Flooding is also common in the lower reaches of the Ganges and Brahmaputra Rivers. The land in the delta area of Bangladesh is only a few meters above sea level; when the rainy season arrives, the area is commonly inundated under 1 meter or more of water. To protect against these annual floods, many of the villages and homes are built upon hills that have been constructed by the local people. The delta area is also hit by cyclonic storms that come out of the Bay of Bengal. These storms cause havoc to the low-lying areas. A cyclonic storm in November, 1970, resulted in catastrophic flooding and caused more than 200,000 deaths.

WATER RESOURCES

The Ganges River is a major water resource for the northern and eastern portions of India. It has been utilized for drinking water, transportation, and waste disposal for thousands of years. The waters of the great river have been diverted for irrigation purposes for hundreds of years. However, the great surge in population growth in India (which many believe will become the most populous country on the Earth by the end of the twenty-first century) and Bangladesh has led to huge demands upon the river system. India's heavy upstream use of fresh water for irrigation has dramatically reduced river flow for irrigation in Bangladesh and through the delta to the ocean. The loss of fresh water flow to the delta has resulted in a landward migration of sea water, leading to the salinization of the coastal lands.

The Ganges River basin is shared by Nepal, India, and Bangladesh, and water rights questions have arisen concerning the share of water to which each country is entitled. The greatest conflict resulted when India built a barrage at Farakka in the early 1970's to divert water to Calcutta. The diversion facilitated transportation to the city, but, in turn, reduced river flow into Bangladesh. The Bangladesh government began a series of negotiations with the Indian government in order to obtain an equal share of the Ganges water. After many frustrating years, the two governments were finally able to craft a water-sharing agreement in 1996 that gave each country a 50 percent share of the water.

Jay R. Yett

CROSS-REFERENCES

Amazon River Basin, 2261; Andes, 812; Deltas, 2312; Floodplains, 2335; Great Lakes, 2270; Lake Baikal, 2275; Mississippi River, 2280; Nile River, 2285; River Bed Forms, 2353; River Flow, 2358; River Valleys, 943; Yellow River, 2290.

BIBLIOGRAPHY

Crow, Ben. *Sharing the Ganges: The Politics and Technology of River Development.* Thousand Oaks, Calif.: Sage, 1995. Crow discusses the political tension that exists between India and Bangladesh regarding the distribution of the flow of the Ganges River. Bibliography and index.

Czaya, Eberhard. *Rivers of the World.* New York: Cambridge University Press, 1983. This is a good, general introduction to the major rivers of the world and contains some information about the Ganges River.

Pollard, Michael. *The Ganges.* New York: Benchmark, 1998. This book, intended for juveniles, traces the course of the Ganges River and describes its physical features, history, and importance. Bibliography and index.

Postel, Sandra. *Pillar of Sand.* New York: W. W. Norton, 1999. This book details the water resource problems of world and gives good detailed information about the water rights issues of India and Bangladesh. Additional information on the salinization of the delta area is also provided.

Shukla, Ashok Chandra, and A. Vandana. *Ganga: A Water Marvel.* New Delhi: Ashish, 1995. The authors focus on the environmental problems faced by the Ganges River, as well as efforts to protect it from further degradation. Bibliography and index.

GREAT LAKES

The Great Lakes represent the largest freshwater lake complex on Earth. Created by continental glaciers over the past 18,000 years, the five lakes provide significant resources for Canadians and Americans occupying the surrounding basin.

PRINCIPAL TERMS

EPILIMNION: a warmer surface layer of water that occurs in a lake during summer stratification; during spring, warmer water rises from great depths, and it heats up through the summer season

GREENHOUSE EFFECT: a natural process by which water vapor and carbon dioxide in the atmosphere absorb heat and re-radiate it back to Earth

ISOSTATIC REBOUND: a tendency of Earth's surface to rise after being depressed by continental glaciers, without faulting

PLEISTOCENE: a geologic era spanning about 2 million years that ended about 10,000 years ago, often considered synonymous with the "Ice Age"

SEICHE: rocking motion of lake level from one end of the lake to the other following high winds and low barometric pressure; frequently, a seiche will follow a storm event

STORM SURGE: a rapid rise in lake level associated with low barometric pressure; the water level is frequently "pushed" above a shoreline on one end of the lake and depressed on the opposite end

THERMOCLINE: a well-defined layer of water in a lake separating the warmer and shallower epilimnion from the cooler and deeper hypolimnion

WETLANDS: areas along a coast where the water table is near or above the ground surface for at least part of the year; wetlands are characterized by wet soils, water-tolerant plants, and high biological production

GEOLOGICAL DEVELOPMENT OF THE REGION

The Great Lakes are superlative features on the North American landscape. They make up the largest freshwater lake complex on Earth and represent about 18 percent of the world's water supply. Covering a total area of 245,000 square kilometers, the Great Lakes have a shoreline length of 17,000 kilometers. Lake Superior (82,100 square kilometers), Lake Huron (59,600 square kilometers), and Lake Michigan (57,800 square kilometers) are among the ten largest lakes on Earth.

The rocks forming the foundation of the Great Lakes date back some 600 million years. On the northern and northwestern shore of Lake Superior are remnants of the Canadian Shield, which is composed of igneous rocks of the Precambrian era, some 1 billion years ago. Following volcanic activity and mountain building during the Precambrian era, the central region of North America was repeatedly covered by shallow tropical seas. At this time, the Paleozoic era (600 million to 230 million years ago), sediments transported by rivers from adjacent eroding uplands were deposited in a shallow marine environment, and lime, salt, and gypsum were precipitated from seawater. All these soft materials were eventually hardened into sedimentary rock layers such as sandstone, shale, limestone, and halite. A multitude of fauna colonized the submarine environment, including corals, brachiopods, crinoids, and several species of mollusks.

As the layers of sediments accumulated over millions of years, the basin began to subside at its center. The Great Lakes basin structure may be compared to a series of bowls, one stacked on top of another. As viewed from above, only the top bowl is completely visible; however, the rims of the progressively deeper bowls are visible as a number of thin concentric rims along the perimeter of the basin.

The Paleozoic era was followed by the Mesozoic era (230 to 63 million years ago), a time of little deposition. In spite of the great age of the rocks making up the foundation of the Great Lakes, the

Lake Michigan. (PhotoDisc)

has a unique history and time of formation, making generalizations difficult. For example, the glaciers repeatedly advanced and retreated from many directions, covering and exposing each lake basin. There is abundant evidence regarding the size, elevation, and precise geographical distribution of each ancestral lake; this historical information is documented by coastal landforms such as higher ancient shorelines and relict wave-cut features. Yet the changing of the lakes' outlets to the ocean and reversals of drainage patterns complicate the sequence of events. Furthermore, because of the weight of the ice, the Paleozoic bedrock subsided. As the glaciers receded, exposing different segments of the basin, the land began to rise. This process, isostatic rebound, is active today, causing elevation changes of many fossil shorelines. Although uplift has slowed since the ice exposed the newly created Great Lakes, the process is continuing.

As the ice began to retreat from the region, glacial landforms were deposited. Along many shorelines, moraines—composed of fragments of rock, sand, and silt—form spectacular bluffs. Along Lake Superior, the ice scraped and removed much of the soil, exposing the bedrock, which now forms high cliffs. Sand eroded from glacial sediment was transported by rivers to the lakes and deposited as beaches. The exposed beach sand was then transported inland by the wind to form coastal dunes.

WEATHER AND CLIMATE

Weather and climate influence several processes occurring in the Great Lakes, including changing lake levels, storm surges and related seiches, and lake stratification and turnover. Through the hydrological cycle, moisture is evaporated from the lake surfaces and is then returned as precipitation over the water and as runoff from the land. During cooler and wetter years, evaporation is retarded, and more water is contributed to

lakes themselves were created in the relatively recent Pleistocene epoch. Between the 220 million years when the basin's bedrock was deposited and the onset of Pleistocene glaciers, the landscape now occupied by the individual lakes was occupied by streams. The streams sought and eroded the softer bedrock. The divides between and parallel to the eroded valleys were more resistant to erosion and are represented by higher elevations.

Streams excavated shales, and weaker limestones now occupied the Lake Michigan and Lake Huron basins. An arc, composed of hard dolomite rock, separates the east side of Lake Huron from Georgian Bay (the Bruce Peninsula). The same structure continues counterclockwise across Michigan's upper peninsula, separating Lake Michigan from Green Bay as the Door Peninsula. The ancient stream channels were favored by the glaciers because they were at lower elevations and composed of more erodable bedrock. The linear shape of the lower Great Lakes is clearly related to the initial erosion by streams followed by the continental ice.

Lake Superior is partly located on the Canadian Shield, and its geologic origin is less obvious. East-west faults underlie Lake Superior, and the rocks form a structural sag, or syncline, oriented along the long axis of the lake in an approximately east-west direction.

The glacial origin and development of the Great Lakes is complex for several reasons. Each lake the lakes by excess precipitation, causing lake levels to rise. In warmer and drier years, evaporation increases, and precipitation is retarded, causing lower lake levels. Such changes in water levels are not cyclic and occur over several years. In 1988, Lake Huron and Lake Michigan had record levels of 177.4 meters above sea level. In 1995, the average water level was 176.3 meters above sea level, a difference of 1.1 meters.

Unstable weather conditions generate storms that pass over the region, generally from west to east. When strong winds persist for several hours from a constant direction over a lake, the water level is "pushed" from one side of the lake to the other. This storm surge, accompanied by low atmospheric pressure, may elevate the water level as much as 2 meters along a shoreline in a matter of a few hours. Gale-force winds on October 30-31, 1996, over Lake Erie raised the lake level 1.25 meters at Buffalo, New York. Concurrently, as the water rose at Buffalo, it was lowered in Toledo, Ohio, at the opposite end of the lake, by 2.25 meters. The total difference of water level was 3.50 meters. Following a storm, the level of a lake rocks back and forth as a "seiche" before settling to its normal level.

In turn, the lake waters dramatically affect the local weather. As winter approaches, lake-effect snows commonly occur. The effect is most common in the fall, before the lakes cool and freeze. Cold winds from the north or west pass over the basin, picking up moisture from the relatively warm lakes. The water vapor is then condensed, forming clouds that, in turn, dump heavy snows in coastal zones, especially along eastern Lake Michigan and southern Lake Superior, Lake Huron, Lake Erie, and Lake Ontario. From November 9 to November 12, 1996, 1.2 meters of snow fell along Lake Erie's south shore, paralyzing local communities.

With the exception of Lake Erie, the lake bottoms were scoured by glaciers to depths below sea level. The lakes thus have variable temperatures from the surface to the bottom. As the water temperature changes from season to season, water density is altered. During winter, as ice forms over the lakes, the water beneath the ice remains warmer. As the ice cover breaks up in spring, the deeper warmer water rises, or "turns over," to the surface. It heats up through the summer months, causing stratification of warmer water (or "epilimnion") above colder, denser water. The contrasting water layers are separated by a thermocline, demarcating a rapid temperature transition between the warmer epilimnion and cooler subsurface water.

An issue of concern to scientists regarding the Great Lakes is the impact of the "greenhouse effect" on the lakes' water levels. The increase of greenhouse gases in the atmosphere, especially carbon dioxide, appears to cause warming at unprecedented rates. Although climatologists differ in their opinions as to the impact of a warmer atmosphere over the Great Lakes, there is general agreement that both evaporation and precipitation will increase and stream runoff will decrease over many years. Based on general circulation models, it appears that lake levels will be from 0.5 meter to 2 meters lower than present levels if the climate continues to warm.

WETLANDS

Wetlands along the shorelines of the Great Lakes are significant ecological zones located at the meeting of the land and lake. Although many

A Lake Superior shoreline. (National Oceanic and Atmospheric Administration)

wetlands around the basin have been lost or degraded, the remaining habitat has multidimensional functions as part of both upland and aquatic ecosystems. The wetlands are exposed to both short-term (storm surges and seiches) and long-term changes in water levels that constantly change the biogeography of these habitats. Because of the state of constant water-level change, or "pulse stability," the distribution and types of wetland plants shift dramatically. Thus, a constant renewal of the flora is occurring. Furthermore, because of the flushing action of the rise and fall of lake levels, peat accumulation in Great Lakes wetlands does not commonly occur, as it does in marine settings.

The coastal wetlands serve significant ecological, economic, and social functions. They provide spawning habitat and nursery and resting areas for many species, including fishes, amphibians, reptiles, ducks, and fur-bearing mammals. Largely because of the sport fishing industry, these habitats contribute significant revenue to the surrounding states and to the province of Ontario. Furthermore, pollution control and coastal erosion protection are additional benefits provided by these habitats. In 1978, the annual value gained by Michigan's coastal wetlands was estimated to be $5,066 per hectare.

STUDY OF THE GREAT LAKES

The creation and development of the Great Lakes have, in terms of geologic time, occurred relatively recently. Also, modifications such as erosion and deposition of coastal features are continual, active processes. To unravel the events leading to changes in the Great Lakes, scientists use techniques that include varve analysis and radiocarbon dating. A common theme of both techniques is that they express time in numbers of years within a reasonable range of accuracy rather than in a relative or comparative way. Varves consist of alternating light and dark sediment layers deposited in a lake. A light-colored mud is deposited during spring runoff; a dark-colored mud is deposited atop the lighter-colored layer during the following winter, as ice forms and there is less agitation of the lake water. One light and one dark band together represent one year of deposition. Numerous layers can be counted, like tree rings, and the number of years that were required for a sequence to be deposited can be determined. To obtain numerous undisturbed varve layers, researchers use a piston-coring device, which consists of a hollow pipe attached to a cable that is released vertically into the lake. As it free falls to the bottom, it plunges into the soft sediment. A piston allows the sediment to remain in the pipe as it is raised by an attached cable. The mud can then be extracted from the tube, cut open, and analyzed.

Radiocarbon dating of carbon-rich material such as peat, lime, coral, and even bone material is useful for absolute dating back to about 50,000 years ago. Carbon's abundance in nature, coupled with the youthfulness of the Great Lakes, makes this tool very useful because many glacial, coastal, and sand dune landforms frequently contain some form of carbon suitable for absolute dating.

To map and detect recent changes in the landscape such as coastal erosion or the rate of dune migration, old maps, navigation charts, and aerial photographs are used. Charts and maps of the coastal zone have been available for more than a century, and aerial photographs of the region have been taken since the 1930's. By observing the position of a shoreline on historical sets of detailed aerial photographs over a ten-year period, for example, the changes in the shoreline can be detected, and the erosion rate per year can be determined.

Geographic positioning systems can accurately locate the latitude and longitude of a point on a shoreline, store the information, and compare the shoreline position with the position at some future time. Satellite pictures help to detect wetland types and determine acreage; this information can then be compared to a later environmental condition, such as a period of higher lake level, to see if species or acreage have changed.

Because of a geological process known as isostatic rebound, fossil shorelines are uplifted. Elevations retrieved from older topographic maps reveal how much uplift has occurred. If the age of a relict shoreline can be determined with radiocarbon analysis, the rate of glacial rebound in millimeters per century can be assessed.

C. Nicholas Raphael

CROSS-REFERENCES

BIBLIOGRAPHY

Biggs, Donald L., ed. *Centennial Field Guide, North-Central Section of the Geological Society of America.* Vol. 3. Boulder, Colo.: Geological Society of America, 1987. Includes twenty-four papers on the geology and landforms of the Great Lakes basin. Features are treated in detail. The discussion is well supported by good maps, photographs, and tables. Suitable for college-level students.

Bolsenga, S. J., and C. E. Herdendorf. *Lake Erie and Lake St. Clair Handbook.* Detroit: Wayne State University Press, 1993. A thorough description of the natural history of Lakes St. Clair and Erie. Included are well-illustrated sections on geology, climate, hydrology, and wetland flora and fauna. A useful reference manual, with numerous maps, tables, and diagrams and a glossary.

Dorr, J. A., Jr., and D. F. Eschman. *Geology of Michigan.* Ann Arbor: University of Michigan Press, 1970. A college-level treatment of the geology of Michigan. Also included are chapters on sand dunes, shorelines, glaciation, and the creation of the Great Lakes. The text includes many photographs keyed to locations on inset maps.

Douglas, R. J. W. *Geology and Economic Minerals of Canada.* Report Number I. Ottawa: Geological Survey of Canada, 1970. Details the regional geology of Canada. Chapters on the Canadian Shield and the geology of the Great Lakes are included. Black-and-white aerial photographs illustrate significant glacial landforms. Intended as a reference work for professionals.

Hough, J. L. *Geology of the Great Lakes.* Urbana: University of Illinois Press, 1966. This 313-page book is a synthesis of research up to the mid-1950's. It includes sections on the present processes in the lakes and their glacial history. The text is jargon-free; it is intended for college students yet suitable for the general reader.

Le Sueur, Meridel. *North Star Country.* 2d ed. Minneapolis: University of Minnesota Press, 1998. This book provides a historical account of the environment, geology, and social aspects of the Great Lakes region. The chapters cover ecology and preservation programs. Appropriate for the high school reader.

Leverett, Frank, and F. B. Taylor. *The Pleistocene of Indiana and Michigan.* Washington, D.C.: U.S. Geological Survey, 1915. This 529-page tome is the classic publication detailing the origin of the Great Lakes. A professional publication, yet suitable for a cross section of audiences.

MacKenzie, Susan H. *Integrated Resource Planning and Management.* Washington, D.C.: Island Press, 1996. Filled with useful maps, this book provides an illustrated description of the ecosystem of the Great Lakes. There is a heavy focus placed on the management of the ecosystem and policies surrounding watershed management.

Paull, R. L., and R. A. Paull. *Geology of Wisconsin and Upper Michigan.* Dubuque, Iowa: Kendall/Hunt, 1977. A readable presentation of the geology of the region. Also included are sections on unique features such as minerals, fossils, and karst topography. Includes a glossary of some three hundred terms. A good introduction to the Great Lakes.

U.S. Environmental Protection Agency. *The Great Lakes: An Environmental Atlas and Resource Book.* Chicago: Great Lakes Program Office, 1995. A free soft-cover publication designed for anyone interested in the lakes. Designed as a public-education tool; the material briefly addresses the physical and cultural environments of the basin. Includes colored maps, tables, and charts.

LAKE BAIKAL

Lake Baikal in southeastern Siberia is the deepest lake in the world and the eighth largest. It has some of the cleanest, coldest fresh water in the world; until recent times, its waters have been among the most pollution free on the Earth.

PRINCIPAL TERMS

BAYKALIA: the geographic region surrounding Lake Baikal

ENDEMIC: found in a particular place and no other

PHYTOPLANKTON: tiny plants that make up the lowest part of the food chain

PLANKTON: forms of microscopic life that drift or float in water

RIFT: a crack or split in the Earth

ZOOPLANKTON: tiny animals that feed on phytoplankton

GEOLOGICAL DEVELOPMENT AND CHARACTERISTICS

Lake Baikal is found in the center of the east Siberian region of Russia. Most of the region is covered with forests of pine and larch trees, though much of the area is also covered with grasslands. The climate is harsh, and there are many areas of permanently frozen subsoil in the region.

Lake Baikal (also spelled "Baykal") is the oldest lake in the world. It is located in southeastern Siberia, about 80 kilometers north of Mongolia. Baikal is the world's eighth-largest lake in area, covering about twelve thousand square miles. It is 640 kilometers long and 50 kilometers in width at its widest part. No lake is deeper than Baikal. It contains about one-fifth of all the fresh water on the Earth's surface and about four-fifths of Russia's freshwater supply.

The lake occupies the deepest continental depression in the world, accounting for its great depth. It is surrounded by mountains that rise to 2,000-meter peaks. The mountain ranges around the lake include the Khamar-Daban, the Baykal, and the Barguzin. Two submerged mountain ranges are found beneath Baikal's surface. They cut across the width of the lake and separate it into three distinct basins. The northern basin depression reaches to 983 meters below the surface, the southern to 1,436 meters; the greatest depths are in the central basin, which goes down to 1,740 meters. The northern edge of the central basin is bordered by the Akademichesky Range, the peaks of which rise nearly 0.6 kilometer from the floor.

Some peaks break through the surface to create islands, the largest being Olkhon, which contains 837 square kilometers and has a mountain peak reaching 795 meters.

Lake Baikal was formed about 80 million years ago in the late Mesozoic era, when dinosaurs still walked the Earth. At that time, the region was made up of a broad basin of shallow lakes and marshes with a subtropical climate. It was then shaken and reshaken by a series of tremendous earthquakes. Earthquake after earthquake pushed the Earth up, forming gigantic mountains and deep valleys. About 25 million years ago, water rushed into the deepest basin and began to fill it up. Earthquakes have continued to strike the region. There have been more than thirty major recorded earthquakes since the eighteenth century. In 1861, a particularly deadly quake killed 1,300 people in a village on the eastern shore. The quake created a new fracture in the Earth, and billions of gallons of water rushed in to create a new part of the lake, Proval Bay.

Baikal has 336 rivers entering it, including the 1,134-kilometer-long Selenga River, which comes from northern Mongolia. The Selenga carries about one-half of all the water received by Lake Baikal. Other major tributaries are the Turka, the Barguzin, the Upper Angara, the Kichera, and the Goloustnaya. The tributaries drain an area of about 540,800 square kilometers. Only one river runs out of the lake, the 1,600-kilometer-long Angara. It eventually reaches the 3,830-kilometer-long Yenisei, which flows all the way north to the Arctic.

Baikal is among the cleanest lakes in the world. The lake's water is clear to about 40 meters because it contains few minerals and very little salt. Baikal's water level moves up and down about 0.6 to 1 meter per year, with the highest water in August and September and the lowest in March and April. Melting rain and snow from the surrounding mountains account for most of the change in water level. Rainfall averages about 30 centimeters per year. Other sources of the waters of Baikal are the vast underground streams that feed into it.

Baikal is usually free of ice only about 110 days a year. The lake is usually frozen over from January to late in May, though in the north the ice usually does not melt until early June. The ice reaches depths of 1 to 2 meters by spring. Water on the surface reaches about 8.8 degrees Celsius by August, though at lower depths the temperature remains a relatively constant 2.75 degrees. Because of its immense size, the lake has a great influence on the climate of Baykalia, as the surrounding area is called. Summers along the lakeshore are usually much cooler than in the surrounding area, while winters tend to be warmer by the lake and colder away from the lake. June temperatures average around 17.6 degrees Celsius in Irkutsk, a city about 58 kilometers from the lake. Irkutsk, on the Angara River, is Baykalia's largest city, with a population of about 600,000 in 1990. Altogether, about 2,500,000 people live in Baykalia.

FLORA AND FAUNA

Despite its generally cold climate, the lake is rich in life, and many forms of plants and animals thrive beneath the ice even in the coldest months of winter. Baikal is home to more than fifteen hundred species of animals, including more than three hundred types of birds and six hundred species of plants. Conditions are unusually superb for aquatic life, and the great age of the lake has allowed more than sufficient time for a great diversity of species to evolve. The lake has some thirteen hundred species of plants and animals that are found only in its waters and nowhere else, including the Baikal seal, a freshwater shrimp, and the golomyanka fish, which gives birth to live young. There is also a fish, the *comechorus baicalensis*, that lives at more than 900 meters below the surface. Other important and unique species include the Baikal oilfish, many varieties of mol-lusks, and the fish found in greatest numbers, the Baikal whitefish, a member of the salmon family.

Plant life is abundant; even in winter, sufficient sunlight breaks through the surface ice to keep plants alive at depths of 10 to 45 meters. Tiny plants, phytoplankton and algae, bloom by the millions at these depths, providing food for rotifers, copepods, and other kinds of freshwater zooplankton, the animal components of plankton. These lowest but important parts of the food chain can usually be seen only through a microscope and are found by the hundreds of millions in lakes and streams.

Baikal is unusual because of the vast amount of life found at its great depths. Most lakes are nearly lifeless below 300 meters, but because of the circulation of Baikal's waters, enough oxygen is carried to depths of 900 meters to support a variety of species. The lake can be divided into three major depth zones. The first are the shallow coastal waters, which extend to depths of from 18 to 21 meters and which are home to hundreds of species of plants and animals familiar to other Siberian lakes. However, several species of caddis flies found in the shallow waters of Baikal are endemic—that is, they are found there but nowhere else. These flies are adapted to bouncing from wave to wave in the lake. They have small wings shaped like paddles, and legs designed for swimming. Their bodies can float on the water. These "flies" have become so well adapted to the water that they no longer have the ability to fly.

The second zone, the intermediate depths, ranges from 22 meters to 200 meters below the surface. Further down is the abyssal zone, where life-forms become even more unusual. In these two zones, 80 percent of the species are endemic. In the 2.2-degree temperatures of the deepest region, there is eternal darkness; hence, many of the species have no eyes or only a minimal sense of sight. One kind of crustacean has a segmented body with seven pairs of legs. This amphipod has small indentations on the sides of its head where its eyes used to be found. The eyes disappeared over millions of years, and now the animal finds its way through the use of long antennae-like objects attached to its body. Many other species in the abyssal zone have made similar adaptations to the absence of light.

Of the fifty species of fish in Baikal, about one-half are found nowhere else. The largest is the lake sturgeon, 3.3 meters in length and weighing up to 230 kilograms. The sturgeon are highly prized by local fishermen as a source of food and also for the caviar, or eggs, that the species produces. Sturgeon were almost fished out of existence, but recent limitations imposed by the Russian government have helped to restore the population.

Other species of fish include the omul, belonging to the salmon family, which reaches more than 0.3 meter in length and can weigh up to 4 kilograms. The omul is a predator, eating other fish. It is usually inactive until water temperatures reach 12 to 15 degrees in the summer. Omuls live in the upper zone, usually not more than 4 meters below the surface, feeding on smaller fish and zooplankton.

The omul is the most important commercial fish in the lake. When it is taken from the icy waters, it sends out a piercing cry as air is expelled from its bladder. The shriek sounds like a bitter complaint; Siberian fishermen say that anyone who whines or complains "cries like an omul."

The golomyanka, endemic to Baikal, has no scales, grows to about 20 centimeters in length, and weighs only a few kilograms. More than one-third of its body weight is oil. The golomyanka feeds at the surface at night, eating zooplankton and algae. The fish has little tolerance for heat, however, and when temperatures reach above 7 degrees Celsius, it will die. When its body washes ashore, its fat melts, leaving behind little but skin and bone. The golomyanka has another unique property: It gives birth to live larvae rather than laying eggs as most fish do, and the female can produce up to two thousand babies during the autumn breeding season. Frequently, the female's belly contains so many larvae that it explodes, killing the mother and most of the offspring.

The only mammal that feeds in the water is the Nerpa seal, one of only two freshwater seals in the world. The Nerpa's closest relatives live more than 3,000 kilometers away in the Arctic. It lives among the rocks on the northeastern coast of Baikal. In the summer, it migrates south, where it sleeps on the shore and feeds on omuls and golomyanka fish. The population of Nerpa seals is estimated at about twenty-five thousand. Scientists believe that the seals originated in the Arctic and made their way to Lake Baikal by way of various tributary rivers about twelve thousand years ago. The Baykal

hair seal is not as numerous as the Nerpa but is also found on the coasts of various islands in the lake.

STUDY OF LAKE BAIKAL

Once among the cleanest lakes on the Earth, Baikal is beginning to be endangered by industrial wastes, despite its great size. There are no major cities right on the lake, but economic development has brought increased pollution to the area. In the 1960's, two paper mills were built on the southern shore, and government officials in the Soviet Union, eager to expand Siberia's economic growth, called for the building of more mills. Protests from the Soviet Academy of Sciences helped stop the construction of two plants and prompted a redesign of another to make it less polluting. Still, in 1977, scientists studying Baikal's biology became alarmed at the decline in zooplankton. Especially serious was the killing of millions of *Epischura*, a tiny shrimplike species that is a major food source for larger fish. Quick action led to some changes in levels of chemical pollutants dumped into the lake by the paper mills. The crisis illustrated how even slight changes in the quality of the lake's water could eventually lead to disaster.

Industrial development began in 1905, after the completion of a railroad along the rocky southern shore. The new train replaced the existing ferry boat that previously was the only way across the lake. The advent of the great 14,580-kilometer trans-Siberian route in 1916 opened the region to contact with Russian territory from Moscow in the west to Vladivostok on the Pacific coast in the east. Still, the region had only a small population of mostly Mongolian peoples. The most important of these were the Buryats, population 353,000, a mostly Buddhist people who had settled in the area of Lake Baikal one thousand years before. It was not until 1938, however, with the completion of the Baikal-Amur trunk line that thousands of farmers and settlers came into the region. Since then, major industries have developed in the area; in addition to paper mills, mining, ship building, fishing, and timber cutting have emerged as local industries.

All the cities and industries of the area lie along the trans-Siberian railroad. Irkutsk is the center of the region and the major industrial city. Ule-Ude, east of Lake Baikal, and Chita are two other cities of more than 100,000 population. The trans-Siberian railroad has been the lifeline of this eastern part of Siberia since its completion. Ule-Ude has a rail link to the Mongolian capital of Ulan Bator. Since 1954, both cities have been linked to Peking and northern China.

Another problem created by the paper mills results from the cutting of huge numbers of trees to make into pulp. The loss of trees dramatically increased soil erosion, which led to severe problems of runoff and contamination of Baikal's once crystal-clear waters. In 1969, the lake became part of a water-conservation area decreed by the Soviet government. Regulations prohibiting logging in certain critical areas were imposed, and strict regulations ordering the treatment of industrial pollutants were issued to reduce the amount of contaminated water entering the lake from the paper mills. Unfortunately, these regulations were not effectively enforced, and economic development was usually placed before environmental concerns by local officials. Tougher regulations and stricter enforcement of existing laws have been promoted by defenders of the environment, but a lack of funding has limited the enforcement of antipollution laws.

Because of the harshness of the climate, agriculture is possible only in limited areas around Lake Baikal. Oats are grown in the region, however, and cattle, sheep, and horses are raised there. Food industries are limited to meatpacking, fish canning, and the processing of some dairy products, chiefly cheese and milk. Mining has a long history in the region, as there is a relatively wide range of metals and minerals such as gold, tin, graphite, mica, and zinc. There are large coal deposits west of Lake Baikal along the Angara River. The Angara is also the home of the giant Bratsk hydroelectric plant, the largest in the world, which supplies power to Irkutsk.

Leslie V. Tischauser

CROSS-REFERENCES

Alpine Glaciers, 865; Amazon River Basin, 2261; Cenozoic Era, 1095; Continental Glaciers, 875; Drainage Basins, 2325; Ganges River, 2265; Glacial Deposits, 880; Glacial Landforms, 915; Great Lakes, 2270; Isostasy, 128; Lakes, 2341; Mississippi River, 2280; Nile River, 2285; Radiocarbon Dating, 537; Waterfalls, 2087; Yellow River, 2290.

BIBLIOGRAPHY

Bogdanov, Uri. *Mysteries of the Deep: From the Depths of Lake Baikal to the Ocean Floor.* Moscow: Progress Publishers, 1989. By a geologist and underwater photographer. Includes illustrations and pictures of some of the more unusual animals that live in the abyssal zone of the lake. Also contains some information of the geological history of the area. No index.

Dapples, Edward C. *Beach Material from Lake Baikal, Siberia.* Miami: Field Research Projects, 1979. Contains photographs, a map, graphs, and drawings of material collected by the geologist and explorer Henry Field on a trip to the lake in 1973. Discusses the formation of the lake and the effect of earthquakes in the region.

Koptyug, Valentin A., ed. *Sustainable Development of the Lake Baikal Region: A Model Territory for the World.* New York: Springer, 1996. This collection of essays and lectures evaluates the economic and industrial development of Lake Baikal and the surrounding environment. Illustrations and bibliographical references.

Kozhova, Olga Mikhaeilovna, and L. R. Izmestseva, eds. *Lake Baikal: Evolution and Biodiversity.* Leiden: Backhuys, 1998. This book examines the evolution and development of the biological systems and the ecosystem of Lake Baikal, with historical context provided. Somewhat technical but suitable for the careful high school student. Illustrations, maps, index, and twenty pages of bibliography.

Matthiessen, Peter. *Baikal: Sacred Sea of Siberia.* San Francisco: Sierra Club Books, 1992. This book combines travelogue-style writing with historical information about the natural evolution and subsequent human-caused deterioration of Lake Baikal. Matthiessen is a popular author who writes from an environmentalist perspective. Color photographs and maps.

Sergeev, Mark. *The Wonders and Problems of Lake Baikal.* Translated by Sergei Sumin. Moscow: Novosti Press Agency, 1989. A brief book with many photographs and wildlife, fish, and plants of the region. Discusses the hazards and costs of industrial pollution and its impact on the lake. Also has a brief history of the geology and history of Baykalia. A detailed map is included.

Symons, Leslie. *The Soviet Union: A Systematic Geography.* Totowa, N.J.: Barnes & Noble, 1983. Contains some useful information on Lake Baikal and its effects on East Siberia. Includes many detailed maps describing the region's mineral, agricultural, and industrial wealth. Also contains a brief description and history of the impact and influence of the trans-Siberian railroad.

MISSISSIPPI RIVER

The longest river in North America, the Mississippi drains a major portion of the continent, with important historical and present-day effects.

PRINCIPAL TERMS

ALLUVIUM: a deposit of soil and mud formed by flowing water

HYDROLOGY: the science that deals with the events of circulation and the properties of water and Earth

SEDIMENT: matter that settles to the bottom of a body of water

GEOLOGICAL HISTORY AND CHARACTERISTICS

The Mississippi River fascinates and amazes both residents and travelers. In North America, its 3,804-kilometer length (of which about 2,592 kilometers are navigable) is surpassed only by those of the Missouri River in the northern United States and the Mackenzie River in western Canada. The Mississippi has often been called the "Nile of North America." The comparison is appropriate; both rivers have deltas that hold fertile land and cliffs that dot their edges. Together with its two main tributaries, the Ohio and the Missouri, the Mississippi drains an immense area of thirty-one states and two Canadian provinces into the Gulf of Mexico. Altogether, the river drains an area 3.2 million square kilometers. It is the fourth-largest drainage basin in the world, exceeded only by the Amazon, Congo, and Nile River basins. The Amazon basin, by far the world's largest, is more than twice as large as the Mississippi basin, draining an area of approximately 7 million square kilometers. The Congo basin is not much larger than the Mississippi's, draining some 3,457,000 square kilometers. The Nile which drains about the same area as the Congo, 3,349,000 square kilometers.

The Mississippi drains lands with greatly differing geologies and environments. The Mississippi and its tributaries flow over a wide variety of igneous, metamorphic, and sedimentary rocks. The sedimentary rocks are easily erodible, and the large amount of material constantly held in suspension by the Mississippi's waters helps explain its muddy appearance. The river flows through lands that have widely differing climates and vege-

tation covers. It drains diverse territories such as the complex mountain ranges in the West, the arid high plains, and the humid southern woodlands.

The geologic history of the Mississippi Valley has developed two distinct basins, the Upper and Lower Mississippi Valleys, with the mouth of the Ohio River as the dividing point. The Lower Mississippi would flood seasonally if it were not regulated by human activity. In the upper valley, the river is a shallower stream enclosed by high, rocky mountains that are geologically unlike the hills that create the borders of the lower valley. The lands over which the Lower Mississippi flows have alluvial surfaces and are only a few thousand years old. This area is marked by inliers, which are outcrops of a formation completely surrounded by another layer, with the older deposits basically engulfing the fresh alluvium of the present lower valley. In the lower valley, the Mississippi is, in fact, a deeper river than it is in the upper valley.

The Pleistocene buildup of ice in the Northern Hemisphere dropped sea levels throughout the world, and the mouth of the Mississippi River was once a few hundred meters lower than it is today. The surface deposits that cover the lower valley are barely one thousand years old, and there is enough diversity in the alluvium of the lower valley to allow archaeologists to determine how long humans have been in the area. The variety of backswamps, natural levees, and crests that create a river sequence look like a winding chain. The prehistory of the very large Lower Mississippi River Valley before the appearance of Europeans is important because the present flora and densely wooded areas are not like those the first human beings in the region saw thousands of years ago.

The Mississippi River system can be categorized into six main basins: the Upper Mississippi, the Missouri, the Ohio, the Arkansas-White, the Red-Ouachita, and the Lower Mississippi. The Lower Mississippi basin includes most of its area, with about 65,000 square kilometers that are flat, while the rest is made up of rolling hills. The basin stretches from the mouth of the Ohio River at Cairo, Illinois, to the Mississippi's mouth below New Orleans. The majority of the Mississippi River floodplain is in the lower valley. Swamplands and timberlands are more common than cleared sections. The Lower Mississippi is a murky stream that is alluvial below Cape Girardeau, Missouri. The lower river receives the discharges of a number of smaller tributary streams, the basins of which are almost wholly within the alluvial valley of the river below Cape Girardeau.

SEDIMENT DEPOSITION

Like several other great rivers of the world, the Mississippi has formed an immense and fertile delta that has been created by the gradual deposition of sediment. Investigators disagree somewhat in their estimates of the river's depositions, but most believe that it amounts to about 260 million tons annually. The sediments largely consist of clay silt and fine sand, with different clays accounting for about 70 percent of the total sedimentary load.

These clays form the subsoils of most of the lower southern states. They vary greatly in their erodibility. The banks and beds of the Mississippi and its tributaries have also contributed greatly to complicating the work of hydrologists and engineers trying to control the flow of these rivers. Since at least Cretaceous times, the drainage basin of the Mississippi River system has been delivering sediments to the Gulf of Mexico. The deposition of

materials was enhanced by the thawing of the glaciers in the Jurassic period. During this stage, sediment-filled streams submerged the Lower Mississippi valleys.

Throughout geologic time, the sites of maximum deposition (depocenters) have shifted along the Gulf Coastal Plain. The bulk of the sediments that make up the Gulf Coast geosyncline, which is a portion of the Earth's crust subjected to downward warping during a large fraction of geologic time, have been derived in part from ancestral Mississippi River drainages. Over long periods of geologic time, the river has constantly fed sediments to the receiving basin, thus building up thick layers of deltaic sediments that have gradually enlarged the coastal plain shoreline seaward.

Deltas form in lobes; in modern times, the seaward progression of the different Mississippi delta lobes has constructed a deltaic plain that has a total area of 28,678 square kilometers. Of this, 23,992 square kilometers are subaerial, or periodically under water. In the past 7,000 years, the site of maximum sedimentation of delta lobes has shifted and has occupied many positions. One of the earliest deltas, the Sale Cypremort, created a deltaic lobe along the western borders of the Mississippi plain. This deltaic lobe was exceptionally widespread and thin. After approximately 1,200

The Mississippi River. (PhotoDisc)

years of build-out, the site of maximum deposition changed to another delta lobe, the Cocodrie system. A similar procedure of unseating developed, and the river abandoned the Cocodrie deposition, beginning a new delta lobe, the Teche. This process continued until the modern delta achieved its present form some 600 to 800 years ago. The result is a complex of deltaic arrangements stretching some 745 kilometers along the Louisiana coast and inland nearly 260 kilometers. The Mississippi Delta is the most recent, geologically speaking.

A tow boat on the Mississippi pushes grain-bearing barges downstream. (© William E. Ferguson)

The soil deposits of the alluvial valley are of uneven thickness. Basins along the river at the north end of the valley hold only few feet of modern river deposits, whereas more than 60 meters of deposits exist beneath the flat-lying coastal marshlands near the mouth of the river. These modern deposits rest on a substratum of heavier sand and gravel. The top layers of land along the Lower Mississippi are extraordinarily fertile because of large amounts of organic material and nutrients contained in the sediment arriving from the north.

MEANDER

Rivers that flow through flat, sedimented deltas rarely have straight channels. In rivers that flow through mountainous areas, the steep slopes and the hardness of the rock force waters to take the shortest path downhill, which generally is a straight line. When rivers cross plains that have gentle slopes, however, the turbulence in the streams is constantly dislodging fragments of soft material and creating new channels, which are usually extremely winding. The channel can then form a sinusoidal pattern that in places comes almost full circle. This process is called "meandering." The meander is the natural response of a river by which it adjusts to the slope of the land through which it flows. Meandering expands stream length and reduces slope. During high flow in a freely meandering stream running over alluvial soil, the exterior concave bank bends erode, and the bed scours and deepens. At the same time, the eroded matter scoured from the bed and banks is dropped on the point bar in the next bend downstream. With the low flows, Earth removal takes place in the bends. A meandering river can change the location of its main channel or split into branches, thus creating islands.

As meandering continues and as concave banks continue to recede, old river bends become extremely elongated; a narrow strip of land is often all that remains between the two bends of the river. When a flood causes a river to exceed the capacity of its channel, water flows over the narrow neck at a steep slope and high velocity, thus eroding a new channel, or cutoff, across the strip of land. Eventually, as the new cutoff channel becomes wider and deeper, the old bendway becomes separated from the river by deposition on bars; it then becomes an oxbow lake. A fully developed stream will maintain much the same length over time, as the decrease in length caused by cutoffs is equal to the added length produced by the growth of other meanders. Each cutoff produces a new meander that climaxes in another cutoff.

The effects of a cutoff on a river and its surroundings can be dramatic. The cutoff shortens the length of the river bottom, which changes both the river-bottom profile and the water-surface profile. Immediately after the cutoff is made, a drawdown curve, or dip, in the water profile occurs in the area of the cutoff. This causes local erosion of the river-bottom upstream of the cutoff to start to take place. The increased erosion upstream will cause a temporary overloading of the river downstream from the cutoff and will result in deposition of the sediment that the river cannot carry. This effect, though temporary, may last for a long time, since the full length of the graded part of the river upstream involves the removal of tremendous quantities of bed material. During this period, the river downstream of the cutoff is overloaded. As a result of the consequent temporary aggravation, river stages and groundwater tables will rise. Because of the reduction in the river's length and loss in storage ability, flood peaks will increase in height. These downstream effects, although temporary, may well outweigh the long-term advantageous effects of the cutoffs that occur upstream.

A successful employment of a human-made cutoff can be found on the Lower Mississippi River, where a channel improvement project that began in 1932 resulted in a lowering of flood levels by approximately 2 meters without creating extreme instability in the river channel. For a cutoff to be permanently successful, however, the hydraulic gradients must be such that the river can preserve them more or less naturally, without requiring periodic human-made adjustments at numerous intervals. To prevent the river from scouring upstream and depositing downstream, the cutoff needs to be planned in such away that the pilot cutoff channel diverts only the swift top water of the river, while the sediment-laden bottom water stays in the old river channel. This is best done by constructing the pilot channel in line with the upstream river channel and making the old river meander the branch channel of the new route.

Delta Wetlands

The flatness of a delta, the meandering of the river channel, and the great fertility caused by the continual deposition of new organic material mean that river deltas are natural wetlands. Marshes and swamps are areas of low, flat-lying wet Earth open to daily or seasonal flooding, although they are not the same things. Physically, marshes and swamps have common traits; they differ pri-

marily in the kind of vegetation that flourishes on the land. Marshes are covered with grasses and other grasslike plants such as cattails. Swamps commonly include hardwood forests. The Mississippi River Delta supports some of the most important wetlands in the world, and it includes about 40 percent of all coastal wetlands in the lower forty-eight states.

The delta wetlands support plant and animal populations that can adjust to shifting water levels. Annual floods supply the system with sediments containing dissolved organic matter and nutrients. Many species of fish feed or spawn in the wetlands. At the mouths of the rivers that drain the wetlands are estuaries that support marine life. The same flooding that brings life-supporting elements to the floodplain forests also flushes organic material of all kinds to the sea. This matter supports the food chains of the estuaries, which are especially complex because of the great gradations of temperature and salinity that exist where fresh river water meets the sea.

A natural part of the life cycle of a river delta is flooding, a phenomenon caused by the vast discharges associated with high-water levels that surpass the capability of the river channel to eliminate water. Floods are caused by melting snow, by rainfall mixed with snow melt, by glacial melts in mountain rivers, or by rainfall alone. Although modern floods produce heavy damage to human life and property, presettlement floods mainly served to rearrange the topography of riparian lands.

During annual floods, river stages will normally top adjacent natural levees. However, flood inundation is not continuous along all portions of a river; in many instances, during a single flood, various crevasse splays will form along low points in the natural levee. These small crevasses will be maintained for a year or more, until enough material is deposited in them to build them back up to the level of the natural levee. They will then cease to be active.

In a natural delta, distributary channels serve as flumes that direct a portion of the water from the parent river system to a receiving basin. With continued enlargement of the distributary channels, a point is reached at which the channel is no longer able to maintain its gradient advantage, and the process of channel abandonment begins.

In the Mississippi Delta, where tidal flooding is not usually a factor, the distributary channels commonly fill with fine-grained material mixed with peats and transported organic debris. One of the major features of the Mississippi River basin is the large amount of silting caused by the formation of crevasses, which break off from the main distributaries and deposit materials in the numerous interdistributary bays. Eventually, deposits form in shallow bays between or adjacent to major distributaries and extend themselves seaward through a system of radial channels, similar in appearance to the veins of a leaf.

In essence, then, it is important to understand that the presettlement Lower Mississippi basin was a system that had functioned well throughout geologic time. The Mississippi River meanders carried runoff from its vast watershed to the Gulf of Mexico, and the sediments brought from the north created vast tracts of land that now form much of the gulf states. The sediments continually enriched this land, which supported an immense diversity of life by depositing new fertilizing organic materials. The river's floods did not destroy land but rather enlarged and enriched it. Once flood waters overflowed the Mississippi's natural levees, they flowed into distributaries or formed crevasses that directed them into the vast area of wetlands that surrounded the main channel.

Loralee Davenport

CROSS-REFERENCES

Abyssal Seafloor, 651; Amazon River Basin, 2261; Earthquake Hazards, 290; Ganges River, 2265; Great Lakes, 2270; Lake Baikal, 2275; Lakes, 2341; Nile River, 2285; Surface Water, 2066; Weather Modification, 1992; Yellow River, 2290.

BIBLIOGRAPHY

Changnon, Stanley A., ed. *The Great Flood of 1993: Causes, Impacts, and Responses.* Boulder, Colo.: Westview Press, 1996. An examination of the flood damage of 1993 and its social and environmental effects. The book also looks at the factors that made the flood so devastating, as well as prevention methods put in place as a result. Illustrations and bibliographical references.

Coleman, James. *Deltas: Processes of Deposition and Models for Exploration.* Baton Rouge, La.: Burguess Publishing, 1981. This book contains an excellent overview of deltas. Provides a particularly good explanation of the changes that take place within a delta.

Dennis, John. *Structural Geology.* 2d ed. Dubuque, Iowa: W. C. Brown, 1987. A good overview of the geology of North America. Useful background for anyone interested in the Mississippi and related rivers.

Foster, J. W. *The Mississippi Valley: Its Physical Geography.* Chicago: S. C. Griggs, 1869. A description of the geography of the Mississippi River in the 1860's. Provides an interesting historical perspective.

Harris, Eddy L. *Flood Control and Water Manage-ment in the Yazoo: Mississippi Delta.* Mississippi: Social Science Research Center, Mississippi University, 1993. A thorough analysis of flood control and prevention policies and management for the Mississippi River. Notable for an excellent bibliography of works on the Mississippi Delta.

_____. *Mississippi Solo: A River Quest.* 2d ed. New York: Henry Holt, 1998. A memoir of canoeing along the Mississippi River. A nice account of the river's ecosystem and social importance. Suitable for all levels. Illustrations.

Miller, A. *The Mississippi.* New York: Crescent Books, 1975. Contains historical maps that allow the reader to trace the changes in the river's course over time.

Petersonn, Margaret. *River Engineering.* Englewood Cliffs, N.J.: Prentice-Hall, 1986. A good overview of the science of river engineering, including explanation of relevant terms.

Tompkins, Frank H. *Riparian Lands of the Mississippi River.* New Orleans, La.: Improvement and Levee Association, 1901. A collection of essays on the river from an engineering point of view.

NILE RIVER

The Nile River is an interesting and important hydrologic system. Its complex geologic evolution began with the drying up of the Mediterranean Sea about 6 million years ago. The Nile carries water northward across the Sahara Desert from high-rainfall regions of equatorial Africa; it nurtured one of the great civilizations of ancient times and sustains millions of people in modern Egypt.

PRINCIPAL TERMS

CATARACT: rough water or a waterfall in a river, generally obstructing navigation

HYDROLOGY: the science that deals with the properties, distribution, and circulation of water on land

MANTLE PLUME: a rising jet of hot mantle material that produces tremendous volumes of basaltic lava

MONSOON: a seasonal air current affecting eastern Africa and southern Asia; in the Northern Hemisphere's winter, dry winds flow from the continents to the ocean, whereas in summer, moist winds flow from the ocean to the continents and cause heavy rains

PYROXENE: a rock-forming mineral commonly found in igneous rocks such as basalt or gabbro

RIFT: a portion of the Earth's crust where tension has caused faulting, producing an elongate basin; rifts fill with sediments and, sometimes, volcanic rocks

TECTONIC: relating to differential motions and deformation of the Earth's crust, usually associated with faulting and folding of rock layers

GEOGRAPHY AND CHARACTERISTICS

The Nile River is by some measures the longest river on Earth, extending 6,650 kilometers from Lake Victoria to the shores of the Mediterranean Sea. Only the Amazon River in South America approaches its length. The river follows a generally south-to-north path; both the source of the White Nile in equatorial Africa and its mouth on the southern shore of the Mediterranean Sea lie within 1 degree of longitude. The Nile crosses 35 degrees of latitude, a distance comparable to the width of the continental United States, and flows across a greater variety of regions than any other river.

In spite of its great length and large drainage basin (approximately 3.4 million square kilometers, or about 10 percent of the area of Africa), the Nile carries relatively little water; during the twentieth century, annual flow ranged from a low of 42 cubic kilometers in the drought year of 1984 to a high of 120 cubic kilometers in 1916. This comparatively small flow results from the fact that no water is added to the river for the final one-half of its journey to the sea. Most rivers merge with increasingly large streams as they approach the sea, joining their waters into an ever-swelling stream.

The Nile flows instead through the Sahara Desert, the largest and most desolate tract of land on Earth, matched only in size and desolation by the icy wastes of Antarctica.

Passage through the Sahara increases the importance of Nile water while reducing its volume. Its greatest value is reached as it nears the sea, where it has for more than five thousand years nurtured agriculture in Egypt. The rich soil, constant sunshine, and abundant water of the Nile Valley in Egypt result in one of the most productive agricultural regions of the world. The ancient civilizations of Egypt were built on this firm economic foundation. Without the Nile, Egypt would be as barren as the rest of the Sahara; with the Nile, Egyptian farms can produce two or even three crops every year. Egyptian civilization was nurtured by the Nile and protected from invasion by the sea to the north and the desert to the east and west. Egypt—and, consequently, much of world culture—thus truly is the "gift of the Nile."

WHITE NILE AND BLUE NILE

Dramatic changes in the amount of water flowing through the Nile occur annually, reflecting the fact that two independent streams, the Blue Nile and the White Nile, join at Khartoum, the capital of Sudan. The White Nile issues from Lake Victoria, the second-largest freshwater lake on Earth. The White Nile tumbles down a series of falls and rapids to Lake Kioga and Lake Albert before spilling out into a huge swamp known as the Sudd in southern Sudan. The Sudd was the great barrier to European explorers seeking the source of the White Nile. The vast swamps caused explorers traveling in boats up the Nile to lose their way, forcing them to turn back or perish. As a result, the source of the White Nile was discovered in 1862 by an explorer, John Speke, who traveled west from Zanzibar to Lake Victoria.

In the Sudd, the White Nile—there known as the Bahr el Gebel—gains several tributaries. From the west flows the Bahr El Arab, and the Sobat flows from the east. The vastness of the Sudd results in huge losses by evaporation; about 50 percent of the water flowing into the Sudd is lost by evaporation. The Sudd acts as a buffer, so that greater water flow into the Sudd causes the swampy area to expand, resulting in increased evaporation. Efforts to plan for expected increases in water demand, especially in Egypt, have focused on reducing these losses by building a 350-kilometer diversion, the Jonglei Canal, around the Sudd. Civil war in southern Sudan stopped its construction, but growing environmental concerns rendered it unlikely that the canal would be quickly finished even if the war ended. The combination of relatively constant rainfall over the Lake Victoria region, the vastness of Lake Victoria, and evaporative losses in the Sudd result in a relatively constant flow of water down the White Nile to Khartoum. North of Malakal, the Nile flows over a flat stretch in a well-defined channel without swamps; the river drops only 8 meters in elevation over its 800-kilometer length from Malakal to Khartoum.

In contrast to the White Nile, the Blue Nile and its little sister, the Atbara River, are seasonal streams. For about one-half of the year they flow as feeble trickles, contributing negligible water to the Nile from January to June. In mid- to late summer, the monsoons sweep large amounts of moisture evaporated from the Indian Ocean toward the Horn of Africa. As the water-laden air flows up over the Ethiopian highlands, it cools, and torrential rains fall. These rains so swell the Blue Nile

that its waters overwhelm the smaller White Nile at Khartoum and cause it to flow back upstream. Between 70 and 80 percent of the Nile's total annual water budget results from the flood phase of the Blue Nile and Atbara Rivers. These raging rivers also erode large amounts of black sand and silt from the basaltic highlands of Ethiopia and carry these sediments north. In his 1899 book *The River War*, Winston Churchill described the Nile in flood: "As the Nile rises its complexion is changed. The clear blue river becomes thick and red, laden with the magic mud that can raise cities from the desert sand and make the wilderness a garden." The historic fertility of the Nile Delta and Valley in Egypt owe as much to the new layer of this rich soil added to the inundated fields as to the deep soaking. This annual flooding and delivery of sediment no longer occurs downstream of the Aswan High Dam, which was constructed in the 1960's to ensure that the water needs of Egypt could be met in spite of drought in the Nile headwaters.

Like the White Nile, the Blue Nile issues from a large lake, Lake Tana. It initially flows to the southwest but progressively turns to the southwest and then northwest as it descends through the Blue Nile Canyon. It leaves the mountains about at the border between Sudan and Ethiopia and continues on a northwestward track across the Sudanese lowlands until it joins the White Nile at Khartoum.

CATARACT NILE, GREAT BEND, AND EGYPTIAN NILE

For a distance of 1,850 kilometers from Khartoum to Aswan, the Nile is defined by two features: the cataracts and the great bend. The cataracts are sections where the river tumbles over rocky outcroppings, creating serious obstacles to navigation. There are six "classical" cataracts, but in reality there are many more, and as much as 565 kilometers of this portion of the Nile—referred to as the Cataract Nile—are affected by cataracts. The cataracts are also significant because these define river segments where granites and other resistant rocks come down to the edge of the Nile. The floodplain in this region is narrow to nonexistent, and opportunities for agricultural development are correspondingly limited. These two factors—navigational obstacles and restricted floodplain—are the chief reasons that this part of the Nile is thinly populated and why the historic border between Egypt in the north and Nubia or Sudan in the south is found not far from the First Cataract at Aswan.

The great bend is one of the most unexpected features of the Nile. For most of its course, the Nile flows inexorably to the north; at one location in the heart of the Sahara, however, it turns southwest and flows away from the sea for 300 kilometers before resuming its northward journey. This deflection of the river's course is the result of tectonic uplift of the Nubian Swell that has taken place during the past few hundred thousand years. This uplift is also responsible for the cataracts; if not for the recent uplift, these rocky stretches would have been worn away long ago by the abrasive action of the sediment-laden Nile.

The northernmost segment is the Egyptian Nile, extending for 1,200 kilometers between Aswan and the Mediterranean Sea. It consists of two parts—the Nile Valley and the Delta. The Nile Valley consists of the broad floodplain that is imprisoned between steep limestone or sandstone hillsides. The boundary between the lush valley floor and the flanking desert is stark and sudden; a visitor can stand with one foot on the black mud, brought 3,000 kilometers from Ethiopia, and the other foot on desert sand or barren limestone. The floodplain widens progressively to the north until it opens up just north of Cairo into the Delta. The mouth of the Nile is where the term "delta" was first applied; the ancient Greeks were impressed by the triangular shape of the land around the Nile's mouth and the similarity that it had to the fourth letter of the Greek alphabet. The Nile splits into two branches at the south end of the Delta, the western, or Rosetta, branch (where the famous Rosetta Stone was discovered) and the eastern, or Damietta, branch. Up to the time that the Aswan High Dam was built, the Delta continued to grow as annual floods laid down their loads of silt. These sediments are now deposited in Lake Nasser, with the result that the sediment-starved Delta is slowly sinking and its shoreline is retreating.

GEOLOGICAL FORMATION OF THE NILE

The evolution of the Nile is an important area of scientific research. The river consists of a series of steeper and flatter segments, and this is thought to indicate that several independent drainage systems existed in the region now drained by the Nile. Much is known about when and how the course of the Nile came into being, but much remains to be learned.

A critical event in the formation of the Nile was the evaporation of the Mediterranean Sea about 6 million years ago. Because of the climatic zone in which it lies, more water evaporates from the Mediterranean than is supplied by the rivers that flow into it. This water deficit requires that replenishing seawater flow into the Mediterranean from the Atlantic. When the Strait of Gibraltar pinched shut as a result of the collision between Africa and Europe, the Mediterranean slowly dried up. The Dead Sea of Israel and Jordan, at about 400 meters below sea level, is presently the deepest spot on the continents, but the Mediterranean seafloor, lying as much as 3,000 meters below sea level, was 0.6 kilometer deeper and one thousand times more vast than the Dead Sea region. This may have been the greatest sea ever to evaporate completely, and the event profoundly affected the streams that flowed into it. A north-flowing river existed in what is now Egypt; as sea level dropped, the stream became steeper and steeper as it cut down into relatively soft limestones. The enhanced erosive power allowed its upper tributaries to extend into the headwaters and capture upstream drainages. The increased water from the captured streams further increased the stream's erosive power, further stimulating the expansion of the drainage system upstream. This led to the development of the "Eonile," which flowed through a huge canyon that was deeper than the Grand Canyon of Arizona and many times longer. This canyon is buried beneath all the Egyptian Nile, but it cannot be traced south of Aswan.

In time, the barrier at Gibraltar ruptured, and a tremendous waterfall brought Atlantic seawater to refill the Mediterranean basin. The "Grand Canyon of Egypt" became a drowned river valley or estuary, similar to the fjords of Norway but very different in origin. Slowly, this estuary filled with sediments brought in by rivers flowing from the south, and a landscape not too different from the present was established by 3 million years ago.

STUDY OF THE NILE

It is much more difficult to discern the development of the Nile upstream from Aswan, but there are some clues. Lake Victoria did not exist prior to about twelve thousand years ago. Before this time, the streams of the Ugandan highlands flowed west to join the Congo, which drains into the equatorial Atlantic. Recent tectonic activity lifted and tilted the region to make the lake and direct its overflow to the north. Similarly, recent tectonic activity caused the Nubian Swell and the Bayuda Uplift to be more or less of an obstacle, with the result that northward flow across Nubia is sometimes permitted and sometimes blocked. Probably the oldest parts of the Nile drainage are those associated with the Sudd. These follow the axes of sediment-filled rifts that formed more 65 million years ago and that have continued to sink and to fill with sediments since that time.

The Ethiopian highlands began to form about 30 million years ago as a result of tremendous volcanic activity, as a mantle plume punctured the crust, but the contribution of the distinctive black sediments from these highlands is not recognized in the Egyptian Nile until about 650,000 years ago. This may be the result of increased rainfall on the Ethiopian highlands accompanying the development and intensification of monsoonal circulation in the recent past. Monsoonal circulation is caused by the change in position of atmospheric low-pressure cells, which lie over the equatorial Indian Ocean during northern winter and over south-central Asia during northern summer. The result is that cold, dry winds blow south from Asia during northern winter, but warm, moist winds blow from the sea toward Asia in northern summer. The westward deflection of summer winds resulting from the Coriolis effect brings part of the moisture-laden air currents over Ethiopia, where the air cools as it rises. Cool air can hold less moisture than warm air, so clouds and then rain form as the monsoon rises over the Ethiopian Plateau. This brings the long, drenching rains in Ethiopia that cause the annual Nile flood. The monsoonal circulation has intensified over the last few millions of years as a result of continued uplift of the Tibetan Plateau. The mystery of the Nile flood

that puzzled the ancient Egyptians can thus best be understood by knowing about mountain-building events occurring thousands of kilometers away. Water from the Ethiopian highlands may not have reached Egypt because the Nubian Swell acted as a barrier, perhaps deflecting the water to the west. It may be that only with the additional water provided as a result of the intensifying monsoon that the upstream Nile was able to erode its way through the Nubian Swell and continue north to the Mediterranean Sea.

The history of the Nile is mostly inferred from sedimentary deposits in the Delta and Egyptian Nile. Scientists know that great river systems carried sediments—preserved today as the Nubian Sandstone—north from central Africa as long ago as the Cretaceous period, about 100 million years ago, but the course of these rivers is poorly known. No link can be established between the Cretaceous rivers and the Nile. The sea invaded Africa from the north toward the end of the Cretaceous period, and much of northeast Africa was a shallow sea during much of the early Tertiary period, about 70 million to 40 million years ago. River deposits from the late Eocene and early Oligocene (about 35 million years ago) are known from west of the present Nile, but these sediments did not travel far, indicating that they came from a relatively small river. This may have been the precursor of the stream that carved the great canyon following evaporation of the Mediterranean Sea about 6 million years ago.

Robert J. Stern

CROSS-REFERENCES

BIBLIOGRAPHY

Al-Atawy, Mohamed-Hatem. *Nilopolitics: A Hydrological Regime.* Cairo, Egypt: American University in Cairo Press, 1996. This is a brief but detailed discussion of the politics and conservation methods surrounding the Nile River watershed, ecology, and water supply. Map and bibliography.

Churchill, Winston. *The River War.* New York: Longmans, Green, 1899. An account of the Mahdi's rebellion in Sudan and the British expedition against it during the late nineteenth century. Accounts of military operations can be skimmed over, but the trials of navigation up the Nile provide the best account of the river upstream from Aswan. Provides a good understanding of why Egypt stops near Aswan. Suitable for the general reader.

Guadalupi, Gianni. *The Discovery of the Nile.* New York: Stewart, Tabori, and Chang, 1997. Complete with fold-out color maps and illustrations, this book allows the reader to explore the entire length of the Nile River. Suitable for all levels.

Hillel, D. *Rivers of Eden.* New York: Oxford University Press, 1994. A wonderful account of the history of water use in the Middle East and the problems that the future holds for the region. Hillel is an environmental scientist and hydrologist, and his account is based on sound technical considerations, but he enlivens the tale with a generous spice of history, religion, and personal experience. Chapter 6 is entitled "The Mighty Nile." Suitable for the general reader.

Howell, P. P., and J. A. Allan, eds. *The Nile: Sharing a Scarce Resource.* Cambridge, England: Cambridge University Press, 1994. Overview of the history of the Nile, with special emphasis on how its flow has changed in the past. Discussion of the future utilization of the river and the problems facing planners. Comprehensive, excellent text, suitable for college students and professionals.

Moorehead, A. *The Blue Nile.* New York: Harper & Row, 1962. Similar to *The White Nile* in its approach. An account of the exploration and development of the Blue Nile and parts of the Egyptian Nile, including the journey of James Bruce to the headwaters of the Blue Nile, Napoleon's invasion of Egypt, the empire of Mohamed Ali, and the collision of Ethiopian and British cultures at the end of the nineteenth century. These stories are told within a context of the land and the people of the Blue Nile basin and Egypt. Suitable for the general reader.

_____. *The Geology of Egypt.* Brookfield, Vt.: A. A. Balkema, 1990. An introduction to the geology of Egypt, including an exploration of the history and significance of the Nile River. Suitable for the layerson.

_____. *The River Nile: Geology, Hydrology, and Utilization.* New York: Pergamon, 1993. Outlines the geologic history of the Nile, with particular emphasis on the record from Egypt. This is the best source for those interested in the history of the Nile Valley and Delta in Egypt. Excellent figures and abundant references. Suitable for geologists and archaeologists.

_____. *The White Nile.* New York: Harper & Row, 1960. An account of the exploration and development of the White Nile, with emphasis of the discovery of Lake Victoria and the source of the White Nile during the mid-nineteenth century by Speke, the search of Henry Stanley for David Livingstone, and the Sudanese revolt and British reaction. An excellent overview of the land and the people, suitable for the general reader.

YELLOW RIVER

The Yellow River—the second largest river in China—flows through three distinct geographic regions in northern China. The river is characterized by varying features of discharge, flooding, and sediment capacity in each region. The easily eroded loess sediments in the central region make the Yellow River the muddiest river in the world. The enormous quantity of sediment deposited by the river has been a primary cause of the catastrophic floods that have destroyed property and killed millions of people in northern China throughout recorded history.

PRINCIPAL TERMS

AGGRADATION: the buildup of sediment in a river channel that causes the level of the river to rise

AVULSION: the process in which a river changes its course because of flooding

DELTA: a landform developed when a river drops sediment as it enters a large standing body of water

FLOODPLAIN: the area covered by water during large flooding; in many areas, the term is used to denote the area between human-made levees

HYPERCONCENTRATED FLOOD: a flood that carries unusually large quantities of sediment

LEVEE: a human-made ridge constructed to contain river waters during flood stage; levees are sometimes breached during large-scale flooding, leading to damage along the surrounding areas

LOESS: wind-blown sediment that can accumulate in great thicknesses and is easily eroded if not stabilized by vegetation

GEOGRAPHY OF THE YELLOW RIVER

The Yellow River (Chinese, Huang He or Hwang Ho) is the second largest river in China; it flows 5,464 kilometers eastward from the Plateau of Tibet to its mouth at the Bohai Sea (Po Hai), a large embayment of the Yellow Sea. The Yellow River, also known as the Huang Ho (or Huang He), has been a major site of civilization for millennia. Because of the tendency of the river to break through its levees during floods and cause death and great damage, the Yellow River is also informally known as "China's sorrow" and the "Ungovernable."

The Yellow River drains a basin of approximately 752,400 square kilometers, the third largest basin in China. The drainage basin can be divided into three distinct regions of different relief in which the river has developed different characteristics. The Yellow River begins in the Payen-k'a-la Mountains and flows for approximately 1,200 kilometers eastward across the Tibetan Plateau highlands, where it drains an area of 124,000 square kilometers. In this upper reach of the river, its channel is incised into hard crystalline rock; in many stretches, the channel is relatively narrow.

The slope of the river channel steepens to more than 2 meters per kilometer as the river leaves the plateau and passes through a number of gorges. The upper basin contributes nearly one-half of the water that flows through the river; however, because of the hard bedrock of the region, it supplies less than 10 percent of the sediment transported by the river.

As the river flows eastward out of the Tibetan Plateau and into the middle basin, it bends and travels northward for 885 kilometers before turning eastward for another 800 kilometers. Finally, the river bends southward for an additional 800 kilometers before it turns eastward and leaves the middle basin. At the southern end of the southward-flowing section of the great loop, the Yellow River is joined by its two largest tributaries, the Wei River and the Fen River. As the Yellow River winds it way around the great bend, the river channel traverses the great Loess Plateau of north-central China and the sandy soils of the Ordos Desert. Loess is an eolian sediment that was deposited by winds blowing southward out of Mongolia after the last ice age. Loess is characteristically a poorly consolidated material that is easily eroded by wind and water. The great sediment load of the Yellow River is primarily contributed by the erosion of the loess in the middle basin. It is the immense quantity of light-colored sediment suspended in the water that gives the river its name in both English (Yellow) and Chinese (*huang* is the Chinese term for yellow). The loess—up to 170 meters thick in some areas—is the thickest soil sequence in the world. The native forests that had long grown on the loess soils had been cut down even before the beginning of the Han Dynasty in the year 206; therefore, there is relatively little vegetation to hold the soil in place. The river easily cuts through this friable material, which has led to the development of deeply incised river channels.

The lower basin extends east of the Loess Plateau from the city of Cheng-chou to the river delta at the Bohai Sea, a distance of about 786 kilometers. The river crosses the North China Plain, a flat alluvial plain built up of sediments deposited over the last 15 million years by the Yellow and other rivers. The sediment dumped into the region filled an ancient embayment of the Yellow Sea, and the Yellow River continues to build its delta seaward today. The river is contained by long human-made levees throughout this heavily populated region, and sedimentation between the levees is often excessive, occasionally leading to catastrophic flooding. The delta of the Yellow River begins approximately 70 kilometers west of the Bohai Sea and covers an area of over 5,440 square kilometers.

General Hydrology

The average annual flow rate of the Yellow River is approximately 1,770 cubic meters per second, but it may range from up to 36,300 cubic meters per second in wet years to below 1,000 cubic meters per second in dry years. The average rainfall for the entire basin is approximately 470 millimeters per year, but it varies significantly between basins and is generally concentrated between July and October, when more than 50 percent of the annual rainfall may occur. Much of the water comes out of the upper (western) basin from the snowfall in the higher elevations of the Tibetan Plateau, but a significant amount of water is added by tributaries in the middle basin. The average an-

nual discharge of the river is estimated to be only 58 billion cubic meters, which is less than the Hanjiang River, a tributary of the Yangtze River. The relatively low discharge of the river is caused by the generally low precipitation rates in the river basin and by the considerable evaporation of water during its journey to the sea. In addition, a large quantity of water is withdrawn from the river for irrigation. The upper basin contributes approximately 48 percent of the water of the river, the middle basin about 37 percent, and the lower basin about 15 percent.

During the winter months, the Yellow River freezes over for as many as fifteen to twenty days per year. The ice is generally broken up with explosives to allow for navigation. In 1972, the lower reaches of the river began to dry up for short periods of time. In 1995, the easternmost stretch of the river was dry for a record 122 days, about one-third of the year. Increased withdrawal of water for irrigation in the upper and middle basin has caused the loss of water in the lower basin.

Sediments

The Yellow River is the muddiest river in the world, carrying an average annual sediment load of about 1.6 billion tons. The maximum recorded sediment load was 3.9 billion tons in 1933, while the minimum load was 488 million tons in 1928. The mean annual sediment concentration is 29.6 kilograms per cubic meter of water. This is about thirty times more concentrated than the Nile River and almost three times the concentration of the Colorado River. The source of most of the sediment transported and deposited by the river is the easily eroded loess of the middle basin. It is calculated that the middle basin contributes approximately 90 percent of the sediment carried by the Yellow River. The arid nature of the middle basin, the lack of soil-stabilizing vegetation, and the friable nature of loess all contribute to extraordinary erosion rates on the Loess Plateau. It is estimated that approximately 10,000 tons of soil per square kilometer per year erode from the plateau. Hyperconcentrated flows often result from such heavy sediment loads on the Yellow River.

The sediment is washed into the Yellow River and carried downstream to be deposited either in the river system or at the coastline. If the sediment reaches the waters of Bohai Sea, it is dropped as the river hits the sea, forming a delta. The modern delta is just the latest of many that developed as the Yellow River and other rivers deposited sediment into a marine embayment west of the present-day Bohai Sea. As the deltas prograded eastward, they filled the embayment with sediment. This process of filling a marine basin has continued for the last 15 million years. More recently, the delta has continued to expand eastward at rather astonishing rates. From 1870 to 1970, the delta prograded eastward approximately 24 kilometers. Some areas experienced even higher rates of growth, with one region of the delta expanding by approximately 25 kilometers between 1949 and 1952. However, beginning in 1972, increased water consumption by the population in the middle and upper basins resulted in a drying up of the lower reaches of the Yellow River such that water and sediment did not reach the coastline for several months during the year. This phenomenon will affect delta growth if it continues and may result in erosion of the coastal sediments and encroachment of marine waters onto the delta.

Much of the sediment eroded from the Loess Plateau is deposited in the river system itself and does not reach the coastal areas. This coarse- to fine-grained sediment is dumped into channels along the lower reach of the river, causing a buildup of the river bed. The aggradation of the bed would normally cause a change in the river channel by avulsion: The river would simply be forced out of its channel during a flood and would establish a new channel in a region of less sediment and lower elevation. This natural process of channel change causes the river to migrate across the alluvial plain that was constructed of earlier deposited sediments. It is clear from geologic studies that the Yellow River has changed its course numerous times over the last four thousand years. Over this time span, the Yellow River has entered the sea at places as much as 700 kilometers apart. The river entered the Bohai Sea along a northern route from about 3000 B.C.E. to 602 B.C.E., then changed its course to a more southerly channel after flooding and entered the sea south of the Shantung Peninsula. In 70 C.E., the river again changed its course to a more northerly channel and entered the Bohai Sea near its present mouth. Changes in the river channel also occurred inland where the Yellow River en-

ters the lower basin. In the last two thousand years, the Yellow River has had 7 major course changes and almost 1,600 levee breaches.

The lower basin is heavily populated, and a course change can affect millions of people. As a result, the Chinese have attempted to control the river by surrounding the river with human-made levees and by the construction in 1960 of a large dam at the city of Sanmenxia. The levees, which were constructed to contain the river during major floods, have generally been successful. However, as they contain the river flood waters, they also trap the huge quantities of silt and sand carried by the waters. As a result, the river bed and surrounding floodplain between the levees continues to aggrade because of sediment deposition. As the elevation of the bed rises, it is necessary to build higher levees to contain future floods. This process has continued for hundreds of years, and now the river is actually contained within the levees at a higher elevation than the surrounding plains. In a few areas, the river level is 30 meters higher than the surrounding land. This means that the river is at the highest elevation in the region and is actually the drainage divide for the area. This situation can lead to catastrophic flooding if a levee is breached.

FLOODING

The Yellow River is called "China's sorrow" because it has caused so many deaths and enormous damage when it has flooded across the North China Plain. Since the river is, in many places, much higher than the surrounding plains and since the alluvial plain is so flat, a breach in the levees will cause widespread flooding. Some of the most catastrophic floods in the history of the world have occurred on the Yellow River. Chinese engineers have constructed thousands of kilometers of levees in an attempt to control the river, but these efforts have only delayed the flooding. There have been nearly 1,600 major and minor levee breaches recorded in the last two thousand years. Between 960 and 1048, thirty-eight major levee breaches occurred, a rate of nearly one break every other year. Better construction and maintenance of levees led a lower rate of levee breaches, but flooding continued to occur. The years of the Taiping Rebellion (1850 to 1864) were marred by a lessening of concern over the levees; as a result, breaches became more common, leading to a major river course change in 1852-1854. The Yellow River broke through its levees near the city of K'ai-feng in 1887, leading to one of the most devastating floods in history. The flood waters covered 130,000 square kilometers of the North China Plain, depositing silts and clays over the entire area. The floods inundated 21,000 hectares of cropland and deposited up to 10 meters of sediment in some areas. In all, 11 cities and 2,000 villages were flooded, and approximately 900,000 people perished. It is estimated that an additional 2 to 4 million people suffered flood-related deaths. Another major flood occurred two years later that destroyed 1,500 more villages. A devastating flood occurred in 1921 in which many villages were destroyed, particularly near the mouth of the river. In 1933, more than 1,500 villages were inundated, and more than 18,000 people died.

One of the most disastrous floods of the Yellow River was directly caused by humans when the levees were intentionally ruptured by Chinese soldiers, who were ordered to dynamite the levees in order to halt the advance of Japanese soldiers in 1938. The ensuing flood successfully stopped the Japanese advance but also destroyed eleven cities, more than four thousand villages, and many thousands of hectares of cropland. More than 1 million Chinese people died, and an additional 11 million people were left homeless. It took the engineers eight years to regain control of the river and route it back into its pre-1938 channel. Another significant flood occurred in 1949.

Jay R. Yett

CROSS-REFERENCES

African Rift Valley System, 611; Amazon River Basin, 2261; Dams and Flood Control, 2016; Drainage Basins, 2325; Earth Science Enterprise, 1752; Floodplains, 2335; Freshwater Chemistry, 405; Ganges River, 2265; Geomorphology of Dry Climate Areas, 904; Geomorphology of Wet Climate Areas, 910; Great Lakes, 2270; Lake Baikal, 2275; Lakes, 2341; Mississippi River, 2280; Nile River, 2285; Paleoclimatology, 1131; Precipitation, 2050; River Bed Forms, 2353; Sand, 2363; Sand Dunes, 2368; Sediment Transport and Deposition, 2374; Surface Water, 2066.

BIBLIOGRAPHY

Czaya, Eberhard. *Rivers of the World.* New York: Cambridge University Press, 1983. A general text on the world's rivers that has valuable information on the fluvial processes and a good description of the changing course of the Yellow River over the last 2,500 years.

Milne, Anthony. *Floodshock.* Gloucester, England: Allan Sutton, 1986. This general reader on flooding throughout the world contains good, readable information on the flooding of the Yellow River. There is a good description of the 1887 Yellow River flood and the dynamiting of the levees in 1938.

Hillel, Daniel. "Lash of the Dragon." *Natural History* (August, 1991): 28-37. This is a well-written description of the problems of soil erosion caused by the Yellow River. The extent of the effects of sedimentation and flooding are also described.

Wan, Chao-hui. *Hyperconcentrated Flow.* Brookfield, Vt.: A. A. Balkema, 1994. This book contains a wealth of information about sediment transport in the Yellow River. Illustrations, bibliography, and index.

Waterlow, Julia. *The Yellow River.* Hove, England: Wayland, 1993. This short book (48 pages), part of the World's Rivers series, is a good introduction to the Yellow River and its characteristics. Includes photographs.

7
WATER AND EARTH: SEDIMENTOLOGY

ALLUVIAL SYSTEMS

Alluvial systems include a variety of different depositional systems, excluding deltas, that form from the activity of rivers and streams. Much alluvial sediment is deposited when rivers top their banks and flood the surrounding countryside. Buried alluvial sediments may be important water-bearing reservoirs or may contain petroleum.

PRINCIPAL TERMS

BRAIDED RIVER: a relatively shallow river with many intertwined channels; its sediment is moved primarily as riverbed material

EPHEMERAL STREAM: a river or stream that flows briefly in response to nearby rainfall; such streams are common in arid and semiarid regions

FLOODPLAIN: the relatively flat valley floor on either side of a river that may be partly or wholly occupied by water during a flood

LONGITUDINAL BAR: a midchannel accumulation of sand and gravel with its long axis oriented roughly parallel to the river flow

MEANDERING RIVER: a river confined essentially to a single channel that transports much of its sediment load as fine-grained material in suspension

OXBOW LAKE: a lake formed from an abandoned meander bend when a river cuts through the meander at its narrowest point during a flood

POINT BAR: an accumulation of sand and gravel that develops on the inside of a meander bend

TRANSVERSE BAR: a flat-topped body of sand or gravel oriented transverse to the river flow

SEDIMENT

Deposits of silt, sand, and gravel produced by the activity of rivers and streams are called alluvial sediments. Sediment in rivers is moved primarily as either suspended load or bed load. The suspended load is the finest portion of sediment—that is, silt and clay—and is carried within the flow itself by fluid turbulence. Material moved along the bottom of the river by rolling, sliding, and bouncing is called the bed load, and it makes up the coarse fraction of a river's sediment. Rivers can be divided into four categories based on their morphology: braided, meandering, anastomosing, and straight. Straight rivers are rare, usually appearing only as portions of one of the other river types, and anastomosing rivers can be considered a special type of meandering river.

Several criteria are used to characterize alluvial systems. They include grain size, dominant mode of sediment transport (suspended load versus bed load), and migrational pattern of the river channel. Alluvial sediments can be broadly divided into three interrelated depositional settings: braided river, alluvial fan, and meandering river.

BRAIDED RIVER

Braided rivers have low sinuosity, which is defined as the ratio of the length of the river channel to the down-valley distance. They are characterized by relatively coarse-grained sediment transported as bed load. Fine-grained sediments make up a minor portion of the deposits. The main channel is internally divided into many subchannels and bars, which give the river a braided pattern. River bars are ridgelike accumulations of sand and gravel formed in the channel or along the banks, where deposition is induced by a decrease in velocity. Transverse bars are flat-topped ridges, oriented transverse to the flow, that grow by down-current additions and migration of sediment. Longitudinal bars are midchannel sand and gravel accumulations oriented with their long axes roughly parallel to current flow. During low-water stages, this braided pattern is very apparent, and water occupies only one or several of the subchannels. It is only during high-water stages that the entire braided channel has water in it. The bars that occur in these rivers form as a result of high sediment loads and fluctuations in river discharge.

A braided stream in Mount McKinley National Park, Alaska; stones were left by the river after a flood. (© William E. Ferguson)

ALLUVIAL FANS

Alluvial fans are deposits that accumulate at the base of a mountain slope. There, mountain streams encounter relatively flat terrain and lose much of their energy, therefore depositing the sediment they were moving. This stream shifts its channel through time, as sediment continues to be deposited, and builds a fan-shaped accumulation of debris that is coarse-grained near the mountain front and becomes progressively finer-grained away from the highland. Alluvial fans are best developed and observed in arid or semiarid climates, where vegetation is sparse and water flow is intermittent. Large quantities of sediment may be moved during short-term flash-flood events. There are also humid-climate fans, such as the enormous Kosi fan in Nepal, which measures 150 kilometers in its longest dimension.

Both arid and humid fans are built at least in part by deposition in a braided stream environment. Stream discharge, and therefore sediment deposition, is discontinuous on arid fans. Braided streams may operate over the fan's entire surface or predominantly on the outer regions, away from the region where the stream leaves the confines of the mountain valley. Arid fans are also built by deposition from mudflows and debris flows. These flows differ from braided stream flows in that they contain less water and much more debris. Humid fans have braided stream systems operating con-

Braided rivers form in regions where sediment is abundant, water discharge fluctuates (but may be high), and, usually, vegetation is sparse. Some braided rivers and streams have flowing water in them sporadically, with long periods of dryness in between. These streams are called ephemeral, or intermittent. Braided rivers tend to have relatively high gradients. As such, they commonly occur as the upper reach of a river that may become a meandering river downstream as the sediment's grain size and the gradient both decrease.

Alluvial fans are deposits that accumulate at the base of a mountain slope. (© William E. Ferguson)

tinuously on their surface. Their overall deposits are similar to braided stream deposits formed in other sedimentary environments.

MEANDERING RIVERS

Meandering rivers have a greater sinuosity than braided rivers and are usually confined to a single channel. They have a lower gradient and therefore are typically located downstream from braided rivers. Sediment in meandering rivers is moved mostly as fine-grained suspended load. Several different types of sedimentary deposit result from the activity of meandering rivers. The coarsest material available to the river is moved and deposited

A low-velocity, meandering stream in Yellowstone National Park. (© William E. Ferguson)

within the deepest part of the channel. These gravelly deposits are thin and discontinuous. Point bars develop on the inside curve of a meander bend and are a major site of sand deposition, although silt and gravel may also be components of point-bar deposits. Deposition takes place on point bars because of flow conditions in the river as water travels around the bend. Erosion takes place on the bank opposite the point bar, and in this way, meanders migrate. When a river floods and tops its banks, water and finer-grained sediment spill out of the channel and onto the surrounding valley floor, which is called the floodplain. On the bank directly adjacent to the river channel, large amounts of sediment are deposited to form natural levees, elongate narrow ridges parallel to the channel; levees help to confine the river but are still topped during major floods. As the river spreads out across the floodplain, silt and clay are deposited.

Meandering rivers are constantly shifting their location within the river valley. In this way, very thick alluvial deposits may accumulate through time. New channels may be created between meanders such that old meander bends are cut off and isolated. These isolated meander bends are termed "oxbow lakes" and tend to fill quickly with sediment.

A special type of meandering river is an anastomosing river. It is characterized by a system of channels that do not migrate as much as meandering river channels and that are separated by large, permanent islands.

TRANSPORT AND DEPOSITION

Alluvial sediments require running water for transport and deposition. This water may be available year-round, such that rivers and streams are constantly active. Sporadic stream discharge produces alluvial sediments in arid and semiarid environments. Alluvial sediments are also associated with glaciers. Large amounts of sediment with a wide range in grain size are deposited directly by glaciers. Streams fed by glacial meltwaters are important and effective agents of transport and deposition of this sediment. Most streams associated with glaciers are bed load streams and therefore have a braided pattern. Their sediment is usually quite coarse-grained.

Alluvial systems form broad, interconnected networks of drainage that feed water and sediment from highlands or mountainous regions to lowlands and eventually to the sea. These drainages form recognizable patterns that are controlled by the type of rock; the type of deformation, if any, that the Earth's crust in the region has undergone; and the region's climate. As river systems age, the landscape changes and evolves. In this way, the surface of the land is sculpted by rivers, in concert with other surface processes.

STUDY OF ALLUVIAL SYSTEMS

The study of alluvial systems can be divided roughly into two categories: the study of modern river systems and the study of sedimentary rocks interpreted to have formed in some type of alluvial system. The study of modern alluvial processes and the sedimentary deposits that they generate is crucial to the understanding of such deposits in the rock record. By understanding modern rivers—the ways in which sediment is moved and deposited, the changes through time in channel shape and location, and the characteristics of the deposits in relation to specific physical conditions—geologists can begin to interpret ancient alluvial sediments.

Modern alluvial systems are studied in a number of ways. It is important to know as much as possible about the flow of water itself, so measurements are made of flow velocity, depth, and width. The channel shape and configuration are also measured. Samples of river sediment, both suspended load and bed load, are collected, and estimates are made of how much sediment is moved by a river. It is also important to look at recent alluvial sediments not now directly associated with the river system. That may be done by digging trenches or collecting samples from floodplains or other alluvial sediments. This type of study looks at the last few hundred or few thousand years of river activity and is critically important to the understanding of these systems. The geologist must study not only what is happening today but also how the system has evolved. In this way, knowledge becomes a predictive tool that greatly enhances the overall understanding of the phenomenon.

It was not until the 1960's that geologists began to realize how large has been the contribution of alluvial systems to sedimentary rocks. This realization came about as a direct result of studies of modern alluvial systems that provided the necessary information to allow the correct identification of ancient alluvial deposits. The study of ancient alluvial deposits takes many forms and provides some information that cannot be gathered from modern deposits. For example, modern deposits, for the most part, have only their uppermost surface exposed, except along the bank and in erosion gullies. Ancient accumulations, in contrast, have been transformed into rock such as shale, sandstone, and conglomerate and commonly are parts of uplifted and dissected terrains, including mountain ranges. These exposures of alluvial deposits allow the three-dimensional architecture of alluvial systems to be studied. Geologists look carefully at vertical changes and associations in the types and abundances of sediment that form as a result of alluvial processes.

It has been found that both meandering and braided rivers commonly form cycles of sedimentation that begin with relatively coarse-grained debris and progress to fine-grained debris. These cycles result mainly from the shifting and migration of the channel system. The coarser-grained base represents the channel and bar deposits, and the finer-grained top represents the overbank floodplain deposits.

The study of ancient alluvial deposits also provides clues to the geologic evolution of a region. The specific mineral composition of sedimentary rocks can indicate the nature of the terrain from which the sediment was derived. This, in turn, contributes to an understanding of the history that led to the generation of the ancient river system that produced the alluvial deposit.

SIGNIFICANCE

Alluvial systems operate over much of the Earth's surface. They move and deposit enormous quantities of sediment each year and are both a boon and a hindrance to humankind. River valleys and floodplains are desirable regions for agricultural development because of the fertile soil often found there. Rivers, however, naturally flood about every 1.5 years. This flooding causes huge losses in property, crops, and sometimes even lives. Flooding associated with alluvial fans can be highly energetic and can occur almost without warning, usually in response to heavy precipitation over a short period. Water levels can build very quickly in narrow valleys with little vegetation and produce a rushing wall of water. Such a flood occurred in 1976 in Big Thompson Canyon, near Rocky Mountain National Park, Colorado.

Many different economically valuable materials are found in alluvial deposits. Because alluvial deposits are relatively coarse-grained, they have spaces between the grains that may contain usable fluids, such as water, oil, and natural gas. Many important aquifers are found in alluvial deposits. Petroleum typically originates in marine deposits, but it commonly migrates and may form reservoirs in alluvial deposits. Most sandy alluvial deposits are composed predominantly of quartz grains; however, concentrations of a number of different minerals and ores, including gold and diamonds, occur in alluvial deposits. Another very important economic resource in alluvial deposits is the sand and gravel itself. This material is used for road construction and in the manufacture of cement and concrete.

The deposition of sediment from river systems represents the wearing away of the land. Much alluvial sediment eventually makes its way to the ocean, where it undergoes reworking by marine processes and deposition on continental shelves or perhaps in the deep sea. Through time, as continents move relative to one another, these sediments may be compressed, folded, and uplifted to become parts of major mountain chains. The Appalachian Mountains in the eastern United States and the Himalaya in northern India and China are only two examples of mountains composed in part of sedimentary rocks, some of which are alluvial in origin.

Bruce W. Nocita

CROSS-REFERENCES

Beaches and Coastal Processes, 2302; Deep-Sea Sedimentation, 2308; Deltas, 2312; Desert Pavement, 2319; Drainage Basins, 2325; Evaporites, 2330; Floodplains, 2335; Lakes, 2341; Reefs, 2347; River Bed Forms, 2353; River Flow, 2358; Sand, 2363; Sand Dunes, 2368; Sediment Transport and Deposition, 2374; Weathering and Erosion, 2380.

BIBLIOGRAPHY

Davis, Richard A., Jr. *Depositional Systems: A Genetic Approach to Sedimentary Geology*. 2d ed. Englewood Cliffs, N.J.: Prentice-Hall, 1992. This college-level textbook has good introductory chapters on alluvial systems and on related subjects, such as sediment transport.

Reading, H. G., ed. *Sedimentary Environments: Processes, Facies, and Stratigraphy*. 3d ed. Cambridge, Mass.: Blackwell Science, 1996. Probably the most comprehensive text available on sedimentary environments. Much of the material is technical, but the text has excellent figures and photographs and will not overwhelm the careful reader.

Schumm, Stanley A. *Active Tectonics and Alluvial Rivers*. New York: Cambridge University Press, 2000. Suitable for college-level readers, this source is packed with information on river systems and tectonics.

Smith, David G., ed. *The Cambridge Encyclopedia of Earth Sciences*. Cambridge, England: Cambridge University Press, 1982. This easy-to-read and thorough source contains a chapter on sedimentation. It covers fluvial, desert, and glacial sediments and includes diagrams and photographs of braided and meandering rivers and an alluvial fan. The production and transportation of sediments are also discussed. For general readers.

Tarbuck, Edward J., and Frederick K. Lutgens. *Earth: An Introduction to Physical Geology*. 6th ed. Upper Saddle River, N.J.: Prentice Hall, 1999. An introductory textbook suitable for high school students. Chapter 9, "Running Water," contains sections on channel deposits (bars), floodplain deposits, alluvial fans, and various types of river. Includes many diagrams and photographs. Review questions and a list of key terms conclude the chapter.

Walker, R. G., ed. *Facies Models: Response to Sea-level Change*. 2d ed. Tulsa, Okla: Society of Economic Paleontologists and Mineralogists, 1994. An excellent compilation on sedimentary systems. Several chapters are devoted to alluvial systems. Suitable for college students.

BEACHES AND COASTAL PROCESSES

The shoreline is the meeting place for the interaction of land, water, and atmosphere, and rapid changes are the rule rather than the exception.

PRINCIPAL TERMS

BEACH: an accumulation of loose material, such as sand or gravel, that is deposited by waves and currents

LONGSHORE CURRENT: a slow-moving current between a beach and the breakers, moving parallel to the beach; the current direction is determined by the wave refraction pattern

LONGSHORE DRIFT: the movement of sediment parallel to the beach by a longshore current

OSCILLATORY WAVE: a wind-generated wave in which each water particle describes a circular motion; such waves develop far from shore, where the water is deep

TSUNAMI: a low, rapidly moving wave created by a disturbance on the ocean floor, such as a submarine landslide or earthquake

WAVE BASE: the depth to which water particles of an oscillatory wave have an orbital motion; generally the wave base is equal to one-half the distance of the length of the wave

WAVE REFRACTION: the process by which the angle of a wave moving into shallow water is changed; the bending that results is also termed wave refraction

BEACH COMPOSITION

The processes that create, erode, and modify beaches are many. Marine processes, such as waves, wave refraction, currents, and tides, work concurrently to modify, create, or erode a beach. This suggests that a beach is a very sensitive landform, and indeed it is. Generally a beach is a deposit made by waves and related processes. Beaches are often regarded as sandy deposits created by wave action; however, beaches may be composed of broken fragments of lava, sea shells, coral reef fragments, or even gravel. A beach is composed of whatever sediment is available. Beaches have remarkably resilient characteristics. They are landforms made of loose sediment and are constantly exposed to wave and current action. On occasion, the coastal processes may be very intensive, such as during a hurricane or tropical storm. Yet, in spite of the intensity of wave processes, these rather thin and narrow landforms, although perhaps displaced, restore themselves within a matter of days. Coastal scientists are inclined to believe that the occurrence and maintenance of beaches are related to their flexibility and rapid readjustment to the varying intensity of persistent processes.

Sediment deposited on beaches is often derived from the continents. Rivers are one source of beach sediment, and the coast is another. As rivers erode the land, the sediment they carry ultimately finds its way to the shore. Because the sediment may be transported several tens or hundreds of kilometers, it is refined and broken down even further to finer-sized particles. Once the river reaches the sea, the sediment is distributed by longshore currents along the shoreline as a beach. Beaches also occur where the shoreline is composed of cliffs, such as along the Pacific coast of North America. Here, waves erode the sea cliffs, and the sediment is deposited locally as a beach. Under these conditions, the beach deposit is most often gravelly because the sediment is transported only a short distance and has not had an opportunity to break up into finer-sized sediment such as sand. Some island beaches are made up entirely of shell fragments.

WAVES

The most obvious energy source working on beaches is waves. Waves approaching a beach are generally created by winds at sea. As wind velocities increase, a wave form develops and radiates

out from the storm. An oscillatory motion of the water occurs as the wave form moves
across the water surface. It is
important to note that the water movement within a wave is
not the same as the movement
of a wave form. In a wave created in deep water that is approaching a beach, the water
particles move in a circular
orbit, and very little forward
movement of the water occurs.
The water at the surface moves
from the top of the orbit (the
wave crest) to the base of the
orbit (the wave trough) and
then back up. Thus the water
particles move in an oscillatory
wave; this motion continues
down into the water. Although

An eroded cliff and sea stack showing tilted bedding in Alaska's Aleutian region. (U.S.
Geological Survey)

the size of the orbits in the water column decreases, motion occurs to a depth referred to as
the wave base. At this point, the depth of the wave
base is less than the depth of the water. As wave
crests and troughs move into shallow water, the
water depth decreases to a point where it is equal
to the wave base. From this point, the orbital motion is confined because of the shallowness of the
water and takes on an elliptical path. The ellipse
becomes more confined as waves enter shallower

water and eventually becomes a horizontal line. At
this point, there is a net forward movement of water in the form of a breaker.

As a wave enters shallow water, many adjustments occur, such as a change in the orbital path
of water particles already described and a change
in the velocity of the wave form. Since the wave
base is "feeling the bottom" in shallower water,
friction occurs, slowing the wave down. As seen
from an airplane, waves entering shallow water do
so at an angle, not parallel to
the shoreline. Therefore, one
part of the wave enters shallow
water and slows down relative
to the rest of the wave. Thus a
part of the wave crest is feeling
bottom sooner than the rest of
the wave. Since the wave crests
and troughs have different velocities, the wave refracts or
bends. In so doing, the wave
crests and troughs try to parallel the shallow bottom topography, which they have encountered.

The wave refraction is seldom completed, and the breaking wave surges obliquely up
the slope of the beach and then

A sandy beach with ripple marks. (© William E. Ferguson)

Near Manati, Puerto Rico, a reverberation, or "echo," bay formed behind a breach in a wind-built ridge of sand. (U.S. Geological Survey)

water, which triggers waves. The waves are low, subdued forms traveling thousands of kilometers over the ocean surface at extremely high velocities, often in excess of 800 kilometers per hour. As a tsunami approaches shallow water or a confined bay, its height increases. There is no method for direct measurement of heights of tsunami waves; however, in 1946, a lighthouse at Scotch Cap, Alaska, located on a headland 31 meters above the Pacific Ocean, was destroyed by waves caused by a landslide-generated tsunami.

The changing character of a beach is very dynamic because the properties of waves are variable. During storm conditions or when more powerful waves strike a beach during the winter season, for example, the beach commonly is eroded, has a steeper slope, and is composed of a residue of coarser sediment, such as gravel, that is more difficult to remove. During fair-weather or summer conditions, however, beaches are redeposited and built up.

returns perpendicular to the shoreline. The result is a current that basically moves water in one direction parallel to the beach in a zig-zag pattern (a longshore, or littoral, current). It is a slow-moving current that is located between the breaking wave out in the sea and the beach. Because longshore-current movement operates in shallow water and along the beach, it is capable of transporting sediment along the shoreline. "Longshore drift" refers to the movement of sediment along beaches. In a sense, the longshore current is like a river moving sand and other material parallel to the shore. Beaches are always in a state of flux. Although they appear to be somewhat permanent, they are constantly being moved in the direction of the longshore current. Along any shoreline, several thousand cubic meters of sediment are constantly in motion as longshore drift. Along the beaches of the eastern United States, about 200,000 cubic meters of sediment are transported annually within a longshore drift system.

A different type of ocean wave is one that is generated on the ocean floor rather than by the wind. Such waves are properly termed tsunamis. Although they popularly have been coined "tidal waves," they are completely unrelated to tides or the movement of planets. Some type of submarine displacement, such as the creation of a volcano, a landslide on the seafloor, or an earthquake beneath the ocean bottom, causes a displacement of

STUDY OF BEACHES AND COASTAL PROCESSES

The study of beaches and coastal processes is not particularly easy because of constant wave motion and changes in the beach shape. Scientists have, however, developed field methods as well as laboratory techniques to study these phenomena. To study wave motion and related current action, several techniques have been devised to include tracers, current meters, and pictures taken from satellites and aircraft. Two types of tracers are commonly used to determine the direction and velocity of sand movement: radioactive isotopes and fluorescent coatings to produce luminophors. (Luminophors are sediment particles coated with selected organic or inorganic substances that glow under certain light conditions.) In the former type, a radioactive substance such as gold, chromium, or iridium is placed on the surface of the grains of sediment. Alternatively, grains of glass may be coated with a radioactive element. The radioactive sediment can then be detected relatively easily and quickly with a Geiger counter. Both techniques can trace the direction and abundance of sediment along the seashore. Bright-colored dyes or current meters can also be used to document the direction of longshore currents. Surveying instruments are used to determine the high and low topography of beaches and adjacent sand bars. By measuring beach topography before and after a storm, for example, scientists can record changes. In this way, they can document the volume and dimensions of beach erosion or deposition that has taken place during a storm. Measurement of wave height and other wave characteristics can be done with varying degrees of sophistication. Holding a graduated pole in the water and visually observing wave crest and troughs is simplest and cheapest. To achieve more refined measurements, scientists place pressure transducers or ultrasonic devices on the shallow seafloor to record pressure differences or fluctuations of the sea surface. These more precise instruments also record wave data for later study and analysis.

All these methods are detailed field techniques. By comparing aerial photographs, detailed maps, satellite pictures, and in some cases government studies in a coastal sector, changes over long periods of time may be detected. Finally, because wave motion cannot be controlled on a shoreline, wave tank studies are used to derive wave theories. Normally, an elongated glass-lined tank with water 0.3 to 1 meter in depth is used. A wave machine creates waves at one end of the wave tank and sediment is introduced at the other end. The heights and other characteristics of the waves can be varied, as can the type of sediment, to form the beach. In this controlled way, the various beach and wave relationships can be studied.

SIGNIFICANCE

More than 70 percent of the population of the United States lives in a shoreline setting, and thus an understand-ing of how waves interact with beaches is important. Shoreline property is highly prized and hence valuable because of the demand for it. On many beaches, the great investments made in hotels and condominiums suggest that beaches are the most sought-after environment on Earth; shoreline frontage is sold by the foot or meter, not by the acre or hectare. Currently, however, such demand and investment are threatened with rising sea levels and continued coastal erosion.

Beaches represent the line of defense against wave erosion. Waves generated in the open sea slow down in shallow water, and the beach deposit absorbs the impact of waves. Beaches are therefore constantly changing and are one of the most ephemeral environments of Earth. Thus a sound knowledge of longshore currents and beach development is necessary prior to nearshore or marine construction. Sea walls and groins, for example, interfere with waves and longshore currents and may cause considerable erosion in selected areas: Sea walls are often constructed perpendicular to a shoreline to slow longshore currents so that a beach can be deposited. Downcurrent, beyond the area of beach deposition, erosion will take place. Similarly, rivers—a major source of the sediment that creates beaches—are sometimes dammed, thus depriving beaches of sediment and resulting in their erosion. Unless planners, developers, and

A barrier sand spit along the South Carolina coast, deposited by storm outwash among other processes. (U.S. Geological Survey)

builders understand these processes, major damage can result: Failure to understand coastal processes has, on occasion, caused a riparian property owner to sue a neighbor who caused beach erosion to take place.

Beaches are in a sense climatic barometers that record changes of sea level. The warmer atmosphere has led to a rising sea level as a result of glacial melting and warmer water temperatures. Beaches do indeed detect changes in the local sand supply, as well as global changes in the atmosphere.

C. Nicholas Raphael

CROSS-REFERENCES

BIBLIOGRAPHY

Bascom, Willard. "Beaches." *Scientific American* 203 (August, 1960).

_____. "Ocean Waves." *Scientific American* 201 (August, 1959). Although older, both of these articles are still current. Well illustrated with photos and diagrams. Technical concepts are explained in language a nonscience layperson can understand. Included is a discussion of tsunami or tidal wave, wave properties, and wave refraction. The more recent article is a continuation of the wave article; it presents aerial photographs of the impact of seawalls and related structures on the beaches.

_____. *Waves and Beaches.* Garden City, N.Y.: Anchor Books, 1964. A good introductory pocket book for nonscientists by an expert. Numerous examples and illustrations have made this softcover book popular. Nonmathematical and nonscience laypersons should have no difficulty with this book.

Bird, Eric. *Beach Management.* New York: John Wiley, 1996. Includes chapters on waves and on beaches. Although most examples are Australian, the book covers fundamental concepts. A good introduction for anyone who has had a high-school-level Earth science course.

Cameron, Silver Donald. *The Living Beach.* Toronto: Macmillan Canada, 1998. This book looks at the ecosystems and biological systems of North American beaches. There is a focus on the relationships between beaches and the environment. Index and bibliography.

Kaufman, Wallace, and Orrin Pilkey. *The Beaches Are Moving: The Drowning of America's Shoreline.* Durham, N.C.: Duke University Press, 1983. This thought-provoking book, written in a nontechnical style, deals with the processes working in the coastal zone, such as winds, waves, and tides. The impact of rising sea levels and the modification and urbanization of the coast are highlighted. A narrative text suitable for all ages.

Komar, Paul D. *Beach Process and Sedimentation.* Upper Saddle River, N.J.: Prentice Hall, 1998. Extensive treatment of waves, longshore currents, and sand transport on beaches. Equations and mathematical relationships are presented and elaborated upon. College-level material. This book is for those interested in the specifics of coastal processes.

Leatherman, Stephen P. *Barrier Island Handbook.* 3d ed. College Park, Md.: Laboratory for Coastal Reasearch, University of Maryland, 1988. Based on actual field studies along the East Coast of the United States. Numerous photographs, diagrams, and tables. Most suitable for coastal managers and government employees; however, very readable and suitable for nonscientists as well as the general scientist. Emphasizes the dynamic nature of beaches, recreation and construction impacts, and nearshore processes.

Lencek, Lena, and Gideon Boskar. *The Beach: The History of Paradise on Earth.* New York: Viking, 1998. A natural history of the environmental and social importance of coastal areas. Illustrations.

Leonard, Jonathan Norton. *Atlantic Beaches.* New York: Time-Life Books, 1972. A regional travel description of the shoreline from Cape Cod, Massachusetts, southward to Cape Hatteras and the Outer Banks, North Carolina. Information is presented in a nonscience narrative form. Color photography is excellent. Useful in planning trips along the East Coast of the United States, as the emphasis is on the scenery of the seascape.

Pethick, John. *An Introduction to Coastal Geomorphology.* Baltimore, Md.: Edward Arnold, 1984. A thorough survey of coastal processes, this 260-page book is divided into three sections: wave energy on the coast and its characteristics, the relationship between currents and the movement of beach material along the shore, and the landforms, such as beaches, mud flats, and estuaries. Most suitable for anyone needing an equation or a technical explanation of selected processes operating in a coastal zone.

Thorsen, G. W. "Overview of Earthquake-Induced Water Waves in Washington and Oregon." *Washington Geologic Newsletter* 16 (October, 1988). A nine-page introduction to the impact of a tsunami on the coasts of Washington and Oregon. An earthquake occurred in March, 1964, in Alaska, and the waves traveled southward along the Pacific coast of North America. The tsunami is discussed in nontechnical language. Damage and economic losses are estimated. Maps and tidal gauge records presented. A good review of wave activity, intended for interested laypersons and teachers.

Walker, H. J. "Coastal Morphology." *Soil Science* 119 (January, 1975). A nontechnical overview of coastal landforms. Discussion includes beaches, deltas, and lagoons. The world view is taken, and maps illustrating processes are included. This fifteen-page article is useful for a nonscientist interested in the causes and distribution of coastal features from a geographical perspective.

DEEP-SEA SEDIMENTATION

Deep-sea sedimentation occurs by the settling of particles to the ocean floor and by the transport of material from shallow to deep water. Knowledge of the distribution of sediments and the processes of sedimentation will help to evaluate the potential of the world's oceans for the mining of natural resources and the storage of waste.

PRINCIPAL TERMS

CALCAREOUS OOZE: sediment in which more than 30 percent of the particles are the remains of plants and animals whose skeletons are composed of calcium carbonate

CARBONATE COMPENSATION DEPTH (CCD): the depth in the oceans at which the rate of supply of calcium carbonate equals the rate of dissolution of calcium carbonate

CLAY: a particle less than 0.1 millimeter in diameter

CLAY MINERALS: a group of hydrous silicate minerals characterized by layers of atoms that form thin sheets that are held together loosely

PELAGIC: meaning "of the deep sea"; it refers to sediments that are fine-grained and are de-

posited very slowly at great distances from continents

SILICEOUS OOZE: fine-grained sediment in which more than 30 percent of the particles are organic remains of plants and animals whose skeletons are composed of silica

TURBIDITY CURRENT: a mass of water and sediment that flows downhill along the bottom of a body of water because it is denser than the surrounding water; common on continental slopes

UPWELLING: an ocean phenomenon in which warm surface waters are pushed away from the coast and are replaced by cold waters that carry more nutrients up from depth

INVESTIGATION OF DEEP-SEA SEDIMENTS

Sedimentation in the deep sea differs substantially from that in any other environment. Deep-sea sedimentation is pelagic. It occurs at great distances from continents at depths in excess of 1,000 meters primarily by the settling of particles through the overlying water. Every 100,000 liters of seawater contains 1 gram of extremely small particles. Settling is extremely slow, and sediments accumulate at rates on the order of a few millimeters per thousand years. Transport of material by turbidity currents along the bottom of the ocean from shallow to deep regions is also a mechanism by which deep-sea sediments are deposited, but this is far less important than settling of particles from the surface. In contrast, current transport dominates sedimentation in all other environments, where sedimentation rates typically exceed several centimeters per year.

The oceans cover 70 percent of the Earth, and, as a result, deep-sea sedimentation may be the most common sedimentary process. Early investigations of the seas were limited, however, to

coastal processes and oceanic circulation because of the lack of techniques available to study the deep ocean bottom. Knowledge of deep-sea sedimentation was profoundly increased by the voyage of HMS *Challenger* from 1872 to 1876, which made the first systematic study of the ocean floor. Prior to the *Challenger* expedition, practical investigations of the oceans concentrated on winds, tides, and water depths in harbors to provide information for commercial navigation and for the laying of submarine telegraph cables. The *Challenger* crew collected data on the temperature and depth of the oceans, marine plants and animals, and sediments on the ocean bottom. Deep-sea sediments were described by John Murray, naturalist to the expedition. He noted that sediments in the deepest parts of the ocean are fine-grained and are composed of particles that settled from the surface. Comparison of rocks exposed on land with those dredged from the ocean led to the assertion that deep-sea sediments could not be found on land. This conclusion caused great debate, and the interpretation of some rocks as deep-sea deposits by a few scientists persisted despite opposition by most of the scientific community.

Investigation of deep-sea sedimentation flourished again during and immediately after World War II. Despite the voluminous amount of data available by 1950, the prevailing doctrine in the first half of the twentieth century was similar to that in the previous century: The continents and oceans were permanent features of the Earth's surface that developed close to their present form near the beginning of geologic time. Thus, the ocean basins were billions of years old; ocean-bottom sediments were tens of kilometers thick, having accumulated from early in the history of the Earth to the present, and deep-sea sediments were not exposed on land, a point that was still severely contested. Then, during the 1950's two discoveries shocked the oceanographic community. First, fossils taken from seamounts (submarine mountains) in the Pacific Ocean at depths of 2 kilometers were only 100 million years old, suggesting that subsidence of the ocean floor was recent. Second, deep-sea sediments were found to be less than 200 meters thick. The advent of the theory of plate tectonics in the 1960's profoundly changed ideas about the ocean basins by providing both an explanation for the two observations of the previous decade and a mechanism for the emplacement of deep-sea sediments onto continents. In addition, the motion and cooling of the oceanic plates predicted by plate tectonics explain the global distribution of deep-sea deposits.

PELAGIC SEDIMENTS

Pelagic sediments are characterized by their fine grain size. Few particles are larger than 0.025 millimeter, and most are smaller than 0.001 millimeter. Particles include terrigenous debris (derived from the continents), such as clay minerals, quartz, and feldspar; volcanogenic grains (derived from volcanoes), such as volcanic glass, pumice, and ash; biogenic material (derived from living organisms), such as fecal pellets (waste produced by organisms inhabiting the surface waters) and skeletons of planktonic plants and animals; and cosmogenic matter (derived from space or the cosmos), such as pieces of meteorites. Pelagic sediments that contain more than 30 percent biogenic material are oozes. If the biogenic debris is composed of calcium carbonate, the sediment is a cal-

careous ooze. If the biogenic debris is composed of silica, the sediment is a siliceous ooze. Most calcareous oozes consist of foraminiferans, which are single-celled animals that secrete calcium carbonate and live in the surface waters. In contrast, siliceous oozes contain diatoms (green, unicellular algae) in cold water and radiolarians (silica-secreting single-celled animals) in warm water. Pelagic clays are sediments that contain less than 30 percent biogenic debris.

The two major controls on pelagic sedimentation are the calcium compensation depth (CCD) and the fertility of the surface waters. The CCD reflects the interplay between the release of carbon dioxide during the decay of a surface organism and the dissolution of its calcium carbonate skeleton as it descends through the ocean. Above the CCD, calcareous oozes are dominant, whereas below the CCD, siliceous oozes and pelagic clays are common. For example, in one region of the ocean, sediments decreased from 90 percent calcium carbonate at a depth of 1,000 meters to 20 percent calcium carbonate at a depth of 5,000 meters. Oozes accumulate below areas of high fertility where upwelling brings nutrient-rich bottom water to the surface. Because upwelling creates increased productivity of both calcareous and siliceous organisms, siliceous oozes form only where the water depth exceeds the CCD. Calcareous oozes, the most abundant biogenic sediments in the ocean, cover only 15 percent of the Pacific Ocean bottom but blanket 60 percent of the Atlantic Ocean floor. This difference reflects the combination of a shallower CCD and greater water depth in the Pacific Ocean than in the Atlantic Ocean. Siliceous oozes are in areas of elevated CCDs such as the equatorial regions, the subarctic and subantarctic zones, and the continental margins of northwestern South America and eastern Africa.

Sedimentation of pelagic clays is restricted to the central portions of the oceans where the fertility is low. This reflects the difference in settling rates between nonbiogenic particles and fecal pellets. Because of its small size, a particle may take many years to sink through the ocean. The concentration of many particles into fecal pellets by organisms feeding at the surface, therefore, is thought to be an important mechanism by which debris settles to the bottom. The greater size of the pellets allows them to fall much faster than other particles. It is only in areas of low productivity, therefore, where biogenic material does not overwhelm the sediment. Wind is the primary means by which nonbiogenic particles reach the sea. Dust occurs in the atmosphere at great distances from continents and constitutes as much as 10 percent of pelagic sediment.

OTHER PROCESSES

Additional processes in deep-sea sedimentation include the rafting of glacial debris by ice, the transport of terrigenous and shallow-water material by turbidity currents along the ocean floor, and the precipitation of nodules on the ocean bottom. Pieces of ice calve off glaciers in the polar regions and float toward warmer waters where they eventually melt, releasing glacial sediment into the ocean. Although this process is most common in the south polar region today, glacial sediments greater than ten thousand years old elsewhere on the ocean floor suggest the process extended over a much greater area in the past. Turbidity currents generated along continental slopes may traverse hundreds of kilometers of ocean basin where ridges and depressions do not occur. Far away from continents, the currents commonly contain only silt but may deposit the grains over great distances. These currents also may locally erode the seafloor, creating breaks in the continuity of the sediment pile by removing the most recent sediments and exposing older sediments at the surface. Chemical reactions of seawater with the surface of the sediment cause the precipitation of ferromanganese (composed of iron and manganese) nodules in areas where sediments accumulate slowly. The nodules initiate by nucleating around manganese in the sediment and by accreting minerals from the seawater. Nodules may cover more than 40 percent of the seafloor.

Sedimentation in the deep sea is a complex interaction between water depth, fertility of the surface waters, and current transport. To illustrate how the depth dependence of sedimentation affects the global distribution of sediments, one must understand the relationship between plate tectonics and the topography of the ocean floor. At a mid-ocean ridge, which rises a few thousand meters above the surrounding ocean basin, molten rock is extruded to form a new piece of seafloor. Calcareous oozes accumulate on the new seafloor because

of its depth above the CCD. As another piece of seafloor forms, the older piece moves away from the mid-ocean ridge toward the deep ocean basin (seafloor spreading), cools, and sinks. The old seafloor gradually descends below the CCD, where siliceous oozes accumulate above the calcareous oozes. Eventually the piece of seafloor passes below a region of low surface productivity such that pelagic clays are deposited above the siliceous oozes. This vertical sequence has been documented in several localities throughout the world's oceans.

STUDY OF DEEP-SEA SEDIMENTATION

Techniques used to study sedimentation in the deep ocean are diverse and include piston coring, seismic reflection profiling, the use of deep-diving submarines, and experiments performed both at sea and in the laboratory. Piston coring is the most important technique because it enables scientists to sample the ocean floor directly. It uses the same kind of rotary-drilling equipment that is standard in petroleum exploration on land to drill into the top few hundred meters of the seafloor. The cores are 7 centimeters in diameter and provide samples that are large enough to preserve features that yield insight into the process by which the sediment accumulated. For example, current-produced features in some samples were the first direct evidence of currents in the deep seas. In addition, piston coring confirmed the vertical sequence of sediments predicted by plate tectonics.

To map the ocean floor using a 7-centimeter drill core is time-consuming and inefficient. Seismic reflection profiling has the advantage of providing information about the thickness and distribution of layers in the sediments on the seafloor relatively quickly. The two major disadvantages are that it cannot determine the composition of the sediment and that it only identifies layers that are tens of meters thick. In seismic reflection profiling, a low-frequency sound source is towed over the bottom. The sound waves are reflected by the seafloor and by layers several kilometers below the boundary between the sediment and the water. The returning signals are displayed on strip charts that reveal the water depth and the thickness and pattern of layers in the sediments. One of the most significant early contributions of seismic reflection profiling was the evidence that sediments on the ocean floor were too thin to have been accumulating since early in the history of the Earth, an observation explained by plate tectonics.

The use of deep-diving submarines has allowed scientists to observe the ocean floor directly. Previously the ocean floor was only photographed; deep-diving submarines allowed scientists to confirm or reject hypotheses based on other indirect techniques. The ranges and endurance times of the submersible vessels, however, are limited. In addition, they frequently cannot be used in areas where current velocities are greater than 1 meter per second.

Simple experiments either at sea or in the laboratory also can be illuminating. For example, the existence of the CCD was verified by lowering spheres of calcium carbonate to various water depths in the Pacific Ocean. When the spheres were examined, it was discovered that the sphere in the deepest water had dissolved the most. Scientists in the laboratory conduct experiments on settling rates by dropping particles of various sizes in large beakers and measuring the time necessary for the particles to reach the bottom. Thus the amount of time represented by the thickness of sediment on the ocean floor was estimated at a few hundred million years.

One of the most important developments in deep-sea sedimentation was the creation in 1965 of the Deep Sea Drilling Project (DSDP), now known as the Ocean Drilling Project (ODP), to drill and investigate the ocean floor. The project began in earnest in 1968 with the maiden voyage of the American ship the *Glomar Challenger*, which was outfitted with state-of-the-art navigational, positioning, and drilling equipment. International teams of scientists boarded the ship in ports all over the world and remained at sea for two months. Since its inception, ODP has undertaken more than sixty-five voyages and contributed significantly to scientists' understanding of the ocean.

SIGNIFICANCE

The oceans cover nearly three-quarters of the surface of Earth. To study deep-sea sedimentation, therefore, is to understand one of the most globally prevalent processes. Understanding deep-sea sedimentation, however, is important for several other reasons. First, sediments on the ocean floor yield insight into bottom currents and water depths, which affect the navigation of submarines

and the installation of underwater communications cables. Second, the sediments preserve the climatic record of the past few hundred thousands of years. For example, glacial deposits approximately ten thousand years old are found over a much greater area of the bottom of the ocean than recent glacial deposits, suggesting that transport of glacial debris by ice was very common ten thousand years ago. This implies that the temperature of the Earth was lower in the past, to allow the formation of significant quantities of ice. Scientists can use this information to gauge present-day and predict future fluctuations in the global climate. Third, the seas receive a significant amount of the garbage produced by modern society. Knowledge of sedimentation rates and currents in the deep sea helps to establish the length of time necessary to bury the waste, the direction the waste will travel prior to its burial, and the effect of dumping waste on the overall health of the ocean.

Fourth, and most important, understanding deep-sea sedimentation and the global distribution of deep-sea sediments provides a framework within which one can estimate the economic potential of the deep sea and minimize the damage of ocean exploitation. As the reserve of natural resources on the continents dwindles because of expanding demand from high-technology industries, exploding population growth, and increasing consumer appetite, attention focuses on the oceans as a possible source of raw materials. Important resources that may be found locally in the deep sea are oil and gas. The decay and accumulation of plants and animals are essential to the generation of hydrocarbons and are processes that occur over a large area of the seafloor. Emphasis also is placed on mining the deep sea for important materials such as manganese, gold, cadmium, copper, and nickel. Similar to most new industries, large-scale exploitation of the natural resources of the deep sea is limited at present by the current state of technology. The incentive to mine the oceans, however, will grow as cheaper and more sophisticated technologies are created.

Pamela Jansma

CROSS-REFERENCES

Bibliography

Andel, Tjeerd Hendrik van. *Tales of an Old Ocean.* New York: W. W. Norton, 1978. A descriptive story of the evolution of thought on the oceans. Chapters discuss deep-sea sedimentation in general terms. Additional material covers the topography of the seafloor and the theory of plate tectonics. Highly enjoyable for anyone interested in the oceans.

Blatt, Harvey. *Petrology: Igneous, Sedimentary, and Metamorphic.* 2d ed. New York: W. H. Freeman, 1996. This textbook describes sedimentary rocks and processes. An excellent introduction to mechanics of sedimentation, classification of sediments, features of sediments that form under different conditions of water flow, environments of sedimentation, and transformation of sediments into rocks. Most of the discussion is purely descriptive. Some chapters assume some quantitative knowledge. Suitable for the college-level student and for the layperson who is technically inclined.

Davis, Richard A., Jr. *Oceanography: An Introduction to the Marine Environment.* 2d ed. Dubuque, Iowa: W. C. Brown Publishers, 1991. This text is written to acquaint students with the principles of physical, chemical, biological, and geological oceanography. Included are sections covering a wide range of subjects and oceanic research, from the origin of the oceans and deep-ocean floor to seawater chemistry and marine resources. Broad in scope and shallow in depth, this book is best suited to a nonscience major or a person entering teaching where a large knowledge base is important.

Grasshoff, Klaus, Klaus Kremling, and Manfred Ehrhardt, eds. *Methods of Seawater Analysis.* 3d ed. New York: Wiley-VCH, 1999. This third edition of a widely used college text is completely revised and extended, devoting more attention to the techniques and instrumentation used in determining the chemical makeup of seawater. Illustrations, bibliography, and index.

Murray, John, and Alphonse J. Renard. *Deep Sea Deposits.* Edinburgh: Neill, 1891. The original volume written by two members of the crew of the *Challenger,* which made the pioneering voyage in deep-sea studies. The classification scheme for deep-sea sediments is still used. An excellent source for the historical framework of deep-sea investigations, the book dramatically emphasizes the importance of observation in science.

Ocean Science: Readings from "Scientific American." San Francisco: W. H. Freeman, 1977. A collection of articles previously printed in *Scientific American.* Articles include discussions of all aspects of the ocean, including circulation, composition of seawater, topography of the oceans, plate tectonics, and deep-sea sedimentation. A good source of information on natural resource exploitation and waste disposal in the world's oceans. Well illustrated.

Press, Frank, and Raymond Siever. *Understanding Earth.* 2d ed. New York: W. H. Freeman, 1998. An introductory text on the evolution of the Earth, designed to accompany a beginning class on geology. Several chapters discuss sedimentary rocks, the oceans, plate tectonics, and deep-sea sediments. Especially suited for high school students.

Reading, H. G., ed. *Sedimentary Environments: Processes, Facies, and Stratigraphy.* 3d ed. Cambridge, Mass: Blackwell Science, 1996. Discusses sediments in terms of the environment (beach, pelagic) in which they occur instead of by type (sand, mud). The chapter on pelagic environments is very detailed and is an excellent text for the comparison of modern and ancient deep-sea sediments. Exceptionally well illustrated; the reference list is thorough. Suitable for college-level students.

Shephard, F. P. *Submarine Geology.* 3d ed. New York: Harper & Row, 1973. A complete description of sediments and the physiographic provinces of the oceans, including sections on continental margins and mid-ocean ridges. Also discusses methods used by geologists to investigate the oceans and the relationship between plate tectonics and deep-sea sedimentation. A good historical treatment of the study of the oceans. Suitable for college-level students.

Thurman, Harold V. *Essentials of Oceanography.* 6th ed. Upper Saddle River, N.J.: Prentice-Hall, 1999. An introductory college text designed to give the student a general overview of the oceans in the first few chapters. This overview is followed by a well-designed, in-depth study of the ocean—involving chemistry of the ocean, currents, air-sea interactions, the water cycle, and marine biology—and a well-developed section on the practical problems resulting from human interaction with the ocean, such as pollution and economic exploitation.

Usdowski, Eberhard, and Martin Dietzel. *Atlas and Data of Solid-Solution Equilibria of Marine Evaporites.* New York: Springer, 1998. This college-level textbook offers the reader an illustrated guide to sea-water composition, including phase diagrams of sea-water processes. The book is accompanied by a CD-ROM that reinforces the concepts discussed in the chapters.

DELTAS

Deltas are dynamic sedimentary environments that undergo rapid changes over very short periods. Found in lakes and shallow ocean waters, they are rich in organic material and provide food and shelter for fish and wildlife.

PRINCIPAL TERMS

CREVASSE: a break in the bank of a distributary channel causing a partial diversion of flow and sediment into an interdistributary bay

DELTA: a deposit of sediment, often triangular, formed at a river mouth where the wave action of the sea is low

DISTRIBUTARY CHANNEL: a river that is divided into several smaller channels, thus distributing its flow and sediment load

GEOARCHAEOLOGY: the technique of using ancient human habitation sites to determine the age of landforms and when changes occurred

INTERDISTRIBUTARY BAY: a shallow, triangular bay between two distributary channels; over time, the bay is filled with sediment and colonized with marsh plants or trees

NATURAL LEVEE: a low ridge deposited on the flanks of a river during a flood stage

PRODELTA: a sedimentary layer composed of silt and clay deposited under water; it is the foundation on which a delta is deposited

SEDIMENT: fragmented rock material composed of gravel, sand, silt, or clay that is deposited by a river to form a delta

WAVE ENERGY: the capacity of a wave to erode and deposit; as wave energy increases, erosion increases

FORMATION AND SHAPE

Deltas contain many valuable resources. Government agencies such as the U.S. Fish and Wildlife Service study the surface properties of deltas because of the enormous wetlands and abundant wildlife that occupy these landforms. Geologists study deltas because they are favored places for the accumulation of oil and gas resources. This low topographical feature serves society in many ways, and that is why it has been the object of intense study.

Deltas are deposits of sediments, such as sand or silt, that are carried by rivers and deposited at the shoreline of a lake, estuary, or sea. As the river meets the water body, its velocity is greatly decreased, causing the river sediment to be deposited. If the accumulated sediment is not removed by waves or currents, a delta will accumulate and continue to extend itself into the lake or ocean. The term "delta" is used to describe this depositional landform because it is often triangular. It is believed that the Greek historian Herodotus coined the term with the shape of the Greek letter *delta* in mind. Herodotus visited Egypt, where the Nile Delta is located, and he correctly defined the shape of that delta; not all deltas, however, are triangular.

The Ganges Delta, the Colorado River Delta, and many other deltas have different shapes. The shape postulated by Herodotus is, in fact, somewhat unusual, but it is applicable to the Nile and the Mississippi River Deltas. The Nile Delta has a smooth but curved shoreline—it is an "arcuate" delta—whereas the Mississippi Delta has spreading channels and resembles the digits of a bird's foot. Occasionally, current and wave action at the shoreline causes sediment to be distributed to the left and right of a river channel, forming smooth beaches on either side. Such a delta is shaped like a cone, the point of which projects toward the sea, and is called a "cuspate" delta. The Tiber River, which empties into the Mediterranean Sea, is a classic example of this type of delta. Rivers such as the Seine, in France, may deposit sediment in elongated estuaries, forming shoals and tidal flats.

Earth scientists have noted that the shapes of deltas are associated with several conditions, such as the character of river flow, the magnitude of wave energy and tides, and the geologic setting. The bird-foot delta of the Mississippi River has extended itself well into the Gulf of Mexico because the river carries and deposits a high volume of sediment on a shallow continental shelf. The wave and tidal forces are low, and the delta deposit is not redistributed along the shoreline or swept away. Conversely, a cuspate delta, such as that of the Tiber, is a product of strong waves moving over a steep continental shelf. Persistent high wave energy redistributes the sediment often, forming beaches and sand dunes along the delta shoreline. The Nile Delta is an arcuate delta characterized by moderately high wave energy and a modest tide range. Occasional high wave conditions deposit beaches and sand dunes along the arc-shaped delta front at the Mediterranean Sea. Tides also play a direct role in the creation of deltas. Deltas in estuaries are formed because of a high tidal range coupled with low wave conditions. The Seine estuary, with its distinctive mud flats exposed at low tide, provides a good example.

LANDFORMS

Although deltas have different shapes that reflect differences in the intensity of river, wave, current, and tidal processes, certain landforms may be identified as characteristic of delta formation. Submarine features are deposited below sea level, and subaerial features form at or just above sea level. As a river empties into the sea, the finest sediments, usually very fine silt or clay, are deposited offshore on the seafloor. This submarine deposit forms the foundation on which the delta sits and is appropriately referred to as a "prodelta deposit." The deposit can often be detected on navigation maps as a relatively shallow, semicircular deposit under the water.

As deposition continues, the prodelta deposit is covered with the extending subaerial delta, which is composed of coarser sediments. Deltaic extension occurs along the distributary channels. During higher river flow, the distributary channels overflow, depositing natural levees along their sides. The digitate distributary pattern of the Mississippi Delta illustrates this process very well. As the distributaries extend to deeper water, the shallow areas between the distributaries are better developed. These areas, known as interdistributary bays, are shallow landforms colonized by aquatic plant life. Over time, deposition occurs in the interdistributary bays through breaches in the nat-

ural levees. As the river mouth distributaries enter a flood stage, the lower regions of a natural levee are broken, and fine suspended sediments are introduced into the interdistributary bay area. Such

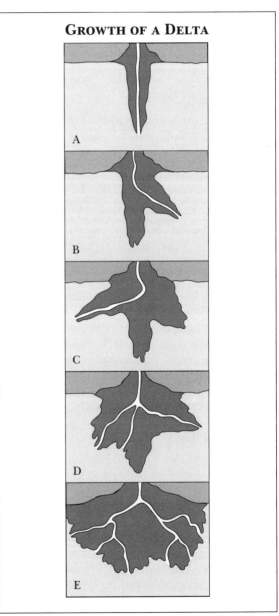

Deltas form at the point where a stream or river terminates in a body of water such as a lake or gulf. As the stream carries sediment into the lake, sand and other debris builds (panel A). Eventually the stream naturally seeks a shorter route to the lake (panels B and C), and as sediment continues to build, several channels may be formed, resulting in an increasing number of routes being etched through to the lake (panels D and E).

overbank splays, or crevasse splays, are primarily responsible for the infilling of a delta. The crevassing is usually a very rapid but short-lived process, occurring during a high river stage and operating over a ten- to fifteen-year period. With the passage of time, the openwater interdistributary bays are silted and colonized. Eventually, however, the marshy bays may subside because of the compaction of the sediment, creating water areas once again.

GEOLOGIC HISTORY

Although the geologic history of large deltas such as the Mississippi is complex, the succession and behavior of shifting deltas have been determined in some detail. Over the past twenty thousand years, the large continental glaciers that occupied much of the upper midwestern United States began to melt. As the climate of the Earth continued to warm, the meltwater was returned to the oceans, and the sea level rose some 100 meters, inundating valleys that had previously been cut by streams. The oceans reached their present approximate level about five thousand years ago. The Mississippi and similar valleys were flooded and became elongated bays. Over time, the shallow water bays were choked with sediments that formed broad floodplains extending down the valleys. Once a depositing river extended beyond the confines of its valley, a delta was deposited in deeper water. Because the river was no longer confined, it was free to shift over great distances. The Mississippi River Delta is actually composed of seven distinct delta lobes extending over an approximate distance of 315 kilometers. The oldest delta, Salé Cypremort, was deposited some 4,600 years ago; the most modern delta was deposited within the past 550 years. Older Mississippi Delta lobes, such as the Teche Delta, have subsided since they were deposited, giving an opportunity for a more recent delta (in this case, the Lafourche) to be deposited on top and more seaward of the older feature. The different delta lobes making up the enormous deltaic plain have resulted from a shifting of the Mississippi River well upstream in its valley. This process is analogous to a hand movement occurring because of a shoulder movement.

Because of significant changes in the shoreline environments, many deltas in marine coastal zones are eroding. The Mississippi River is at the edge of the continental shelf and cannot build out into deep water. Also, subsidence and a slight rise in sea level are causing the delta to erode. The Nile Delta is eroding as well. With the construction of the High Aswan Dam upstream, there has been a decrease in the sediment supplied to the Nile Delta. This lack of sediment, along with a slight rise in sea level, has led to erosion. Some Earth scientists have suggested that the wave action in coastal Egypt is cutting back the Nile Delta at a rate of 15 to 30 meters per year in some areas.

STUDY OF DELTAS

Deltas are difficult to study because many of the features are very flat, marshy, or under water. Since deltas change rapidly over only a few decades, however, maps are an important tool with which to determine changes. Navigation maps and maps that illustrate the topography of coastal areas around the world have been made for generations. By comparing the size and location of a delta on old maps and new maps, changes can be analyzed. Also, aerial photographs and pictures taken from satellites aid in identifying the erosion and deposition of delta landforms.

Often, older delta lobes were settled by ancient peoples. Through the science of geoarchaeology, it can be determined when changes occurred. As deltas such as the Mississippi shift from side to side over time, the human population follows the deltas from place to place. By examining the location of archaeological sites over the past fifteen hundred years, scientists have determined the minimum age of the several deltas forming the Mississippi deltaic plain. The Indian pottery found there reveals that the delta framework was deposited very recently. Cultural remains indicate other changes—in salinity, subsidence, and delta deterioration.

By boring holes into the soft sediments of a delta, geologists can decipher its subsurface aspects. Because oil and natural gas are often associated with deltas, oil companies have bored holes in many delta landforms. Information derived from this method of study reveals the composition of the thick delta sediments and the rate of delta accumulation. In fact, boreholes in some deltas have indicated that older deltas once existed and are now buried beneath younger deltas.

Finally, deltas can be created in the laboratory.

In nature, deltas are often very large and complex. To make the study of deltas less difficult, scientists use tanks filled with water and sediments. Experiments can be performed that, for example, control the amount of sediment used to build deltas. Relationships between sediments and current velocities may be studied to gather information on such properties as the rate of delta growth. By controlling the phenomena that cause deltas to form, geologists can gain an overview of the behavior and processes of delta development.

Significance

Deltas, with their marshes and bogs, are not aesthetically pleasing; however, depositional landforms have been useful to prehistoric and historic populations in many ways. Deltas, along with estuaries, are among the most biologically productive areas on the Earth's surface. Most deltas are colonized with wetland swamps or marshes, which are breeding areas for wildlife. In the United States, for example, duck hunting is a popular sport associated with wetlands. In marine deltas where there is tidal influence, fresh water and salt water mix. The river brings oxygen and nutritive substances into the delta, and the result is an enormous production of sea life. High biological productivity attracted humans to this land feature. Deltas have often been centers of civilization; the deltas of the Nile, in Egypt, and the Tigris-Euphrates, at the head of the Persian Gulf, have supported important societies. Soils in delta regions are nourished through seasonal flooding, and water tables are high, guaranteeing adequate water with which to irrigate crops, even in the dry season. Food and crop production from tropical deltas is significant because most tropical soils are not very productive. Deltas such as those of the Mekong and the Ganges are outstanding examples.

Deltas are transition zones between the land and the sea, between river and marine processes. Their rivers are also links between ocean and continent. Cities such as New Orleans, Venice, Amsterdam, and Rotterdam owe their prosperity to their delta geography. Such cities, known as *entrepôts,* thrive on marine traffic entering a country or on overland traffic exiting the country.

Since deltas are areas of vast accumulations of sediments, they generate building material for future mountains. The young mountains of the world, such as the Alps and the Himalayas, parallel coastal areas and are composed of sedimentary rocks. Marine fossils frequently found in such rocks reveal that they not only are composed of sediment but also were once deposited under water, later to be thrust upward to great heights.

C. Nicholas Raphael

Cross-References

Bibliography

Bird, Eric. *Beach Management.* New York: John Wiley, 1996. An introductory text on coastal zones and processes. Most examples are taken from Australia. The chapter on deltas is well illustrated with maps and diagrams. The text is nontechnical and comprehensible to readers with little scientific background.

Coleman, J. M. *Deltas: Process of Deposition and Models for Exploration.* Boston: International Human Resources Development Corporation, 1982. A detailed review of deltaic processes, including an overview of the Mississippi River Delta and discussions of other deltas and their variability. Numerous maps and diagrams illustrate specific points. For readers with some background in the subject, this text can be a useful supplement.

Davis, R. A., Jr. *Oceanography: An Introduction to the Marine Environment.* 2d ed. Dubuque, Iowa: W. C. Brown Publishers, 1991. A book on deposition in coastal areas. Topics include deltas, beaches, marshes, and estuaries. The treatment of deltas is generally narrative, and equations are sparingly used. A comparative presentation of deltas is instructive and not difficult to understand; some background in physical geology is useful but not necessary.

Harrison, Robert W. *Flood Control and Water Management in the Yazoo: Mississippi Delta.* Mississippi: Social Science Research Center, Mississippi University, 1993. A thorough analysis of flood control and prevention policies and managemnt for the Mississippi River. Notable for an excellent bibliography of works on the Mississippi Delta.

LaBlanc, R. J., ed. *Modern Deltas.* Tulsa, Okla.: American Association of Petroleum Geologists, 1976. A college-level text describing some of the world's deltas. The treatment is generally nonmathematical and descriptive, but it will be most useful to those who have had a course in geology. Well illustrated with diagrams, maps, and pictures.

Morgan, J. P. "Deltas: A Resume." *Journal of Geologic Education* 18 (1970): 107-117. An excellent introductory article on deltas. The emphasis is on the Mississippi River Delta, which the author studied for many years. Different processes and their influence on delta development are covered. An excellent and not overly technical paper. Includes maps and tables.

_____, ed. *Deltaic Sedimentation Modern and Ancient.* Tulsa, Okla.: American Association of Petroleum Geologists, 1970. An old but not outdated series of chapters by different authors on deltas around the world. Thorough in its treatment but not overly technical,

the book emphasizes sedimentary differences and the biological and physical character of deltas. Well illustrated with maps and aerial photographs. Some knowledge of geology or Earth science would be useful.

Oti, Michael N., and George Postma, eds. *Geology of Deltas.* Brookfield, Vt.: A. A. Balkema, 1995. This is a detailed yet easily understood account of the geology of deltas and their function in the environment and in river systems. Illustrations, maps, and bibliographical references.

Peterson, J. F. "Using Miniature Landforms in Teaching Geomorphology." *Journal of Geography* 85 (November/December, 1986): 256-258. This paper discusses small-scale landforms and their advantages in the classroom. Deltas and related features, such as alluvial fans, are highlighted. The nontechnical text is supplemented with photographs. A good article for those interested in constructing a delta on a small scale. Suitable for high school students.

Raphael, C. N., and E. Jaworski. "The St. Clair River Delta: A Unique Lake Delta." *The Geographical Bulletin* 21 (April, 1982): 7-28. A delta in the Great Lakes is described and compared with the Mississippi River Delta. The paper suggests that although many deltas look the same, different processes are at work in them. Good aerial photographs, maps, and cross sections. A nontechnical treatment of delta processes, forms, and vegetation.

Thornbury, W. D. *Principles of Geomorphology.* 2d ed. New York: John Wiley & Sons, 1969. A well-written introductory textbook on landforms. The various landforms of deltas and related river features are discussed in detail. The book emphasizes form rather than process and is nontechnical in its presentation. Suitable for those who want a rapid, nonmathematical introduction to deltas and related features, such as floodplains and coasts.

DESERT PAVEMENT

Desert pavements are concentrations of stones on the land surface of arid areas, produced by wind and running water erosion and upward movement of stones through the soil. Stone pavements may signal serious soil erosion that must be addressed before the land can be irrigated for agriculture.

PRINCIPAL TERMS

CREEP: the slow, gradual downslope movement of soil materials under gravitational stress

DEFLATION: the sorting out, lifting, and removal of loose, dry, silt- and clay-sized soil particles by turbulent eddy action of the wind

DIRT CRACKING: a process in which clays accumulate in rock cracks, take on water, and expand to rupture the rock

EXPANSION-CONTRACTION CYCLES: processes of wetting-drying, heating-cooling, or freezing-thawing, which affect soil particles differently according to their size

SALT WEATHERING: the granular disintegration or fragmentation of rock material affected by saline solutions or by salt-crystal growth

THERMAL FRACTURE: the formation of a fracture or crack in a rock as a result of temperature changes

VENTIFACT: any stone or pebble that is shaped, worn, faceted, cut, or polished by the abrasive action of windblown sand, generally under desert conditions

OCCURRENCE OF DESERT PAVEMENT

Desert pavements are extensive stony surfaces in arid areas that occur not only on slopes but also on a range of lowland surfaces, including water-eroded and deposited areas. The stones tend to be closely packed on flat or moderately inclined plane surfaces. Stone pavements are rendered prominent by their lack of vegetation but, in any case, they form most readily by processes typical of arid regions where plant roots do not bind the soil very much. The pavements are most striking in areas of low relief, where it is the largely flat, stony surface itself, rather than the contours of the land, that impresses the observer.

Desert stone pavements range from rocky or boulder-strewn surfaces to smooth plains of the finest gravel. Terminology for different types commonly derives from Arabic, as one might expect, given the prevailing aridity across North Africa and the Middle East, where that language is spoken. Thus the term "hamada" (Arabic for "unfruitful") describes a bouldery terrain, and "reg" (meaning "becoming smaller") indicates a finer pavement of small stones. The term "serir" is a synonym for "reg" in the central Sahara.

Most hamada pavements are residual, consisting of stones derived from the bedrock beneath, or constitute boulders transported only short distances. Most regs consist of transported stones. These distinctions of size and transport distance become rather blurred over time, as the larger rock fragments are progressively weathered to finer sizes and transported farther by wind and water. Nevertheless, a residual origin of many hamada pavements is clearly evident in a number of places by the angularity and lack of sorting of the rock fragments and by the similarity to the bedrock beneath. The water-deposited nature of a reg pavement can be indicated by sorted and rounded gravels of mixed composition and distant origin and also by the pavement's occurrence close to dry stream channels. Residual regs, which are usually closely associated with hamadas, commonly consist of angular flakes of local bedrock or of a less weatherable residue. Reg pavements can be composed of two elements in varying proportions: a compact mosaic of small stones embedded in soil and a more uneven component of larger fragments lying loose on that surface or protruding through it.

WIND AND WATER EROSION

Desert pavements are polygenetic; that is, there is more than one process by which they can be produced. The most traditional explanation, however, is that the stones are concentrated by means of wind erosion or deflation of the fine particles. Deflation may eventually settle the pebbles into such stable positions that they fit together almost like the blocks of a cobblestone street.

The effectiveness of deflation on desert pavements is less, however, than is commonly claimed. Under natural conditions, soils that have been tested by scientists can resist wind erosion because they are silty and cohesive and tend to form crusts as a result of repeated wetting and drying. A large grain size of the soil makes it resistant to wind erosion. Soil cohesiveness is an important factor; even where particles are spherical, the magnitude of the cohesive forces between particles less than 0.1 millimeter in diameter is greater than the weight of the particle. Small particles, however, are usually more irregular in shape, or even platy, a factor that increases their cohesion. Since cohesion is so important between small particles, silts and clays should be quite resistant to wind erosion. That is true, however, only where the surfaces are smooth, the clays and silts are uniformly fine, and no other material is blown onto them. Otherwise, large grains are dislodged from fine soils as aggregates of particles or knocked loose by the impact of other large sand grains. In any case, the amount of lowering and stone concentration by the wind is limited; deflation diminishes markedly as the protective stone cover increases, and deflation becomes virtually ineffective when stones cover 50 percent of the surface. Undisturbed stone pavements are among the most windstable of desert surfaces.

Rain and water flow appear to be more effective than wind in eroding fine-textured soils on sloping desert pavements. In test plots cleared of stones in one experiment on 5-degree slopes, water wash accounted for most of the 5- to 50-centimeter surface lowering that occurred in a period of five years. During this time, the stone pavement was renewed with stones from below this differential erosion.

UPWARD DISPLACEMENT

Where subsoils are so clay-rich that they do not erode easily by either wind or water and the subsoils are also largely stone-free, the formation of stone pavements by deflation is more difficult to understand. In these situations, the mechanism of formation seems to have been forces of expansion and contraction within the soil that cause upward displacement and concentration of stones on the surface. Soils that exhibit this phenomenon contain expansive clays in alkaline chemical conditions and are subject to swelling and heaving upon wetting and to shrinkage and cracking upon drying. Periods of heating followed by cooling and of freezing followed by thawing also contribute to the expansion-contraction cycles, which cause stones to move upward and concentrate on the surface.

The exact mechanism for upward stone movement in deserts is not precisely known, although by analogy to known stone movement in areas of intense freeze and thaw, some details are understood. It is possible, for example, that the stones shift upward as the underlying

Desert pavement in Arizona: Wind blows away sand and clay until only small stones remain. (© William E. Ferguson)

soils swell in wet periods and that as the soil shrinks in dry periods, fine particles fall into cracks beneath the stones and prevent the return of the rock fragments. Stones may also induce differential swelling in the soil by speeding the downward advance of a wetting front around and over them. The stones are then likely to be displaced upward, away from the dry zone beneath them, and held tightly by the wet and sticky soil above. Soil may squeeze into the space left and thus prevent a return movement of the stones. These processes also resemble those postulated for certain types of sorted, patterned ground, where stones become arranged in polygons or striped zones. In these cases, it is clear that no process other than upward movement along certain zones could produce such patterns.

Stone pavements of the upward-movement type occur on and in the topsoils of weakly salty soils on the stony tablelands of arid Australia, where the subsoils are almost stone-free. The pavement stones are silica-rich, originally precipitated within the soils by mobilization and concentration of silica in unusual chemical reactions. For those stone pavements to have originated as a residue from erosion would have required a stripping of more than 1.5 meters of erosion-resistant, clay-rich soil. Such a process is unlikely; an upward movement is far more probable, given the swelling and cracking potential of salty clays. Similar pavements have been noted in deserts in Nevada and California as well as in the Atacama desert of Peru.

An alternative form of displacement is possible where the soil material has been transported, particularly by the wind. For example, in South Australia, a stone pavement occurs on a layer of clay rich in gypsum (calcium sulfate), which is thought to have been blown into place by the wind. The stone pavement on the surface resembles a buried pavement. It is possible that the original pavement first trapped a small amount of windborne dust among the stones. Some of the rock fragments were then displaced upward, little by little, through wetting and swelling of the eolian (wind-deposited) clays during the course of their accumulation, and thus the stones were never deeply buried.

Concentration of stone pavement through winnowing by wind or wash can be relatively rapid, but the contribution by movement through the soil may take longer. Once formed, a pavement is relatively stable. The closely spaced stones act as a drag on the surface wind, restrict the entrainment of finer intervening materials, and so limit deflation. On moderate slopes, runoff water is spread over the surface by the stone mantle and thus does not tend to cut a deep gully. Selective erosion is countered further as the stone concentration increases. Whether residual or transported, pavements naturally consist of materials resistant to weathering and therefore serve as "armor" for deserts.

SURFACE WEATHERING

Another manner in which stone pavements can be generated is by relative concentration through surface weathering. Rock fragments in the relatively moist environment of a desert soil are more susceptible to weathering than those on the arid surface, particularly where the soil is impregnated with salt or gypsum. Consequently, a stone pavement may survive above a soil that itself has few stones because they disintegrate at depth over time. The phenomenon is pronounced in granitic gravels; for example, some terrace sediments are known to have larger-sized stone pavements above horizons of small granite fragments formed by the weathering breakdown of boulders beneath the surface. These subsurface zones of fine particles are deepest and most free of subsurface stones on the highest and oldest terraces, where they may be more than 50 centimeters thick. In such cases, it is thought that stone pavements first formed from the abundant rock fragments in the area, but as subsurface weathering progressed all the buried stones were destroyed, while those on the surface remained relatively unweathered.

In spite of their status as a protective armor on desert surfaces, stone pavements are subject to further evolution over thousands of years as the stones weather and their secondary products are redistributed. This evolution is generally toward an increasingly even surface of small grain size and greater compaction. Also common is progressive darkening by surface weathering and the formation of rock or desert varnishes of manganese and iron stains, precipitated on the stone surfaces by microbial and weathering action.

Weathering of pavement stones occurs not only beneath the stones, where they are in contact with the protected soil layers, but also on their exposed surfaces. Pavement stones can be wetted frequently by dew, which also contributes salts to both surface and subsurface in the weathering process. Because pavements are generally unchanneled and provide little runoff, the features are particularly subject to episodic or seasonal cycles of shallow wetting by rainfall and evaporative drying, through which weathering is activated. Bare and generally dark-colored pavement, among the most strongly heated surfaces of the desert, are areas of considerable evaporation, as (mostly saline) moisture is drawn upward through narrow cracks. As a result, thin salt crusts are widespread, the soil itself is impregnated with chlorides and sulfates of calcium and sodium, and salt weathering is significant. In this process, the growth of various salt crystals in the pores of the stones causes forces sufficient to disrupt the rocks.

Pavement stones also trap windborne dust in fissures and cracks, and dirt cracking can result from expansion of the dust particles when they are wetted. Lichens and algae also exploit the shaded and somewhat more moist environments under the stones, adding an organic element of chemical decomposition to the weathering process.

In the breakdown of pavement stones, there tends to be a further selective concentration of resistant fine-grained material—for example, siliceous flint or chert pebbles that accumulate on the surface as soft limestones are weathered. Such stones can eventually break down by incorporation of water into their microcrystalline atomic structures, by spalling (chipping), by blocky fracturing (crazing), or by complete cleavage and radial splitting. Coarser-grained stones undergo granular disintegration and pitting. Fracturing of pavement stones also has been attributed to differential expansion and contraction caused by solar heating, but the idea is controversial. Such thermal fracturing may be only a partial cause; much of the broken stone is superficially altered chemically, and dirt-cracking expansion may be more important. In some cases, stones below the surface, where solar heating is not a possibility, can be seen to have been pried apart in this fashion.

Wind abrasion is a form of natural sandblasting. Its effectiveness is related to wind velocity, the hardness of the sand and dust carried by the wind, and the hardness of the rock fragments being eroded. It is probably most effective in certain polar areas where cold, dense air can carry large particles at very high velocities and where, at winter temperatures, even ice has a hardness of some minerals. The "dry valleys" of Antarctica, for example, have extensive stone pavements that have been affected in this way.

BOULDER AND PEBBLE FORMS

Boulders and pebbles in stone pavements may be fluted, scalloped, and faceted by wind abrasion. Flute and scallop forms vary in length from a few millimeters to several meters, but large flutes are not common in hard rocks. It is thought that they are produced by turbulent helical (helix-shaped) flows carrying dust and sand and that the scallops grow downwind. Where they are cut on large, immobile boulders, they are clear indicators of the strongest (dominant) wind directions. For this reason, they are most commonly reported from places where wind directions have remained largely unchanged for thousands to hundreds of thousands of years.

Some of the best-known wind-eroded stones in pavements are ventifacts (a general term for wind-faceted pebbles and boulders). A rock face abraded by wind may be pitted if there is a range of hardness in the minerals of the face, or it may be smooth where the rock is fine-grained or composed of only one mineral. Thus, rocks such as coarse-grained granite have pitted surfaces, but fine-grained quartzite generally produces only smooth, polished surfaces. Ventifacts have a great range of surface shapes, with plane and curved faces and two or more facets. The German term "dreikanter" is used for ventifacts with three facets, and "einkanter" is used for two-faceted stones. Multiple facets indicate either that there was more than one wind direction or that the stones have been turned over through time. As the sizes of the stone pavements' fragments are progressively reduced, a matrix of increasingly fine particles is supplied to the pavement. The proportion of such secondary material reveals the maturity of development of a desert stone pavement, although in practice it may be difficult without sophisticated analysis to distinguish between a new pavement in the process of being formed and an old one being degraded.

The fine particles in a degraded pavement are redistributed by wash and rain, a process that contributes to the smoothing and compaction of the pavement. Any exposed soil is puddled and sealed by heavy rainfall or runoff water so that a saturated layer flows into hollows between the stones. Bare soil interspersed through a pavement is generally crusted above a bubbly or vesicular horizon, 1 to 3 centimeters thick, that also extends around and beneath the pavement stones. The bubbles result from the escape of entrapped air. Equally important in compaction is the gravitational settling of the stones during the expansion and contraction of the surface upon heating and cooling and (especially in saline and expansive clay soils) upon wetting and drying. Pavements of this type are generally soft and puffy after rain, but, upon drying, the stones become ever more firmly embedded in a tight mosaic.

The slow downhill creep of water-saturated surface materials can also assist in either the smoothing or the roughening of a sloping stone pavement, particularly where dispersal of the matrix is accentuated by salinity. Microrelief on a stone pavement can be reduced by the differential flow of the fine sediments, and the pavement stones can thereby become more evenly distributed and further embedded. In some cases the stone pavements can move into a series of rough steps, aligned along the contours of the slope.

John F. Shroder, Jr.

CROSS-REFERENCES

Alluvial Systems, 2297; Beaches and Coastal Processes, 2302; Biostratigraphy, 1091; Coal, 1555; Continental Shelf and Slope, 584; Dams and Flood Control, 2016; Deep-Sea Sedimentation, 2308; Deltas, 2312; Drainage Basins, 2325; Evaporites, 2330; Floodplains, 2335; Floods, 2022; Geoarchaeology, 1028; Lakes, 2341; Petroleum Reservoirs, 1728; Reefs, 2347; River Bed Forms, 2353; River Flow, 2358; River Valleys, 943; Sand, 2363; Sand Dunes, 2368; Sea Level, 2156; Sediment Transport and Deposition, 2374; Turbidity Currents and Submarine Fans, 2182; Weathering and Erosion, 2380.

BIBLIOGRAPHY

Chouha, T. S. *Desertification in the World and Its Control.* Jodhpur, India: Scientific Publishers, 1992. Chouha deals with desertification on a global scale but focuses much attention on India. Several sections deal with tactics that have been used and proposed to control the problem. Maps and index.

Climate, Drought, and Desertification. Geneva, Switzerland: World Meteorological Organization, 1997. This brief booklet, prepared by the World Meteorological Organization, deals with the climatic factors, such as drought, that lead to desertification worldwide. Includes color illustrations.

Cooke, Ronald U. "Stone Pavements in Deserts." *Annals of the Association of American Geographers* 60 (1970): 560-577. One of the most comprehensive nontechnical journal articles available on desert pavements. The emphasis is on polygenesis, with examples mainly from California and Chile. The author notes that deflation may be a relatively unimportant process of pavement formation, a view decidedly in contrast with the ideas of many in the field. He does note that the relative importance of the different processes may vary greatly between sites.

Cooke, Ronald U., and Andrew Warren. *Desert and Geomorphology.* London: UCL Press, 1996. One of the chief English-language sources on dry climate geomorphology. The text has a long section on the formation of desert pavements as well as on the many other collateral processes and landforms that occur with or around such features.

Mabbutt, J. A. *Desert Landforms.* Cambridge, Mass.: MIT Press, 1977. This book has a chapter on stony deserts that is a superb exposition on desert pavements. Their polygenetic origin is well explained, and numerous examples and illustrations aid the discussion. The remainder of the book is replete with material on other arid landforms and processes that occur in the same context as do desert pavements.

McGinnies, William G., B. J. Goldman, and P. Paylore, eds. *Deserts of the World.* Tucson:

University of Arizona Press, 1968. This survey of research into the physical and biological environments of the world's deserts is one of the most comprehensive studies of arid regions ever attempted. Gives valuable information specific to many particular deserts and has an extensive reference list. The section on desert pavements is short but useful and contains many references not generally cited elsewhere.

Twidale, C. R. *Australian Landforms: Structure, Process, and Time*. Adelaide, Australia: Gleneagles, 1993. A book on general geomorphology that has a short section on desert pavements, known as "gibber" in Australia, where such features are most common. Twidale's extensive experience with Australian desert landforms makes this a useful text for one seeking to understand the context of desert pavements.

DRAINAGE BASINS

Drainage basins reflect the operation of physical laws affecting water flow over the Earth's surface and through rocks. They define natural units that concentrate flow into rivers, which, in turn, remove both water and sediment, the latter eroded from the surface of the basin. They become a focus for human activities whenever water is exploited as a resource.

PRINCIPAL TERMS

BASIN ORDER: an approximate measure of the size of a stream basin, based on a numbering scheme applied to river channels as they join together in their progress downstream

CHANNEL: a linear depression on the Earth's surface caused and enlarged by the concentrated flow of water

EROSION: the removal of sediment from the Earth's surface

GROUNDWATER: water that sinks below the Earth's surface and that slowly flows through the rock toward river channels; it keeps rivers flowing long after rainstorms

HYDROLOGICAL: relating to the systematic flow of water in accordance with physical laws at or close to the Earth's surface

LIMESTONE: a rock composed primarily of the mineral calcite or dolomite, which can be dissolved by water that is acidic

CHARACTERISTICS OF DRAINAGE BASINS

A drainage basin is an area of the Earth's surface that collects water, which accumulates on the surface from rain or snow; its slopes deliver the water to either a channel or a lake. Normally, the channel that collects the water leads eventually to the ocean. In this case, the drainage basin is defined as the entire area upstream whose slopes deliver water to that channel or to other channels tributary to it. Thus, strictly speaking, drainage basins are defined as natural units only when streams enter bodies of water such as lakes or the ocean or when two streams join; one speaks of the Mississippi River drainage basin or the Ohio River drainage basin—this latter basin lies above the Ohio River's confluence with the Mississippi.

Less often there is no exit to the ocean: This type of basin is called a basin of inland or interior drainage. Notable examples are the basins containing the Great Salt Lake in Utah, the Dead Sea in Israel and Jordan, and the Caspian Sea in Asia. The Basin and Range province in the Rockies (an area of about 1 million square kilometers extending from southern Idaho and Oregon through most of Nevada, western Utah, eastern California, western and southern Arizona, southwestern New Mexico, and northern Mexico) has at least 141 basins of inland drainage. The center of these bas-ins is usually marked by a playa—a level area of fine-grained sediments, often rich in salts left behind as inflowing waters evaporate. At certain times in the geological past, when the annual rainfall regime was wetter, some of these basins completely filled with water to the point of overflowing, at which point the drainage system may have connected to another interior basin or connected to a river system that drained to the sea. The basin now containing the Great Salt Lake (known to geologists as Lake Bonneville) overflowed at Red Rock Pass about 15,000 years ago. The overflowing waters discharged into the Snake River system, and thus to the Columbia River and the Pacific. Drainage basins may change in character over relatively short periods of geological time. There is some evidence that the entire Mediterranean Sea was a basin of inland drainage for a period about 3 to 5 million years ago: Substantial salt deposits are found on its bed, and traces of meandering rivers have been seen in certain geological sections.

GROUNDWATER AND EROSION

Although the term "drainage basin" is normally thought of as applying to the surface, a very important component of the basin as a hydrological unit is the rock beneath the surface. Much of the water that arrives on the surface sinks into the soil and the underlying rocks, where it is stored as soil water and as groundwater. Soil water either sinks farther to become groundwater, or it may flow through soil and back out onto the surface downslope when and where the soil is saturated. Groundwater moves very slowly through the rock (millimeters to centimeters per day), but it eventually seeps into stream channels and so keeps water flowing in rivers long after rain has finished falling.

This characteristic of groundwater can lead to a circumstance that alters the definition of a drainage basin when the rocks are primarily composed of limestone or any rocks susceptible to solution. Because limestone is soluble in acidic water (natural rain is slightly acid; acid rain accentuates the acidity), over thousand of years, percolating groundwater dissolves substantial volumes of rock and causes a system of underground channels to develop, which may sometimes be enlarged to caverns. Yugoslavia is famous for its underground cave systems; in the United States, Kentucky, Florida, and New Mexico (with its famous Carlsbad Caverns) are well known for their limestone terrain and underground drainage systems. In this case, the route of the water underground may bear little or no relation to the pattern of channels and slopes seen on the surface, some or all of which may have become completely inactive. The determination of the drainage basin is then very difficult as various types of tracer (colored dyes or tracer chemicals) have to be placed in the water in order to track it, and the pattern of water flow to any given site will also depend on the location of the storm waters causing the flow. The very slow solution of limestone to form underground river systems and caves emphasizes the fact that water moving through the basin removes solid rock. As the rivers dissolve their way downward, they leave some caves "high and dry" above the general level of the underground water (the water table), which points up the fact that rivers work down through the rock with the passage of time.

This aspect of drainage basins—that solid rock is slowly removed—is harder to observe in areas of less soluble rock, even though water flowing out of the drainage basin carries sediment (small particles of soil and rock) and has been doing so for long periods of geological time. Thus, in the long run, the surface of the Earth is lowered, and even in the short run, enormous amounts of sediment may be removed from the basin every year. The Mississippi removes a total of about 296 million metric tons per year, or 91 tons per square kilometer, though this is small compared to the Ganges, which takes out 1,450 million metric tons per year, or 1,520 tons per square kilometer. If there were no corresponding uplift of the drainage basins or other interference, this removal would lead to the leveling of entire basins within 10 to 50 million years, depending on the lowering rate and the mean altitude of the basin. If the Ganges basin has a mean relief of 4,500 meters, it would be completely lowered in about 7 million years at present erosion rates. If the Mississippi has an average relief of 1,000 meters, the time needed is on the order of 40 million years. These figures, however, are unrealistic because other processes cause compensating uplift, but they do indicate that drainage basins are being actively eroded within reasonable spans of geologic time.

The process of erosion proceeding at different rates in adjacent basins may cause the drainage divide (the line separating different flow directions for surface waters) to migrate toward the basin with the lower erosion rate. This is most common in geologically "new" terrain when stream systems are not deeply incised into the rocks. Drainage diversions may be simulated, as in the Snowy Mountain diversion in Australia, where waters are diverted across a divide by major engineering works in order to provide irrigation water for the Murray Darling river basin and also for hydroelectric power generation.

FLOODS

When winter snow melts (the late spring and early summer peak in Mississippi River flow is caused by melting snow in the Rocky Mountains) or severe storms bring heavy rain to large areas, the water that falls flows over the surface to channels and generates floods in the rivers. Floods are not abnormal; they are an expectable occurrence in drainage basins. It is easy to understand that when basin relief is high and slopes are steep (in the Rockies or the Appalachians), floods tend to generate higher flood peaks than when slopes are very gentle. A basin that is round in shape tends to concentrate floodwater quickly because the streams tend to converge in the middle, whereas a long, narrow stream has the effect of attenuating the flow peak, even when the total amount of water falling on the basin may be the same. Similarly, a forest tends to attenuate flood peaks and to promote higher river flows between flood peaks than does open farmland. With the latter, there is a tendency for water to flow very rapidly off the surface into channels, whereas with a forest much water is intercepted by the leaves of trees and the impact of rain on the surface is weaker, in part because leaf cover protects the soil. Because the soil is not so well protected, sediment loss from the surface into streams is greater from farmland than from forests and is higher again from land disturbed by major building projects.

CLASSIFICATION AND MEASUREMENT OF DRAINAGE BASINS

Various methods have been devised to classify basins according to size. The most common method depends on a numbering system applied to the streams that drain them. All the fingertip streams are labeled with 1. When two of these tributaries meet the channel, it is termed a second order channel and is labeled with 2. Subsequently, the order of a stream increases by one only when two streams of equal order join. Otherwise, if two streams of unequal order join, the order given to the downstream segment is that of the larger of the two orders. The order of the drainage basin is then the order of the stream in the basin. In this type of numbering system (called Horton/Strahler ordering), the Mississippi drainage basin is an eleventh or twelfth order basin; the exact number depends on the detail (map scale) with which the fingertip streams are defined. The larger orders are rare because if the Mississippi basin is twelfth order, it would take another river of roughly similar magnitude to join it to make a thirteenth order basin.

Basin order may be used as a relatively natural basis for the collection of other data about

the basin. The simplest measure is of the area in square kilometers. In addition, the basin relief, the height difference between the lowest and the highest points, and the mean relief, or the average height of the basin above the outlet, may be recorded. The most precise method of recording basin relief is by computing the hypsometric (height) curve for the basin, which requires an accurate topographic map. When constructed, it shows, for any altitude, the proportion of the basin area above that particular altitude and, for comparative purposes, it may be produced in a dimensionless form by dividing both the height and the area measures by their maximum values or by the difference between the maximum and minimum heights if zero is not the minimum height.

Basin shape and basin dimensions (length and width) may also be recorded, although the notion of basin shape suffers from the problem that no completely unambiguous numerical measure exists that can be used to define the shape of an area in the plane (that is, on a map), and the problem is especially intractable if there are indentations in the edge of the basin. All measures are dependent to a considerable degree on the accuracy of source maps; in mountainous terrain, such maps may often be much less than perfect, if they exist at all. Even with automated drafting aids and

CHARACTERISTICS OF SELECTED MAJOR DRAINAGE BASINS

River	Outflow	Length	Area	Average Annual Suspended Load
Amazon	180.0	6,300	5,800	360
Congo	39.0	4,700	3,700	—
Yangtze	22.0	5,800	1,900	500
Mississippi	18.0	6,000	3,300	296
Irawaddy	14.0	2,300	430	300
Brahmaputra	12.0	2,900	670	730
Ganges	12.0	2,500	960	1,450
Mekong	11.0	4,200	800	170
Nile	2.8	6,700	3,000	110
Colorado	0.2	2,300	640	140
Ching	0.06	320	57	410

NOTE: Rivers are ordered by outflow; outflow is multiplied by 1,000 cumecs (a cumec is 1 cubic meter of water per second); length is measured in kilometers; area is measured in square kilometers multiplied by 1,000; average annual suspended load is measured in millions of metric tonnes.

digitizers (which automatically record positions on maps and save them for a data file), the measurement of basin properties is a tedious and time-consuming process. Unless there are pressing reasons for a new analysis, it is common to rely on data tabulations made by hydrological or environmental agencies whenever possible.

Measurements are made of drainage basin properties because they are often used in statistical analyses together with the known flow of the few gauged rivers in order to predict flow characteristics for rivers that have not been metered. The direct measurement of stream flow, while straightforward in principle, is time-consuming, especially in the early stages, and a flow record is not very useful for predictive purposes until it has recorded at least twenty years of flow (preferably much more). Because of the high capital and maintenance costs involved in collecting river records, there has been an understandable emphasis on records for large rivers; the economic benefits from prediction (and eventual control) of the flow are more obvious, and measured flow records can sometimes be supplemented by anecdotal evidence of historic large floods—those flows that are often of most interest in land-use planning (for example, in the zonation of land for residential use). It has been acknowledged that the hydrological behavior of low order basins is less well understood, and more information has been collected on them, especially for urban areas where the routing of the large quantities of water that run off from impermeable surfaces in the city (roofs and roadways) has been recognized as a serious planning problem, especially when the flow systems connect with urban sewage systems.

SIGNIFICANCE

The control of water outflow from drainage basins is necessary in some regions in order to promote irrigation, to supply domestic and industrial water, to generate power, and to implement flood control. The Hoover Dam on the Colorado was originally conceived as a control dam, but hydroelectrical generators were also built in order to help defray costs by selling power. There are nineteen major dams in the Colorado basin. Difficulties (aside from the legal technicalities of water ownership and redistribution) arise from the fact that to control substantial amounts of water, large areas of the basin have to be regulated; in addition, there are economies of scale in large projects, particularly in the construction of large dams and reservoirs. A single control dam strategically placed may regulate flow downstream for hundreds of kilometers, whereas it would require hundreds of small dams on first- and second-order streams to achieve the same effect.

Large control dams do generate problems. The reservoirs trap sediment coming from upstream, which will eventually fill them, at which point they will be useless; small dams may fill within a few years. An original estimate for the Hoover Dam suggested that it would take only four hundred years to fill Lake Mead; after only fourteen years, surveys revealed that the water capacity had been reduced by 5 percent and that sediment in the lake bottom reached a maximum of 82 meters where the upstream river entered the still waters of the lake. Downstream of a dam, the reduced sediment content and the regulated water flow often seriously affect riparian environments. There may be a variety of channel responses, often unpredictable, to the interference in the river regime caused by the dam. The stream may cut into its bed, it may change the dimensions of its channel, or it may even aggrade its bed. In the case of the Hoover Dam, the water downstream, deprived of its sediment by the dam, had an increased ability to remove fine sediment from the river bed but left coarser rocks behind because the flood peaks that would remove them normally were now controlled—that is, much reduced. The result is an "armoring" of the stream bed with coarse rocks, an effect that extends 100 kilometers downstream in the case of the Colorado River below the Hoover Dam. In the Colorado system as a whole, the net effect of controlling flood peaks has been for rapids to stabilize and to increase in size as sediment becomes trapped in them. A corollary of the "winnowing" of fine material has been the disappearance of river beaches and an increased propensity to pollution as sediment becomes much less mobile and more concentrated in space.

Keith J. Tinkler

CROSS-REFERENCES

tion, 1473; Evaporites, 2330; Floodplains, 2335; Geomorphology of Dry Climate Areas, 904; Lakes, 2341; Mars's Valleys, 2452; Reefs, 2347; River Bed Forms, 2353; River Flow, 2358; Sand, 2363; Sand Dunes, 2368; Sediment Transport and Deposition, 2374; Soil Erosion, 1513; Weathering and Erosion, 2380.

BIBLIOGRAPHY

Chorley, Richard J. "The Drainage Basin as the Fundamental Geomorphic Unit." In *Water, Earth, and Man*, edited by Richard J. Chorley. London: Methuen, 1969. Treats the modern geologic and geometric approaches to the measurement of the physical characteristics of the basin: stream numbering and ordering techniques, relief measures, and the relations of basin size, shape, and relief with stream flow behavior. Excellent bibliography.

Graf, William L. *The Colorado River: Instability and Basin Management*. Washington, D.C.: Association of American Geographers, 1985. An excellent, well-written study of the particular management problems and practices associated with this large and famous river. Focuses on the way the river has adjusted to a variety of changes caused by climatic change, rangeland management, the building of large dams, and the extraction of water for irrigation. Easily understood by the layperson; sound bibliography.

Gregory, K. J., and D. E. Walling. *Paleohydrology and Environmental Change*. New York: John Wiley, 1998. A comprehensive academic textbook aimed at the serious undergraduate or a well-prepared, scientifically minded layperson. Examples from around the world; well-illustrated with photographs, maps, and diagrams. Detailed information on instrumentation, and the implications for socioeconomic management are carefully considered. Extensive bibliography.

Henry, Georges. *Geophysics for Sedimentary Basins*. Translated by Derrick Painter. Paris: Editions Technip, 1997. This book looks at the geological characteristics of sedimentary basins. Suitable for the college-level reader. Color illustrations, index, and bibliographical references.

More, Rosemary J. "The Basin Hydrological Cycle." In *Water, Earth, and Man*, edited by Richard J. Chorley. London: Methuen, 1969. Lays out in considerable diagrammatic detail how water circulates in and through the basin and the theoretical framework possible for modeling and studying it, particularly with a view to interfering with basin flows in an optimal manner. Excellent bibliography.

Parnell, John, ed. *Geofluids: Origin, Migration, and Evolution of Fluids in Sedimentary Basins*. London: Geological Society, 1994. This book looks at the evolution of geofluids and their dynamics and migration in relation to sedimentary basins. Intended for the college student, the book is filled with helpful illustrations and maps. Index and bibliography.

Smith, C. T. "The Drainage Basin as an Historical Basis for Human Activity." In *Water, Earth, and Man*, edited by Richard J. Chorley. London: Methuen, 1969. Explains how the drainage basin has long been a natural unit for the focus of human economic activity, with examples from China, Europe, and the Americas. The importance of the basin declined somewhat as the Industrial Revolution progressed, but the need for large-scale planning of water use may be reversing this trend. A helpful guide to the historical perspective.

EVAPORITES

Evaporites provide a large variety of minerals, some of which are absolutely essential to most living organisms. Although a few evaporites are very abundant and widely used, others are relatively rare; it is the rarer evaporites that are necessary for most manufacturing and industrial uses. Their widespread use, from individual consumption to far-reaching scientific applications in Earth and space research, makes evaporites indispensable to modern industrial society.

PRINCIPAL TERMS

BRECCIA: a coarse-grained clastic rock composed of angular broken fragments held together by a mineral cement or fine-grained matrix

BRINE: warm to hot, highly saline seawater containing calcium, sodium, potassium, chlorine, and other small amounts of free ions

DISSOLVED SOLIDS: a term that expresses the quantity of dissolved material, such as minerals, in a sample of water

RELATIVE HUMIDITY: the ratio, expressed as a percentage, of the actual amount of water vapor in a given volume of air to the amount that would be present if the air were saturated at the same temperature

SALTS: ions that have been removed from minerals and are free to move while in a solution; they are electrovalent or ionic compounds when crystalline

SUPRATIDAL: the shore area marginal to shallow oceans that are just above high-tide level

OCCURRENCE OF EVAPORITES

Evaporites are deposits formed by precipitation from solutions or brines containing concentrated salts. Concentrations of salts are brought about by evaporation; therefore, the deposits have been termed evaporites. Common evaporite minerals include gypsum, anhydrite, halite, and the carbonate minerals calcite and possibly dolomite. Although these carbonates may be precipitated, they are generally defined and described with limestone and are therefore excluded here. In addition, more than eighty evaporite minerals have been reported from marine evaporite deposits. Most of them are categorized into chlorides, sulfates, carbonates, and borates. In nonmarine evaporite deposits, the minerals already listed are precipitated, as well as many that are usually unique to this environment of deposition and include the well-known mineral species trona and some potash salts.

In general, modern evaporites occur in the subtropical zones between 15 and 40 degrees latitude, but they are also present in both arctic deserts and high equatorial plateaus. As such, they are known from all continents. They are thin deposits and cover relatively small areas in modern settings. Yet modern evaporite environments provide insight into the origin of very widespread and much thicker deposits from the geologic past. Evaporite deposits may form in those areas of the Earth where the amount of total rainfall is exceeded by the total amount of evaporation and are generally restricted to regions having arid climates. These conditions may occur in marine or nonmarine environments.

Marine settings may be either large or small bodies of water that have physically restricted access to large bodies of open seawater and from very little to no contribution from freshwater streams. Such restrictions may be caused by sedimentary deposits that, over a period of time, partially close the access or shallow depths at the access that are caused by other geologic phenomena. Ocean marginal areas slightly above maximum high tide, or supratidal zones, may be the sites for saline deposits resulting from flooding during abnormally high tides or wind-driven high water or both. Nonmarine or continental settings are associated with inland lakes of various sizes. Many continental lakes that contain evaporites are in inland basins on the lee side of mountains that obstruct the paths of the prevailing winds. Most moisture present in the wind is precipitated on the opposite side of the mountain from the basin, creating an arid climate on the downwind side of the mountain or a rain-shadow desert. These nonmarine evaporitic basins may contain permanent brine lakes, such as the Dead Sea along the border between Israel and Jordan and the Great Salt Lake of Utah. Others may be smaller and have ephemeral lakes or playas, such as Little Salt Lake and Siever Lake of western Utah, Carson Lake and Ruby Lake of Nevada, and Death Valley and Searles Lake of California. Some lakes, such as those found at Saline Valley, California, have only an efflorescent crust of a depth of 1 meter or less of accumulated salts, while others may have salt-rich muds and sands. Many inland accumulations are not composed of the common minerals stated previously; rather, they are sodium-rich carbonates such as trona or sodium and calcium-rich sulfates such as glauberite. Ancient lakes that formed the world-famous Green River formation in Wyoming precipitated these types of minerals as part of the basin fill.

CHARACTERISTICS OF EVAPORITES

Of importance is the major difference between evaporites and most other sedimentary rocks, that is, the extent to which evaporites may be altered after deposition. Evaporite minerals are easily dissolved in rock fluids and are commonly replaced by other evaporite minerals or nonevaporite minerals such as barite, calcite, and those rich in silica. They may be totally dissolved and leave only residues and collapsed breccias that are produced as the roofs of the solution voids collapse, such as the Kirschberg Collapse breccia of the Edwards Plateau in south-central Texas. During the burial process of successive deposits on top of an evaporite zone, the evaporites may become deformed, producing brecciated beds, contortions, and pearcement features. Pearcement salt bodies (diapirs), or salt domes, are typical of the ancient Louann salt deposits (160 million years old) of the subsurface of the Mississippi, Louisiana, and Texas Gulf coastal plain and similar ancient deposits of the Salt Mountains of Iran. Most other sedimentary

rocks are not as easily altered or removed from their site of deposition.

Evaporite deposits accumulate with great rapidity, as much as one hundred times more rapidly than most other sedimentary deposits. Rates of up to 100 meters per 1,000 years have been observed in solar salt ponds, yet it is not expected that such restricted conditions as exist in these ponds can be maintained through prolonged periods of geologic time. Yet depositional rates of 3 to 4 meters per 1,000 years have been demonstrated by geologists for the evaporitic episode of the Mediterranean Messinian, south of Greece, where the interval of time for a thickness of 1.5 to 2.0 kilometers of halite (rock salt) and gypsum to be deposited was no more than 500,000 years.

A major control on the evaporite minerals precipitated in arid climates is the relative humidity of the region of deposition; this is the limiting factor in water evaporation. In order to produce brines from which halite, sodium chloride, or rock salt can be precipitated, the mean relative humidity must be less than 76 percent and for potassic salts

A salt crust on a floodplain near Cottonball Flat, Inyo County, California, the result of evaporation. (U.S. Geological Survey)

less than 67 percent. Relative humidities between 70 percent and 80 percent and higher are present at most low-latitude coastal regions, with values of less than 67 percent occurring over landmasses. Therefore, gypsumanhydrite minerals will be the principal precipitates along many arid coastlines—in fact, the only minerals precipitated if the relative humidity is above 76 percent. Potassic and chloride salts have such high-solubility characteristics that the ideal locations for their precipitation are areas in which there is a relatively small marine basin surrounded by land with a restricted opening to normal seawater or an inland basin.

Although known occurrences are rare, evaporites appear to be among some of the Earth's earliest known sedimentary rocks, such as those in the Isua Belt of Greenland with an approximate age of 3.5 billion years and those in the Strelly Pool chert of the Pilbara Block, Western Australia, that are roughly 3.4 billion years old. Occurrences become more abundant in rocks 1 to 2 billion years old and very common in rocks less than 600 million years old. The distribution of evaporites in time and location is very irregular. Geologists have found that a combination of world climatic constraints, plus the rather small areas of flooded continental crust having shallow seas, produces evaporitic deposits mainly in continental deserts and along arid coastlines, the largest deposits occurring in nonmarine settings such as Ch'inghai, China. Ancient extensive shallow to deep marine basins have large deposits of evaporites, the most recent forming roughly 5 to 6 million years ago in the Mediterranean region.

Seawater composition varies very little throughout the oceans of the world. Although almost all the natural elements of Earth have been found to be present in seawater, only a dozen constituents are present in concentrations greater than one part per million. Total dissolved solids in seawater have a mean of 35 parts per 1,000; of this, sodium and chlorine account for 85 percent. If a 1,000-meter column of seawater were to be evaporated and the salts precipitated, it would yield a deposit 4.6 meters thick, of which 0.8 meter (17.4 percent) would be calcium sulfate, 3.7 meters (80.4 percent) would be sodium chloride (halite), and the remaining 0.1 meter (2.2 percent) of precipitate would be complex magnesium-rich and potassium-rich salts and a very minor amount of iron-bearing limestone. Therefore, ancient salt deposits hundreds of meters thick required the evaporation of a very large volume of seawater. Geologists generally consider that the composition of seawater has varied very little over the past 2 billion years. If that is true, it does not necessarily follow that great thicknesses of salt were deposited in deep basins; rather, shallow water deposits probably occurred in a gradually subsiding basin.

CONTINENTAL DEPOSITS

Almost one-fourth of the continental areas of the Earth are underlain by evaporites; in 60 percent of the occurrences the evaporitic sequence contains chlorides. Most of the deposits are in the Northern Hemisphere and are commonly associated with petroleum-bearing rocks in large sedimentary basins. Such well-known areas are found in the Permian Basin of New Mexico, west Texas, and northern Mexico; the Persian Gulf region; the Zechstein province of northern Europe; the Northern Gulf of Mexico Basin in Texas, Louisiana, Mississippi, and offshore Gulf of Mexico; and smaller sedimentary basins known for New York State and for Michigan State.

The New York Salina evaporites have a subsurface areal extent of some 25,000 square kilometers in western New York, Pennsylvania, eastern Ohio, and northern West Virginia. Individual layers are 12 to 24 meters thick. At least seven salt beds interbedded with shale have a cumulative thickness of 76 meters between depths of 580 and 952 meters. In the Michigan Basin, which includes all of Michigan and part of adjacent Ontario, salt beds have a maximum aggregate thickness of 488 meters.

North America has two of the most spectacular evaporite deposits that are currently known on Earth and that have been studied for many years, the Permian Basin and the Northern Gulf of Mexico Basin. In the Permian Basin of west Texas, New Mexico, and northern Mexico, the Castile formation, roughly 260 million years old, underlies an area of almost 325 kilometers in diameter and is composed of more than 95 percent salts. The Castile formation and the associated Salado formation combined comprise a maximum thickness of 1,220 meters, of which up to 457 meters are laminated anhydrite, calcium sulfate, which is one of the thickest known evaporite deposits on Earth.

The lower portion is mostly very thin laminated anhydrite in which each lamina averages some 1.6 millimeters in thickness. The laminae of anhydrite alternate with dark-colored bitumen-rich calcareous layers that are slightly thinner than the anhydrite. Individual layers have been correlated over distances of 113 kilometers, demonstrating the contemporaneity of uniform environmental characteristics at the time of deposition. At the surface, the anhydrite has hydrated to form gypsum. In the upper portion of the Castile-Salado sequence, the principal mineral is halite or rock salt, with lesser amounts of potash-bearing salts, those rich in calcium, potassium, and magnesium. The ratio of halite to gypsum is 1:1 rather than the 22:1 that would exist if simple evaporation of seawater took place. In most marine evaporite sequences the ratio is approximately 3:1, commonly less at 1:1, or sometimes as low as 1:100.

The 260-million-year-old evaporitic deposits of the Zechstein province of northwestern Europe are in the subsurface of a basin that includes much of the North Sea and adjacent land areas of the northeastern part of England, most of Denmark and the Netherlands, and the north German plain, extending through Poland and into Lithuania. Evaporites underlie an area of at least 250,000 square kilometers, and in the region of Stassfurt,

East Germany, they exceed 1,000 meters in thickness. These evaporites include the typical evaporitic sequence of carbonates, gypsum, anhydrite, halite, and the more soluble minerals, particularly potash salts.

A very large evaporite deposit in the subsurface of the Gulf coastal plain of Texas, Louisiana, Mississippi, and the associated offshore continental shelf extends more than 650,000 square kilometers, and sodium chloride, rock salt, had an estimated original thickness of 1,500 to 2,000 meters. Total amount of other salts is uncertain. This deposit is roughly 160 million years old and is closely associated with the production of petroleum.

James O. Jones

CROSS-REFERENCES

Alluvial Systems, 2297; Beaches and Coastal Processes, 2302; Dams and Flood Control, 2016; Deep-Sea Sedimentation, 2308; Deltas, 2312; Desert Pavement, 2319; Drainage Basins, 2325; Floodplains, 2335; Groundwater Movement, 2030; Hydrologic Cycle, 2045; Lakes, 2341; Land-Use Planning, 1490; Reefs, 2347; River Bed Forms, 2353; River Flow, 2358; Sand, 2363; Sand Dunes, 2368; Sediment Transport and Deposition, 2374; Weathering and Erosion, 2380.

BIBLIOGRAPHY

Friedman, Gerald M., and John E. Sanders. *Principles of Sedimentary Deposits: Stratigraphy and Sedimentology.* New York: Macmillan, 1992. An excellent college-level text for the serious undergraduate, with several sections covering various aspects of evaporite sedimentology.

Guilbert, John M., and Charles F. Park, Jr. *The Geology of Ore Deposits.* 4th ed. New York: W. H. Freeman, 1985. One of the best of the available college-level texts on ore deposits, this book has a descriptive section concerning evaporite minerals. Widely available in university and large public libraries.

Jensen, Mead L., and Alan M. Bateman. *Economic Mineral Deposits.* 3d ed. New York: John Wiley & Sons, 1979. A very good beginning-level college text on ore deposits. Includes a useful section concerning statistical data on

production and mineral usage. Widely available in university and large public libraries.

Jones, James O. "Marine Dominated Coastal Sabkha and Tidal-Flat Deposition of the Blaine Formation, Texas." In *Proceedings of the International Geological Congress.* Vol. 2. Moscow: International Geological Congress, 1984. The article is part of the proceedings from the International Geological Congress of 1984. The congress meets at four-year intervals in a different host country and seeks to enhance international exchange of scientific data and cooperation. This article concerns a well-known and major evaporite-producing region. Available in university and major libraries.

Jones, James O., and Tucker F. Hentz. "Permian Strata in North-Central Texas." In *Centen-*

nial Field Guide: South-Central Section. Vol. 4. Boulder, Colo.: Geological Society of America, 1988. This article is part of the multivolume set done by the Geological Society of America as part of the celebration of its one-hundredth anniversary to summarize what is known about the geology of all North America. Concerns field aspects of evaporite deposition.

Logan, Brian W. *The MacLeod Evaporite Basin, Western Australia*. Memoir 44. Tulsa, Okla.: American Association of Petroleum Geologists, 1987. This reference is widely available in university and major libraries. It contains detailed coverage of a site of evaporite deposition that was researched in the 1980's. It also includes general and statistical data.

Pettijohn, F. J. *Sedimentary Rocks*. 3d ed. New York: Harper & Row, 1975. A good college-level text for beginners, with a major section concerning evaporite deposits and their characteristics. Widely available in university and large public libraries.

Reading, H. G., ed. *Sedimentary Environments: Processes, Facies, and Stratigraphy*. 3d ed. Cambridge, Mass: Blackwell Science, 1996. Probably the most comprehensive text available on sedimentary environments. Much of the material is technical, but the text has excellent figures and photographs and will not overwhelm the careful reader.

Schreiber, B. C. "Arid Shore Lines and Evaporites." In *Sedimentary Environments and Facies*, edited by H. G. Reading. 2d ed. Oxford, England: Blackwell Scientific, 1986. The best of the available college-level texts with a chapter on evaporites. The evaporite sedimentology of deposits from diverse world locations is covered with abundant references to technical papers. Available in university and major libraries.

Usdowski, Eberhard, and Martin Dietzel. *Atlas and Data of Solid-Solution Equilibria of Marine Evaporites*. New York: Springer, 1998. This college-level textbook offers the reader an illustrated guide to seawater composition, including phase diagrams of seawater processes. The book is accompanied by a CD-ROM that reinforces the concepts discussed in the chapters.

Walker, R. G., ed. *Facies Models: Response to Sea-level Change*. 2d ed. Tulsa, Okla: Society of Economic Paleontologists and Mineralogists, 1994. An excellent compilation on sedimentary systems. Several chapters are devoted to alluvial systems. Suitable for college students.

Warren, John. *Evaporites: Their Evolution and Economics*. Malden, Mass.: Blackwell Science, 1999. This college-level textbook looks at the geology of evaporites, their role in the environment, and their economic use to humans. Illustrations, maps, index, and fifty-page bibliography.

FLOODPLAINS

Floodplains historically have been the locations of dense human population, providing fertile land and water that enabled early human settlements to attain some degree of permanence. They have also been the locations for recurring damage from floods.

PRINCIPAL TERMS

ALLUVIUM: sediment that is deposited by a stream

BLUFF: the edge of the remnant higher-elevated land that marks the margin of the floodplain

DISCHARGE: the volume of water that is transported by a stream, normally stated as cubic meters per second or cubic feet per second

FLUVIAL: pertaining to running water; for example, fluvial processes are those in which running water is the dominant agent

GEOMORPHOLOGY: the study of the origins of landforms and the processes of landform development

LOCAL BASE LEVEL: that elevation below which a stream of equilibrium will not degrade

MEANDER: a large sinuous curve or bend in a stream of equilibrium on a floodplain

ONE-HUNDRED-YEAR-FLOOD: a hypothetical flood whose severity is such that it would occur on an average of only once in a period of one hundred years; equates to a 1 percent probability each year

OXBOW: a lake that is a remnant of a cutoff stream meander

STREAM OF EQUILIBRIUM: a stream that is carrying its maximum load of sediment; it will not erode its channel any deeper but instead will establish a floodplain

CHARACTERISTICS

Floodplains are well named. They are some of the most level surfaces on Earth, and they are known for frequent floods. They also are probably the most agriculturally productive lands on Earth. The earliest concentrations of civilization are associated with the floodplains of historical rivers whose names are well known: the Nile, Ganges, Euphrates, Yangtze, and Huang. These floodplains are still among the most populous areas on Earth.

A floodplain is a landform, a physical feature that is studied in the discipline of geomorphology. Its origin is linked with frequently recurring floods. One of the most studied floodplains is that of the Mississippi River. Extensive levees and flood walls have been built in an effort to restrict floodwaters to certain areas where they will do the least damage. A large flood, the so-called superflood or one-hundred-year-flood, will still do tremendous damage. A large river cannot be prevented from flooding; it is a part of the natural fluvial process. It can only be altered to some extent so that its floodwaters will do a minimum of damage.

The variables of climate and topographic relief will result in floodplain variation around the world, but all will possess certain identifiable characteristics. The foremost distinguishing characteristic is the very low relief, or its almost "flat" appearance. Of course, the land is not in fact level, or the river would not flow. The floodplain possesses a very slight gradient to the sea. It is an expanse of sediment that was deposited by the river itself. Thousands of years of flooding have spread layer upon layer of river-borne sands, silts, and clays, referred to as alluvium. As the waters receded after each flood, the river returned to its normal channel.

FORMATION

In order to understand the floodplain, it is necessary to understand the processes that brought it into being. The originating river did not always possess the floodplain, and many rivers today do not have floodplains. A stream with a floodplain may have—perhaps millions of years ago—originally flowed in a narrow channel with higher land along either side. There was no level land adjacent to the stream—no floodplain; its formation would require eons of time of erosion and deposition. As the stream eroded, its channel deepened and also attempted to erode laterally, but the vertical

downcutting was dominant and the lateral cutting ineffective. At some point in the fluvial process of downcutting, the stream reached an elevation, referred to as local base level, beyond which the stream could not downcut. The local base level is not some hard, resistant rock that stops downcutting, but rather is a particular elevation in respect to that stream, which in relation to the stream's mouth at sea level determines the gradient of the stream.

When the stream downcuts its valley, the stream's gradient, or slope of flow, is lowered. As the gradient is lowered, so is the stream's velocity, and consequently the stream's ability to carry sediment load is lowered. At some gradient, the stream's ability to carry its sediment load equals the sediment load that has been brought to the master stream by its tributaries. When that happens, the stream is said to be in equilibrium. Its sediment-load carrying capacity is then equal to the sediment load. This "stream of equilibrium" condition is a prerequisite for the development of a floodplain. The Mississippi River is a classic example of such a stream. When the stream attains equilibrium and stops downcutting, its lateral cutting becomes the dominant stream activity.

Thus, the processes for floodplain development are set in motion. The lateral cutting by the stream results in a widening of the valley, but without any deepening. A broader-level valley floor then begins to develop, as the stream swings side to side, without downcutting. Over many years, the floor of the valley becomes wider than the channel of the stream itself. This level land on the valley floor adjacent to the stream, small at first, is the floodplain. It will be enlarged over time. The stream's volume of water, or discharge, will naturally vary with seasonal precipitation. During a flood, the discharge exceeds the channel capacity, and the excess overflows onto the level adjacent land. The term "floodplain" is now appropriate.

The floodplain is continually widened as the stream cuts laterally. The "bluffs" that mark the edge of the floodplain are driven farther back from the river as more of the adjacent higher land is eroded. When wide

TYPES OF FLOODPLAINS

Wide Channel:
Free Meander

Narrow Channel:
Confined Meander

Braided Channel

A low-velocity stream, the Serpentine River meanders through a floodplain on the Seward Peninsula of Alaska. An oxbow lake is visible in the center. (© William E. Ferguson)

enough, the stream channel develops great curving, sinuous "meanders." Meanders are perhaps the most prominent identifying characteristics of a stream of equilibrium, and true meanders are always on a floodplain. Meanders may subsequently be cut off from the main channel by its persistent migrations or wanderings over the floodplain. Such cutoff meanders may persist for hundreds of years as lakes, called oxbows. The oxbows will gradually be silted in and become oxbow swamps for a period of time. Even after thousands of years, the oxbow lakes and oxbow swamps may survive as meander "scars" on the floodplain, showing up as vegetation and soil differences.

These curving scars are even visible in fields of cropland. The river continues to flood periodically, frequently on an annual basis. Each time the floodwaters spread over the floodplain, an additional layer of silt is deposited. The entire floodplain is a compilation of layer upon layer of alluvial deposits from flood after flood, over thousands of years. The silts that are deposited on the plain have been eroded earlier from soils upstream. They tend to be fertile topsoils from upstream drainage basins, and thus the floodplain tends to receive almost annual increments of fertile topsoil. The recurring flooding and silting shortens the life span of the oxbows. Older ox-

bows are slowly obliterated, while new ones are cut off by the meandering stream. Floodplains of large rivers can attain impressive dimensions. The Mississippi floodplain is 60 to 100 kilometers wide and some 800 kilometers long, extending from the confluence of the Ohio River in southern Illinois to the Gulf of Mexico.

FLOOD CONTROL

The Mississippi floodplain is superb agricultural land. Great floods in 1927 and 1937, however, wrought such havoc that massive flood-control measures were initiated to prevent further occurrences. The controls are not complete and probably never will be, as it is a never-ending battle to stay abreast of the river. During the great Mississippi flood of 1973, floodwaters were largely confined, and a flood of catastrophic proportions was avoided. In 1993, the Mississippi River once again flooded. Big cities, such as St. Louis, were protected by floodwalls, either for aesthetic or economic reasons. However, Des Moines, which had no floodwall, was severely affected: Its water-treatment plant flooded, leaving the community without a safe water supply for one month. The exceedingly costly flooding of 1993 once again threw into question many policies regarding levee and floodwall construction, which protect some farmland and communities but worsen flooding elsewhere. The U.S. Army Corps of Engineers breached at least one agricultural levee to "save" a town downstream during the 1993 flood. Also problematical is the federal flood insurance program, which "rewards" people who live in the floodplain by giving them benefits after each flood. The damage that occurred in 1993 prompted calls to change the program so that a recipient would receive benefits only for the first insured flood loss.

A major aspect of all floodplains is the relationship between the rivers' natural tendencies to flood and the human occupants' efforts to protect themselves. There are a number of flood-control measures and devices. Principal among them are levees. Levees are human-made earthen ridges constructed to parallel the channel about 0.5 kilometer back from the riverbanks. The function of the levees is to confine the overbank floodwaters to a narrow flood zone along the river and thus protect the greater portion of the floodplain. Without the levees, the entire floodplain could be flooded in a major flood all the way to the bluffs. A levee failure would be disastrous. Another major flood-control measure is the construction of reservoirs on tributary streams, where water can be held back when the master stream is in flood. These reservoirs are not on the floodplain itself and may be many miles upstream in the tributaries. An area-wide, integrated plan must be adopted to ensure efficient flood control.

STUDY OF FLOODPLAINS

Floodplains have been studied for almost as long as human civilization has existed. Most of the earliest permanent human settlements were in fact associated with floodplains, where fertile land and water were available. Early concerns were focused on two issues: distribution of irrigation water for agriculture and protection from excess water during floods. Modern concerns are similar. In the Mississippi floodplain, the concern is primarily flood control. The area's climate is humid, and river water has never been important for irrigation in the region. Research and study, therefore, have been directed toward developing effective measures and devices for control of overbank waters. An additional area for study, and for considerable expenditure as well, has been channel management for navigation purposes. The maintenance of an efficient channel is necessary for both navigation and flood control. Flood control, however, is concerned with maximum flow. Channel management for navigation purposes is primarily concerned with minimum flow, that is, maintaining minimum shipping channel depths and widths during dry seasons and droughts.

Studies and data collection are directed toward the mechanics of stream flow. Stream-gauging stations are established in which measurements are taken to record discharge, velocity, and flow turbulence. Many of the gauging stations are automatic, producing data for long-term analysis and also alerting warning monitors to flash-flood conditions. Gauging networks are established for trib-

utary streams as well as for the master stream on the floodplain. The morphology of the channel is studied. The width, depth, and cross-sectional areas of the streams are analyzed for their capabilities to pass the flood flows. Great effort and expense are applied in river-engineering works to improve channel flow and to increase capabilities for high water discharge. Some of the adjustments are channel straightening, bank clearing, dredging and sandbar removal, and the construction of levees and dikes. Models of the stream channels, floodplain, and levees are constructed to precise scale. Water is passed through the models so as to determine the best configurations for efficient flow. Dikes are placed in critical locations in the model, and the flow is studied to observe the effect of the dikes on bank erosion and sandbar formation. While the river conditions cannot be perfectly replicated in a scale model, its use does aid in the selection of sites for river-engineering works.

SIGNIFICANCE

Floodplains will continue to play an important role in human activities. They have always been areas of concentrated population and will be even more so in the future as the world population increases. Floodplains will acquire even more significance for agriculture. More people will live on the wide level surfaces that were once the undisputed domains of the rivers, yet the rivers can no longer be allowed to flood the plains. It is natural for a stream to flood the plain periodically. It is not natural to confine the flood to a narrow strip along the channel as has been done with the Mississippi. More than 3,000 kilometers of levees have been constructed along the lower Mississippi River. The levees have been a success in preventing disastrous floods such as those that occurred in 1927 and 1937. They have brought protection to the valuable farmlands and to the towns and settlements on the plains. However, losses along the upper Mississippi were enormous during the 1993 flood, which caused fifty-four deaths and $12 billion in property damage.

The protection offered by the levees serves as an invitation for continued settlement and development, which proceed today on the world's floodplains as if there were no threats of flooding. A levee failure today can be more catastrophic than one in the past, even with the same amount of flooding, because so much more lies in the path of the flood. It is probably correct that the greatest flood is yet to come.

Adding to the threat of flooding on the floodplain is the fact that human habitation of floodplains increases the likelihood of floods. The floodplain can be thought of as the bottom of a funnel. Runoff from all basins upstream merges on the floodplain. Many flood-control measures that are applied to tributary streams, such as channel straightening, actually increase the likelihood of flooding downstream by speeding up the runoff. Additionally, changes in land use today exacerbate flooding. Deforestation and the drainage of wetlands increase runoff and intensify flooding. The increase in urbanization, with its growing expanse of asphalt paving and rooftops, accelerates runoff onto areas downstream. As the drainage basins undergo change, so does the potential for flooding on the floodplain. The floodplain is a dynamic and changing environment, and it will require continued study and adjustment. Humans are now settled densely on the world's floodplains. The plains have become more significant as food-producing areas in a world of growing demands. Great engineering works have been constructed and have doubtless prevented many floods, yet the potential for disastrous floods has only increased. Continued vigilance is thus essential.

John H. Corbet

CROSS-REFERENCES

Alluvial Systems, 2297; Beaches and Coastal Processes, 2302; Deep-Sea Sedimentation, 2308; Deltas, 2312; Desert Pavement, 2319; Diagenesis, 1445; Drainage Basins, 2325; Evaporites, 2330; Geochemical Cycle, 412; Geomorphology of Dry Climate Areas, 904; Karst Topography, 929; Lakes, 2341; Minerals: Physical Properties, 1225; Reefs, 2347; River Bed Forms, 2353; River Flow, 2358; Rocks: Physical Properties, 1348; Sand, 2363; Sand Dunes, 2368; Sediment Transport and Deposition, 2374; Sedimentary Mineral Deposits, 1637; Sedimentary Rock Classification, 1457; Weathering and Erosion, 2380.

BIBLIOGRAPHY

Bloom, Arthur L. *Geomorphology: A Systematic Analysis of Late Cenozoic Landforms.* 3d ed. Upper Saddle River, N.J.: Prentice-Hall, 1998. A thorough text on geomorphology. Four chapters are devoted to the processes and landforms of streams. Diagrams illustrate the stages of floodplain development. The book assumes some knowledge on the part of the reader but is not difficult. An extensive list of references is provided with each chapter.

Changnon, Stanley A., ed. *The Great Flood of 1993: Causes, Impacts, and Responses.* Boulder, Colo.: Westview Press, 1996. An examination of the flood damage of 1993 and its social and environmental effects. The book also looks at the factors leading up to the flood and prevention methods put in place as a result. Illustrations and bibliographical references.

Chorley, Richard J., ed. *Introduction to Fluvial Processes.* London: Methuen, 1975. A paperback that is a technical treatment of the various aspects of fluvial processes. Includes diagrams, graphs, and formulas explaining stream hydraulics and channel morphometry.

Harrison, Robert W. *Flood Control and Water Management in the Yazoo: Mississippi Delta.* Mississippi: Social Science Research Center, Mississippi University, 1993. A thorough analysis of flood control and prevention policies and management for the Mississippi River. Notable for an excellent bibliography of works on the Mississippi Delta.

Lobeck, A. K. *Geomorphology: An Introduction to the Study of Landscapes.* New York: McGraw-Hill, 1939. A well-known text on geomorphology. The diagrams are particularly helpful in understanding floodplains. The text is easy to read and is supplemental to the illustrations. Although dated and out of print, it is considered a classic in the field.

Morisawa, Marie. *Streams: Their Dynamics and Morphology.* New York: McGraw-Hill, 1968. A paperback that is a more detailed and technical treatment of stream processes than are the other listed references. Includes data on graphs, as well as maps and diagrams. The hydraulics of stream flow are explained.

Smith, Keith, and Graham Tobin. *Human Adjustment to the Flood Hazard.* New York: Longman, 1979. A paperback that deals specifically with flood hazards. Provides information on the nature of floods, planning strategies, and techniques of control. This specialized book is for the more advanced student.

Strahler, Arthur N., and Alan H. Strahler. *Introducing Physical Geography.* 2d ed. New York: John Wiley & Sons, 1998. A general introductory text on physical geography. It gives a good explanation of fluvial processes and the development of floodplains. Well illustrated with maps and photographs. Recommended as a first source of information. Accompanied by a CD-ROM.

Tarbuck, Edward J., and Frederick K. Lutgens. *Earth: An Introduction to Physical Geology.* 6th ed. Upper Saddle River, N.J.: Prentice Hall, 1999. A general beginning text for college Earth science, easily readable by the layperson. The book employs quality color diagrams to explain the development of a floodplain, plus excellent illustrations throughout. Recommended as an initial source.

Thornbury, William D. *Principles of Geomorphology.* 2d ed. New York: John Wiley & Sons, 1969. A geomorphology textbook for the more advanced student. Three chapters deal with fluvial processes. Includes diagrams of floodplain features and the Mississippi Delta. References are listed.

LAKES

Lakes are geologically short-lived features, and sediments deposited in lakes (called lacustrine sediments) constitute only a tiny part of the sedimentary rocks of the Earth's crust. Nevertheless, lake sediments are important sources of information about past climates. Several important economic resources—including oil shales, diatomaceous Earth, salt and other evaporites, some limestones, and some coals—originate in lakes.

PRINCIPAL TERMS

ALLOGENIC SEDIMENT: sediment that originates outside the place where it is finally deposited; sand, silt, and clay carried by a stream into a lake are examples

BIOGENIC SEDIMENT: sediment that originates from living organisms

CLASTIC SEDIMENTS: sediments composed of durable minerals that resist weathering

CLAY: a mineral group that consists of structures arranged in sandwichlike layers, usually sheets of aluminum hydroxides and silica, along with some potassium, sodium, or calcium ions

CLAY MINERALS: any mineral particle less than 2 micrometers in diameter

ENDOGENIC SEDIMENT: sediment produced within the water column of the body in which it is deposited; for example, calcite precipitated in a lake in summer

MINERAL: a solid with a constant chemical composition and a well-defined crystal structure

MINERALOID: a solid substance with a constant chemical composition but without a well-ordered crystal structure

PLANKTON: plant and animal organisms, most of which are microscopic, that live within the water column

SESTON: a general term that encompasses all types of suspended lake sediment, including minerals, mineraloids, plankton, and organic detritus

GEOLOGICAL ORIGIN OF LAKES

Several geologic mechanisms can create the closed basins that are needed to impound water and produce lakes. The most important of these mechanisms include glaciers, landslides, volcanoes, rivers, subsidence, and tectonic processes.

Continental glaciers formed thousands of lakes by the damming of stream valleys with moraine materials. Glaciers also scoured depressions in softer bedrock, and these later filled with water to form lakes. Depressions called kettles formed when buried ice blocks melted. Mountain glaciers also produce numerous small, high alpine lakes by plucking away bedrock. The bowl-shaped depressions that occur as a result of this plucking are called cirques; lakes that occupy cirques are called tarns. Sometimes a mountain glacier moves down a valley and carves a series of depressions along the valley that, from above, look like a row of beads along a string. When these depressions later fill with water, the lakes are called paternoster lakes, the name coming from their similarity to beads on a rosary.

Landslides sometimes form natural dams across stream valleys. Large lakes then pond up behind the dam. Volcanoes may produce lava flows that dam stream valleys and produce lakes. A volcanic explosion crater may fill with water and make a lake. After an eruption, the area around the eruption vent may collapse to form a depression called a caldera. Some calderas, such as Crater Lake in Oregon, fill with water. Rivers produce lakes along their valleys when a tight loop of a meandering channel finally is eroded through and leaves behind an oxbow lake, isolated from the main channel. Sediment may accumulate at the mouth of a stream, and the resulting delta may build, bridging across irregularities in the shoreline to create a brackish coastal lake.

Natural subsidence creates closed basins in areas underlain by soluble limestones or evaporite deposits. As the underlying limestone is dissolved

2341

away, the Earth above collapses to form a cavity (sinkhole), which later fills with water. Finally, large-scale (tectonic) downwarping of the Earth's crust produces some very large lakes. Large basins form when the crust warps or sinks downward in response to deep forces. The subsidence produces very large closed basins that can hold water. A few immense lakes owe their origins to tectonic downwarping.

SEDIMENTATION

With few exceptions, most lakes exist in relatively small depressions and serve as the catch basins for sediment from the entire watershed around them. The natural process of sedimentation ensures that most lakes fill with sediment before very long periods of geologic time have passed. Lakes with areas of only a few square kilometers or less will fill within a few tens of thousands of years. Very large lakes, the inland seas, may endure for more than ten million years. Human-made lakes and reservoirs have unusually high sediment-fill rates in comparison with most natural lakes. Human-made lakes fill with sediment within a few decades to a few centuries.

Lake sediments come from four sources: allogenic clastic materials that are washed in from the surrounding watershed; endogenic chemical precipitates that are produced from dissolved substances in the lake waters; endogenic biogenic organic materials produced by plants and animals living in the lake; and airborne substances, such as dust and pollen, transported to the lake in the atmosphere.

Allogenic clastic materials are mostly minerals; they are produced when rocks and soils in the drainage basin are weathered by mechanical and chemical processes to yield small particles. These particles are moved downslope by gravity and running water to enter streams, which then transport them to the lake. Clastic materials also enter the lake via waves, which erode the materials from the shoreline, and via landslides that directly enter the lake. In winter, ice formed on the lake can expand and push its way a few centimeters to 1 meter or so onto the shore. There, the ice may pick up large particles, such as gravel and cobbles. When spring thaw comes, waves can remove that ice, together with its enclosed particles, and float it out onto the lake. The process by which the large particles are transported out on the lake is called ice-rafting. As the ice melts, the large clastic particles drop to the bottom; they are termed dropstones when found in lake sediments. A landslide into a lake or a flood on a stream that feeds into the lake can produce water heavily laden with sediment. The sediment-laden water is more dense than clean water and therefore can rush down and across the lake bottom at speeds sufficient to carry even coarse sand far out into the lake. These types of deposits are called turbidite deposits.

Endogenic chemical precipitates in freshwater lakes commonly consist of carbonate minerals (calcite, aragonite, or dolomite) and mineraloids that consist of oxides and hydroxides of iron, manganese, and aluminum. In some saline and brine lakes, the main sediments may be carbonates, together with sulfates such as gypsum (hydrated calcium sulfate), thenardite (sodium sulfate), or epsomite (hydrated magnesium sulfate), or with chlorides such as halite (sodium chloride) or more complex salts. Of the endogenic precipitates, calcite is the most abundant. Its precipitation represents a balance between the composition of the atmosphere and that of the lake water.

Diatoms are distinctive microscopic algae that produce a frustule (a kind of shell) made of silica glass that is highly resistant to weathering. When seen under a high-powered microscope, diatom frustules appear to be artwork—beautiful and highly ornate saucer- and pen-shaped works of glass. A tiny spot of lake sediment may contain millions.

A lake's sediment may contain from less than 1 percent to more than 90 percent organic materials, depending upon the type of lake. Most organic matter in lake sediments is produced within the lake by plankton and consists of compounds such as carbohydrates, proteins, oils, and waxes that are made up of organic carbon, hydrogen, nitrogen, and oxygen, with a little phosphorus. Plankton, with an approximate bulk composition of 36 percent carbon, 7 percent hydrogen, 50 percent oxygen, 6 percent nitrogen, and 1 percent phosphorus (by weight), includes microscopic plants (phytoplankton) and microscopic animals (zooplankton) that live in the water column. Lakes that are very high in nutrients (eutrophic lakes) commonly have heavy blooms of algae, which contribute much organic matter to the bottom sediment. Terrestrial (land-derived) organic material such as leaves, bark, and twigs form a minor part of the organic matter found in most lakes. Terrestrial organic material is higher in carbon and lower in hydrogen, nitrogen, and phosphorus than is planktonic organic matter.

Airborne substances usually constitute only a tiny fraction of lake sediment. The most important material is pollen and spores. Pollen usually constitutes less than 1 percent of the total sediments, but that tiny amount is a very useful component for learning about the recent climates of the Earth. Pollen is among the most durable of all natural materials. It survives attack by air, water, and even strong acids and bases. Therefore, it remains in the sediment through geologic time. As pollen accumulates in the bottom sediment, the lake serves as a kind of recorder for the vegetation that exists around it at a given time. By taking a long core of the bottom sediment from certain types of lakes, a geologist may look at the pollen changes that have occurred through time and reconstruct the history of the climate and vegetation in an area.

Volcanic ash thrown into the atmosphere during eruptions enters lakes and forms a discrete layer of ash on the lake bottom. When Mount St. Helens erupted in 1980, it deposited several centimeters of ash in lakes more than 160 kilometers east of the volcano. Geologists have used layers of ash in lakes to reconstruct the history of volcanic eruptions in some areas. Although dust storms contribute sediment to lakes, such storms are usually too infrequent in most areas to contribute significant amounts.

WATER CIRCULATION

Lake waters are driven into circulation by temperature-induced density changes and wind. Most freshwater lakes in temperate climates circulate completely twice each year; they are termed dimictic lakes. Circulation exerts a profound influence on water chemistry of the lake and the amount and type of sediment present within the water column. During summer stratification, the lake is thermally stratified into three zones. The upper layer of warm water (epilimnion) floats above the denser cold water and prevents wind-driven circulation from penetrating much below the epilimnion. The epilimnion is usually in circulation, is rich in oxygen (from algal photosynthesis and diffusion from the atmosphere), and is well lighted. This layer is where summer blooms of green and blue-green algae occur and calcite precipitation begins. The middle layer (thermocline) is a transition zone in which the water cools downward at a rate of greater than 1 degree Celsius per meter. The bottom layer (hypolimnion) is cold, dark, stagnant, and usually poor in oxygen. There, bacteria decompose the bottom sediment and release phosphorus, manganese, iron, silica, and other constituents into the hypolimnion.

Sediment deposited in summer includes a large amount of organic matter, clastic materials washed in during summer rainstorms, and endogenic carbonate minerals produced within the lake. The most common carbonate mineral is calcite (calcium carbonate). The regular deposition of calcite in the summer is an example of cyclic sedimentation, a sedimentary event that occurs at regular time intervals. This event occurs yearly in the summer season and takes place in the upper 2 or 3 meters of water. On satellite photos, it is even possible to see the summer events as whitings on large lakes, such as Lake Michigan.

As the sediment falls through the water column in summer, it passes through the thermocline, into the hypolimnion, and onto the lake bottom. As it sits on the bottom during the summer months, bacteria, particularly anaerobic bacteria (those that thrive in oxygen-poor environments), begin to decompose the organic matter. As this occurs, the dissolved carbon dioxide increases in the hypolimnion. If enough carbon dioxide is produced, the hypolimnion becomes slightly acidic, and calcite and other carbonates that fell to the bottom begin to dissolve. The acidic conditions also release dissolved phosphorus, calcium, iron, and manganese into the hypolimnion, as well as some trace metals. Clastic minerals such as quartz, feldspar, and clay minerals are not affected in such brief seasonal processes, but some silica from biogenic material such as diatom frustules can dissolve and enrich the hypolimnion in silica. As summer progresses, the hypolimnion becomes more and more enriched in dissolved metals and nutrients.

Autumn circulation begins when the water temperature cools and the density of the epilimnion increases until it reaches the same temperature and density as the deep water. Thereafter, there is no stratification to prevent the wind from circulating the entire lake. When this happens, the cold, stagnant hypolimnion, now rich in dissolved substances, is swept into circulation with the rest of the lake water. The dissolved materials from the hypolimnion are mixed into a well-oxygenated water column. Iron and manganese that formerly were present in dissolved form now oxidize to form tiny solid particles of manganese oxides, iron oxides, and hydroxides. The sediment therefore becomes enriched in iron, manganese, or both during the autumn overturn, the amount of enrichment depending upon the amount of dissolved iron and manganese that accumulated during summer in the hypolimnion. Dissolved silica is also swept from the hypolimnion into the entire water column. In the upper water column, where sunlight and dissolved silica become present in great abundance, diatom blooms occur. The diatoms convert the dissolved silica into solid opaline frustules.

As circulation proceeds, the currents may sweep over the lake bottom and actually resuspend 1 centimeter or more of sediment from the bottom and margins of the lake. The amount of resuspension that occurs each year in freshwater lakes is primarily the result of the shape of the lake basin. A lake that has a large surface area and is very shallow permits wind to keep the lake in constant circulation over long periods of the year.

As winter stratification comes, an ice cover forms over the lake and prevents any wind-induced circulation. Because the circulation is what keeps the lake sediment in suspension, most sediment quickly falls to the bottom; sedimentation then is minimal through the rest of winter. If light can penetrate the ice and snow, some algae and diatoms can utilize this weak light, present in the layer of water just below the ice, to reproduce. Their settling remains contribute small amounts of organic matter and diatom frustules. At the lake bottom, the most dense water (that at 4 degrees Celsius) accumulates. As in summer, some dissolved nutrients and metals can build up in this deep layer, but because the bacteria that are active in releasing these substances from the sediment are refrigerated, they work slowly, and not as much dissolved material builds up in the bottom waters.

When spring circulation begins, the ice at the surface melts, and the lake again goes into wind-driven circulation. Oxidation of iron and manganese occurs (as in autumn), although the amounts of dissolved materials available are likely to be less in spring. Once again, nutrients such as phosphorus and silica are circulated out of the dark bottom waters and become available to produce blooms of phytoplankton. Spring rains often hasten the melting, and runoff from rain and snowmelt in the drainage basin washes clastic materials into the lake. The period of spring thaw is likely to be the time of year when the maximum amount of new allogenic (externally derived) sediment enters the lake.

Spring diatom blooms continue until summer stratification prevents further replenishment of silica to the epilimnion. Thereafter, the diatoms are succeeded by summer blooms of green algae, closely followed by blooms of blue-green algae. Silica is usually the limiting nutrient for diatoms; phosphorus is the limiting nutrient for green and blue-green algae.

DIAGENESIS

After sediments are buried, changes occur; this process of change after burial is termed diagenesis. Physical changes include compaction and dewatering. Bacteria decompose much organic matter and produce gases such as methane, hydrogen sulfide, and carbon dioxide. The "rotten-egg" odor of black lake sediments, often noticed on boat anchors, is the odor of hydrogen sulfide. After long periods of time, minerals such as quartz or calcite slowly fill the pores remaining after compaction.

One of the first diagenetic minerals to form is pyrite (iron sulfide). Much pyrite occurs in microscopic spherical bodies that look like raspberries; these particles, called framboids, are probably formed by bacteria in areas with low oxygen within a few weeks. In fact, the black color of some lake muds and oozes results as much from iron sulfides as from organic matter. Other diagenetic changes include the conversion of mineraloid particles containing phosphorus into phosphate minerals such as vivianite and apatite. Manganese oxides may be converted into manganese carbonates (rhodochrosite). Freshwater manganese oxide nodules may form in high-energy environments such as Grand Traverse Bay in Lake Michigan.

STUDY OF LAKES

Scientists who study lakes (limnologists) must study all the natural sciences—physics, chemistry, biology, meteorology, and geology—because lakes are complex systems that include biological communities, changing water chemistry, geological processes, and interaction among water, sunlight, and the atmosphere.

Modern lake sediments are collected from the water column in sediment traps (cylinders and funnels into which the suspended sediment settles over periods of days or weeks) or by filtering large quantities of lake water. Living material is often sampled with a plankton net. Older sediments that have accumulated on the bottom are collected with dredges and by piston coring, which involves pushing a sharpened hollow tube (usually about 2.5 centimeters in diameter) downward into the sediment. Cores are valuable because they preserve the sediment in the order in which it was deposited, from oldest at the bottom to youngest at the top. Once the sample is collected, it is often frozen and taken to the laboratory. There, pollen and organisms may be examined by microscopy, minerals may be determined by X-ray diffraction, and chemical analyses may be made.

Varves are thin laminae that are deposited by cyclic processes. In freshwater lakes, each varve represents one year's deposit; it consists of a couplet with a dark layer of organic matter deposited in winter and a light-colored layer of calcite deposited in summer. Varves are deposited in lakes where annual circulations cannot resuspend bottom sediment and therefore cannot mix it to destroy the annual lamination. Some lakes that are small and very deep may produce varved sediments; Elk Lake in Minnesota is an example. In other lakes, the accumulation of dissolved salts on the bottom eventually produces a dense layer (monimolimnion), which prevents disturbance of the bottom by circulation in the overlying fresher waters. Soap Lake in Washington State is an example. Because each varve couplet represents one year, a geologist may core the sediments from a varved lake and count the couplets to determine the age of the sediment in any part of the core. The pollen, the chemistry, the diatoms, and other constituents may then be carefully examined to deduce what the lake was like during a given time period. The study is much like solving a mystery from a variety of clues. Eventually, the history of climate changes of the area may be learned from the study of lake varves.

Edward B. Nuhfer

CROSS-REFERENCES

Alluvial Systems, 2297; Beaches and Coastal Processes, 2302; Dams and Flood Control, 2016; Deep-Sea Sedimentation, 2308; Deltas, 2312; Desert Pavement, 2319; Drainage Basins, 2325; Evaporites, 2330; Floodplains, 2335; Floods, 2022; Hydrologic Cycle, 2045; Land-Use Planning, 1490; Reefs, 2347; River Bed Forms, 2353; River Flow, 2358; River Valleys, 943; Sand, 2363; Sand Dunes, 2368; Sediment Transport and Deposition, 2374; Surface Water, 2066; Weathering and Erosion, 2380.

BIBLIOGRAPHY

Bailey, Ronald. *Rivers and Lakes.* New York: Time-Life Books, 1985. A book on lacustrine (lake) and fluvial (river) environments, suitable for general readers. Part of the Time-Life Planet Earth series.

Håkanson, Lars, and M. Jansson. *Principles of Lake Sedimentology.* New York: Springer-Verlag, 1983. Though this book is a reference for professionals in the field of lake sedimentology, parts of it may be understood by high school students. Most books on limnology focus on lake water; this reference is one of the few to focus on lake sediments in detail. Provides methods of sampling and discusses the influence of lake type and shape on the sediments formed in the lake, the circulation of lake waters, the chemistry of sediments, and the pollution of lakes.

Hutchinson, G. Evelyn. *A Treatise on Limnology.* 3 vols. New York: John Wiley & Sons, 1957-1975. A highly comprehensive reference about lakes. The set derives its information from worldwide sources and is well indexed by subject as well as by specific lakes and their geographic locations. Volume 1 (in two parts: *Geography and Physics of Lakes* and *Chemistry of Lakes*) discusses the geologic formation of lakes, lake types, the interaction of sunlight with lake waters, water color, heat distribution, and water circulation. The chemistry of lake waters is discussed in detail. Volume 2, *Introduction to Lake Biology and Limnoplankton*, covers plankton and the factors that influence their growth. Volume 3, *Limnological Botany*, covers larger aquatic plants (macrophytes) and attached algae. Most of the treatise is readable by college undergraduates and high school seniors, but a few parts will be well understood only by specialists.

Imberger, Jeorg, ed. *Physical Processes in Lakes and Oceans*. Washington, D.C.: American Geophysical Union, 1998. This extensive volume details the origins, processes, and phases of oceans, lakes, and water resources, as well as the ecology and environments surrounding them. Illustrations and maps.

Lerman, Abraham, ed. *Lakes: Chemistry, Geology, Physics*. New York: Springer-Verlag, 1978. This book fills a gap in the Hutchinson treatise by focusing on the geologic processes of sedimentation in freshwater and brine lakes. The book is actually a compilation of chapters, each written by specialists. Particular attention is given to carbonate sediments, clastic and endogenic minerals, human influence on natural lakes, and organic compounds. The chapters on lake sediments and carbonate sedimentation are well illustrated. Most parts of these chapters are accessible to college undergraduates and high school students. Other chapters will be well understood only by specialists.

Lerman, Abraham, Dieter M. Imboden, and Joel R. Gat, eds. *Physics and Chemistry of Lakes*. New York: Springer-Verlag, 1995. This book offers a nice introduction to limnology by examining the geological, chemical, and physical properties of lakes. Suitable for the nonscientist. Illustrations, index, and bibliography.

Stumm, Werner, ed. *Chemical Processes at the Particle-Water Interface*. New York: John Wiley & Sons, 1987. A highly technical reference book, designed for specialists and graduate students. Chapters are authored by a variety of experts, who focus on the chemical interactions that occur between sediment particles and the surrounding lake waters, the process of clotting of particles, and the role of particle surfaces in removing trace metals from water. Chapter 12, by Laura Sigg, focuses specifically on lake sediments and may be understood by undergraduates who have had rigorous courses in introductory chemistry and geology.

U.S. Environmental Protection Agency. *The Great Lakes: An Environmental Atlas and Resource Book*. Chicago: Great Lakes Program Office, 1995. A free soft-cover publication designed for anyone interested in the lakes. The material briefly addresses the physical and cultural environments of the basin. Includes colored maps, tables, and charts.

Wetzel, R. G., ed. *Limnology*. 2d ed. Philadelphia: Saunders, 1983. A very well-written textbook typical of those used by undergraduates and graduates in their first limnology courses. Covers physical, biological, and chemical aspects of lakes. High school algebra, chemistry, physics, and biology courses will be necessary prerequisites to understand most of the text.

REEFS

Reefs are among the oldest known communities, existing at least 2 billion years ago. They exert considerable control on the surrounding physical environment, influencing turbulence levels and patterns of sedimentation. Ancient reefs are often important hydrocarbon reservoirs.

PRINCIPAL TERMS

CALCAREOUS ALGAE: green algae that secrete needles or plates of aragonite as an internal skeleton; very important contributors to reef sediment

CARBONATE ROCKS: sedimentary rocks such as limestone, which is composed of the minerals calcite or aragonite, or dolostone, which is composed of the mineral dolomite

CORALLINE ALGAE: red algae that secrete crusts or branching skeletons of high-magnesium calcite; important sediment contributors and binders on reefs

RUGOSE CORALS: a Paleozoic coral group also known as "tetracorals"; sometimes colonial, but more often solitary and horn-shaped

SCLERACTINIAN CORALS: modern corals or "hexa-corals," different from their more ancient counterparts in details of the skeleton and the presence of a symbiosis with unicellular algae in most shallow-water species

STROMATOLITES: layered columnar or flattened structures in sedimentary rocks, produced by the binding of sediment by blue-green algal (cyanobacterial) mats

STROMATOPOROIDS: spongelike organisms that produced layered, mound-shaped, calcareous skeletons and were important reef builders during the Paleozoic era

TABULATE "CORALS": colonial organisms with calcareous skeletons that were important Paleozoic reef builders; considered to be more closely related to sponges than to corals

"TRUE" REEFS VS. REEFLIKE STRUCTURES

Reefs or reeflike structures are among the oldest known communities, extending back more than 2 billion years into the Earth's history. These earliest reefs were vastly different in their biotic composition and physical structure from modern reefs, which are among the most diverse of biotic communities and display amazingly high rates of biotic productivity (carbon fixation) and calcium carbonate deposition, despite their existence in a virtual nutrient "desert." Reefs are among the few communities to rival the power of humankind as a shaper of the planet. The Great Barrier Reef of Australia, for example, forms a structure some 2,000 kilometers in length and up to 150 kilometers in width.

It is necessary to distinguish between "true," or structural, reefs and reeflike structures or banks. Reefs are carbonate structures that possess an internal framework. The framework traps sediment and provides resistance to wave action; thus, reefs can exist in very shallow water and may grow to the surface of the oceans. Banks are also biogeni-cally produced but lack an internal framework. Thus, banks are often restricted to low-energy, deep-water settings. "Bioherm" refers to moundlike carbonate buildups, either reefs or banks, and "biostrome" to low, lens-shaped buildups.

REEF CLASSIFICATION

Modern reefs are classified into several geomorphic types: atoll, barrier, fringing, and patch. Many of these may be further subdivided into reef crest or flat, back-reef or lagoon, and fore-reef zones. Atoll reefs are circular structures with a central lagoon, thought to form on subsiding volcanic islands. Barrier reefs are elongate structures that parallel coastlines and possess a significant lagoon between the exposed reef crest and shore. These often occur on the edges of shelves that are uplifted by faulting. Fringing reefs are elongate structures paralleling and extending seaward from the coastline that lack a lagoon between shore and exposed reef crest. Patch reefs are typically small, moundlike structures, occurring isolated on shelves or in lagoons. The majority of fossil reefs would be classified as patch reefs, although many examples of extensive, linear, shelf-edge trends are also known from the geologic record.

Reefs form one of the most distinctive and easily recognized sedimentary facies (or environments). In addition to possessing a characteristic fauna consisting of corals, various algae, and stromatoporoids, they are distinguished by a massive (nonlayered) core that has abrupt contacts with adjacent facies. Associated facies include flat-lying lagoon and steeply inclined fore-reef talus, the latter often consisting of large angular blocks derived from the core. The reef core is typically a thick unit relative to adjacent deposits. The core also consists of relatively pure calcium carbonate with little contained terrigenous material.

REEF ENVIRONMENTS

Modern reefs are restricted to certain environments. They occur abundantly only between 23 degrees north and south latitudes and tend to be restricted to the western side of ocean basins, which lack upwelling of cold bottom waters. This restriction is based on temperature, as reefs do not flourish where temperatures frequently reach below 18 degrees Celsius. Reef growth is largely restricted to depths greater than 60 meters, as there is insufficient penetration of sunlight below this depth for symbiont-bearing corals to flourish. Reefs also require clear waters lacking suspended terrigenous materials, as these interfere with the feeding activity of many reef organisms and also reduce the penetration of sunlight. Finally, most reef organisms require salinities that are in the normal oceanic range. It appears that many fossil reefs were similarly limited in their environmental requirements.

Some of the most striking features of modern reefs include their pronounced zonation, great diversity, and high productivity and growth rates. Reefs demonstrate a strong bathymetric (depth-related) zonation. This zonation is largely mediated through depth-related changes in turbulence intensity and in the quantity and spectral characteristics (reds are absorbed first, blues last) of available light. Shallow (1- to 5-meter) fore-reef environments are characterized by strong turbulence and high light intensity and possess low-diversity assemblages of wave-resistant corals, such as the elk-horn coral, *Acropora palmata*, and crustose red algae.

With increasing depth (10-20 meters), turbulence levels decrease and coral species diversity increases, with mound and delicate branching colonies occurring. At greater depths (30-60 meters), corals assume a flattened, platelike form in an attempt to maximize surface area for exposure to ambient light. Sponges and many green algae are also very important over this range. Finally, corals possessing zooxanthellae, which live in the coral tissues and provide food for the coral host, are rare or absent below 60 meters because of insufficient light. Surprisingly, green and red calcareous algae extend to much greater depths (100-200 meters), despite the very low light intensity (much less than 1 percent of surface irradiance). Sponges are also important members of these deep reef communities.

REEF COMMUNITIES

Coral reefs are among the most diverse of the Earth's communities; however, there is no consensus on the mechanism(s) behind the maintenance of this great diversity. At one time, it was believed that reefs existed in a low-disturbance, highly stable envi-

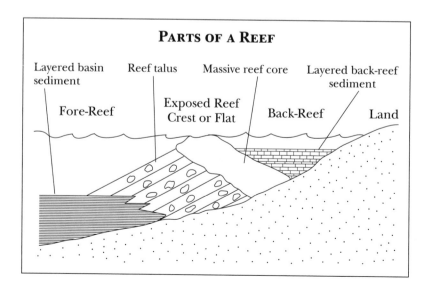

PARTS OF A REEF

Layered basin sediment   Reef talus   Massive reef core   Layered back-reef sediment

Fore-Reef   Exposed Reef Crest or Flat   Back-Reef   Land

Jaluit Reef, Marshall Islands. (U.S. Geological Survey)

ronment, which allowed very fine subdivision of food and habitat resources and thus permitted the coexistence of a great number of different species. Upon closer inspection, however, many reef organisms appear to overlap greatly in food and habitat requirements. Also, it has become increasingly apparent that disturbance, in the form of disease, extreme temperatures, and hurricanes, is no stranger to reef communities.

Coral reefs exhibit very high rates of productivity (carbon fixation), which is a result of extremely tight recycling of existing nutrients. This is necessary, as coral reefs exist in virtual nutrient "deserts." Modern corals exhibit high skeletal growth rates, up to 10 centimeters per year for some branching species. Such high rates of skeletal production are intimately related to the symbiosis existing between the hermatypic or reef-building scleractinian corals (also gorgonians and many sponges) and unicellular algae or zooxanthellae. Corals that, for some reason, have lost their zooxanthellae or that are kept in dark rooms exhibit greatly reduced rates of skeleton production.

In addition to high individual growth rates for component taxa, the carbonate mass of the reefs may grow at a rate of some 2 meters per 1,000 years, a rate that is much higher than that of most other sedimentary deposits. This reflects the high productivity or growth rates of the component organisms and the efficient trapping of derived sedi-

ment by the reef frame. Although the framework organisms, most notably corals, are perhaps the most striking components of the reef system, the framework represents only 10-20 percent of most fossil reef masses. The remainder of the reef mass consists of sedimentary fill derived from the reef community through a combination of biosynthesis (secretion) and bioerosion (breaking down) of calcium carbonate. An example of the relative contributions of reef organisms to sediment can be found in Jamaica, where shallow-water, back-reef sediment consists of 41 percent coral, 24 percent green calcareous algae, 13 percent red calcareous algae, 6 percent foraminifera, 4 percent mollusks, and 12 percent other grains. The most important bioeroders are boring sponges, bivalves, and various "worms," which excavate living spaces within reef rock or skeletons, and parrot fish and sea urchins, which remove calcium carbonate as they feed upon surface films of algae.

TYPES OF REEF COMMUNITIES

A diversity of organisms has produced reef and reeflike structures throughout Earth history. Several distinct reef community types have been noted, as well as four major "collapses" of reef communities. The oldest reefs or reeflike structures existed more than 2 billion years ago during the Precambrian eon. These consisted of low-diversity communities dominated by soft, blue-green algae, which trapped sediment to produce layered, often columnar structures known as stromatolites. During the Early Cambrian period, blue-green algae were joined by calcareous, conical, spongelike organisms known as archaeocyathids, which persisted until the end of the Middle Cambrian. Following the extinction of the archaeocyathids, reefs again consisted only of blue-green algae until the advent of more modern reef communities in the Middle Ordovician period. These reefs consisted of corals (predominantly tabulate and, to a much lesser extent, rugose corals), red calcareous algae, bryozoans (moss animals), and the spongelike stromatoporoids. This community type persisted through the Devonian period, at which time a global collapse of reef communities occurred. The succeeding Carboniferous period largely lacked reefs, although algal and crinoidal (sea lily) mounds are common. Reefs again occurred in the Permian period, consisting mainly of red and green calcareous algae, stromatolites, bryozoans, and chambered calcareous sponges known as sphinctozoans, which resembled strings of beads. These reefs were very different from those of the earlier Paleozoic era; in particular, the tabulates and stromatoporoids no longer played an important role. The famous El Capitan reef complex of West Texas formed during this interval. The Paleozoic era ended with a sweeping extinction event that involved not only reef inhabitants but also other marine organisms.

After the Paleozoic extinctions, reefs were largely absent during the early part of the Mesozoic era. The advent of modern-type reefs consisting of scleractinian corals and red and green algae occurred in the Late Triassic period. Stromatoporoids once again occurred abundantly on reefs during this interval; however, the role of the previously ubiquitous blue-green algal stromatolites in reefs declined. Late Cretaceous reefs were often dominated by conical, rudistid bivalves that developed the ability to form frameworks and may have possessed symbiotic relationships with algae, as do many modern corals. Rudists, however, became extinct during the sweeping extinctions that occurred at the end of the Cretaceous period. The reefs that were reestablished in the Cenozoic era lacked stromatoporoids and rudists and consisted of scleractinian corals and red and green calcareous algae. This reef type has persisted, with fluctuations, until the present.

STUDY OF MODERN REEFS

Modern reefs are typically studied by scuba (self-contained underwater breathing apparatus) diving, which enables observation and sampling to a depth of approximately 50 meters. Deeper environments have been made accessible through the availability of manned submersibles and unmanned, remotely operated vehicles that carry mechanical samplers and still and video cameras. The biological compositions of reef communities are determined by census (counting) methods commonly employed by plant ecologists. Studies of symbioses, such as that between corals and their zooxanthellae, employ radioactive tracers to determine the transfer of products between symbiont and host. Growth rates are measured by staining the calcareous skeletons of living organisms with a dye, such as Alizarin red, and then later collecting and sectioning the specimen and measuring the amount of skeleton added since the time of staining. Another method for determining growth is to X-ray a thin slice of skeleton and then measure and count the yearly growth bands that are revealed on the X-radiograph. Variations in growth banding reflect, among other factors, fluctuations in ocean temperature.

Reef sediments, which will potentially be transformed into reef limestones, are examined through sieving, X-ray diffraction, and epoxy impregnation and thin-sectioning. Sieving enables the determination of sediment texture, the relationships of grain sizes and abundance (which will reflect environmental energy and the production), and erosion of grains through biotic processes. X-ray diffraction produces a pattern that is determined by the internal crystalline structure of the sediment grains. As each mineral possesses a unique structure, the mineralogical identity of the sediment may be determined. Thin sections of embedded sediment or lithified rock are examined with petrographic microscopes, which reveal the characteristic microstructures of the individual grains. Thus, even highly abraded fragments of coral or algae may be identified and their contributions to the reef sediment determined.

STUDY OF FOSSIL REEFS

Because of their typically massive nature, fossil reefs are usually studied by thin-sectioning of lithified rock samples collected either from surface exposures or well cores. Reef limestones that have not undergone extensive alteration may be dated through carbon 14 dating, if relatively young, or through uranium-series radiometric dating methods.

NATURAL LABORATORIES AND ECONOMIC RESOURCES

Modern reefs serve as natural laboratories, enabling the geoscientist to witness and study phenomena, such as carbonate sediment production, bioerosion, and early cementation, that have been responsible for forming major carbonate rock bodies in the past. The study of cores extracted from centuries-old coral colonies shows promise for deciphering past climates and perhaps predicting future trends. This is made possible by the fact that the coral skeleton records variations in growth that are related to ocean temperature fluctuations. The highly diverse modern reefs also serve as ecological laboratories for testing models on the control of community structure. For example, the relative importance of stability versus disturbance and recruitment versus predation in determining community structure is being studied within the reef setting.

Modern reefs are economically significant resources, particularly for many developing nations in the tropics. Reefs and the associated lagoonal sea-grass beds serve as important nurseries and habitats for many fish and invertebrates. The standing crop of fish immediately over reefs is much higher than that of adjacent open shelf areas. Reef organisms may one day provide an important source of pharmaceutical compounds, such as prostaglandins, which may be extracted from gorgonians (octocorals). In addition, research has focused upon the antifouling properties exhibited by certain reef encrusters. Reefs also provide recreational opportunities for snorkelers

and for scuba divers, a fact that many developing countries are utilizing to promote their tourist industries. Finally, reefs serve to protect shorelines from wave erosion.

Because of the highly restricted environmental tolerances of reef organisms, the occurrence of reefs in ancient strata enables fairly confident estimation of paleolatitude, temperature, depth, salinity, and water clarity. In addition, depth- or turbulence-related variation in growth form (mounds in very shallow water, branches at intermediate depths, and plates at greater depths) enables even more precise estimation of paleobathymetry or turbulence levels. Finally, buried ancient reefs are often important reservoir rocks for hydrocarbons and thus are important economic resources.

W. David Liddell

CROSS-REFERENCES

Alluvial Systems, 2297; Beaches and Coastal Processes, 2302; Carbonates, 1182; Continental Structures, 590; Deep-Sea Sedimentation, 2308; Deltas, 2312; Desert Pavement, 2319; Drainage Basins, 2325; Earth Resources, 1741; Evaporites, 2330; Floodplains, 2335; Fossil Plants, 1004; Fossilization and Taphonomy, 1015; Lakes, 2341; Radiocarbon Dating, 537; River Bed Forms, 2353; River Flow, 2358; Sand, 2363; Sand Dunes, 2368; Sediment Transport and Deposition, 2374; Weathering and Erosion, 2380.

BIBLIOGRAPHY

Bathurst, Robin G. C. *Carbonate Diagenesis.* Boston: Blackwell Scientific Publications, 1990. Provides an excellent general reference on carbonate sediments, from reef and other environments, and their diagenesis.

Darwin, Charles. *The Structure and Distribution of Coral Reefs.* Berkeley: University of California Press, 1962. Darwin's book, originally published in 1851, is replete with observations on coral reefs from around the world. In addition, the theories presented on the formation of reef types such as atolls have withstood the test of time.

Davidson, Osha Gray. *The Enchanted Braid: Coming to Terms with Nature on the Coral Reef.* New York:

Wiley, 1998. This is a beautifully illustrated introduction to the ecosystems of coral reefs. Sections deal with coral reef biology, ecology, and endangered reef ecosystems. Suitable for nonscientists. Color illustrations and maps.

Frost, S. H., M. P. Weiss, and J. B. Sanders, eds. *Reefs and Related Carbonates: Ecology and Sedimentology.* Tulsa, Okla.: American Association of Petroleum Geologists, 1977. Includes a broad array of papers covering such aspects of the ecology and geology of modern and ancient reefs as bioerosion, diagenesis, paleoecology of ancient reefs, the role of sponges on reefs, and sedimentology of basins adjacent to modern reefs.

Goreau, Thomas F., et al. "Corals and Coral Reefs." *Scientific American* 241 (August, 1979): 16, 124-136. A good overview of the ecology of modern coral reefs. The discussion of coral physiology and the symbiotic relationship with zooxanthellae is particularly valuable.

Jones, O. A., and R. Endean, eds. *Biology and Geology of Coral Reefs*. 4 vols. New York: Academic Press, 1973-1977. This series of volumes encompasses both biological and geological aspects of coral reefs. Of particular value are review chapters covering, for example, the reefs of the western Atlantic.

Kaplan, Eugene H. *A Field Guide to Coral Reefs of the Caribbean and Florida, Including Bermuda and the Bahamas*. Boston: Houghton Mifflin, 1988. In addition to providing descriptions and illustrations (many in color) of common reef organisms, this book provides an excellent overview of modern reef community structure, zonation, and environments.

Laporte, Leo F., ed. *Reefs in Time and Space: Selected Examples from the Recent and Ancient*. Tulsa, Okla.: Society of Economic Paleontologists and Mineralogists, 1974. Includes papers on reef diagenesis, reef geomorphology, and the distribution of carbonate buildups in the geologic record.

Newell, Norman D. "The Evolution of Reefs." *Scientific American* 226 (June, 1972): 12, 54-65. Provides an overview of the composition of reef communities throughout the Earth's history, including the various collapses and rejuvenations.

Smith, F. G. *Atlantic Reef Corals: A Handbook of the Common Reef and Shallow-Water Corals of Bermuda, the Bahamas*. Rev. ed. Baltimore, Md.: University of Miami Press, 1971. Smith provides taxonomic keys, descriptions, illustrations, and zoogeographic distributions for the Atlantic reef corals. In addition, he provides much general information on the distribution of coral reefs and their ecology.

Stoddart, D. R. "Ecology and Morphology of Recent Coral Reefs." *Biological Reviews* 44 (1969): 433-498. This article provides a review of the global distribution of coral reefs with emphasis on their ecology and geomorphology.

Vacher, Leonard, and Terrence M. Quinn. *Geology and Hydrogeology of Carbonate Islands*. New York: Elsevier, 1997. This book looks at the geology, hydrology, geochemistry, and evolution of reefs and carbonate masses. Suitable for the careful high school-level reader. Illustrations, maps, index, and bibliography.

Wood, Rachel. *Reef Evolution*. New York: Oxford University Press, 1999. This detailed account follows the geophysical and geochemical phases of reefs. Special attention is paid to reef ecology and the animal communities that make up these environments. Appropriate for the nonscientist. Illustrations, maps, index, and bibliographical references.

RIVER BED FORMS

Bed forms, produced by flows of water or air in natural environments and in artificial channels, are a distinctive aspect of the transport of granular sediment along a sediment bed. They have a strong effect on the magnitude of bottom friction felt by the flow. Because of their great variety as a function of flow conditions, bed forms are valuable in interpreting depositional conditions of ancient sedimentary deposits.

PRINCIPAL TERMS

ANTIDUNE: an undulatory upstream-moving bed form produced in free-surface flow of water over a sand bed in a certain range of high flow speeds and shallow flow depths

BED CONFIGURATION: the overall geometry of a sediment bed molded by sediment transport by a flowing fluid

BED FORM: an individual geometrical element of a bed configuration

COMBINED FLOW: a flow of fluid with components of both unidirectional and oscillatory flow superposed on one another to produce a more complex pattern of fluid motion

CURRENT RIPPLE: a small bed form, oriented dominantly transverse to flow, produced at low to moderate flow speeds in unidirectional water flows

DUNE: a large bed form, oriented dominantly transverse to flow, produced at moderate to high flow speeds

EOLIAN DEPOSITS: material transported by the wind

FLUME: a laboratory open channel in which water is passed over a sediment bed to study the nature of the sediment movement

OSCILLATION RIPPLE: a small to large bed form, oriented dominantly transverse to flow, produced at low to moderate flow speeds in oscillatory water flows

OSCILLATORY FLOW: a flow of fluid with a regular back-and-forth pattern of motion

PLANE BED: a bed configuration without rugged bed forms produced in both unidirectional and oscillatory flows at high flow speeds

UNIDIRECTIONAL FLOW: a flow of fluid oriented everywhere and at all times in the same direction

CHARACTERISTICS

A striking feature of the transport of loose granular sediment over a bed of the same material by a turbulent flow of fluid such as air or water is that in a wide range of conditions of flow and sediment size, the bed is molded into topographic features, called bed forms, on a scale ranging from hundreds of times to even a million times larger than the grains themselves. Examples of such bed forms are sand ripples at the seashore or on a dry riverbed, sand dunes in the desert, and (less apparent to the casual observer) large underwater sand dunes in rivers and in the shallow ocean. The overall geometry of a sediment bed molded by a flow of fluid is called the bed configuration; bed forms are individual elements of this configuration. The term "bed form" includes both the over-

all geometry and the individual elements of that geometry.

The enormous range of bed-form-producing flows, together with the complex dynamics of the response of the bed, makes for striking variety in the scale and geometry of bed forms. Scales span a range of five orders of magnitude, from a few centimeters to more than 1,000 meters in spacing. Bed forms may appear as long ridges or circumscribed mounds, and their crests may be rounded or sharp. Most bed forms are irregular in detail, but elongated bed forms tend to show a more or less strong element of regularity in their overall arrangement, some even being perfectly regular and straight-crested. Elongated bed forms tend to be oriented transverse to flow, although flow-parallel forms are produced under certain conditions, and forms with no strongly preferred orientation are produced in some flows. Most bed forms are approximately wave-shaped and are often likened to waves, but they are waves only in a geometrical sense, not in a mechanical sense.

The most common bed forms are in sands (sediments with mean size lying between about 0.1 millimeter and 2 millimeters), but bed forms are produced in silts (sediments with mean size lying between about 4 micrometers and 0.1 millimeter) and gravels (sediment with mean size greater than 2 millimeters) as well. Bed forms produced by flows of air or water over mineral sediments in natural flow environments are of greatest interest to geologists, but a far wider range can be produced by flows of fluids with other densities and viscosities over sediments less dense or more dense than the common mineral sediments, which have densities mostly in the range 2.5 to 3.0 grams per cubic centimeter.

Types of Fluid Flow

Bed forms are made by unidirectional flows of air or water, as in rivers and tidal currents and under sand-moving winds, and by oscillatory flows, as on the shallow seafloor beneath wind-generated surface waves, which cause the water at the bottom to move back and forth with a period the same as that of the waves and with horizontal excursion distances of a few centimeters to a few meters. Bed forms are also made by what are called combined flows: superpositions of unidirectional flows and oscillatory flows. Such combined-flow bed forms are not as well understood as those made by unidirectional and oscillatory flows, but they are common in the shallow ocean. Bed forms made under wind are called subaerial or eolian bed forms, and bed forms made under water are called subaqueous bed forms.

At first thought, it might seem that the natural mode of sediment transport would be over a planar bed surface. In certain ranges of flow, a planar transport surface is indeed the stable bed configuration; technically, such a plane bed is a bed configuration with no bed forms. In reality, plane-bed transport is the stable configuration only under certain conditions; rugged bed forms cover the transport surface over a wide range of conditions in both oscillatory and unidirectional flows. Why such bed forms develop at all on transport surfaces is poorly understood. In certain ranges of flow, the planar transport surface is unstable in the sense that small bed irregularities of the kind that can be built at random by the plane-bed sediment transport become amplified to grow eventually into bed forms rather than being smoothed out again. The physics behind this instability is complex and is still not clearly understood. An essential element of this complexity is that there is a strong interaction or feedback between the bed configuration and the flow: The flow molds the bed configuration, but the bed configuration in turn affects the nature of the flow.

In unidirectional flow, current ripples are formed in fine sands as soon as the current is strong enough to move sand, and they persist to moderate currents of about 0.5 meter per second. In a vertical cross section parallel to the flow, these current ripples are triangular in shape, with downstream slopes about equal to the angle of repose of sand under water (about 30 degrees) and with gentler upstream slopes. In top or horizontal view, they are oriented mostly transverse to flow and are irregular in detail. Their spacings are 10 to 20 centimeters, and their heights are a few centimeters; this characteristic size changes little with flow strength or sediment size. With increasing unidirectional-flow speed, ripples give way to dunes, which are geometrically fairly similar to ripples but are of much larger scale—meters to thousands of meters in spacing and tens of centimeters to tens of meters in height, depending in a complex and poorly understood way on conditions of flow as well as sediment size. With further increase in flow speed to about 1 meter per second, dunes are replaced by a plane-bed configuration. The sequence of bed configurations with increasing unidirectional flow speed over coarse sands is different from that over fine sands: Transport is over a plane bed at flow strengths just above the threshold for sediment movement, and then dunes develop with further increase in flow strength. In sediments coarser than about 0.6 millimeter, current ripples are not formed in any range of flow speeds.

Current ripples and dunes move upstream or downstream at a speed far lower than the flow speed, by erosion of sediment on the upstream side of the bed form and deposition on the downstream side. When an individual ripple or dune can be watched carefully for a time, it is seen to change its size, shape, and speed irregularly, eventually to disappear by being absorbed into a neighboring bed form. Offsetting this loss of bed forms is the production of new ones by a kind of subdivision of one larger form into two smaller ones. In flows of water with a free upper surface, like rivers and tidal currents, undulatory bed forms called antidunes make their appearance at high flow speeds and shallow flow depths. Antidunes, so named because they tend to move slowly upstream by erosion of sediment on the downstream sides and deposition on the upstream sides, come about by a complex effect of standing surface waves (waves that move upstream about as fast as the flow is moving downstream) on the sediment bed.

In oscillatory flow, regular straight-crested bed forms called oscillation ripples, with sharp crests and broadly rounded troughs, are formed as soon as the sediment begins to be moved. At their smallest, their spacing is a few centimeters, but they grow in size to more than 1 meter in spacing as the period and amplitude of the oscillatory motion increases. In fine sands, these larger ripples become irregular and mound-shaped, but observations are not yet adequate for a detailed description. When the maximum flow speed during a single oscillation reaches about 1 meter per second, the bed forms are washed out to a plane-bed mode of transport. Combined flows produce a whole range of ripples intermediate between current ripples and oscillation ripples.

If the flow changes with time, as is the rule rather than the exception in natural flows, the bed configuration adjusts in response. Usually the bed configuration lags behind the change in the flow, with the result that the bed configuration is more or less out of equilibrium. Dunes formed by reversing flow in a tidal channel are a good example of such disequilibrium; commonly the weaker of the two flows (whether ebb or flood) modifies the shape of the dunes but does not reverse the asymmetry. Bed forms on a riverbed during passage of a flood also tend to lag behind changes in the flow.

STUDY OF BED FORMS

Bed forms can be observed and studied in natural flow environments and in laboratory tanks and channels. Each of these approaches has it advantages and disadvantages. In nature, observations on bed configurations are limited by various practical and technical difficulties. In laboratory tanks and channels, the bed forms can be studied much more easily, but for the most part water depths are unnaturally shallow.

A laboratory open channel in which a flow of water is passed over a sediment bed is called a flume. Flumes range from a few meters to more than 100 meters long, from about 10 centimeters to a few meters wide, and from several centimeters to about 1 meter deep. The water is usually recirculated from the downstream end to the upstream end to form a kind of endless river. The sediment may also be recirculated, or it may be fed at the upstream end and caught in a trap at the downstream end. In a flume, it is fairly easy to measure the profile of the bed configuration with a mechanical pointer gauge or with a sonic depth sounder, and the bed forms can be studied visually and photographed through the sidewalls and perhaps from the top. Time-lapse motion pictures of bed-form movement are also instructive. Oscillatory-flow bed forms can be studied either in long open tanks, in which water waves are passed over a sand bed to produce oscillatory flow at the bed, or in closed horizontal ducts, in which water is pushed back and forth in a regular oscillation between tanks at the ends of the duct.

In some natural flow environments, such as rivers and tidal currents, the geometry of the bed forms can be studied when they are exposed at low water. Observations of the bed forms when they are being molded by the flow, however, are more difficult. If the flow is not too strong, divers can make direct observations. Profiling of the bed geometry along lines or even across wide areas is usually possible by means of various sonar techniques, whereby the travel times of sound pulses reflected from the bed are converted to water depths. If the water is clear enough, bottom photographs of small areas can be taken. Current velocities can be measured with current meters anchored on the bed. Movement of large bed forms is difficult to measure because bed-form speeds are slow and usually no fixed reference points are available.

The status of observations on bed configurations leaves much to be desired. Even in laboratory channels and tanks, where the major outlines are by now fairly well known, there is much room for further work for two reasons: The narrowness of the flow tends to distort the three-dimensional aspects of the bed geometry, and little work has been done on bed forms in combined flows. There is a need for more observations on the geometry and movement of bed forms in natural flows as a function of flow strength, as well as on the effect of disequilibrium.

SIGNIFICANCE

Bed forms are ubiquitous in natural flow environments. They are most apparent to the casual eye in fields of sand dunes in deserts and in certain coastal environments where the wind molds available loose sand into dunes. They are widely present but less obvious in rivers and the shallow ocean. Apart from their intrinsic scientific interest as a widespread natural phenomenon, bed forms are of importance in both engineering and geology for various reasons. Large underwater bed forms many meters high in rivers and marine currents can be obstacles to navigation, and their movement can be a threat to submarine structures. Also, the inexorable movement of desert dunes or coastal dunes can bury roads and buildings.

The rugged topography of ripple and dune bed forms leads to a pattern of flow over each bed form in which the pressure on the upstream surface is relatively high and the pressure on the downstream surface is relatively low, much like an unstreamlined motor vehicle on the highway or a house in a strong wind. This pressure difference adds greatly to the force the flow exerts on the bed and, conversely, to the force the bed exerts on the moving flow. Hydraulic engineers have expended much effort on the effect of this resistance force on the depth a river assumes when it is given a particular rate of flow to carry. A river with a planar bed can pass a given flow rate at a shallower depth and greater velocity than can a river with a bed roughened by large dunes, which exert a large resistance force on the flow and make the velocity smaller and the depth greater.

Geologists have given attention to bed forms partly because of their effect on the geometry of the stratification that develops as a sediment bed is deposited while bed forms are active on the sediment surface. It is often possible to tell the kind of bed form that was present just by examining the stratification in a sedimentary rock like a sandstone. If the flow conditions responsible for making that kind of bed form are already known—from laboratory experiments or from observations in modern natural flow environments—the depositional conditions of that sedimentary rock (which may be geologically very ancient) can be interpreted. Such interpretations are one of the tools used in mapping the geometry of subsurface petroleum reservoirs in sedimentary rocks.

John Brelsford Southard

CROSS-REFERENCES

BIBLIOGRAPHY

Allen, J. R. L. *Principles of Physical Sedimentology.* London: Allen & Unwin, 1985. A lucidly written introduction to the movement and deposition of sediment for sedimentology students at the college level and beyond. The chapter on bed forms is moderately mathematical, but there is some good descriptive material and illustrations of bed forms. The treatment of bed forms is concise but fairly comprehensive.

Collinson, J. D., and D. B. Thompson. *Sedimentary Structures.* 2d ed. London: Allen & Unwin, 1989. This book, written for beginning college-level students in sedimentology, presents a nonmathematical treatment of bed forms in the context of their sedimentological significance. Three chapters are devoted to bed forms and the sedimentary structures they produce. Numerous illustrations of the various kinds of bed forms.

Klingeman, Peter C., et al., eds. *Gravel-Bed Rivers in the Environment.* Highlands Ranch, Colo.: Water Resources Publications, 1995. This somewhat technical book describes the sediment transport and ecology of rivers and riverbeds, as well as their interaction with surrounding environments. Intended for the college-level reader. Illustrations, maps, index, bibliography.

Leeder, M. R. *Sedimentology and Sedimentary Basins: From Turbulance to Tectonics.* Malden, Mass.: Blackwell, 1999. An introductory college-level text on sedimentology, with a well-illustrated and mostly nonmathematical chapter on bed forms. Suitable for high school readers who are willing to do some preparatory reading in the earlier parts of the book.

Middleton, G. V., and J. B. Southard. *Mechanics in the Earth and Enviromental Sciences.* New York: Cambridge University Press, 1994. This book offers a long and mostly nonmathematical chapter on bed forms for sedimentologists, with material on both the observational characteristics and the basic hydrodynamics of bed forms. Emphasis on underwater bed forms. No photographs.

Reineck, H. E., and I. B. Singh. *Depositional Sedimentary Environments.* 2d ed. Berlin: Springer-Verlag, 1980. This almost entirely nonmathematical book, designed as a monograph on modern depositional environments for sedimentologists and marine geologists, has several long, nonmathematical, and extensively referenced sections on bed forms in a great variety of flow environments. Suitable for high school readers. Unusually well illustrated.

Vanoni, V. A., ed. *Sedimentation Engineering.* New York: American Society of Civil Engineers, 1975. This engineering reference manual, aimed at students and practicing engineers, deals with the fundamentals of sediment-transport mechanics from an applied standpoint. Parts of two chapters deal with bed forms, mainly in the context of rivers. The best and most authoritative source for information on the effect of bed forms on the behavior of rivers. Fairly heavily mathematical.

Yalin, M. S. *River Mechanics.* New York: Pergamon Press, 1992. This college-level textbook, designed for engineering students, devotes a long chapter to an unusually fundamental treatment of the mechanics of bed forms. Heavily mathematical but with numerous qualitative insights into bed-form behavior. No illustrations.

RIVER FLOW

Worldwide, rivers are the most important sources of water for cities and major industries. Hydroelectric power is a major source of electrical power, and transport of heavy, bulky goods by river barge is a vital link in most transportation systems. Understanding and predicting low, high, and average flows of rivers is therefore important to the people and industries that depend on them.

PRINCIPAL TERMS

DISCHARGE: the total amount of water passing a point on a river per unit of time

EVAPOTRANSPIRATION: all water that is converted to water vapor by direct evaporation or passage through vegetation

HYDRAULIC GEOMETRY: a set of equations that relate river width, depth, and velocity to discharge

HYDROGRAPH: a plot recording the variation of stream discharge over time

HYDROLOGIC CYCLE: the circulation of water as

a liquid and vapor from the oceans to the atmosphere and back to the oceans

HYDROLOGY: broadly, the science of water; the term is often used in the more restricted sense of flow in channels

RATING CURVE: a plot of river discharge in relation to elevation of the water surface; permits estimation of discharge from the water elevation

TURBULENT FLOW: the swirling flow that is typical of rivers, as opposed to smooth, laminar flow

TYPES OF WATER FLOW

There are two very different fundamental types of flow of water. Laminar flow is a smooth flow, in which two particles suspended in the water will follow parallel, nearly straight paths. Turbulent flow is a complex, swirling flow in which the paths of two suspended particles have no necessary relation to each other. Turbulent flow may have velocity components that are up or down from, sideways or upstream of the average flow direction, although the average flow direction is always in the downstream, or downslope, direction. Flow near the bottom and sides of a river or stream is always laminar, but the zone of laminar flow is very narrow. Turbulent flow dominates throughout the cross section of stream flow.

Because flowing water exerts a shear stress, or viscous shear, along the bottom and sides of the river, average flow velocity is least at the bottom and increases upward into the body of the river. Because of energy losses to surface waves, the average turbulent downstream velocity (hereafter referred to simply as velocity) at the river's surface is slightly below the maximum velocity. The maximum velocity occurs at about 0.6 of the river depth above the bottom of the river.

Because of viscous shear between flowing water and the bottom materials, the small (and sometimes large) sediment particles that make up most river bottoms are moved by the flowing water. Once in motion, small particles whose settling velocity is less than the upward component of turbulent flow move downstream with the water, settling out only in backwaters, such as the still water behind a dam, where flow velocities are very low. This continuously moving sediment is referred to as suspended load and is restricted, in most cases, to clay- and silt-sized material. Sand-sized and larger sediment grains are moved during periods of high flow velocity and dropped where, or when, flow velocity decreases. This coarse material is the "bed load." Bed-load transport normally occurs only near the river bed. It is the sediment in transport that does the work of the river: erosion and transport. Erosional features of river valleys—such as canyons and potholes—are produced by the "wet sandblast" of the suspended bed load, a process that is much more effective during periods of high flow velocity.

During periods of rainfall or snowmelt, direct precipitation into streams and runoff from adjacent land provides stream flow. Between periods of rainfall or snowmelt, stream flow comes from the slow seepage of groundwater into surface streams. Small streams, the smallest of which flow only during wet periods, join to form larger streams, which join to form rivers. Because rainfall or snowmelt occurs only occasionally in an area drained by a river (the river's drainage basin), the quantity of water flowing past any point on a river, or any point on any of its tributary streams, varies with time.

DISCHARGE

The quantity of water passing a point on the river is measured in cubic meters per second (or, in common North American and British practice, cubic feet per second) and is called the discharge. For a period of time after a rainfall or snowmelt event—for example, a flood—discharge increases at all points in the drainage basin. If the flood event occurs in only one part of the drainage basin, say the higher part of the basin with the smaller streams, the increase in discharge will occur first in the higher part of the basin and will occur in the larger streams lower in the basin at some later time.

If the discharge at a given point on a stream is measured continuously over a period of days and the discharge is plotted against time, with dis-

charge on the vertical axis of the graph and time in days on the horizontal axis, the increase in discharge related to a storm or snowmelt (a flood) will appear as a hump in the discharge curve. A graph of discharge with time is called a hydrograph, and a hydrograph that shows a flood-related hump is a flood hydrograph. Flood hydrographs tend to be more pronounced—that is, the curve is higher and has steeper sides—on streams near the source of the flood water and longer and less pronounced—that is, lower and with more gently sloping sides—farther downstream. Putting it another way, the flood hydrograph attenuates, or dies out, downstream. The low, flat portion of the hydrograph that measures flow between flood events is called baseflow. Experience with flood hydrographs for a river enables hydrologists to predict the effect of future rainstorms and snowmelts of varying intensity and to predict low flow during prolonged dry periods.

An increase in discharge is accompanied by an increase in velocity. For rivers with sandy bottoms (sandy streambeds), the higher water velocity causes erosion, or scour, of the bottom; that is, the sand begins to move along the bottom with the water, and the river channel becomes deeper. The elevation of the water surface relative to some fixed point on the riverbank also increases. As a rule, the banks are not vertical but sloping, and the increase in elevation of the water surface causes an increase in width. In summary, an increase in discharge is accompanied by increases in flow velocity, stream width, water surface elevation, and depth of channel (the last two items add up to an overall increase in depth). A reduction in discharge has just the opposite effect, including the deposition of new sand, arriving from upstream, in the channel as the velocity decreases.

HYDRAULIC GEOMETRY

In 1953, Luna Bergere Leopold and Thomas Maddock, Jr.,

River flow eroded a pothole through this ledge below a dam at Blackstone quadrangle in Worcester County, Massachusetts. (U.S. Geological Survey)

introduced a concept that describes the relationships among the variables that change when discharge changes. The concept is called hydraulic geometry, and it is embodied in the following three equations: $w = aQ^b$, $d = bQ^f$, and $v = kQ^m$, where w is stream width, d is average depth of the stream, v is the flow velocity, and Q is discharge. The coefficients a, b, and k and the exponents b, f, and m are determined empirically—that is, by measurement in the field. Since the discharge is equal to the cross-sectional area of the stream (wd) times the flow velocity (that is, $wdv = Q$), $wdv = aQ^b \times bQ^f \times kQ^m = Q$. For this to be true, the product of a, b, and k must be 1 ($abk = 1$), and the sum of the exponents b, f, and m must also be 1 ($b + f + m = 1$). Many field studies have shown these equations to be good approximations of actual variation in width, depth, and velocity with variation in discharge.

The coefficients of the hydraulic geometry equations have little effect, relative to the powerful exponents, and are usually ignored. The exponents of the hydraulic geometry equations depend on the physical characteristics of the drainage basins and stream channels involved, but for a given point on a given stream, average values are $b = 0.1$, $f = 0.45$, and $m = 0.45$, which means that the increase in discharge during a flood event is expressed primarily in increases in depth and flow velocity.

The discharge of rivers increases downstream because of the larger length of stream receiving baseflow and the contribution from many tributaries, and hydraulic geometry equations may also be applied to downstream changes in discharge. Average values for the exponents in the downstream hydraulic geometry equations for width, depth, and velocity are $b = 0.5$, $f = 0.4$, and $m = 0.1$. This phenomenon is a paradox. Casual observation would suggest that small streams in the higher parts of the drainage basins flow faster than the large rivers in the lower parts of the drainage basins. Yet, appearances are deceptive: Water in the wider, deeper channels of the larger streams actually flows faster than does water in the smaller headwater streams.

When a flood flow exceeds the capacity of the river channel, the surface elevation of the river rises above the elevation of the riverbanks, and flooding of the surface adjacent to the stream oc- curs. This is what has generally been called a flood, or an overbank flood. Overbank floods do serious damage to homes, businesses, industrial facilities, and crops on the flooded areas, and much time and money are devoted to flood prevention. The principal method of flood prevention is the construction of levees, large Earth embankments or concrete walls along the stream bank, at locations where the potential economic loss because of flooding justifies the expense of their construction and maintenance.

STUDY OF RIVER FLOW

Research on flow in rivers, or hydrology, almost invariably involves the determination of flow velocity and discharge. Because discharge (Q) equals stream width (w) times average depth (d) times average flow velocity (v), velocity and discharge are closely related. As implied by the world "average" in the preceding sentence, depth and velocity vary across the width of a river or stream. Velocity is lowest in the shallower parts of the river. Therefore, the cross section of the stream is divided into sections, and the discharge is taken as the sum of discharges of all the sections.

For a relatively small river, a rope or strong cord is stretched across the river and marked at regular intervals, typically 2 meters. The depth is measured at each marked point, and a current meter is used to measure the velocity at each point. The flow in rivers and streams is turbulent, and the current meter actually measures the average downstream component of velocity. Average downstream velocity varies with depth of the channel and also through the vertical section at any point on the stream, the lowest velocities occurring at the bottom and at the surface. Average velocity at any given vertical section of flow occurs at about $0.6\,d$ above the bottom, which is where the current meter is positioned. Alternatively, two velocity measurements may be taken, at 0.2 and $0.8\,d$ above the bottom, and averaged. The equation $Q = wdv$, where w is the interval at which measurements were taken, is computed for each section, or interval, across the stream; all the resultant discharges are summed to obtain the total discharge at the point on the stream, which is now called a station.

This process is laborious. If discharge at a station must be more or less continuously monitored, a quicker method is desirable, which is accomplished by developing a rating curve for the station. Discharge is measured in the manner already described at several different times, with as wide a range of discharges as practical. A staff gauge (a board mounted vertically in the stream and marked in units of length, usually feet in American and British practice) is erected at the station, and the level of the water surface, called the stage of the river, is determined each time the discharge is measured. The rating curve consists of a plot of stage against discharge. Once the rating curve has been determined, discharge is estimated from river stage by use of the rating curve. Because of the scour and fill of the river bottom that occur with each increase and decrease in discharge and velocity, the rating curve is not a straight line. Moreover, because the scour and fill may, over time, change the character of the river channel at the station, it is necessary actually to measure the discharge periodically to check the validity of the rating curve. If large changes in the channel occur, it is necessary to establish a new rating curve.

The U.S. Geological Survey maintains a large number of rating stations, or gauging stations, and periodically reports stream discharges. At most stations operated by the agency, the staff gauge is replaced by a vertical pipe driven in the streambed and perforated near the bed. Water is free to flow in and out of the pipe as river stage changes, but the water surface inside the pipe is not disturbed by surface waves. A cable with a weight at the end is attached to the float, and the cable is passed over a wheel near the top of the pipe. The float is free to move with the water surface inside the pipe and, as it moves, it turns the wheel. Sensitive instruments monitor the position of the wheel and therefore the river stage as well. In this way, the stage is periodically and automatically reported to a central office by telephone line or radio.

Robert E. Carver

CROSS-REFERENCES

Alluvial Systems, 2297; Beaches and Coastal Processes, 2302; Deep-Sea Sedimentation, 2308; Deltas, 2312; Desert Pavement, 2319; Drainage Basins, 2325; Evaporites, 2330; Floodplains, 2335; Floods, 2022; Lakes, 2341; Reefs, 2347; River Bed Forms, 2353; Sand, 2363; Sand Dunes, 2368; Sediment Transport and Deposition, 2374; Weathering and Erosion, 2380.

BIBLIOGRAPHY

Dingman, S. L. *Physical Hydrology.* New York: Macmillan Publishers, 1994. A thorough introduction to hydrology, this book offers a clear development of all the equations that are essential to fluvial hydrology. For full understanding, a year of college calculus is necessary, but about 80 percent of the material can be mastered by a student with knowledge only of algebra and trigonometry. An excellent treatment of the subject, with many answered problems and a helpful annotated bibliography. Accompanied by a computer optical laser disk.

Dunne, Thomas. *Fluvial Geomorphology and River-Gravel Mining: A Guide for Planners.* Sacramento, Calif.: California Department of Conservation, 1990. A thorough look at fluvial systems and geomorphology. Emphasis is on design to avoid environmental problems associated with water in all its manifestations. Many worked problems, most involving only basic mathematics.

Kindsvatter, C. E. *Selected Topics of Fluid Mechanics.* U.S. Geological Survey Water Supply Paper 1369-A. Washington, D.C.: Government Printing Office, 1958. Develops in an exceptionally clear way the fundamental concepts of fluid mechanics that underlie fluvial hydrology. Less than 10 percent of the material requires a knowledge of calculus. Many aspects of the approach are much superior to more recent texts.

Leopold, Luna B. *A View of the River.* Cambridge, Mass: Harvard University Press, 1994. A lucid explanation of rivers and their processes. Clearly written and easy to follow. Highly recommended for both beginners and professionals.

_____. *Water, Rivers, and Creeks.* Sausalito, Calif.: University Science Books, 1997. A brief, very readable introduction to hydrology and the study of rivers. Highly recommended for all newcomers to the field and most professionals.

Manning, J. C. *Applied Principles of Hydrology.* 3d ed. Westerville, Ohio: Charles E. Merrill, 1997. An excellent short introduction to the principles of surface water and groundwater hydrology. Incorporates very little of the mathematics involved in the two fields, but the descriptions and illustrations of methods of making field measurements are clear and informative.

Richards, Keith. *Rivers, Form, and Process in Alluvial Channels.* London: Methuen, 1982. A college-level text requiring some familiarity with calculus for complete understanding. Its great strength is the very large number of research papers cited in the text. Conveys a sense of the quantity and type of research that has been done.

SAND

Sand is the most continental of all sediments. It is important as a source of precious gems and ores, as an abrasive, and as a reservoir for the storage of valuable fluids.

PRINCIPAL TERMS

ALLUVIUM: sediment deposited by flowing water

CATACLASTIC: those formative processes of sand that relate to crushing

DIAGENESIS: all physical or chemical changes after deposition

ENDOGENETIC: those formative processes of sand that relate to chemical and biochemical precipitation

EPICLASTIC: those formative processes of sand that relate to weathering

PALEOCURRENT: the current system at the time of deposition

PRECIPITATE: to condense from a solution

PROVENANCE: all the factors relating to the production of sand

PYROCLASTIC: those formative processes of sand that relate to volcanic action

STRATIGRAPHY: the internal fabric and structures, the external geometry, and the nature of the basal contact of sand bodies

TURBIDITY CURRENT: the movement under gravity of a stream of fluids under, through, or over another fluid

FORMATION AND COMPOSITION

Sand is any Earth material that consists of loose grains of minerals or rocks that are larger than silt but smaller than gravel. Sand includes those grains that are less than 2.12 millimeters but greater than 0.06 millimeter in diameter. Materials meeting this definition are of widely diverse origins.

There are five processes that lead to the formation and release of sand-sized grains. The first of these processes is weathering, including both disintegration and decomposition. Some rocks crumble by the action of the air, rain, or frost. Decomposition is probably the origin of the bulk of the quartz sands. Most of the rock is converted to fine-grained, clay-sized material from which the inert, undecomposed quartz grains are liberated. Sands produced in this manner are called epiclastic sands. The second source of sand is explosive volcanism. The explosive action of volcanoes yields vast quantities of sand-sized debris: glass, crystal fragments, and lava particles. Sands produced in this manner are called pyroclastic sands. Sand-sized materials may also be produced by crushing action, which, as opposed to ordinary abrasion, produces a significant volume of sand. The impact from meteors will shatter rocks, and much of the

material that falls back into the crater is sand-sized. Earth movements may crush rocks but do not produce a significant deposit; glacial crushing, on the other hand, does produce a considerable body of sand-sized material. Sands produced in this manner are called cataclastic sands. Much sand-sized material is also produced by chemical and biochemical precipitation. They are generated within the basin and are not the products of the wastage of landmass. Biochemical and chemical precipitation form oölitic sands that may accumulate in a significant deposit. Sands produced in this manner are called endogenetic sands. The fifth process that may produce sand is pelletization; it involves the sand-sized pellets produced by organisms and the pellets of other origins, such as those that are blown into "clay dunes." Like endogenetic sands, these sands are often found in basins.

Because quartz is the principal product of rock decomposition, it is the principal component of sand. Quartz may be classified in three ways: igneous, including volcanic quartz; metamorphic, including pressure quartz; and sedimentary, including new crystals and overgrowth. One important type of metamorphic quartz is polycrystalline quartz, which includes those grains that are composed of one or more crystal units. Next to quartz, feldspar is the most common mineral of sand. Feldspar is released by the granular disintegration of acidic rock. Several types of feldspar are usually present in sand, although the alkali-rich feldspars seem to be more abundant than the calcic feldspars. Although mica is conspicuous in some sandstones, it is never a major component. Mica is derived from igneous rocks. In general, abundant mica points to a metamorphic provenance for sand.

In addition to quartz and feldspar, sandstone usually contains rock particles. These rock particles, most of which are fine-grained, carry their own evidence of provenance. Most common are the shales, which tend to be molded about the more resistant quartz. Because they are among the most informative of all the components of sand, any attempt to trace the origin of sand usually begins with an examination of these constituents.

DISTRIBUTION

Sand is distributed over the Earth. The deep oceanic basins are practically devoid of sand because the grains are too large to be blown or washed far off the continents; therefore, the principal environments of sand are on the continents. Sand is produced on the continent and is shifted from the higher elevations to the lower ones. Only a small amount is carried to the deep sea by currents that transport sand down the continental slopes to the plains. The absence of any sand probably results more from an absence of supply than from conditions unfavorable to its accumulation.

The most obvious places to find sand are the rivers and beaches and, to a lesser extent, glacial outwash plains, dunes, and shallow shelf seas. Sand is transported to the floodplains and to the deltas. A little sand also escapes the river channel and ends up in swamps and bayous. Shoreline sand includes that which is found in beaches, in lagoons, and on tidal flats. Although windblown dunes are closely associated with beaches and major rivers, the most impressive are found in the dune fields of some desert basins. Marine sands are primarily shelf sands.

Sand is transported primarily by water. It moves mostly along the bottom of a stream as "bed load." Sand does not move continuously; instead, sand

grains travel in periodic, short jumps along the bottom. The alluvial transport of sand creates forms ranging from small ripples only a few centimeters high to belts many kilometers high and more than 50 meters thick. A wide range of dunes and bars occur in between.

Sand dispersal on beaches depends on the interaction of shoreline configuration and underwater topography with the energy supplied by waves and currents. As a deep-water wave moves toward a beach, it pushes against the bottom so that the wave tends to parallel the shore. As the water shallows, wave steepness increases, causing loose grains to move back and forth on the bed. This process is called surge transport. In response to passing waves, fluid particles follow a circular path except near the bottom, where the particles move back and forth above the bed, perpendicular to the shoreline.

Wind transports vast quantities of sand in deserts and carries inland the sands of beaches that are supplied by currents. When the wind reaches a certain velocity, dry grains of sand begin to roll and accelerate. The round shape of grains of desert sand can be attributed to this rolling action. Unlike silt and clay, which travels long distances suspended in the wind, sand travels by successive jumps on the ground. The grains finally return to the surface in a long, parabolic path.

Turbidity currents are a unique type of transport system for sand. These currents are turbid because of the suspension of fine particles of sand within them. Turbidity currents form as muddy rivers enter freshwater lakes or reservoirs, in which the turbid water passes beneath the clear water above. The river current slows down as it enters the lake and is maintained as a separate layer of water with a significantly higher density than the clear lake water. Under the force of gravity, the turbid layer then moves along the floor of the lake until it loses momentum, at which time the suspended material gradually settles.

Diagenesis includes all the changes that sand undergoes after it has been deposited. The major direct evidence of diagenesis in sandstones is the nature of the textural relations between mineral grains and crystals. During diagenesis, the original material is preserved in a medium that is not original. The most frequently occurring types of replacements are those that preserve faint outlines of the original. The most obvious diagenetic modification is the introduction of cementing agents. Calcite is the most common carbonate mineral cementing sandstone, although silica cementation also occurs frequently. Not all sands undergo diagenesis at the same rate. Sands that are millions of years old may be incoherent, while relatively recent sands may be already bound together.

STUDY OF GRAINS

The earliest method of studying the external form of the grains, which is still used today, involves separating the particles by means of a sieve. Before sifting begins, the sand must be well washed in water, using a small, stiff brush to detach any mud still sticking to the grains. When dry, the coarsest particles are separated by one sieve and the smallest particles by another so that the sizes of the particles may be compared. The coarse sieve allows grains of 0.36 millimeter in diameter to pass through, and the smaller sieve keeps back all grains that are larger than 0.25 millimeter in diameter. After a sample of medium quality is obtained, it is mixed thoroughly and spread on a horizontal glass plate so that the rounded grains do not separate from the flat and angular grains. The characteristics of the individual grains are then brought into sharp relief with the assistance of a polarizing microscope.

Before the invention of powerful microscopes, the thin-section technique was devised in the nineteenth century to examine the mineral composition of sandstone. A smooth cross section of the stone is then obtained so that the natural history of the formation may be traced in the various layers of sediment. The thin-section technique has also been applied to igneous rock.

In recent years, sedimentation tubes have been developed. They are gaining in popularity because they provide a more rapid analysis than do sieves. Sedimentation tubes operate on the principle that the sample is introduced at one end of the tube and settles to the other. This method categorizes grains according to their settling velocity, not their size.

The question of provenance is one of the most difficult for the sedimentary petrographer to solve because sands are derived from preexisting sands and because the source areas may have changed with time. Sedimentary petrographers examine both internal and external evidence as they attempt to trace the "birth" of sediment. Internal evidence is provided by examination of a single grain of sand. A tourmaline grain, for example, may show a secondary growth on a rounded core. This outgrowth may imply weathering and release of the grain, followed by transportation and abrasion, or it might suggest deposition in a new deposit of sand. A more complete analysis of provenance can be made from a sample than from a single grain. If more than one sample is obtained, it is possible to map sedimentary petrologic provinces.

The external evidence relating to provenance is of two types. Regional stratigraphy will contribute to the analysis of provenance by establishing the relative ages of the stratigraphy. In order to study the stratigraphy of an area, investigators use a paleogeologic map, which indicates the formations exposed and subject to erosion. Paleocurrent analyses are another important approach to the problem of provenance. They are especially helpful in the study of sandstones of alluvial origin because the up-current direction of these alluvial sands is in the direction of the source.

Once investigators have examined both the internal and external evidence, they present their conclusions as a kind of "flow sheet" or provenance diagram, of which there are two types. The first type is based solely on what can be seen in the rock itself through thin-section analysis. The second type of provenance diagram is based on a study of the thin section and on a knowledge of regional geology and stratigraphy. To construct this type of diagram, any and all manner of geologic data must be pieced together.

SIGNIFICANCE

Sand is economically important because of the mineral content of certain shore and river sands. As the lighter components are removed by the current, the heavier components become concentrated. Some of these deposits, which are called placers, yield diamonds and other gemstones, gold, platinum, uranium, tin, monazite (containing thorium and rare-earth elements), zircon (for zirconium), rutile (for titanium), and other ores. During the California Gold Rush of 1849, millions of dollars worth of gold was taken from placers. In modern times, rare metals for jet engines come from placers in Florida, India, and Australia. The greensands, which are found over the ocean floor, are widely sought after because their green color indicates the presence of potash-bearing material. These sands have been used for land dressing and water softening. In addition, potash has been successfully extracted from them. The search for these sands has been refined to an art called alluvian prospecting.

Sands derived from specific minerals are indispensable to certain industries. Very pure quartzose sands are used as a source of silica in the pottery, glassmaking, and silicate industries. Similar sands are required in making the lining for the hearths of acid-steel furnaces. Sands utilized in foundries for making the molds in which metal is cast are those that have a clayey bond uniting the quartz grains. Quartz sands and garnet sands are also used as abrasives in sandpaper and sand blasting because of their hardness and poor cleavage. They are employed in the grinding of marble, plate glass, and metal. Some sands are used as soil conditioners or fertilizers. Ordinary sands find a multitude of other uses. Sand is an essential ingredient of mortar, cement, and concrete. Sand is also added to clays to reduce shrinkage and cracking in brick manufacture and to asphalt to make "road dressing." Additionally, it is used in filtration and as friction sand on locomotives.

Aside from its usefulness as a mixer, sand serves as a reservoir for the storage of valuable fluids. The pore systems of sand and sandstones are capable of containing large systems of fresh waters, of brines, and of petroleum and natural gas. Sand strata are also conduits for artesian flow. Fluids may also be injected into the sands. Before fluids can be extracted, geologists must become familiar with the shape and porosity of these sand reservoirs. A working knowledge of diagenesis is essential for geologists who are searching for petroleum deposits.

Geologists and geomorphologists concerned with shore erosion and harbor development are concerned with sand production, movement, and deposition. In order to solve the problems of shore engineering, some understanding of sand supply and sand deficit or removal is necessary. Geologists must also have a solid understanding of sand as they try to prevent the encroachment of sand on cultivated lands and forests or on roads and other structures. Finally, sand is involved in many of the problems of river management.

Alan Brown

CROSS-REFERENCES

Alluvial Systems, 2297; Aquifers, 2005; Beaches and Coastal Processes, 2302; Deep-Sea Sedimentation, 2308; Deltas, 2312; Desert Pavement, 2319; Drainage Basins, 2325; Evaporites, 2330; Floodplains, 2335; Floods, 2022; Geomorphology of Dry Climate Areas, 904; Geomorphology of Wet Climate Areas, 910; Hydrologic Cycle, 2045; Lakes, 2341; Land-Use Planning, 1490; Precipitation, 2050; Reefs, 2347; River Bed Forms, 2353; River Flow, 2358; River Valleys, 943; Sand Dunes, 2368; Sediment Transport and Deposition, 2374; Surface Water, 2066; Weathering and Erosion, 2380.

BIBLIOGRAPHY

Allen, J. R. L. *Physical Processes of Sedimentation.* Winchester, Mass.: Allen & Unwin, 1977. Sand is examined in the form of shallow marine deposits. Because the emphasis is on the formation of these structures, the author assumes that the reader has a thorough knowledge of the properties of sand. The technical nature of the material requires more illustrations than were provided.

Blatt, Harvey, and Robert J. Tracy. *Petrology: Igneous, Sedimentary, and Metamorphic.* New York: W. H. Freeman, 1996. This book brings together the major theories of sedimentation. The formation of sand is discussed in terms of modern physics and chemistry. As a result, the work is suitable for college students and geologists. The references that are included at the end of each chapter are very useful.

Houseknecht, David W., and Edward D. Pittman, eds. *Origin, Diagenesis, and Petrophysics of Clay Minerals in Sandstones.* Tulsa, Okla.: Society for Sedimentary Geology, 1992. A collection of essays written by leading experts in their respective fields, this book examines the geochemical and geophysical properties of sand, sandstone, clay minerals, and other sedimentary rocks. Filled with illustrations and includes an index and bibliographical references.

Kukal, Zdenek. *Geology of Recent Sediments.* Prague: Academia Publishing House of the Czechoslovakia Academy of Sciences, 1971. The chapter on the formation and structure of beach sediments explains every facet of sand deposits in great detail. The illustrations, charts, and references are useful, though a bit technical. Written for college students and geologists.

Pettijohn, F. J., Paul Edwin Potter, and Raymond Siever. *Sand and Sandstone.* 2d ed. New York: Springer-Verlag, 1987. An in-depth analysis of sand. This lengthy book is amply supplemented with photographs, illustrations, and charts. Includes a glossary at the end of each chapter. For college students with a solid background in geology.

Prothero, Donald R., and Fred Schwab. *Sedimentary Geology: An Introduction to Sedimentary Rocks and Stratigraphy.* New York: W. H. Freeman, 1996. A thorough treatment of most aspects of sediments and sedimentary rocks. Well illustrated with line drawings and black-and-white photographs, it also contains a comprehensive bibliography. Chapters 11 and 12 focus on carbonate rocks and limestone depositional processes and environments. Suitable for college-level readers.

Reading, H. G., ed. *Sedimentary Environments: Processes, Facies, and Stratigraphy.* Oxford: Blackwell Science, 1996. A good treatment of the study of sedimentary rocks and biogenic sedimentary environments. Suitable for the high school or college student. Well illustrated, with an index and bibliography.

Reineck, H. E., and I. B. Singh. *Depositional Sedimentary Environment.* New York: Springer-Verlag, 1973. This book is concerned primarily with the various types of sand deposit. Diagrams and photographs help to clarify the technical language. Useful for college students majoring in geology.

Sorby, H. C. "On the Structure and Origin of Non-calcareous Stratified Rocks." In *Sedimentary Rocks*, edited by Albert V. Carozzi. New York: Halsted Press, 1975. Written by one of the earliest and most authoritative sedimentary petrologists. The ten-page section on sand is a concise introduction to the different types of sand. Although it is written for scientists, the chapter avoids highly technical terms, making it accessible to high school and college students.

Tucker, Maurice E. *Sedimentary Rocks in the Field.* New York: John Wiley & Sons, 1996. Presents a concise account of biogenic sedimentary rocks and other sedimentary rocks. Classification of sedimentary rocks is well covered. Depositional environments are only briefly discussed. References are well selected. Suitable for undergraduates.

SAND DUNES

Sand dunes form when wind or water deposit sand. Unlike many other geological processes that produce no visible change in a single human life span, the deposition of sand as dunes occurs on a short time line. Moving dunes overrun cropland and forests, supply and deny barrier islands and mainland beaches, and act as a barrier to prevent inland floods.

PRINCIPAL TERMS

BARCHAN DUNE: a crescent-shaped sand dune of deserts and shorelines that lies transverse to the prevailing wind direction

CROSS-BEDDING: layers of rock or sand that lie at an angle to horizontal bedding or to the ground

ECOSYSTEM: in the environment, the unit of the interactions between living elements and non-living factors in a sustaining system

GEOMORPHOLOGY: the branch of Earth science that interprets surface landforms

LONGITUDINAL DUNE: a long sand dune parallel to the prevailing wind

SALTATION: the hopping, jerking, and jumping movements of sand grains by wind or water action

SLIP FACE: the downwind or steep leeward front of a sand dune that continually stabilizes itself to the angle of repose of sand grains

STAR DUNE: a starfish-shaped dune with a central peak from which three or more arms radiate

DISTRIBUTION

Perhaps the most familiar sand dunes are those of the great flat deserts such as the Sahara and Kalahari in Africa, the Mojave in North America, and the Gobi in Asia. It is these mounds of sand that come to the minds of most people when they think of deserts, yet less than 12 percent of arid lands are covered by dunes. Moreover, dunes are not solely a product of desert dynamics; they also form on the beaches of lakes, rivers, and oceans, on barrier islands and river floodplains, under water, and on Mars. Wherever there is a ready supply of sand and an agent of transportation, such as wind or water, a mound of sand with a crest—the dune—will form.

Nevertheless, most of the sand dunes of the world are found in the giant deserts. The 2,340,000-square-kilometer Arabian Desert bordering the Red Sea and the 728,000-square-kilometer Rub' al-Khali farther south appear to be ridge after endless ridge of dunes and blowing sand. Other arid regions, such as China's Takla Makan Desert (crossed by Marco Polo); Australia's Simpson Desert; Death Valley and the deserts of Colorado, Chihuahua, and Sonora in North and Central America; and India's Thar Desert contain sand dunes.

Less familiar are coastal dunes, on the Bay of Biscay and the Dutch Frisian Islands in Europe; Bermuda Island; the eastern United States barrier islands from which rise Miami, Florida, and Atlantic City; the south shore of Lake Michigan; the Pacific coast in Oregon; the Namib coastal desert in southwestern Africa; and the Atacama Desert in coastal Peru. Nor are dunes limited to ocean shores; the lee side of channels along the Mississippi River and the beaches of other large rivers, such as the Koyukuk and Kobuk in Alaska, are windblown dunes. Shoreline dunes are confined to a belt just inland from the beach. Otherwise they are similar in characteristics and design to those of the desert. Where valleys of inland rivers, such as the Platte, Arkansas, and Missouri Rivers, cross the Great Plains, dunes form. The stable dunes of the Sand Hills region of western Nebraska cover about 57,000 square kilometers. The Juniper Dunes area of southwestern Washington is a remnant of the largest known flood in Earth history, the Spokane flood. Nor are sand dunes confined to the Earth. In 1976, the Viking 1 Mars probe discovered crescent-shaped dunes and other desert phenomena on the plains of Memnonia on Mars. Sand dunes also form under water.

Beneath tidal inlets, such as Cook Inlet in Alaska, are ripples and dunes formed as water swiftly flows into and out of the basin. Dunes require neither dry land nor hot air to form.

FORMATION AND MOVEMENT

The sand dune is the mature form, the "adult," of the "infant" sand pile. Both shoreline and desert sand dunes begin with an obstacle—a large rock, pebbles, a small sand heap, or vegetation—around which sand builds up. On deserts, loose sand is supplied to the wind by the weathering of sand-rich bedrock. On beaches, longshore currents and waves supply sand. As the stream of windblown sand deflects and separates around an obstacle, a wind shadow zone forms between the forks. The wind shadow is an island of sheltered air in which the wind blows much more slowly than in adjacent streams. Zones of slower wind speed provide a resting place for some of the sand, which piles up into a drift that then presents a larger barrier to additional blowing sand.

With constant wind and sand supply, dunes continue to build in size. Some Sahara dunes may reach 100 meters in height, although dunes of 30 meters are more common. The typical dune shape is distinctive: a long, low-angled windward slope rising to a peak and a steeper leeward slope. Sand grains pushed over the crest eventually settle to a constant 30- to 35-degree angle on the downwind slope. This angle of repose is the maximum slope of the intersection of the leeward dune side and the ground. Sand moves either by surface creep or saltation. Sand grains driven forward by the wind hit resting sand grains that either are forced forward from the impact like a billiard ball (creep) or are launched into the air like a Ping-Pong ball (saltation). A saltating grain of sand can bounce about 1 meter high and with such force that it can move a grounded grain six times larger. Usually larger sand particles roll along the surface, while finer particles saltate and sandblast anything

in their path. Like a ground fog, saltation layers may be dense enough to obscure the ground beneath yet form a sharp edge with clear air above. Saltating sand makes a dune appear to be "smoking."

Dunes are extremely mobile. During the process of dune growth, as the saltating grains spill over and pile up on the leeward side, or slip face, the entire dune moves downwind. The windward slope, then, is constantly eroding while the sand grains are deposited on the leeward slope. Dunes march forward at a barely detectable 3 meters per year, although dunes in areas of strong, constant winds can move more than 120 meters in one year, or about 0.3 meter every day. On low-lying coasts with continuous supplies of sand, such as those in the Netherlands and southwestern Africa, dunes migrate inland unless anchored with stabilizing vegetation.

CLASSIFICATION

Different methods of classifying sand dunes have been developed. The most commonly used system distinguishes dunes by shape, sand supply, and orientation to the wind. Another classification scheme factors in the presence of vegetation as well as crosswinds in determining dune form. Earth scientists usually categorize a sand dune as either a transverse dune, a longitudinal dune, or a star dune.

One type of transverse dune is the crescent-shaped barchan that lies transverse to—that is,

Sand dunes in the Namib Desert, Namibia. (© William E. Ferguson)

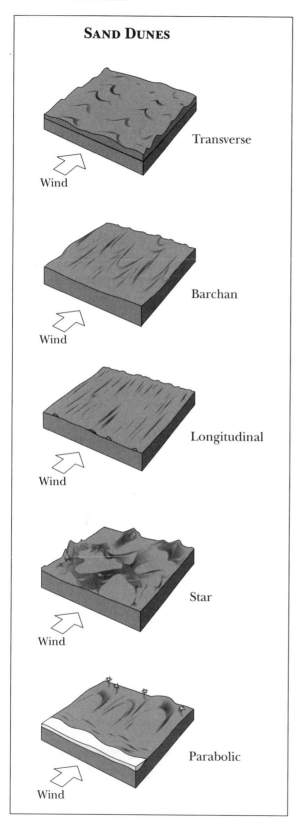

SAND DUNES

Transverse

Wind

Barchan

Wind

Longitudinal

Wind

Star

Wind

Parabolic

Wind

across—the predominant wind direction. The tips of the crescent grow forward on the downwind side of the dune. The gently sloping leeward side is concave, and the steeper windward slip face is convex. When sand supply is limited and the wind direction is fairly uniform, a barchan dune will form. The barchan is much broader than it is tall; a maximum-height dune reaching 33 meters would be about 400 meters wide. The steady northeast trade winds supply sand to the barchan dunes of the North African desert. When sand supply is more abundant, transverse dunes form— not sharp-edged crescents but more amorphous, wavy ridges still perpendicular to the direction of the wind. Typical shoreline dunes, supplied with beach sand, belong in this category, as do some dunes of the Saudi Arabian desert.

Longitudinal dunes, such as the sinuous line of ridges called seifs in the Sahara Desert, lie parallel to the average direction of prevailing winds. When wind direction varies within a 90-degree pie-shaped wedge, the moderate supply of sand is forced from slightly different angles to form a straight-lined dune with a "peak and saddle" pro-file. An aerial survey of dunes in the world's most famous deserts would find primarily linear sand ridges. The southeast trade winds have built spec-tacular parallel dunes in the Great Sandy, Simp-son, and Victoria Deserts of Australia. In Egypt's Western Desert, continuous 70- to 100-meter-tall ridges, snaking 30 to 40 kilometers in length, are not uncommon. In southern Iran, 225-meter-tall longitudinal dunes are known.

The star dune, formed in a confined basin by variable winds blowing from radically different di-rections, is the most complex and the least studied of the three types. Star dunes have been described from the Namib Desert in southwestern Africa and the Mojave Desert in California, although only about 5 percent of documented dunes be-long to this category. The typical dune form of this type resembles a starfish with a central peak and three or more radiating arms. Arms of star dunes in the northern Mojave correspond to the dominant wind directions over the seasons. As soon as any one wind direction stabilizes, the star dune grades into a less distinctive composite shape. Less stable, star dunes often are flanked by and grade into both barchan and linear dunes.

As sand dune deposits are buried to become

part of the geologic record, their overall shape, such as crescent or linear, is disturbed. What remains in desert-formed dunes is the conspicuous slip face, the downwind slope, which appears as a distinctive cross-bedded wedge. Wedges group together and intersect one another from different directions, with no evidence of normal horizontal layering. The approximately 100-million-year-old Navajo sandstone of the Colorado Plateau shows spectacular cross-bedding. An older dune formation, the Permian Coconino sandstone, and the younger Lower Bunter sandstones near Birmingham, England, are also ancient sand dunes. Since beach dunes are less frequently preserved, few have been unequivocally identified in the geologic record. Sections of the St. Peter sandstone of the Mississippi River Valley are thought to be sand dune deposits. Sand deposited in sand dunes has distinct characteristics that remain in the fossil record. Windblown quartz grains, limited to a narrow range of small sizes, are extremely round, with a frosted glass surface resulting from the constant collisions of saltation.

Parabolic dunes occur where vegetation partly covers the sand, often along shorelines. Here, the shape of the dune is similar to the barchan but is oriented in the opposite direction, pointing into the wind rather than away from it. A break in the vegetation causes the wind to "blow out" the sand and deposit it in a curved rim formation that grows higher and higher.

SAND DUNES AND HUMANKIND

The vastness and inhospitability of sand dunes in the great deserts, such as the Arabian and Egyptian, evoked awed respect from early Arab traders and later European explorers, such as the famous T. E. Lawrence ("Lawrence of Arabia"), Charles Montague Doughty, and Carsten Niebuhr. Adding to the mysterious reputation of sand dunes are the incredible booming or roaring noises heard by travelers. Native tales attributed the drumlike rumbles to cantankerous spirits, such as Rul, jinni of the dunes, or to a legend of tolling bells buried beneath the sands. Crossing the dunes of the Takla Makan Desert in Mongolia, Marco Polo reported similar unexplained sounds: "Often you fancy you are listening to the strains of many instruments, especially drums and the clash of arms." The loud noises are attributed to the endless movement of

sand on a dune's slip face. As the sand on the steep forward slope avalanches to readjust to its angle of repose, grains vibrate and produce sounds.

Humankind's primary interaction with the sand dunes of ocean beaches, saltwater estuaries, and, to a lesser extent, the deserts has been an attempt to restrain their constant migration. A line of dunes is a natural barrier against flooding from exceptionally high tides or violent storm-driven waves. Inland migrating dunes may, however, overpower inhabited beachfronts. The usual way to stabilize dunes and reduce migration is to plant vegetation such as sea oats, marram, and other grasses or to erect slatted fences. Vegetation must then be protected from the intense use of dune buggies and excessive visitation. In the oil-abundant Iranian desert, a petroleum product has been sprayed on the shifting dunes to provide a water-holding mulch in which shrubs and trees can root. Another factor important to long-term dune survival is an assured supply of sand. If dunes are cut off from ocean-supplied sand, they eventually wither. Sea walls and jetties erected by engineers attempting to reduce beach erosion also prevent waves from supplying sand to the dunes. Bay Ocean, an unsuccessful resort development on Tillamook Spit of the Oregon coast, erected jetties to harbor ships in Tillamook Bay. The jetty also, however, prevented sand, typically carried by longshore currents up and down the coast, from supplying the dunes. First, the dunes and the beach itself disappeared, followed, a few years later, by the spit, which was breached by a severe storm.

One of the most famous programs of dune stabilization was the United States government's attempt to stabilize 360 kilometers of barrier island beaches off the Atlantic coast. During the Depression of the 1930's, a sand fence was installed to encourage the development of exceptionally large, artificial dunes. Beach grasses, shrubs, and trees were then planted on the offshore slope. Forty years later, the human-made dunes on Hatteras Island continued to rise in height, but the original beach, 220 kilometers wide, had receded to less than 33 meters, and the marshes behind the towering dunes were drying up. The unnaturally high dunes had prevented sand deposition on the beach and transport of seawater to the marsh at the back of the island. In 1973, the federal government changed its policy of dune stabilization to allow nature to take its course.

STUDY OF SAND DUNES

Specialists from several disciplines of Earth science contribute to the interpretation of sand dunes. The geologist in the specialty of geomorphology, or the study of land features, in addition to tedious personal observations on foot, uses remote-sensing technologies, such as aerial photography and satellite imagery, to piece together a cohesive view. The sedimentologist, who studies processes of rock formation from sediments, uses data collected from both field observations and laboratory environments such as a flume or wind tunnel. Engineers have contributed inadvertently to understanding the behavior of sand dunes by a failure to recognize the dune as a phenomenon dependent on at times unidentified but essential processes.

Like other natural landscape phenomena, sand dunes were first studied by local inhabitants, explorers, and adventurers traveling overland. Earliest European explorations, such as the thirteenth century trek across a portion of the Mongolian Gobi and Marco Polo's journey across the high, frigid Takla Makan Desert, yielded general observations about the arid lands. Using the traditional transportation mode, the camel, Europeans touched the vast tracts of the Arabian Desert and the Rub' al-Khali in the late nineteenth and early twentieth centuries. Later, using automobiles, explorers and scientists such as Roy Chapman Andrews in Mongolia and Ralph Bagnold in Egypt's Western Desert contributed careful observations about dune shapes and sand movement. Dune classifications have been proposed by Bagnold from his studies in the Egyptian Sand Sea, by John T. Hack in his studies of Arizona dunes, by C. T. Madigan describing the sea of dunes in the immense Australian deserts, and by others. Field studies of the relationship between wind regime—that is, variation of wind direction and wind velocity—and dune morphology, or form, continue today. Where geologists can wet down a dune face, they use shallow trenches and take core samples to study airflow, sedimentary processes, and internal dune structure and deposits.

In the 1930's, Bagnold pioneered studies of sand transport by wind with experiments in the wind tunnels of the Imperial College of Science and Technology in London. Controlled laboratory experiments such as these provide data on the relationship between wind speed and sand grain size; sand grains of a specific size do not move until the wind reaches a certain speed. On one type of sand tested by Bagnold, wind speed reached a velocity of about 18 kilometers per hour before the grains began to roll. Models of sand grain transport and saltation are continually being updated by field studies and laboratory experiments. Sedimentologists use laboratory flume studies to provide data on the formation of dunes under water. Dunes in a water environment form and remain stable only at a specific stream velocity; dunes do not form if the velocity is too slow and will disappear at higher velocities. Flow of water and eddies around streambed ripples and dunes resembles airflow in wind-formed dunes.

Technological advances in remote sensing, the field of data collection that uses aerial photography and images from satellites, provide insights unavailable to earlier scientists. Geologists use color and infrared aerial photographs taken from multiple, overlapping flights of specialized airborne cameras to map sand dunes. Detailed maps showing airflow and sand-grain flow and ripple migration direction combine data gathered from aerial photographs and field studies. Photographic images of topographic forms such as sand dunes are produced by a type of sensing equipment called side-looking airborne radar. Computer-enhanced photographs taken over the same area on repeating dates by Landsat and National Oceanic and Atmospheric Administration (NOAA) satellites have revealed time-lapse views of dune migration, such as the westward advance of dunes of the African Namib Desert into the Atlantic Ocean. Space exploration has added to the technology of studying sand dunes. Experiments carried by the *Columbia* space shuttle revealed previously unknown characteristics of the Selima sand sheet in the Libyan Desert, and the Viking 1 Mars probe provided close-ups of the Martian sand dunes, which greatly resemble those of the Egyptian Desert.

Geological and civil engineers have discovered previously little-understood features of sand dune migration and relationship to beach sand sources by attempting to stabilize dunes with vegetation and fences. The ocean-fronting beaches and continent-facing marshes of barrier islands on the eastern United States seaboard were reduced considerably by forty years of a federally appointed dune stabilization program. Relationships between the continuing existence of the barrier islands and sand dune mobility were understood only after "stabilized" islands such as Hatteras began to disappear. In many cases, the use of vegetation to anchor shoreline dunes and to inhibit their advance has been more successful. If road or dwelling construction or a natural increase of sand supply does not disturb established vegetation, the balance between sand and vegetation will hold the dune in place.

Cory Samia

BIBLIOGRAPHY

Bascom, Willard. *Waves and Beaches: The Dynamics of the Ocean Surface.* Rev. ed. Garden City, N.Y.: Doubleday, 1980. A classic text by a well-recognized oceanographer, this book gives many specific examples of ocean and onshore processes. Written in easily understood style, with diagrams.

Dunbar, Carl O., and John Rodgers. *Principles of Stratigraphy.* New York: John Wiley & Sons, 1966. A basic text in stratigraphy, or the study of layered rocks, written for the serious student of geology. Excellent sections on recognition of ancient dunes and processes of sedimentation.

Holmes, Arthur. *Holmes' Principles of Physical Geology.* 4th ed. New York: Chapman and Hall, 1993. A revised version of an often-used general geology text that details processes of land formation and topography as well as tectonic and crustal phenomena. Many examples are provided.

Ittekko, Venugopalan, et al., eds. *Particle Flux in the Ocean.* New York: John Wiley and Sons, 1996. This volume contains descriptions of the chemical cycles of the oceans, currents, and marine particles, as well as their their relationship to coastal areas such as beaches. Suitable for the high school reader and beyond. Illustrations, index, bibliography.

Packham, John R., and Arthur John Willis. *Ecology of Dunes, Salt Marsh, and Shingle.* New York: Chapman and Hall, 1997. This is a clearly written introduction to coastal ecology. Sections focus on the geochemical and geophysical phases of sand dunes, salt marshes, and other coastal areas. Intended for the nonscientist. Illustrations, maps, index, and twenty-five pages of bibliographical references.

Page, Jake. *Arid Lands.* Alexandria, Va.: Time-Life Books, 1984. A well-written and photographed volume of a series of Earth science topics (called the Planet Earth series) for the general reader. This issue describes and locates deserts of the world and adjacent lands and relates the dynamic processes of their formation.

Press, Frank, and Raymond Siever. *Understanding Earth.* 2d ed. New York: W. H. Freeman, 1998. If a student newly introduced to the field of geology were to buy a single text, this book should be the one. A geology text for the beginning college student, it covers with thoroughness major topics and includes methods of study as well as current information.

Sackett, Russell. *The Edge of the Sea.* Alexandria, Va.: Time-Life Books, 1983. Like other volumes in this series for the general Earth science reader (the Planet Earth series), this issue presents easily read information accompanied by excellent graphics. Particularly pertinent is the section on barrier islands and dune stabilization.

Thornbury, William D. *Principles of Geomorphology.* 2d ed. New York: John Wiley & Sons, 1969. This frequently used text for college-level geomorphology courses gives thorough coverage to landscape features of all kinds. Good section on applied geomorphology and geological engineering.

SEDIMENT TRANSPORT AND DEPOSITION

Flowing water, wind, and glaciers move sediment from where it is produced by rock weathering to sites of deposition in river basins, lakes, and the oceans. Much of the Earth's landscape is shaped directly or indirectly by the movement of sediment. Interpretation of sedimentary deposits, modern and ancient, rests on understanding of sediment transport and deposition.

PRINCIPAL TERMS

BED LOAD: sediment in motion in continuous or semicontinuous contact with the sediment bed by sliding, rolling, or hopping

BED SHEAR STRESS: the force per unit area exerted by the flowing fluid on the sediment bed, averaged over an area that is large compared to individual bed particles

COMPETENCE: a concept that expresses the size of the largest sediment particles that can be moved by a given fluid flow

DEBRIS FLOW: a flowing mass consisting of water together with a high concentration of sediment with a wide range of sizes, from fine muds to coarse gravels

SALTATION: a mode of sediment movement in air wherein relatively large, heavy particles are lifted from the bed at a substantial angle to the horizontal and take an arching trajec-

tory downwind, little affected by the turbulence in the air

SEDIMENT DISCHARGE FORMULA: a formula or equation designed to predict the sediment discharge that would be observed for a given combination of flow conditions and sediment characteristics

SEDIMENT DISCHARGE: the rate of transport of sediment past a planar section normal to the flow direction, expressed as volume, mass, or weight per unit time; also called sediment transport rate

SUSPENDED LOAD: sediment in motion well above the sediment bed, supported by the vertical motions of turbulent eddies

THRESHOLD OF MOVEMENT: the conditions for which a flow is just strong enough to move the sediment particles at the surface of a given sediment bed

MOVEMENT OF SEDIMENT

Weathering of bedrock exposed on the continents produces solid particles of mineral or rock, ranging in size from the finest clay to large boulders. In most places, sediment moves slowly downslope toward stream channels, largely by the direct or indirect effects of gravity. Once in stream channels, the sediment particles are moved efficiently by flowing water.

A flow of water or air exerts a force on a solid particle resting on a loose bed of similar particles. This force, which arises both from the friction of the flowing fluid and from the existence of relatively high fluid pressure on the upstream side of the particle and relatively low pressure on the downstream side, tends to move the particle in the direction of the flow. This force commonly has an upward component, called the lift, as well as a

downstream component, called the drag. When the fluid force on the bed particle, which is counteracted by the weight of the particle, is sufficient to lift the particle up from its underlying points of support or to rotate the particle downstream around its points of support, the particle begins to move downstream. This condition is called incipient movement, and the overall force per unit area, or stress, that the flow exerts on the bed under those conditions is called the critical or threshold bed shear stress. Another way of looking at incipient movement is in terms of competence: What is the largest particle that can be moved by a flow that exerts a given bed shear stress?

WATER AND AIR TRANSPORT

Once a sediment particle is set in motion by a flow of water, it is likely to move by some combination of sliding, rolling, or hopping close to the bed. The material in this kind of motion is called bed load. If turbulent eddies in the flow have upward speeds greater than the downward settling speed of the particles relative to the fluid in their immediate vicinity, some of the moving particles are swept up into the flow to travel long distances downstream before returning to the bed. The material in this kind of motion is called the suspended load. As the strength of the flow increases, a larger percentage of the load travels in suspension, but bed load is always present near the bed even in strong flows. Suspended particles are not really suspended, in that they are continuously settling downward relative to surrounding fluid and ultimately are redeposited on the bed after traveling distances ranging from less than 1 meter (for coarser particles) to hundreds or even thousands of kilometers (for the finest particles).

The sediment particles transported by water range in size from the finest clay sizes (of the order of 1 micrometer in size) through silts (a few tens of micrometers) to sands (of the order of 1 millimeter) and gravels (coarser than a few millimeters). Clays and silts are carried mostly in suspension and are deposited only where current velocities are very small. Sands are transported both as bed load and in suspension, depending on the strength of the flow, and gravels are transported mainly as bed load in flows prevalent on the Earth's surface.

In air, the most prominent mode of sediment particle movement is saltation, in which the particles are lifted off the bed to move in a regular arching trajectory upward at a fairly large angle to the bed and then downward at a fairly small angle to the bed. Saltating sand grains commonly rise no more than 1 or 2 meters above the sand surface while traveling as much as several meters downwind. Sands and even small gravel particles are transported in saltation; finer particles are put directly into suspension by the wind.

AGENTS OF SEDIMENT TRANSPORT AND RESULTS

Means	*Deposit*	*Result*
Breakers	Beach	Sandstone
Currents	Beach	Conglomerate, Sandstone
Glaciers	Moraines	Tillite
Gravity	Avalanches	Conglomerate
Gravity	Galus	Breccia
Gravity	Landslides	Conglomerate
Groundwater	Stalactite	Dripstone
Lakes	Varved clay	Shale
Organisms	Reefs	Shell limestone
Rivers	Alluvial fans	Conglomerate
Rivers	Till	Sandstone
Seawater	Evaporites	Rock salt, anhydrite
Seawater	Precipitates	Manganese oxide concretions, cherts
Springs	Varved clay	Travertine
Swamps	Peat	Coal
Wind	Ash	Tuff
Wind	Dunes	Sandstone
Wind	Dust	Loess

In the oceans, sediment is moved not only by unidirectional currents but also by oscillatory flow resulting from the passage of wind-generated waves at the sea surface. Moreover, unidirectional flows and oscillatory flows can be superposed to produce combined flows, in which the water at the bed has an oscillatory motion but undergoes net movement in some direction as well. The concepts of threshold, bed load, and suspension apply to oscillatory and combined flows as well as to unidirectional flows.

DISCHARGE AND DEPOSITION

The time rate at which sediment is carried across some planar section, real or imaginary, normal to the flow direction is called the sediment transport rate, or sediment discharge. It is expressed as either mass, weight, or volume of sediment per unit time. The transport rate of the bed load and of the suspended load can be considered either separately or together as the total transport rate. The sediment transport rate, expressed per unit width normal to the flow direction, is a steeply increasing function of the bed shear stress or the flow velocity. The great mathematical complexity of turbulent flow carrying discrete solid particles has hindered development of theories to

Sediment from the Mississippi River (right) flows into the Gulf of Mexico. (U.S. Geological Survey)

predict the sediment transport rate as a function of sediment characteristics and flow conditions. A large number of formulas or equations, often called sediment discharge formulas, have been developed to predict the sediment transport rate. All have been built around one or another physically plausible mechanism that provides the general mathematical form of the equation. The specific form of the equation is then found by fitting or adjusting coefficients in the equation so that the equation conforms to some set of actual measured data on transport rates. None of these sediment discharge formulas is significantly better than any other, and there can be differences by as much as a factor of ten in predicted transport rates.

The volume concentrations of suspended sediment in most flows of water in rivers or in the oceans, as expressed in volume of sediment per unit volume of water-sediment mixture, is usually no more than a few percent. In certain situations, however, water-saturated masses of sediment can begin to flow even on a gentle slope of 1 or 2 degrees by liquefaction, either spontaneous or induced by earthquake shocks. Such flows, called debris flows, may have sediment concentrations of up to 70 percent by volume. Debris flows can be formed either on the land surface or under water.

Glaciers are locally responsible for the transport of significant volumes of sediment. Glaciers derive most of their sediment load from erosion of the bedrock or from preexisting sediment beneath the glacier, although valley glaciers can carry on their upper surface large quantities of sediment that falls from the valley walls. Glaciers are far less selective of the sediment sizes they carry than are flows of water or air.

Transported sediment is ultimately redeposited in some way. Deposition is always associated with one or both of two kinds of changes in the flow. One of these is temporal: There is everywhere a net loss of load from the flow to the bed with time because the flow becomes weaker everywhere with time. An example is redeposition of sediment picked up by a river flood as the flood subsides. The other, which is usually more important in building thick deposits of sediment, is spatial: The flow becomes weaker in the downstream direction, causing the sediment transport rate to decrease in the downstream direction. The only way the sediment transport rate can decrease downstream is for sediment to go into storage at all points on the bed, thus building up the bed. An example is the expansion and weakening of flow in a river delta, where a river meets a large lake or the ocean.

STUDY OF SEDIMENT

Sediment transport is studied in natural flows, in laboratory tanks, and by computer modeling. Laboratory studies have the advantage that the conditions of sediment transport can be closely controlled, so that the various factors thought to be important in determining the mode and rate of sediment movement can be varied independently. The disadvantage of laboratory work, aside from necessarily small physical scales of the flow, is that the phenomenon may be too simplified to simulate natural flow environments well. In the laboratory, sediment transport and deposition are studied mostly in open channels, called flumes, in which a flow of water is passed over a sediment bed. Flumes range from a few meters to more than 100 meters long, from about 10 centimeters to a few meters wide, and from several centimeters to about 1 meter deep. The water is usually recirculated from the downstream end to the upstream end to form a kind of endless river. The sediment may also be recirculated, or it may be fed at the upstream end and caught in a trap at the downstream end. In laboratory flumes, the sediment movement may be observed visually or photographed either through a transparent sidewall or from the water surface if suspended sediment is not abundant.

Measurement of sediment transport rates is notoriously difficult, not only in natural flows but also in the laboratory. In flumes and in some specially instrumented rivers, it is possible to pass the flow over or through a section where all of the sediment in transport as bed load is extracted for measurement. In general, however, bed load must be measured using traps of various designs that are lowered to the bottom, opened for a certain time interval, and then brought back to the surface. The problem is that such samplers tend to distort the flow and therefore the sediment movement in their vicinity, and there is usually no good way of estimating and correcting for that effect.

In both the laboratory and nature, suspended load is usually sampled by extracting samples of the suspensate (water plus sediment) at several levels in the flow by sucking or siphoning through small-diameter horizontal tubes with their opening facing upstream. Care must be taken to match the extraction speed to the local flow speed to minimize overcatching or undercatching. Both in rivers and in shallow marine environments such as beaches, the direction and rates of sediment movement have been estimated by emplacing plugs of sediment tagged with short-lived radioisotope tracers and taking closely spaced sediment samples in the surrounding area at some later time, after the tagged sediment has been dispersed by transport. Measurements of this kind have the advantage of integrating the transport over a long time period.

SIGNIFICANCE

Because most of the Earth's surface topography is produced by erosion, transport, and deposition of particulate material derived from weathering of bedrock, consideration of sediment transport is essential in any attempt to account for how the landscape of the Earth develops through time. Even in the driest of deserts, most geological work is accomplished by running water: On the rare occasions of heavy rains, the abundant loose sediment produced by weathering is entrained and transported by floods. Erosion and deposition of sediment in rivers and estuaries in the processes of channel shifting lead to great changes in river geometry on time scales that range from days to decades. These changes often make it difficult to maintain navigable channels in rivers and harbors. The useful water-storage life of reservoirs is determined by the rate of transport of sediment from upstream in relation to the reservoir capacity; because of saltation, reservoir life is typically limited to decades rather than to hundreds of years. Rivers downstream of reservoirs commonly experience a substantial lowering of the level of the riverbed as the now sediment-free flow seeks

A glacier transported this very large piece of sediment, leaving it to stand alone after the ice melted, in this case almost precariously balanced. Such boulders are called "erratics" because they seem unrelated to the surrounding landscape. (U.S. Geological Survey)

to pick up new sediment, leading to what is called degradation.

In the coastal zone, sediment is transported by tidal currents in shifting tidal channels and by nearshore currents of various kinds that flow parallel to open shorelines. Sediment movement along open shorelines is augmented greatly by the effect of waves: The strong oscillatory bottom flows produced by waves tend to suspend the sediment, which then can be carried for some distance even by unidirectional currents too weak to move sediment by themselves. Understanding of the rates of sediment erosion, transport, and deposition is essential for dealing with problems of shoreline changes in the coastal zone.

The sedimentary record of the Earth is the outcome of sediment transport and deposition, in the same ways observed presently. Understanding modern processes of sediment movement and deposition is essential in interpreting the ancient depositional environments in which the Earth's sedimentary record was produced. Understanding of the controls on the complex geometry of sediment bodies, ultimately a matter of sediment transport and erosion, plays an important role in petroleum exploration because petroleum is commonly found in porous sedimentary rocks that were formed by particle-by-particle deposition of sediments in river systems and in the oceans.

John Brelsford Southard

CROSS-REFERENCES

Aerial Photography, 2739; Alluvial Systems, 2297; Beaches and Coastal Processes, 2302; Dams and Flood Control, 2016; Deep-Sea Sedimentation, 2308; Deltas, 2312; Desert Pavement, 2319; Desertification, 1473; Drainage Basins, 2325; Evaporites, 2330; Floodplains, 2335; Geomorphology of Dry Climate Areas, 904; Lakes, 2341; Land-Use Planning, 1490; Landsat, 2780; Reefs, 2347; River Bed Forms, 2353; River Flow, 2358; Sand, 2363; Sand Dunes, 2368; Weathering and Erosion, 2380.

BIBLIOGRAPHY

Allen, John R. *Principles of Physical Sedimentology.* Winchester, Mass.: Allen & Unwin, 1985. A lucidly written introduction to the movement and deposition of sediment for sedimentology students at the college level and beyond. The chapters on sediment transport and deposition are moderately mathematical but contain unusually clear qualitative discussions of the physical effects.

Blatt, Harvey, and Robert J. Tracy. *Petrology: Igneous, Sedimentary, and Metamorphic.* New York: W. H. Freeman, 1996. A textbook on sedimentology aimed at the upper college level. The chapter on sediment movement, a self-contained treatment on fluid flow and sediment transport, is somewhat mathematical but is an excellent starting point for someone looking for fundamentals.

Leeder, Mike R. *Sedimentology and Sedimentary Basins: From Turbulence to Tectonics.* Oxford: Blackwell Science, 1999. A college-level text on sedimentology, with separate chapters on fluid flow and sediment transport. Some mathematics is used, but the treatment remains coherent.

Middleton, Gerard V., and John B. Southard. *Mechanics in the Earth and Enviromental Sciences.* New York: Cambridge University Press, 1994. This book contains several chapters on the fluid dynamics of sediment-transporting flows, written for nonspecialists who have at least an elementary knowledge of calculus. Three later chapters deal with sediment transport, with emphasis on important physical effects.

Prothero, Donald R., and Fred Schwab. *Sedimentary Geology: An Introduction to Sedimentary Rocks and Stratigraphy.* New York: W. H. Freeman, 1996.. A thorough treatment of most aspects of sediments and sedimentary rocks. Well illustrated with line drawings and black-and-white photographs, it also contains a comprehensive bibliography. Chapters 11 and 12 focus on carbonate rocks and limestone depositional processes and environments. Suitable for college-level readers.

Pye, Kenneth. *Aeolian Dust and Dust Deposits.* London: Academic Press, 1987. An excellent reference about airborne transport and sediment deposition.

Reading, H. G., ed. *Sedimentary Environments: Processes, Facies, and Stratigraphy.* Oxford: Blackwell Science, 1996. A good treatment of the study of sedimentary rocks, sedimentary transport, and biogenic sedimentary environments. Suitable for the high school or college student. Well illustrated, with an index and bibliography.

Vanoni, V. A., ed. *Sedimentation Engineering.* New York: American Society of Civil Engineers, 1975. This engineering reference manual, aimed at students and practicing engineers, deals with the fundamentals of sediment-transport mechanics from an applied standpoint. The best and most authoritative source for information on the engineering aspects of sediment movement. Emphasis is on engineering practice, but without slighting basic understanding of the physical effects. Fairly heavily mathematical.

Yalin, M. S. *Mechanics of Sediment Transport.* 2d ed. Oxford, England: Pergamon Press, 1977. This college-level textbook, designed for engineering students, is devoted entirely to quantitative treatment of sediment movement. Theoretical and heavily mathematical, but with numerous qualitative insights into the physics of sediment movement.

WEATHERING AND EROSION

The weathering process breaks down the rocks of the Earth's surface into soluble material and into the particles that form the basis for soils and agriculture. These weathered products are then swept away by the various erosional agents, such as rivers and glaciers, that shape the planet's rocky surface.

PRINCIPAL TERMS

ABRASION: the wearing away of rock by frictional contact with solid particles moved by gravity, water, ice, or wind

ACID RAIN: rain with higher levels of acidity than normal; the source of the high levels of acidity is polluted air

CHEMICAL WEATHERING: the chemical decomposition of solid rock by processes that change its original materials into new chemical combinations

EROSION: the general term for the various processes by which particles already loosened by weathering are removed

GRANITE: a coarse-grained, igneous rock composed primarily of the minerals quartz and feldspar

LIMESTONE: a sedimentary rock composed of calcium carbonate

MECHANICAL WEATHERING: the physical disintegration of rock into smaller particles having the same chemical composition as the parent material

MINERAL: a naturally occurring, inorganic, crystalline material with a unique chemical composition

SANDSTONE: sedimentary rock composed of grains of sand cemented together

WEATHERING: the general term for the group of processes that break down rocks at or near the Earth's surface

DESTRUCTIVE FORCES

Although the landscape rarely appears to change, constructive and destructive forces are at work within the Earth, building the crust up and breaking the rocks down and carrying the resulting debris away. The destructive forces are known as weathering and erosion. "Weathering" refers merely to the mechanical disintegration and chemical decomposition of rocks and minerals at or near the Earth's surface. No movement of these materials is implied. Exposure to weather causes rocks to change their character and either crumble into soil or become transformed into even smaller particles that are readily available for removal. "Erosion" refers to the processes by which particles already loosened by weathering are removed. This process involves two steps: First, the loose materials must be picked up, or entrained; second, the materials must be physically carried, or transported, to new locations. The major ways by which Earth materials are eroded are by means of rivers, underground water, moving ice, waves, wind, and landslides.

Weathering is a near-surface phenomenon because it involves the response of Earth materials to the elements of sunlight, rain, snow, and the like. It does not affect rocks that are deeply buried within the Earth's crust. Only after these rocks are exposed at the Earth's surface, after a long period of uplift and removal of overlying material, does weather begin to affect them. Now they are in a changed environment, subject to the comparatively hostile action of acid rain, subfreezing temperatures, and high humidity. The resulting transformations that take place in the rock are what is called weathering.

MECHANICAL AND CHEMICAL WEATHERING

Scientists recognize two different types of weathering: mechanical and chemical. Although the two types are generally discussed separately, it is important to keep in mind that in nature they generally work hand in hand. Mechanical weathering (also known as disintegration) involves the physical breakdown of the rock into smaller and smaller grains, usually because of the application of some kind of pressure such as the expansion of water during freezing or the growth of plant roots in rock crevices. The chemical composition of the rock, however, remains unchanged. The end result of mechanical weathering is smaller pieces of rock that are identical in composition and appearance to the original larger rock mass.

Chemical weathering (also known as decomposition) involves a complex alteration in the materials that compose the original rock. These materials are chemically changed into new substances by the addition or removal of certain elements, usually through the action of water. The familiar rusting of iron is an example of chemical weathering; there is a total change in the composition and appearance of the original material. At first, there is a hard, silvery metal; afterward, all that remains is a soft, yellowish-brown powder.

Consider the effects of mechanical weathering and chemical weathering on a cube of rock that measures 6 centimeters on a side. Assume that by means of mechanical weathering, the cube is broken down into 216 cubes measuring 1 centimeter on a side. Now, much more surface area is available for chemical attack. For this reason, chemical weathering proceeds much faster when a rock is first broken into smaller pieces by mechanical weathering.

In nature, mechanical weathering can proceed in a variety of ways. The best-known example involves the action of freezing water. Because water increases about 9 percent in volume as it freezes, enormously large outward-directed pressures develop within a rock when water freezes in its pore spaces and cracks. The result is the angular rock fragments found scattered about most mountain tops and sides. Soil contains water, too, and horizontal lenses of ice may form within the soil when water freezes in it. These create bumps in lawns and the familiar "frost heaves" of mountain roads. When heavy trucks rumble over these heaved pavements during thaws, the pavement gives way to create potholes.

In deserts, soil water is drawn upward through the rock and evaporates at the hot upper surface, leaving its dissolved salts behind as crystals growing in the pore spaces of the rock. These growing salt crystals also exert powerful pressures within the rock, so that porous rocks, such as sandstone, undergo continuous grain-by-grain disintegration in desert climates. Mechanical weathering can also be produced when the extreme heat from a forest fire or a lightning strike causes flakes to chip off a rock or when a growing plant or tree extends its root system into cracks and splits a rock apart. Another type of rock splitting is known as exfoliation; it is caused by the spontaneous expansion of rock masses when they are freed from the confining pressures of overlying and surrounding rock. This process produces large, dome bedrock knobs with an onionlike structure. Stone Mountain, Georgia, and Half Dome, in Yosemite National Park, are examples.

Chemical weathering is a more complex process than mechanical weathering because the original rock material is actually transformed into new substances. The rusting of iron has already been mentioned as an example of chemical weathering. Many common rocks and minerals contain iron.

An example of mechanical weathering and rock uplifted by tree roots. (U.S. Geological Survey)

During chemical weathering, the iron in these substances combines with oxygen from the air to form various compounds known as iron oxides.

Another way in which chemical weathering attacks rocks is by dissolving them. Large areas of the Earth's surface are underlain by a rock type known as limestone. Limestone is readily dissolved by water containing small quantities of acid. All rainfall is weakly acidic as a result of its dissolving carbon dioxide from the air to produce dilute carbonic acid. Rains that originate in areas of high air pollution are even more acidic, a condition known as acid rain. When rainfall that contains carbonic acid comes in contact with limestone bedrock, the acid reacts with the calcium carbonate in the rock to produce calcium bicarbonate, a soluble substance that is readily carried off in solution.

A final example of chemical weathering involves the weathering of granite, a hard igneous rock composed primarily of feldspar and quartz. When granite undergoes chemical weathering, each mineral is affected differently. The feldspar is gradually transformed into a new mineral, clay, which is soft and easily molded when wet. Clay offers very little resistance to erosion. Quartz, on the other hand, is highly resistant to chemical attack and is left behind when the clay is removed. Some of the quartz grains remain in the soil, but most will be carried off by rivers, becoming rounded as they tumble along. Eventually they form the sands

of beaches and, in time, the sedimentary rock known as sandstone.

EROSION

The term "erosion" refers to those processes by which the loose particles formed by weathering are picked up and carried to new locations. Erosion is a highly significant phenomenon at the Earth's surface. Examples of it range from small gullies in a farmer's field to a catastrophic landslide in a high mountain valley. Nevertheless, the general principle involved in all types of erosion is the same: Weathered Earth materials move downslope from their place of formation to a new location, with gravity as the driving force. The materials may simply slide downhill as a landslide, or they may be carried down the hill by an erosional agent, such as running water. Worldwide, running water in the form of streams and rivers is probably the single most important erosional agent. Locally, other erosional agents may be highly significant, including underground water flow, glaciers, waves, and wind action.

The downhill movement of weathered materials under the influence of gravity alone results in landslides if the downslope movement is rapid but in "creep" if the movement is imperceptibly slow. When large quantities of water are present in the weathered material, the downslope movement is called a mudslide. Running water can erode material from its channel banks in four different ways: Soluble material can be dissolved by weakly acidic river water, bedrock can be worn smooth as a result of abrasion by sand and gravel carried along the streambed, unconsolidated bank and bed materials can be swept away by a strong current (resulting in bank caving), and upwardly directed turbulent eddies in the water may lift small particles from the bottom and entrain them in this fashion. Underground water erodes bedrock primarily by dissolving it, whereas glaciers act more like rivers, abrading the underlying rock by means of rock frag-

A frost-split granite boulder in Sequoia National Park, Tulare County, California. (U.S. Geological Survey)

A block of granite that has been hollowed out by windblown sand at Llanode Caldera, Atacama Province, Chile. (U.S. Geological Survey)

ments frozen in the ice. Glaciers are also able to pluck rock masses from their channel walls when these rock masses have frozen to the main ice mass; the rocks are torn loose as the ice moves on.

Waves erode shorelines, wearing rock surfaces smooth by means of the sand and gravel they carry. Waves can also dislodge particles from a cliff face. Cracks quickly open in cliffs, seawalls, and breakwaters, and when water is forced into these cracks, the air in the cracks becomes highly compressed, exerting still further pressure on the rock. Wind erosion, on the other hand, relies on the abrasive action of sand grains transported by the wind and on the lifting power of eddies, which are able to entrain finer-grained soil particles.

STUDY OF WEATHERING AND EROSION

Scientists have analyzed the rate at which rocks weather and have found that the important factors are rock type, mineral content, amount of moisture present, temperature conditions, topographic conditions, and amount of plant and animal activity. A rock type may be highly resistant to

weathering in one climate and quite unresistant in another. Limestone, for example, which forms El Capitan, the highest peak in the desert region of southwest Texas, underlies the lowest valleys in the humid climate of the Appalachian Mountains.

Using field observations and laboratory experiments, scientists have studied the rate at which different minerals are attacked by chemical decomposition. Among the minerals formed by igneous activity, quartz is least susceptible to chemical attack, whereas olivine, a greenish-colored mineral rich in iron and magnesium, is one of the most susceptible. The reason is that olivine forms at high temperatures and pressures when melted rock first begins to cool and is consequently unstable at the lower temperatures and pressures that prevail at the Earth's surface. Quartz, on the other hand, forms late in the cooling process, when the temperatures and pressures are more similar to those encountered at the Earth's surface; therefore, quartz more readily resists attack by chemical weathering. Scientists have concluded that the more the conditions under which a mineral forms are akin to those at the Earth's surface, the more resistant to chemical attack the mineral will be.

Numerous observations have been made relating to the rapidity with which weathering takes place. The eruption of Mount St. Helens in Washington State on May 18, 1980, has provided a natural laboratory for such study. During the eruption, vast quantities of volcanic ash were hurled into the air and deposited to depths exceeding several meters in the vicinity of the volcano. Scientists have carefully analyzed the changes that are taking place in the ash because of mechanical and chemical weathering and the rate at which this ash is being converted into a productive soil for the growth of vegetation. Scientists also study the rate at which tombstones and historic monuments of known age are attacked by weathering. For marble tombstones in humid climates, the amount of weathering within a single lifetime may amount to several millimeters.

The rate at which Earth materials are moved from place to place on the Earth's surface by the various agents of erosion is also of interest. One way to approach this problem is to measure the quantity of sediment being carried by a river each year and then to calculate how much of a loss this amount represents for the entire area drained by the river. Data from various locations in the United States suggest that the overall rate of erosion amounts to approximately 6 centimeters per 1,000 years. Corroborating evidence comes from another source: Photographs made in scenic areas and compared with photographs made from the same vantage point one hundred years ago or more show surprisingly little erosional modification of the Earth's surface. Once humans occupy an area intensively, erosion rates increase significantly.

SIGNIFICANCE

Weathering affects not only bedrock outcrops but also human-made structures. Unless they are continually repaired and restored, all structures become weather-beaten and, in time, weaken and fall into ruin. Beginning in the early 1970's, people also became aware of the harmful consequences of acid rain. Concrete, limestone building blocks, and the marble used for statuary are all susceptible to the dissolving action of acid rain. In fact, many statues adorning public buildings in Europe have become unrecognizable because of this process.

The Earth materials produced by weathering are of great significance. The larger grains are known as regolith and, by the addition of partly decomposed organic matter, are turned into soil, the basis for agriculture. Other grains are carried by rivers into the sea to become the raw materials from which beaches are made. Residues of weathered materials are sometimes left behind, such as clay and ores of iron, aluminum, and manganese, which may form valuable mineral deposits.

Erosional processes also affect human life.

Gravity's influence may bury small villages in catastrophic landslides, or it may trigger the imperceptible downslope movements, known as creep, that cause structures on hillsides to collapse. When the Alaska Pipeline was being built, construction of all kinds was hampered by problems caused by soil flowage when the permafrost in the ground thawed. The erosional activity of rivers shapes the landscape, cutting gorges and supplying sediment to alluvial fans, floodplains, levees, and deltas. Sometimes the erosional activity of a river gets out of hand, causing devastating floods. Even where no stream channel is present, farmlands can be seriously damaged by soil erosion.

The erosional power of moving ice sculptures some of the world's most spectacular mountains. Along coastlines, wave erosion creates cliffs or threatens human-made structures such as lighthouses, seawalls, and breakwaters. Coastal currents may carry sand away, causing severe beach erosion. The wind contributes to erosion when it moves sand grains to create a sandstorm. This blinding cloud can sandblast the paint off a car or break a telephone pole in two. Even dust-sized material, when lifted from the ground in the form of a dust storm, can have a devastating effect. During the 1930's, an area known as the Dust Bowl developed in the Great Plains region of the United States. A prolonged drought and unwise agricultural practices resulted in severe dust storms that blew away valuable topsoil, lowering the ground level by nearly 1 meter in some places.

Donald W. Lovejoy

CROSS-REFERENCES

Alluvial Systems, 2297; Beaches and Coastal Processes, 2302; Deep-Sea Sedimentation, 2308; Deltas, 2312; Desert Pavement, 2319; Drainage Basins, 2325; Evaporites, 2330; Floodplains, 2335; Glacial Deposits, 880; Lakes, 2341; Reefs, 2347; River Bed Forms, 2353; River Flow, 2358; Sand, 2363; Sand Dunes, 2368; Sediment Transport and Deposition, 2374; Surface Ocean Currents, 2171.

BIBLIOGRAPHY

Bertin, Léon. *The Larousse Encyclopedia of the Earth.* New York: Prometheus Press, 1965. This reference book has well-written sections on the weathering processes of disintegration and decomposition. There are also lengthy sections on mass wasting, transportation by the wind, subsurface or groundwater, running water, wave erosion, and the work of glaciers. The text is copiously illustrated with excellent black-and-white and color plates. Suitable for general readers.

Birkeland, P. W. *Soils and Geomorphology.* 3d ed. New York: Oxford University Press, 1999. This book on soils is one of the best written from the perspective of a geologist. Many examples from around the world are given. More than one-half of the book is devoted to soil formation, development, and erosion. The final chapter on applications of soil formation to geology is excellent. Suitable for advanced high school and university-level readers.

Bland, Will, and David Rolls. *Weathering: An Introduction to the Scientific Principles.* New York: Arnold, 1998. This introductory college textbook provides a clear look at weathering and erosion processes. The authors explore the relationships among climate, weather, and the environment. Illustrations, references, and index.

Cain, J. A., and E. J. Tynam. *Geology: A Synopsis.* Vol. 1, Physical Geology. Dubuque, Iowa: Kendall/Hunt, 1980. A condensed treatment of physical geology that gives the reader a quick overview of the various weathering and erosion processes. There are many helpful diagrams, tables, and photographs that illustrate the principles involved. An excellent introduction to geology for the nonscientist. Suitable for high school readers.

Jensen, M. L., and A. M. Bateman. *Economic Mineral Deposits.* 3d ed. New York: John Wiley & Sons, 1979. An outstanding economic geology text, containing detailed information on the processes of formation and the occurrence of residual deposits of iron, manganese, aluminum, and clay as a result of weathering. There are cross sections of individual deposits. Suitable for college-level readers and the interested layperson with some technical background.

Judson, S., M. E. Kauffman, and L. D. Leet. *Earth: An Introduction to Geologic Change.* Englewood Cliffs, N.J.: Prentice Hall, 1995. This introductory text discusses the various types of weathering and the methods for studying the rates of weathering and erosion. The text is illustrated with photographs, diagrams, and tables that provide important data. Written at a level suitable for undergraduates.

Plummer, Charles C., and David McGeary. *Physical Geology.* 5th ed. Boston: McGraw-Hill, 1999. An unusually readable text. Extended discussions of weathering and the various erosion processes are supplemented by excellent photographs and line drawings. Each chapter has an extended list of supplementary readings, and there is an excellent glossary. Suitable for college-level readers and interested nonspecialists.

Shelton, J. S. *Geology Illustrated.* San Francisco: W. H. Freeman, 1966. This book contains some of the finest black-and-white photographs ever taken of geologic features, along with explanatory text. Part 3, "Sculpture," has numerous aerial views showing how the Earth's surface has been modified by the erosion processes of landslides, streams and rivers, groundwater, glaciers, waves, and wind. Suitable for laypersons.

Tarbuck, Edward J., and Frederick K. Lutgens. *Earth: An Introduction to Physical Geology.* 6th ed. Upper Saddle River, N.J.: Prentice Hall, 1999. This popular text offers an introduction to weathering and erosional processes. It is concisely written and generously illustrated with color photographs and color line drawings. Key terms are in boldface in the body of the text, and there is a helpful glossary. Suitable for high school readers.

EARTH SCIENCE

Alphabetical List of Contents

Categorized List of Contents

VOLCANOES AND VOLCANISM

WATER AND WATER PROCESSES

WEATHER AND METEOROLOGY